Understanding Data

Alan Garfinkel · Yina Guo

Understanding Data

A 21st Century Approach to Statistics and Data Science

Alan Garfinkel
University of California
Los Angeles, CA, USA

Yina Guo
Los Angeles, CA, USA

ISBN 978-3-032-18599-0 ISBN 978-3-032-18600-3 (eBook)
https://doi.org/10.1007/978-3-032-18600-3

This Springer imprint is published by the registered company Springer Nature Switzerland AG
The registered company address is: Gewerbestrasse 11, 6330 Cham, Switzerland

Preface

This book is highly unusual, if not unique, among statistics books. Virtually all statistics texts teach the student formulas that are recipes for calculating "statistical significance" for various types of data structures.

But in 2019, there were over 800 signatories to an article in *Nature* entitled "Retire Statistical Significance" ("We call for the entire concept of statistical significance to be abandoned"). Their reasons were compelling: bad statistical practices and concepts have led to the *crisis of irreproducibility* that is the current nightmare of the biomedical sciences. These bad practices include the concept of "statistical significance" itself, and the phenomenon of "p-hacking", which includes sifting through thousands of tests of significance, in the hope of finding a few chance hits.

These practices have led to a situation where many if not most published results are irreproducible and unreliable, resulting in a crisis of confidence in science ("Weak statistical standards implicated in scientific irreproducibility", *Nature*, 2013). The crisis of irreproducibility has undermined public trust in science.

In our opinion, any statistics course or text that does not begin with this crisis is now obsolete. There needs to be a radical re-thinking of what statistics is, how it is used, and how it is taught.

A major culprit behind the crisis is the fact that statistics is taught to the student as a list of formulas to memorize for various cases. What little justification is given for the formula is based on assumptions that everything in sight is a perfect bell curve, that sample sizes are very large, and that we use the mean value to describe all distributions.

This book takes a diametrically opposite path. The famous formulas were derived as approximations, made under harsh and unrealistic assumptions, assumptions that were necessary because, *at that time*, the true realization of statistics couldn't be done by hand! But with modern computing, we can find out how probable something is by simulating the situation many thousands of times, in a few seconds, and simply counting how often it occurred.

Freed from the need to calculate gobbledygook formulas, the student can experience statistics directly. If we see a difference between two groups and want to know if it could have happened just by chance, we simulate the situation by throwing the two groups into a box and "resampling" two new groups from the box, and seeing how much *they* differ. How often something happens in thousands of simulations gives the student a deeper understanding of probability and statistics.

Another major drawback of the formula-based approach is that the formulas are applicable only to a very narrow range of concepts, such as the use of the mean to describe all distributions, and the restriction to perfectly bell-shaped ("Normal") distributions. This worship of the mean (which has been dubbed "the flaw of averages") imprisons our thought, and constrains it to thinking only in terms of means, standard deviations, and other concepts that pure mathematics can prove theorems about. We can only look for the car keys under the lamppost.

The resampling-based approach, by contrast, gives us a method, valid in virtually any situation, to calculate critical concepts like confidence intervals, without making harsh and unrealistic assumptions.

Instead of forcing scientists to fit their questions into the formulas, the resampling-based approach frees scientific inquiry to ask more relevant questions, knowing that the statistics can always be done by resampling.

The goal of this book is to raise a new generation of scientists that will not be committing the statistical errors of previous generations.

The audience The book is aimed at first-year undergraduate students in the biological, biomedical, and social sciences. It has no mathematical prerequisites whatever, and is perfectly suited to introductory courses in Data Science.

Contents

Chapter 1

The Problem with Statistics

1.1 Statistical significance

Traditional statistics courses are routine, procedural and uncontroversial. You memorize a bunch of formulas that you don't understand, and whose assumptions are rarely mentioned and are hidden from you, and you plug-and-chug by calculating numbers to plug into those formulas.

The purpose of the formulas, their output, is something called a "p-value", which determines the all-important notion of "statistical significance".

A p-value is a probability, namely, the probability of obtaining, by pure chance, a result as extreme or more extreme than your observed result. If that probability is less than some pre-established standard (say, 0.05), then your result is said to be statistically significant.

Almost all statistics courses aim to teach you the concept of "statistical significance". Statistical significance is held to be important because it is supposed to tell you whether you have a publishable result. It's the goal of the formulas, it's what's taught in courses, it judges the final product of what you do as a scientist. Traditional statistics is all about calculating statistical significance, because traditional science is judged by whether or not you have achieved statistical significance.

It is therefore significant that in March of 2019, the leading journal *Nature* published an editorial entitled "Retire statistical significance" (Amrhein *et al.*, 2019). The editorial was signed by over 800 leading statistics professionals. (They only had a week to get signatures, and they apologized to everyone that didn't have a chance to sign it.)

COMMENT **Retire statistical significance**

Valentin Amrhein, Sander Greenland, Blake McShane and more than 800 signatories call for an end to hyped claims and the dismissal of possibly crucial effects.

21 MARCH 2019 | VOL 567 | NATURE | 305

We ... call for the entire concept of statistical significance to be abandoned.

Over 800 serious professionals have called for the abandonment of what was said to be the main purpose of statistics! How can this be possible?

A. Garfinkel and Y. Guo, *Understanding Data*,
https://doi.org/10.1007/978-3-032-18600-3_1

1.2 Is there a problem? Yes!

Unreliability crisis The reason why all those researchers were prompted to call for the retirement of statistical significance is because there is a major problem: many published scientific findings cannot be trusted. A major factor responsible for those untrustworthy results is the concept of statistical significance itself.

Science writer Sharon Begley called attention to the crisis in a 2011 article.

Newsweek CULTURE

Why Almost Everything You Hear About Medicine Is Wrong

Sharon Begley

She wrote: "First garlic lowers bad cholesterol, then — after more study — it doesn't. Hormone replacement reduces the risk of heart disease in postmenopausal women, until a huge study finds that it doesn't (and that it raises the risk of breast cancer to boot). Eating a big breakfast cuts your total daily calories, or not — as a study released last week finds....a major study concluded there's no good evidence that statins (drugs like Lipitor and Crestor) help people with no history of heart disease".

She continues: "Two 1993 studies concluded that vitamin E prevents cardiovascular disease; that claim was overturned by more rigorous experiments, in 1996 and 2000. A 1996 study concluding that estrogen therapy reduces older women's risk of Alzheimer's was overturned in 2004. Numerous studies concluding that popular antidepressants work by altering brain chemistry have now been contradicted (the drugs help with mild and moderate depression, when they work at all, through a placebo effect), as has research claiming that early cancer detection (through, say, PSA tests) invariably saves lives".

A 2005 article was entitled "Why most published research findings are false", which is quite a title (Ioannidis, 2005). The article uses mathematical reasoning, which we will develop in detail in section 12.5 *Bayes' Theorem and the problem of unreliable results*, to argue that the standardly-used statistical methods, taught in traditional books and courses, produce a very large number of "False Positives," that is, statistically significant "findings" that are in fact erroneous. Add that to another effect: "The greater the financial and other interests and prejudices in a scientific field, the less likely the research findings are to be true," and several other similar factors, and the overall effect accounts for the article's title.

Essay

PLOS | MEDICINE

Why Most Published Research Findings Are False

John P. A. Ioannidis

Irreproducibility crisis Poor statistical techniques, and the unreliable "results" that they produce, have created a phenomenon called **the crisis of irreproducibility**.

A 2015 article in the Boston Globe detailed the crisis of irreproducibility with some shocking documentation. Attempts to replicate well-known published studies were often unsuccessful, leading to a loss of confidence in science and medicine.

Companies like Bayer and Amgen are practical, and their interest is not in "getting published"; their interest is in finding out what is true, before they commit hundreds of millions of their own dollars. Therefore, this lack of reproducibility must be taken very seriously.

The Boston Globe

In science, irreproducible research is a quiet crisis

Carolyn Johnson

Scientists at Bayer HealthCare reported that when they tried to reproduce 67 published discoveries in oncology, women's health, and cardiovascular disease, only a quarter of the time were their in-house results completely consistent with what had been reported.

A scientist at Amgen disclosed that, despite concerted effort, the company had successfully repeated only six of 53 major cancer findings.

Last year, neuroscience researchers tried to repeat five studies that showed 17 links between certain brain structures and behaviors — for example, a finding that people with more gray matter in particular brain regions also have more Facebook friends. The researchers could not replicate any of the connections and found evidence for what scientists call the "null hypothesis" — the idea that there was no relationship.

The lesson has not been lost on some leading figures in medical research, who have been warning about this for some time. Douglas Altman was for many years the head of medical statistics at the Imperial Cancer Research Fund in Britain. His 1994 article "The scandal of poor medical research", says that "Huge sums of money are spent annually on research that is seriously flawed through the use of inappropriate designs, unrepresentative samples, small samples, incorrect methods of analysis, and faulty interpretation".

Believability crisis Poor statistical methods, as taught in traditional courses, have led to a believability crisis. One recent article looked at dozens of published studies about common foods and their association with cancer. Each study calculated a "relative risk" for that particular food; a relative risk < 1 says that the food is associated with lower cancer, while a relative risk > 1 says that the food is associated with increased cancer.

The results of this meta-analysis are sobering: most popular foods have a number of studies that claim an increased risk of cancer and also have a number of studies that claim a decreased rate of cancer! (Figure 1.1).

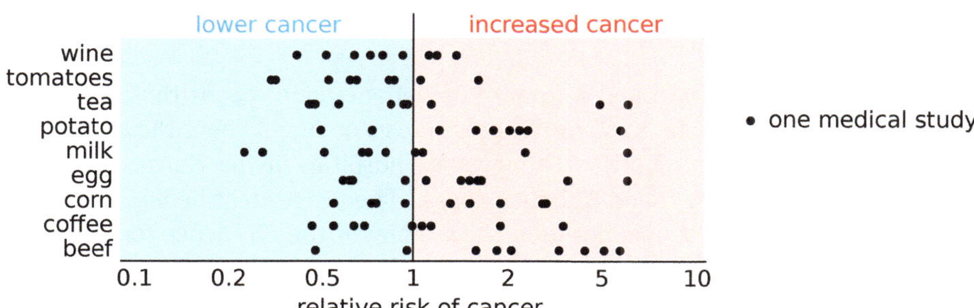

Figure 1.1 A meta-analysis on dozens of published studies about common foods and their association with cancer. Each dot represents one medical study. In each case, the finding was "statistically significant" (Schoenfeld and Ioannidis, 2013).

Science writers and science professionals are becoming aware of the crisis in research in biology and medicine. A deep skepticism is growing in response to the outpouring of results that are simply erroneous.

One of the prime causes of the problem is the use of poor and inappropriate statistics, especially the use of p-values to declare "statistical significance".

1.3 #1 cause of the crisis of unreliable and irreproducible results: bad statistics

"genes for"

The idea of "genetic bases for disease" has captivated a lot of recent interest.

Such "genes for X" studies are extremely popular, and are often featured on news channels on TV and the internet, with headlines like "genes for cancer found" and claims of a "genetic basis of schizophrenia", "genes for heart disease", or "genes for depression".

However, many if not most of these claims are spurious. For example, an article in the *Journal of the American Medical Association* followed up on the claims of genetic bases for Acute Coronary Syndrome (ACS), a common cardiac pathology (Morgan *et al.*, 2007).

ORIGINAL CONTRIBUTION

Nonvalidation of Reported Genetic Risk Factors for Acute Coronary Syndrome in a Large-Scale Replication Study

Thomas M. Morgan, MD
Harlan M. Krumholz, MD, MS
Richard P. Lifton, MD, PhD
John A. Spertus, MD, MPH

Context Given the numerous, yet inconsistent, reports of genetic variants being associated with acute coronary syndromes (ACS), there is a need for comprehensive validation of ACS susceptibility genotypes.

Objective To perform an extensive validation of putative genetic risk factors for ACS.

Conclusions Our null results provide no support for the hypothesis that any of the 85 genetic variants tested is a susceptibility factor for ACS. These results emphasize the need for robust replication of putative genetic risk factors before their introduction into clinical care.

JAMA 2007

Their literature review found a total of 85 variants in 70 genes that had been claimed to be "significant susceptibility factors" for atherosclerosis or ACS. Then, they set up their own study: they studied all patients reporting to university hospitals in the Kansas City area and meeting their study criteria. They found 811 such cases. Then they identified 650 age- and sex-matched controls, and asked the simple question: how many of the 85 "genes for" variants were, in fact, over-represented in the patient group vs. the matched controls?

The answer is astonishing: only 1 of the 85 variants showed a statistically significant difference between the two groups, a result, as the authors say, that could easily have happened by chance given 85 randomly chosen genetic variants.

This is a stunning rebuke of the statistical methods involved in the "genes for" literature. It was published in the prestigious *Journal of the American Medical Association*, and yet it is simply ignored, as scientists engage in the frenzied hunt for "genes for" diseases, a push for headlines and grant funding that has infected modern physiology.

Their study was written up in *Science News* in an article entitled "Odds are, It's Wrong: Science fails to face the shortcomings of statistics" (Siegfried, 2010).

Odds Are, It's Wrong

Science fails to face the shortcomings of statistics

By Tom Siegfried

Nowhere are the problems with statistics more blatant than in studies of genetic influences on disease. In 2007, for instance, researchers combing the medical literature found numerous studies linking a total of 85 genetic variants in 70 different genes to acute coronary syndrome, a cluster of heart problems. When the researchers compared genetic tests of 811 patients that had the syndrome with a group of 650 (matched for sex and age) that didn't, only one of the suspect gene variants turned up substantially more often in those with the syndrome — a number to be expected by chance.

How could so many studies be wrong? Because their conclusions relied on "statistical significance," a concept at the heart of the mathematical analysis of modern scientific experiments.

SCIENCE NEWS | 2010

Weak statistical standards

The #1 cause of the crisis of irreproducibility? Bad statistics. A news article in *Nature* in 2013 was titled "*Weak statistical standards implicated in scientific irreproducibility*".

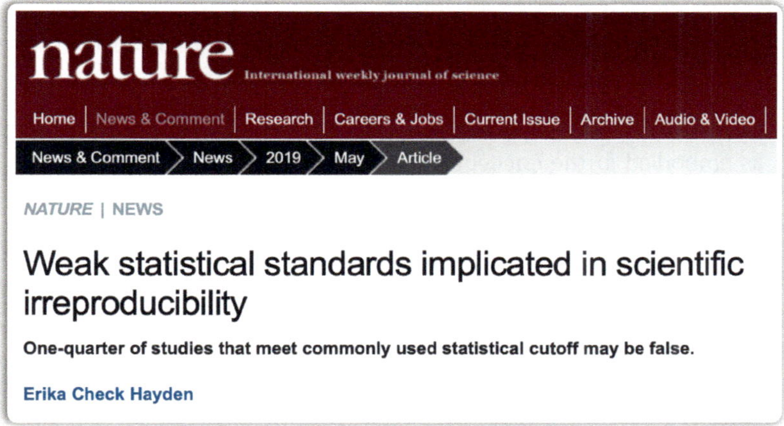
nature International weekly journal of science

Home | News & Comment | Research | Careers & Jobs | Current Issue | Archive | Audio & Video |

News & Comment > News > 2019 > May > Article

NATURE | NEWS

Weak statistical standards implicated in scientific irreproducibility

One-quarter of studies that meet commonly used statistical cutoff may be false.

Erika Check Hayden

The article discusses a study (Johnson, 2013) that uses mathematical reasoning to show that a single factor, namely, loose standards for declaring statistical significance, can already account for an error rate of 17%–25%, due to that factor alone, making the claim that "most published research is false" chilling but plausible.

The fact that scientists continue to use debunked methods, and statistics departments continue to teach them, is distressing, especially in the light of very clearly written and well-published articles calling attention to the problem.

We cannot continue with statistics-as-usual, plugging numbers into invalid formulas and becoming the next generation of bad scientists. But that does mean that the thrust of this book is against 90% of the rest of the bioscience world on the subject of statistical methods.

p-values and statistical significance

So what are these "p-values"? What is the concept of "statistical significance"? That is the subject of Chapter 3, but we will preview it here. "p" stands for "probability", and the p-value of an observed result is the probability of a result that extreme or more extreme happening under "the Null Hypothesis". Typically, the Null Hypothesis is that there is no relationship, and all observed differences are the result of random sampling. The idea is that if we have obtained a result that has a very low probability of happening under the Null Hypothesis (a low "p-value"), then that result is "statistically significant".

It is this method of declaring "statistical significance" that has brought us to the present crisis of irreproducibility (Halsey *et al.*, 2015).

nature methods COMMENTARY

The fickle *P* value generates irreproducible results

Lewis G Halsey, Douglas Curran-Everett, Sarah L Vowler & Gordon B Drummond

The reliability and reproducibility of science are under scrutiny. However, a major cause of this lack of repeatability is not being considered: the wide sample-to-sample variability in the *P* value. We explain why *P* is fickle to discourage the ill-informed practice of interpreting analyses based predominantly on this statistic.

Take-home messages

- There is a crisis of reliability in scientific publications.
- The principal cause of this crisis is the (mis-)use of the concept of "statistical significance" as embodied in the concept of "p-value".

One recent paper (Colquhoun, 2014) said:

> "If you use p = 0.05 to suggest that you have made a discovery, you will be wrong at least 30% of the time. If, as is often the case, experiments are underpowered, you will be wrong most of the time."

What is the remedy?

The remedy for the crisis has to go very deep. We need to completely re-think how we approach data.

We need a revolution in statistics, beginning with the very ways in which we approach data from the outset, and in the most basic ways that we describe and present data.

That is the subject of the next two chapters.

Chapter 2

Describing and Presenting Data

LEARNING OBJECTIVES

After reading this chapter, you will be able to

- choose what types of plots to make to show the distribution of the data, and to know for what kinds of data each type of plot is appropriate.
- decide when a typical value (mean, median, etc.) can be used to summarize a data set.
- measure the variability of data within a data set.

Let's say we have made a measurement on every member of a sample population. If the population is people, this could be their blood pressures, or their heights, or their survival times on some drug; if the population is neurons, it could be their firing rates or their sizes or response times. We then have a data set consisting of those measurements.

The first decision we have to make is: how to present our data? As we will see, the first and most important question is: ***What is the distribution of the data?*** In order to decide which methods are best for a given data set, we need to see the shape of the distribution.

There are several methods for showing the overall distribution of the data, depending on the size of the data set, and the type of data that make it up.

After studying the distribution, we may then choose to summarize the data set by "typical values" like the mean or median. We may also try to summarize the variability of the data set by a number. What we do depends completely on the shape of the data.

One of the founders of modern statistics, JW Tukey, said it best: "There is no excuse for failing to plot and look" (Tukey *et al.*, 1977).

For example, we often hear about "the average" value, or, to use the fancier term, "the mean" value. We talk about the "average class size" or the "mean firing rate" of a collection of neurons.

As we will soon see, there are many cases where the mean is a poor measure of a data set. If we report such a data set only by its mean value, we are committing statistical malpractice, and have drawn a conclusion that is, at best, highly misleading.

As a symptom of this problem, last year we gave a poll question to 272 students on the first day of a freshman statistics for biologists class. The students were shown a headline:

A. Garfinkel and Y. Guo, *Understanding Data*,
https://doi.org/10.1007/978-3-032-18600-3_2

CNN BUSINESS Markets Tech Media Success Perspectives Videos

Average tax refunds up $40 for 2019

By Donna Borak, CNN Business
Published 4:36 PM EST, Thu February 28, 2019

The headline said "Average Tax refunds up $40 for 2019", and the question asked you to say what is likely to happen to you.

Based on the news headline, which of the following is true?

A. I can expect around $40 more this year.

B. I can expect to get a bigger refund, amount uncertain.

C. Many people will get approximately $40 more, but, as they say, "Individual results may vary".

D. A typical refund will be $40 more.

E. None of the above.

The class answers are shown below:

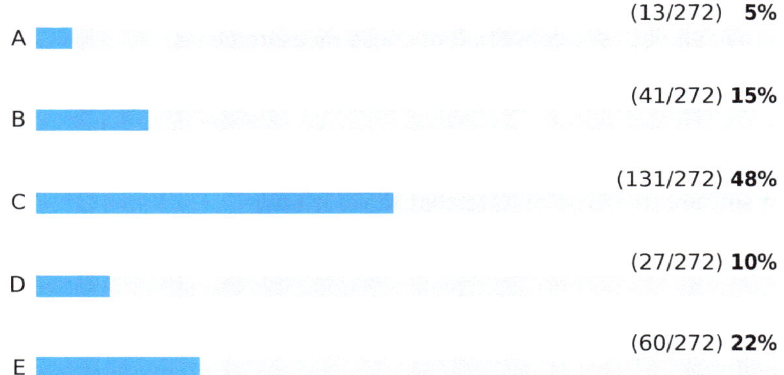

The only even remotely correct answer is E, "None of the above", because given the information in the headline, you can say absolutely nothing about your likely outcome. It could be $40 more, or $40,000 more, or $40,000 less than your last year's return.

Notice that 78% of the class, over three quarters, gave a completely incorrect answer to the question.

How could this happen? How can 78% of the population get that wrong? The answer is that we've been led to believe that a certain shape, the "bell curve", holds almost automatically for all distributions in nature. And if you were unconsciously assuming that tax returns follow a bell curve distribution, also known as a "Normal" distribution, then you would be excused for assuming that the average (= mean) value is also the most likely value, because that happens to be true for bell-shaped distributions.

But are all or most distributions in nature shaped like bell curves? No! As we will see later in this chapter and in Chapter 5 (Normal Distributions), the assumption that most distributions are bell-shaped is the source of a great deal of error in science.

And that's the reason why we aren't beginning with calculating numbers like the mean. We have to first take a step back and ***ask whether the mean or any other numerical summary is justified, at all, for this data.*** In this particular case, the average is a poor descriptor of the data!

So we are not going to just "Take the average" because we're not ready to take the average!

And then there is a second point: even if the mean *is* appropriate for a given data set, if we choose to simply summarize the data set by a single number, we are choosing to ignore a great deal about that data. The most important fact we are missing is the *variability* of the data set. Is the data set highly variable, so that a "typical number" is a poor representative?

So, before we start calculating any summary numbers for a data set, like means or anything else, we first have to look at the data set. Step 1 is to visualize the data set in some way that makes clear what its distribution is. Only then, as step 2, can we decide what numerical summary measures to use.

2.1 Visualizing data: dot plots and beeswarm plots

What is data? It is any set of facts that describe a state of affairs. In this book, data will usually be expressed in numbers. But data could also be a picture, or a fingerprint at the scene of a crime.

The most typical situation is when we have a set of subjects 1, 2, . . . , n and we measure a quantity X for each subject. We will call these numbers x_1, x_2, . . . , x_n. In this case we speak of a *data set* $\{x_1, x_2, \ldots, x_n\}$ describing the group.

In statistics, the data set is often referred to as a sample, and n is then called the sample size. In this book, we will always use this convention that n is the number of data points in a sample.

For example, we might have a group of people, and we measure a systolic blood pressure (SBP) for each of them. Blood pressures are generally measured in units of millimeters of mercury, mmHg, which is a standard of pressure.

In this case the data set might look like:

Subject No.	Name	Blood Pressure (mmHg)
1	Sally E	$x_1 = 108$
2	Trisha K	$x_2 = 126$
⋮	⋮	⋮
n	Steve W	$x_n = 118$

Table 2.1 A sample data set, consisting of blood pressures from n participants.

The first thing to do with any data set is: look at it! This may seem obvious, but it is surprising how often this is not done, with serious consequences.

How to look at data

Dot plot In the case where the data set is small to middling, the most important visualization is the **dot plot**: we simply plot the data values as points on a single axis, horizontal or vertical (Figure 2.1).

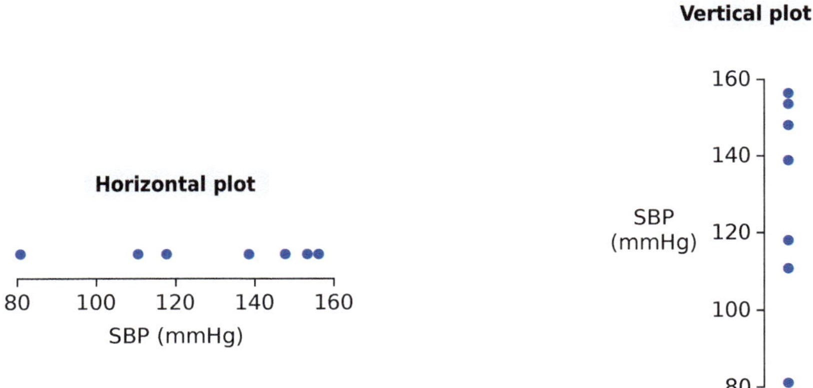

Figure 2.1 Two presentations for dot plots, horizontal and vertical.

In this chapter, we will use the horizontal presentation, but in later chapters, when we are comparing several groups, we will use the vertical presentation.

When the data set gets larger than 20 or so, there is a real chance that data points might overlap, giving us a misleading picture of the data. Consider the four different data sets in Figure 2.2. If the data set is small (n = 10) (first row) there is no problem with the dot plot, and it shows the data nicely. For n = 20 (second row) there is already some overlapping of data points: only 17 dots can be seen; the other 3 are overlapped. By the time n = 50 (third row) the degree of overlap is substantial, so the plot mis-represents the data, and when n = 100 (bottom row) the mis-representation is much worse.

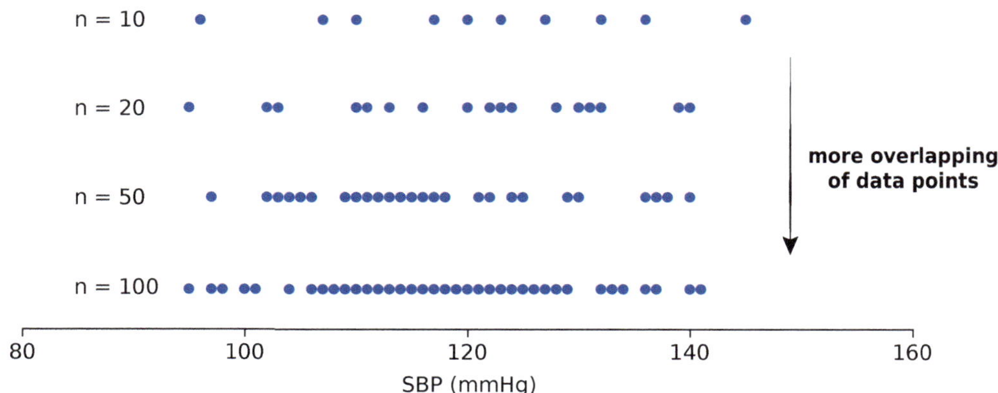

Figure 2.2 Dot plots for four different data sets of increasing size n. Note the overlap of data points in the larger sets.

For these larger data sets, we need a method for preventing overlap of the data points.

Beeswarm The best method for avoiding overlapping is the "beeswarm" plot. Since the data is one-dimensional, we can use the second dimension of the page to spread the data out. We create a fictitious second axis, and then add a small fixed quantity to each data point in this (meaningless) axis. This stacks up data points that are equal or close so they don't overlap (Figure 2.3).

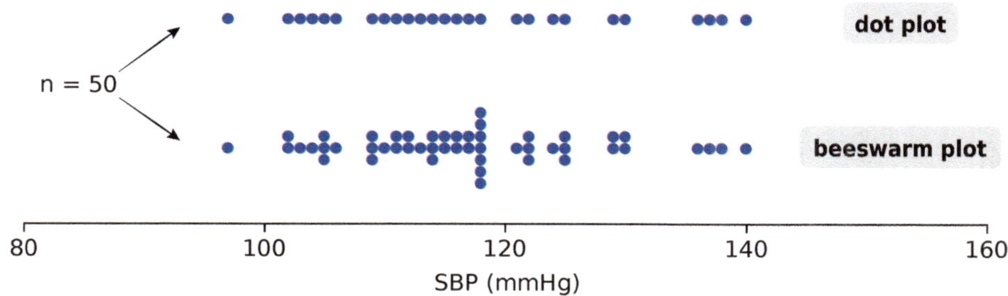

Figure 2.3 Dot and beeswarm plots for the n = 50 data set in Figure 2.2.

Looking at the two forms of presentation, it is clear that the simple dot plot is fine for small data sets, but that for larger ones, the beeswarm plot gives a better picture of the data (Figures 2.3 and 2.4).

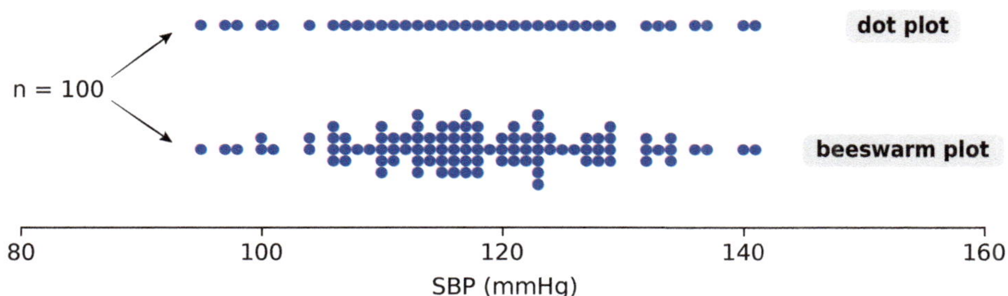

Figure 2.4 Dot and beeswarm plots for the n =100 data set in Figure 2.2.

What to look for in a dot plot or beeswarm plot

The most important first step in looking at a data set is to *look at the data set*, that is, to make a picture of its distribution.

The true picture of the data set is its distribution, and we must always begin by looking at it and seeing what it is trying to tell us. Some important features to look for are:

Bunching up of data (symmetry vs. skew) Are the data fairly evenly distributed over their range, or do they "bunch up"? If they do bunch up, are the more popular values in the middle, making the distribution fairly symmetric? Or do they bunch up on the left, with a long tail to the right, which is called "skewed to the right", or bunch up on the right, with a long tail to the left, which is called "skewed to the left"? (Think of the term "skewed to the..." as meaning "sticking out asymmetrically to the....") For example, the data in Figure 2.5 are skewed to the right. Of course, this can only be seen in the beeswarm plot.

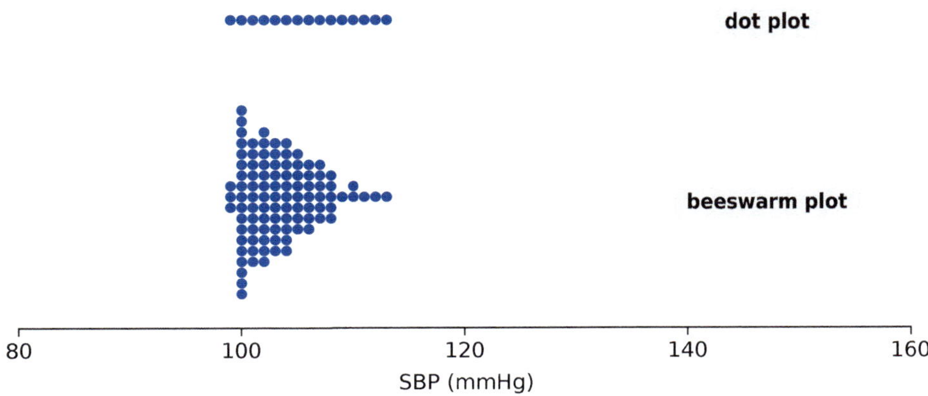

Figure 2.5 Data skewed to the right.

Outliers Are all the data in the same general range, or are there some points that are "far away" from the others? These extreme values are often called "outliers" (Figure 2.6).

How we deal with outliers is an important scientific question, for which there is no quick or automatic answer.

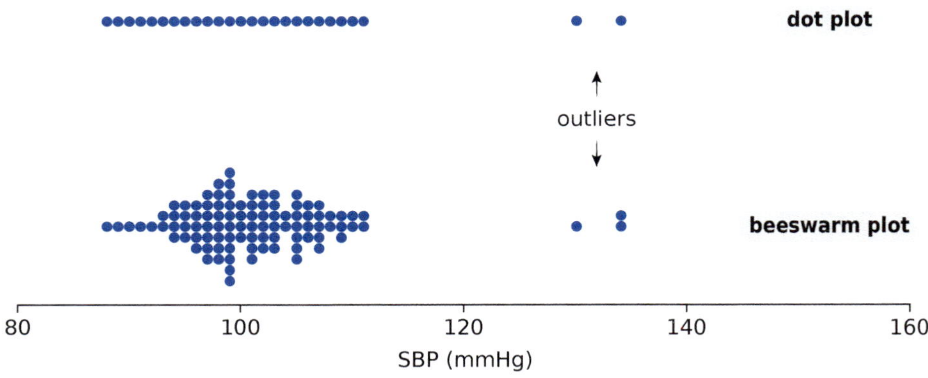

Figure 2.6 Data that has a few outlying values.

When we see outliers, our first duty is to figure out why that data point is showing such extreme values. Outlying values could represent a mistaken measurement (for example, a blood pressure of 1200 mmHg is a typo or a defective instrument), but they could also represent important extreme cases.

Outliers can be excluded only if you are reasonably sure that there is a causal explanation for this value that shows it to be produced by a defective mechanism, one other than the mechanism producing the rest of the data. Certainly data entry errors or defective instruments count as defective mechanisms, but consider also the data in Example 1 on page 14. Here we are plotting the month-by-month cancellation rates for commercial airlines from 2001 to 2003.

We note an outlier: there is one point whose monthly cancellation rate is 20%. When we examine the data, we see that the outlying month is September 2001, when the attacks on the World Trade Center in New York caused massive airline cancellations.

In contrast to this policy, some sources suggest that there can be purely mathematical criteria to exclude points that are, say, more than two standard deviations from the mean, or that pass a certain "critical ratio test". It is very important to understand that **there is no mathematical criterion that justifies "removing" outlying values just because they are outlying!** Removing outlying points simply because they are outlying is scientific fraud.

It is *mandatory* to exclude data that is from a defective Data Generating Process (like a broken meter, a mis-assigned subject, etc.) whether the data is small or large or typical. Thus, we should investigate data that is unusual. It is *never* correct to exclude "outliers" on some mathematical criterion ("two standard deviations" or "critical ratio test").

Tails Note the left and right tails of the data distribution. Are the tails "thin", with most of the data bunched up somewhere in the middle and distributed sparsely in the tails? Or are the tails "thick", implying a more even distribution, with more extreme values well-represented?

Subgroups Are there noticeable gaps between bunches of data? In other words, does the data set seem to group into distinct subgroups, like a "low group" and a "high group"? Or does it seem to be varying continuously, with no noticeable gaps?

If a data set has distinct subgroups (as in Figure 2.7, for example), then the average would not work well! Instead of mindlessly carrying out and applying statistics to the data set, we should first be asking ourselves questions such as: what causes this? What's driving it?

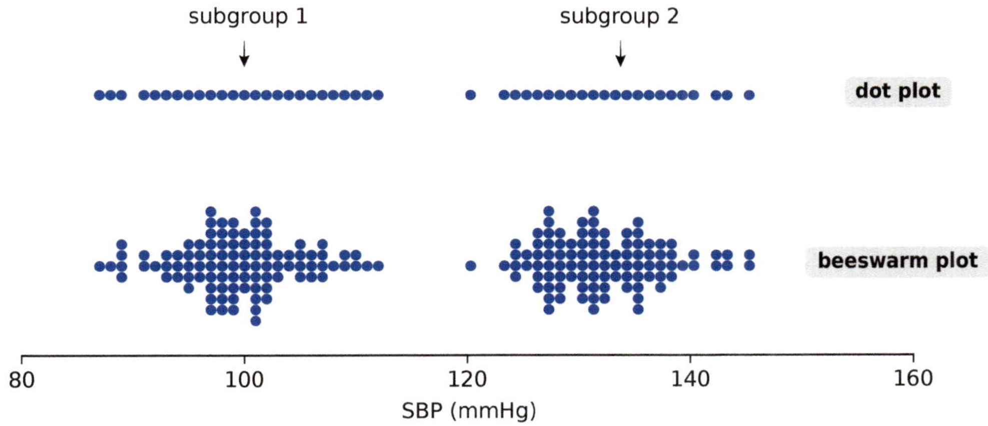

Figure 2.7 Data displaying subgroups.

Notice that the language of this section has been very visual and qualitative. You are being asked to form a visual impression of some qualitative features of the data set. This absolutely critical step is often skipped over, partly out of a prejudice that these impressions are "subjective". **In fact, they are the most important features of the data set, and our judgments of the distribution of the data are critical in making decisions about how to analyze it.**

Later on, we learn methods for turning all these visual impressions into precise mathematical concepts. But it will always be true that the visual impression is the first step.

Example 1 US airline cancellation rates

The United States Department of Transportation monitors and reports to the general public many types of information, including the monthly cancellation rates of airline flights. Cancellation rates are calculated as the ratio of cancelled flights to originally scheduled fights.

Here are the monthly flight cancellation rates (in percent) from 2001 to 2003.

	Jan	Feb	Mar	Apr	May	Jun	Jul	Aug	Sep	Oct	Nov	Dec
2001	3.8%	3.7	3.4	2.2	1.8	3.0	2.1	2.4	20.2	1.5	1.1	1.0
2002	1.7	1.1	1.3	1.0	1.0	1.7	1.3	1.1	0.9	1.0	0.9	1.7
2003	1.5	4.0	1.7	1.2	0.8	0.8	1.4	1.6	1.5	0.9	1.4	2.1

Question What is the distribution of the flight cancellation rates?

Answer To determine the distribution, the tabular form obviously doesn't help much. Therefore, we will make a beeswarm plot.

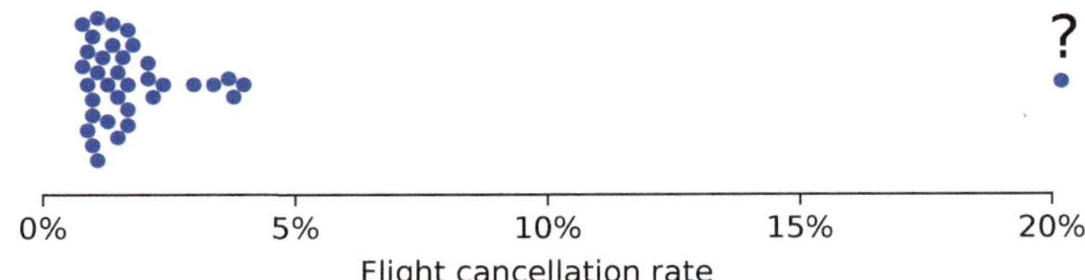

We see an extreme value: one month had a 20% cancellation rate. Is that point an excludable "outlier"? When we verify that that month was September 2001, the month of the World Trade Center attacks, we are justified in removing this data point. It does not give us any insight into the typical behavior of airlines. Here is the beeswarm plot excluding September 2001.

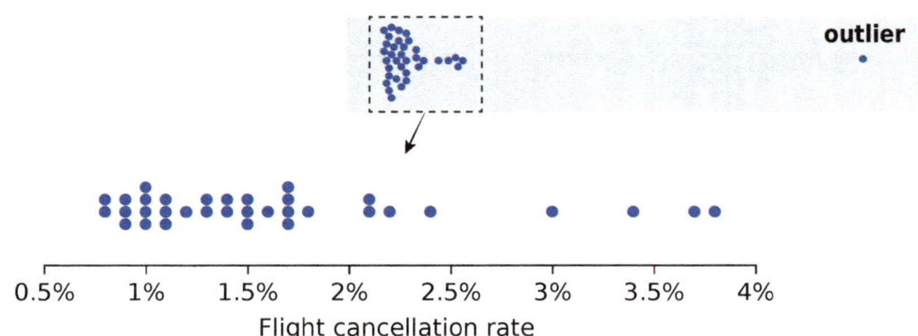

After removing the outlier, we see that the distribution is asymmetric. The data points bunch up on the left, with a long thin tail to the right, so the distribution is skewed to the right.

Question Are there subgroups?

Answer There are no obvious subgroups.

Example 2 Housing prices in San Francisco

From real estate transaction records, there were 886 3-bedroom houses sold in San Francisco in 2018. Here's a beeswarm plot of the sale prices.

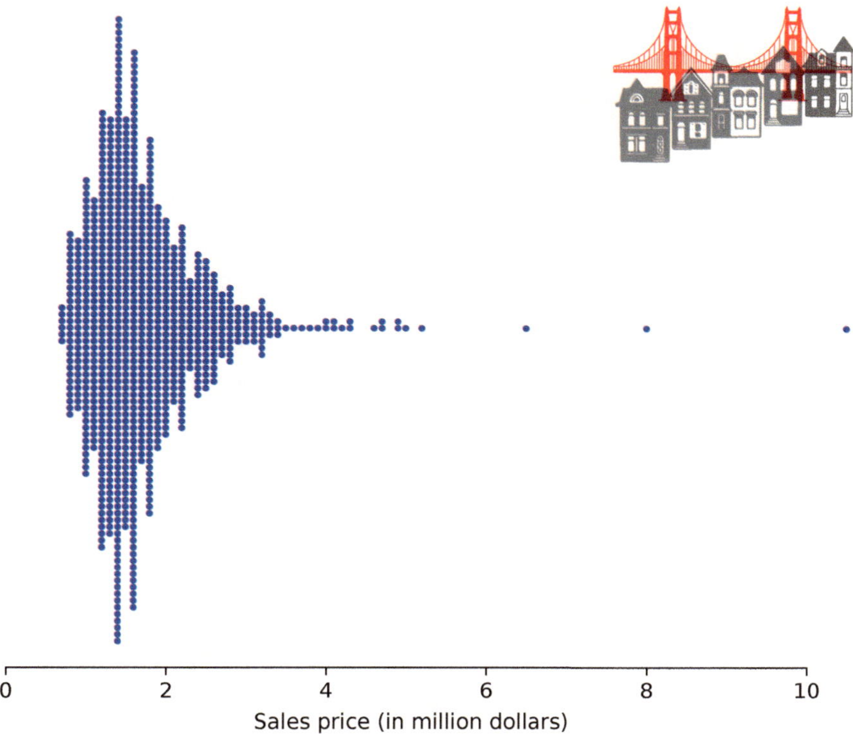

Question Are there outliers?

Answer While there are some large or even extreme values, they cannot be excluded as outliers because they aren't the result of some sort of anomaly; they are just very expensive homes! There is no mathematical criterion that permits the exclusion of these large values just because they are outlying.

Question How would you describe the distribution?

Answer It is clearly asymmetric. The data points bunch up on the left, with a long tail to the right, so the distribution is skewed to the right.

Question What do the tails of the distribution look like?

Answer There is a long thin tail on the right, and a short, thick tail on the left.

Question Are there subgroups?

Answer No subgroups are obvious.

Example 3 Vaccination rates in the US

As of late September 2021, 6 months after rollout of the COVID vaccine, the US state-by-state vaccination rates vary widely. Here is a beeswarm plot of their values.

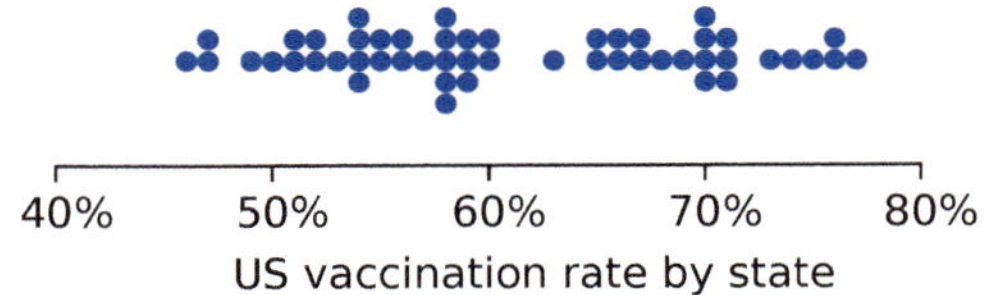

Question Are there outliers?

Answer No, there are no outliers.

Question How would you describe the distribution?

Answer The distribution has a high variability. It does not have a prominent peak, and the data points are scattered widely in a fairly uniform manner.

Question What do the tails of the distribution look like?

Answer The left and right "tails" are hardly tails at all, because the data is distributed fairly uniformly.

Question Are there subgroups?

Answer There do appear to be two subgroups, a higher vaccination rate group and a lower vaccination rate group.

FURTHER EXERCISES 2.1

1. **Random number generator.** A random number generator was set to produce whole numbers from 10 to 25. The plot below shows the results from 52 trials. Which of the following descriptions of this beeswarm plot is appropriate?

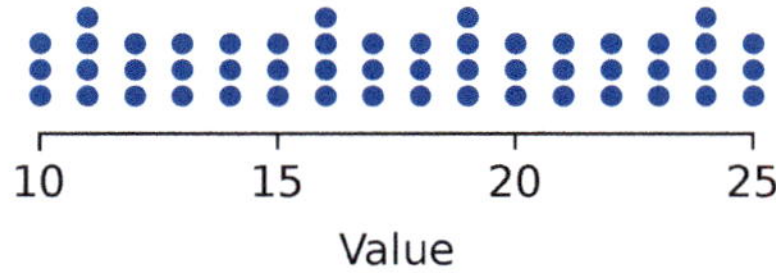

a. Right-skewed
b. Left-skewed
c. Symmetric with thick tails
d. Symmetric with thin tails
e. Uniform

f. Bimodal

2. **Daily screen time.** Four people tracked their daily screen time for 5 weeks.

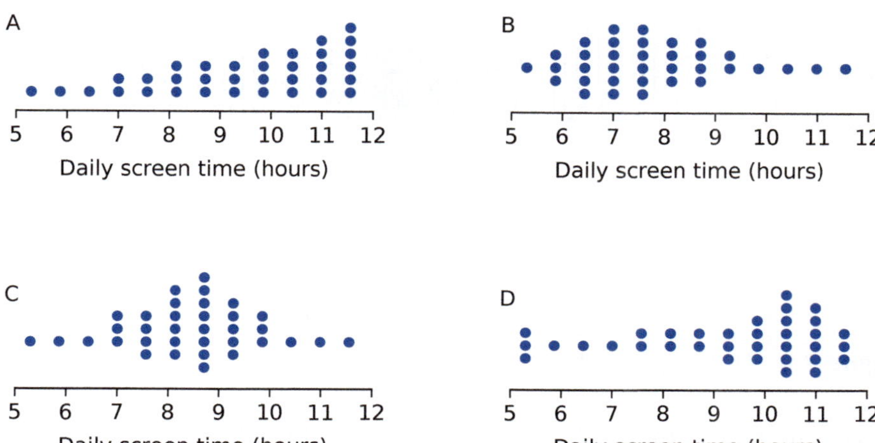

a. Describe the distribution for each person. Comment on center, spread, shape, and any unusual days.
b. Which person likely has a schedule requiring high, steady screen time?
c. Identify which plot is right-skewed and explain what that could mean in a real-life context.
d. Who has the greatest day-to-day variability in screen time?
e. Which person likely has high screen use most days, with some days of even higher usage? This pattern might describe a content creator, competitive gamer, or someone with many online meetings.
f. Which person's data shows the most outliers or extreme values compared to their typical pattern?
g. If someone wanted to reduce their screen time, which person would find it easiest to identify specific patterns to change?

2.2 Looking at the distribution: histograms and kernel density plots

Histograms

As the data sets begin to get larger, another form of data visualization comes into play, the **histogram**. In a histogram, we no longer show the individual data points. Rather, we divide the data axis into evenly spaced intervals, called **bins**, and sort the data values into the bins. Then we show the number of data points in each bin as a vertical bar erected over the interval that defines that bin (Figure 2.8).

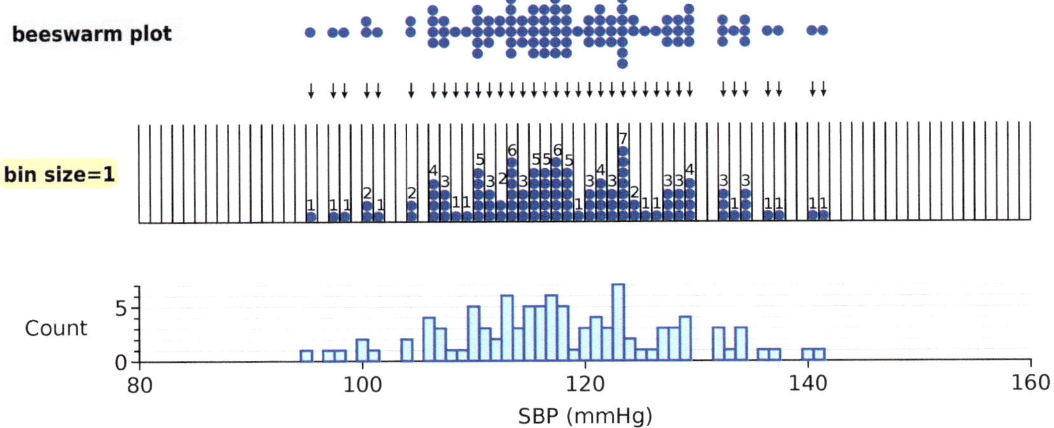

Figure 2.8 Histogram of the n = 100 data set in Figure 2.2, using a bin width = 1.

In order to make a histogram, we have to make two choices:

- a choice of cutoff points
- a choice of bin width

Choice of cutoff points The cutoff points are the left and right extremes of the histogram. The most common rule is to use the minimum and maximum values of the data as the left and right cutoff points, respectively.

An exception to this rule would be when we want to compare two histograms, which is frequently done. In that case, we might want to insist that the two histograms have the same cutoff points, so the differences between the two distributions are obvious. In Figures 2.8 to 2.12, for example, we are using constant cutoff points of 80 and 160. The choice of cutoffs has a large effect on the shape of the histogram.

Choice of bin width The bin width is the other critical choice. It has a radical effect on the shape of the histogram. It is useful to experiment with a variety of bin widths; each choice of a bin width brings out some features of the data and suppresses others. For example let's begin with a histogram with bin width = 1 (Figure 2.8). Notice that there is a lot of fine structure in the histogram. There are gaps in the histogram at 102 and 131, as well as other places, a dip at 119, and a tall peak at 123. But do these mean anything? Or are they just accidents, due to the fact that any sample of 100 numbers between 80 and 160 will have some gaps, dips and peaks? For this reason, it is not a good idea to have *too many* bins; here we have 100 data points and 80 bins.

When we double the bin width to 2, there are still "too many" bins, which produces a number of random gaps and peaks (Figure 2.9). There is no reason to think that these gaps and peaks have any meaning.

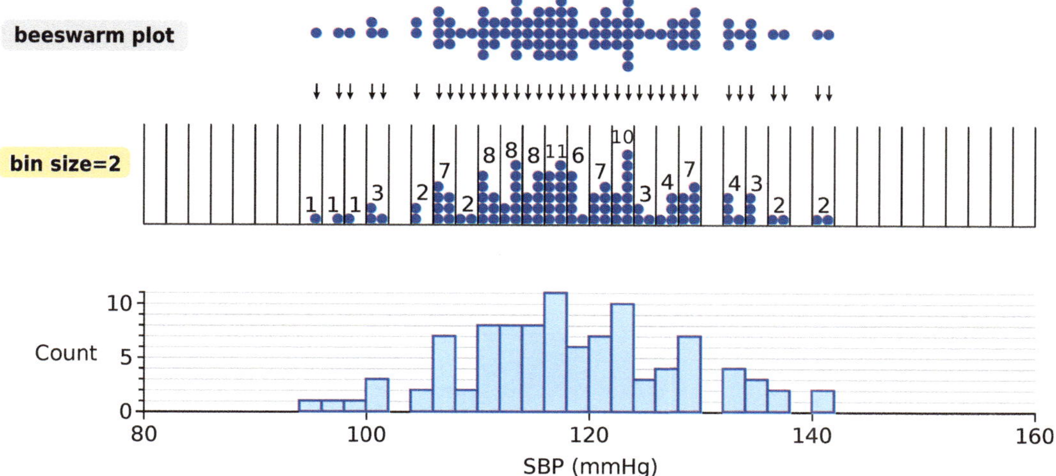

Figure 2.9 Histogram of the n = 100 data set in Figure 2.2, using a bin width = 2.

It is only when we get to a bin size = 5 that these random fluctuations are suppressed, and the resulting bins are large enough to give us a better picture of the data (Figure 2.10).

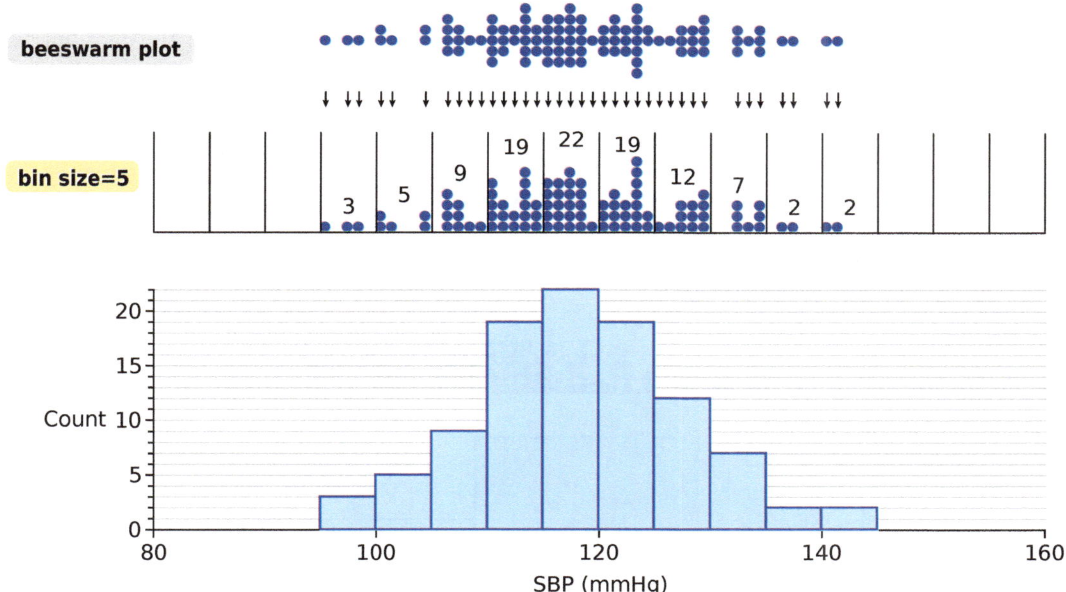

Figure 2.10 Histogram of the n = 100 data set in Figure 2.2, using a bin width = 5.

We see that the distribution is *unimodal* (single-humped) and roughly symmetric. We also see that the left and right tails are tapered and thin. (We should note that we are dismissing the thin peaks and gaps as "random" as we smooth out the histogram. We could eventually discover that there really *was* a peak at that point, but we would need a much larger data set to argue that.)

A bin size = 10 confirms these observations (Figure 2.11).

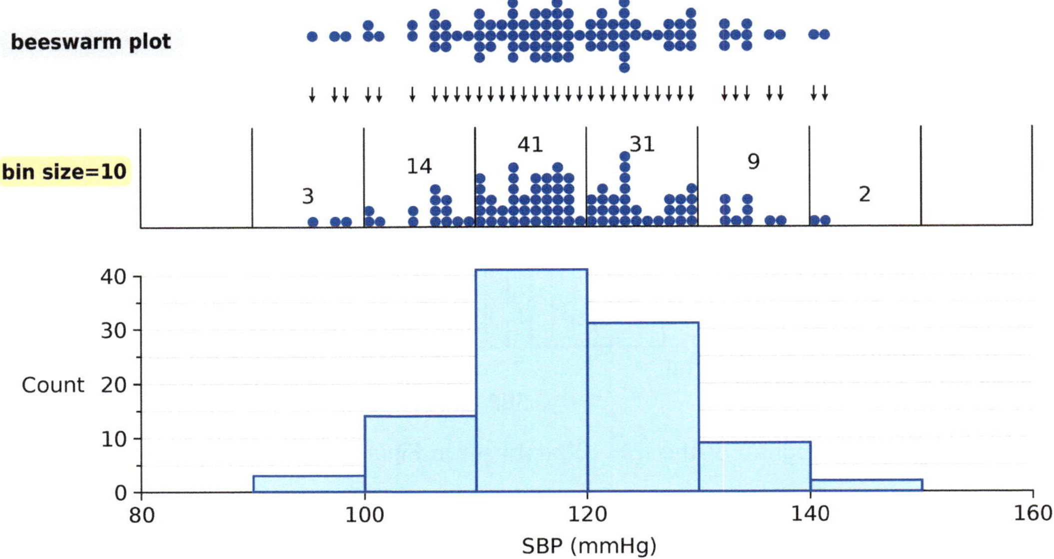

Figure 2.11 Histogram of the n = 100 data set (Figure 2.2), using a bin width = 10.

But when we take our bin size = 20, we are beginning to lose key features of the distribution. The gentle tapering from the center to the tails is now lost, and virtually all data are in the two middle bins. We have lost information about the distribution (Figure 2.12).

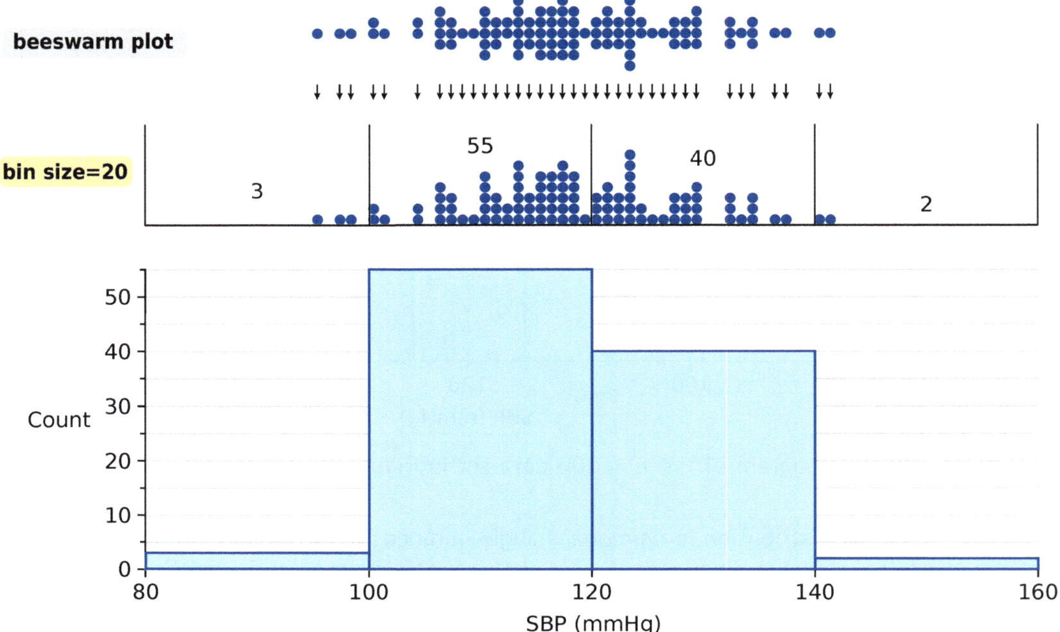

Figure 2.12 Histogram of the n = 100 data set (Figure 2.2), using a bin width = 20.

The choice of bin width depends on the number of data points. There are some simple rules of thumb for choosing bin width. Generally, instead of talking about bin width, we will talk about the inverse, which is the number of bins. Obviously,

$$\text{bin number} = \frac{\text{data}_{\text{max}} - \text{data}_{\text{min}}}{\text{bin width}}$$

The two most popular rules for choosing bin number are:

- **square root rule**: choose the smallest whole number bigger than or equal to $\sqrt{n}$, where n is the number of data points.
- **Rice rule**: choose the smallest whole number bigger than or equal to $2 \times (n^{1/3})$.

In this case, where $n = 100$, the square root rule gives a bin number of $\sqrt{100} = 10$, while the Rice Rule gives a bin number of $2 \times (100^{1/3}) \approx 9.3$, also requiring 10 bins.

There are other more sophisticated rules, that take into account the variability of the data.

Example 4 Earthquakes in or around California

California sits on the San Andreas Fault, and earthquakes have long been a problem. The state monitors and records seismic activity.

Here is a histogram of earthquakes recorded in the 7 day period ending on September 30, 2021. There were 976 earthquakes recorded during this time, measured in Richter scale. We chose the bin size to be 0.5 on the Richter scale for the histogram. The count of earthquakes that fall into each bin is shown on top of each bin.

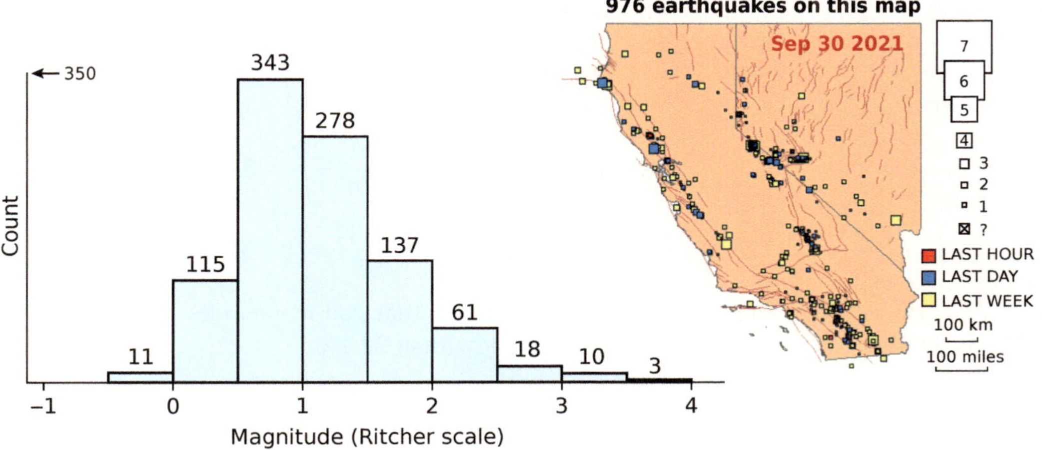

Question In this data set, what percentage of the earthquakes had magnitude between 0.5 and 2.0 in Richter scale?

Answer From the histogram, we see that there are 3 bins between 0.5 and 2.0:

- bin [0.5, 1)
- bin [1, 1.5)
- bin [1.5, 2)

The count of earthquakes that fall into those 3 bins are 343, 278, and 137, respectively. Adding them together, we get the total number of earthquakes between magnitude 0.5 and

2.0, which is 343 + 278 + 137 = 758, while the total number of earthquakes is 976. Therefore, the percentage of the earthquakes had magnitude between 0.5 and 2.0 on the Richter scale is 77.7%.

$$\% \text{ of earthquakes between 0.5 and 2.0} = \frac{343+278+137}{976} = 77.7\%$$

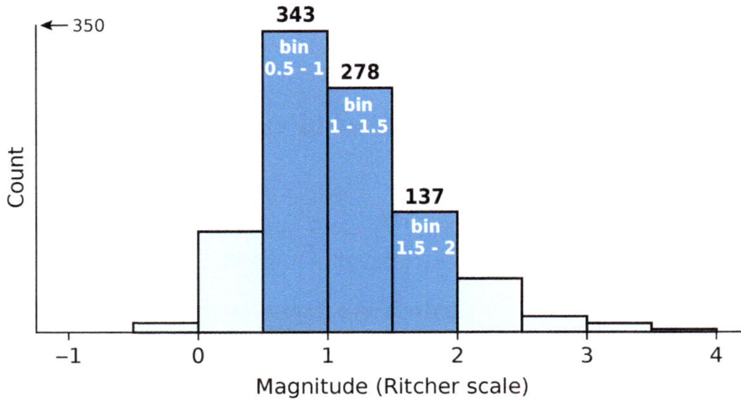

Question From this histogram, can you tell the exact percentage of earthquakes that had magnitudes between 1.8 and 3.0 on the Richter scale?

Answer No, you cannot tell the exact percentage of earthquakes that had magnitudes between 1.8 and 3.0. The reason is that the lower end of the range in question, 1.8, doesn't lie exactly on the edge of any bins. This is one of the shortcomings of histograms.

The most we can get out of this histogram is by rounding 1.8 up and down to 1.5 and 2.0, and calculate the percentages of earthquakes between range 1.5 – 3.0 and range 2.0 – 3.0 on the Richer scale.

$$\% \text{ of earthquakes in bin } [1.5, 3) = \frac{137+61+18}{976} = 22.1\%$$

$$\% \text{ of earthquakes in bin } [2, 3) = \frac{61+18}{976} = 8.1\%$$

We can state that the percentage of earthquakes that had magnitudes between 1.8 and 3.0 on the Richer scale is more than 8.1% and less than 22.1%.

Question Can you tell the min and max magnitude of earthquakes of this data set from the histogram?

Answer No, you can't. This is another shortcoming of histograms.

Question As you continue to update the histogram every time there is a new earthquake, and the next has a magnitude of 2.3. Which bin would need to be updated?

Answer Based on the binning we chose for this histogram, an earthquake with a magnitude of 2.3 falls into the bin [2, 2.5). Hence, bin [2, 2.5) needs to be updated by increasing the count number by 1, from 61 to 62.

Exercise 2.2.1 You are given the following data set:

$$\{100, 97, 114, 95, 100, 86, 104, 113, 102, 125, 109, 106, 105, 106, 99\}$$

a. Construct a histogram for the data set using the provided bins (thick lines are the bin limits).

b. The data are length measurements in mm. Correctly label both axes.

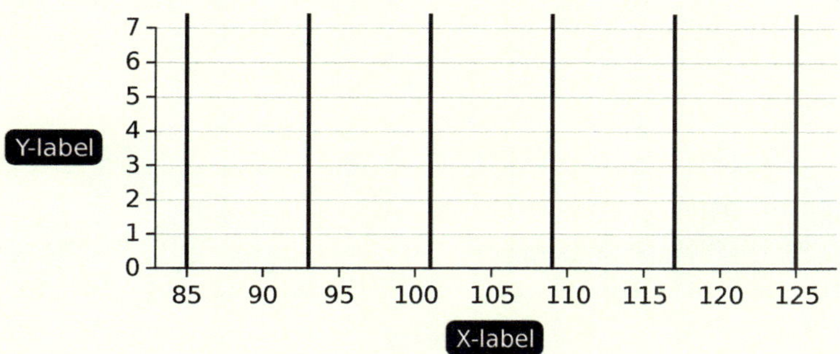

Exercise 2.2.2 The following four histograms are based on the recorded time intervals between eruptions of the Old Faithful geyser, with different bin widths.

a. Which would you choose to represent the data and why? What does the bin rule say? Do you agree?

b. Based on your choice, describe the data set.

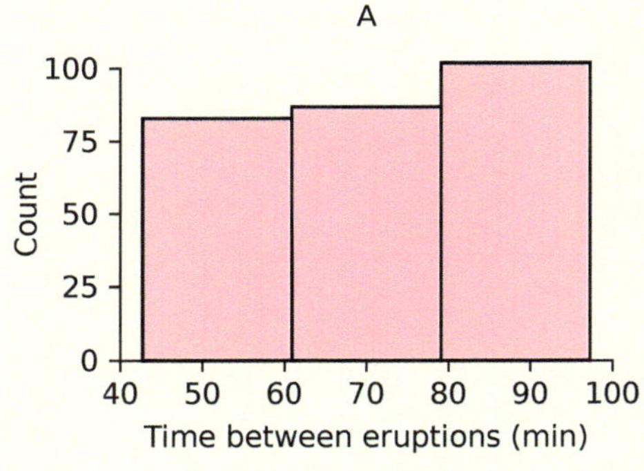

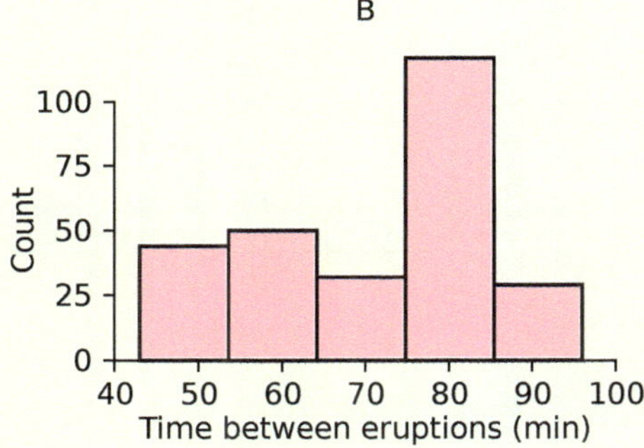

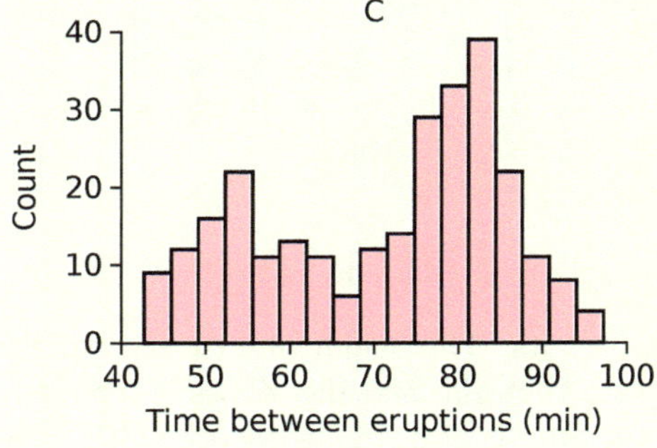

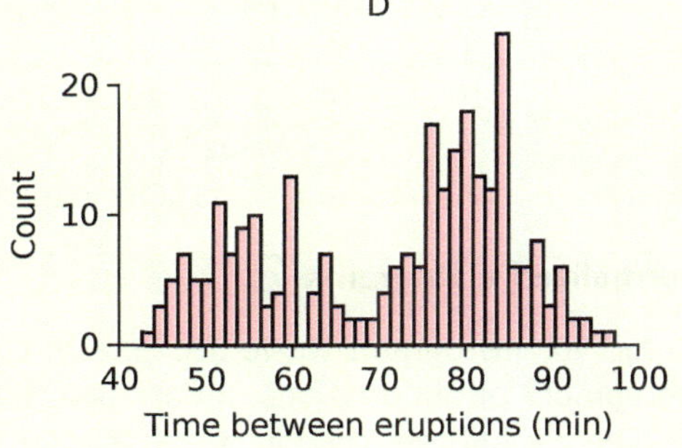

Exercise 2.2.3 Which histogram is right-skewed? Which histogram is bimodal?

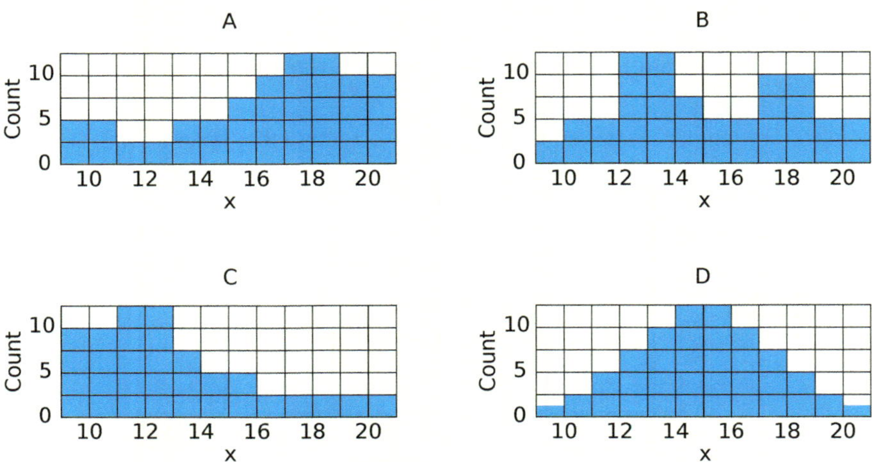

Exercise 2.2.4 Which histogram matches the beeswarm plot?

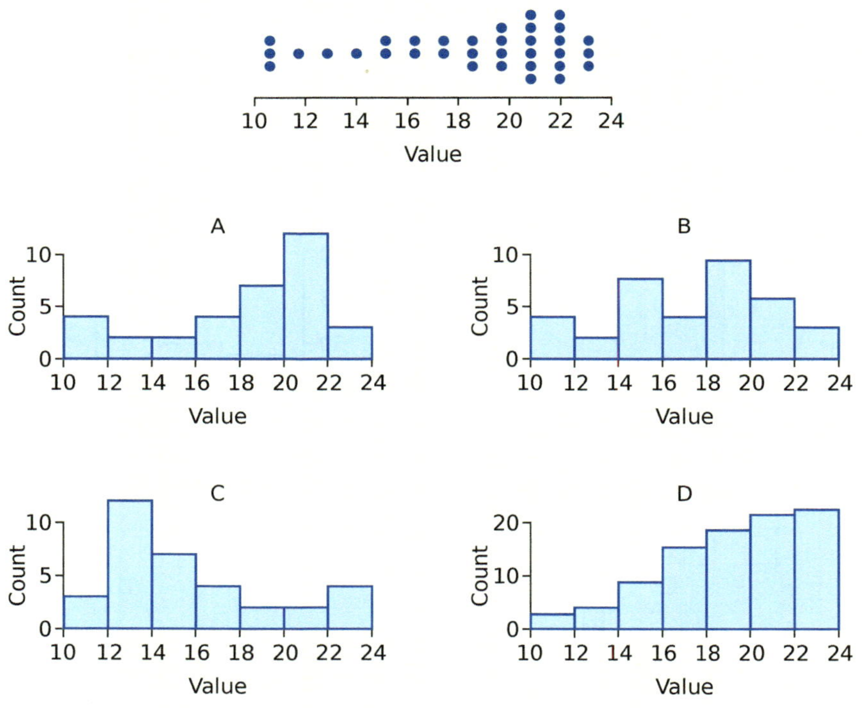

Normalized histograms

So far, the histograms we've been making are of raw counts. The height of each bin represents the number of observations or data points that fall into that bin, and the Y-axis is "counts".

There is another variant of this standard histogram, that is also frequently used. It's called a

normalized histogram. In a normalized histogram, the height of each bin represents the fraction of the total observations that fall into that bin. The Y-axis is therefore "fraction of total counts". Note that the shape of the histogram doesn't change, just the units on the Y-axis (Figure 2.13).

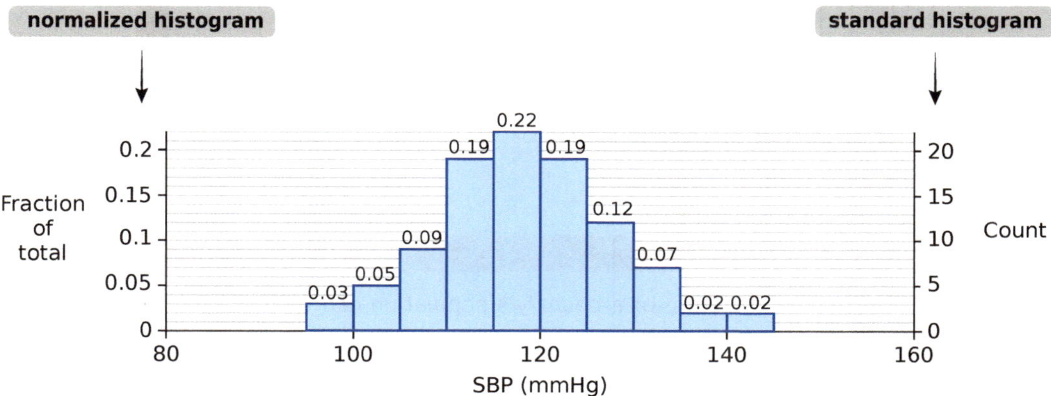

Figure 2.13 Turning a standard histogram into a normalized histogram.

Since the bin heights represent fractions of the total observations, it follows that the sum of all the heights must equal to 1.

Example 5 Distribution of generations: comparing the US and the UK

Here are two histograms showing the age distributions of the US and the UK in 2020, with a bin size of 10 years. By looking at the histograms, we can see that the population distributions by age of the two countries are very similar, even though there are almost five times as many Americans as Britons. Standard histograms like those show the raw count of people fall into each bin. For example, we can read out the number of 70 year olds in the US (24.7 million), and the number of 70 year olds in the UK (5.8 million).

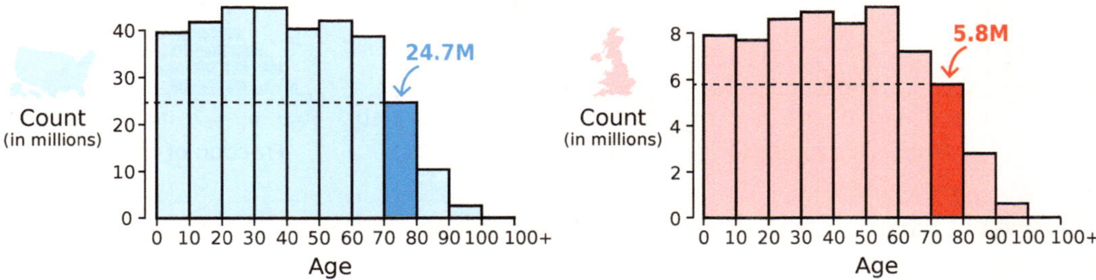

Question Which country has relatively more 70 year olds?

Answer This question is hard to answer from the histogram of counts. It is best answered by constructing a normalized histogram, in which we show not the numbers in each age group, but the fractions of the total population that each age group claims.

From the normalized histogram, we easily read off that that people in their 70s constitute 7.5% of the US population and 8.6% of the population in the UK.

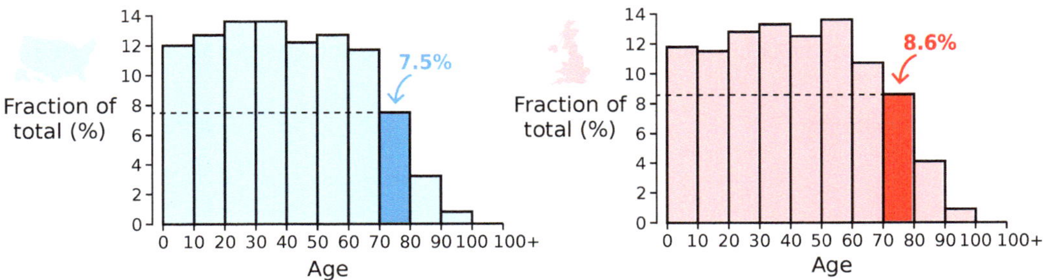

Example 6 Population pyramid

The age and gender structure of a country's population can be illustrated graphically by a population pyramid. A population pyramid is essentially two vertical histograms, the male and female age distributions, put side by side with a shared Y-axis. In a vertical histogram, the horizontal and vertical axes are switched; the X-axis is the count or fraction of total, and the Y-axis is the age. The left side of the pyramid is usually the male age distribution and the right side is the female. The shape of a country's population pyramid tells a story about the history of its population growth.

Here are three population pyramids A, B and C for three different societies.

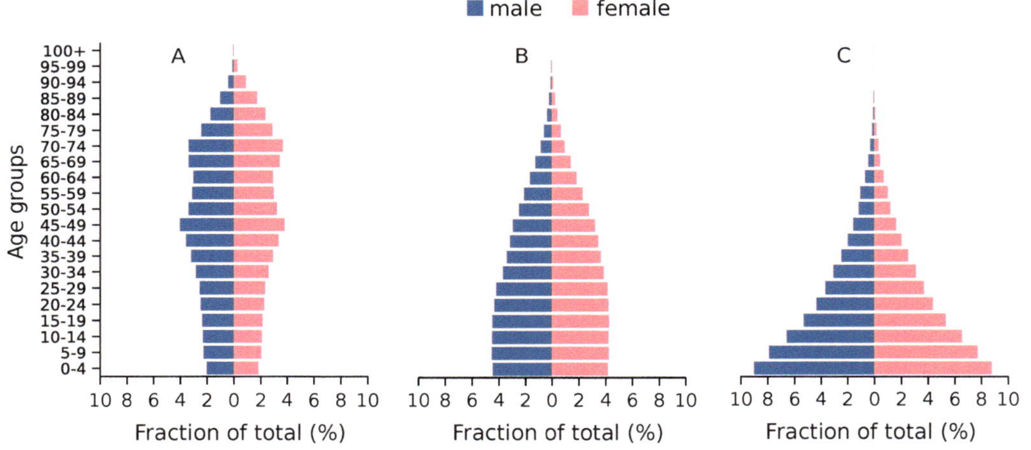

Question Comment on the shape of each distribution, and what does each one tell you about its demographical trend?

Answer Judging from the shape of each population pyramid, we can see that:

Society A has roughly equal male and female populations. It is mostly composed of middle-age people, with almost equal percentages across all ages. The bottom is getting narrower and narrower, meaning there are fewer and fewer children born every year than the year before, and therefore lower and lower fertility rate. It tapers off gently towards the top, meaning people live to a long age, and the society's mortality rate is very low as well.

Society B also has roughly equal male and female populations. Its population pyramid is wide at the bottom and narrow at the top. It is largely composed of young people, teenagers, and babies. The fertility rate is high and the mortality rate is high as well. There are more and more children born every year than the year before. People die early in this society.

Society C also has roughly equal male and female populations. Its age structure graph is extremely wide at the bottom, showing that its population is mostly composed of young people, teenagers, and babies. For example, by eyeballing, we can see that the under 5 year olds in Society C are about 9 percent of the total population. The population pyramid gets wider and wider in each younger generation means that every year more children have been born than the year before. The population pyramid tapers sharply at the top, much faster than both A and B, implying that people die early and few can make to middle age. In Society C, both the fertility and mortality rates are extremely high.

Cumulative histograms

Since the histogram can be very sensitive to the choice of bin width, a more robust presentation is sometimes used: the **cumulative histogram**, in which each bin is given a height which is not the number of data points in that bin, but rather the number of data points in all bins less than or equal to it.

Thus it is keeping a cumulative sum, and the height of the rightmost bin will be equal to the total number of data points (Figure 2.14).

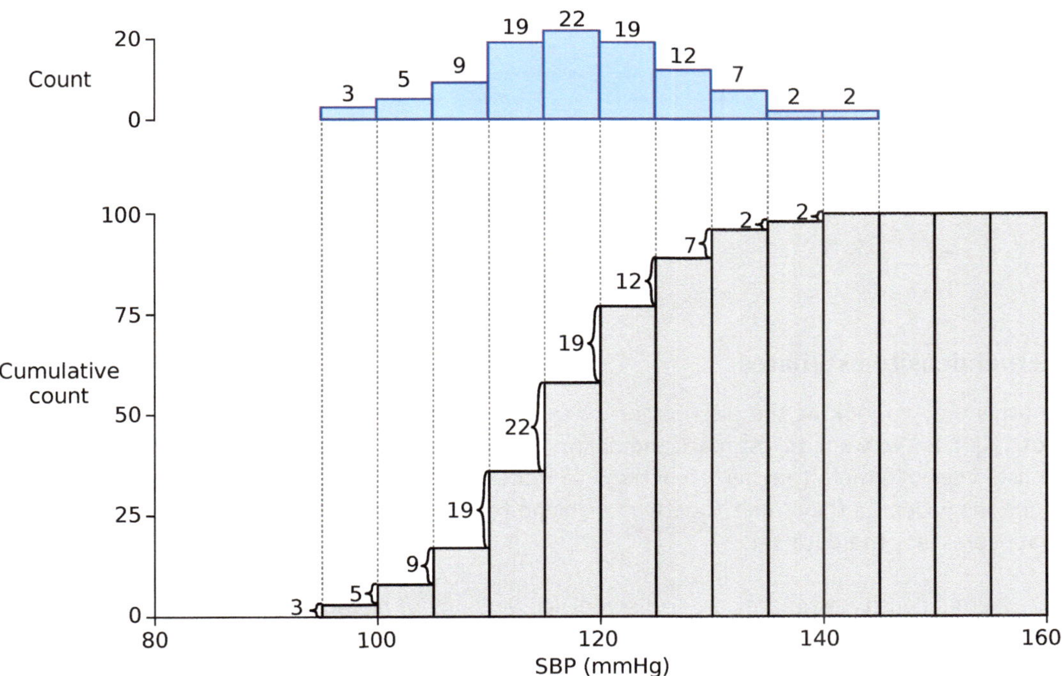

Figure 2.14 Turning a standard histogram (upper) into a cumulative histogram (lower).

Cumulative normalized histograms

Just as we can make cumulative histograms of counts, we can also make cumulative histograms of normalized fractions. In a **cumulative normalized histogram**, each bin is given a height which corresponds to the fraction of the data values in all bins less than or equal to it. Thus it keeps a cumulative sum of normalized fractions, and the height of the rightmost bin will be equal to 1 (or 100%).

Exercise 2.2.5 Here is a histogram of waiting times of 800 calls to a hospital in a week.

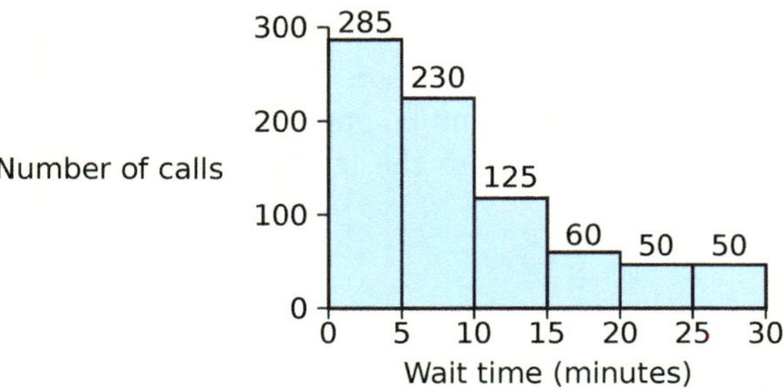

a. Make a cumulative histogram using the left Y-axis as the cumulative total. Fill the axis ticks as needed.

b. Make a cumulative normalized histogram using the right Y-axis as the cumulative total. Fill the axis ticks as needed.

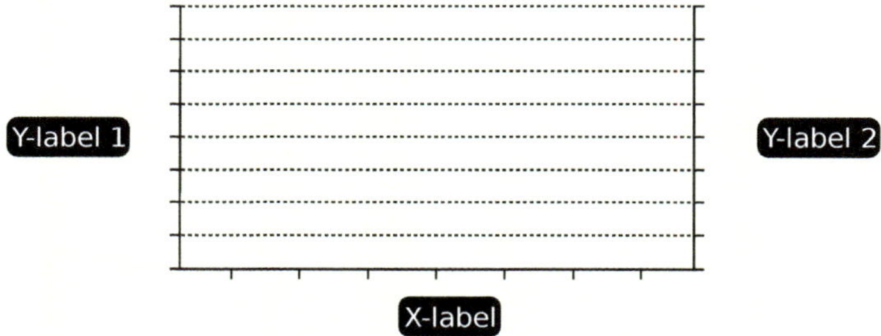

Kernel density estimates

Another way to look at the distribution of the data is to make what is called a **kernel density plot** (KDE). We want to estimate the shape of the distribution, and a good way to do that is to use a set of simple ("kernel") functions to represent each data point, and then let the kernel functions widen so they "melt together" or blend to make a picture of a continuous distribution that represents the data set.

Kernel density estimates are like smoothed versions of histograms.

There are several different kernel functions we could choose: they could be little triangles, little rectangles, or little smooth curves like the "bell curve" $Y = e^{-X^2/\sigma}$. The key is that each of these functions has a parameter that controls its width. In the case of little rectangles or triangles, the parameter is the width of the base, and in the case of the bell curves, the parameter is σ: the bigger the σ, the wider the curve (Figure 2.15).

To form a kernel density estimate, we start with narrow kernel functions, so narrow that each kernel surrounds one data value. Then we let the width get slowly bigger and bigger until the many little curves have merged into a single smooth curve. As σ gets bigger, the narrower kernels 'melt' into a smoother function. This is called a **kernel density estimate** of the distribution.

Here we will use bell curves as our kernel functions (Figure 2.15).

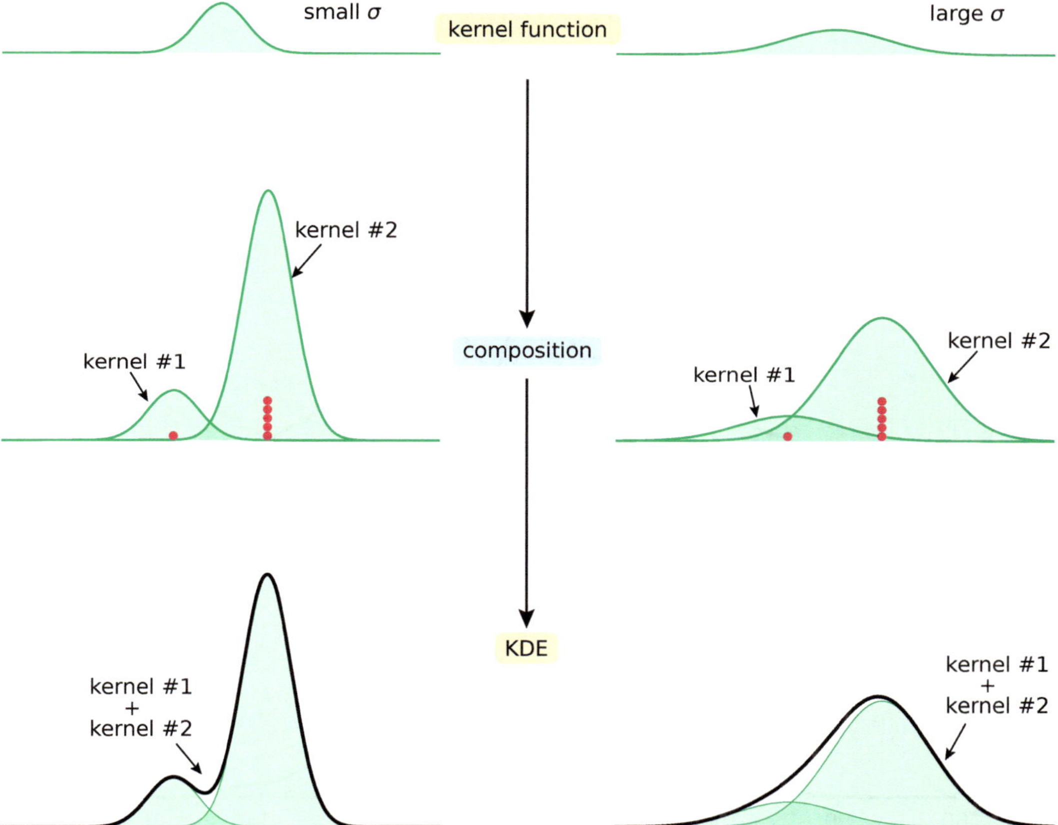

Figure 2.15 Two kernel density estimates of a data set (red dots, $n = 6$). Left: a smaller σ gives a narrow kernel function. Right: a larger σ gives a wider kernel function. Upper row: kernel function shape. Middle row: separate kernel functions. Lower row: sum of kernel functions gives the kernel density estimate (black line).

If we carry out kernel density estimates for the $n = 100$ data set, we see that the shape of the overall estimate depends strongly on the choice of kernel width (Figure 2.16).

For $\sigma = 0.4$, the kernel functions are so narrow that even when added together, they hardly overlap, and so produce a sum that has many jagged peaks that carry little information. A similar statement can be made about $\sigma = 1$. Even $\sigma = 2$ has a waviness to it whose information value is problematic: is it truly a property of the distribution, or is this because any sample of 100 points and $\sigma = 2$ will have some waviness to it? Only for $\sigma = 3$ do we begin to see a smoother distribution.

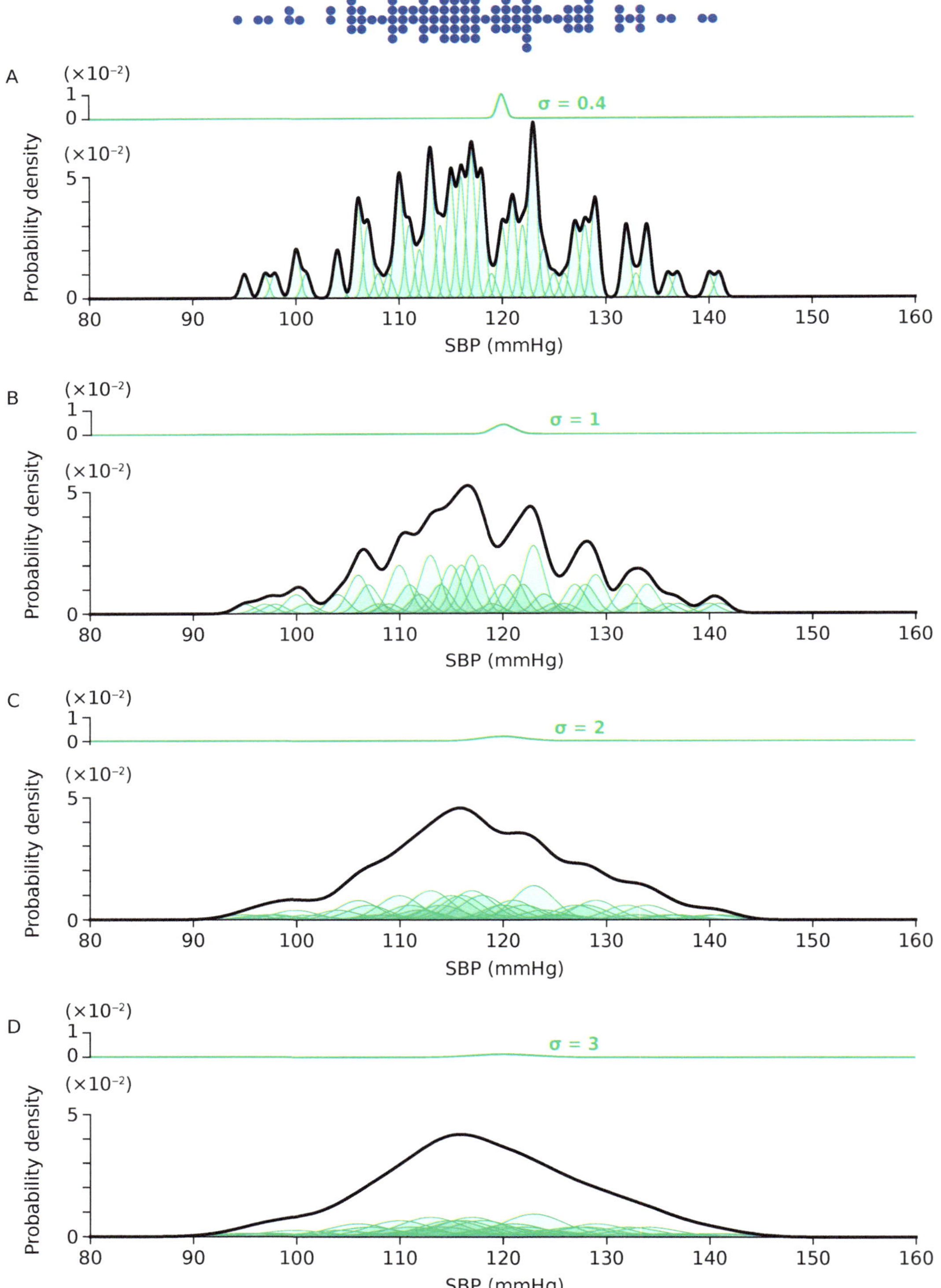

Figure 2.16 Kernel density estimates (black lines) for the same data set (blue dots), using four different widths for the kernel function.

Exercise 2.2.6 Consider the following data set

The left column shows four kernel functions with four different choices of σ. The right column shows kernel density estimates for the same data set (red dots on the horizontal axis). All 8 plots have the same scale.

a. Rank σ_A, σ_B, σ_C and σ_D.

b. Match the kernel function with the corresponding kernel density estimate.

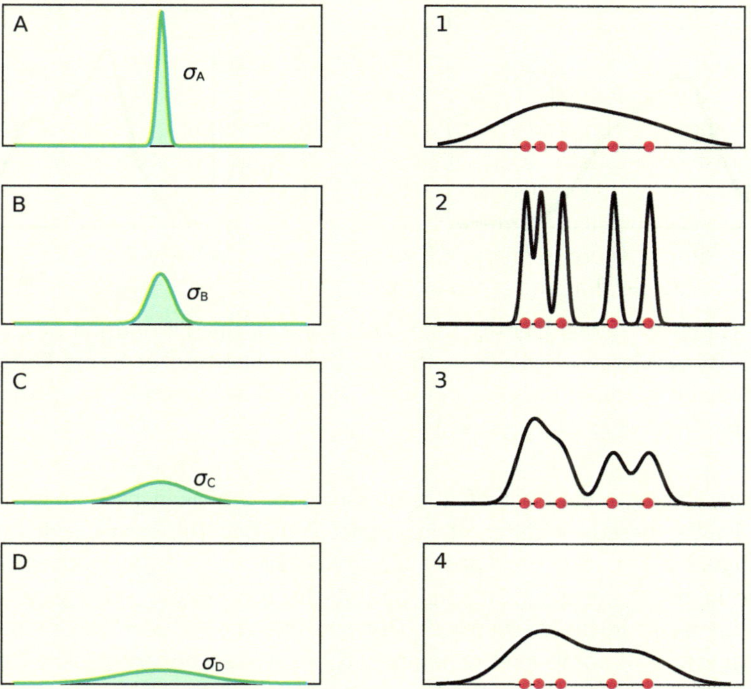

Obviously, the choice of kernel width is crucial to the appearance of a KDE; a very narrow kernel width will just give us many little bumps, one around each data point. Not very helpful! At the other extreme, choosing too large a kernel width will just give us a big bell curve, regardless of the original distribution. This is also misleading.

So how to choose a kernel width? There are many technical research papers debating different strategies for choosing kernel width. We will use the strategy that is built into the Python library `scipy`, called `gaussian_kde`. But please note that, according to scipy.org:

> "`gaussian_kde` ... includes automatic kernel width determination. The estimation works best for a unimodal distribution; bimodal or multi-modal distributions tend to be oversmoothed."

As a glimpse into the problem of KDE oversmoothing, consider the interesting example in Jones *et al.* (1996) "A brief survey of bandwidth selection for density estimation".

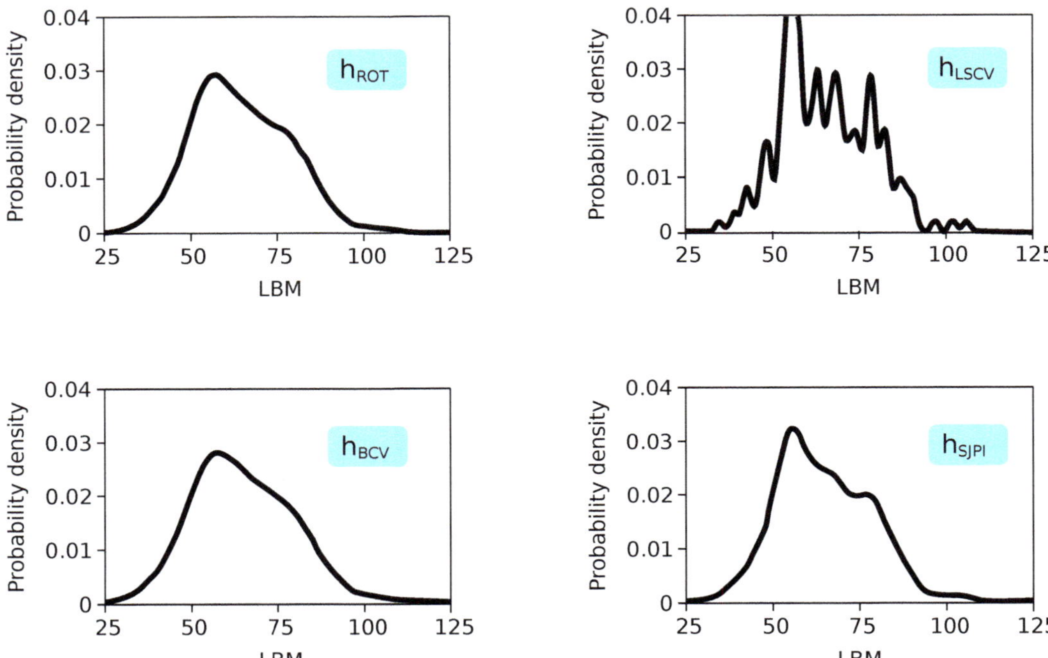

Figure 2.17 Gaussian Kernel Density Estimates showing the distribution of $n = 202$ measurements of Lean Body Mass (LBM) from the Australian Institute of Sport Data, using various data-driven bandwidths: h_{ROT}, h_{LSCV}, h_{BCV} and h_{SJPI}. The strongest suggestion of bimodality is given by h_{SJPI}. See Jones *et al.* (1996).

In their study, a single data set of 202 measurements of lean body mass is rendered into four different KDEs, based on different ways of calculating the kernel width (also called the bandwidth) (Figure 2.17). See their paper for the details of how the methods differ. The two on the left are heavily smoothed, and the one at the top right is very spiky, indicating what could be spurious peaks in the distribution. But consider the KDE on the bottom right. There is a strong suggestion of some kind of bimodality, with a secondary peak on the right. As it turns out, the authors tell us, this is the best presentation of the data, because the actual data is not a single distribution, but rather, is *two* distributions, one for females and one for males, superimposed on each other (Figure 2.18).

Since the `scipy` KDE function tends to oversmooth, we should definitely experiment with some hand-set values of the kernel width.

As we experiment, we have to be very careful about not getting too excited by bumps in the KDE until we have satisfied ourselves that they are meaningful, and not just artifacts of the smoothing procedure.

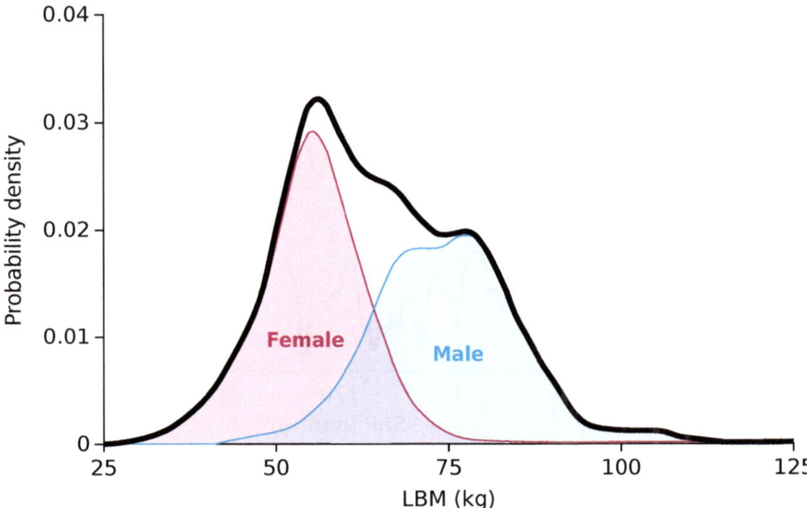

Figure 2.18 Kernel Density Estimates using h_{SJPI} (for full population), based on subpopulations of n = 102 females (red line) and n = 100 Males (blue line). The black line represents the combined data set. This verifies that the bimodal structure in Figure 2.17 was an important feature of the data worth deeper investigation.

Example 7 Vaccination rates in the US

In Example 3 on page 16, we looked at a dot plot of US state-by-state vaccination rates. There was an impression that they fell into two subgroups. This impression is now confirmed by a KDE, which is clearly bimodal.

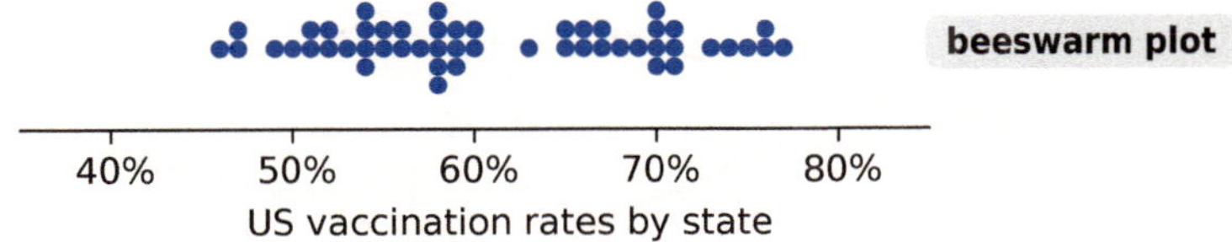

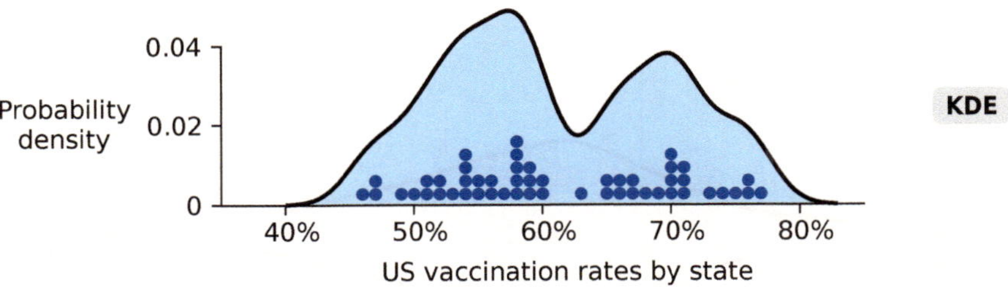

Violin plot Finally, another version of the kernel density plot that is widely used is the **violin plot**. To make a violin plot, take a kernel density estimate, form its mirror image, and then glue the two together along the flat sides to make a symmetric figure (Figure 2.19). Violin plots help to visualize the distribution of the data.

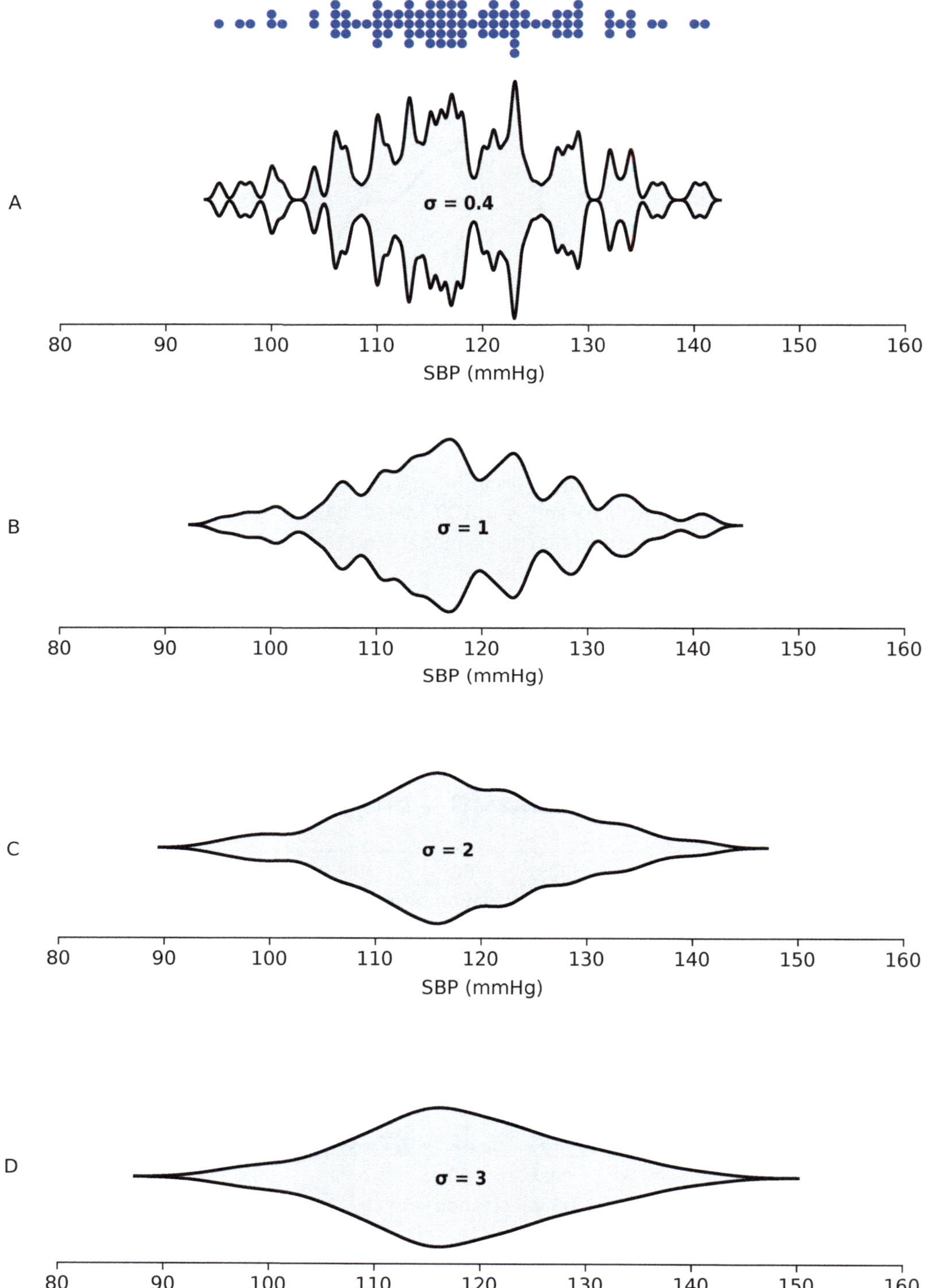

Figure 2.19 Violin plots for the same data set (top blue dots), using four different widths for the kernel function.

Violin plots are especially useful when we need to compare two or more groups, represented by different distributions, a subject that will be discussed in Chapter 6 *Comparing Two Groups* and Chapter 7 *Comparing Three or More Groups*.

As a sneak peek into group comparisons, consider three groups, which we will call A, B and C. They all have the same mean value, namely, 10. But they are very different. The differences become clear when we present them as three violin plots side-by-side (Figure 2.20).

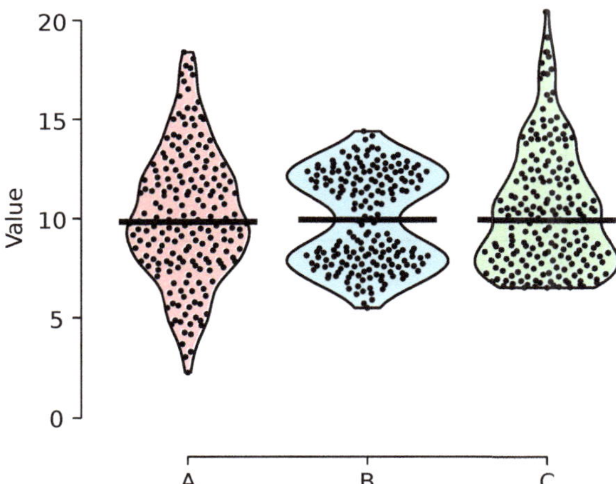

Figure 2.20 Violin plot presentation of three groups side by side.

KDE surfaces in 3D So far, we have been thinking about a single data set $\{x_1, x_2, \ldots, x_n\}$. If we visualize these data as points in X-space, we can then put kernel functions around those points on the X-axis and build up our KDE curve. But what if we have not just one measurement X of the subjects 1, 2, ..., n, but another one Y, so that we now have a data set which is n pairs of (X-value, Y-value) (Figure 2.21).

subject i	x_i	y_i
1	x_1	y_1
2	x_2	y_2
3	x_3	y_3
⋮	⋮	⋮
n	x_n	y_n

Figure 2.21 Bivariate data set presented in a tabular format.

Data sets with two values for each subject are called *bivariate* data sets. We will study some measures of them in Chapter 9 .

For now, we will try to visualize the distributions of these data sets with KDE functions. First of all, we have to realize that in 1D (univariate) data, our kernel functions were bell-shaped *curves*, now in 2D bivariate data sets, our kernel functions have to be bell-shaped *surfaces*.

These KDE surfaces can give us very insightful pictures of bivariate data sets. Let's take the data set shown in Figure 2.22.

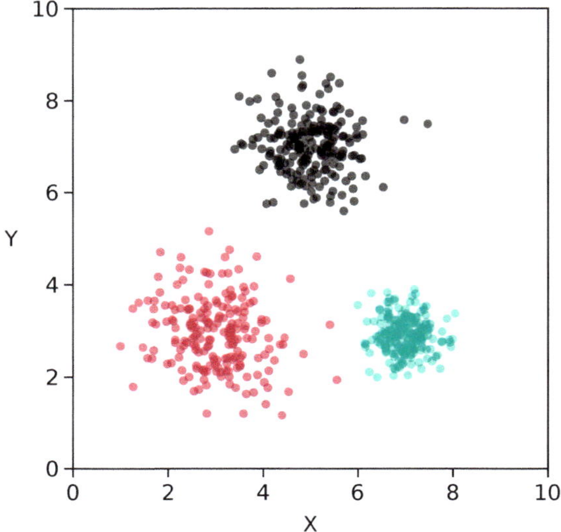

Figure 2.22 An artificial example of an (X, Y) data set that has three clusters.

Then applying `scipy`'s `gaussian_kde` to the bivariate data set gives us the KDE in 3D (Figure 2.23). The Z-axis is the probability density; the height of the KDE graph shows the probability density at any given (X, Y) pair. (Because this a probability density function, the total volume under the surface is 1.)

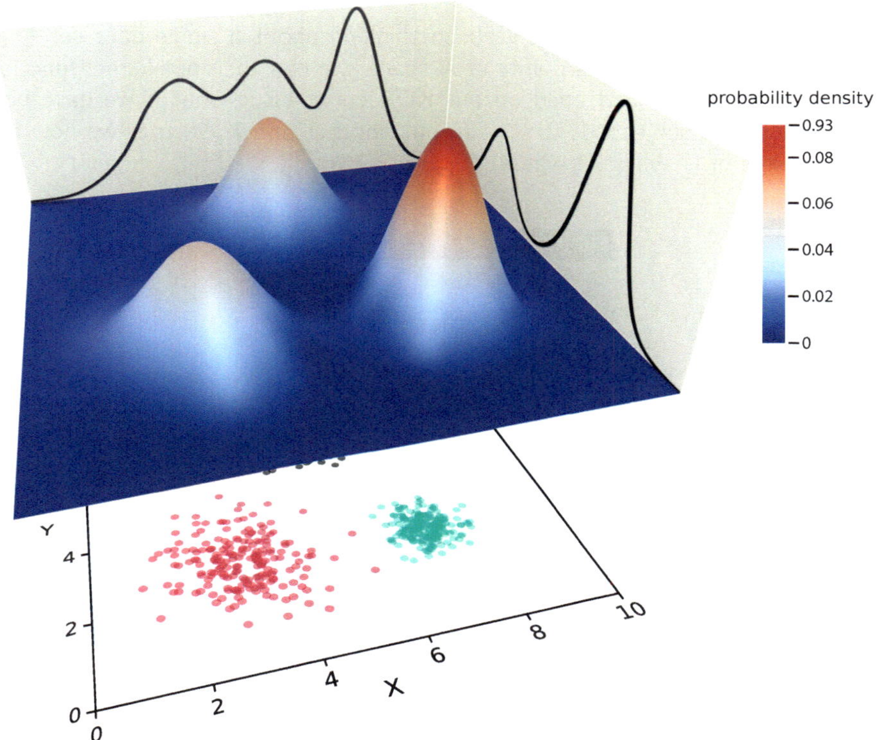

Figure 2.23 KDE plot of the data in Figure 2.22.

2D KDE using color instead of height Our first real world example is one in which the X- and Y-axes are measurements done on plants: the "Iris Dataset"[1], a collection of data originally presented in *The use of multiple measurements in taxonomic problems* (Fisher, 1936). It actually has four different measured values for each of 50 plants of three different species. Here we show a KDE estimate of just two of the measurements, for just two of the three species, shown in red and blue (Figure 2.24).

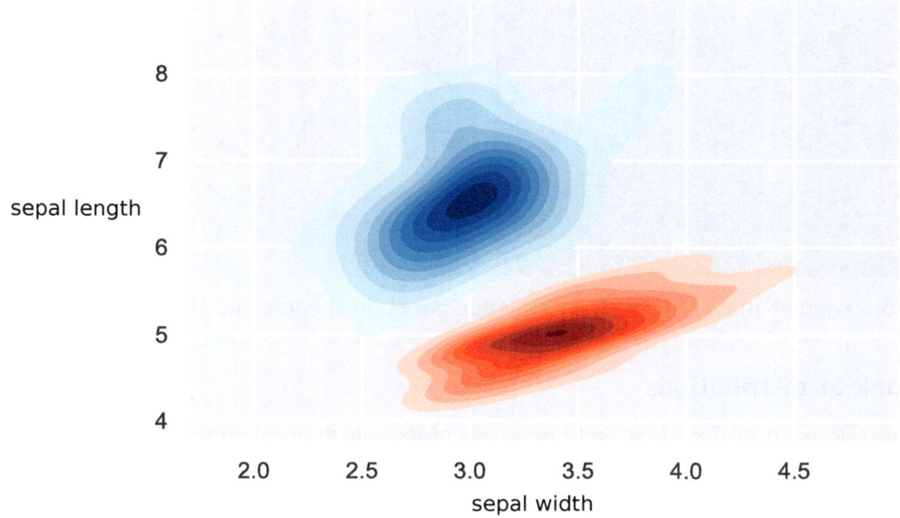

Figure 2.24 2D KDE of sepal width and sepal length from the Iris Dataset. Red and blue are two different species. Darker colors denote higher values.

Notice how we can distinguish the two species clearly using the bivariate descriptor, a task that would be much harder using only one measurement alone.

Another important example of 2D KDE plots is in the rapidly growing subject called **Geospatial Analysis**. The use of Geographic Information Systems (GIS) gives us many high-quality data sets, where the X and Y measurements are the longitude and latitude of particular points of interest.

We can then make a KDE of that data to show the density of whatever feature we can map, using color or height or both to show the density of the objects of interest. A nice example can be found in "A web-based geospatial toolkit for the monitoring of dengue fever" (*Applied Geography*, 2014). The authors created first a point map of locations of dengue cases (Figure 2.25 left), and then a KDE of the density of dengue cases, using color to show the third dimension. Notice an important fact: the KDE, on the right, very clearly shows the existence and location of "hotspots" of dengue, which can be important clues to the epidemiology of the disease.

A note to students: Two-dimensional KDE estimates are a highly advanced topic, something you might expect to see in a second or even third course in data science. We are introducing this very early because it's an important method of data visualization for bivariate data. You will now be in a position, when you go interview for that lab research position, to

[1] `https://archive.ics.uci.edu/ml/datasets/Iris`

say "Oh yes, I can do bivariate kernel density estimates in Python". This is impressive.

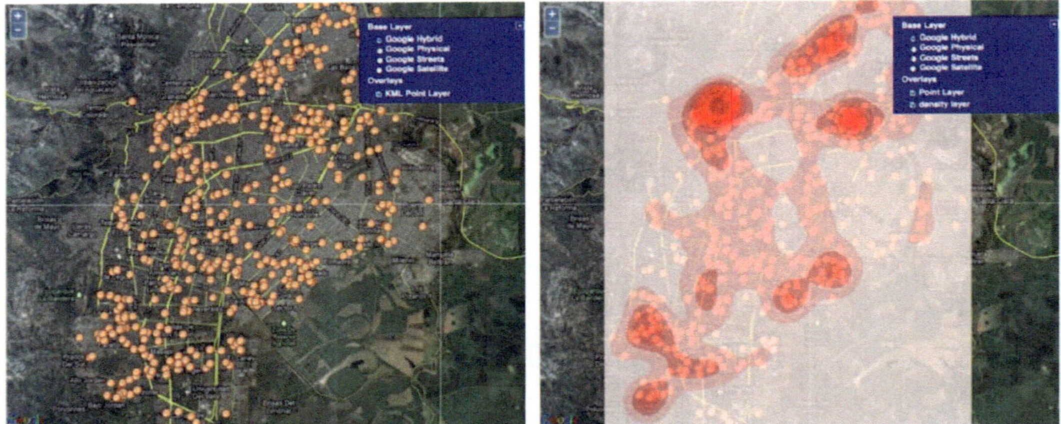

Figure 2.25 A point map of locations of dengue cases on the left and the KDE on the right.

How to look at distributions

When we were looking at dot plots and beeswarm plots, which are directly plotting all the data, we noted a number features to look for in those pictures:

- Skewness vs. symmetry
- Thick tails vs thin tails
- Outliers
- Subgroups

Now that we are looking at pictures of the distribution, whether histograms or kernel density estimates, the same features are important in these distribution plots:

- **Skewness vs. symmetry:** can still be seen in the distribution plots.
- **Thick tails vs. thin tails:** can also still be seen in the distribution plots.
- **Outliers:** Outliers make problems for distribution plots. If the distribution estimate includes the outlying data points, it may distort the shape of the plot. On the other hand, it is not correct to just reject the outlier point or points because they complicate the plot.
- **Subgroups:** We talked about the importance of recognizing subgroups when we looked at a dot plot. Similarly, when we look at a distribution plot like a histogram or Kernel Density Estimate, we should look for subgroups, now in the form of somewhat distinct and separate peaks. For example, in the middle data set in Figure 2.20, the violin plot shows two distinct peaks. These peaks are also called *modes*, and so this distribution would be called *bimodal*. Other distributions may have multiple peaks; these distributions are called *multi-modal distributions*.

Defining bell-shaped Please note that in this entire section, we have been speaking only in descriptive terms, based on our visual impressions of the data distribution. No numbers have been invoked or used. All talk has been in terms of shapes: symmetry, size of tails, peaks, etc.

Now we are ready to go on, and talk about putting summary numbers on distributions. However, there is one, all-important definition we need to make. We have already, in several places, referred to certain curves as "bell-shaped", without further definition. We will now make a semi-rigorous definition of bell-shaped: we will say that a distribution is **bell-shaped** if it is

- Symmetric
- Thin tailed
- Unimodal

As we will see, the choice of descriptive statistics is relatively easy when the data distribution is bell-shaped. However, we should never assume that a distribution is bell-shaped just because that would make it easy to analyze! We must always look at the data.

Spatial mapping: John Snow and the invention of epidemiology

The example of the dengue map is an excellent modern-technology throwback to the most famous spatial analysis map in medical history, which is the map made by John Snow of the spatial pattern of cholera in London (Snow, 1855). In 1854, the city of London was struck by an epidemic of cholera, an often fatal disease that is still around today. The prevailing theory about the cause of cholera was that it was caused by foul air ("miasma"). Snow was a doctor with good mathematical skills, and he was convinced that the "miasmatists" were wrong. So he went out and actually gathered data on people who had contracted cholera.

He had the idea of plotting where the cases were on a local map (Figure 2.26), and immediately realized that a lot of cases clustered near a public water source called the Broad Street pump. The Broad Street pump, as was the practice back then, took its water directly from the Thames River. In that period, the Thames was also used as a public sewage dump. Raw sewage, including untreated human excrement, was dumped directly into the Thames upriver of the pump.

Snow concluded that cholera was being transmitted by something in the foul water of the Thames downstream from the sewage plant. This was the first great victory of epidemiology, made possible by spatial analysis.

Figure 2.26 Left: section of John Snow's original 1855 map. Black squares ■ are cholera cases. Right: a modern day GIS rendering of the same data.

The introduction of data visualization: the story of Florence Nightingale

Most school children are taught about Florence Nightingale as a caring nurse. Very few people are aware that she played a pioneering role in the introduction of data visualization methods, as told in a recent *Scientific American* article (Figure 2.27).

AUGUST 1, 2022 | 4 MIN READ

How Florence Nightingale Changed Data Visualization Forever

The celebrated nurse improved public health through her groundbreaking use of graphic storytelling

BY RJ ANDREWS

Figure 2.27

While leading a medical expedition to the Crimean war in 1855 (Figure 2.28), Nightingale became concerned about the number of soldiers who were dying due to non-combatant deaths, especially preventable diseases due to poor sanitation.

Figure 2.28 Nightingale, on horseback, surveying a battlefield in Crimea.

She gathered extensive data, but immediately realized that her audience, which was army generals, politicians, and ultimately Queen Victoria, were not able to follow the data presentations that were standard at the time, namely, tedious tables (Figure 2.29 left). The politicians and senior officers accepted high death rates for soldiers as something inherent in soldiering. But Nightingale showed through an impressive visualization, that the overwhelming majority of soldier deaths were from preventable diseases, far outnumbering deaths due to battle wounds (Figure 2.29 right).

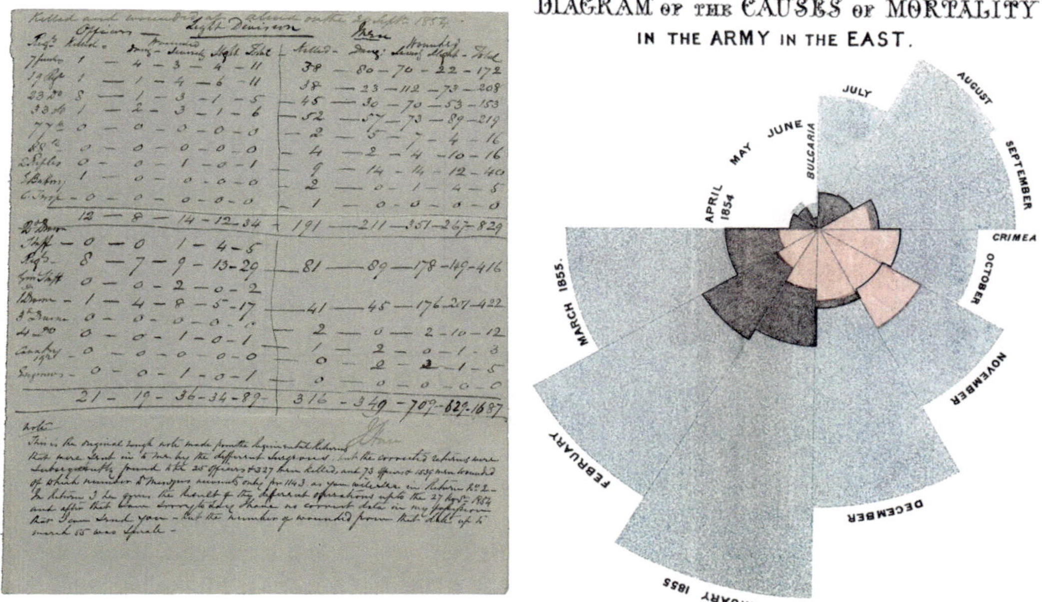

Figure 2.29 Left: tables of mortality data, typical at the time. Right: Nightingale's presentation of the data distribution by month. In each wedge, the pink area in the center represents battlefield-related deaths, the blue area represents deaths due to preventable diseases, and the gray area represents deaths due to other causes.

As a result of her efforts in using data visualization to facilitate storytelling, the article tells us that "mortality from preventable disease among soldiers declined to less than that in the comparable civilian population". Moreover, "the reforms Nightingale fought for were finally codified in the British Public Health Act of 1875".

FURTHER EXERCISES 2.2

1. Given a data set with 15 data points. They are height measurements in feet.

$$\{2.5, 8.9, 4.7, 11.1, 1.7, 6.3, 10.5, 0.3, 14.8, 13.2, 1.5, 2.9, 1.2, 7.4, 5.9\}$$

   a. Draw a histogram representing the data set. Here we choose bin width to be 3 and the bin limits are solid lines.

   b. Label both axes on the histogram.

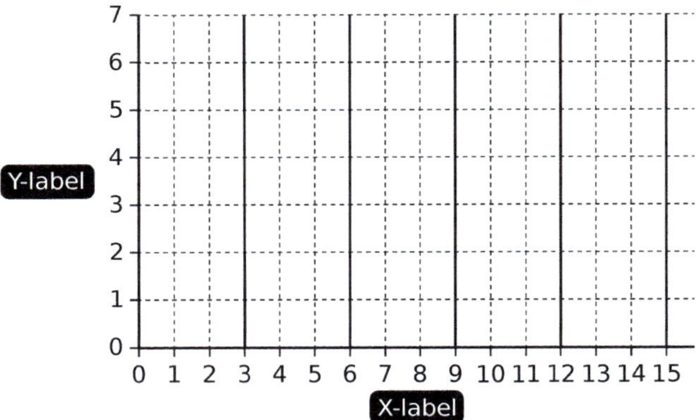

2. The age and gender structure of a country's population can be illustrated graphically by a population pyramid. Here are three population pyramids of three different societies A, B and C. For each population pyramid, the left histogram is the male population and the right is the female population. Comment on the shape of each distribution, and what you can learn about each society's demographical trends.

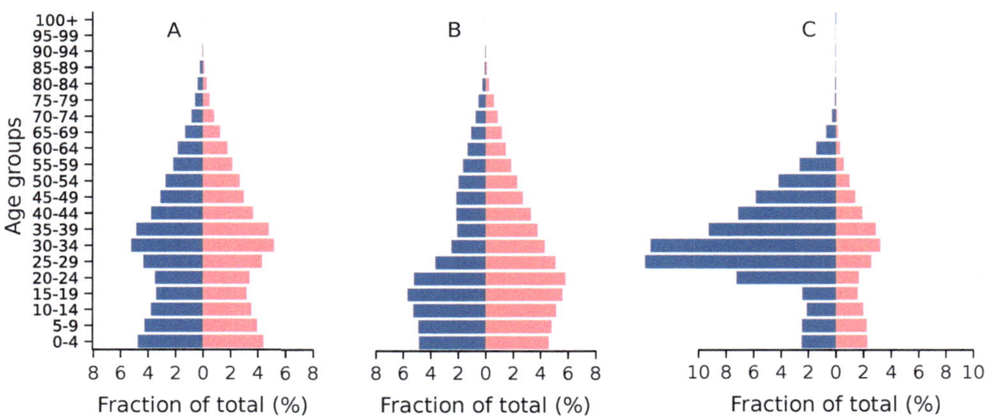

3. Figures in the left column show the daily new cases of Covid-19 in countries A, B, C, D and E from the beginning of the pandemic till August 2021. Figures in the right column show the total confirmed cases of those five countries during the same time period. Match each daily new cases graph on the left with the corresponding cumulative total cases graph

on the right. (The countries A, B, C, D, E are China, New Zealand, South Africa, India, Poland.)

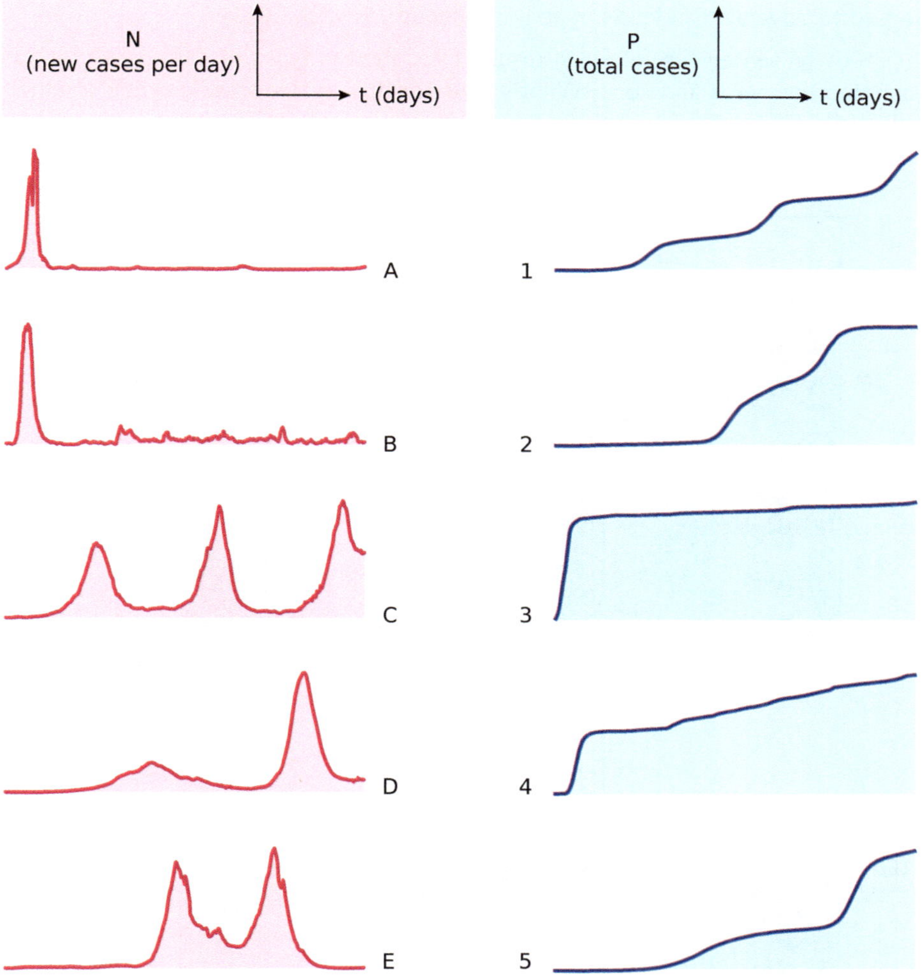

4. **Water supply and cholera epidemic.** During the London cholera epidemic in 1854, John Snow's investigation identified the following data between water supply companies and cholera occurrence.

   a. Based on the data, which company is more likely to be transmitting "morbid matter" and causing cholera?

   b. Choose a data visualization to present your finding.

Water Supply Company	Number of Houses	Cholera Deaths	Cholera Deaths per 10,000 Houses
Southwark and Vauxhall	40,046	1,263	315
Lambeth	26,107	98	38
Rest of London	256,423	1,422	55

5. **Lead exposure in early childhood.** Lead is a developmental neurotoxin. A study was carried out to understand the extent of the US population exposed in early life to high levels of lead, and the true social costs and benefits of lead regulation (see article "Half of US population exposed to adverse lead levels in early childhood", *PNAS*, 2021).

   The study presented the following result. Describe the types of data presentation in each graph and their pros and cons. What's your conclusion from reading their results?

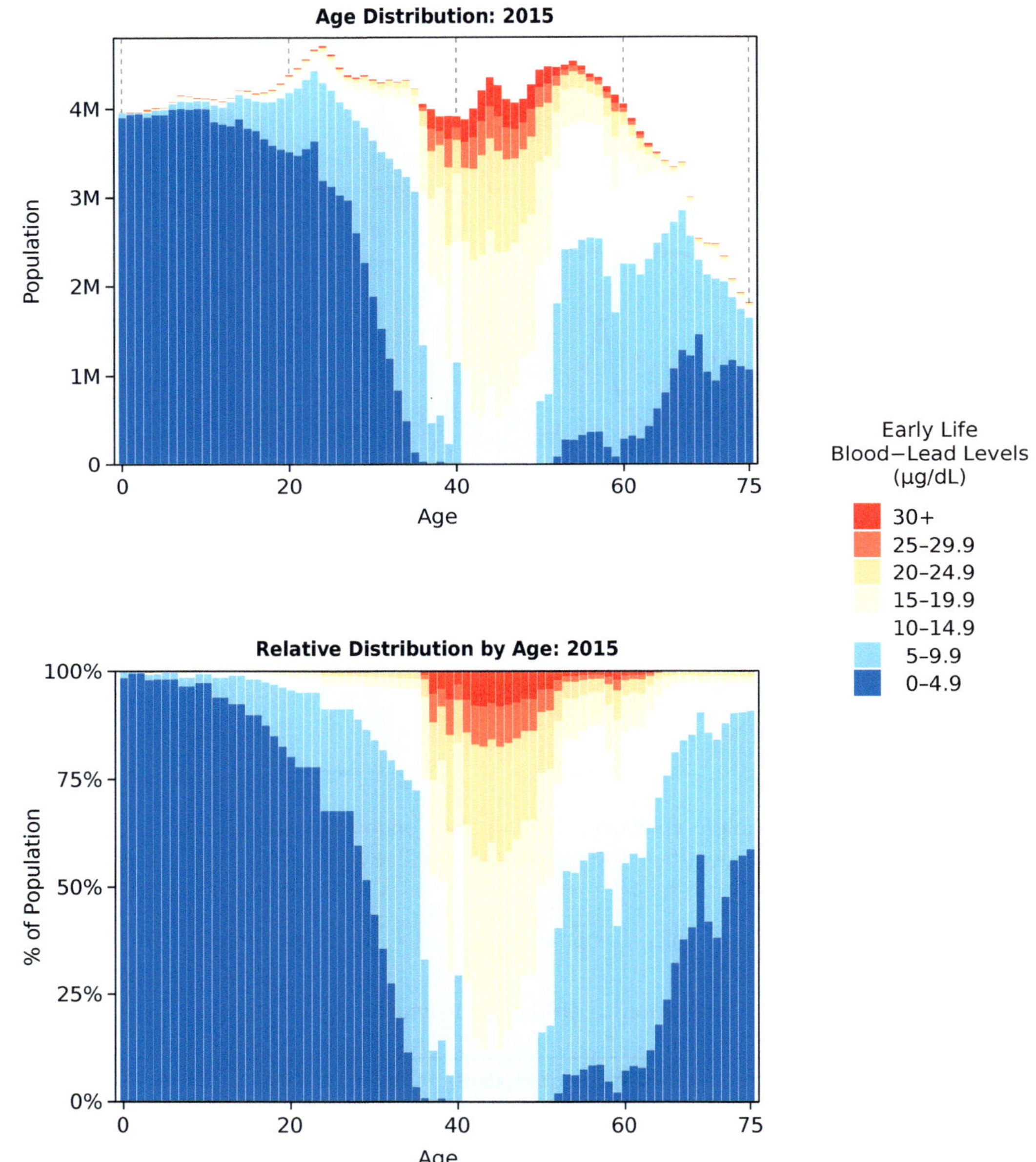

2.3 Summarizing a data set: "measures of central tendency"

Suppose we want to describe the whole data set, that is, the whole set of numbers, by one summary (or representative) number. There are a variety of reasons why we might want to do this. A single number helps us to characterize the group, and makes it easier to compare groups. *But we have to be very careful about using a single "typical" number to characterize a group.* There are many cases where it's a bad idea to describe the group with a particular single number, and other cases where it's a bad idea to use any single number at all!

The decision to describe the data set by a single number, and which number to use, is a scientific choice that must be justified by the shape of the data distribution and the purposes for which the number is needed.

The mean or average value

The most common choice of a typical number to describe a data set is the **average** or **mean** value of the data. Suppose X is a data set that has n data points x_1, x_2, ..., x_n, then

$$\overline{X} = \text{mean}\{x_1, x_2, \ldots, x_n\} = \frac{1}{n}(x_1 + x_2 + \cdots + x_n) = \frac{1}{n}\sum_{i=1}^{n} x_i$$

The mean is often written as μ or μ_X or $\overline{x_i}$, or $\overline{X}$. All four symbols mean the same thing.[2]

Exercise 2.3.1 What number would you divide by to calculate the mean of 3, 2, 4, and 5?

Exercise 2.3.2 The mean of four numbers is 15. If three of the numbers are 10, 15, and 15, what is the value of the fourth number?

Exercise 2.3.3 What is the mean of the following data sets?

data set A: 40, 40, 40, 40, 40

data set B: 0, 20, 40, 60, 80

data set C: 0, 0, 0, 0, 200

data set D: −10000, −10000, −10000, −10000, 40200

Exercise 2.3.4 Invent a data set ($n = 5$) whose average is 40.

The use of the average value is so widespread that it is done almost as a reflex. But it is important to realize that **there are many kinds of data distribution for which the average or mean is a poor choice for a "typical value".**

Recall the example of the tax refunds, from the beginning of this chapter. We saw the word 'average', and got conned into believing that the "average" must be "typical". But this is *only* true for bell-shaped distributions, which the distribution of tax refunds is most definitely not. (Why not?)

---

[2] **The mean of the data and the "population mean".** We are using the terms μ_X and $\overline{X}$ to mean the same thing, namely, the average value of the sample $X = (x_1, x_2, \ldots, x_n)$. In some statistics texts, these mean different things; $\overline{X}$ is the sample average, and μ_X is the "true" mean of the underlying distribution from which X was drawn. We do not make this distinction because, typically, we have no idea what "true" distribution the data are drawn from. We have only our data.

Most distributions are not Normal

Far from the standard belief that "most distributions are bell-shaped", the fact is that most distributions of biological and social quantities are *not* bell-shaped. In a review paper by (Bono *et al.*, 2017), called "Non-normal Distributions Commonly Used in Health, Education, and Social Sciences: A Systematic Review", the authors began: "The data obtained in many fields of health, education, and the social sciences . . . clearly deviate from those of the normal distribution." They described one study that looked at 693 distributions from real data and found that "most distributions were non-normal", indeed, "only 5.5% of the distributions were close to expected values under normality".

We frequently encounter distribution shapes for which the mean is a poor descriptor. Here are some warning signs. If you see them, carefully consider whether to use a mean as a descriptor.

Danger: bimodal data When data are clearly separated into two subgroups, the choice of the mean value to represent them is a poor choice, since the mean value will be typical of no one, and therefore mis-represents the data. Suppose, for example, we are making a graph of the distribution of political opinions in some group. It turns out that this particular group is heavily polarized, so there are lots of people on the Left and lots of people on the Right. Does it make scientific sense to describe this group by saying "the average opinion is Centrist"?

There are many cases where the data are *bimodal* (two peaks), and it is generally a bad idea to use the average to describe those groups. For example, the disease Hodgkin's lymphoma tends to strike two different age groups: people in their 20s and people in their 70s (Figure 2.30). It would be a poor idea to say "the average age of lymphoma onset is about 45".

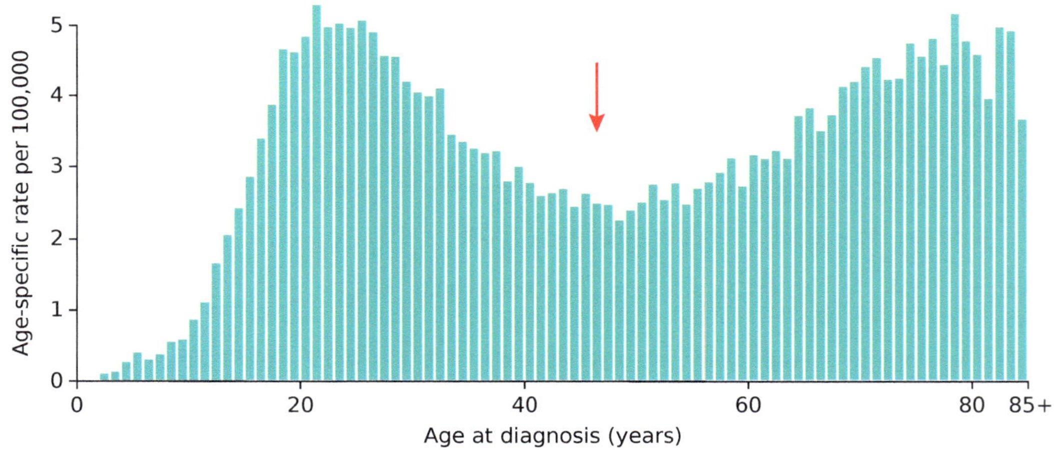

Figure 2.30 Age-specific rates of Hodgkin's lymphoma. Data obtained from the Surveillance, Epidemiology, and End Results Program Research Data (1973-2014).

For another example, freeway use depends on time of day: there are many users in the morning and evening rush hours, making two peaks in the graph of "number of cars" vs. "time of day" (Figure 2.31).

Would it make sense to say "The average time of freeway use is 1 pm"? This would be very misleading. Indeed, any summary number would be un-helpful; we would do better to note the number and magnitude and location of peaks, or lack thereof.

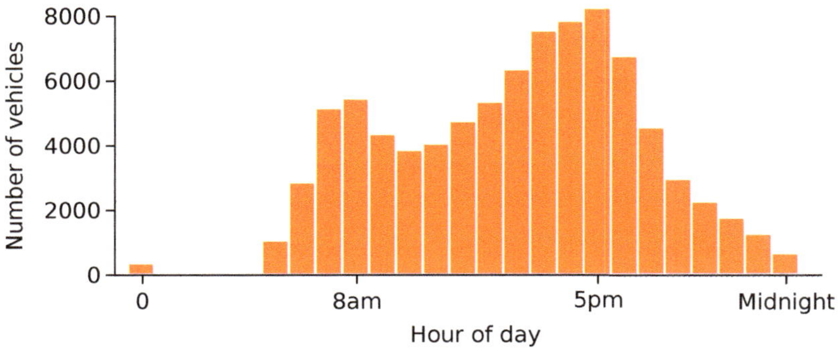

Figure 2.31 Histogram of number of vehicles at any given time of day.

For another example, consider a histogram of traffic speeds of vehicles driven on a major highway in a day (Figure 2.32). The speed limit at this location is 60 miles/h.

However, note the significant number of vehicles traveling at a speed below 30 miles/h, indicating congestion!

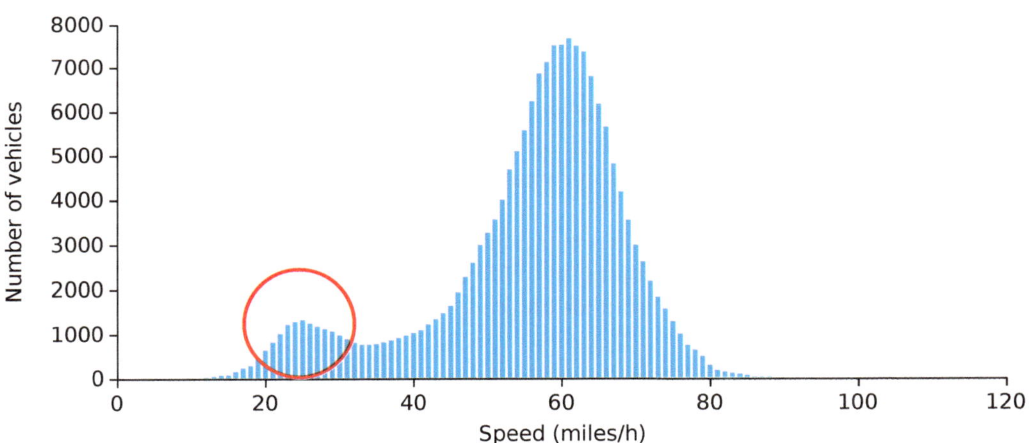

Figure 2.32 Histogram of traffic speed of vehicles driven on a major highway in a day. The speed limit at the location is 60 miles/h.

The histogram gives us a valuable information about the data. We should not rush to summarize histograms by single numbers! Stare at the histograms, they are telling you important things.

Danger: skewed data When the data are skewed to the left or right, it is also a bad idea to use the average value to summarize it. This is because, in the process of averaging, the more extreme data values, out in the tail, will tend to pull the average out. Therefore, in a skewed distribution, the mean tends to lie not in the main body of the data, but is pulled out toward the tail.

This can be seen in the distribution of US household incomes (Figure 2.33). It is strongly skewed to the right. As a consequence, the mean household income, $88,439, is well into the top 50% of the distribution, that is, well above the median. Clearly, it would be misleading to present the mean income to describe this data set. To say "the average US household income is around $90,000" presents a very different picture of the US economy than to say "Half of US households earn less than $67,000".

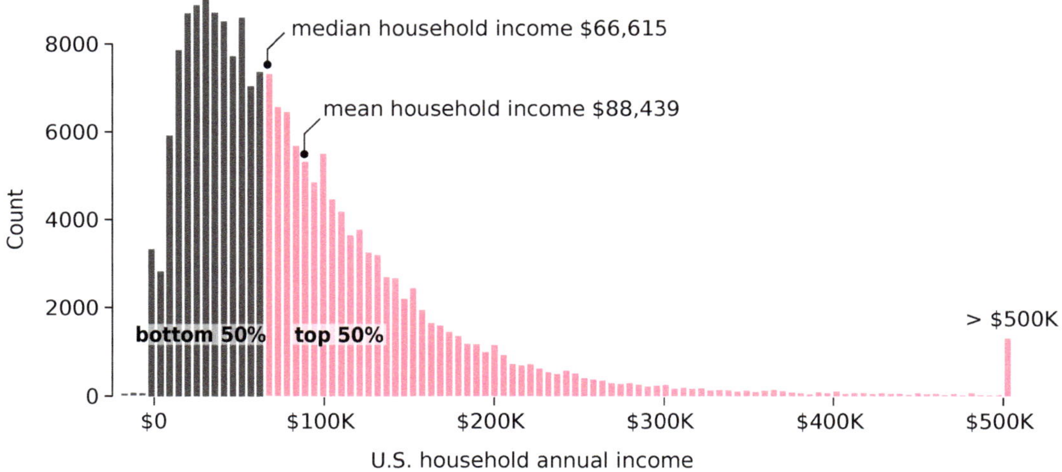

Figure 2.33 Distribution of US Household incomes in 2015.

Danger: outliers This is especially true of the extreme form of skew that is seen when there are outliers. If a billionaire happens to be sitting in a classroom of 100 typical students, then can we really say that the average personal wealth in that classroom is $10 million? This is not a useful or representative number.

For example, consider the distribution of life expectancies across the 44 countries in the Organisation for Economic Co-operation and Development (OECD). It is obviously skewed to the left, and, in addition there is an outlier on the low end, namely South Africa (Figure 2.34).

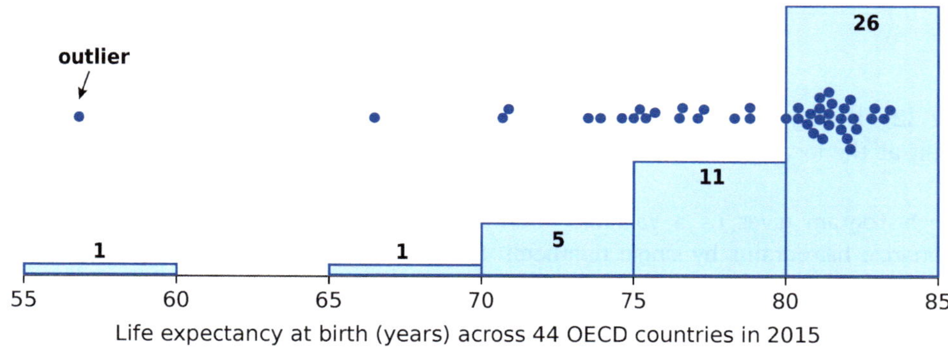

Figure 2.34

Clearly, it would be a poor idea to construct an "average life expectancy" of all the OECD countries, since the South Africa data point will strongly influence that average, and pull it down. But it is also important that the simple use of a single number also does not do justice to the case of South Africa or India (the second most extreme outlier).

When we dive deeper and look at life expectancy in South Africa, we are immediately struck by another fact: the very idea of "life expectancy" is itself questionable! That's because life expectancy is defined as an average: the average life span. But the actual distribution of lifespans in South Africa is markedly bimodal, with a large peak at 0-1 year, representing substantial infant mortality. This has pulled down the average life span, hence the life expectancy (Figure 2.35).

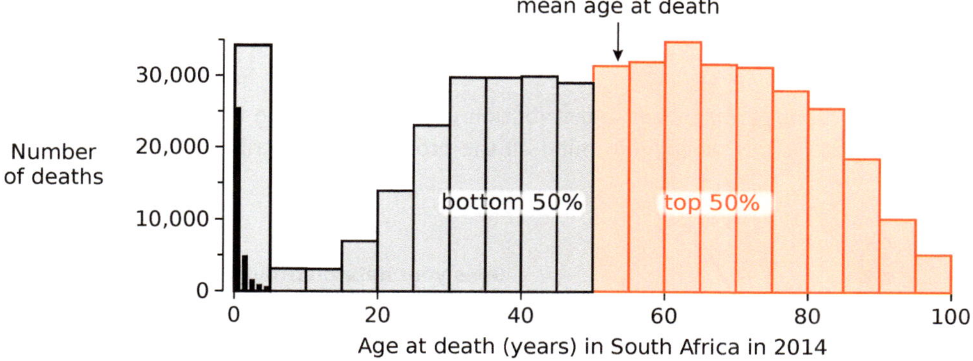

Figure 2.35 Histogram of reported age at death in South Africa, 2014. The bin width is 5 years. Within the first bin, we have shown a further breakdown of the data, with a bin width of 1 year. (Data from World Health Organization Mortality Database.)

Another striking feature of the histogram for South Africa is that there is a large increase in deaths beginning at age 30, due to the HIV/AIDS epidemic.

Exercise 2.3.5 Download other countries' demographics data from the WHO Database Website and compare them to each other and South Africa.

Danger: uniform data When the data is distributed fairly uniformly over its range, the average value is also not typical in any sense. Suppose we look at the age distribution of the US (Figure 2.36). We see that every decade of age (0–10, 10–20, etc.) has about the same number of people, up to the age of 60. Would it make sense to say "the average age of this group is 40?" That statement is true, of course, but is it good science to describe the group in this way?

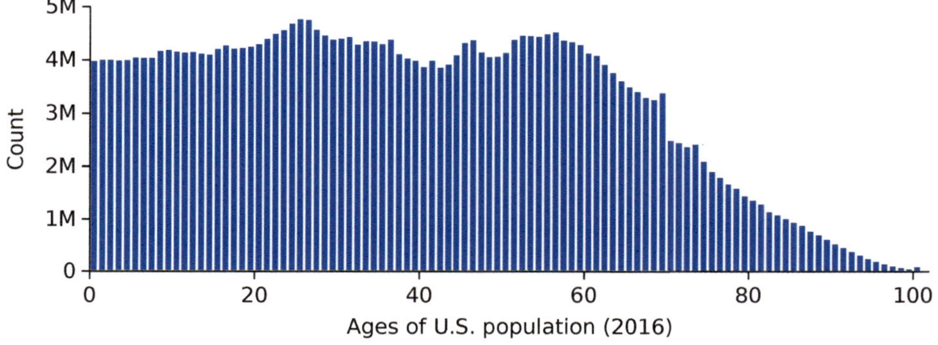

Figure 2.36 Histogram of the ages of the US population in 2016.

When is it OK to use the mean?

When is it OK to use the mean to describe a data set? We can set out a rule of thumb. Let's say that a distribution is "bell-shaped" if it is

- symmetric
- single humped
- thin tailed

then our rule of thumb is that when a distribution is bell-shaped, the use of the mean value to describe the data is OK, **keeping in mind all the problems of describing a data set by any single number.**

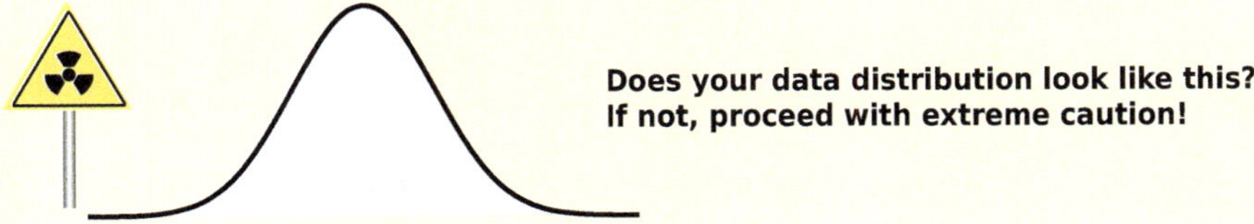

The "Flaw of Averages"

We have just discussed for which types of distributions it is "not OK" to use the mean value, and we've sort of implied that for the others, such as bell-shaped distributions, it is "OK" to use the mean.

Why a single number at all? But that presupposes a deeper question: why are we using a *single number* at all to represent a whole population? Why are we suppressing the variability of the data and insisting that it be represented by a single number?

This is a deep philosophical question, that has an interesting history. Why do we need to summarize the data in this way? Why a single number?

One factor that has contributed to the growing use of single number summaries is that the use of single number summaries, and the mean in particular, grows out of the social and corporate need to process large populations mechanically (Ziliak and McCloskey, 2008). With the rise of industrial science and medicine after WW2, the need for summary measures was created by the need for evaluation of large programs and populations. Managers say "Give me a single number", for reasons of expedient processing:

1. we need a single number because our Medical School receives ~15,000 applications, out of which only ~200 will be accepted
2. we need a single number because we have to decide which 15% of professors should get their grants funded
3. we need a single number because we have to evaluate hundreds of programs

It is important to note that these are not scientific considerations. They are industrial or programmatic or managerial or social, but they are not scientific requirements.

The obsession with the average These non-scientific pressures created a national obsession, peaking in the 1950s, about the "average" value of populations. The story of the obsession was remarkably documented in Todd Rose's article "When US Air Force discovered the flaw of averages" (Rose, 2016).

He tells the story of cockpit design in US Air Force planes. In the early 1950s, the USAF was suffering an appallingly high rate of pilot deaths, all attributed to "pilot error". "After multiple inquiries ended with no answers, officials turned their attention to the design of the cockpit itself." They took multiple measurements of 4,063 pilots, took their averages, and found that *no single pilot had the average measurements!*

One pilot would have a longer-than-average arm length, while another had a shorter-than-average upper body length, the distance from seat to shoulder. The variability was impressive. Researchers concluded that the "average size fits all" approach could not work. When they suggested the idea of adjustable controls for pedals, levers, etc., it was dismissed as absurdly impractical by the manufacturers. But when the Air Force insisted on it, the airplane manufacturers developed adjustable seats and control surfaces, which of course we have in every car on the road today.

"Norma": the perfect woman? In Rose's account, the obsession with the average became a kind of national psychosis. "A doctor in Cleveland took the averages of dozens of body measurements on thousands of women, and proclaimed that the averages represented the ideal woman. They even had a sculptor make a statue, called "Norma", based on these average measurements. . . .Norma was featured in Time magazine, in newspaper cartoons, and on an episode of a CBS documentary series, This American Look, where her dimensions were read aloud so the audience could find out if they, too, had a normal body." (Figure 2.37).

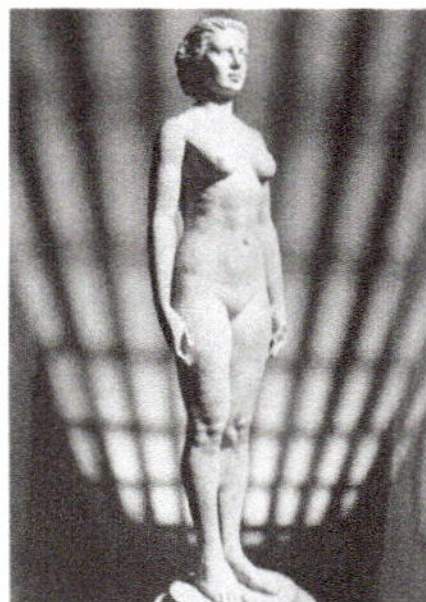

Figure 2.37 Norma, based on average measurements of size data collected from 15,000 young adult women.

Of course, we find this laughably stupid today, but maybe not so laughable when you realize that the doctor in question was Dr. Robert L. Dickinson, president of the American Gynecological Society and chairman of obstetrics at the American Medical Association.

TORONTO STAR **The Flaw of Averages**

In addition to displaying the sculpture, the Cleveland Health Museum began selling miniature reproductions of Norma, promoting her as the "Ideal Girl," launching a Norma craze. A notable physical anthropologist argued that Norma's physique was "a kind of perfection of bodily form," artists proclaimed her beauty an "excellent standard" and physical education instructors used her as a model for how young women *should* look, suggesting exercise based on a student's deviation from the ideal. A preacher even gave a sermon on her presumably normal religious beliefs.

Todd Rose

The story of Norma should be read and studied as a warning against the use of the "average" to describe a population.

The psychosis of "the average" lent itself to what is probably the worst statistical idea of the 20th century: the idea that population variability is a kind of "noise" or "error", and that we can use the average value as representative of the distribution.

In traditional statistics courses, you will even see the formula

$$\text{data} = \text{mean} + \text{error}$$

which embodies two major fallacies (1) the idea that the mean is representative of the data set, which of course is often false and (2) the idea that population can be represented by a single number, with variability being "error".

21st century statistics, by contrast, sees population variability as an important factor. Biological variability is not "noise" and it is not "error". It is not a bug in the program, it's a design feature. Population variability and diversity play many important causal roles in evolution, for example, as well as in the susceptibility to diseases. We have to resist the temptation to look only at average values, and instead, we must focus on distributions. We'll discuss measures of the variability of a data set in section 2.4 *Summarizing a data set: measures of variability*.

Situations where averages shouldn't be used

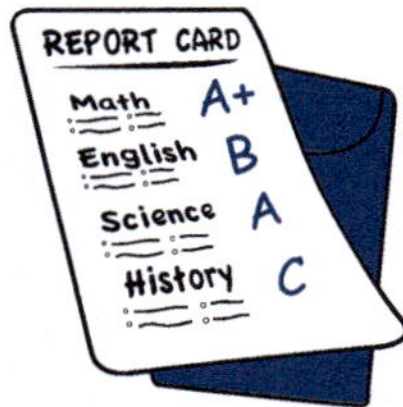

- **Batting averages.** In baseball and softball, a player's batting average is the average value of the player's performance in their at-bats, with 1 meaning "got a hit" and 0 meaning "made an out". So if a player had 120 at-bats and got 30 hits, their batting average is 30/120 = .250.

  Are batting averages a good measure of a hitter? Although this statistic was used for many years to evaluate players, when data analytics became popular around the year 2000, it was realized that it was a poor measure of the player, because it doesn't distinguish between critical situations and uncritical situations.

- Probably the worst average we can think of is the **grade point average**. First of all, the distribution of one person's grades is far from a Normal distribution: all values lie between F and A+, with a "hard wall" at both ends. This makes the distribution highly non-Normal: if you get a C in one class, you can't get a grade beyond A+ to balance the C.

Treating the average patient The flaw of averages, the idea that a population can be characterized by an average value plus "error", has disastrous consequences for clinical treatment. A telling demonstration of this fact was shown by Kent and Hayward (2007), in their article "When Averages Hide Individual Differences in Clinical Trials". They showed the work of artist Chris Dorley-Brown, who took 2,000 photographs of people living in a small town in England,

and created a digital "average face" (center photograph in Figure 2.38). The average person is neither male nor female, neither young nor old, and neither sick nor healthy. Medical therapy tailored to the average person will miss many important clinical dimensions.

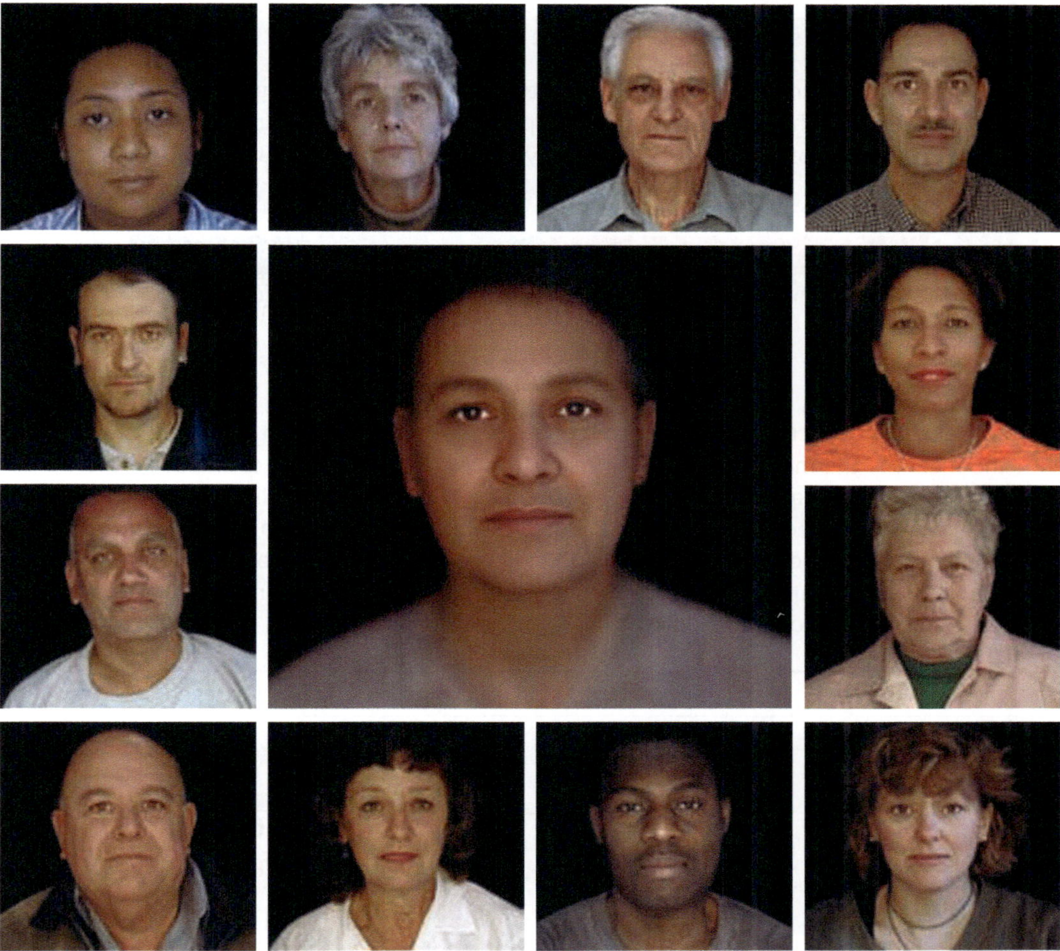

Figure 1. In the year 2000 artist Chris Dorley-Brown took 2,000 digital photos of people living in the small town of Haverhill in Suffolk, England, and then used software to merge these photographs, step by step, until all 2,000 had been blended into one image. This illustration shows 12 double portraits (two photographs morphed together) and a blend made up of all of the double portraits. (Dorley-Brown has left out the original portraits in order to protect the privacy of the participants.) In the modern clinical trial, the responses to treatment of thousands of individuals are typically summarized in a single number in the same way the center photograph represents all the other individuals. As the data are averaged, important individual differences are lost. The authors propose ways to better examine clinical-trials results to guide medical practice.

Figure 2.38 Reprint of Figure 1 from the article "When Averages Hide Individual Differences in Clinical Trials" by Kent and Hayward (2007).

Exercise 2.3.6 The R_0 of an infectious disease is the mean number of new people infected by a patient with the disease.

a. Suppose two diseases have the same R_0. What might this fact hide?

b. Speculate on the possible significance of this for disease control.

c. During the COVID epidemic, U.S. states closely monitored their calculated R_0, in order to decide when and how to re-open non-essential businesses. Given your answer to (a), what might be a problem with this approach for a larger state like California (pop. 40 million) versus a smaller state like Rhode Island (pop. 1 million).

Exercise 2.3.7 Universities often emphasize that they have a really small average class size. Is that consistent with your experience or impression? If not, why do you think that's so? Here is a histogram of the class sizes at a typical university. The average class size is 18 people.

47.8%	29.3%	22.8%
classes with less than 20 students	20-49	50 or more

The median value

Another important descriptor of a data distribution, which is unfortunately sometimes confused with the mean, is the **median value**. The median value is the midpoint of the data, the "middle number": 50% of the data lie below the median, and 50% of the data lie above it.

To calculate the median $\widetilde{X}$ of a set of n data points, rank the data from the lowest to the highest value. If n is odd, then the median value is the middle value on the list. If n is even, there is no "middle value" in the data set, so we use the standard convention and report the median as the average of the middle two data values (Table 2.2).

	n is odd	n is even
data set	{15, 8, 21, 4, 10}	{15, 8, 21, 5, 4, 10}
mean	$\frac{15+8+21+4+10}{5} = 11.6$	$\frac{15+8+21+5+4+10}{6} = 10.5$
ranked data set	{4, 8, 10, 15, 21} (↑ at 10)	{4, 5, 8, 10, 15, 21} (↑ at 8 and 10)
median	10	$\frac{8+10}{2} = 9$

Table 2.2 How to calculate the median value of a data set.

Exercise 2.3.8 What is the median of the following data sets?

data set A: {76, 38, 4, 42, 41, 25, 39, 72}

data set B: {3, 7, 3, 2, 3, 4}

data set C: {3, 7, 3, 2, 3, 4, 1000}

Exercise 2.3.9 The front row in a movie theater has 19 seats. If you were directed to sit in the seat that occupied the median position, which seat would you have to sit in?

Note that there is a big difference between the mean and median in terms of how they are defined. There is a formula for the mean (see page 45). But there is no *formula* for the median;

rather, there is a *procedure* for calculating it. This may seem like a minor difference, but it is important: it means that it is relatively easy for mathematicians to prove theorems about the mean, by operating on the formula, but it is much harder to prove theorems about the median.

Nevertheless, and despite the difficulty of proving theorems about it, the median has a number of very helpful properties. One big difference between the mean and the median is that the median is resistant to influence by outliers. If you put a billionaire into a room of 100 students, the average income changes dramatically, but the median income changes very little; from the median's point of view, you are just adding one more large value, and it wouldn't matter to the median if it was $10 more than the previously highest value, or $10 million more. **Thus it is often essential to use the median in the presence of data with outliers.**

Indeed, whenever a distribution is skewed, the median may be a more informative measure than the mean.

It follows from the definition of median that if the distribution is skewed to the right, then the mean will lie to the right of the median. And if the data are skewed to the left, the mean will lie to the left of the median (Figure 2.39). How far away they are is a measure of how skewed the data are. And if the data are symmetric (one of the requirements for bell-shaped), then the mean and median are close.

Figure 2.39

For example, consider the histogram of US household incomes, which is skewed to the right (Figure 2.33). The *mean* household income is $88,439. The *median* income is much less, at $66,615, meaning that half of US households earn less than that.

Reading out the median from a cumulative histogram We can directly read out the median from the cumulative histogram, just as with Figure 2.40: we find the 50% point on the "count" (vertical) axis, and draw a line across to see where it intersects the cumulative curve, in this case that is the point $66,615.

The large difference between mean and median indicates that the distribution is heavily skewed.

Percentiles

The median value is the midpoint of the data; 50% of the data lie below it and 50% of the data lie above it. It is also called the 50th percentile. The kth percentile is the value in the data set that has k% of the data values at or below it and (100 – k)% of the data values at or above it.

From the cumulative histogram, we can conveniently obtain any percentiles. For example, to read out the 25th percentile, we follow the same procedure, and need only to find the 25% point on the "count" (vertical) axis, so we draw a line across to see where it intersects the cumulative curve and read out the corresponding X-value (as illustrated in Figure 2.40).

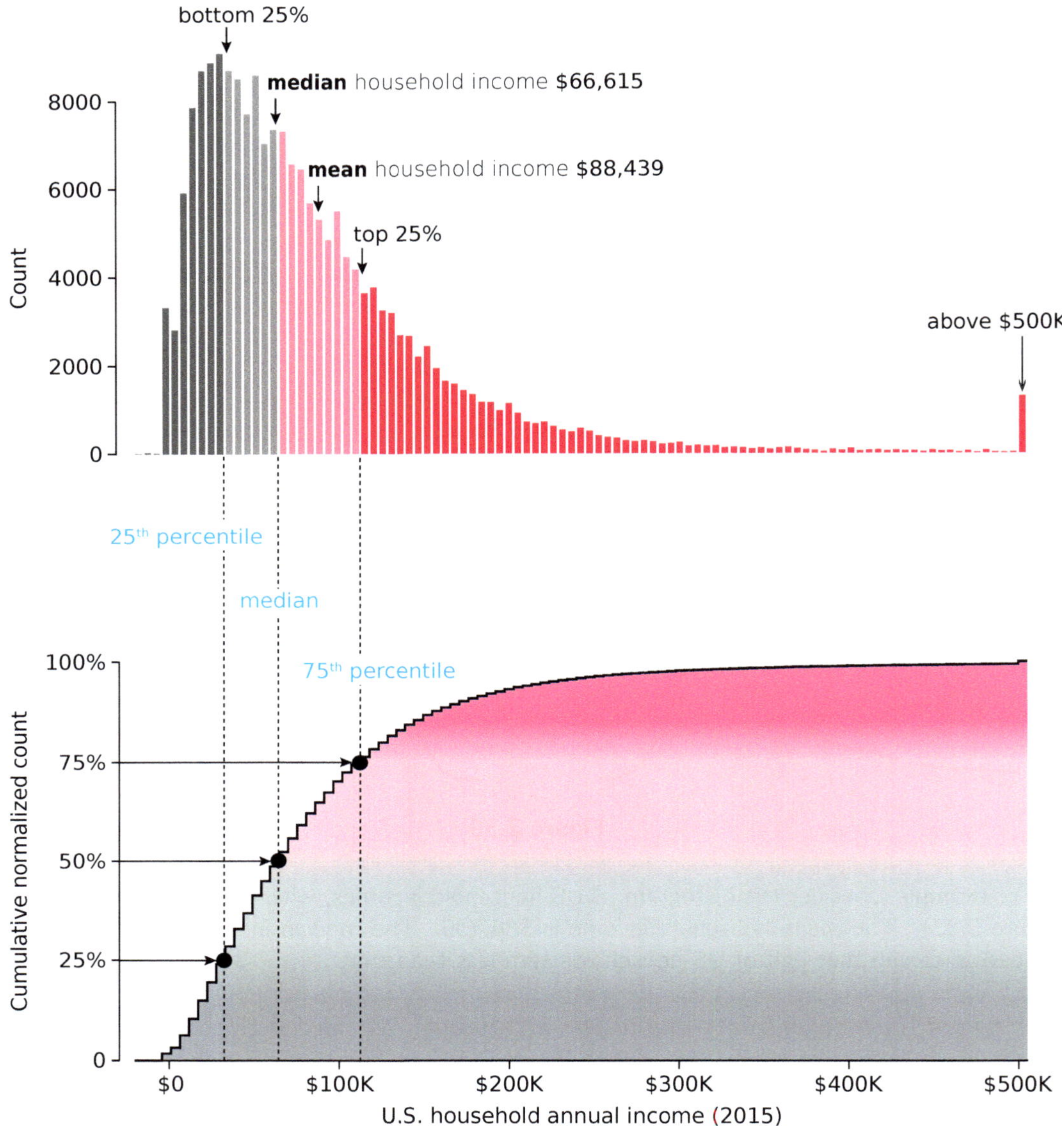

Figure 2.40 Upper: Histogram of US household annual incomes. Lower: cumulative histogram of the same data. Note markings for mean and median, as well as the cutoffs for the 25th and 75th percentiles.

Exercise 2.3.10 In Ex 2.3.7 on page 54, we were given the distribution for a University's class sizes and the average class size.

a. What can you say about the median class size? How does it compare to the average class size?

b. What can you say about the skewness of the distribution?

The meaning of the mean and the meaning of the median

Whether to use the mean or the median as a descriptor is a choice that depends partly, but only partly, on the shape of the distribution. It also depends on the purpose of our inquiry, and on the very meaning of the terms "mean" and "median".

It is certainly true that, for most purposes, if a distribution is skewed, then the mean is a poor "typical value".

There are many cases in which the median of a skewed distribution is a more informative measure than the mean, because the median is directly interpretable as the "50% chance" point: a data point has a 50% chance of falling below (or above) it.

Perhaps the best example would be survival data. A presentation that is frequently seen is called a survival curve, which shows the number of people still alive after time t. It's the inverse of a normalized cumulative histogram showing death before time t. In a normalized cumulative histogram, as we go from the left to the right along the X-axis, we start from 0% and add up each bin and the rightmost bin has a value of 100%. Whereas in a survival curve, as the survival time increases, the fraction of survivors decreases from 100% to 0.

For example, think about the distribution of survival times for cancer patients given a particular drug. As is frequently the case with survival data, the data are skewed to the right, with a few patients surviving a long time. The long-surviving patients pull up the average survival time.

The biologist Stephen J. Gould was diagnosed with cancer, and asked his doctor what was his likely survival. The doctor said "well, the average survival with this type of cancer is x years." Gould went home and thought about it, and he realized that the mean survival time told him nothing of value, no fact he could use (just like the case where you learned that the average tax refund will be $40 more). If, for example, most patients die within months but a handful live 10 or more years, what you want to know is the *median* survival time, which is directly interpretable: it tells you that you have a 50% chance of living this long.

Consider the survival curves for Hodgkin's lymphoma, broken down into distinct age groups (Figure 2.41). The median survival time decreases dramatically with increasing age: for the 20–40 age group (red), the median survival time is 13 years, whereas for the 60–80 age group (green), it is less than 5 years. Note how much the means are pulled towards the right due to the right skewed distributions.

The choice of mean or median can also reflect your interest in this particular data set. For example, if you are applying to work for a company, what you want to know is not the mean salary. One reason for this is that salaries are notoriously skewed to the right, so if the CEO makes hundreds of millions and everyone else makes minimum wage, the average salary would be pulled very high. But there's another reason for wanting the median, which has to do with your purpose in looking at this salary data. If you are thinking about working for the company, the median salary would be much more informative to you: it's some guide to what you're likely to be paid. But from the point of view of the company's financial officer, what is of interest is the average salary, because that's what the company has to pay out. *If the payroll is thought to be too high, then what must be reduced is the mean salary, not the median.*

Exercise 2.3.11 Suppose we have daily rainfall data for 365 days in a certain area. What would the mean of that data tell you? What would the median tell you?

Exercise 2.3.12 For each scenario below, state whether mean or median would be more appropriate and why. 1) Describing typical family size in a country. 2) Reporting typical recovery time after a medical procedure.

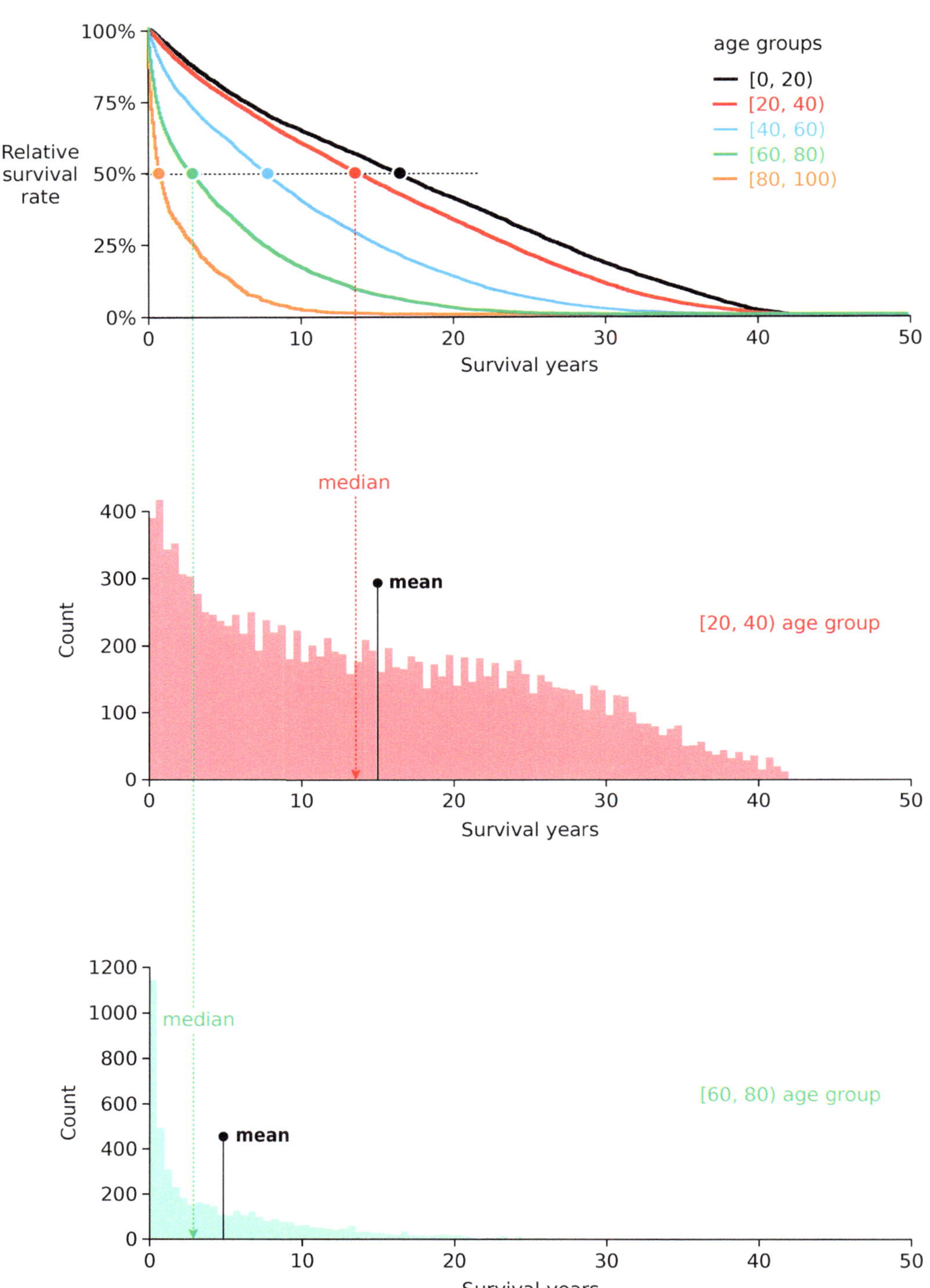

Figure 2.41 Upper: survival curves (survival rate as a function of survival time) for Hodgkin's lymphoma, broken into distinct age groups. Middle: histogram of survival times for the [20, 40) age group. Lower: histogram of survival times for the [60, 80) age group.

Mode

The mode of a data set is its most popular value, the highest point in the histogram, the "peak". Its value in characterizing data is limited. It's the "most popular" value for the data, but how popular are other far-away values? Is it a good idea to neglect substantial peaks, that may be far away, because they are ever-so-slightly lower than the mode? For the same reason, the mode is very unstable: when the data are changed slightly, the mode can change significantly. And importantly, the notion of "the highest point in the histogram" depends strongly on our choice of bin width and choice of cut-offs for the histogram. The mode is not much used, and is mentioned here only for completeness, and because it lent its name to some useful words. We say that a distribution is **unimodal** if it has a single prominent peak, **bimodal** if it has two peaks, and multi-modal if it has more than two peaks.(Figure 2.42).

Figure 2.42

Look at the distribution, not the summary number

When we view a data set, it is critical to realize that we need to see the distribution of the data before calculating summary numbers. Consider the figure below. Suppose we are looking at the effects of a drug or an environmental condition (here, hot weather). To look only at the average or median effect is to miss the fact that the condition had a drastic effect on one participant.

FURTHER EXERCISES 2.3

1. Calculate the mean and median of the data set: {10, 15, 19, 13}.
2. What number would you divide by to calculate the mean of 3, 4, 19, 5 and 6?
3. What measure of central tendency is calculated by adding all the values and dividing the sum by the number of values?
4. You measure the weights of 10 randomly chosen shells at the beach. The average weight from the shells you collected is 200 grams. Does this imply that at least 4 of the shells weigh less than 200 grams? If so, explain why. If not, using statistics terminology, explain why and provide a counter example.
5. Sketch a distribution for which a mean is appropriate and two distributions for which a mean is not appropriate (each in a different way). For each, explain why a mean is or is not appropriate.
6. **General shape of a distribution.** Each of the following KDEs represents a data set. Choose the best descriptor for each distribution:

   left skewed, right skewed, symmetrical, uniform and multi-modal.

7. What term is used to describe the distribution of a data set that has a single prominent peak?

   a) Multimodal b) Unimodal c) Nonmodal d) Bimodal

8. Consider the histogram below, which shows how many mammalian species have a given longevity (life span). For mammals, how does average longevity compare to median longevity? How do you interpret the mean and the median?

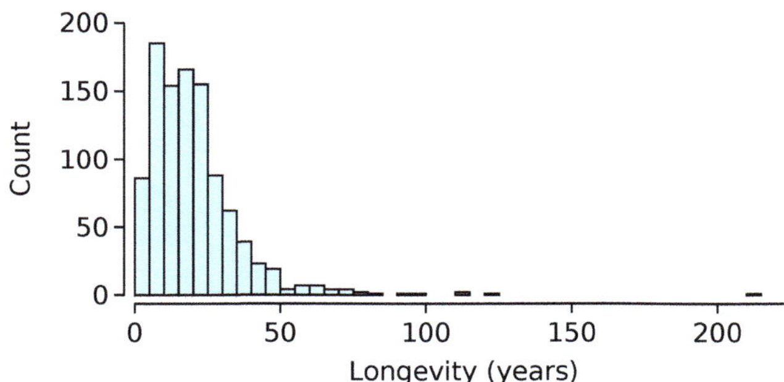

9. Briefly describe three situations in which it would be inappropriate to describe the central tendency of a set of data with a mean. For at least one of them, sketch it using the appropriate visualization.

2.4 Summarizing a data set: measures of variability

Measures like the mean and median are often called "measures of central tendency": they report a single number that "represents" the data. By using a summary measure of central tendency, like the mean or median, we are losing a lot of information about the data set.

The most important thing we are losing is any sense of the **variability** of the data. Look at the two distributions in Figure 2.43. They have the same mean value, but one has a lot wider variation than the other.

This is a useful thing to know about a distribution or a data set, and so we would like to find a measure that reflects the variability of a data set.

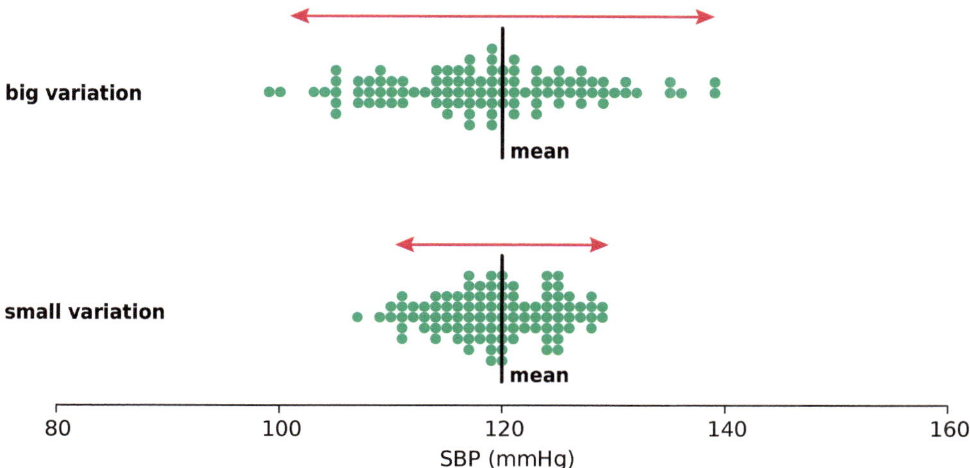

Figure 2.43 Two different distributions have the same mean, but differing variabilities, a fact that would be lost by simply presenting the means.

Variance

The best-known measure of variability is the **variance**. The idea behind variance is basically to add up the distances from each point to the mean value of the data. If there is a lot of variability in the data set, then there will be a lot of large distances to the mean, and they will add up to a large variance.

There are a few technical issues that have to be faced. The most important is that we used the concept of "distance". Distances, as we normally think of them, are always positive. The distance from Chicago to LA is the same as the distance from LA to Chicago, around 2000 miles. We don't think of one distance as "+2000" and the other as "−2000".

So how are we to define the "distance" between two data points x_1 and x_2, which we visualize as two points on the X–axis? If we use "$x_1 - x_2$" then we have the problem that the distance from x_1 to x_2 is not the same as the distance from x_2 to x_1. Also, because one is the negative of the other, if one point lies 100 units to the left of the mean μ, (so $x_1 - \mu = -100$ is negative) and another point x_2 lies 100 units to the right of μ, (so $x_2 - \mu = 100$ is positive), then, if we add the two "distances", we would get zero.

So how can we prevent "positive distances" and "negative distances" from canceling each other out? The answer that is standard in the statistics literature is that we *square* all the positive and negative distances. Since squares are always positive (or 0), we can add them up without some canceling others. The **variance** is then defined as the average value of these squared distances from the mean:

$$\text{variance of data set } \{x_1, x_2, \ldots, x_n\} = \frac{1}{n}\sum_{i=1}^{n}(x_i - \overline{X})^2 = \sigma^2$$

where $\overline{X}$ is the mean of data set $X: \{x_1, x_2, \ldots, x_n\}$. The variance is often written as σ^2 which is read "sigma squared".

Statistical theorists and mathematicians like the concept of variance because it is relatively easy to prove theorems about squared quantities, especially when the distributions are bell-shaped. But we are less interested in mathematical formalisms here, and more interested in the question: how good an estimator of variation is the variance?

One problem with the concept of variance is that the variance of a set has units that are difficult to interpret: if the data set is incomes in dollars, then the variance is in "dollars squared". This creates a problem for interpretation: whatever can the notion of "dollars squared' (or "years squared") possibly mean? How can we report the variability of a data set in a unit that we can't make sense of?

Another drawback that the variance suffers from comes from the very nature of squaring: squaring a large quantity makes it gigantic, and squaring a small quantity makes it negligible. Thus, the variance will be dominated by a few outlying or skewed values. This makes the variance a less desirable estimator in situations where there is substantial skew.

Standard deviation

Statistical theorists have tried to "undo" the distortions due to squaring the differences by then taking the square root of their sum. That leads to the concept of **standard deviation (s.d.)**, which is the square root of the variance:

$$\text{standard deviation of data set } \{x_1, x_2, \ldots, x_n\} = \sqrt{\frac{1}{n}\sum_{i=1}^{n}(x_i - \overline{X})^2} = \sigma$$

The standard deviation is often written as σ.

The standard deviation has the right units, which are the same as the original data units. However, it still suffers from the same domination by extreme values.

And yet another drawback of the notion of variance or standard deviation is that their definitions contain the concept of the "mean" *twice*: we take the mean distance (or the mean squared distance) from the mean of the data set, **Therefore, the concepts of variance and standard deviation are dependent on the mean, and if the mean is not a good measure of the data set (for example, skewed distributions) then the variance and standard deviation will be poor descriptors of the variability.**

Exercise 2.4.1 Consider the following three lists of numbers:

- List X contains 80 random numbers generated (with uniform probability) from the interval [15.3, 25.3]
- List Y contains the number 99 repeated 80 times
- List Z contains 80 random numbers generated (with uniform probability) from the interval [101, 105]

- List W contains 80 random numbers generated (with uniform probability) from the interval [20.5, 21]

Answer the following questions:

a. What is the standard deviation of List Y?

b. Is it true that the standard deviation of List Z is more likely to be larger than the standard deviation of List X?

c. Rank the standard deviations: σ_X, σ_Y, σ_Z, σ_W.

Standard error of the mean

Finally, just for completeness, we have to mention one more measure of variability. The **standard error of the mean (s.e.m.)** is defined as the standard deviation of the data divided by $\sqrt{n}$, where n is the size of the data set. The concept of the s.e.m. is supposed to cast light on the variability of the mean value of the data, as opposed to just the variability of the data, which is measured, for better or worse, by σ. The standard error of the mean purports to tell us how variable the estimate of the mean might be. Think of a single chemistry class. The grades within that class will be highly variable, with some A$^+$s and some Ds. But the *average* grade in the class is probably going to be around B$^+$, and if we looked at many chemistry classes, **the variability in the average grade across the classes is going to be much less than the variability of grades within a single class.**

There are theorems that tell us that *if* the data are perfectly bell-shaped, then the variability of the mean values of a lot of samples will be given by the standard error, but that is, as always, a big "if".

Absolute deviation

Statistical theorists and mathematicians like the concepts of variance and standard deviation, which involve squared distances.

But there is nothing wrong with our original concept of distance, as in "the distance from Chicago to LA". The distance we are talking about here would be expressed as the absolute value: $|x_1 - x_2| = |x_2 - x_1|$.

To apply this concept to data, we begin with the median of the data, which we will call m. Then we will take all the absolute values of the distances from every point to m, and then take the median of those distances. The result is called the **Median Absolute Deviation (MAD)**. We define the **MAD** as

$$\text{MAD of data set } \{x_1, x_2, \ldots, x_n\} = \text{median}\Big\{|x_1 - m|, |x_2 - m|, \ldots, |x_n - m|\Big\}$$

where m is the median of the set $\{x_1, x_2, \ldots, x_n\}$.

It gives us another measure of variability, one that does not suffer from the distortions of squaring and un-squaring.

MAD

Notice that in the definition, the term "MAD" uses the median *twice*. It's the median of the distances from the median.

There are more definitions of "MAD": in the literature, you will sometimes see the term

"MAD" used to refer to the median absolute deviation around the mean, or the mean of the absolute deviations around the median, or the mean of the absolute deviations around the mean. This can be very confusing, and it's important to be clear about which version is being used.

- Whether to use the mean or the median as the group measure depends on whether the data distribution meets the strict requirements for the use of the mean.
- Whether to use the mean or the median as the "M" in "MAD" depends on the distribution of the absolute differences.

As usual, if the distribution of the absolute differences meet the requirements for the use of the mean, then we could be justified in using the mean of the differences, but if it isn't, the mean might be misleading, and we might do better by using the median. Each method has strengths and weaknesses, for example, using the median of the absolute distances would become unstable and highly variable when the sample size is small.

Interquartile Range

Another measure of the "spread" or variability of a data set is the **Interquartile Range** (IQR), the difference between the cutoff for the 25th percentile and the cutoff for the 75th percentile. When the Interquartile range is large, that means that there is a lot of variability in the data.

More specifically, a large IQR means that there is a lot of variability *in the central part of the data*. Consider the house price data in Figure 2.44.

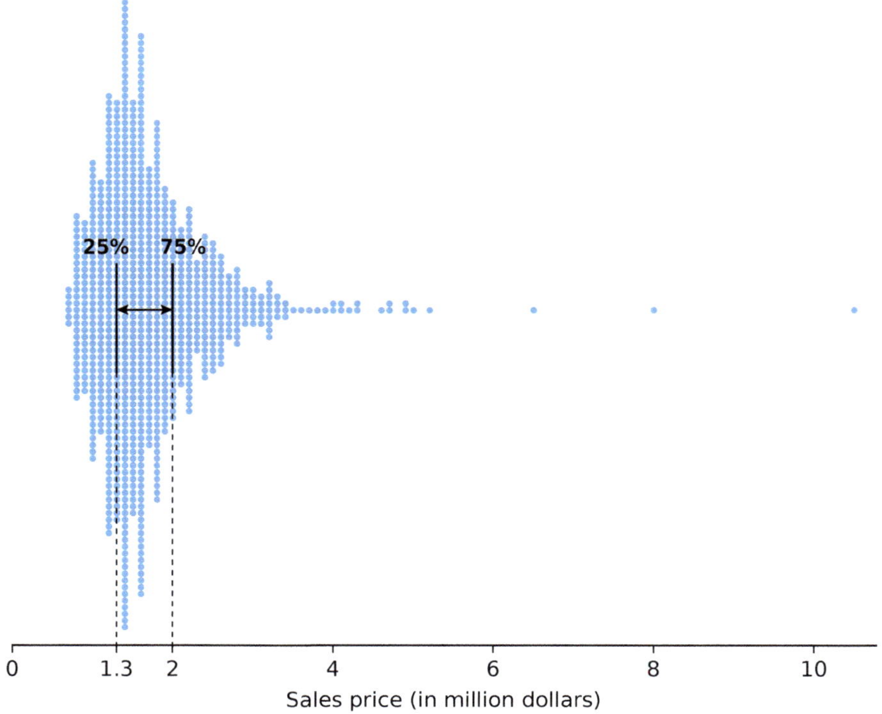

Figure 2.44 Interquartile range for all the 3-bedroom houses sold in San Francisco in 2018.

Note that there are many high-end outlying values, but they are not being captured by the IQR. The IQR is looking at the central portion of the data, and note that it is not very large. There is not a great deal of variability in the middle part of the housing market, however much variability there is in the high-end.

Range

Finally, another measure of the variability of a data set is its **range**, which is simply the maximum value of the data minus the minimum value.

$$\text{range} = \text{max} - \text{min}$$

It can give us a useful insight into the overall variability of the data set. However, note that it depends on only two data points out of the whole set, so it is a highly unstable measure of the set. It can change drastically if one of those two data points is changed.

So, to summarize, the definition of variability as variance, standard deviation or standard error of the mean all suffer from the urge to square everything in sight, and all of them depend on the mean being a representative number. Thus they are questionable when the data is not bell-shaped, and the mean is a poor descriptor.

We can calculate all 6 definitions of variability for a given data set. **Which definition we should use in a given case depends strongly in the shape of the data distribution.**

If the data lies in a bell-shaped distribution (symmetric, unimodal and thin-tailed), then we can consider using the mean-based concepts of variance and standard deviation. We can also consider using the standard error of the mean, while keeping in mind that the s.e.m. is not a measure of the variability of the data, it's a measure of the variability of the average value of the data.

But if the data distribution is not bell-shaped, for example if the data are skewed, then we can't use the mean-based measures, and it is better to use a median-based measure like MAD.

Range and IQR do not depend on the shape of the data, but keep in mind that they give *very* limited information.

Exercise 2.4.2 By looking at the distribution of the following data sets, which measure of variability should you use?

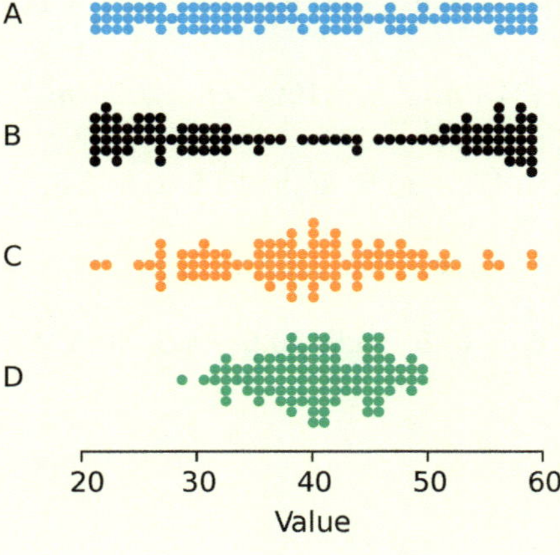

Example 8 Arsenic in rice

In 2013, the US Food and Drug Administration (FDA) released a report on rice products available in the United States. Among those many rice types, here are the arsenic levels (ppb or parts per billion) in a total of 12 samples of Brown Basmati rice that were included in the report: {86, 84, 178, 161, 82, 118, 124, 127, 134, 146, 199, 200}.

Question How would you describe the distribution of this data set?

Answer The first step is always to plot the data. The beeswarm plot shows that the data points are spread out with no prominent peaks (modes). The data is almost uniformly distributed.

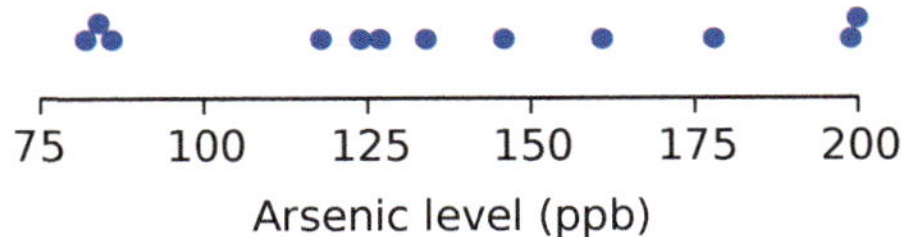

Question What is the appropriate measure of central tendency?

Answer Since the data does not fit the criteria for a "bell-shaped" curve, namely, symmetric, single humped and thin tailed, we can only opt for the median as the measure for central tendency.

In order to find the median value, sort the 12 numbers from smallest to largest. Since the sample size 12 is an even number, the median value is the average of the middle two numbers.

$$82,\quad 84,\quad 86,\quad 118,\quad 124,\quad \overset{\downarrow}{127},\quad \overset{\downarrow}{134},\quad 146,\quad 161,\quad 178,\quad 199,\quad 200$$

$$\text{median: } m = \frac{127+134}{2} = 130.5 \text{ ppb}$$

Question What is the appropriate measure of variability?

Answer Since the data does not fit the criteria for a "bell-shaped" curve, we should not be using variance, standard deviation or sem. Both the range and MAD do not depend on the shape of the data, therefore we should report range and MAD as the appropriate measure of variability for this data set.

$$\text{Range: max} - \text{min} = 200 \text{ ppb} - 82 \text{ ppb} = 118 \text{ ppb}$$

$$\begin{aligned}
\text{MAD} &= \text{median}\left\{|86-m|,\ |84-m|,\ \ldots,\ |199-m|,\ |200-m|\right\} \\
&= \text{median}\left\{44.5,\ 46.5,\ 47.5,\ 30.5,\ 48.5,\ 12.5,\ 6.5,\ 3.5,\ 3.5,\ 15.5,\ 68.5,\ 69.5\right\} \\
& \qquad \Downarrow \text{sort from smallest to largest} \\
&= \text{median}\left\{3.5,\ 3.5,\ 6.5,\ 12.5,\ 15.5,\ \mathbf{30.5},\ \mathbf{44.5},\ 46.5,\ 47.5,\ 48.5,\ 68.5,\ 69.5\right\} \\
&= \frac{30.5+44.5}{2} \\
&= 37.5 \text{ ppb}
\end{aligned}$$

FURTHER EXERCISES 2.4

1. What is MAD a measure of? a) Central tendency b) Variation
2. Why is the MAD a better measure for most data sets than other measures (such as standard deviation, standard error of the mean)? Give at least two reasons.
3. Find the MAD of the data set {14, 9, 7, 19, 3}.
4. What is the interquartile range of the data set {2, 5, 7, 9, 15, 15, 17, 19, 22, 24, 29, 30}?
5. **Heart rates.** The histograms below show recorded heart rates (in beats per minute) collected from four different individuals over a 24-hour period using fitness trackers. Interpret what each histogram might tell us about the individual's activity level and fitness.

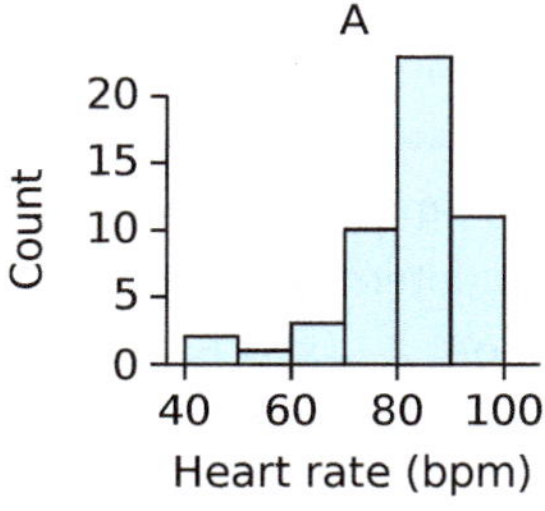

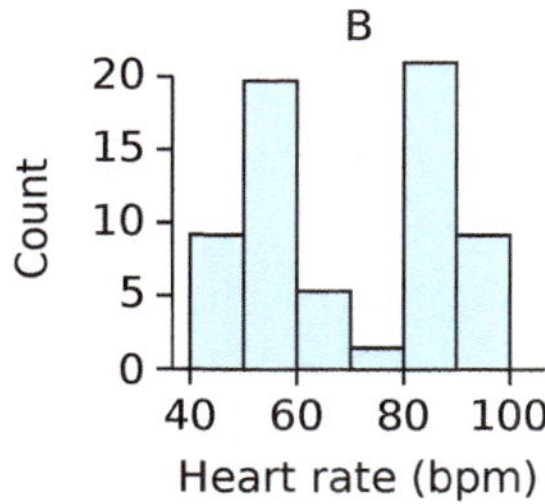

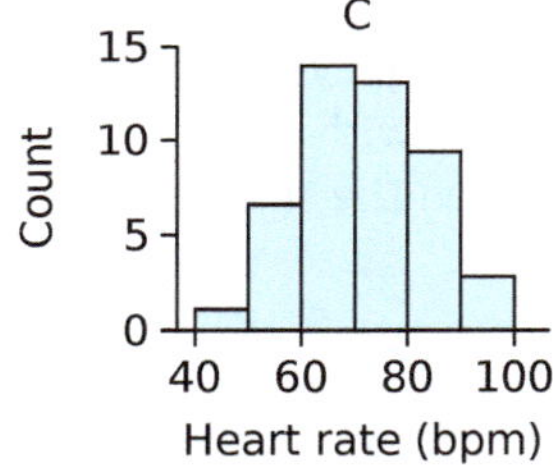

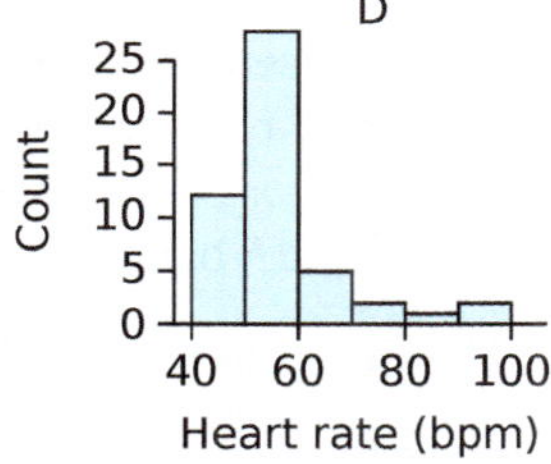

a. For which individual(s) is the mean a representative value for their typical heart rate? Briefly explain at least two features of the data that led to your choice.
b. For which individual(s) should you use a median? Why?
c. For which individual(s) would the mean be greater than the median? Why?
d. Which person appears most physically fit? Which person likely exercised during the monitoring period?
e. What does Histogram A suggest about this person's daily routine? Consider the peak around 80-90 bpm. What measure of variability would you use for individual A? Give at least two reasons why this measure is better than other possibilities.
f. Histogram D is right-skewed with most measurements at 50-60 bpm. What are two possible explanations for such a low resting heart rate?
g. Compare Person A's and Person D's typical heart rates.
h. Why should we visualize the distribution of our data before doing any statistical analysis and present that visualization as part of our analysis?

2.5 Presenting data: show the data

We have now discussed several ways of presenting data sets: dot plots, beeswarm plots, histograms, Kernel Density Estimates, and violin plots. We used these presentations to get a sense of the overall distribution of the data.

Based on the distribution of the data, we made decisions on how to represent central tendency and variability. We discussed measures of central tendency using "typical values" such as the mean and the median, and we discussed ways of measuring variability, including standard deviations, variance, standard error of the mean, MAD, interquartile range, and overall range.

measures of a data set

central tendency	variability
• mean • median	• variance • standard deviation • sem (standard error of the mean) • MAD (median absolute deviation) • IQR (interquartile range) • range

Now it is time to ask: how do we present these findings for publication or further analyses? When it's time to present our data, the overall rule is: **show the data**.

1. **Always begin with a presentation of the data that displays its distribution**: either a dot plot or a beeswarm plot to show the data, together with a histogram or kernel density estimate if the data is more numerous. Comment on the shape of the data distribution: is it bell shaped? skewed? bimodal?

2. **Depending on the outcome of (1)**, choose measures of central tendency and/or variability, as appropriate for this distribution. For example, be wary of the mean when the distribution is not bell-shaped.

Box plots

There is a standard form of data presentation that falls short of the injunction to show the data, yet is frequently used. It's called a **box plot**. A box plot is a graphic presentation of the data using its 5-number summary (Figure 2.45).

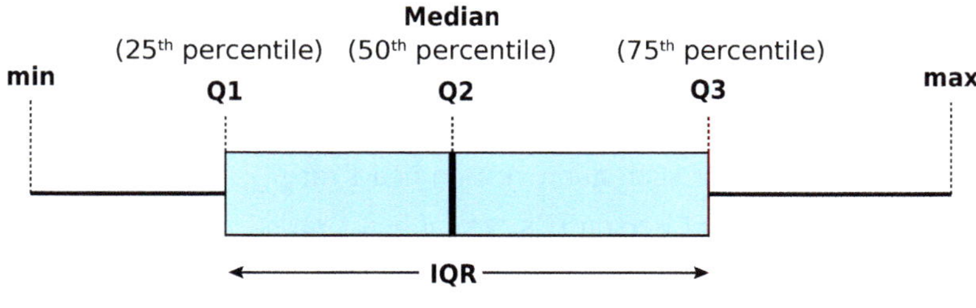

Figure 2.45 A box plot.

It shows the median of the distribution, which is the 50th percentile of the data, the cut-offs for the 25th and 75th percentiles of the data, called Q1 and Q3, respectively, and the max and min of the data.

A common form of box plot is to show only the middle two quartiles, and to calculate the width of those middle two quartiles, that is, the Interquartile Range, the difference between the 75th and 25th percentiles.

A box plot covers a few, but certainly not all, of the "measures of a data set". It shows the median and the quartile cutoffs as well as the min and the max. From these, we can also derive the IQR and the range of the data.

Sometimes, if there are outlying values that would distort the box plot, a choice is made to present the outlying values as free-standing data points, excluding them from the box plot summary.

The box plot should be thought of as the minimum summary description of a data set. It gives us, at the least, 5 pieces of information about the location and shape of the distribution.

Exercise 2.5.1 Here is a box plot of the student exam scores for a class.

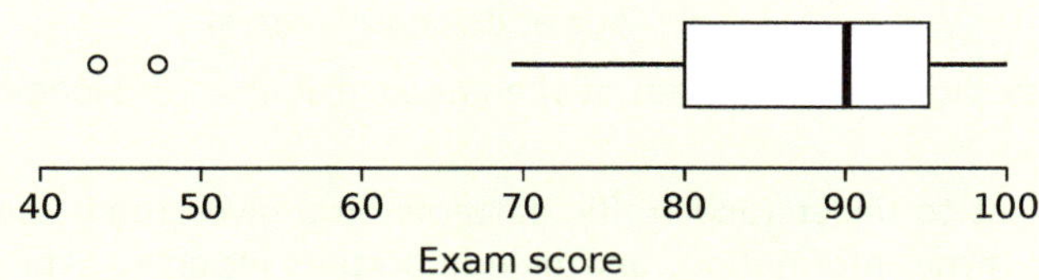

a. What information can we learn from this box plot about the student exam scores?

b. Interpret the box plot. Make sure you discuss symmetry/skew, modality, tails, and outliers.

c. Based on this plot, was the exam easy or hard?

Exercise 2.5.2 Sketch a box plot for the flight cancellation data.

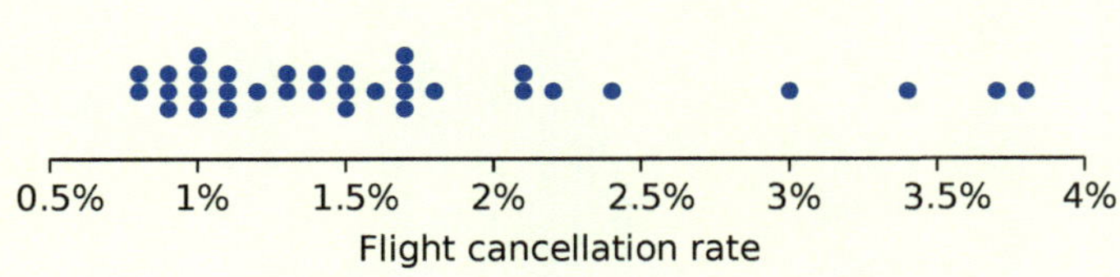

Beyond the box plot: show the distribution

However, a box plot is a poor substitute for a presentation of the actual distribution.

A distribution plot reveals key features Consider the data on the age at diagnosis of Hodgkin's Lymphoma (Figure 2.46). The box plot is shown above, and the distribution of the data is shown below. Notice how much more information is conveyed by the histogram compared to the box plot! The histogram shows us that while there is a substantial peak in the mid-to-late 20s, the overall count of Hodgkin's Lymphoma cases continues at a fairly steady pace throughout the years 40-70.

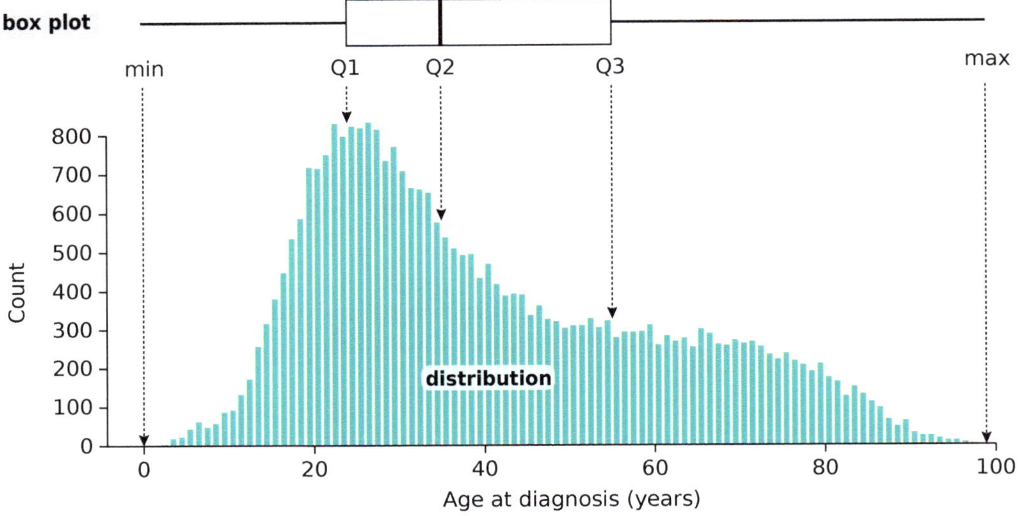

Figure 2.46 Box plot and histogram of the age at diagnosis of Hodgkin's Lymphoma.

Similarly, if we want to understand traffic patterns on a given road, summary data like the box plot give us very little information, and hide important insights. The distribution plot of vehicle speeds reveals that, in addition to the expected peak around the speed limit, there is also a second peak at a much lower speed. This tells us that there are traffic jams on this road (Figure 2.47).

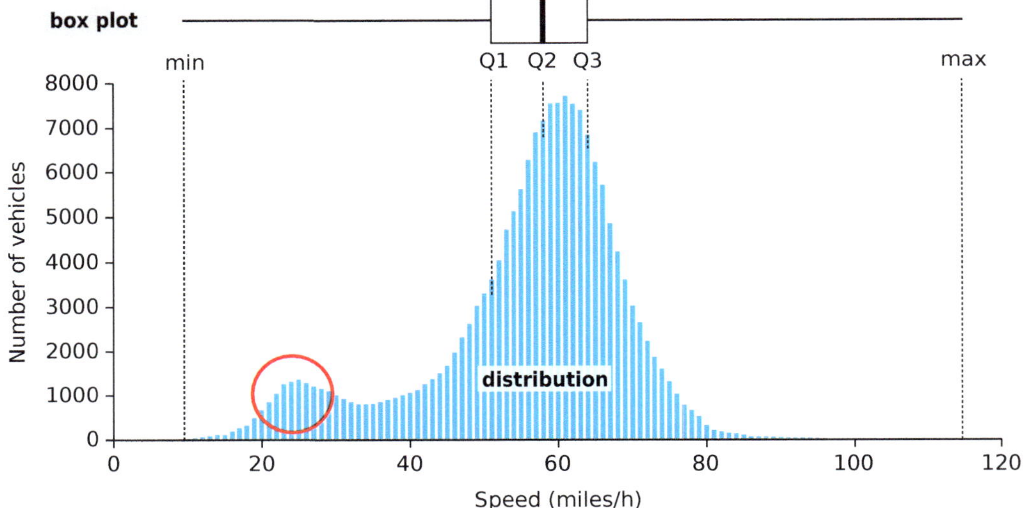

Figure 2.47 Box plot and histogram of traffic speed of vehicles driven on a major highway in a day. The speed limit at the location is 60 miles/h. Note the significant number of vehicles traveling at a speed below 30 miles/h, totally lost in the box plot.

The presentation depends on the data We always have to be flexible about how we are presenting data, and let the shape of the distribution determine the presentation. For example, here we have the data on the 2020 compensation of 200 leading CEOs (Figure 2.48).

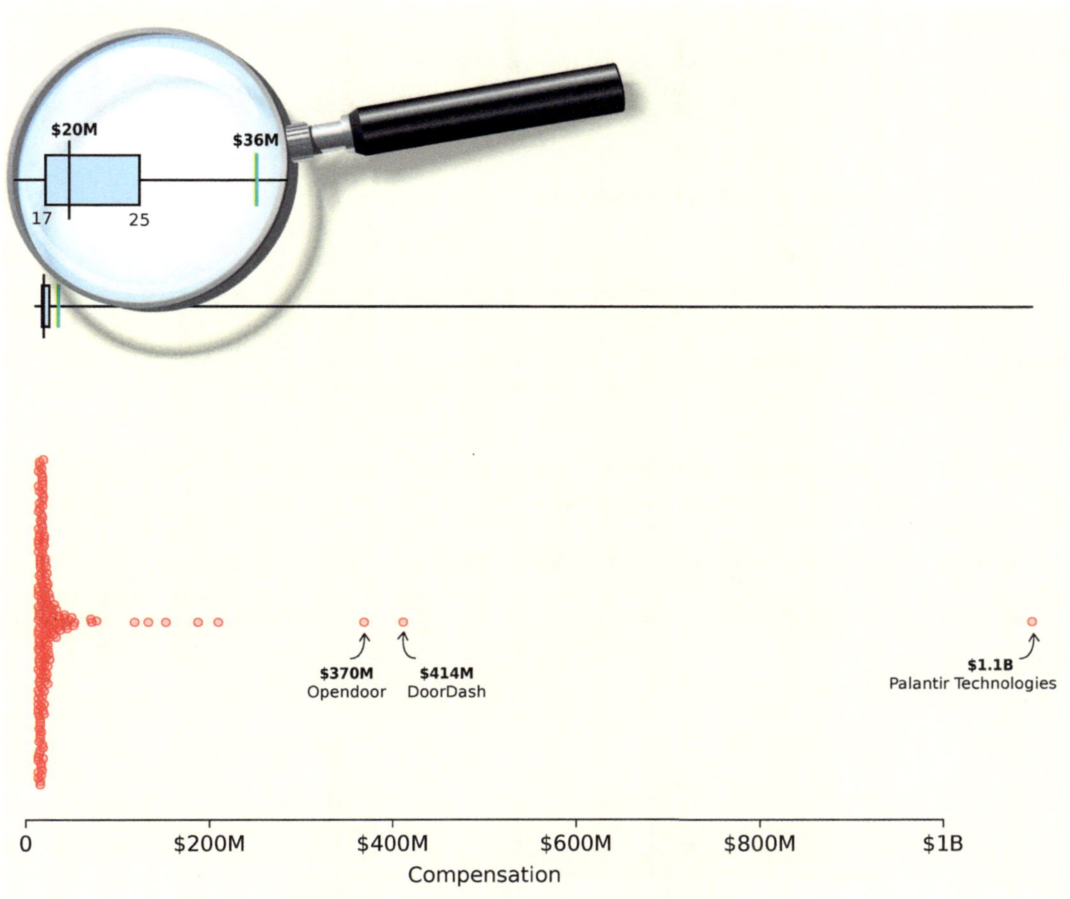

Figure 2.48 Box plot and beeswarm plot of Equilar 200 CEO compensation in 2020.

The data is extremely skewed, so the box plot is a poor choice. Virtually all the data bunches up to the extreme left, so much so that we need a magnifying glass to see what's going on there! A histogram wouldn't do that much better either. But the beeswarm plot is more helpful; it shows the overall distribution and captures the skewness of the distribution, all without losing sight of the extreme values.

Exercise 2.5.3 Try to plot a histogram on Equilar 200 CEO compensation data set and see how it compares to a beeswarm.

Look at the distribution! As another example of the limitations of box plots, let's consider three groups A, B, and C. Their box plots are shown in Figure 2.49. From this presentation, they don't look very different.

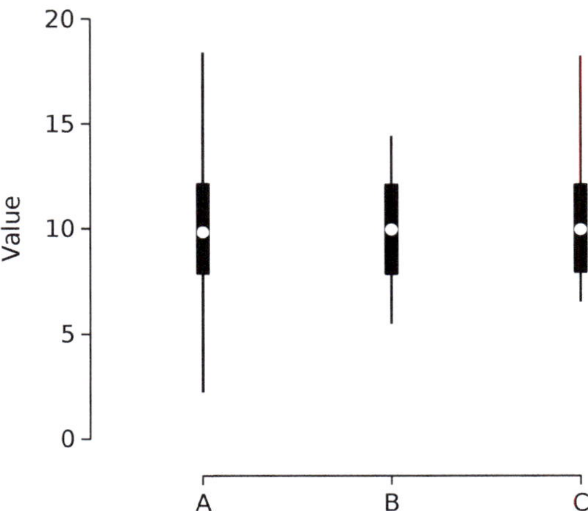

Figure 2.49 Box plots for three data sets. Median values are shown as white dots, the interquartile range (between 25th and 75th percentiles) is shown by the thick bar.

Note that all three groups have the same median (the white dots in the center), the same 25th and 75th percentiles (the thick black bars). There are some differences among their max and min values.

Although the three groups A, B and C have similar box plots, nevertheless, they are very different distributions, as is clear when we show a violin plot of the distributions: group A more or less bell-shaped, group B is markedly bimodal, and group C is significantly skewed to the right (Figure 2.50).

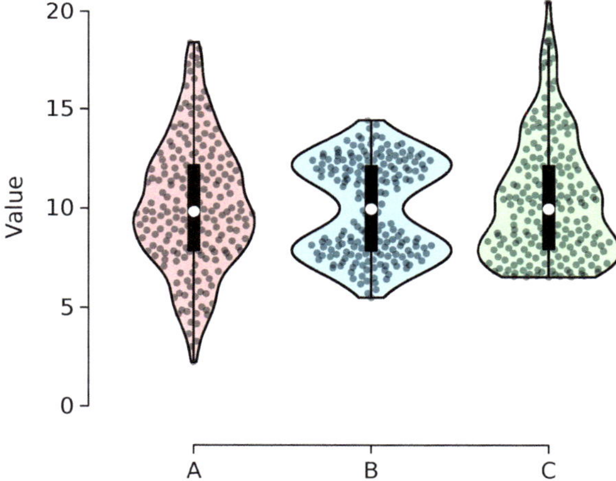

Figure 2.50 Adding a violin plot to the presentation reveals three markedly different distributions.

Exercise 2.5.4 The left panel shows the box plot of a data set. The right panel shows three data sets as violin plots overlayed with actual data points. Which of the three data sets could fit the box plot?

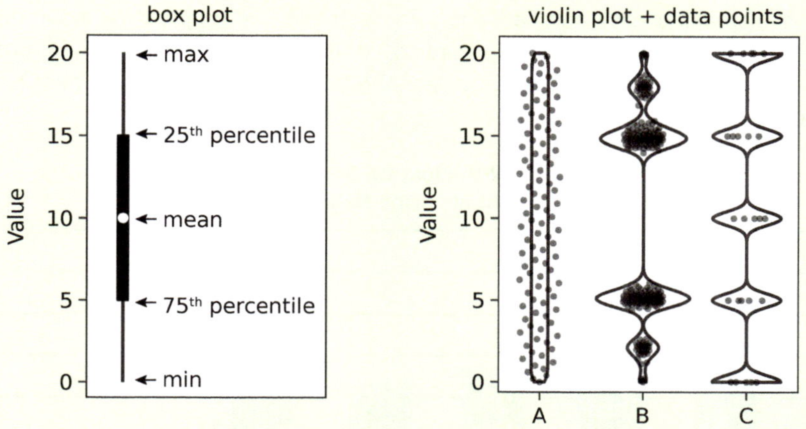

Exercise 2.5.5 Is it appropriate to create a box plot for the date set of US vaccination rate by state?

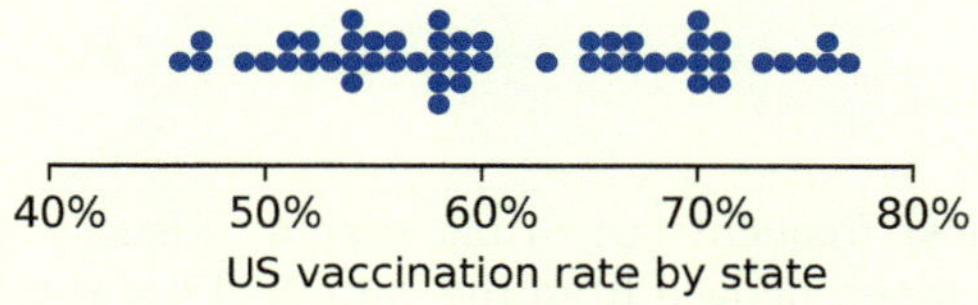

Bar graphs

Probably the form of data presentation that is most common is the "bar graph". Bar graphs are very popular, but, as we will soon see, 90% of the time **they are deeply wrong, and must be avoided at all costs.**

"OK" bar graphs

Bar graphs are no problem at all when the bar is just used to represent a single number, not a summary (Figure 2.51).

Figure 2.51 OK use of a bar graph to show a single raw number.

Of course, we are spilling a lot of ink just to denote a single number. In lieu of this graph, we could have just shown the number "35". All those green pixels convey zero additional information.

But there is nothing wrong with using the bar for visualization purposes. It can even serve a useful purpose of making easy comparisons. For example, we might want to compare the oil reserves of a number of countries, using bars to compare them (Figure 2.52). In this kind of case, the "bar chart" is really just reporting raw numbers, not summaries. There is no problem with this kind of presentation.

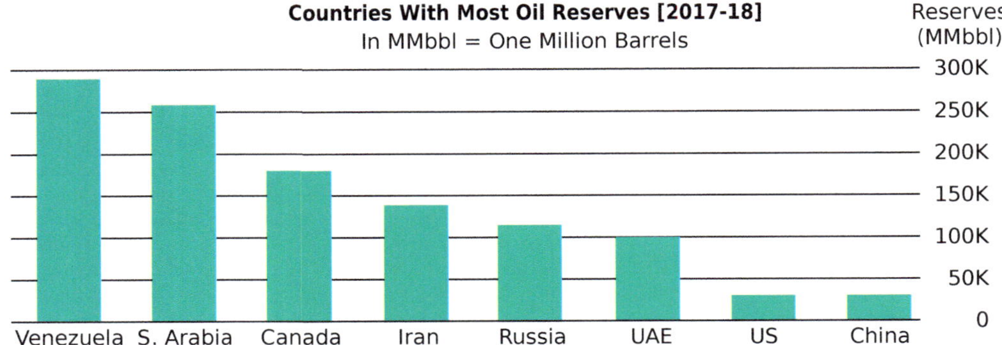

Figure 2.52 Using many bars to represent raw numbers can facilitate a comparison of those numbers.

"Not OK" bar graphs

However, this is not the most frequent use of bar graphs. Usually, the bar is being used to represent some *summary* number, derived from the data, and not the data itself. This is where the problems begin.

The use of "bar graphs" to represent summary data is extremely widespread in scientific publications and presentations. It is very common to read something like "one group was placed on the special diet, while the control group ate as they normally do. The weight loss of the two groups is shown in the Figure", and we are referred to a bar graph, like Figure 2.53.

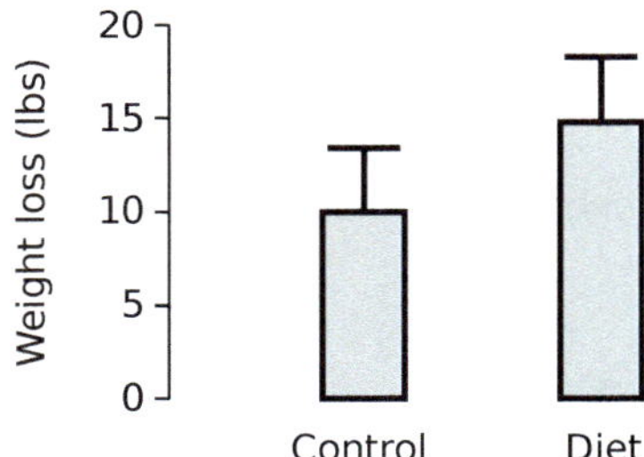

Figure 2.53 Exemplary bar graphs representing findings you would often see in research papers and presentations.

Now we have some big problems.

What do the heights of the gray bars represent? Most of the time, if this is a figure in a research paper or a slide in a presentation, we are not told and we have to guess. Sometimes, but not always, if it is a research paper there will be a "Statistical Methods" paragraph or two, that will say something like "data are shown as mean values, ..." But in research talks, this information is often omitted, and we don't know if it's a mean or a median or . . . .

"Error bars" The other part of the standard "bar graph" is the "error bar" on top of each gray bar, generally in the shape of a capital T (Figure 2.54).

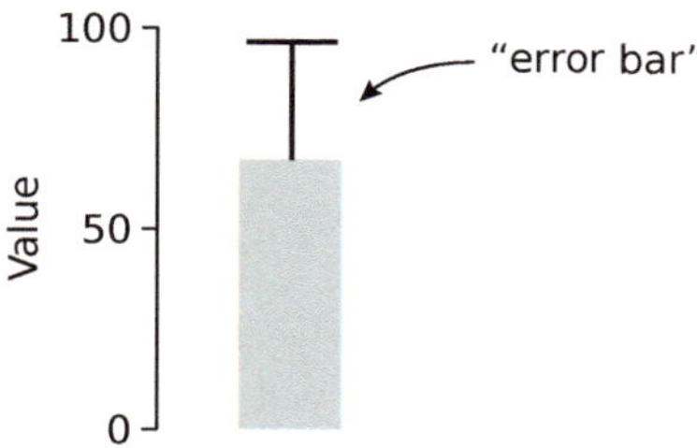

Figure 2.54 A bar graph with an error bar. Some number (the height of the bar) is approximately 65 and some other number (the height of the T) is approximately 30, *but what do those numbers represent?*

The idea behind the error bar is to show the variability of the data in addition to just a summary value. As we've seen, measures of central tendency (if appropriate) should be supplemented by, at the very least, a measure of the variability of the data. The error bar is supposed to indicate some measure of the variability of the data. But what measure? We are usually not even told.

Error bar: "standard deviation"? If we go back to the Statistical Methods section of the paper (if any), we might see that it's the standard deviation of the data set.

But if the standard deviation is being used as the measure of variability, then we are back in all of the problems we encountered with this concept in section 2.4 (*Summarizing a data set: measures of variability*). The standard deviation, we saw, is the square root of the average squared distance-from-the-average-value-of-the-data. Consider that sentence: it uses the concept of average twice. It first finds the average of the data set, squares the distances from each data point to the average, then finds the average of the squared distances, and takes the square root of that average. In section 2.4 (*Summarizing a data set: measures of variability*) we discussed all the problems with using the standard deviation as the measure of variability, so all of those objections apply to standard deviations when they are used as "error bars". In particular, variance and standard deviation are only valid for bell-shaped distributions, where the use of the average is justifiable.

Another, related, problem is that the error bar is frequently shown as rising above the top of the bar. So it is only showing the part of the error above the average. How about the variability below the average? We are forced to assume that those two are equal. But that is only true in a symmetric distribution, and is false if the distribution is skewed. So the one-sided error bar is invalid if the distribution is skewed, and yet that has not been checked.

Error bar: "standard error of the mean"? Another common use for the error bar is to represent the standard error of the mean. Recall that the s.e.m. is the standard deviation of the data, divided by the square root of n. Why would we do that? Why would we use a concept like s.e.m.? What meaning can it have to divide the standard deviation by the square root of n?

The answer is buried very deeply in the mathematical theory of bell-shaped, so-called Normal distributions.

Recall that the idea behind the s.e.m. is that while there might be a lot of variability within a data sample, there is a lot less variability in the average values of data samples. For example, within a given chemistry class, there will be some As and some Ds, but if you look at the average grade in the class, it's typically not going to be an A or a D. So there's much less variability among average grades across classes than there is within a class.

How much less? There is a theorem that says that **if a distribution is truly Normal** (that is, Gaussian, see Chapter 5) and we sample n from it, producing a set with mean m, then we can calculate something called a "95% confidence interval" as

$$m \pm 1.96 \times \text{s.e.m.}$$

and, assuming that the distribution is Normal, we can then conclude that 95% of the time, confidence intervals calculated in this manner will contain the true mean of the underlying Normal distribution.

For that reason, the interval $[m - 1.96 \times \text{s.e.m.}, m + 1.96 \times \text{s.e.m.}]$ is called the 95% **confidence interval** for the mean m. So the s.e.m. is an estimate of how variable the average is (if the distribution is Normal). It can be thought of as telling us what the precision is of the sample mean m as an estimate of the mean of the underlying Normal distribution. So the idea behind the notion of confidence interval is that we would like to have some estimate of the precision of this mean as an estimate of the mean of the whole, unseen, underlying population.

The sentiment behind confidence intervals is valid: it would be informative to know how precise the average value of our sample m is as an estimate of the mean of the unseen population. However, the idea that the confidence interval can be calculated by this formula depends entirely on the assumptions that the distribution of the underlying universe is Normal (Gaussian), and that we are using the mean value as our descriptor.

We present a resampling-based approach to confidence intervals in Chapter 4. This resampling approach does not depend on assumptions about the shape of the distribution, and is valid for any summary measure, not just the mean.

What bar graphs hide

The most significant criticism of bar graphs is that they hide important facts about the data and the analysis. Indeed, some writers have suggested that sometimes, the very purpose of using bar graphs is to hide embarrassing facts about the data.

How large is the "n"? Notice that the n, the size of the data set, isn't shown in the bar, which therefore hides that fact about the data. The forms of data presentation we learned in this chapter, such as dot plots, beeswarm plots, histograms, etc, make it easy to see how large the n is, and therefore the quality of the study. But a bar graph simply hides this fact, which is a poor practice. It's an act of deception, whether intentional or not.

What's the distribution? And then the most important and most pressing question is: **what is the distribution of the data? is it skewed, etc.?**

After all, as we have just learned, **if the data distribution is not bell-shaped, summarizing the data by the mean value can be a bad idea.** Therefore, anyone who is showing you an average value is asking you to trust them that the data distribution is bell-shaped. But this is often not the case, and the average is being used wrongly. **Using bar graphs is a way of hiding the data: the investigator does not want you to see the size or the distribution of the data. We have to ask "why?"**

These are very good questions:

- what is the "n"?
- what is the distribution of the data?
- does the distribution justify the measures used?

None of these questions is answered by the bar graph. In fact, the bar graph is designed to avoid them. It is impossible to overstate the damage that has been done by investigators using bar graphs to hide data, whether it is hiding a small "n" or hiding highly skewed data, by just showing the bar for the average, and leaving the rest unsaid. Trying to hide unattractive features of our data by using bar graphs is dishonest and poor science. We should always use data presentation methods that force a discussion of the size and shape of the data, and do not suppress the variability of the data, its range, and even the number of data points we have! Unfortunately, this is widely done. It is very common to see a data set presented as just a bar graph representing the mean value, possibly with an "error bar" representing the standard deviation or the standard error of the mean. Using this form of presentation, hiding the distribution of the data, hiding the "n", etc. is unhelpful at best, and dishonest at worst.

It's important to re-state the fundamental principle: **measures that depend for their validity on the distribution being bell-shaped cannot be used without first establishing that the distribution is, in fact, bell-shaped.**

It is malpractice to use these measures in the absence of this verification. The use of s.d. (or s.e.m.) is only valid if the distribution is bell-shaped.

Under no circumstances should we present only a summary measure, like a mean, as a "bar graph" to represent the data set. This practice suppresses the variability of the data, its range, and even the number of data points we have!

"Dynamite Plunger Plots"

This is the basis of the "error bar": in the typical bar-graph-with-error-bar, we are seeing the average value as the gray bar and the standard deviation or the s.e.m. as the T-shaped bar on top.

The respected statisticians Gordon Drummond and Sarah Vowler wrote an editorial in the *Journal of Physiology* in 2011, whose title was "Show the data, don't conceal them" (Drummond and Vowler, 2011). Note the ethical tone of the title; they sharply criticized the practice of showing means and standard deviations as bar graphs with "error bars"; they are accusing bar graphs, quite rightly, of "concealing the data". We agree.

The Journal of Physiology STATISTICAL PERSPECTIVES

Show the data, don't conceal them

G. B. Drummond and S. L. Vowler

In that paper, Drummond and Vowler coined a derisive name for these data-hiding bar-graphs-with-error-bars. They called them "dynamite plunger plots" (Figure 2.55).

Figure 2.55 Wile E. Coyote has his hands on a dynamite plunger, preparing to kill the elusive Roadrunner bird. It typically does not go well for the coyote.

(For those that didn't grow up watching American cartoons, Figure 2.55 shows the devious Wile E. Coyote from the Roadrunner cartoons, with his hands on a dynamite plunger, in another vain attempt to kill the Roadrunner.)

Their derogatory name for these bar graphs, "Dynamite Plunger Plots" has caught on.

An editorial in the prestigious *Journal of Clinical Investigation* took the same tone, saying

> **JCI** The Journal of Clinical Investigation
>
> "To maintain the highest level of trustworthiness of data, we are encouraging authors to display data in their raw form and not in a fashion that conceals their variance. Presenting data as columns with error bars (dynamite plunger plots) conceals data. We recommend that individual data be presented as dot plots shown next to the average for the group with appropriate error bars."

We strongly support this movement away from "bar graphs," with or without "error bars". The bar graph suppresses the shape of the distribution.

A good illustration of why this is important can be seen in the data in Figure 2.56. There are three different groups: A, B and C. The distribution of their data is as shown. In each data set, the average value is 100 and the standard deviation is 35. Therefore, **all three data sets have the exact same dynamite plunger plot, which is shown at the top.**

If you had used the dynamite plunger presentation, you would have completely missed the interesting differences among the data sets. Suppose that A was the control group, and B and C are two different drugs. Did the drugs have any effect? Yes, certainly. Is there a result here? Absolutely! Drug B eliminates all the variation, increasing small values and decreasing large ones, it drastically tightened the distribution to a value below the control group mean, with the exception of one poor (or lucky) individual who had an even higher value than was seen in the control group. Drug C, on the other hand, has a different type of effect: it has separated the population into 2 distinct subgroups, high responders and low-responders, which of course leads us to ask questions like: who are the 2 subgroups? are they men and women? old and young? etc.

This entire discussion would be lost if we just used dynamite plunger plots. Bar graphs suppress all this, because each data set has the same bar graph and "error bar"! This is why we avoid bar graphs.

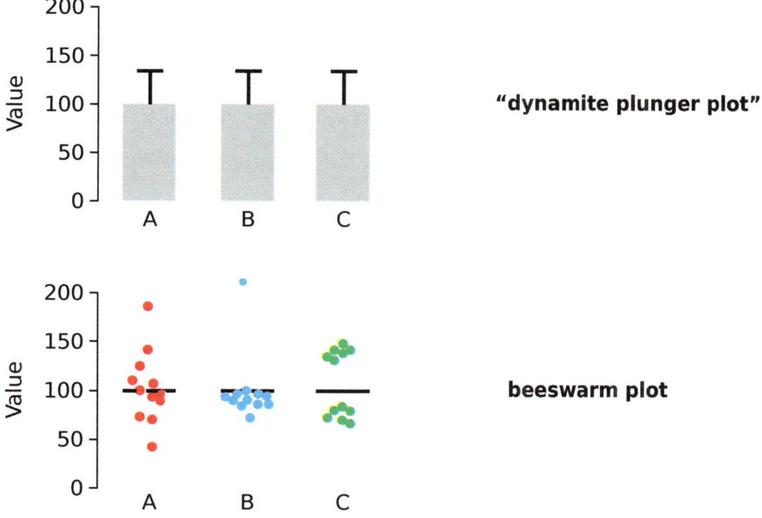

Figure 2.56 Bar graphs conceal the data.

Avoid making Dynamite Plunger Plots!

We stress here the approach that asks us first of all to look at the data visually, and to take special note of its distribution, looking for key features like variability and skew. The graphic presentations that best help with this key task are dot plots, beeswarm plots, histograms, and kernel density estimates/violin plots.

We have not said yet what the best candidate is for the "error bar". For now, we can use the variance or MAD. But what it is really calling for is a *confidence interval*, the subject of Chapter 4.

In summary, we have seen good and bad ways of describing and presenting data.

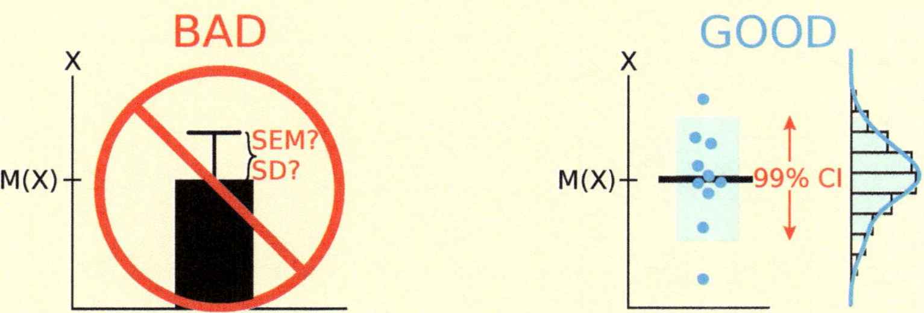

On the left are the bad ways: dynamite plunger plots.

- use of the mean without justification
- unexplained error bars, that were in fact calculated using formulas that depend on the data being bell-shaped
- hiding the n
- concealment of the distribution

On the right are the good ways:

- dot plots showing the data and their distribution
- a histogram and a KDE also showing the distribution of the data
- a 99% confidence interval (shaded bar) for the measure M of the data set X.

These are the good and bad ways to present data. The only unanswered question is: how can we calculate confidence intervals if our data set is not bell-shaped, and/or we don't wish to use the mean as our descriptor? Is there a general method for calculating confidence intervals? The answer is yes, and we'll dive into that in Chapter 4 (*Confidence Intervals: one measure, one group*).

Plotting the data vs. summary numbers

There's a well-known figure that illustrates the perils of only calculating summary numbers to describe data. It's called "The datasaurus" (Figure 2.57), and it was developed by Alberto Cairo. There are 13 wildly different data sets, *but every one of them has the same summary numbers.* Each data set has the same X-mean, the same Y-mean, the same standard deviation in X, the same standard deviation in Y, and the same correlation coefficient (approximately = 0) (see Chapter 9 *Bivariate Data: Correlation and Beyond*).

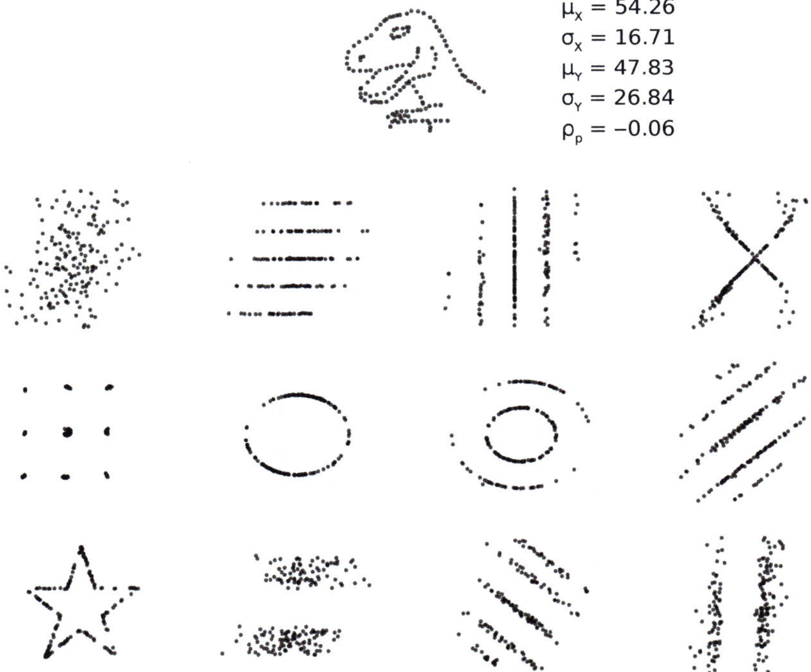

Figure 2.57 The datasaurus. All 13 data sets have the same summary numbers, shown on top.

If a researcher were to obtain one of these data sets, and only calculated summary numbers, they would completely miss the structure of the data, and end up throwing out a valuable finding!

FURTHER EXERCISES 2.5

1. What are we looking for in a data plot?

   a) Skewness vs. symmetry
   b) Thick tails vs. thin tails
   c) Outliers
   d) Subgroups (multi-modality)
   e) Variability

2. Give two examples of information that is lost when only a measure of central tendency and a measure of variation are used to describe a data set, as in a bar chart with error bars.

3. Which of the following types of graphs hides information about modality?

   a) Plunger plot
   b) Dot plot
   c) Box plot

4. What is the interquartile range?

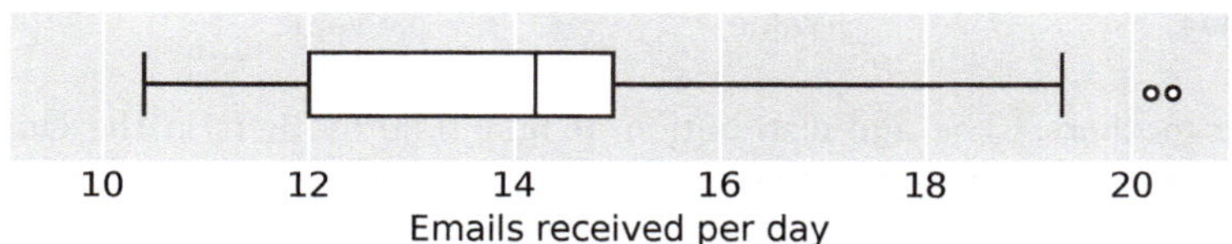

5. **What Plunger Plots miss.** Two data sets, A and B, show the daily number of power outages in a solar farm, before and after implementing a grid-balancing system. In the plunger plots, the bars represent the mean number of outages, and the error bars show the SEM. Compare the two types of plot and describe what information is lost in the plunger plots.

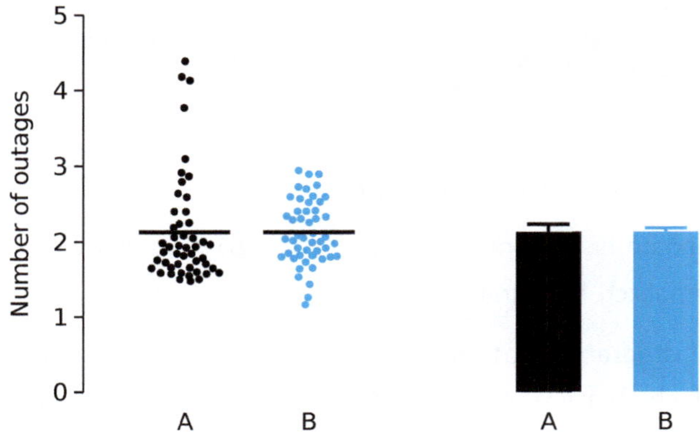

6. **Binning your data.** When we use histograms to represent a data set, it is essential to choose a proper number of bins. The three histograms below are based on the same data set.

   a. Which histogram would be the appropriate presentation of the data set? Explain why.
   b. What is the rule of thumb when it comes to selecting the number of bins?

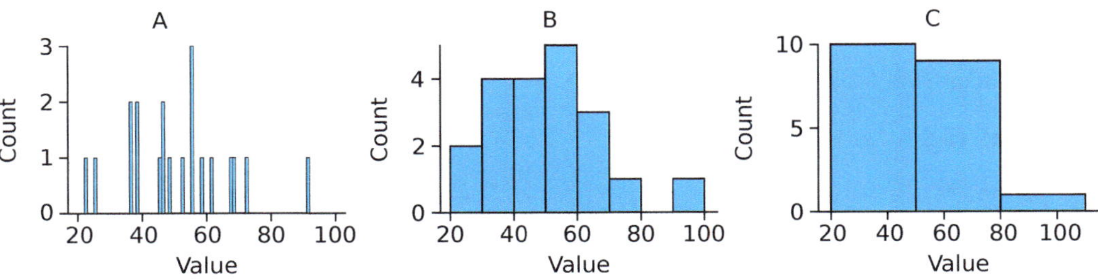

7. Which box plot is right skewed?

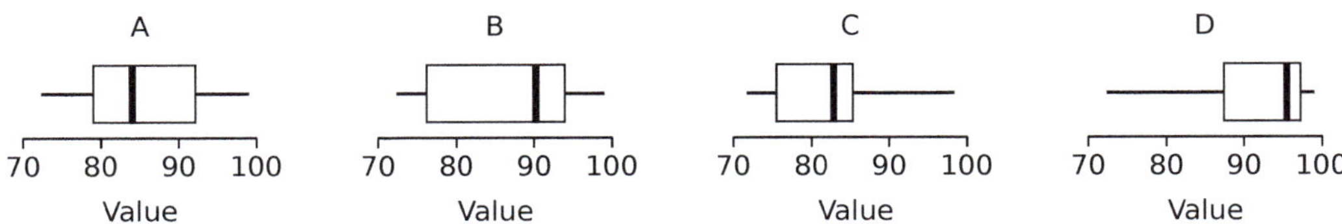

8. **First time mothers.** The age distribution of first time mothers in the United States has changed greatly over the last forty years (*New York Times*, 2018). Here are two histograms showing the ages of first time mothers in 1980 and 2016.

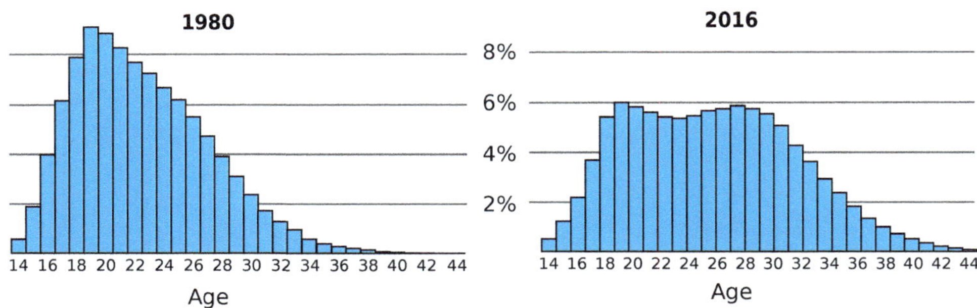

   a. What is the type of the histograms?

      a) Standard histogram
      b) Cumulative histogram
      c) Normalized histogram
      d) Cumulative normalized histogram

   b. For each histogram, what are the sums of the heights of all bins?
   c. Decide which measure of central tendency (if any) you would present to an audience and explain your reasoning.
   d. Describe/interpret each histogram. Discuss symmetry/skew, modality, tails, outliers.

9. **Grade distribution.** Consider the following relative frequency distribution, which details the results of a midterm exam taken by 120 students.

Grade	Less than 50	50–59	60 –69	70–79	80–89	90–100
Relative frequency	15%	10%	30%	20%	12%	13%

a. The median score is in which grade range?
b. How many students got at least a 70 on the exam?
c. What is the interquartile range?
d. How would you describe the grade distribution?
e. We have seen that problematic results can occur due to outliers. Rank the following measures in the order of "least affected by outliers" to "most affected by outliers".

   a) mean, median, range
   b) median, mean, range
   c) range, median, mean
   d) median, range, mean
   e) range, mean, median

10. **Emergency Room visits.** The distribution of Emergency Room wait times is shown below.

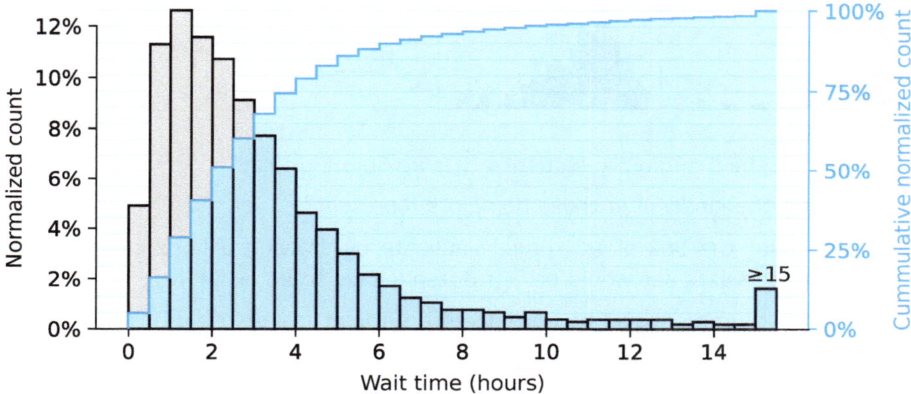

a. What types of graphs are shown?
b. What is the probability that a patient will have to wait more than 2 hours?
c. What is the probability that a patient will have to wait more than 4 hours?
d. What is the longest time a patient will have to wait?
e. If the Board of Health requires an ER to process 75% of the patients within 2 hours, is the hospital meeting this requirement?
f. Estimate the median and roughly sketch a box plot for ER wait times.

11. Write the type of visualization that is appropriate for each situation. The options are: bar chart, beeswarm plot, box plot, and violin plot. Some may be used more than once or not at all.

a. Our data is the number of times A,B,C,D,E were the correct response on a multiple choice exam with 100 questions. We want to know if C is the most common correct response.
b. Our data is 100 data values for each of two treatments, and we want to know the values at 25%, 50%, and 75% of the data for each group.
c. We have 200 data points for each of three treatments, and we want to know if the distribution of each group is bimodal or unimodal.

12. Match the histogram with the corresponding box plot.

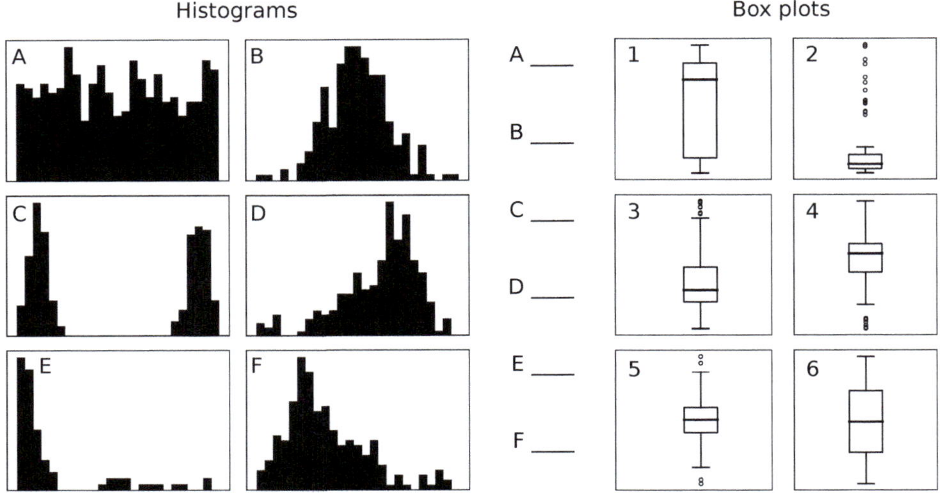

a. What are the 5 summary statistics that box plots display?
b. What do the circles ∘ in those box plots represent?
c. On each of the box plots, roughly indicate by drawing a horizontal line where the mean of the data set would be. (It doesn't have to be exact.)
d. Which histograms would be considered a symmetrical distribution?
e. The data in plot F can best be described as

a) Right-skewed b) Bimodal c) Bell-shaped d) Left-skewed

f. Select all distributions for which the mean would be a useful and informative measure of central tendency.

13. **Daily exercise time.** Answer the following questions based on the figure below:

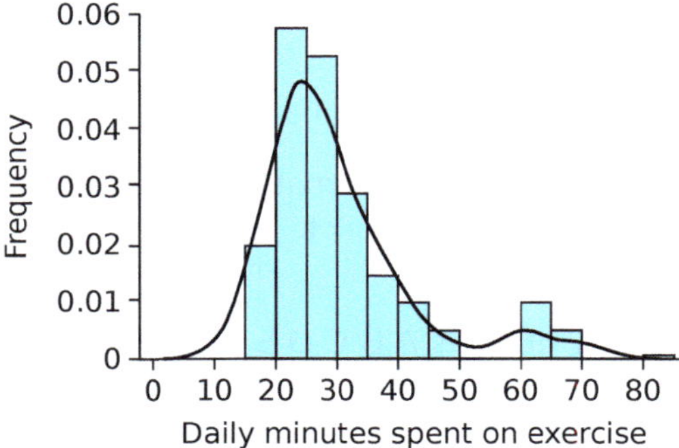

a. The type of plot the bars represent is a(n) ________.
b. The curve is called a(n) ________.
c. What measure of variation would be best suited to this data? Give one reason why.

d. Briefly describe the data in terms of shape, skew, and modality (including approximate values of any modes).

14. **Longevity.** Here is a histogram representing the longevities of mammals.

    a. Comment on the shape of the data. Discuss symmetry/skew, modality, tails, and outliers.

    b. Would a beeswarm plot be better to represent this data set? Explain why.

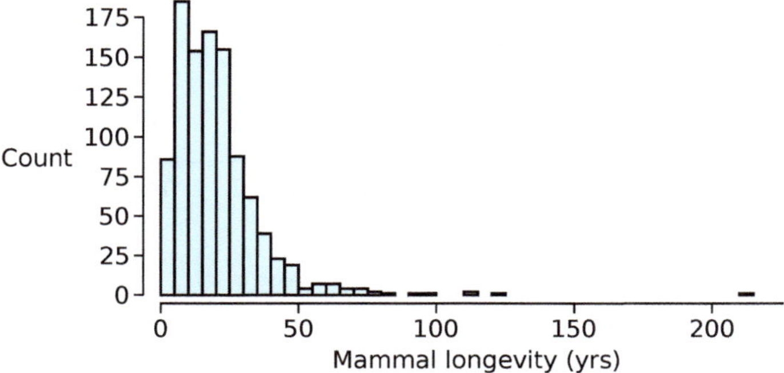

15. **Air pollution.** The figure below shows the results of a study tracking air pollution levels (in $\mu g/m^3$) on weekdays versus weekends across three different cities. Describe three things that should be improved.

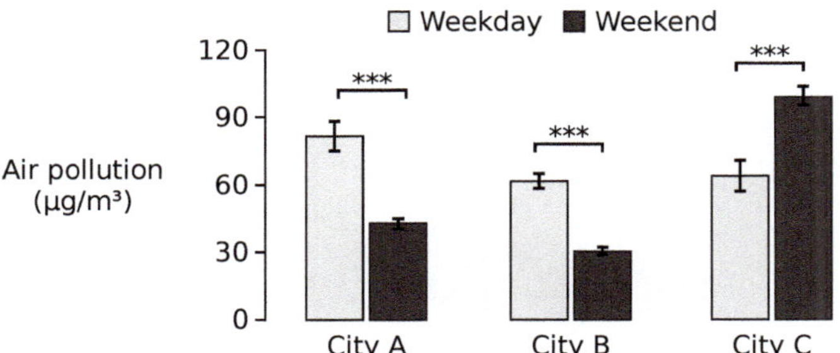

16. **Traffic.** The plot below shows the average traffic congestion level on a Friday in Los Angeles. What type of plot is this? Is it appropriate to use this type of plot?

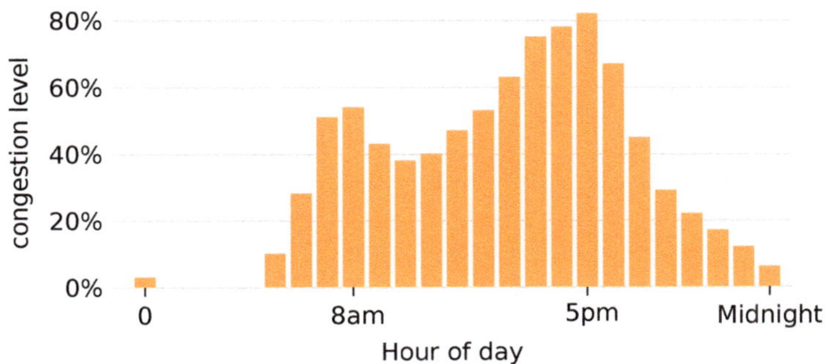

17. **Clinical trials.** Would you recommend using this type of graph in the figure below to depict this data? Provide three reasons for your recommendation based on this specific figure.

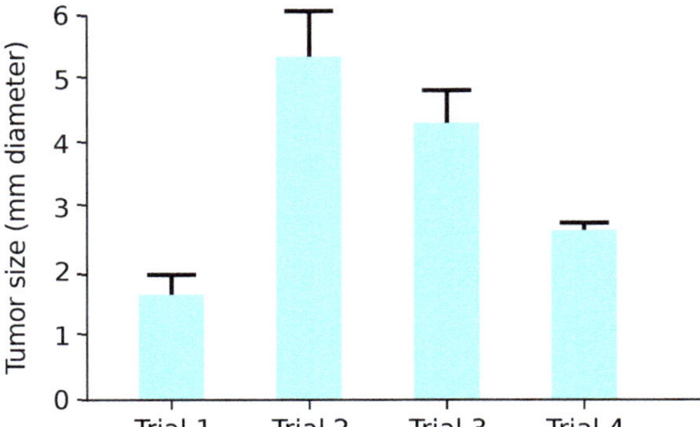

Chapter 3

Introduction to Probabilities and Hypothesis Testing

LEARNING OBJECTIVES

After reading this chapter, you will be able to

- calculate the probabilities of various kinds of events by making box models and sampling from them.
- determine the degree of statistical significance of an observed result as compared to theoretical expectations.
- state clearly five reasons why "Null Hypothesis Significance Testing" is problematic.
- explain the difference between "Type I" and "Type II" errors in hypothesis testing.

3.1 Modeling and simulating probabilities

What is probability?

How do we estimate the probability of something occurring? Sometimes, estimating probabilities can be very hard. If we are asked to estimate the probability of a female US president, or of peace in the Middle East, or even of your favorite team winning the championship, probability estimates are difficult and may be controversial.

But sometimes, estimating probabilities can be easy, because we know in advance exactly how many distinct, equally likely events are possible, and we just have to count how many of them satisfy our criterion. Estimating the probability of drawing an ace in poker is easy: there are 52 cards, and 4 of them are aces, so the probability of drawing an ace is $\frac{4}{52}$ or around 8%.

In this case, the *probability* of drawing an ace is interpreted as the *frequency* of aces in the deck. This is the "frequentist" interpretation of probability, and it works well when the event in question can be put in a standardized form like drawing from a box whose contents are known, or flipping a coin.

If we are interested in calculating the probability of more complex events, there are often helpful formulas. For example, the probability of drawing a pair of aces is the probability of drawing one ace from a deck of 52, which is $\frac{4}{52}$, and then drawing another ace from the

A. Garfinkel and Y. Guo, *Understanding Data*,
https://doi.org/10.1007/978-3-032-18600-3_3

remaining cards, which is $\frac{3}{51}$. We use an elementary law of probability that if X and Y are independent, then the probability of having both X and Y is equal to (the probability of X) × (the probability of Y). Drawing an ace from a deck of 52 cards is independent of drawing an ace from a deck of 51 cards that has 3 aces in it. Using the multiplication rule, we calculate the probability of drawing two aces from a poker deck of 52 cards as $\frac{4}{52} \times \frac{3}{51}$, which is around 0.005, or one half of one percent.

So far, so good. But what happens when we want to calculate more complex events? For example, a straight in poker is a hand of 5 cards that can be arranged into a sequence of 5 consecutive ascending numbers, such as (7, 8, 6, 4, 5) or (J, K, Q, A, 10) (Figure 3.1).

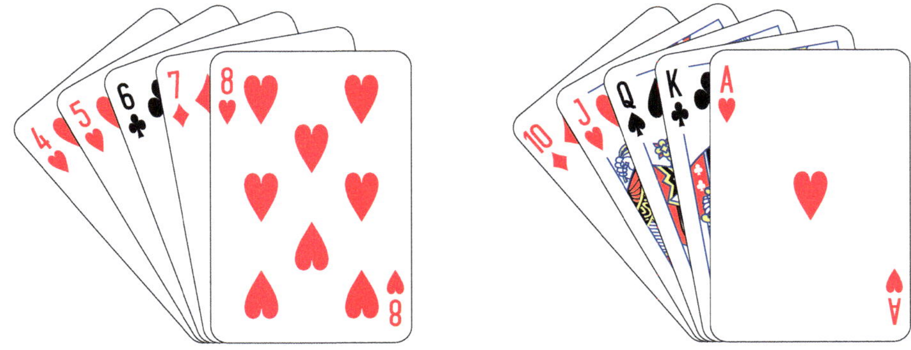

Figure 3.1

What is the probability of being dealt a straight in poker? There is no easy formula. And yet there are only finitely many possibilities, and these can be worked out one by one in a brute force way. But that is a long calculation, with many special cases.

There are also cases where there might be a formula, somewhere, but we don't remember it.

What is the probability of a family with 5 children having 4 or more of them being girls? (Figure 3.2).

Figure 3.2 What is the probability of a family with 5 children having 4 or more of them being girls?

There is a theoretical way to calculate this, in terms of a formula for "the number of combinations of n things taken k at a time" with lots of factorials in it, but what if we don't have that handy and/or are unsure of the assumptions that went into the formula?

Exercise 3.1.1 Find the probability of getting exactly two heads when flipping three coins.

Exercise 3.1.2 What is the probability of throwing one die and getting a number greater than 4?

Exercise 3.1.3 What is the probability of throwing two dice and getting "snake eyes" (two 1s)?

Exercise 3.1.4 What is the probability of throwing two dice and getting a total of 11?

Exercise 3.1.5 What is the probability that a single card chosen from a poker deck is not an ace?

Exercise 3.1.6 Three cards are chosen at random from a poker deck. What is the probability of choosing a five, a four and a three, in order?

Simulating probabilities

There is another way to answer these questions of probability. In the case of drawing a straight in poker, we could deal out millions of 5-card hands, shuffling carefully each time, and just count how many of them were straights. In the case of the "4 or more girls problem", we could let a black card mean "is a girl" and a red card mean "is a boy", and then deal out millions of 5-card hands, and count how many of them had 4 or more black cards.

These are what we might call **physical simulation** approaches to calculating probabilities.

It's a lot easier, and much faster, to write a computer program to simulate the same process. Languages like Python all have random number generators in them, and therefore can carry out simulations of random choices from a list, the equivalent of shuffling a deck of cards and dealing from it. "Deal a million hands" is no problem for modern computing.

Box models We formulate a **box model**, in which we set out the list of possible choices, as it were in a box, and then set out rules for randomly choosing from the box, to model any particular problem. The box consists of a set of elementary objects. The objects could be numbers, or they could be symbols for the colors (Red, Blue, Green), or any set of symbols that can be put in a list. The "outcome" is modeled as some combination of elementary objects, such as "drawing 2 Reds and a Green". Then we simulate the outcomes of the random model by choosing randomly from the box to replicate the more complex condition, such as drawing a straight in poker or picking 2Rs and a G from the box (R, B, G). We do this thousands of times, and then we just count what percentage of the simulations had the property we are interested in.

When the problem can be put in the form of "choosing from a box model", we can estimate probabilities by simulation of the box model.

Simulating the "4 or more girls" problem For the "4 or more girls" problem, we would have a box model with 2 elementary objects: a symbol for "girl" and a symbol for "boy". For each family, to form a set of 5 children, we go to the box model 5 times, each time randomly selecting one of the two elements (Figure 3.3).

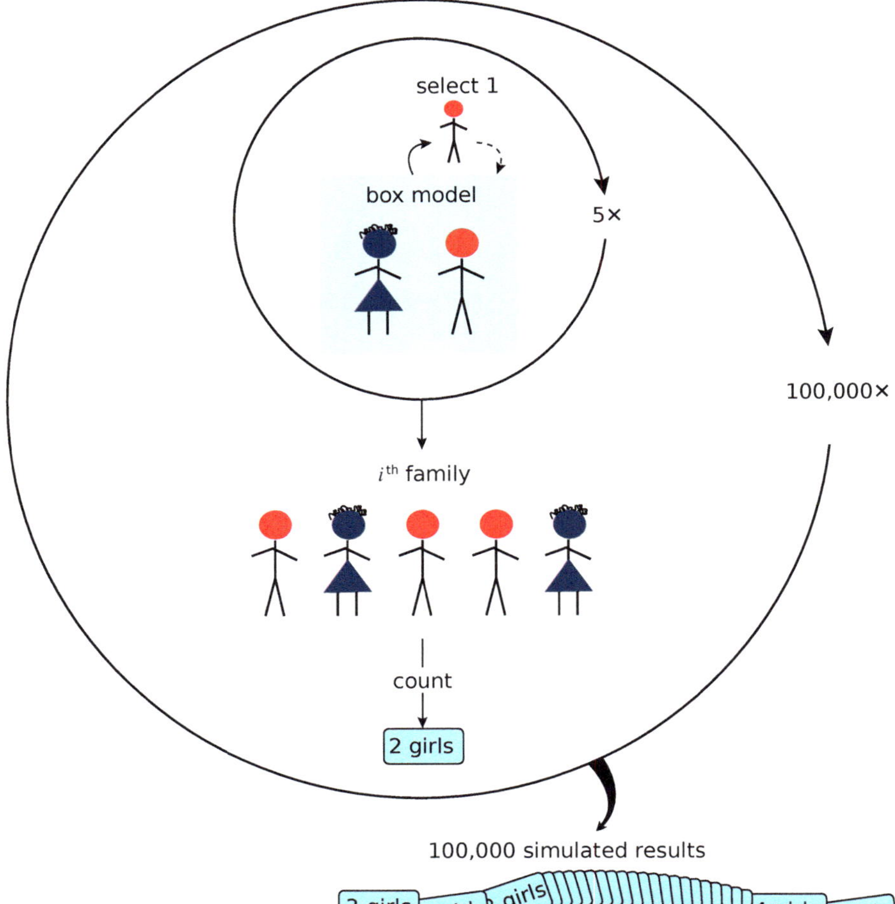

Figure 3.3 How to simulate the probability of having 4 or more girls in a family with 5 children.

"randomly selecting one of the two elements"

When we say "randomly selecting one of the two elements", we have to further specify the probability of selecting each element. In the simplest model, the two might have equal probability, but we can also specify unequal probabilities. For example, we might learn that in a particular region, female births outnumber male births 55:45. We can formulate a box model that has a 55:45 probability built into it (say, a box with 55 girls and 45 boys), to determine the likelihood of 4 or more girls in that particular scenario.

Then, for each simulated family, count the number of girls and record that number. We repeat the process of forming a family and counting the number of girls many times, such as 100,000 times. Eventually, we would have simulated the formation of 100,000 families and have 100,000 values of the number of girls.

Then, to answer the question, what is the probability of having 4 or more girls in a family with 5 children, we count the number of the 100,000 families that have 4 or 5 girls, divided by the total number of simulations, 100,000.

Here's the pseudo-code:

3.1.1 Probability for 4 or more girls in a family with 5 children

```
Make a box called gender
Put 2 elements in it:  boy and girl
gender = [boy, girl]

Define a family as a sample of 5 (with replacement) from the gender box.
For each family, let numGirls be the number of girls in that family.

for i = 1, 2, ..., 100,000 do
    Draw a family from the gender box, that is, a sample with replacement
     of size 5
    This generates the i^th family
    Calculate the number of girls in this i^th family, numGirls_i
    (For example, family_i = [girl, girl, boy, girl, boy], and the number
     of girls in this i^th family, numGirls_i = 3)
end

We now have 100,000 values of numGirls
Count how many have numGirls = 4 or numGirls = 5
Call that number count
```

The estimated probability is $\frac{\text{count}}{100{,}000}$

We could also make a histogram of the 100,000 values. The histogram from our simulation looked like this (Figure 3.4).

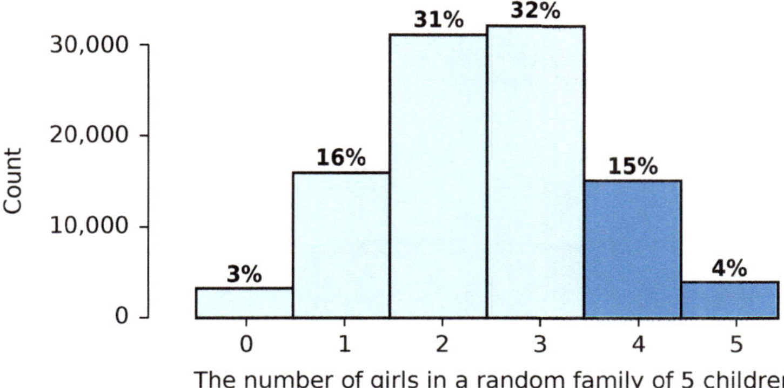

Figure 3.4 Histogram showing the distribution of the number of girls in 100,000 simulated families with equal probabilities of boys and girls.

So that's the answer: the probability of a family of five children having 4 or more girls is estimated to be around 19% (the actual value in our simulation was 18.79%). **What formula did we use? We didn't use a formula, we carried out many simulations.**

Long run pattern of simulating probabilities

How many simulations to carry out? How many is many?

First let's take a look at a simple case of flipping one coin. If the coin is fair, then we know, by theory, that the probability of Heads is 50%.

When using simulation to calculate the probability, the result evolves as we increase the number of simulations. In our first run, the first iteration of the simulation gives out a Head, so the probability of Heads from this one simulation gives us a result of 100%. But then we got more Tails and the probability came down below 50%, but then went above 50% again.... It fluctuates with smaller and smaller amplitude and stabilizes at 50% by 5000 iterations.

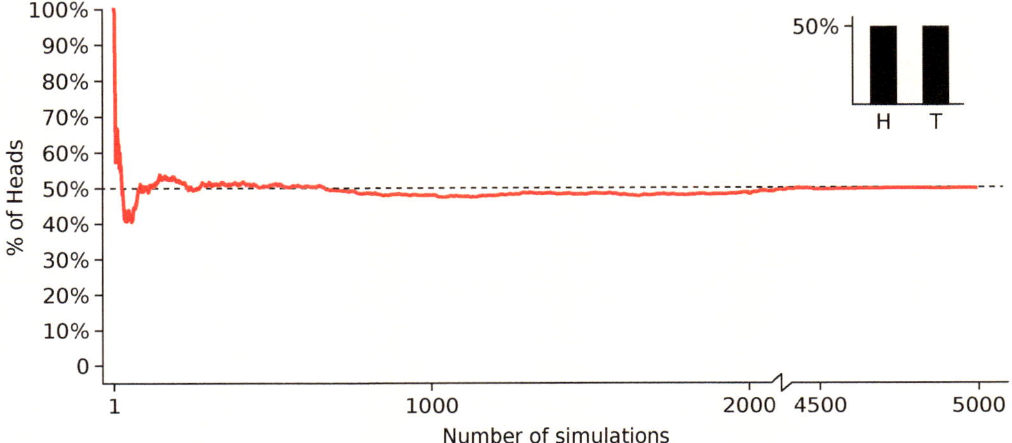

Not every run has to follow the same route, here we show two more runs (green and blue lines). The overall pattern is that the simulated probability evolves and approaches the theoretical solution as the number of simulations increases.

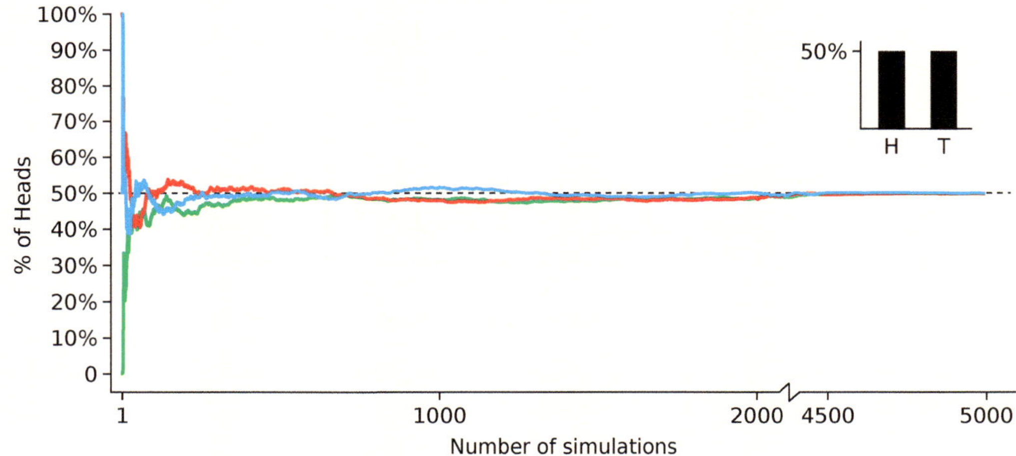

So when we say many, thousands of iterations will give us a relatively stable result. In this text, we'll typically use 10,000.

As for the probability of 4 girls in a random family of 5, we showed the final result of 100,000 simulations in Figure 3.4. Below is the actual route that our simulated probability has taken. Yours would look different, but the final results are all approaching the same theoretical solution.

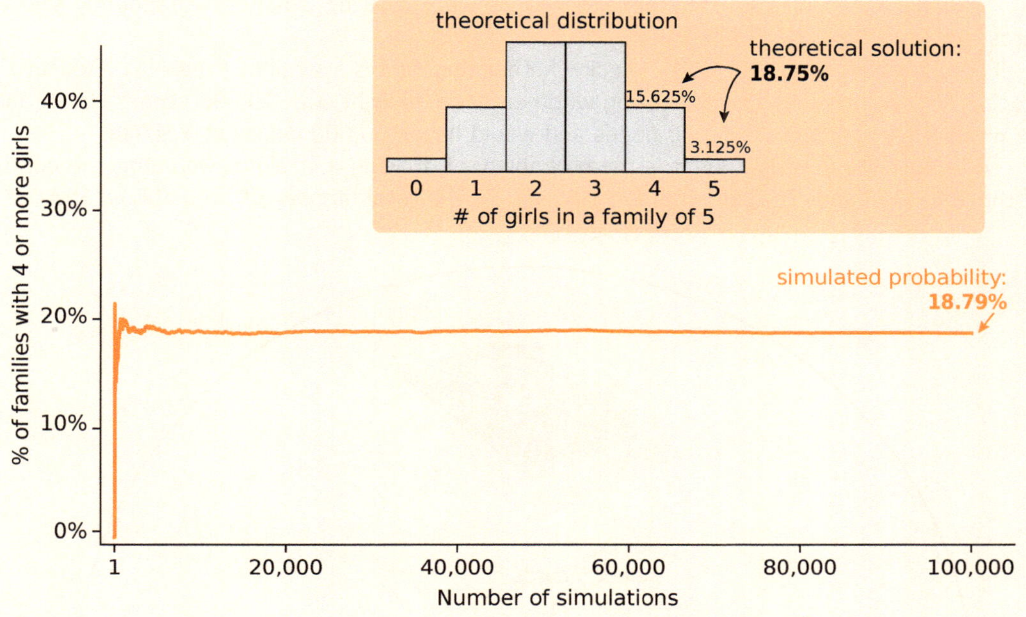

As we can see, 100,000 simulations are more than enough. The simulated probability stabilizes around 10,000 and doesn't change much afterwards.

"The formula"

We have to answer an important objection. Someone might say: But there really ***is*** a formula for this, and mathematicians worked hard to produce it, and I had to memorize it in high-school or college math, although I can't quite remember it right now. Something about "factorials". Shouldn't we be using it? How can we disregard the math?

The answer is: yes, there is a formula for this case. It produces the answer 0.1875, very close to our estimate of 0.1879. But then we have to learn some abstract probability theory, which may or may not be interesting to us, who really just want an answer to our question.

But we are going to resist the formula approach here.

- First of all, are we really sure that the correct formulation of the "4 or more girls" problem is correctly put as "Choose n from k", and that the factorials are all set up right? Or are we in danger of committing what is sarcastically called "Type III error: using the wrong formula"?
- Second, and more important, ***we need to calculate probabilities for cases where there is no formula.***

Simulating the probability of drawing a straight In poker, a straight is a hand that consists of five cards in sequential rank, regardless of suit. For example, 7, 8, 9, 10, J is a straight, whatever suits the cards have.

Let's calculate the probability of drawing a straight in poker. There is no formula for this!

In the simulation approach, we would start with a box model with 52 elementary objects: 1, 2, ..., 13 with 4 copies of each number. To simulate drawing a hand, we randomly select 5 cards from this box model (Figure 3.5).

Then, for each simulated hand, check whether the hand is a straight (Yes/No). Repeat this process of drawing a hand and checking whether it is a straight, say, 100,000 times. Eventually, we would have simulated 100,000 hands and would have 100,000 values of Yes/No.

To answer the question, what is the probability of drawing a straight, we count the number of the 100,000 hands that are straights, divided by the total number of simulations, 100,000.

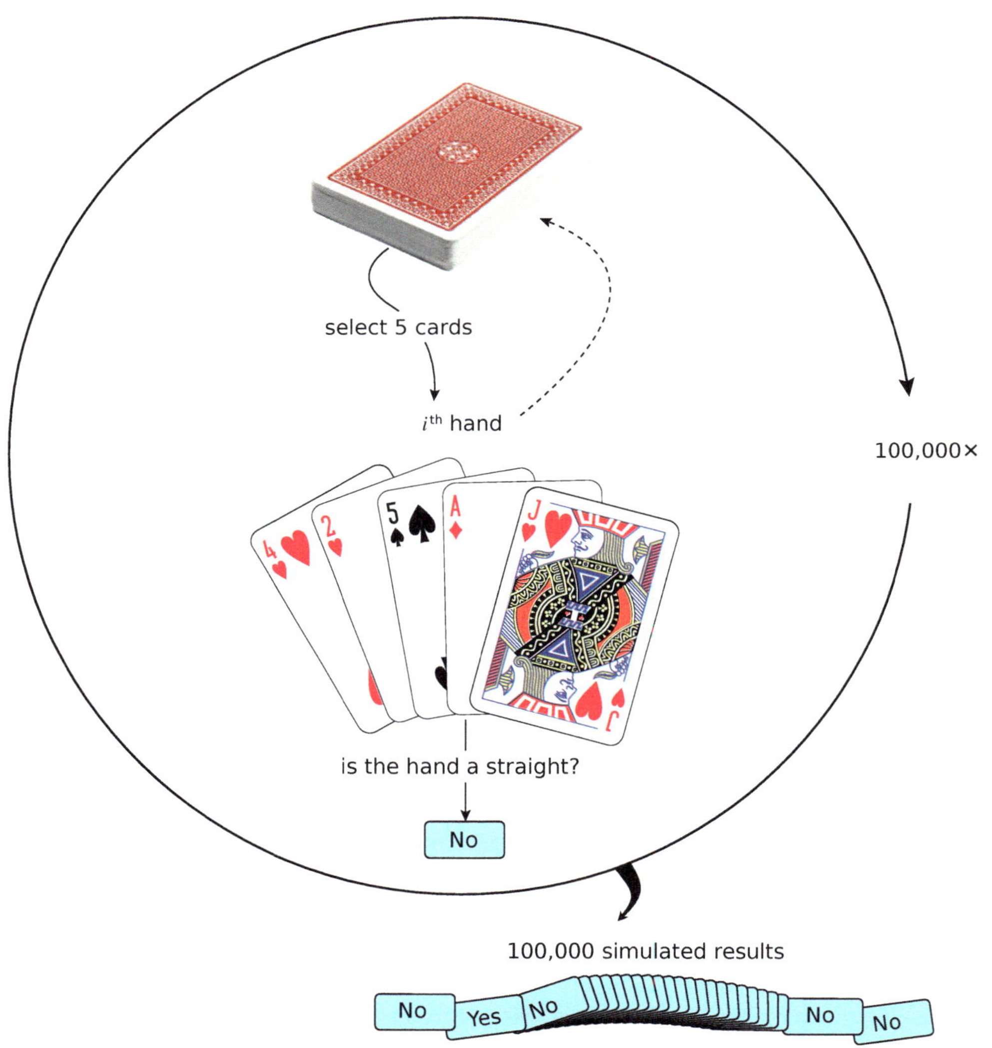

Figure 3.5 Procedure to simulate the probability of drawing a straight.

We will do it here in pseudo-code:

3.1.2 Simulating the probability of drawing a straight

```
Make a box with 13 numbers 1, 2, ..., 13, containing 4 copies of each
 number
Call each of the 52 elements a card
box = [1, 1, 1, 1, 2, 2, 2, 2, ..., 13, 13, 13, 13]

for i = 1, 2, ..., 100,000 do
    Randomly draw 5 cards out of the box (without replacement) to make a
     hand
    Check whether the hand is a straight
    if the newly drawn hand is a straight then
        add 1 to the count of straights
    end
end

Probability of drawing a straight = count of straights / 100,000
```

In our simulation, we estimated that the probability of drawing a straight is around 0.39% (Figure 3.6). What formula did we use? **We didn't use a formula, we carried out many simulations.**

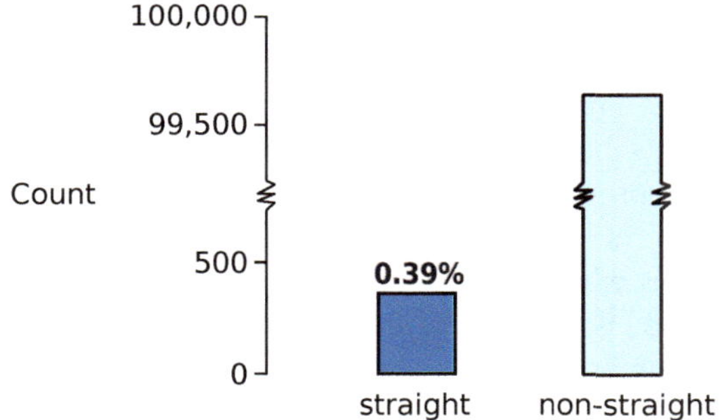

Figure 3.6 Histogram showing the distribution of drawing a straight in 100,000 simulated hands.

So, to summarize, we have developed a method for using simulations to answer questions of probability, **provided that the question can be put in the form of selection from a box model.**

Monte Carlo simulation The process of randomly selecting items from a list resembles gambling: for example, in the casinos of Europe, a roulette wheel is used to randomly choose a number between 0 and 36. This is equivalent to choosing one from a box model that has the elements {0, 1, 2, ..., 36}. (In the US, roulette wheels have a 38th slot called "00".)

The archetype of gambling casinos is the casino at Monte Carlo, which has led to these simulation methods being called **Monte Carlo methods**.

FURTHER EXERCISES 3.1

1. You flip 5 fair coins. What is the probability of observing 4 heads?
2. Three cards are chosen at random from a poker deck with replacement. What is the probability of choosing a five, a four and a three, in order?
3. A jar contains 4 blue, 3 red, 2 green and 5 yellow marbles. A marble is chosen at random from the jar. After replacing it, a second marble is chosen. What is the probability of choosing a blue and then a red marble?
4. **Fair ticket price.** You are offered the opportunity to play the following game: your opponent rolls 2 regular 6-sided dice. If the difference between the two rolls is at least 4, you win $20. Otherwise, you get nothing. What is a fair price for a ticket to play this game once? In other words, what is the expected earnings of playing the game?
5. **Four letter words.** How many different 4-letter "words" can you make using the 26 letters of the alphabet? For example, two of them are aaaa, cbgb, etc.
6. **Free throws.** A guard on the women's basketball team made an amazing 92.2% of her free throws this season, including a record-breaking streak of 33 in a row. What is the likelihood of her making 33 free throws out of 33, given her record this season? Use simulation to find this value and create a histogram of your simulation results.

3.2 Null Hypothesis Significance Testing

Now we can apply the concept of probability to the practice that is called **Null Hypothesis Significance Testing**, or **NHST**. This is a practice that is very widely followed, and accounts for the majority of scientific results being published in journals today. Unfortunately, it is also a major cause of the crisis of irreproducible results, and can be a recipe for doing bad science.

The thinking behind NHST is this: let's say you obtained some result. For example, in a sample of 300 births in families who live near power plants, there were 52% females and 48% males. Now the question arises: is this an "interesting" finding? Should we try to publish this? Should we be alarmed that something in the environment near power plants is affecting the development of the fetus and producing unequal sex ratios?

We can imagine an inner dialogue, with one voice saying "yes, 52/48 is a discrepancy, and it should be noted and published". What is the other, "no", voice saying? Probably something like "no, this is only a minor discrepancy, and discrepancies like this happen by chance all the time". So the key is the question "*how often* do discrepancies like this happen by chance?" If the answer is "30% of the time" then we would be hesitant to publish this result, because of the significant possibility that it might be just due to chance. After all, *any* sample will have some chance variations. On the other hand, if the answer is "results this big happen only 0.1% of the time" then we would have some reason to believe that the result might not just be a random occurrence.

Surprise value One way to think of this is to ask for the "surprise value" of the finding. If the finding is likely to have happened by chance, then the surprise value of the finding is low, because "it is to be expected". On the other hand, if the finding is *unlikely* to have happened by chance, then the surprise value of the finding is high, and the result merits further attention.

Therefore, the question becomes "**how** likely is a given result to have happened by chance?" **This question is answered by seeing how often a result this extreme happens in the chance model, that is, in the box model.**

We now introduce the central concept of traditional statistics, the notion of **statistical significance**. A finding is said to be statistically significant if it is highly unlikely to have occurred by chance, that is, if it has high surprise value.

"statistical significance"

- Did my result happen by random chance?
- How surprising would it be if it did?
- If the surprise value is high, then my result is statistically significant.

The Null Hypothesis But what is the "box model"? It's really a *Hypothesis*, the hypothesis that the process that is generating this data is no different from random sampling from a world in which there is no difference between girls and boys in birth frequency, and in which each successive birth is independent of every other.

This hypothesis, that "everything is random" is called the **Null Hypothesis**. In the case of the sample of 300 births near power plants, the Null Hypothesis is that boys and girls have an equal probability of birth and that successive births are independent. By contrast, an **Alternative Hypothesis** is any hypothesis that assumes some non-random effect.

So when we see, for example, a 52/48 result, when we ask for the surprise value for this result, we are asking how likely is the result to have happened "under the Null Hypothesis".

In any given situation, the Null Hypothesis is the hypothesis that the result in question is purely random, a statistical fluke. Therefore, *the box model embodies the Null Hypothesis.*

Simulating the Null Hypothesis

We can then estimate how likely or unlikely various results are under the Null Hypothesis.

Let's go back to our original finding that in 300 births near a power plant, we saw a 52%/48% split of girls vs. boys.

Box models of the Null Hypothesis One way to formulate the Null Hypothesis is as a box model: we make a box with the two elements: "boy" and "girl" with equal probability of being chosen. Then we select one randomly from the box model, record its value, and put it back in the box. We repeat this process 300 times to simulate the Null Hypothesis (Figure 3.7).

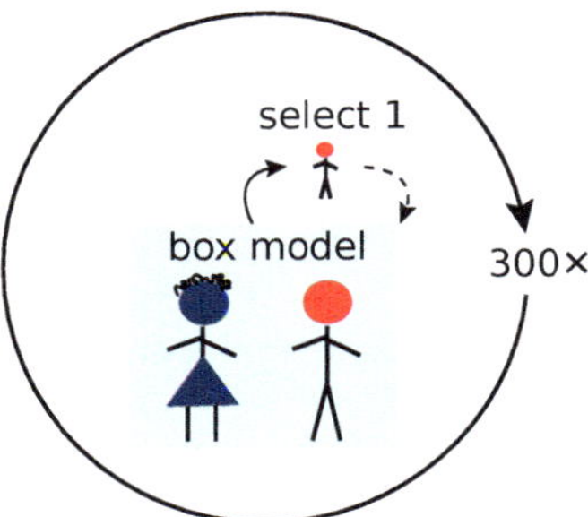

Figure 3.7 The box model that formulates the Null Hypothesis that boys and girls have an equal probability of birth and that successive births are independent.

Sampling with and without replacement. Note that when we sampled from the box model, we immediately replaced the sampled element in the box before making our next selection. This is called **sampling with replacement**.

We could also model the process of 300 independent draws (with replacement) from a box model by sampling randomly **without replacement** from a box containing a large number of red balls and an equal large number of blue balls (Figure 3.8). It's important to realize why we say "large".

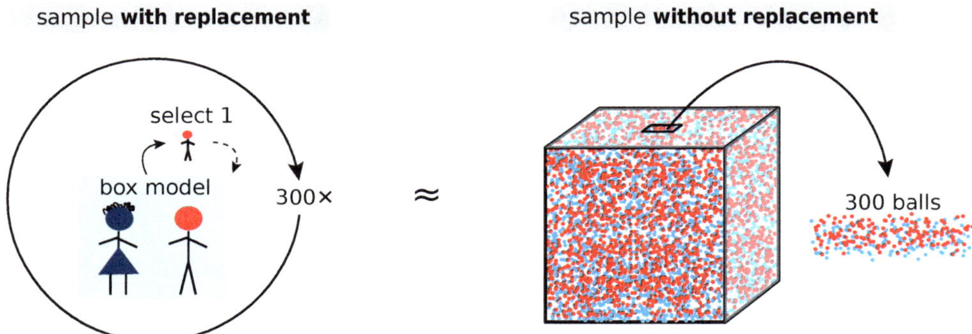

Figure 3.8 Two different sampling methods: with replacement and without replacement. They are approximately equivalent when there the number of red balls and blue balls are equal and large.

As we saw, the probability of drawing an ace in a 52-card poker deck is 1/13, or about 7.7%. Assuming the first draw is an ace, and we do not replace it in the deck, what is the chance of a second draw to also be an ace? It is not also 7.7%. Because we have already removed one of the four aces, the probability of drawing a second ace, given that we have already removed one ace, is 3/51, or about 6%. In this example, the probabilities change because the events are not independent: taking one ace out of the deck changes the probability of drawing any kind of card from the remaining deck.

Therefore, if we tried to model the Null Hypothesis in the male/female birth problem by an experiment drawing red and blue balls from a box, we would have a problem: drawing one ball from the box changes the probability of the next ball, so the draws are not independent!

There are two ways to solve this problem.

The first way to make the draws independent is to sample as before, but to assume that the number of red and blue balls is very large. If the number is very large, then the act of removing one ball still changes the probabilities on the next draw, but not by very much. If there are 10,000 red balls and 10,000 blue ones, then drawing a red ball on the first draw (exactly 50% chance) means that the probability of drawing a red ball on the second draw is not 50%, but 49.99%).

Thus, making the assumption that the number of balls is very large gives us an approximation to "independent" draws, and the larger the number, the better the approximation. However, this approach has limitations; for example, it had better be the case that the number of draws is small relative to the number of balls, or else we would again be running out of balls, and facing limited and therefore non-independent choices.

The second, and better, strategy is to sample with replacement. In sampling with replacement, we draw a ball from the box, write down its color, and then put it back in the box before the second draw. Now the draws are fully independent. As a consequence, we could even draw the very same ball twice. This is why we use sampling with replacement: to make every draw equal and independent.

Sampling without replacement in casino Blackjack

The problem of independence came up in a practical way in the management of blackjack in large gambling casinos. In blackjack, each player plays against the dealer. The dealer and each player is dealt two cards, and the player can optionally take additional cards to get closer to a total of 21, but if the player goes over 21, that's an automatic "lose". The Dealer has no options and plays by algorithm: draw another card if the total is less than or equal to 16 and stand pat on 17 or higher.

It is clear (or rather, it became clear to certain people in the 1970s) that high value cards (10, J, Q and K are all worth 10) are "bad" in the sense that they are likely to put you over 21 if you drew them as optional additional cards.

Clearly, then, if a player knew that there were lots of high cards still unplayed in the deck, they would make a larger bet because the deck is unfavorable to the dealer. The reason is that the player can opt not to take an additional card, but the Dealer must keep drawing cards until they have a total of at least 17. Thus, a large number of high cards still unplayed mean that the Dealer is more likely to lose by going over 21.

This is where the finiteness of the deck comes in: it means that successive cards are not independent. If there were relatively few high cards played so far, that implies that the remaining deck is rich in high cards.

Using this non-independence forms the basis of the blackjack system called "card counting", in which well-trained players can exploit the non-independence by keeping track of the relative frequency of high and low cards already played. It had been estimated that, with card counting, the usual house advantage shifts to a 2% advantage to the player.

Casinos wanted to defeat this advantage, which was cutting into profits. They could have sampled with replacement, that is, returned the used cards to the deck and re-shuffled them on every hand. But that takes time, and therefore costs money. Instead, the casinos hit on the idea of simply using many decks instead of one: the typical Las Vegas blackjack table uses 7 decks at once. The draws are still non-independent, but the deviation from independence is very small. Recently, the casinos have introduced Continuous Shuffle Machines, which are essentially sampling with replacement, making card-counting obsolete. (The classic source for card counting is Thorp's *Beat The Dealer*.)

Coin flipping model of the Null Hypothesis Another way to formulate the random process of the Null Hypothesis is to think of it as a **coin-tossing problem**.

For example, in the case where we are estimating the probability of getting 4 or more girls in a family with 5 children, we could set up the random model as a case of coin flipping: we flip a coin, and let Heads = is a boy, and Tails = is a girl. Then we flip the coin 5 times and count how many Tails we got. If the number of Tails is 4 or 5, we write down 'yes', if not we write down 'no'.

Similarly, in the case of a sample of 300 births in families who live near power plants, we would just flip the coin 300 times and generate a sample of 300 births (Figure 3.9).

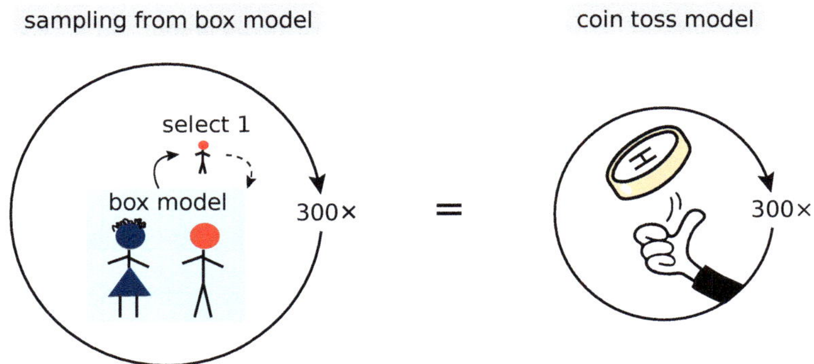

Figure 3.9 The box model and coin toss model are equivalent in formulating the Null Hypothesis.

Of course, we need to specify the coin: we need to stipulate in advance that it is a "fair", 50/50 coin, and we need to stipulate something that is automatic with actual coin tosses, that each toss is "independent" of every other: whether one flip comes out Head or Tail has no effect on subsequent tosses. If the probabilities are not 50/50, as in the case where we believe that girl births outnumber boy births 55:45, then we have to toss an "unfair" coin, where the probabilities of Heads and Tails are not equal, but stand in the ratio 55:45. This is easy to do in programming.

Now we have done one simulation of the Null Hypothesis.

10,000 simulations In order to calculate the probabilities of the observed result happening under the Null Hypothesis, we have to simulate the Null Hypothesis many times, say, 10,000 times. For the example of the sex ratio in 300 births around power plants, the procedure is:

Simulate the Null Hypothesis 10,000 times for the sex ratio near power plants study

Step 1 ► Flip the coin 300 times.

Step 2 ► Calculate the fraction of Heads.

Step 3 ► Repeat Steps 1 and 2 for a total of 10,000 times.

Step 4 ► Write the 10,000 fractions to an array.

The results of the 10,000 experiments are shown in the histogram (Figure 3.10). The histogram of outcomes under the Null Hypothesis is called the **null distribution**.

So we see that a result of 52/48 happened roughly 10% of the time in the null distribution.

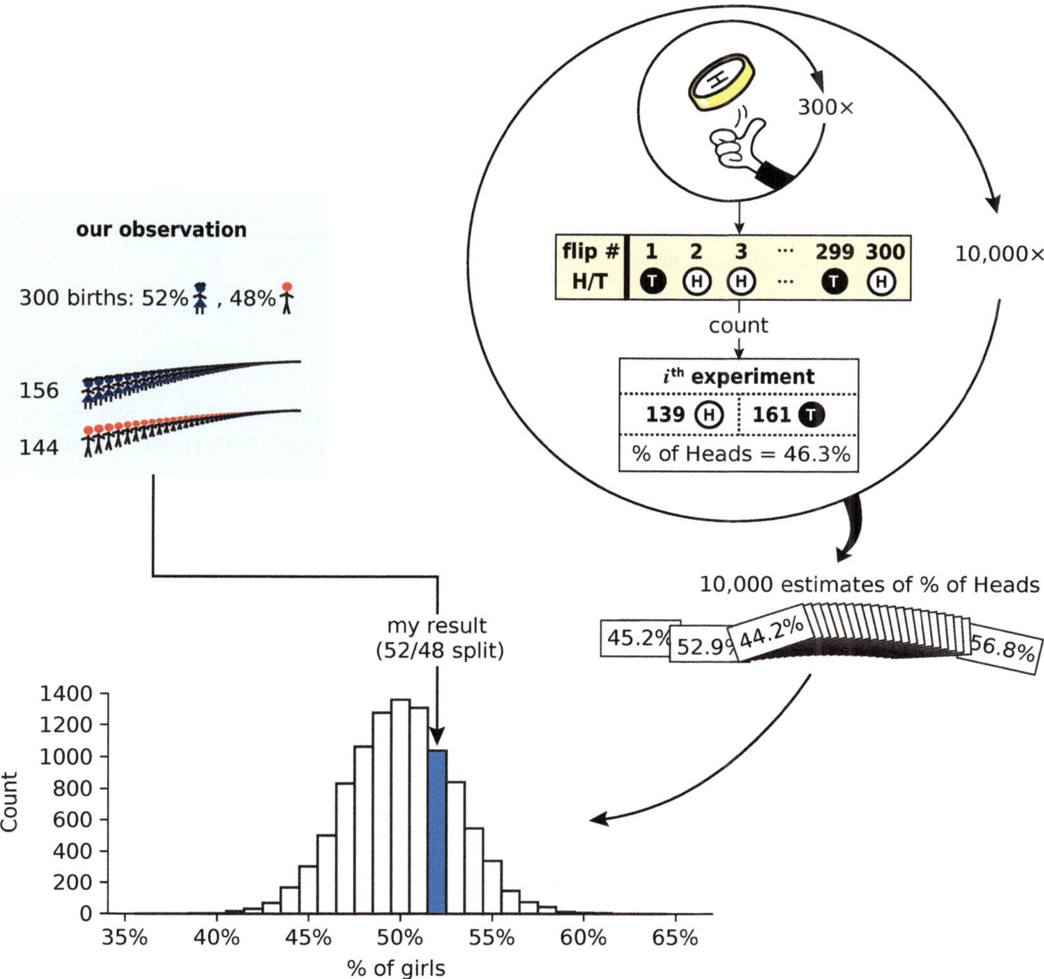

Figure 3.10 Comparing the observed result to the null distribution, which is the histogram of results of 10,000 simulations of flipping a fair coin 300 times and counting the number of Heads.

Surprise value

But we are not quite done. We have to re-think why we were asking this question in the first place. We wanted to know how likely this result was under the Null Hypothesis because we wanted to know how surprising it is; how much should we be impressed with the strength of this finding? When we put it that way, we have to ask: what about results that are ***even more extreme*** than ours? The frequency of even more extreme events is relevant because all of those even more extreme events are decreasing the "surprise" factor of our result. For example, if 55/45 occurs frequently under the Null Hypothesis, then that should count against the surprise factor of a 52/48 finding.

Therefore, the full count of all simulations that resulted in our ratio (52/48) *or higher* must count against the surprise value of our finding (Figure 3.11). Adding up the blue rectangles, we see that around 27% of the time, the result was as extreme or more extreme than ours.

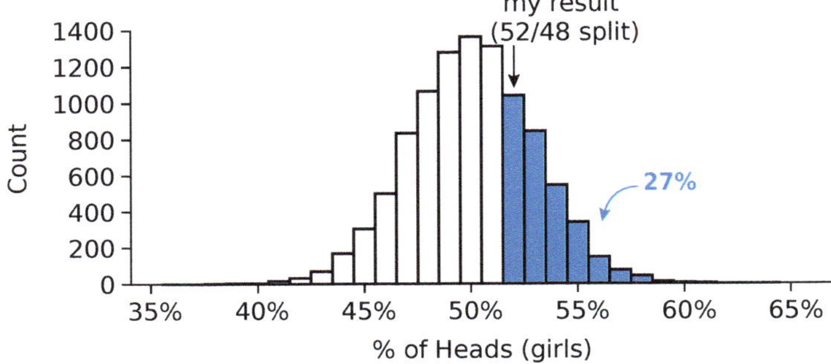

Figure 3.11

Two-sided tests Then there is a second factor: we've been talking about a 55/45 split as "more extreme" than our finding. But what about a 45/55 split, that is, 45 Heads and 55 Tails?

You will probably agree that the occurrence of a 45/55 split should decrease the surprise factor of a 52/48 split. Really, the question we are asking, or should be asking, is not "what is the likelihood of 52% females and 48% males" but rather "what is the likelihood of one gender having 52% representation and the other 48%?"

If we take the question that way, then the "number of results as extreme as or more than yours" has to be drawn from both sides of the null distribution, cases where Heads outnumber Tails and cases where Tails outnumber Heads.

The resulting histogram looks like Figure 3.12.

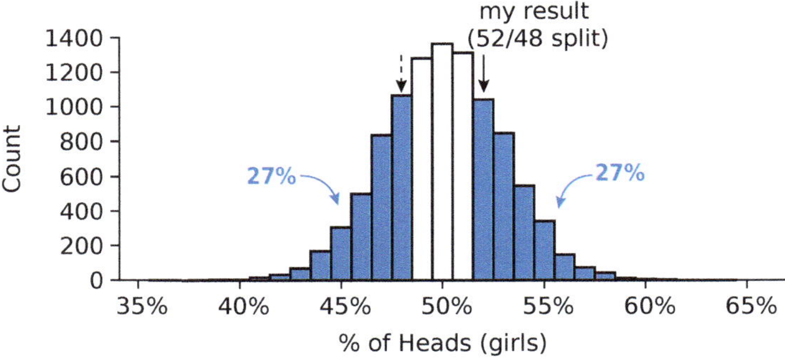

Figure 3.12

This is what is called a "2-sided" probability. Results as or more extreme than our result (52/48) or the corresponding flipped result (48/52) count against the surprise factor of the 52/48 result.

Obviously, using a 2-sided probability calculation decreases your "surprise factor". It makes the probability of our result (or more extreme) happening by pure chance to be roughly twice as large. Here, for example, it makes the real surprise value of our result to be very small, because results as extreme or more extreme as ours occurred around 54% of the time!

Is it ever correct to use a one-sided test? Usually not. But there are exceptions: one website says "a one-tailed test is appropriate if the estimated value may depart from the reference value in only one direction". For example, this applies when the quantities in question are distances, which can only be positive. In Chapter 8 *Independence, Proportions, and Relative Risk*, the effect size is the distance between two histograms, which can only be positive, and the null value is zero, so a one-sided p-value calculation should be used and is the correct method.

But there are other arguments for one-sided testing which are not appropriate. For example, another website says "So, if you are only interested in determining if Group A scored higher than Group B, and you are completely uninterested in possibility of Group A scoring lower than Group B, then you may want to use a one-tailed test." This seems completely wrong to us and to many other investigators. How can someone announcing what possibilities they are "completely uninterested in" *ever* affect the outcome of a scientific calculation?

What's even worse is that the weaker standards involved in the one-sided test invite a dishonesty in the scientist. Since the one-sided test is twice as lenient as the two-sided, you can cheat, and give yourself a standard that is twice as easy just by declaring that you have "no interest" in the opposite result. This is a serious temptation to dishonesty, and it appears from surveys of the literature that many investigators have given in to this method of cheating.

one-sided or two-sided?

In a well-written survey, "When should we use one-tailed hypothesis testing?" Ruxton and Neuhäuser (2010) write "we very rarely find ourselves in a position where we are comfortable with using a one-tailed test." And, in an article entitled "The Use of One- Versus Two-Tailed Tests to Evaluate Prevention Programs", Ringwalt *et al.* (2011) wrote "almost all authorities now agree that one-tailed tests are rarely (if ever) appropriate". The authors conclude that

> "one-tailed tests should never be used because they introduce greater potential for [False Positive] errors and create an uneven playing field when outcomes are compared ... In conclusion, it is our position that investigators should abandon one-tailed tests altogether and that they should rely exclusively on two-tailed tests to reject their null hypotheses."

The surprise value of our result Returning to our finding of 52% girls out of 300 births near power plants, we can say that a result this extreme or more occurred by chance a huge 54% of the time, so **it would be mistake to think we had found something important.**

Exercise 3.2.1 Pepsi vs. Coke: can anybody really tell the difference? A student participates in a Coke versus Pepsi blind taste test. He correctly identified the brand of the beverage 7 out of 9 tries. He therefore claims that he can tell the difference between the two soft drinks.

You have just learned the concept of null hypothesis significance testing and the resampling approach, and decided to determine the probability of anyone getting at least 7 right out of 9 tries just by chance. Which of the following would be your course of action in order to provide an accurate estimate of that probability?

a. Have the same student repeat this experiment a very large number of times and calculate the percentage of times that he correctly identified the brand of the beverage.

b. Call upon all your friends (you have lots and lots of friends), asking each of them to repeat this experiment, and calculate the percentage of people who make at least 7 correct guesses out of 9 tries.

c. Use a computer, and repeat a very large number of simulations, each simulation consisting of a set of 9 tries with a 50% chance of guessing the correct beverage on each try, and calculate the percentage of simulations that resulted in 7 or more correct guesses out of 9 tries.

d. All of the methods listed above would provide an accurate estimate of this probability

e. Using a computer, repeat a very large number of simulations, with each simulation consisting of a set of 9 tries with a 7/9 chance of guessing the correct beverage on each try, and calculate the percentage of simulations that resulted in 7 or more correct guesses out of 9 tries.

p-values

We have now shown how to calculate the probability, under the Null Hypothesis, of obtaining a result as or more extreme than the one that was observed.

We can now make an important definition:

The **p-value** of the observed result is the probability of a result as extreme or more extreme happening purely by chance, that is, happening "under the Null Hypothesis".

The p-value is also called the "degree of statistical significance" of the observed result.

We began this discussion by talking about the surprise value of the observed finding. We can now say that the intuitive concept of surprise value can now be formally defined: the lower the p-value of the observed result, the higher the surprise value.

Resampling method to calculate p-values

Step 1 ► Make a "random" model (this is the Null Hypothesis).

Step 2 ► Simulate the Null Hypothesis, say, 10,000 times.

Step 3 ► Count how many of the simulations had a result as extreme or more extreme than the actual observation.

Step 4 ► Divide that number by 10,000.

Step 5 ► That's the p-value for the observed result.

Exercise 3.2.2 When we do a test of significance, the p-value is:

a. The probability, assuming the Null Hypothesis is true, that the test statistic will result in a value that is equally or more extreme than that actually observed.

b. The probability, assuming the Null Hypothesis is false, that the test statistic will result in a value that is equally or more extreme than that actually observed.

c. The probability that the Null Hypothesis is true.

d. The probability that the Null Hypothesis is false.

From p-values to Null Hypothesis Significance Testing

The point of calculating the p-value of the result was to put a number on the question "how surprising is this?" If the p-value is very low, then the likelihood of its happening by chance (that is, under the Null Hypothesis) is very low, and this strengthens our belief that this result is worthy of our notice.

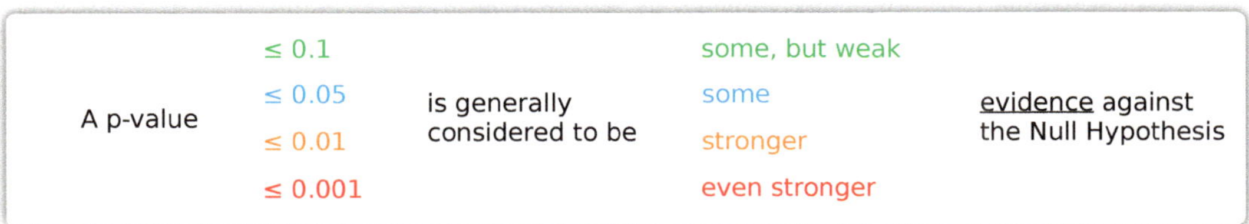

What is NHST? We now have a numerical p-value. The doctrine of Null Hypothesis Significance Testing was developed to answer another question: suppose we had to make a Yes/No decision about some study. For example, suppose we are editors of a journal, and we have to make the decision to publish or reject a given study. In order to satisfy this need for a Yes/No decision, the authors have to convert the numerical p-value into a Yes/No conclusion. Converting the p-value to a Yes/No conclusion is **Null Hypothesis Significance Testing (NHST)**.

Ronald Fisher, writing in 1925, proposed to set a threshold p-value, call it α (alpha), and then declare any $p < \alpha$ to be "statistically significant" and any $p \geq \alpha$ as "not statistically significant".

The logic of NHST In the doctrine of NHST, the idea is then that if an observed result produces a p-value $< \alpha$, then observing this result enables us to "reject" the Null Hypothesis. And since we have rejected the Null Hypothesis, that is taken to be evidence in favor of the Alternative Hypothesis. In most cases, the Alternative Hypothesis is simply that the Null Hypothesis is false, that is, that there must be some violation of the Box Model, some nonrandom factors at work.

For the sex ratio near power plants study, we would state the Null Hypothesis as "boy/girl births are equally likely and sequentially independent", so the Alternative Hypothesis is simply that "either boy/girl births are not equally likely or successive births are not independent".

However, we'll see that the basic argument of NHST is based on a fallacy (see *The p-value is the answer to the wrong question* on page 127).

Where to draw the line? The procedure of NHST is then to choose α, the threshold cutoff for the p-value, and declare that any p-value that is less than α is "statistically significant". But how to choose α?

In the scientific literature, a widely adopted choice for α is the value 0.05, originally suggested by Ronald Fisher in 1925. (He wrote: "It is convenient to take this point as a limit in judging whether a deviation is to be considered significant or not". A year later, he wrote "Personally, the writer prefers to set a low standard of significance at the 5 per cent point".) There is absolutely no scientific basis for this choice.

As we will see in section 3.3 (*What's wrong with Null Hypothesis Significance Testing?*), "$p < 0.05$" turned out to be a highly problematic idea, and has resulted in a huge amount of bad science. Nevertheless, this number has persisted in spite of dozens of strenuous objections, including the article in *Nature* in 2019, with over 800 signatories, entitled "Retire Statistical Significance" (see Chapter 1).

NHST

Step 1 ► Carry out a study and calculate the observed result.

Step 2 ► Form a box model of the Null Hypothesis of the experiment.

Step 3 ► Use the box model to simulate the Null Hypothesis many times. Calculate the probability of obtaining a result as extreme or more extreme than the observed result. This probability is the p-value.

Step 4 ► Choose a value for α as the cutoff for statistical significance (0.05 has been a common choice, but this is a major cause of false positives).

Step 5 ► If the p-value is less than α, declare the observed result to be "statistically significant".

Example 1 Measles outbreak in 2019

Measles is one of the most contagious diseases in the world. One person with measles is expected to infect between 12 and 18 others, spread by virus-laden droplets released when an infected person coughs or sneezes. The only way to prevent an outbreak of measles is to rely on "herd immunity". In order to achieve herd immunity for measles, the vaccination rate of a population needs to be at least 95%.

However, in recent decades, measles cases have been on the rise. In 2019, from January 1 to October 1, a total of 1249 cases of measles were reported in the United States. The number and vaccination status of measles cases are shown in the table below.

Measles cases	Vaccination status count and %		
	unvaccinated	vaccinated	unknown
1,249	872 (70 %)	142 (11%)	235 (19%)

Given that the vaccination rate of the general public at the time in the United States is 92%, we'd like to understand if the measles vaccine prevents infection.

Question How would you translate the question "does the measles vaccine prevents infection" into math?

Answer We can reason like this: if the measles vaccine had no effect whatsoever, then the proportion of the vaccinated in the measles cases ought to be the same as the proportion of vaccinated in the overall population. And if the vaccine actually prevented measles, then the proportion of vaccinated in the measles cases ought to be much lower than the proportion of vaccinated in the overall population. Therefore, mathematically, what we are asking is: is the proportion of vaccinated in the measles sample statistically significantly different from its proportion in the general population?

Question What is the Null Hypothesis H_0? What is the Alternative Hypothesis H_a?

Answer

H_0: the proportion of vaccinated in the measles sample is **similar to** its proportion in the general population.

H_a: the proportion of vaccinated in the measles sample is **different from** its proportion in the general population.

Question What is the data we need to collect?

Answer From the way we framed the Null Hypothesis and the Alternative Hypothesis, we need to know 1) the proportion of vaccinated in the measles sample and 2) the proportion of vaccinated in the general public.

- **The proportion of vaccinated in the measles sample is 14%.** In the table above, there are three categories: unvaccinated, vaccinated and unknown. However, for our purpose, only the first two categories are relevant, and we would need to discard the third category, the unknown, and recalculate the percentages.

Measles cases	Vaccination status count and %	
	unvaccinated	vaccinated
1,014	872 (86%)	142 (14%)

- **The proportion of vaccinated in the general public is 92%.**

Question Write the pseudo-code for NHST.

Answer Since flipping a coin that has a 92% chance of getting heads is the equivalent of randomly sampling a person from a general public that has a 92% vaccination rate, the pseudo-code for NHST can be written as:

1. Formulate the Null Hypothesis.
   - Flip a 92% (Heads) coin 1014 times, record the number of Heads.
   - Calculate the percentage of Heads, call it c_i.
2. Simulate the Null Hypothesis 10,000 times.
   The percentages of heads for each experiment are $\{c_1, c_2, c_3, \ldots, c_{10,000}\}$.
3. Count the number of experiments in which the percentage of Heads ≤ 14%.
4. Calculate the p-value.

$$\text{p-value} = \frac{\text{the number of experiments in which the percentage of Heads} \leq 14\%}{\text{total number of experiments } (= 10{,}000)}$$

Question What is the calculated p-value? How to interpret this?

Answer $p < 0.0001$. The observed value, 14%, is so far out of the range of the simulations that we had to make a break in the axis to show it. In fact, 14% or less never occurred in our 10,000 simulations. So we can say that $p < 0.0001$.

In the case of zero simulations giving results as or more extreme than the observed, all we can say is that p is less than 0.0001. This is all we can say given 10,000 simulations; we can't give an exact numerical p-value. If, for some reason, we need a more precise p-value, we can increase the number of simulations.

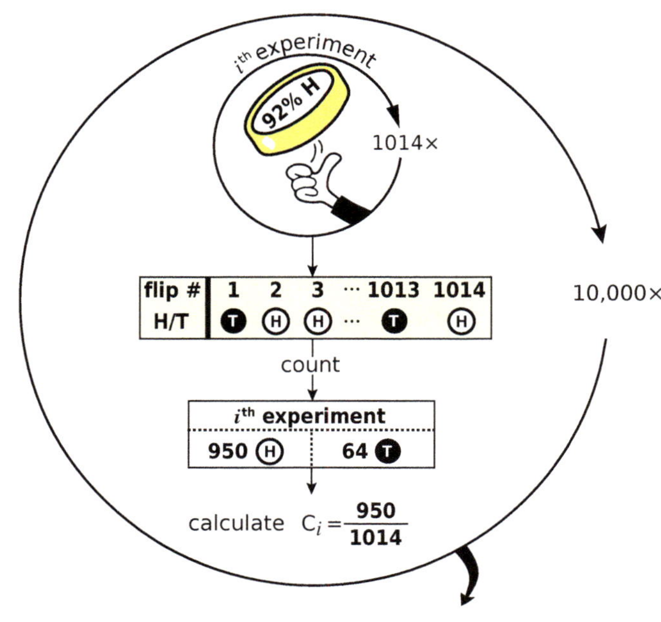

10,000 estimates of vaccination rate under the Null Hypothesis

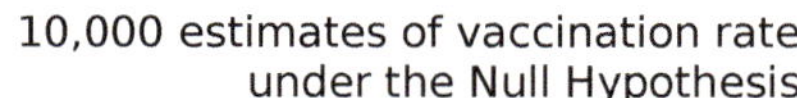

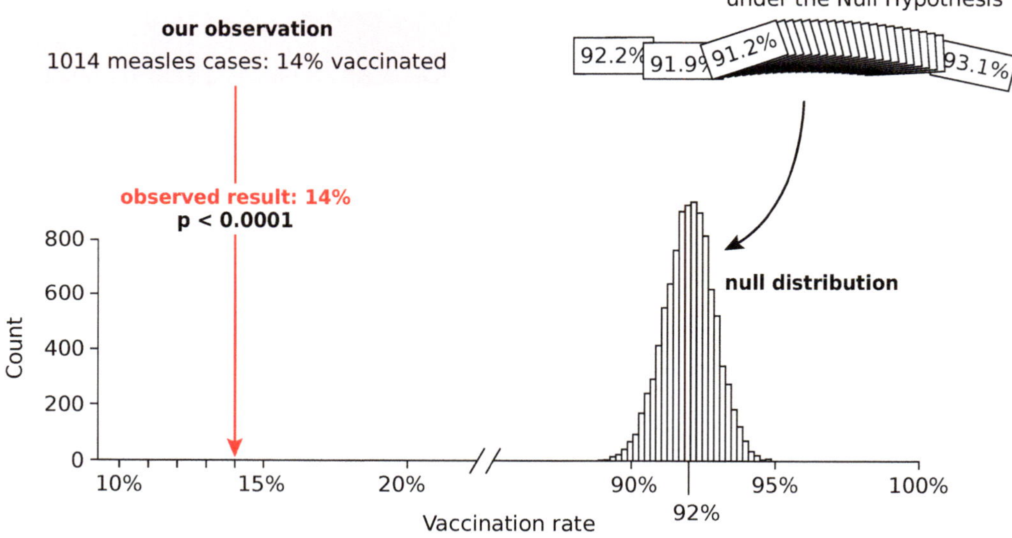

This very small p-value says that the observed result (14% vaccinated) in the sample is very unlikely to happen under the Null Hypothesis. We conclude that the percentage of vaccinated in the measles groups was statistically significantly lower than the percentage of vaccinated in the general population, therefore, the vaccine was effective.

Exercise 3.2.3 COVID oral antiviral pill. In early October 2021, data was released from a clinical study of molnupiravir, a five-day course oral antiviral medicine, and its effectiveness in treating COVID-19.

From the press release, we learn that, in the placebo group, 53 patients, or 14.1% were hospitalized or died. For those who received molnupiravir, 28, or 7.3% were hospitalized or died. This can be summarized in the table below. The company claimed that their drug "significantly reduced the risk of hospitalization or death".

Groups	n	Hospitalized or died cases (%)
Placebo	377	53 (14.1%)
Molnupiravir	385	28 (7.3%)

a. Given this data, we'd like to evaluate the company's claim that the antiviral pill significantly reduces hospitalization or death of COVID-19 patients. How to translate this question into math?

b. What is the Null Hypothesis H_0? What is the Alternative Hypothesis H_a?

c. What data do we need to collect?

d. Write the pseudo-code for NHST.

e. What is the calculated p-value? Does the result back the company's claim?

The press release also gave some information on the side effects, claiming that "the incidence of any adverse event was comparable in the molnupiravir and placebo groups (35% and 40%, respectively)". This can be summarized as the table below.

Groups	n	Adverse event cases (%)
Placebo	377	132 (35%)
Molnupiravir	385	154 (40%)

f. Given this data, we'd like to evaluate the company's claim that the incidence of adverse events was comparable in the molnupiravir and placebo groups. How to translate this question into math?

g. What is the Null Hypothesis H_0? What is the Alternative Hypothesis H_a?

h. What data do we need to collect?

i. Write the pseudo-code for NHST.

j. What is the calculated p-value? Does the result back the company's claim on "adverse events"?

Factors that influence the p-value

Effect size Let's do an example, based on our power plant finding.

As we saw, the p-value of the finding "52% females in 300 births" was hardly surprising, since it (or a more extreme event) happened by pure chance 54% of the time, so its p-value was 0.54.

Now let's suppose we did another study, and found a different result. Let's say that this time, we found 56% girls in a sample of 300 births, a larger effect size.

By simulating the Null Hypothesis many times, we can assemble the histogram of results, that is, the null distribution. From the null distribution, we can say in advance how big a ratio we would need in a sample of 300 to be statistically significant at the 0.05 level by finding the top 2.5th percentile and the bottom 2.5th percentile cutoffs. The answer is a ≥ 55.7% girl birth rate. Our new result (56% girl birth rate) has exceeded that by a whisker (Figure 3.13). In fact, when we calculated the p-value, $p = 0.0453$.

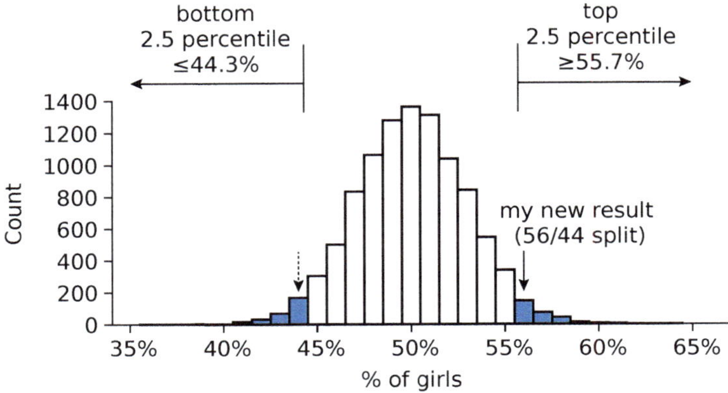

Figure 3.13 Null distribution of the girl birth ratio in 300 births, under the Null Hypothesis that both genders are equally likely, and successive births are independent. The cutoffs for $p < 0.05$ are shown, as well as the new observed girl birth rate of 56% ($p = 0.0453$).

Then, using a $p < 0.05$ standard, we declare that the result is "statistically significant".

In summary, in the original power plant study, we found a 52/48 sex ratio in 300 births near power plants. The result was considered not surprising at all (under the Null Hypothesis). It had a p-value of 0.54. In the new study with an increased effect size, 56/44 sex ratio, found $p = 0.0453$. Since the p-value was just under 0.05, we declared our finding "statistically significant".

The resulting paper is called "Altered birth gender ratios in the vicinity of power plants". The key sentence in the paper is "Since the p-value of a 56/44 split among equally likely outcomes in a population of 300 is less than 0.05, the result is statistically significant."

This is NHST. It's important that you know and understand this method. NHST is the most frequently used kind of statistical argument in the scientific literature. However, it is also the methodology that is responsible for the false "results" that are plaguing science.

Other things being equal, the larger the effect size, the smaller the p-value, hence the easier to find statistical significance. This relationship is an overall trend, but we should not assume that the relationship is linear: doubling the effect size does not necessarily halve the p-value.

Sample size What if the sex ratio of families near power plants was derived from a study of 150 births instead of 300 births? And on the other hand, what if we obtained the same sex ratio from a study of 600 births (twice as many)?

Do you think that increasing the sample size would:

- increase the strength of evidence against the Null Hypothesis?
- decrease the strength of evidence against the Null Hypothesis?
- or have no impact on the strength of evidence?

Intuitively, it seems reasonable to think that as we increase the sample size (the "n" of our study), keeping the same sex ratio, then the strength of evidence against the Null Hypothesis will increase.

To find out how sample size impacts p-value, we carried out 10,000 simulations for each sample size, n = 150, 300, and 600 (Figure 3.14).

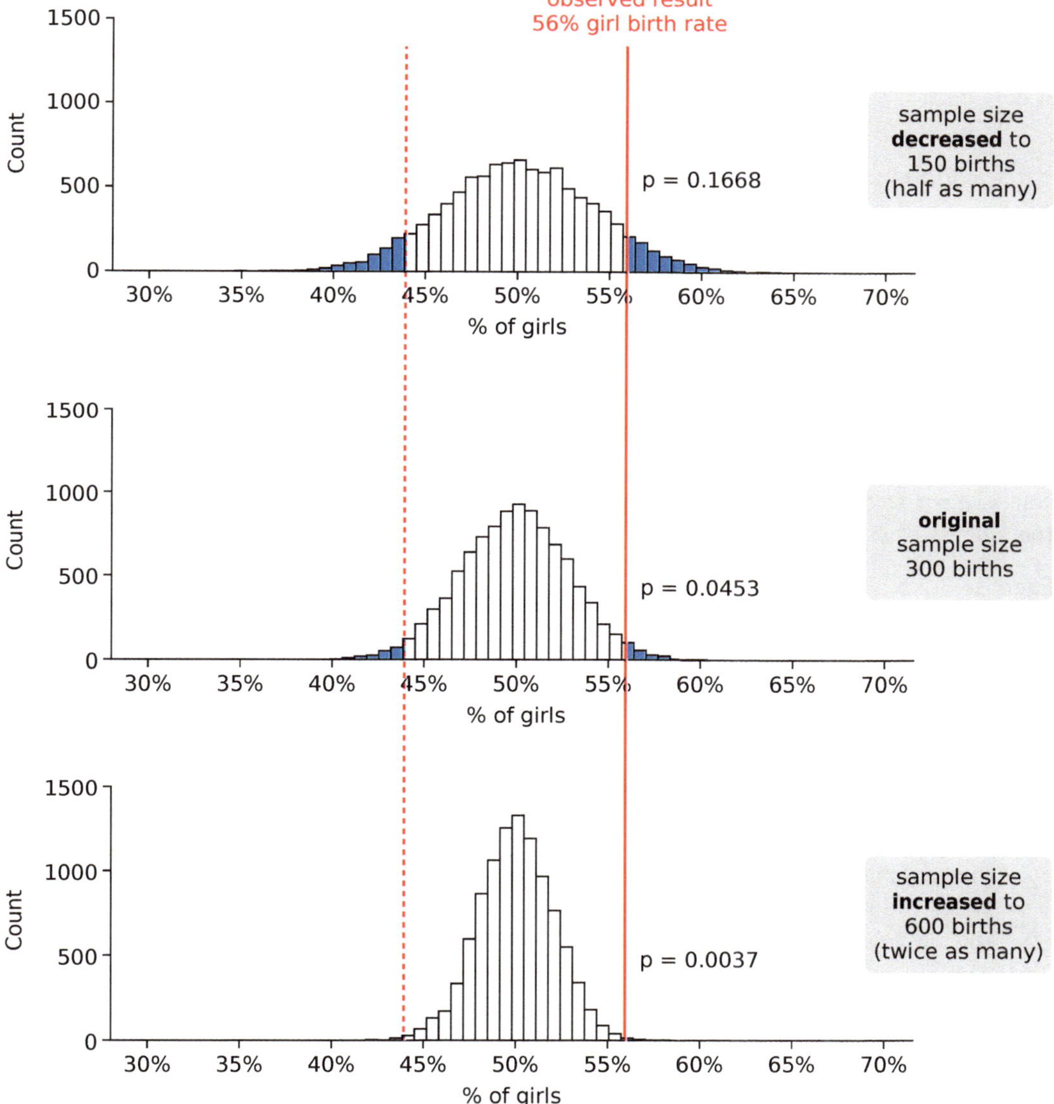

Figure 3.14 The greater the sample size, the less variability in the null distribution.

We find that as the sample size increases, the null distribution becomes narrower and narrower. Therefore, assuming the observed sex ratio stays the same (56/44), the narrower the null distribution, the fewer simulations there would be as or more extreme than 56/44, yielding a smaller p-value, indicating more surprise for 56/44 to occur under the Null Hypothesis. The strength of evidence against the Null Hypothesis increases with increased sample size.

In general, the larger the sample size, the less variability we will see in the statistic of interest from sample to sample, and therefore, the less variability we will see in the null distribution. This implies that a given observed result will be judged to be more extreme, hence a lower p-value, since it is being compared to a tighter null distribution.

However, this is a double-edged sword: if the sample size is very large, then even a trivial, medically irrelevant difference in effect size will be declared to be highly statistically significant. This has happened, for example, in large scale clinical trials of the effects of aspirin on heart attacks, a small risk reduction was declared to be highly significant, leading to physicians recommending daily aspirin to their patients, a practice which has since been discontinued (see "aspirin and heart attacks" on page 126).

Statistical power A larger sample size or an increased effect size both make it easier to find statistical significance when a result is really "there". When a result is really there and the study is negative, that is, failed to find statistical significance, we call it a Type II error. A Type II error is also called a False Negative. The effect was really there, but the study failed to see it. In general, the ability of a test to avoid False Negatives is called "power" (think about the power of a microscope). We have just seen that increasing sample size and increasing effect size directly increase the power of a study, by producing lower p-values. We'll discuss power in Chapter 11.

Exercise 3.2.4 You have been invited to engage in a basketball skills contest with the legendary LeBron James. The contest involves shooting free throws. Because he challenged you, you get to specify the exact terms of the contest. What (fair) contest design gives you the highest probability of winning?

Exercise 3.2.5 Someone claims they have a psychic ability to influence coin flips. They flip a coin 10 times and get 9 heads.

a. **Logic of NHST.** The Null Hypothesis (H_0) is: They're not psychic, and the coin is fair (50/50). Now we're asking: "If the coin were fair, how often would we see something *this* extreme in *either* direction?" The 2-sided p-value is 0.02. What does p = 0.02 mean?

b. **What can we conclude?** Since 2% is rare, we reject the Null Hypothesis at $\alpha = 0.05$, and conclude that if the coin is fair, it's unlikely to have 9 out 10 heads. But did we prove that psychic powers exist? or have we only showed that our observed result is unlikely under the Null Hypothesis of "pure chance"? Could NHST tell us which alternative is true: psychic powers, a weighted coin, a rare but genuine chance outcome, or sleight of hand?

FURTHER EXERCISES 3.2

1. Given a box model of {1, 2, 3}, with each element having equal probability of being selected.

   a. Which of these are possible outcomes of 3 resamples with replacement?

      a) {2, 1, 3}　b) {2, 2, 2}　c) {1, 3, 1}
      d) {2, 1, 3, 2}　e) {3, 2, 1}

   b. Which of these are possible outcomes of 3 resamples without replacement?

a) {2, 1, 3}　　b) {2, 2, 2}　　c) {1, 3, 1}
d) {2, 1, 3, 2}　　e) {3, 2, 1}

2. **Measles, mumps, rubella vaccination.** The local paper reported measles, mumps, and rubella (MMR) vaccination rates at local kindergartens. An elementary school with 122 kindergarten students had an MMR immunization rate of 79%. The U.S. national MMR vaccination rate for children of this age at the time was 91%. Describe how you would determine if the vaccination rate for this school is statistically significantly different from the U.S. as a whole. Include your study's parameter of interest, Null Hypothesis, box model, and sample size.

3. **Measles outbreak in Disneyland.** During the winter of 2014–2015, there was an outbreak of measles among Disneyland visitors. Of the 110 visitors who fell ill, 49 were unvaccinated and 61 had received the measles vaccine. At first glance this seems puzzling, as vaccinations are supposed to prevent infection, leading some to hypothesize that the measles vaccine does not work, or worse, causes infections ("The majority of measles victims were vaccinated!").

   A closer look at the numbers tells a different story: roughly 91% of the general population had already been vaccinated. Given this, is it actually surprising that 61 out of 110 infected visitors were vaccinated? If the vaccine offered no protection at all, how likely would we be to see 61 or fewer vaccinated individuals in a random group of 110?

   a. Before writing any code, prepare the following as comments. This is the first step toward writing pseudo-code, which is always good practice.

      a) Summarize the information given and state your goal.
      b) State the Null Hypothesis.
      c) Describe the box model for the Null Hypothesis.
      d) Describe the procedure for calculating the p-value.

   b. Use simulation to find the p-value. Show the null distribution with the observed result indicated.
   c. Write a sentence or two interpreting the p-value.
   d. In cases where the calculated p-value is 0, does this really mean that such a result is impossible under the Null Hypothesis? If not, how to get a more precise p-value?

4. **Can people and dolphins communicate?** In the 1960s, the U.S. Navy started a series of experiments to understand interspecies communication between people and dolphins. One experiment that Dr. Javis Bastian conducted in 1964 was called "The Transmission of Arbitrary Environmental Information Between Bottlenosed Dolphins."

   In the experiment, two dolphins, Buzz and Doris, were trained to push one of the two buttons in a pool. If the experimenter shined a constant light, the dolphins were trained to push the right button to get a fish. If he shined a blinking light, the dolphins were trained to push the left button to get a fish.

   To see if they could communicate, the dolphins were separated by placing a curtain down the middle. While Buzz had access to the buttons, only Doris could see the light. Doris needed to let Buzz know which button to push. Both dolphins got the reward only when the correct button was pushed.

In the study, the experimenter shined the light a total of 16 times, each time randomly selecting whether to make the light shine constantly or blink, which only Doris could see. Buzz pushed the correct button 15 times out of 16.

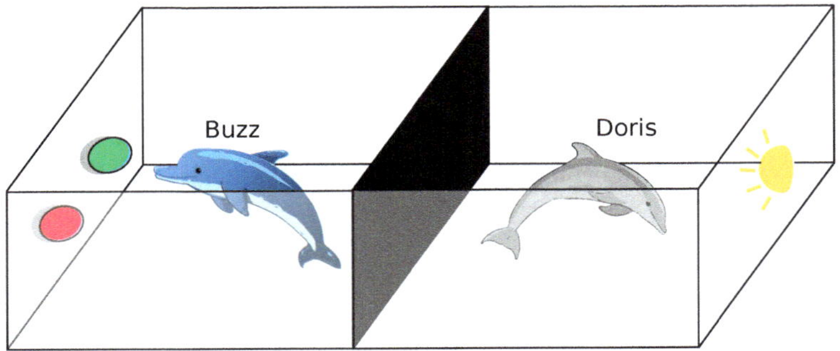

Based on these data, do you think Buzz somehow knew which button to push or could Buzz have just been guessing? Is this result unlikely to happen due to random chance?

a. What are the two possible explanations for Buzz pushing the correct button 15 out of 16 times.
b. Formulate the Null Hypothesis.
c. How do you use a coin flipping model to formulate the Null Hypothesis?
d. How do you use a box model to formulate the Null Hypothesis?
e. If Buzz was just guessing which button to push each time, what would the distribution of the number of correct choices look like? Either write a pseudo-code or graphical flowchart to illustrate how you would obtain this distribution.
f. What is the probability for the observation "Buzz pushed the correct button 15 times out of 16" to occur if Buzz was just randomly guessing?
g. What's your conclusion?

5. **Voter support.** Assume 35% of registered voters in a state support a new climate policy. At a town hall meeting, all 160 attendees were surveyed and 40% supported the policy. Is this town hall representative of the state?

   a. Which is an appropriate Null Hypothesis?

      a) The proportion of supporters at this town hall is less than 35%.
      b) The proportion of supporters at this town hall equals 35%.
      c) The proportion of supporters at this town hall is greater than 35%.

   b. To formulate the Null Hypothesis using a box model, which would you choose? Let S = "supports the policy" and N = "does not support."

      a) Box model = 65 Ss and 35 Ns
      b) Box model = 3 Ss and 1 Ns
      c) Box model = 1 Ss and 3 Ns
      d) Box model = 35 Ss and 65 Ns
      e) Box model = 40 Ss and 120 Ns

c. Alternatively, to formulate the Null Hypothesis using a coin flipping model, what would it be? Let H = "supports the policy" and T = "does not support".

d. Describe the procedure to calculate the p-value.

e. Choose the distribution that best describes the proportion of supporters at this town hall under the Null Hypothesis.

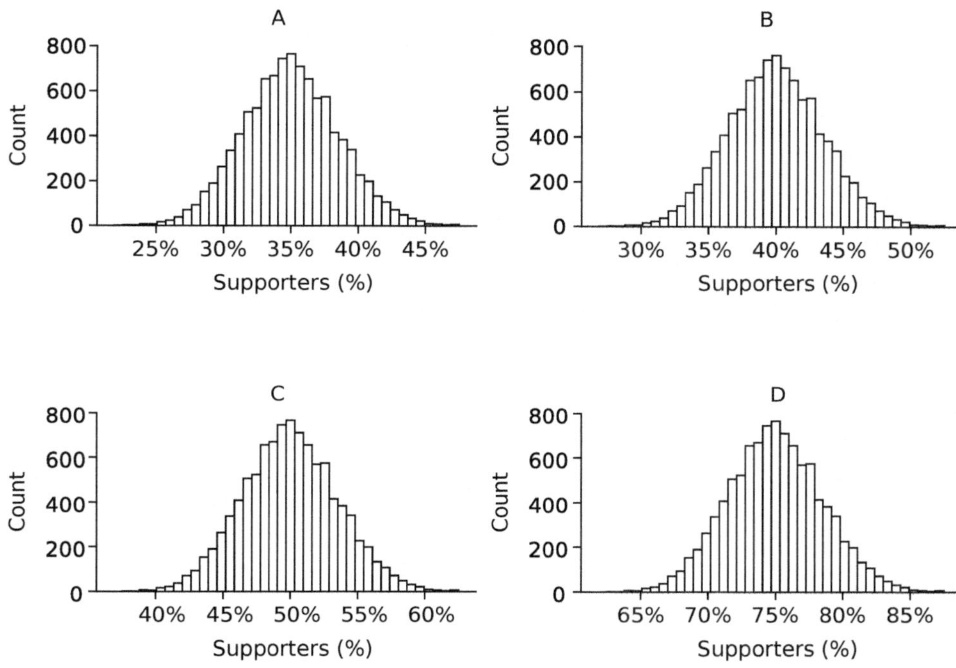

f. To calculate the p-value, should you use a one-sided or two-sided test?

g. Which statement is the correct interpretation of a p-value of 0.22?

a) 22% of voters at this town hall support the policy.

b) There is a 22% probability that this town hall is a random sample from the state.

c) In a random sample of 160 voters from the state, there is a 22% probability of observing 40% or more supporters.

d) If this town hall is not a random sample from the state, there is a 22% probability of observing 40% or more supporters in a group of 160 voters.

6. **Just guessing.** If you are trying to pass a true/false exam but you know absolutely nothing about the subject, would you rather have more questions or fewer questions on the test?

7. **Lung infections.** A researcher is working on a new drug to clear lung infections. Using the standard treatment it typically takes 120 hours to clear the infection. In an early trial of the new drug, 4 subjects had their infection cleared in an average of 50 hours. Although this length of time is less than half that of the standard treatment, the results are not statistically significant. How would you explain this?

a. The placebo effect is present, which limits statistical significance.
b. The early trial was probably not double-blind.
c. There was probably a mistake in the calculation. These results are definitely significant.
d. The sample size is small.

8. **Dog training.** A dog is trained to detect the scent of explosives. When presented 3 bags of which only one contains explosives, 15 out of 21 times the dog successfully detected the bag containing the explosives. Did the training have an effect on this dog or could this result be explained by random chance?

9. **Sample size and p-value.** As the sample size increases and the statistic of interest stays the same, the strength of evidence against the Null Hypothesis would

   a) increase b) decrease

10. **Effect size and p-value.** If we are comparing the difference between the means of two groups, as the means get farther apart, the strength of evidence against the Null Hypothesis would

    a) increase b) decrease

11. **Meaning of 0.05.** You're scrolling through social media and see a viral article with the headline: "Low-Fiber Diet Raises Cholesterol — Only a 5% Chance This Is Wrong!" The article describes a study where researchers tracked the diets and cholesterol levels of 200 volunteers over 8 weeks.

    Your friend, who has never taken a statistics course, is alarmed and says: "5% sounds really small, so there's a 95% chance the finding is true, right?" They ask you what the "5% chance of being wrong" actually means.

    a. Explain where the 5% figure likely comes from and what it actually means. Your explanation should be understandable to someone without any statistics background.
    b. Should you automatically believe this finding? What additional information would you want to know before changing your diet based on this study?

3.3 What's wrong with Null Hypothesis Significance Testing?

$\alpha = 0.05$ is not strong enough; use $\alpha = 0.01$ or better 0.001

One of the biggest problems created by NHST with $p < 0.05$ is the very large number of False Positives that this method creates. Recall Chapter 1, where we were discussing the crisis of false results in the scientific literature. We mentioned the concept of False Positives, that is, "results" that are claimed but are actually false.

When we do Null Hypothesis Significance Testing, there are four distinct possibilities: the Null Hypothesis can be true or false in reality, and our study result can either reject the Null Hypothesis or not (Figure 3.15).

		In reality	
		Null Hypothesis *really* false	Null Hypothesis *really* true
Study result	study rejects Null Hypothesis	**True Positive (TP)**	**False Positive (FP)**
	study does not reject Null Hypothesis	**False Negative (FN)**	**True Negative (TN)**

Figure 3.15 Four possible outcomes of NHST.

In the context of NHST, using a $p < 0.05$ criterion for statistical significance, that is, $\alpha = 0.05$, guarantees that we will falsely declare significance 5% of the time when the Null Hypothesis is true. If we define the **False Positive Rate** as the likelihood of falsely rejecting the Null Hypothesis and declaring significance, then α is the False Positive Rate.

The False Positive Rate can be defined as in Figure 3.16.

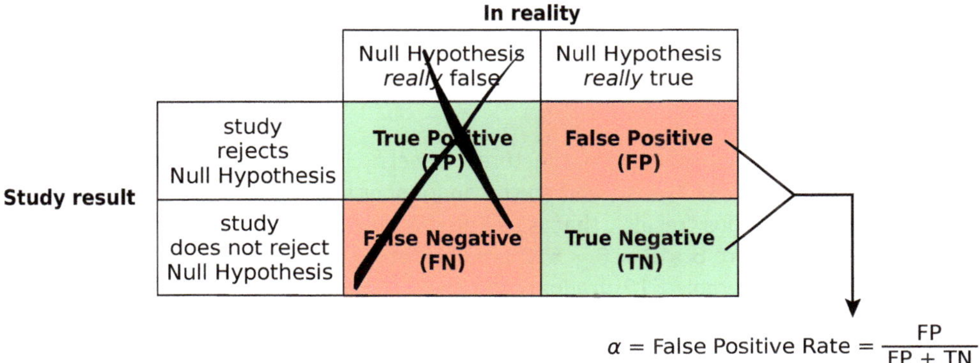

Figure 3.16 Definition of False Positive Rate (α).

After all, to say that "results this extreme (or more) happen by chance less than 5% of the time" means that "results this extreme or more happen by chance as much as 5% of the time". By using $p < 0.05$ as our standard, we are accepting the fact that 5% of the time, we will declare random numbers to be "statistically significant". We are accepting a 5% failure rate in our science. Perhaps you think this isn't too bad, although a carmaker or a drug manufacturer with a 5% failure rate would not last very long. We wouldn't even buy a toaster that had a 5% failure rate. Is that failure rate acceptable for our science?

But the False Positives problem due to NHST is actually much worse than that. When Ronald Fisher first proposed "$p < 0.05$" as a standard for statistical significance, he did not anticipate that if this method were applied tens of thousands of times, by automated procedures and massive amounts of data, the result would be hundreds or even thousands of False Positives.

Problem of multiple testing The Jim Borgman cartoon below shows 3 wheels of fortune, each with 8 possibilities. The first wheel lists 8 possible causal factors ("stress", "coffee" etc.), the second wheel lists 8 medical conditions ("heart disease", "hypothermia", etc.) and the third wheel lists 8 possible reference populations ("children", "twins", etc.) (Figure 3.17).

Figure 3.17

The idea is that scientists spin the three wheels. In this particular case, they landed on "coffee", "depression" and "twins", and so the announcer says "coffee causes depression in twins". Of course this is a bogus "result", that is, a False Positive, but consider this: suppose you had a large budget. There are $8 \times 8 \times 8 = 512$ possible "studies" that could be done. Now, with your large budget, you assemble 512 graduate students, and assign each one of them one of these studies. However, you have no actual data to give them, so instead, you make up completely random numbers as data for each student project.

What will happen? If each grad student applies a "$p < 0.05$" standard, then the 5% False Positive Rate times 512 studies equals 26 bogus results, on average. That means that 26 of our grad students will stand up and announce that they have found a statistically significant result, and we would have put another 26 false studies into the scientific literature.

p-hacking It's important to understand that this cartoon is no joke: **it's a plausible model for how so many false results have made it into the literature.** The unfortunate consequence of Null Hypothesis Significance Testing is that it has created a phenomenon called **p-hacking**, which includes the practice in which researchers just do lots and lots of tests, and then keep and publish only the ones that were significant. The practice of p-hacking is widespread in science today, and is responsible for a lot of the false results in the literature (Head *et al.*, 2015).

Zodiac signs and p-hacking

An entertaining rebuke and satire of p-hacking was provided by a group from the University of Toronto, who had access to the medical records of over 10 million residents of the Province of Ontario. They assigned each person their astrological sign, based on their month and day of birth. The investigators then "searched through 223 of the most common diagnoses for hospitalization until we identified two for which subjects born under one astrological sign had a significantly higher probability of hospitalization compared to subjects born under the remaining signs combined ($p < 0.05$)". By testing multiple hypotheses, they found, for example, that "Sagittarians had a higher probability of humerus fracture ($p = 0.0123$) compared to all other signs combined." Their objective was "To illustrate how multiple hypotheses testing can produce associations with no clinical plausibility", and they certainly achieved that.

Journal of Clinical Epidemiology

Testing multiple statistical hypotheses resulted in spurious associations: a study of astrological signs and health

Peter C. Austin, Muhammad M. Mamdani, David N. Juurlink , Janet E. Hux

Objectives: To illustrate how multiple hypotheses testing can produce associations with no clinical plausibility.

Study Design and Setting: We conducted a study of all 10,674,945 residents of Ontario aged between 18 and 100 years in 2000. Residents were randomly assigned to equally sized derivation and validation cohorts and classified according to their astrological sign. Using the derivation cohort, we searched through 223 of the most common diagnoses for hospitalization until we identified two for which subjects born under one astrological sign had a significantly higher probability of hospitalization compared to subjects born under the remaining signs combined ($P < 0.05$).

Results: We tested these 24 associations in the independent validation cohort. Residents born under Leo had a higher probability of gastrointestinal hemorrhage ($P = 0.0447$), while Sagittarians had a higher probability of humerus fracture ($P = 0.0123$) compared to all other signs combined. After adjusting the significance level to account for multiple comparisons, none of the identified associations remained significant in either the derivation or validation cohort.

Conclusions: Our analyses illustrate how the testing of multiple, non-prespecified hypotheses increases the likelihood of detecting implausible associations. Our findings have important implications for the analysis and interpretation of clinical studies.

Thus, one problem with using 0.05 as the cutoff for statistical significance is that it simply isn't low enough, and thereby allows in too many false positives.

Where does "$p < 0.05$" come from? We've been mentioning the idea of "$p < 0.05$" as the criterion for "statistical significance" but we haven't really asked the key question: where did the number "0.05" come from? What led us to choose that number, and what is its validity? These are good questions.

The answer is that the number "0.05" was originally suggested by Ronald Fisher, based on nothing more than his personal judgment. He just asked himself "What would a good number be?" and answered himself that he thought that a 5% chance of being wrong was small enough.

We are therefore justified in asking about the sanctity of this number.

Let's distinguish between the p-value, which is the calculated probability for the observed data under the Null Hypothesis, and the threshold α that we set for p-values, in order to declare that a result is "statistically significant". So NHST can be put in the general form: "if $p < \alpha$, then declare the observed result to be statistically significant".

So our question can be put as: what determines the choice of α?

As we already saw, raising α (say, to 0.1) makes it easier to declare statistical significance (even if it's wrong), and lowering it (say, to 0.01) makes it harder to declare significance (even if there really is a result there).

Type I and Type II error The choice of α is really a special case of a much more general problem of striking a balance between two types of errors (Figure 3.18).

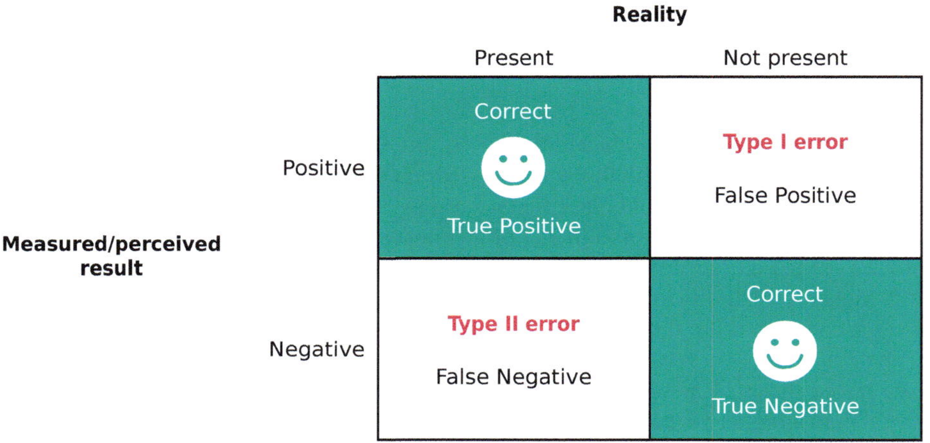

Figure 3.18 Type I and Type II errors.

Let's say we have a Reality, in which some condition is either there or not.

Reality = True: condition is there
Reality = False: condition is not there

Then in addition, we have a Result, in which we declare that the condition is there or that it isn't.

Result = Positive: we declare that the condition is there
Result = Negative: we declare that the condition is not there

We can then make two important definitions.

- A **Type I error** is a (Reality = False, Result = Positive) error, in other words, a False Positive.
- A **Type II error** is a (Reality = True, Result = Negative) error, which is a False Negative.

The tradeoff between Type I and Type II error is found throughout science and medicine. In general, we can always lower the Type I error by tolerating a higher Type II error and vice versa.

For example, the MRI shown in Figure 3.19 has a tiny smudge in a kidney (red arrow). Does that smudge indicate kidney cancer?

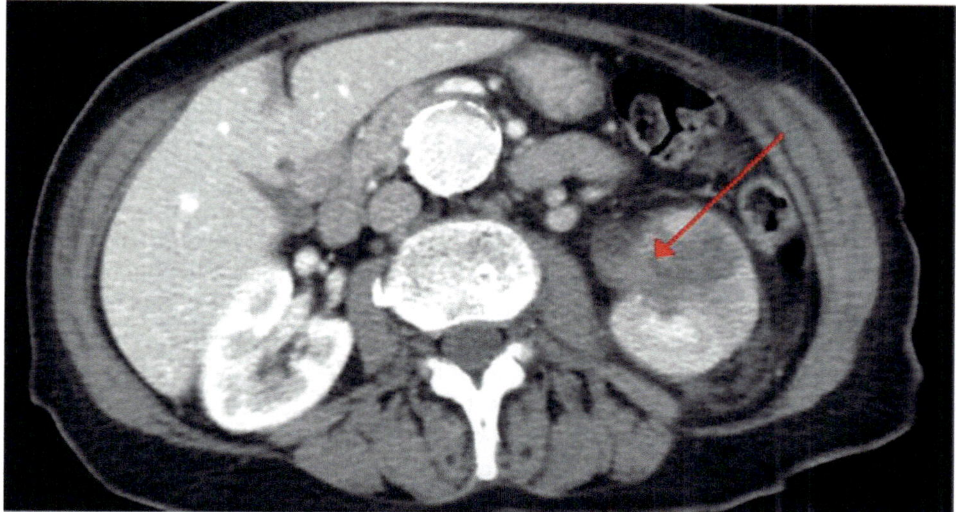

Figure 3.19 A tiny smudge in a kidney MRI scan.

In answering this question, there are two types of mistakes we, the radiologist, could make:

- **Type I error** (False Positive) we could say it's kidney cancer when it isn't.
- **Type II error** (False Negative) we could say it's not kidney cancer when it is.

We can obviously set our threshold for declaring cancer to be as high or low as we like. We can make it hard to declare cancer and easy to say not cancer, by saying something like "it must have all of the following 6 features present in order to declare cancer". In that case, Type I error (saying it's cancer when it's not) will be rare, but Type II error (saying it's not cancer when it really is) will be more common. At the other extreme, we could declare that any unusual smudges are likely to be cancer; then we will make more Type I errors and fewer Type II errors.

There is clearly a tradeoff between these two types of errors. We can always decrease the number of false negatives by increasing the number of false positives, and vice versa. If we just said "cancer" to every image, our Type II error rate would be zero. We would have no False Negatives because we would have no negatives at all! Of course, our Type I error (False Positive) rate would be very large.

Exercise 3.3.1 For each scenario, which type of error is being committed?

a. The weather forecast predicted rain today but there was no rain today.

b. The weather forecast predicted no rain today but it rained today.

It's clear that we could, theoretically, draw the line anywhere we want. What factors influence our choice? What determines the right trade-off?

Social costs of the two types of error The answer is that the trade-off point is determined by the *social costs* of the two types of error. If we make the mistake of not declaring cancer when it is really there (Type II error), the patient continues undiagnosed, and will not receive treatment until the disease has progressed further, which is obviously not good. But what is the cost of a Type I error (saying it's cancer when it really isn't)? First, there is the psychological stress of receiving the phone call that says "please come in for further testing". Then the additional social costs are defined by what the "further testing" is. If it's just another image or another blood test, then the social cost is limited to the cost of the additional scans or other tests, in addition to the worry factor. In cases like this, where the social cost of Type I error is modest, we set the trade-off point to tolerate False Positives in the desire to avoid False Negatives.

But what if the only further tests were invasive, like a needle biopsy or even an exploratory surgery? Now the costs of a False Positive have gone way up.

In criminal law, we also have to set a bar between False Negatives and False Positives. In the US, the evidence to convict on a criminal charge must be "beyond a reasonable doubt". This is a high threshold for declaring "guilty" (equivalently, a low threshold for declaring "not guilty"). It reflects our social value that False Positives (innocent people declared guilty) are much worse than False Negatives (guilty people declared not guilty). Of course, if we changed that standard of evidence to the weaker "more likely than not", then we would correctly convict many more truly guilty people, but we would also wrongly convict many more innocent people.

The choice of different standards of evidence in the law is equivalent to the choice of varying levels of α in NHST. It reflects how many False Positives (innocent people declared guilty) the society is willing to tolerate to avoid having too many False Negatives (guilty people set free). By setting $\alpha = 0.05$, we are saying that we will tolerate 5% False Positives in order to avoid having too many False Negatives.

Exercise 3.3.2 In the 1980s, a drug (Bendectin), given to pregnant women for nausea and vomiting, was suspected of causing birth defects. In a study, partly funded by the drug company, the authors claimed that there was "no significant association" between the drug and birth defects (see "Antenatal exposure to doxylamine succinate and dicyclomine hydrochloride (bendectin) in relation to congenital malformations, perinatal mortality rate, birth weight, and intelligence quotient score", *American Journal of Obstetrics and Gynecology*, 1977).

If we analyze the data given in the paper, the drug increased "musculoskeletal defects" by about 40%. If we calculate the likelihood of finding this level of increase under the Null Hypothesis, the p-value is 0.14. Therefore, the authors have taken this finding, that there is a 40% increase in musculoskeletal defects ($p = 0.14$), and repackaged it as "the studies are reassuring in that they suggest that Bendectin does not appear to be harmful to the fetus".

a. In the context of this study, what are Type I and Type II errors?
b. Which type of error is worse?
c. State the Null Hypothesis.
d. Is it appropriate for this study to choose $\alpha = 0.05$? Explain why or why not.

Exercise 3.3.3 You're developing a test to detect a strain of the bacterium *E. coli* on lettuce. If the test is positive, all the lettuce grown in that area that year must be removed from stores and destroyed. Symptoms of this particular *E. coli* strain include violent cramps, severe diarrhea, and vomiting. 95% of people who ingest it will eventually recover, but 5% will not. The standard α used for NHST by the Food and Drug Administration (FDA) is 0.01. Should you adjust α in your test? If so, how should you adjust it? Briefly explain your decision.

Why are we seeing so many False Positives?

If we choose an alpha of 0.05, then we are allowing in 5% False Positives. **This is an alarmingly high rate of error, which is compounded by the phenomenon of multiple testing using automated programs on "Big Data" sets, so the actual number of False Positives becomes quite high, and the literature is cluttered with false results.** This results in the crisis of confidence in published results that we are seeing today. "$p < 0.05$" is simply too lax a standard, and has resulted in the pollution of the scientific literature with false results.

Something that "only" happens by chance 5% of the time will happen by chance 5% of the time.

A lower value for alpha? A number of respected sources have called for a lowering of the alpha value for published research, that is, for higher standards to publish. In a paper called "Redefine Statistical Significance", over 70 authors said "We propose to change the default p-value threshold for statistical significance from 0.05 to 0.005 for claims of new discoveries" (Benjamin *et al.*, 2018).

Why are they calling for a lowering of alpha? Because "In biomedical research, 96% of a sample of recent papers claim statistically significant results with the $p < 0.05$ threshold. *However, replication rates were very low for these studies*" (Italics added; see their paper for the references for these claims).

This is a step in the right direction, but just lowering alpha is not enough. Lowering alpha to 0.005, as they proposed, does not solve the multiple testing problem. If we do thousands of tests, then we will find some that meet the $p < 0.005$ cutoff, but are false positives.

Recall the study from the University of Toronto (page 118) that took the 12 Zodiac signs, and searched through 223 medical diagnoses to find significant associations. For each medical diagnosis, they carried out pairwise comparisons of each zodiac sign against others, making for a total of 66 pairwise comparisons. 223 diagnoses × 66 pairwise comparisons per diagnosis makes for a total of 14,718 tests in all.

A number of these tests returned p-values less than 0.005, for example, the association of Taureans with the diagnosis "Diverticula of intestine" had $p = 0.0006$, and people born under Pisces had a higher diagnosis of Heart Failure, with $p = 0.0013$.

Therefore, merely lowering the alpha value, even by an order of magnitude, doesn't protect against false positives if we are doing several orders of magnitude more tests.

Brain activity in a dead salmon?

Another group of scientists created their own satire of the multiple testing practice. They took a dead salmon, and put it under an fMRI machine, the technology that produces those pictures of colored areas of the brain that "light up" during various tasks or conditions. They held up pictures of people to the dead fish, "asked" it what emotion the picture was demonstrating, and recorded an fMRI.

Then, when they subjected the results of the fMRI scans to standard statistical testing without correction for false positives, they found 16 voxels (that is, 3D pixels) out of 8064 that showed "significant" differences even at the 0.001 level.

Neural correlates of interspecies perspective taking in the post-mortem Atlantic Salmon: An argument for multiple comparisons correction

Craig M. Bennett Abigail A. Baird Michael B. Miller George L. Wolford

INTRODUCTION

With the extreme dimensionality of functional neuroimaging data comes extreme risk for false positives. Across the 130,000 voxels in a typical fMRI volume the probability of a false positive is almost certain. Correction for multiple comparisons should be completed with these datasets, but is often ignored by investigators. To illustrate the magnitude of the problem we carried out a real experiment that demonstrates the danger of not correcting for chance properly.

RESULTS

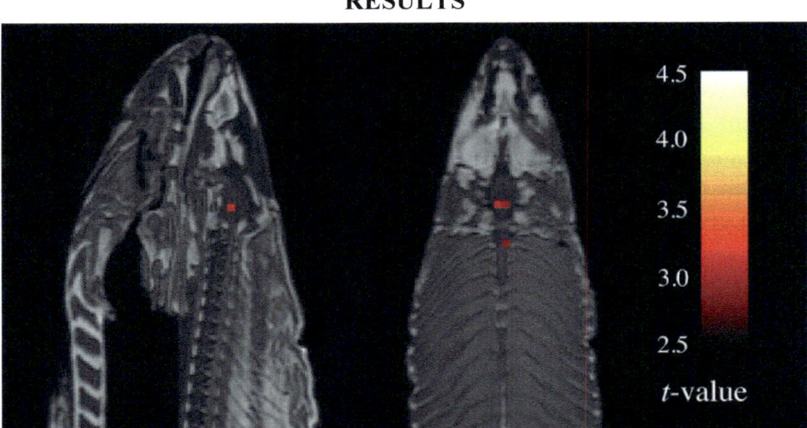

A t-contrast was used to test for regions with significant BOLD signal change during the photo condition compared to rest. The parameters for this comparison were t(131) > 3.15, p(uncorrected) < 0.001, 3 voxel extent threshold.

Several active voxels were discovered in a cluster located within the salmon's brain cavity (see above figure). The size of this cluster was 81 mm3 with a cluster-level significance of p = 0.001. Due to the coarse resolution of the echo-planar image acquisition and the relatively small size of the salmon brain further discrimination between brain regions could not be completed. Out of a search volume of 8064 voxels a total of 16 voxels were significant.

Identical t-contrasts controlling the false discovery rate (FDR) and familywise error rate (FWER) were completed. These contrasts indicated no active voxels, even at relaxed statistical thresholds (p = 0.25).

As the authors observed, what was needed was a correction to control the number of False Positives that will likely occur overall, given the number of tests we are performing, whereas alpha, the False Positive Rate is a per-test rate and doesn't take the number of tests into account.

What is needed to combat the multiple testing problem is not just a fixed lowering of the alpha value, as the zodiac and dead salmon authors clearly explain, but rather a correction for the tendency to produce false positives in any multiple testing.

Family-wise Error Rate (FWER) The simplest correction for false positives in multiple testing is called the Bonferroni correction. It seeks to control the **Family-wise Error Rate**, that is, the chance that multiple testing will produce at least one false positive.

The Bonferroni correction says that if we want an *overall* FWER of α, and we do m tests, then our actual p-value cutoff must be the adjusted cutoff α^*, which is the per-test α divided by the number of tests m. So in the case of zodiac signs, if we wanted an overall FWER of 0.05 and we did 14,718 tests, then we would need to set $\alpha^* = \frac{0.05}{14{,}718} = 0.0000034$. None of the tests would have made this cutoff (We'll learn more about the Bonferroni correction in Chapter 7, see *Bonferroni correction* on page 302).

False Discovery Rate (FDR) A more sophisticated method for controlling false positives in multiple testing focusses on maintaining an overall **False Discovery Rate**, which is the **overall probability of a positive finding being a false positive.**

We can define the False Discovery Rate (FDR) for a multiple testing procedure to be

$$\text{FDR} = \frac{\text{FP}}{\text{TP} + \text{FP}}$$

where FP is the number of False Positives, and TP is the number of True Positives (Figure 3.20).

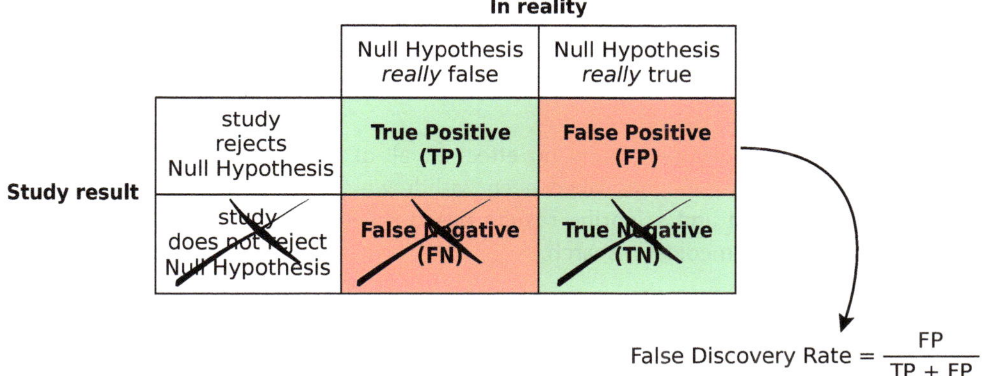

Figure 3.20

The best known method for controlling the False Discovery Rate is the Benjamini-Hochberg procedure. Instead of using a constant threshold for all tests, it uses a sliding threshold. We'll learn the details in Chapter 7 (see *Benjamini-Hochberg correction* on page 303).

"p < anything" is just 'yes' or 'no'. Give the p-value, not just "< 0.05"

In a paper called "Sifting the evidence — what's wrong with significance tests?", in the journal *Physical Therapy*, Sterne and Smith wrote:

> An arbitrary division of results, into "significant" or "non-significant" according to the p-value, was not the intention of the founders of statistical inference.
>
> ...
>
> A P value of 0.05 need not provide strong evidence against the null hypothesis, but it is reasonable to say that P < 0.001 does. In the results sections of papers the precise P value should be presented, without reference to arbitrary thresholds.

This is generally good advice, but there is also a problem with it. Many times in science, we *have to* make yes/no decisions. The editor has to decide: do we publish this result or not? The public has to decide: is there a problem here or not?

In these cases, especially the decision to publish or not, we have to draw an "arbitrary threshold" in addition to providing a numerical p-value.

NHST gives us no clue as to effect size

A deep problem lying at the heart of NHST is that the Null Hypothesis is either rejected or it is not rejected. We are told that a certain treatment, say, lowers heart disease, and that the lowering is statistically significant. Maybe we are also given the numerical p-value, but we still want to know: lowers heat disease *by how much*? This *how much* is called the **effect size** of the treatment.

Aspirin and heart attacks For example, a few decades ago, the Physicians Health Study of the use of aspirin found a highly statistically significant reduction in heart attacks in people taking daily aspirin. They studied more than 22,000 subjects over an average of 5 years and reported that aspirin was associated with a reduction in myocardial infarction (MI, also known as heart attack) that was highly statistically significant: $p < 0.00001$.

But, as observed by Gail Sullivan and Richard Feinn ("Using effect size — or why the p-value is not enough", Sullivan and Feinn (2012)), "the effect size was very small: a risk difference of 0.77%." So there was a lowering of risk that was less than one percent. And, "as a result of that study, many people were advised to take aspirin who would not experience benefit yet were also at risk for adverse effects".

Then, more recent studies found no effect at all of aspirin on heart disease, and the recommendation to use aspirin has since been withdrawn. **This is an important example of the unreliability of medical and scientific results based on NHST, and one of the reasons for the crisis of confidence in medical research.**

The take home message is simple: don't just say "statistically significant increase"; say how much the increase was. Give an effect size.

- **never** say "Treatment X significantly improved survival ($p <$ whatever)".
- You **must**, at the very least, say "Treatment X improved survival by Y% or Z years ($p <$ whatever)".

Confidence intervals

Even an effect size and a p-value does not give you any sense of the *precision* of the estimated effect size (the "plus or minus"). Say our study finds that Treatment X improves survival by 30%, let's say, $p < 0.01$. We need to know how precise that 30% estimate was. Could it have been 60%? Could it have actually worsened survival? To put it another way, suppose we were to redo the experiment; how variable would our effect size be?

In traditional statistics, this question is approached through the notion of a **confidence interval** for that estimated effect size. We will discuss confidence intervals in depth in Chapter 4.

The p-value is the answer to the wrong question

The fifth problem with NHST is perhaps the deepest and the most important. First of all, it is critical that we clear up a major confusion about the interpretation of p-values. Suppose we have done a study and have a result with $p < 0.05$. What are we justified in saying? As we saw, the correct interpretation of that statement is "The probability of obtaining a result that extreme or more, under the Null Hypothesis, is less than 0.05".

However, there is a very seriously mistaken interpretation of that result, a mistake we should never make, but which is found thousands of times in the literature. The mistake is to say "There is a 95% chance the Null Hypothesis is false" or "There's a 95% chance the observed result is true". You *can* say "under the Null Hypothesis, this happens less than 5% of the time" but you *may not* say "there's a 5% chance that the Null Hypothesis is true".

The p-value fallacy The confusion about the interpretation of p-values is a big problem. The cause of that confusion is that the mistaken interpretation is *what we would really like to know.* We would really like to be able to say "there's a 95% chance the Null Hypothesis is false" because that statement is completely interpretable and understandable. Nevertheless, it is precisely what you *may not* say!

The mistaken interpretation is called the **p-value fallacy**. In order to explain the fallacy, we need to introduce a little bit of terminology, called **conditional probability**, which we will use again later in this book, especially in the chapter on Bayesianism. The conditional probability of "A given B", that is, the probability of A given that B has occurred, will be written as "$p(A \mid B)$".

$$p(A \mid B) = \text{prob}\,(A \mid B) = \text{probability of A given B} = \text{probability of A “on B”}$$

We will see a formal definition of $p(A \mid B)$ in the chapter on Bayesianism (Chapter 12). Right now we can think of it simply as the fraction of Bs that are also As. That is, it's the answer to the question: given B, what is the probability of also being an A?

$$\mathbf{p(A \mid B) \neq p(B \mid A)}$$

So for example, the probability of being tall, given that the person is a pro basketball player, is very high, because most pro basketball players are tall. But the probability of being a pro basketball player, given that the person is tall, is much lower, since few tall people are pro basketball players.

$$\text{prob}\left(\begin{array}{l}\text{person}\\ \text{is tall}\end{array}\middle|\begin{array}{l}\text{person is a pro}\\ \text{basketball player}\end{array}\right) \neq \text{prob}\left(\begin{array}{l}\text{person is a pro}\\ \text{basketball player}\end{array}\middle|\begin{array}{l}\text{person}\\ \text{is tall}\end{array}\right)$$

Exercise 3.3.4 Give a real-world example where $p(A \mid B)$ is not the same as $p(B \mid A)$. Include both a verbal description and symbolic notation for your example.

Using the concept of conditional probability, we can now say clearly what is wrong with saying that "$p < 0.05$ means that there's a 95% chance that the Null Hypothesis is false".

The p-value fallacy can be summarized as equating the two conditional probabilities, $p(A \mid B)$ and $p(B \mid A)$.

- The statement "there's at least a 95% chance that the Null Hypothesis is false" is saying "given our observed result, the probability of the Null Hypothesis being true is less than 5%".

$$\text{prob}\,(H_0 \mid \text{our observed result}) < 5\%$$

- But the p-value statement "$p < 0.05$" is not that, rather, it's

$$\text{prob}\,(\text{data at least as extreme as our observed result} \mid H_0) < 5\%$$

  That is, that given that the Null Hypothesis is true (that is, "under the Null Hypothesis"), the probability of obtaining results at least as extreme as our observed result is < 5%.

  But what is prob(data at least as extreme as our observed result | H_0)? It's the sum of a number of terms, all of which have the form $p(D \mid H_0)$, where D ranges over all outcomes that are at least as extreme as our observed result.

And are those two statements equivalent? No, they are not!
If we let H_0 = Null Hypothesis and D = our observed data, we see that

p-values are an answer to the wrong question

p-values are an answer to a question in the form:

How likely is this data given the Null Hypothesis? = $\text{prob}(D \mid H_0)$

But what we really want to know is whether the data supports my hypothesis or not:

How likely is the Null Hypothesis given this data? = $\text{prob}(H_0 \mid D)$

Importantly,

$$\text{prob}(D \mid H_0) \neq \text{prob}(H_0 \mid D)$$

A really bad philosophical argument A number of papers and books on NHST try to argue in favor of the wrong interpretation with a philosophical argument that is completely fallacious. The argument begins with the valid point that there is a principle in logic called the **argument from the contrapositive**. It says that

if	**"if A, then B"** is true
and	**not-B** is true,
then	we can conclude that **not-A** is true

This is a valid principle. An example is:

- if it's raining, then the streets are wet
- the streets are not wet
- therefore it's not raining

The mistaken argument that $p(H \mid D)$ is really equal to $p(D \mid H)$ then goes like this: we are just going to make a probabilistic version of the argument from the contrapositive, saying, "if H, then *probably* not D, but D, therefore *probably* not H".

However this "principle" is *completely fallacious*, as a simple example will illustrate.

- if a person is an American, they are *probably* not a US Senator (probability $\frac{100}{300{,}000{,}000} = 0.0000003$)
- but (my Senator) is a Senator
- therefore (my Senator) is *probably* not an American

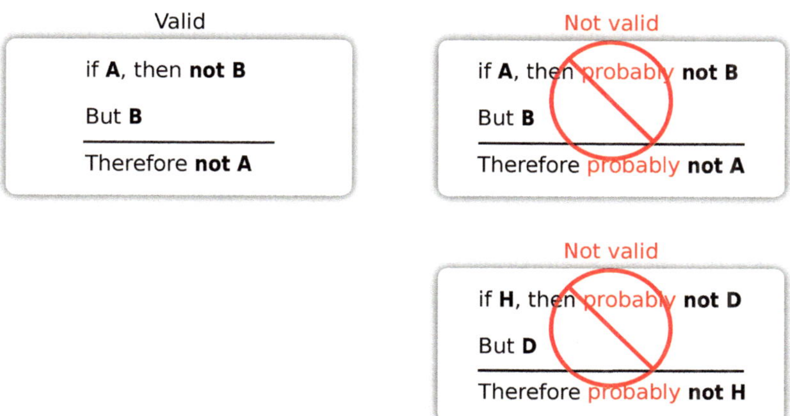

The engine behind the p-value fallacy is the mistake of thinking that $\text{prob}(D \mid H_0)$ is the same as $\text{prob}(H_0 \mid D)$ when in fact,

$$\underbrace{\text{prob}(H_0 \mid D)}_{\text{what you really wanted}} \neq \underbrace{\text{prob}(D \mid H_0)}_{\text{p-value}}$$

The p-value fallacy

It is surprising that many papers, books and lectures continue to assert this fallacy. The fallacy was excellently diagnosed and described in detail in an important paper by the statistician Jacob Cohen in 1994 ("The earth is round ($p < .05$)", *American Psychologist* (Cohen, 1994)). He calls it "an error in elementary logic" and cites other researchers who call it an "illusion" and "wishful thinking".

We will discuss the p-value fallacy further in Chapter 12, on Bayesianism.

California vs. Collins In the meantime, we will mention briefly a famous example of this fallacy, the 1968 case of *People of the State of California v. Collins*. The facts are these: a violent robbery was committed in the city of San Pedro. An eyewitness at the scene described the couple committing the robbery by a number of characteristics (races, genders, color and make of car, physical characteristics, presence of a dog, etc.).

Days later, a couple answering this description was stopped by law enforcement miles away and charged with the crime, based on their answering the description of the wanted couple. At trial, the prosecutor argued that the chance of a random couple meeting this description was 1 in 10 million[1]. Therefore, he argued, there is a 1 in 10 million chance that they are innocent.

[1] For clarity of exposition, we have rounded off some of these numbers. The overall conclusion is not changed.

At trial, the jury convicted the couple, accepting the 10 million to 1 odds as "proof beyond a reasonable doubt".

The prosecutor argued:

If a couple is innocent, then the chance of them matching the description is 1 in 10 million.
But this couple matches the description.
Therefore, they are probably not innocent (1 in 10 million chance of being innocent).

The jury convicted.

But the verdict was reversed on appeal to the California Supreme Court, who explained the fallacy clearly.

On appeal, the courts reversed the finding.

The judge reasoned:

- You arrested this couple miles away.
- Within a driving radius of the crime, there are 30 million people.
- Assuming that the odds of a random couple matching the description is 1 in 10 million, there will be, on average, 3 such couples in the area.
  These 3 couples are false positives.
- In addition, there is the truly guilty couple, so there will be 4 couples in this area that match the description: 3 false positives plus the 1 true positive.
  Therefore, the odds of any couple matching the description being the guilty one is 1/4.
  There's a 3/4 chance they are innocent!!!

We can see that this is an example of the p-value fallacy.
Let

$$H_0 = \text{"the couple is innocent"}$$
$$D = \text{"matches the description of the guilty couple"}$$

Then

$$\text{prob}(D \mid H_0) = 1 \text{ in } 10 \text{ million}$$
$$\text{prob}(H_0 \mid D) = 3/4$$

The prosecutor's fallacy is to equate $\text{prob}(H_0 \mid D)$ and $\text{prob}(D \mid H_0)$.

A similar case was seen in the UK, when Sally Clark was wrongly accused of murdering her two sons, based on similar reasoning.

Californians flipping coins Suppose we were highly concerned about people who cheat at gambling by using unfair coins. We define a *fair coin* as one that has a 50% chance of Heads. We then define a criterion, and say that a coin will be considered unfair if it produces 25 heads in a row in a trial of 25 tosses.

$$p(25 \text{ heads in } 25 \text{ tosses} \mid \text{fair coin}) < \frac{1}{33 \text{ million}} \approx 0.00000003$$

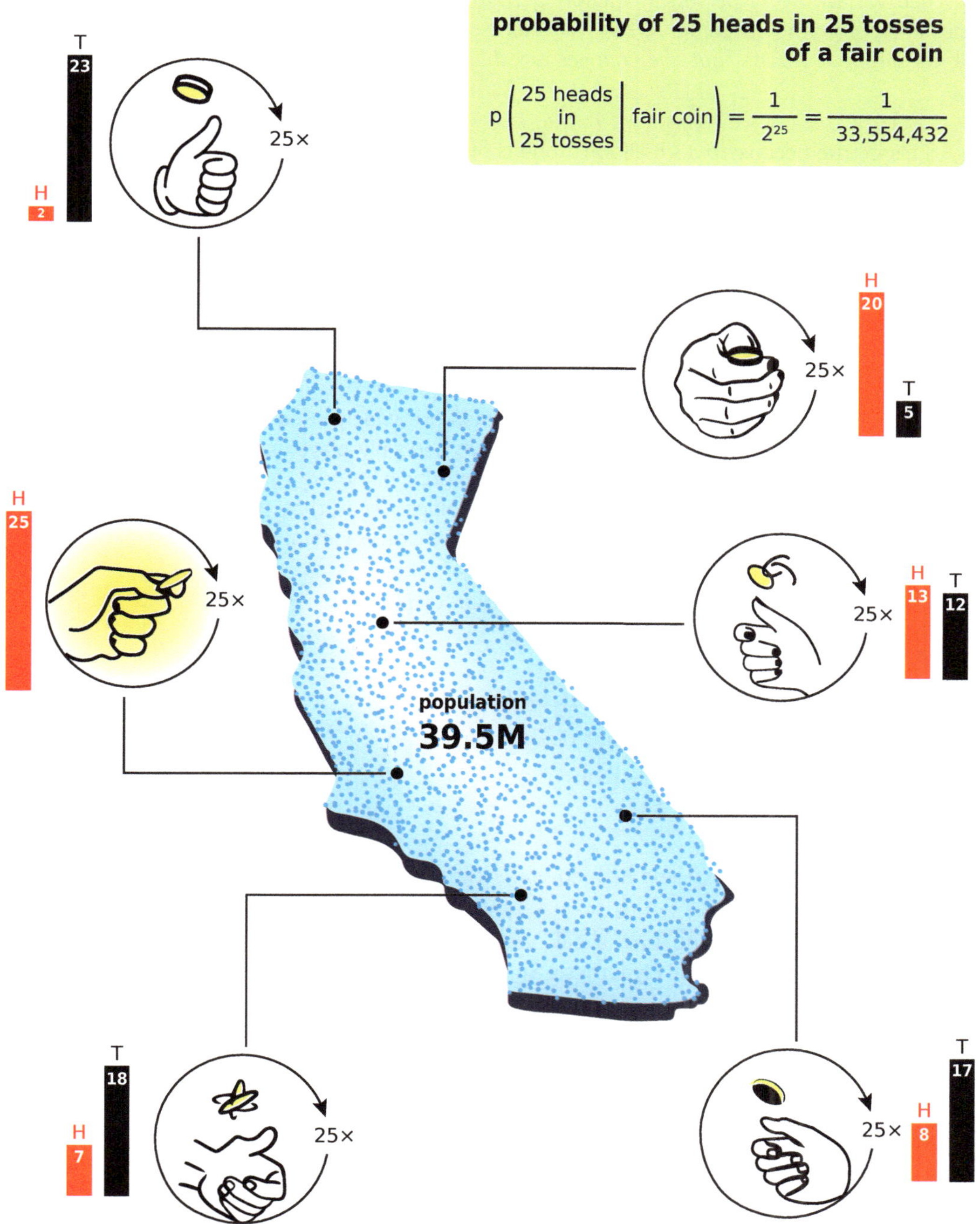

Figure 3.21 If 39.5 million Californians each flip a fair coin 25 times, the chances are that at least one of them will get 25 heads. Here the "unlucky" person lives near Bakersfield and is highlighted in yellow.

Note that the probability of 25 Heads in 25 tosses is 1 in 2^{25}, or about 1 in 33 million. Thus, a coin that produces 25 Heads in 25 tosses has 33 million to 1 evidence against it being fair.

Then we ask every person in California (population around 39.5 million) to flip a coin 25 times and report the outcome. What will the results be? As Figure 3.21 shows, different people will get different results, *but the chances are that at least one person will get 25 heads in their trial!*

This is true because a 33 million-to-one shot is an excellent bet (70% chance of success) if you have 39.5 million chances at it!

So, there is a 70% chance that there will be at least one person who will be the lucky (or unlucky) person. Suppose that in this case it is someone living near Bakersfield. *Are they guilty of possessing an unfair coin?*

Conclusion: problems with p-values

Null Hypothesis Significance Testing and p-values are widely used in the scientific literature. Their use is not completely wrong, but they have certainly played a critical role in generating the false results that have led to the crisis of irreproducibility.

In reviewing the problems with NHST and p-values, recall that the deepest problem lies in the tricky interpretation that we have to make of the statement "$p < 0.05$".

The interpretation of p-values

CANNOT SAY:

- The Null Hypothesis is probably false.
- The result is probably not due to chance.
- The result is probably real.

CAN ONLY SAY:

- Results this extreme happen under the Null Hypothesis less than 5% of the time.

When we go further, and turn a calculation of a p-value into a judgement of "statistical significance", we are doing Null Hypothesis Significance Testing. NHST faces a host of additional problems.

What's wrong with Null Hypothesis Significance Testing? ("$p < 0.05$")

1. 0.05 is not strong enough, giving us too many false positives.
2. "$p <$ anything" is binary, just yes or no. What is the actual p-value?
3. p-values give us no clue as to effect size. We need an estimate of effect size.
4. Even if an effect size is given, there is no sense of the precision of that estimate of the effect size: we need a measure of the uncertainty.
5. The p-value is the answer to the wrong question. What you'd like to know is $p(H_0 \mid D)$, but the p-value gives you $p(D \mid H_0)$.

Problems 1-4 are addressable, and we will deal with methods to answer them in this book, such as methods for controlling the False Discovery Rate, and methods for constructing confidence intervals. Problem 5 goes deeper, to the very core of NHST, and questions its meaning and relevance. The answer to it is in Chapter 12 *Bayesian Statistics*.

Example 2 No difference?

Question "We compared two groups and calculated p = 0.9, so there is no difference between the two groups." Is this interpretation correct?

Answer This interpretation is incorrect because all we can say is "we were not able to detect a statistically significant difference" or "we fail to reject the Null Hypothesis" at $\alpha = 0.01$ or whatever. Consider the difference between "our study did not find life on Mars" and "our study found that there is no life on Mars". (Recall the saying "Absence of evidence is not evidence of absence.") If we had a larger sample size or a different study design, we might find a statistically significant difference.

FURTHER EXERCISES 3.3

1. Consider each of the following scenarios. Which type of error is being committed?
   a. The fire alarm in the building went off; firefighters came but there was no fire.
   b. A pregnancy test indicates that a woman is not pregnant, but she actually is.
   c. A car alarm went off but it was triggered by a loud motorcycle going by.
2. Are the following statements true or false?
   a. "$p < 0.05$" means that at least 1 in 20 "statistically significant" results is likely to be false.
   b. "$p < 0.05$" means the False Discovery Rate is at least 5%.
   c. The "alpha" value, α, is the cutoff that you choose for a p-value to be "statistically significant".
   d. "$p < 0.05$" means "there is a 95% chance that the Null Hypothesis is false".
   e. "$p < 0.05$" means "under the Null Hypothesis, the probability of getting an outcome as extreme or more than the observed result is less than 5%".
   f. "$p < 0.01$" means "the probability that the Null Hypothesis is false is less than 1 percent".
3. If we use a random data generating machine, and produced 1000 studies with random numbers, and use $\alpha = 0.05$ as the cutoff for declaring statistical significance, how many false positives would we expect?
4. Describe a real-world example of Type I and Type II error. Explain what the two types of error mean in the context of your example.
5. Which is worse, Type I error (false positives) or Type II error (false negatives)?
6. What types of errors could a diagnostician make when evaluating a scan for the presence of cancer? What are the social costs of these errors?

7. **Car alarms.** Identify which type of error is each of the following.

   a. Sound the alarm when there is no real threat.
   b. Don't sound the alarm when there is real threat.

   99% of car alarms are false alarms. What judgement of the relative costs of Type I and Type II errors is being made?

8. **Cancer cluster.** A neighborhood near a chemical factory has experienced 13 new cases of leukemia in the past year, in a total population of 5,000 adults. The number of new cases of leukemia in the United States for the same period was 14 per 100,000 adults. Describe how you would determine if this neighborhood is a "cancer cluster" where the rate of new cases of leukemia is statistically significantly higher than the background rate for the U.S.

9. **Thigh vs. arm.** A researcher wants to know whether children are more likely to have a severe reaction to the DTaP (diphtheria, tetanus, and pertussis) vaccine if it is administered in the thigh or the arm. They collected data on severe reactions to this vaccine in 100,000 children aged 3 to 6 years old. 0.63% of children who have the vaccine in the thigh have a severe reaction, and 0.65% of children who have the vaccine in the arm have a severe reaction. Although these rates of severe reactions are not very different, the results are statistically significant at the $\alpha = 0.01$ level ($p = 0.007$).

   a. How is it possible for such a small difference to be statistically significant? Explain your reasoning.
   b. The p-value alone does not tell the whole story. What else should the researcher report, and why is it important to report?

10. **US criminal law.** In American criminal law, we think Type I error is much worse than Type II error, so we set the standard very high; "proof beyond a reasonable doubt". Here are the four outcomes of a jury trial,

		TRUE STATE OF AFFAIRS	
		INNOCENT	GUILTY
JURY VERDICT	NOT GUILTY	INNOCENT with NOT GUILTY VERDICT	GUILTY with NOT GUILTY VERDICT
	GUILTY	INNOCENT with GUILTY VERDICT	GUILTY with GUILTY VERDICT

   a. Identify which are Type I error and Type II error.
   b. Explain how the phrase "proof beyond a reasonable doubt" manifests in terms of the social cost and the balancing between the two types of error.

11. You're a doctor and you read a headline in a newspaper that says a particular dietary supplement decreases blood pressure, with $p = 0.005$. The supplement is generally safe but sometimes causes unpleasant side effects. Assuming this is all you know, answer the questions below.

a. What does this headline mean? Incorporate the definition of a p-value into your answer.

b. What would the Null Hypothesis be for the blood pressure supplement study described above?

c. Should you recommend the supplement to patients with high blood pressure? Why or why not? Give at least two distinct reasons to support your decision.

12. **Mini-pooling blood samples.** Blood donations are tested for transfusion-transmitted diseases and other contaminants in order to keep the blood supply safe. Since the majority of blood donations are safe, under certain circumstances, it saves time and cost to mini pool samples together into one batch and conduct only one test rather than test each individual sample. Assume a test for a certain toxin is to be performed, and the entire batch needs to be discarded if the toxin is detected.

    a. Classify each of the following scenarios into the four possible outcomes: true positive, false positive, true negative, false negative.

        a) The batch is kept and entered into the blood supply when it actually contains the toxin.

        b) The batch is kept and entered into the blood supply when it does not contain the toxin.

        c) The batch is discarded when it actually contains the toxin.

        d) The batch is discarded when it actually does not contain the toxin.

    b. What would be the consequence of a Type I error (false positive)? What would be the consequence of a Type II error (false negative)?

13. **How accurate is an at-home COVID test?** Some of the at-home antigen tests have an overall sensitivity of roughly 85 percent, which means that the test catches roughly 85 percent of the time when the person is actually infected with the virus, and misses 15 percent. They have a specificity of roughly 90 percent, which means that the test will show negative results 90 percent of the time when the person is not infected with the virus. The table below shows the four possible outcomes of the test.

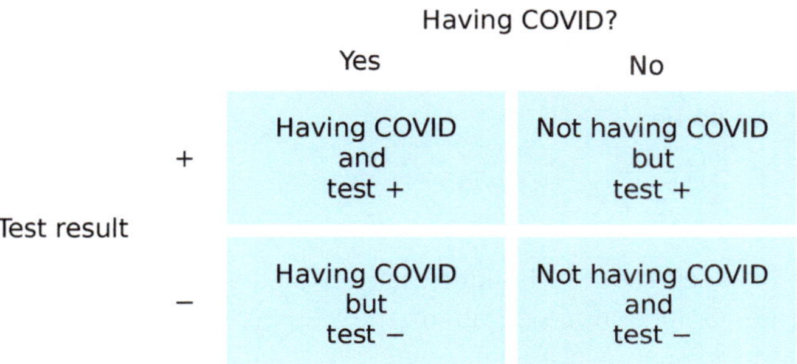

    a. What is type I error and type II error?

    b. How are sensitivity and specificity expressed in terms of the entries in the table?

c. In a town of 10,000 people where 10% of the population have COVID, suppose a mass testing is carried out. Of all the people who tested positive, how many of them actually have the disease?

d. Imagine a mass testing in a different town of 10,000 people, where 1% of the population have COVID. Of all the people who tested positive, how many of them actually have the disease?

14. **Emergency auto-braking.** Uber's fatal self-driving car crash in March 2018 was the first recorded case of a pedestrian fatality involving a self-driving car. According to the report from National Transportation Safety Board, Uber's system detected the victim 6 seconds before impact. The recorded telemetry showed that Uber's self-driving system classified the victim as an unknown object, then a vehicle, then finally a bicycle. (The victim was pushing a bike.) Only 1.3 seconds prior to the impact, the system determined that emergency braking was required. However, from Uber's perspective, "emergency braking maneuvers are not enabled while the vehicle is under computer control, to reduce the potential for erratic vehicle behavior". Which type of error were Uber's engineers trying to avoid by not letting the car auto-brake? Which type of error did they end up committing?

15. **Self driving cars.** In early testing, auto engineers coded a semi-autonomous software to try to avoid braking every time its sensors spot a highway sign, bridge, or the word "stop" on a billboard. However, sometimes, it also failed to identify stopped firetrucks, overturned vehicles, parked police cars or concrete highway barriers, and ended up driving into those stationary objects at highway speed. Which type of error did the engineerrs commit?

16. **Cure rate vs. side effects.** We are evaluating a drug to cure a disease, but the drug has some serious potential side effects. We are comparing the number of patients cured by this drug to the number cured by the current alternative (which has fewer side effects) to determine whether the new drug has a higher cure rate. The standard α used for NHST by NIH is 0.05. Would you recommend using the standard α or adjusting it higher or lower? Why?

17. **Oximeters.** Medical staff often use pulse oximeters clamped to a person's finger to measure the oxygen level of their blood and determine whether they need supplemental oxygen or a ventilator. What types of errors could occur with this tool? What are the associated costs to the patient in each case?

18. Devise an example of a situation in which you might want to use a higher-than-usual alpha (α) value. Explain why this would be appropriate in this situation and what the disadvantages would be.

Chapter 4

Confidence Intervals
one measure, one group

LEARNING OBJECTIVES

After reading this chapter, you will be able to

- understand how to draw inferences from a sample population to make conclusions about a larger population.
- given a measure on a data set, quantify the uncertainty of that measure using the concept of confidence intervals, and explain the meaning of the confidence interval for that measure.
- understand how to use computer resampling as a method for determining the confidence interval for any measure on any data set.
- use confidence intervals as "error bars" to make more informative graphical presentations of data.

4.1 The idea of a Confidence Interval

Inference to a larger population In Chapter 2, we talked about data sets and how to describe them: what summary statistics to use (like the mean or the median or the MAD or ...), and how to describe their distributions. This is the subject generally called **descriptive statistics**, because we have limited ourselves to only describing the data set at hand.

Now we want to ask a larger question. In most cases, we are interested in our data set not just for its own sake, but because we think it can tell us something about the larger population from which it was drawn. We study a sample of patients with, let's say, pancreatic cancer. But we are not only interested in those particular people; we want to make an inference about *all* pancreatic cancer patients. Because we are trying to make an inference about the larger population, this subject is often called **inferential statistics**.

In inferential statistics, we begin with a universe U that we are interested in, but which we can't examine in total. So we take a sample X from U and try to infer properties of U from measurements on X (Figure 4.1).

A. Garfinkel and Y. Guo, *Understanding Data*,
https://doi.org/10.1007/978-3-032-18600-3_4

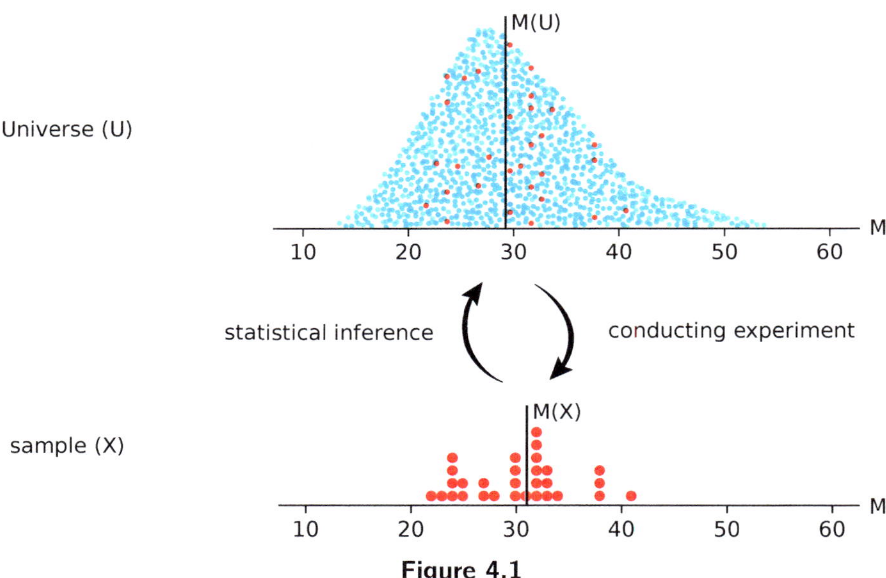

Figure 4.1

Quantifying uncertainty Let's suppose we have a sample X, from the universe U, and an effect size M(X). *Now we want more.* We want to make an inference about M(U).

We want to know *how uncertain* that effect size is as an estimate of M(U). For example, let's say the observed effect size, M(X), is a 40% improvement. Could M(U) have been 60%? Could it have been 10%? Could the true effect even have been negative?

It's important to realize that when we talk about *how uncertain* is the effect size M(X), we must be talking about our uncertainty about M(U), the true value in the universe U. If we are only interested in X, there is no uncertainty in M(X). When we ask "how uncertain is the effect size M(X)", we mean "how uncertain is M(X) as an estimate of M(U)".

The general idea of *uncertainty quantification* has entered into popular discourse: TV reports of the results of political polls will often cite a "margin of error", as in "Candidate A is predicted to receive 53% of the vote, with a margin of error of ±3%." The "margin of error" is a form of uncertainty quantification, the "plus or minus", for the actual percentage that candidate A will receive in the election.

This idea of quantifying uncertainty is an important development in the use of statistics to report scientific results. Scientific journals are increasingly asking for not just "point estimates" of a parameter (like "M = 5.35") but also a range of probable values for that point estimate.

Confidence intervals In traditional statistics, the concept of the margin of error is approached by something called a "confidence interval". For example, the *New England Journal of Medicine*, in their "Guidelines for Statistical Methods", says

> Significance tests should be accompanied by confidence intervals for estimated effect sizes, measures of association, or other parameters of interest.

In fields ranging from medical clinical trials to studies of evolution, scientists are being required to provide confidence intervals. For example, in a widely cited article in *Biological Reviews*, Nakagawa and Cuthill (2007) say

> We advocate presentation of measures of the magnitude of effects (i.e. effect size statistics) and their confidence intervals (CIs) in all biological journals.

What *is* a confidence interval? Intuitively, we would like to find an interval [a, b] that has a 95% probability of containing the true value M(U) (assuming we are calculating a 95% confidence interval). In this chapter, we will develop a resampling-based approach to calculating confidence intervals.[1]

But first, we have to address a key concept in inferential statistics, which is the quality of samples and sampling.

Samples and sampling

From sample to population Suppose we take a sample from a population, make some measurements on the members of the sample, and calculate a summary statistic from that sample.

This could be a sample of people, and our measurements could be a set of blood pressures. Then we take that set of blood pressures from our sample population, and we calculate some summary measure, let's say a median.

What we would like to know is: knowing the median blood pressure from our study, **what could we possibly say about the median blood pressure in the larger population?** What we need is some way of **quantifying the uncertainty** of our sample measurement. If the measure of our observed sample X is M(X), which we will also call M_{obs}, how different could the true measure of the Universe, M(U), be? (Figure 4.2)

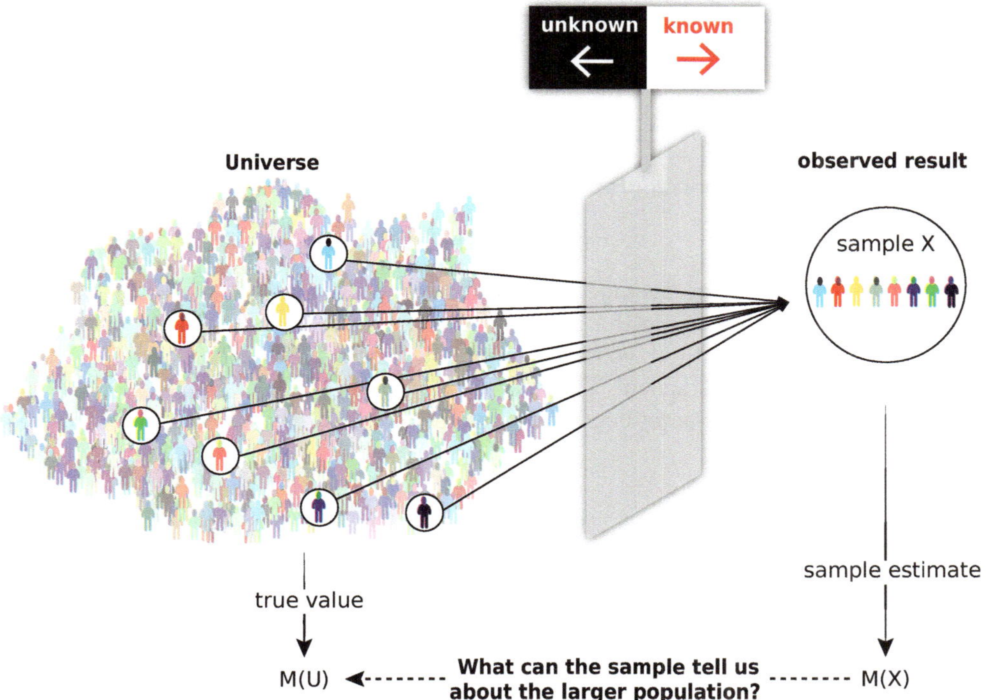

Figure 4.2 We would like to know the value of M(U), the value of our measure M in the whole universe U. But we cannot look at all of U. So we sample n people from U, giving us the sample X, and we calculate M(X) for our sample. Given M(X), what can we say about M(U)?

[1] As we will see later in this chapter (section 4.9 *The interpretation of the confidence interval*), the interpretation of the statement "the 95% confidence interval for M(U) is [a, b]" is "if this experiment was repeated many times, 95% of the confidence intervals calculated in this manner will contain M(U)".

The confidence interval is the answer to the question: given the summary measure M(X), derived from the sample X, how much can we say about M(U), the true measure of the universe U? What is the "margin of error" for M(X) as an estimate of M(U)?

Exercise 4.1.1 The universe U is all entities that we're interested in. However, it's almost never possible or practical to study the whole population U, so we take a sample X (preferably randomly) to help us understand and represent the whole population. For each of the research goals below, what are U, M(U), X and M(X)?

a) A team wants to estimate the insect diversity of a specific 500-hectare rainforest. They spend one week trapping and identifying 450 individual beetles.

b) A nutritionist suspects the sugar content of US breakfast cereals varies far more than consumers realize. She tests 40 boxes from a single grocery store.

c) Antibiotic resistance in soil bacteria is poorly understood. To shed light on this, a researcher isolates 50 bacterial colonies from a single gram of forest soil and tests each one for antibiotic resistance.

Samples As we said, the concept of uncertainty only makes sense if the data set X is a **sample** from some larger population U. If the data set X *is* the whole population, not a sample, then the concept of uncertainty makes no sense and does not apply.

For example, suppose our set is the set of US States

$$\{\text{Alabama, Alaska, } \ldots, \text{ Wyoming}\}$$

and our measure is the population of each state, giving us the data set

$$\{P_1, P_2, \ldots, P_{50}\}$$

We could certainly form the average or the median population for this set. There is no uncertainty or variability whatever in the outcome, because there is no "sampling": the average is what it is and is known exactly (given the exactness of the state population counts). There is no new uncertainty due to sampling.

But if the sample is from a larger population, we can legitimately ask: **how uncertain are we of M_{obs} as a measure of U?**

Sampling We are now talking about sampling a data set from some larger universe. We get our sample by choosing a small subset of the larger population. How we choose that sample is critical. Everything depends on how "representative" it is. This is the subject called "study design."

If we are using the sample to infer about the population, we have to take precautions to **ensure that our sample is representative of the population.**

There are some famous historical examples of sampling failures, due to poor study design.

For the 1936 presidential election, the prestigious magazine *Literary Digest* conducted a sample poll to predict the winner (Figure 4.3). Based on the staggeringly large sample size of around 2.4 million people, the poll predicted that the challenger, Republican Alfred Landon, would win in a landslide, 57% to 43%. The actual result was a bigger landslide in the opposite direction: the Democrat, Franklin Roosevelt, won 62% to 38%.

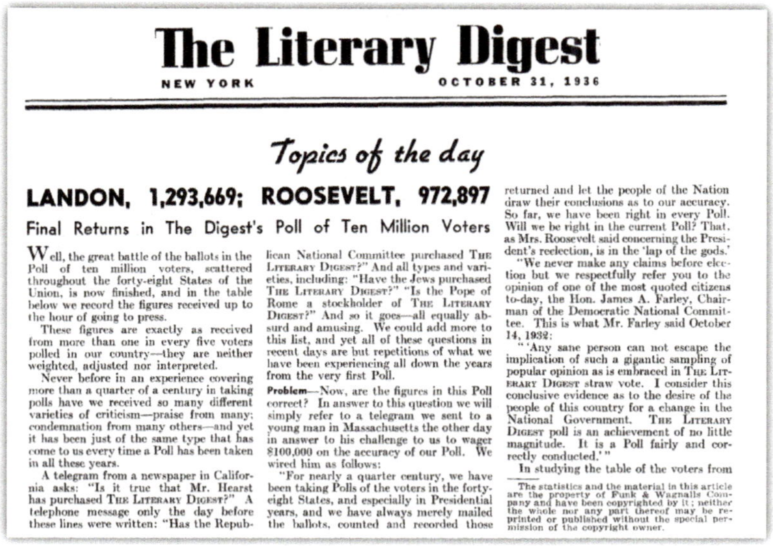

The Literary Digest

NEW YORK OCTOBER 31, 1936

Topics of the day

LANDON, 1,293,669; ROOSEVELT, 972,897

Final Returns in The Digest's Poll of Ten Million Voters

Well, the great battle of the ballots in the Poll of ten million voters, scattered throughout the forty-eight States of the Union, is now finished, and in the table below we record the figures received up to the hour of going to press.

These figures are exactly as received from more than one in every five voters polled in our country—they are neither weighted, adjusted nor interpreted.

Never before in an experience covering more than a quarter of a century in taking polls have we received so many different varieties of criticism—praise from many; condemnation from many others—and yet it has been just of the same type that has come to us every time a Poll has been taken in all these years.

A telegram from a newspaper in California asks: "Is it true that Mr. Hearst has purchased The Literary Digest?" A telephone message only the day before these lines were written: "Has the Republican National Committee purchased The Literary Digest?" And all types and varieties, including: "Have the Jews purchased The Literary Digest?" "Is the Pope of Rome a stockholder of The Literary Digest?" And so it goes—all equally absurd and amusing. We could add more to this list, and yet all of these questions in recent days are but repetitions of what we have been experiencing all down the years from the very first Poll.

Problem—Now, are the figures in this Poll correct? In answer to this question we will simply refer to a telegram we sent to a young man in Massachusetts the other day in answer to his challenge to us to wager $100,000 on the accuracy of our Poll. We wired him as follows:

"For nearly a quarter century, we have been taking Polls of the voters in the forty-eight States, and especially in Presidential years, and we have always merely mailed the ballots, counted and recorded those returned and let the people of the Nation draw their conclusions as to our accuracy. So far, we have been right in every Poll. Will we be right in the current Poll? That, as Mrs. Roosevelt said concerning the President's reelection, is in the 'lap of the gods.'

"We never make any claims before election but we respectfully refer you to the opinion of one of the most quoted citizens to-day, the Hon. James A. Farley, Chairman of the Democratic National Committee. This is what Mr. Farley said October 14, 1932:

"'Any sane person can not escape the implication of such a gigantic sampling of popular opinion as is embraced in The Literary Digest straw vote. I consider this conclusive evidence as to the desire of the people of this country for a change in the National Government. The Literary Digest poll is an achievement of no little magnitude. It is a Poll fairly and correctly conducted.'"

In studying the table of the voters from

The statistics and the material in this article are the property of Funk & Wagnalls Company and have been copyrighted by it; neither the whole nor any part thereof may be reprinted or published without the special permission of the copyright owner.

Figure 4.3 Poll by *Literary Digest* 1936.

The problem with the poll was certainly not its size; it was their sampling method, producing a non-representative sample. What the *Literary Digest* did was take as its source every telephone directory in the United States, as well as lists of magazine subscribers, rosters of clubs and associations, etc. Out of that data base, a mailing list of about 10 million names was created. Every name on that list was then mailed a mock ballot and asked to return the marked ballot to the magazine.

What's wrong with this method? Two things: **sampling bias** and **nonresponse bias.**

Sampling bias refers to the fact that the sample that was sent the mock ballot did not represent the voting population. Criteria like "has a phone", "subscribes to magazines", "belongs to clubs", "owns a house", etc. tend to select a wealthier segment of the population, which tended to vote Republican.

Nonresponse bias refers to the fact that of the 10 million people sent mock ballots, only 2.4 million, less than a quarter, responded. People who are more engaged in the civic process will tend to respond, while people who are disillusioned or disengaged will do so much less. This is obviously a biasing factor in the sample.

Survivorship bias A well-known example of sampling bias occurred during World War II. The leadership of the U.S. Army Air Forces was concerned about the number of their bomber planes that were being shot down. In order to find out which areas on the aircraft were the most vulnerable, they surveyed the bombers as they returned to base, and carried out a statistical survey of the locations of bullet holes. They found relatively fewer hits in the engine/midwing area, and relatively more hits in the fuselage and wing tip areas. Decades later, this finding was summarized in a famous graphic (Figure 4.4).

The Air Forces leadership drew a conclusion from this data, to put more armor on the fuselage area compared to the engine area. But statistician Abraham Wald pointed out that the sample

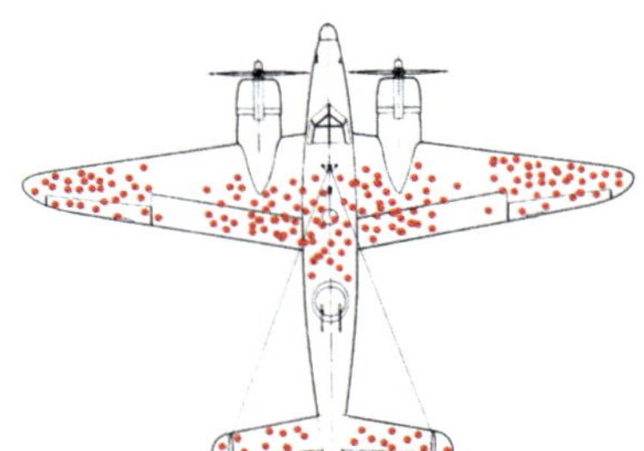

Figure 4.4

was biased. The survey was, obviously, only considering the airplanes that had successfully returned; the airplanes that had taken hits in the engine/midwing area did not survive to be sampled!

Brexit A more recent example was seen in Britain, where the polls in advance of the 2016 Brexit referendum predicted the vote to be against Brexit. However, Brexit won, and afterwards, in the debate over "how could the polls have been wrong?", it emerged that the polling methodology included many young people, who tended to be against Brexit, but who also tend not to vote!

AI A similar problem is faced by applications of Artificial Intelligence, in which an algorithm makes a prediction about a future case, based on a "training set" of data. The validity of such a prediction obviously depends heavily on whether the training set is representative of the population in which the AI is trying to make predictions. For example, in healthcare applications, a training set that partially omits an underrepresented population could well lead AI to dangerous medical misdiagnoses. Similarly, the algorithm that runs a self-driving car must be trained on a fully representative training set, including examples such as pedestrians in wheelchairs, etc.

Study design There are many courses and texts in the subject of "study design" that cover methods to ensure that a sampling process is fair or representative. In representative sampling, the sampling process must have no causal relation to the properties that might affect the outcome. For example, having an odd-numbered social security number has no conceivable relation to voting choice. In the 1936 poll, the property of owning a telephone (part of the causal chain in the sampling process) has a direct causal relation to being wealthy. At that time, this implied a tendency to vote Republican.

This very short discussion will have to suffice for the subject of sampling methods, which are beyond the scope of this book. From now on we will simply assume that our sampling process was well-designed, and that our samples are fair and "representative".

No statistical method helps when your sample is not representative.

Calculating the confidence interval: computer simulation vs. formulas

As we will see at the end of this chapter (section 4.7 *Are there formulas for confidence intervals?*), there are certain kinds of measures, and certain kinds of distributions, for which there are formulas for confidence intervals. For example, there's a formula for the confidence interval for the mean value of a bell-shaped distribution. However, there are two major limitations with the application of this formula: the fact that it is for the mean value only, and only for bell-shaped distributions.

What are we to do when we want CIs for measures other than the mean, say, the median or the ratio of highest to lowest, and/or for data distributions other than bell-shaped? That is the subject of this chapter.

Freedom of measure We want "freedom of measure". We do not want to be confined to what statistics theoreticians can prove theorems about! So what are we to do when we want, for scientific purposes, to use a measure that is not on the list of measures for which theorems can be proved? And what are we to do when our distributions are markedly not bell-shaped?

The general answer is that computer simulation methods can provide answers to these questions.

In 1991, Efron and Tibshirani published their groundbreaking article "Statistical Data Analysis in the Computer Age". Note the title: they are suggesting that our whole approach to statistics is different because of computers.

Science **Statistical Data Analysis in the Computer Age** ARTICLE

Bradley Efron and Robert Tibshirani

Most of our familiar statistical methods, such as hypothesis testing, linear regression, analysis of variance, and maximum likelihood estimation, were designed to be implemented on mechanical calculators. Modern electronic computation has encouraged a host of new statistical methods that require fewer distributional assumptions than their predecessors and can be applied to more complicated statistical estimators. These methods allow the scientist to explore and describe data and draw valid statistical inferences without the usual concerns for mathematical tractability. This is possible because traditional methods of mathematical analysis are replaced by specially constructed computer algorithms. Mathematics has not disappeared from statistical theory. It is the main method for deciding which algorithms are correct and efficient tools for automating statistical inference.

In the abstract of their paper, they say "Modern electronic computation has encouraged a host of new statistical methods that require fewer distributional assumptions than their predecessors and can be applied to more complicated statistical estimators."

By the phrase "more complicated statistical estimators" they mean measures other than the mean. By "require fewer distributional assumptions" they mean that the distributions no longer have to be bell-shaped; they can have any shape whatever.

We want the scientific freedom to devise measures that reflect the scientific hypothesis at hand, and then know that we can construct confidence intervals for them, regardless of the underlying distribution. This will be given to us by resampling-based simulations.

FURTHER EXERCISES 4.1

1. If you're running for governor and the polls say you're getting 52% of the vote, what is the likelihood that you will win?
2. Taking a random sample from a population allows us to:

a. Determine cause and effect between variables.
b. Construct a null distribution for hypothesis testing.
c. Generalize findings from the sample to the broader population.
d. Guarantee a narrow confidence interval for our estimate.

3. Describe a real-world scenario where a sample fails to represent the population being studied. Identify the population, the sample, and describe why the sample is not representative.
4. When reporting the results of a study, why is it more informative to report an effect size alongside a confidence interval than to report the effect size alone?

4.2 Confidence intervals by resampling

Having obtained M(X) (also known as M_{obs}) for our sample X, we want to be able to make an inference about M(U), the true value "out there" in the Universe. In particular, what we would really like to have is a confidence interval for M(U).

In order to get to a confidence interval for M(U), we first need to understand how reliable is our sample statistic M(X)? How variable could our sample observation be?

So we begin with a hypothetical question: what if we could somehow go back into the universe U and do another, similar study Y? How different could M(Y) be from M(X)?

"10,000 (actual) friends" go back into U

Imagine, then, that a friend went out and did the experiment again: went out into the universe U, and sampled from U, choosing the same number of people, and got a new data set Y. When we apply our measure M to it, we get M(Y). Clearly, M(Y) is not going to be exactly equal to M(X). Now suppose we did that many times, and got a large number of observations of M. What would the variability of those Ms be?

10,000 studies In the first replica experiment, we go back into the universe U and choose a fresh sample, call it Y_1, from which we calculate $M(Y_1)$. Suppose we did 10,000 of these replica experiments, and got estimates

$$M(Y_1),\ M(Y_2),\ \ldots,\ M(Y_{10,000})$$

Then we could make a histogram of these 10,000 values of $M(Y_i)$ (Figure 4.5).

From this histogram, we could then find the cutoffs for the 0.5^{th} percentile and the 99.5^{th} percentile. We can then say that there is an approximately 99% probability that a new study Y of U will produce an M(Y) lying in the interval [0.5^{th} percentile cutoff, 99.5^{th} percentile cutoff].

$$\text{prob}\left(0.5^{th} \text{ percentile cutoff} \leq M(Y) \leq 99.5^{th} \text{ percentile cutoff}\right) \approx 99\%$$

However, this interval does not give us information about M(U), which we will also call M_{true}. But that was our original question: what can we say about M(U)?

The 10,000 $M(Y_i)$ values can be viewed as 10,000 *estimates* of the true measure M(U) of the universe U. The 0.5^{th} percentile cutoff and the 99.5^{th} percentile cutoff are telling us that another trip into the universe U would, with 99% probability, produce an estimate lying in this interval.

But that is a different question from: what is a 99% confidence interval for M(U)?

The histogram of the 10,000 $M(Y_i)$ values gives us no information as to where M(U) lies. What we really want is a 99% confidence interval for the true value, M(U).

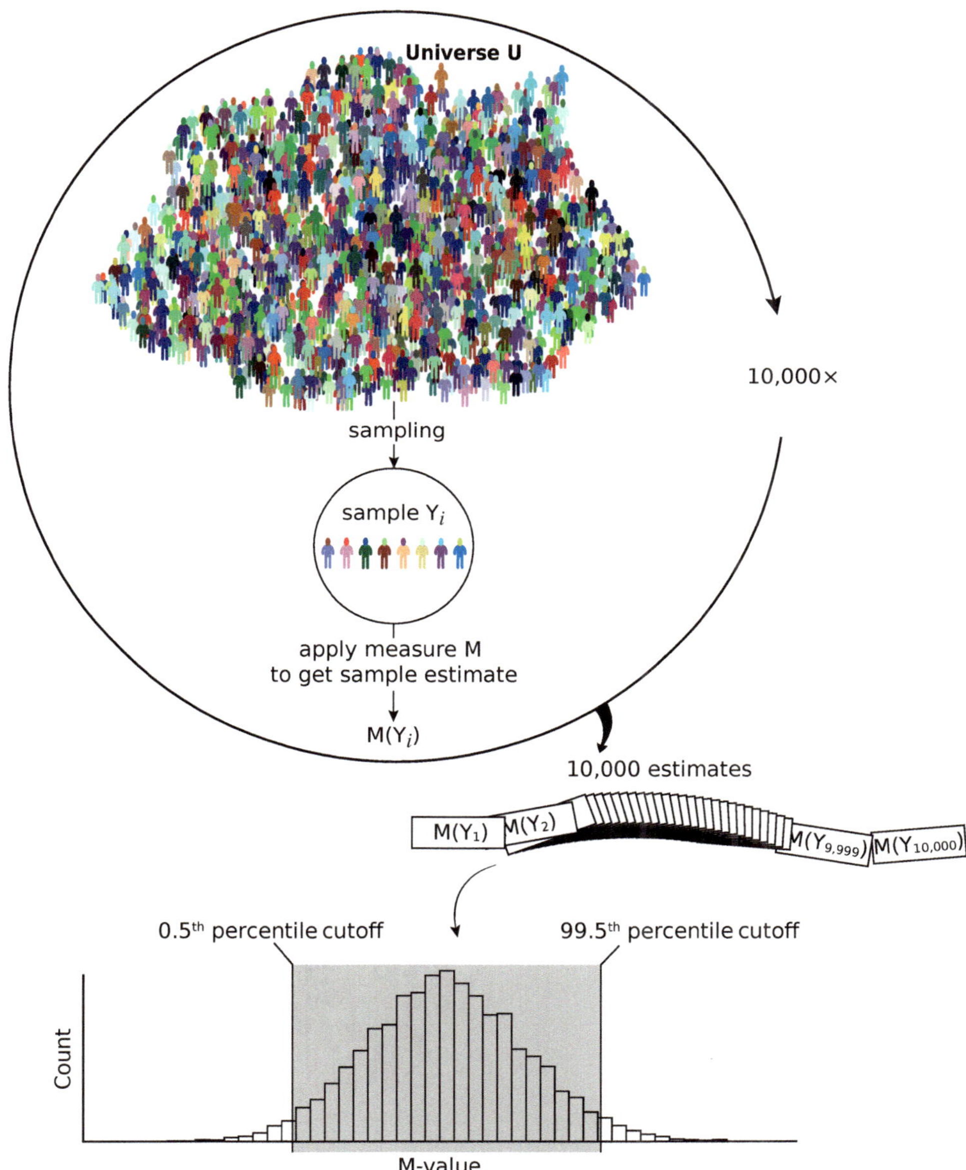

Figure 4.5 The ideal (but absurd) experiment: we have 10,000 friends, each going back into the universe U and generating fresh samples Y_1, Y_2, ..., $Y_{10,000}$. For each sample, we apply measure M, generating 10,000 measurements $M(Y_1)$, $M(Y_2)$, ..., $M(Y_{10,000})$, which we then collect into a histogram. The shaded gray area represents the central 99% of the 10,000 sample estimates.

"10,000 (virtual) friends" go back into X

Of course, the biggest problem with "10,000 friends go back into U" is that it is ridiculous to suggest doing 10,000 additional studies. We don't *have* 10,000 friends, and the cost in money and time of doing this is completely absurd.

Because it is absurd to talk about going back into the universe U, we will make an important assumption about what U must look like to our best knowledge. All we know is our observed data set X. **The only assumption we are justified in making is that the unknown reality U must resemble our data, which was drawn from that reality! Therefore we will use our data set X as a surrogate for the unknown underlying population U** (Figure 4.6).

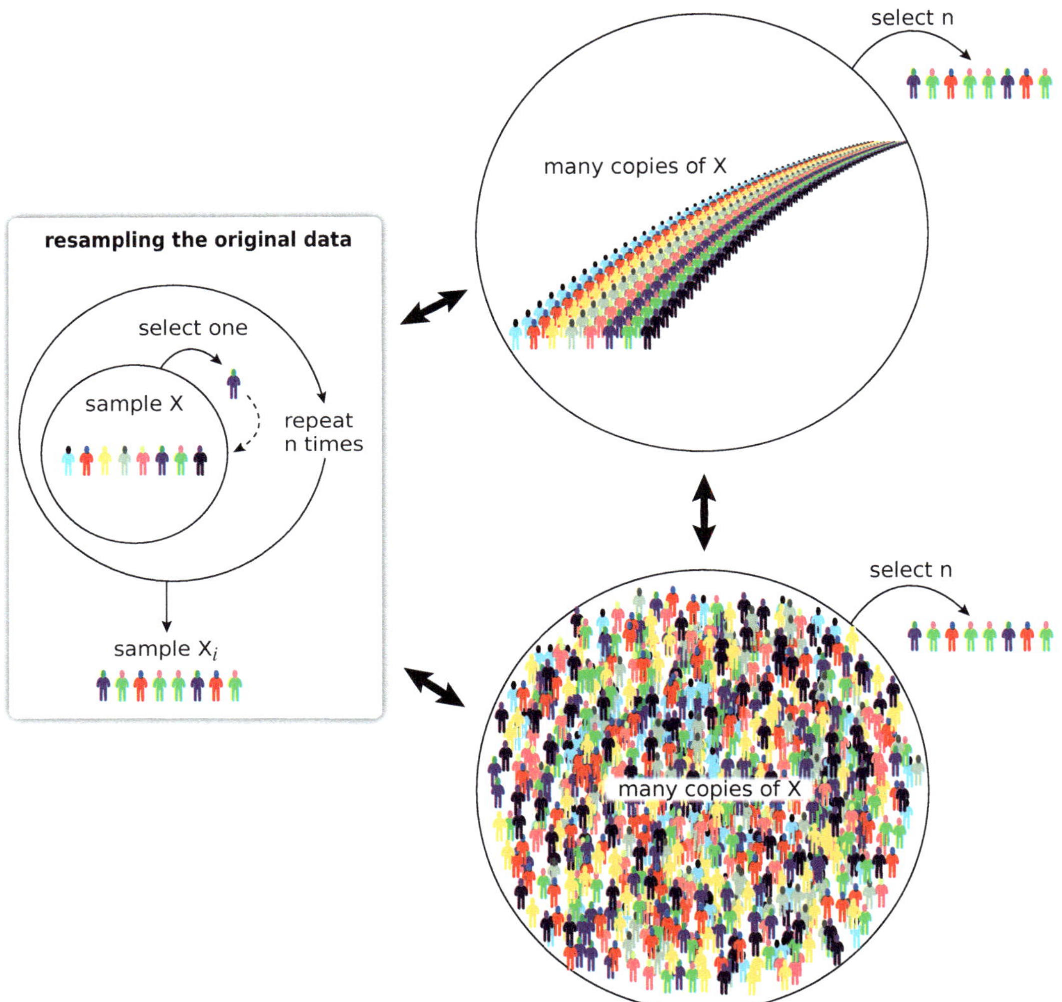

Figure 4.6 Sampling with replacement from X is equivalent to sampling without replacement from a set, called the "resampling universe", consisting of many copies of X.

We can't go back into U! We can, however, go back into our own data X, and resample it. The key idea here is that we can understand U by resampling from X, aka "resampling methods" or "bootstrapping", namely, random sampling with replacement.

Therefore, what we will do to create a virtual study is "resample" our original data. So virtual study #1 consists of going back into the original data set X = $\{x_1, x_2, \ldots, x_n\}$, and randomly sampling n elements with replacement from X. We call this "resampling" the original data.

Resample 10,000 times When we do this resampling, we sample "with replacement". As before (see *Sampling without replacement in casino Blackjack* on page 99), what this means is that we put the data points $\{x_1, x_2, \ldots, x_n\}$ into a box. To generate our first virtual study, we reach blindly into this box and select a data point. Let's say we chose x_5. Then we put x_5 back into the box and choose a second data point. This second choice might have been x_3, but it also might have been x_5 again. We put that data point back and then choose a third, until we have chosen n in this way. This constitutes our first virtual study X_1, and we calculate its M-value and call that $M(X_1)$ or M_1. In this way, we generate 10,000 virtual studies, and calculate the corresponding 10,000 M-values $M_1, M_2, \ldots, M_{10,000}$. The resulting histogram is called the **resampling distribution** (Figure 4.7).

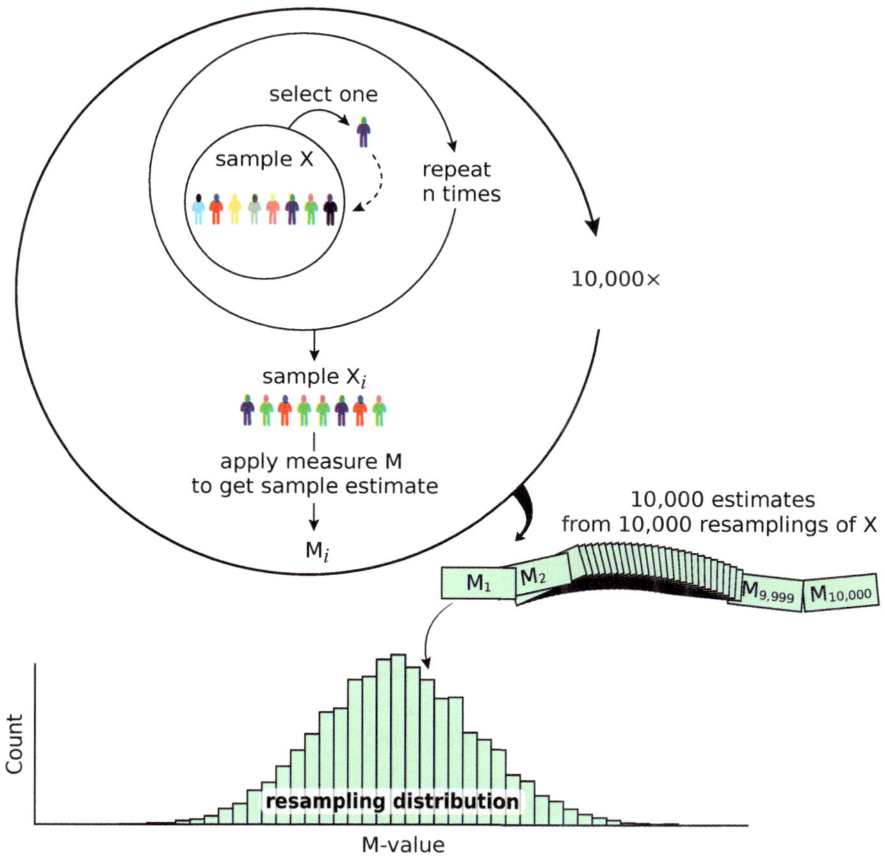

Figure 4.7 Generate the resampling distribution of the measure M from 10,000 virtual studies. Each study is created by resampling X.

There are several reasons why we sample with replacement. The most important reason is that a fundamental assumption of *all* statistics is that every data point is statistically independent of every other. In order to satisfy this requirement, we assume that the world "out there" is just the same as our sample, only much larger. We have one value x_1 in our sample, so we assume that in the larger universe, there are many copies of x_1. Similarly, there are many copies of x_2 in the larger universe, many copies of x_3, ...

So when we sample with replacement from X, we are in effect sampling from the "resampling universe", which we created out of our original sample X and consists of many copies of X. The number of copies of X must be much larger than the original sample size n, so that the probability of drawing each data point is practically unchanged by the previous draws (Figure 4.6).

Resampling works for any measure M Note that the overall resampling procedure does not specify what measure we can or cannot use. In fact, we can use *any* measure M, as long as it's appropriate for our data set, whether it's a mean, median, MAD, variance, absolute-value-averaged, or any measure that is appropriate for our scientific question.

Generating the resampling distribution

Step 1 ► Generate virtual studies by resampling from the original study, using the same sample size as in the original study.

Step 2 ► Compute the statistic of interest M for each of the virtual studies.

Step 3 ► Collect the M-values from the virtual studies to create a resampling distribution.

Once we have this resampling distribution, we are ready to construct the confidence interval.

Constructing the confidence interval

Our original problem was: what can we infer from our sample X about the population U that it was drawn from? In other words, we know the data set X and $M_{obs} = M(X)$. What we would really like to achieve is to go from M_{obs} to some kind of quantification of the uncertainty in M_{true} (= M(U)), the true (and unknown) value for the total population.

In order to make this connection from M_{obs} to M_{true}, we need to make a fundamental assumption about the resampling process.

Assumption about the resampling process

Suppose that we went back into U many times, and obtained samples $Y_i = Y_1, Y_2, \ldots, Y_{10,000}$, yielding a distribution of $M(Y_i)$-values. **Each of these $M(Y_i)$-values can be thought of as an M_{obs} for that hypothetical experiment.**

We assume that the distribution of those hypothetical M_{obs}-values around M_{true} (whatever it is) is well approximated by the distribution of the resampled values $M(X_i)$ around our actual M_{obs}.

$$\underbrace{\text{distribution of } (M_{true} - M_{obs})}_{\text{sampling distribution}} \approx \underbrace{\text{distribution of } (M_{obs} - M_i)}_{\text{resampling distribution}}$$

Note that, on the left hand side, the "distribution of $(M_{true} - M_{obs})$" is the variation of those hypothetical values for M_{obs} around M_{true}. We call this the **sampling distribution**. Here, M_{obs} is a variable. On the right hand side, the "distribution of $(M_{obs} - M_i)$" is the variation of the resampling estimates M_i around our actually observed M_{obs}. Here, M_{obs} is a number. We call this the **resampling distribution**.

By this assumption that those two distributions are similar, we can use the distribution of resampling results around our M_{obs} to tell us how close our actual result M_{obs} is likely to be to M_{true} if we were to redo the experiment many times.

We will use this one assumption to generate a confidence interval for M_{true}. Later, we will validate this assumption experimentally (section 4.5 *Validating the resampling principle*).

How to construct the confidence interval

Step 1 ► **Generate 10,000 sample estimates by resampling from X.**
Resample from the observed data set X to generate 10,000 resamples

$$X_1, X_2, \dots, X_{10{,}000}$$

and calculate their M-values

$$M(X_1), M(X_2), \dots, M(X_{10{,}000})$$

which we will call

$$M_1, M_2, \dots, M_{10{,}000}$$

These are the 10,000 sample estimates M_i for $i = 1, 2, \dots, 10{,}000$.

Step 2 ► **Find M_{lower} and M_{upper}.**
Rank the 10,000 sample estimates (M_i) from smallest to largest, calling them

ranked $MR_1, MR_2, \dots, MR_{50}, \dots, MR_{9951}, \dots, MR_{9{,}999}, MR_{10{,}000}$
smallest ⟶ largest

so that MR_i is the ith smallest value of the 10,000 sample estimates. Let M_{lower} be the 50th smallest value, which is MR_{50}, and let M_{upper} be the 50th largest value, which is $MR_{9{,}951}$.

Step 3 ► We know, by definition, that the probability of M_i lying between M_{lower} and M_{upper} is exactly 99%

$$\text{prob}(M_{lower} \le M_i \le M_{upper}) = 99\%$$

We then can subtract M_{obs} from each of the three terms to get

$$\text{prob}\Big((M_{lower} - M_{obs}) \le (M_i - M_{obs}) \le (M_{upper} - M_{obs})\Big) = 99\%$$

Now multiply each of the three terms in this inequality by -1, which of course requires us to reverse the $\le$ sign to $\ge$. This gives us,

$$\text{prob}\Big((M_{obs} - M_{lower}) \ge (M_{obs} - M_i) \ge (M_{obs} - M_{upper})\Big) = 99\%$$

which is the same as

$$\text{prob}\Big((M_{obs} - M_{upper}) \le (M_{obs} - M_i) \le (M_{obs} - M_{lower})\Big) = 99\%$$

Step 4 ► But what we are really looking for is information about M_{true}, the unknown value for the universe U. We would like to know how close our observed result M_{obs} is likely to be to the true measure $M_{true} = M(U)$. We assume that if we repeated our experiment many times and obtained many M_{obs} values, they would cluster around M_{true} like the M_i-values cluster around our actual M_{obs} (see *Assumption about the resampling process* on page 148). We can now say

$$\text{prob}\Big((M_{obs} - M_{upper}) \le (M_{obs} - M_i) \le (M_{obs} - M_{lower})\Big) = 99\%$$
$$\Downarrow$$
$$\text{Assuming distribution of } (M_{obs} - M_i) \approx \text{distribution of } (M_{true} - M_{obs})$$
$$\Downarrow$$
$$\text{prob}\Big((M_{obs} - M_{upper}) \le (M_{true} - M_{obs}) \le (M_{obs} - M_{lower})\Big) \approx 99\%$$

Step 5 ► Then, to get a useful piece of information about M(U), we add M_{obs} to each term, getting

$$\text{prob}\Big((2\times M_{obs} - M_{upper}) \le M_{true} \le (2\times M_{obs} - M_{lower})\Big) \approx 99\%$$

which is telling us that

$$99\% \text{ confidence interval}^{a} \text{ for } M(U) = \Big[2\times M_{obs} - M_{upper},\ 2\times M_{obs} - M_{lower}\Big]$$

[a] Please note that there are *many* 99% confidence intervals. For example, instead of using a symmetric 0.5% and 99.5% cutoffs as M_{lower} and M_{upper}, we could have used the 0.2% cutoff as M_{lower}, and 99.2% cutoff as the M_{upper}, or we could have used the 0.6% cutoff as M_{lower}, and 99.6% cutoff as the M_{upper}. Here we use the symmetric definition because our resampling distributions are roughly symmetric and we are equally concerned about over- and under-estimation. However, if your resampling distributions are skewed or you care more about one type of error, asymmetric confidence intervals — which place different probabilities in each tail — may be more appropriate.

The process of constructing a 99% confidence interval is shown in Figure 4.8.

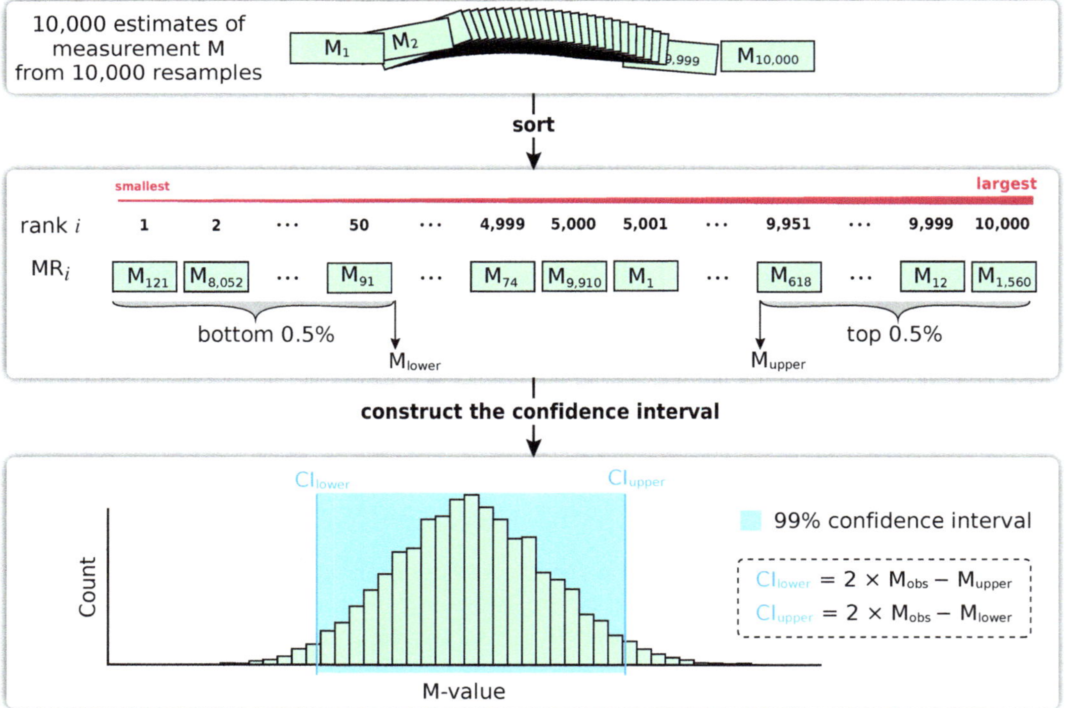

Figure 4.8 Constructing a confidence interval for M_{true}.

Thus, by making the assumption that the variation of the bootstrap samples around the observed value fairly approximates the variation of observed values around the true value, we have used M_{obs} to give us a very useful piece of information, namely, a 99% confidence interval for the true value $M_{true} = M(U)$.

However, it's important to remember that the "99% probability" must be interpreted as a property about the confidence interval generating procedure, not about *this* particular confidence interval. It is saying that if we repeated our experiment many times, and each time generated a 99% confidence interval, then those confidence intervals would contain M_{true} approximately 99% of the time.

Power of the resampling method

- The resampling procedure does not require any assumption about the distribution of our data or the distribution of the underlying population. We assume that our data is the best approximation to the underlying population.
- The resampling procedure can be applied to any measure we are interested in (median, mean, ratio, correlation coefficient, etc.).
- The resampling procedure is the gold standard: for any statistical test that has a formula, when data sets meet the requirements for applying the formula, the resampling-based calculation will agree with the formula result; when the data sets don't meet the requirements, resampling is the only valid method.

FURTHER EXERCISES 4.2

1. Briefly explain how the steps for computing a 95% confidence interval are different from those for computing a 99% confidence interval.
2. A sample of 20 households was drawn randomly from a neighborhood. A 95% confidence interval was calculated to estimate the median household income for the neighborhood. Which of the following values will definitely fall into the range of this confidence interval?
   a. The median household income of the neighborhood
   b. The mean household income of the neighborhood
   c. The p-value
   d. The median household income of the sample
   e. The standard deviation of the sample
   f. Both the median and mean household income of the sample.
3. What does the 95% confidence interval mean?
   a. We are 95% sure that the true value is in this interval.
   b. We are 95% sure that this specific interval contains the true value.
   c. 95% of the data is in this interval.
   d. If we sampled this population again many times, 95% of the confidence intervals that we calculated would contain the true value.

4.3 Example of a confidence interval

Let's test the confidence interval procedure on an example. In this example, we are going to create an artificial situation, in which we actually do know the universe.

The universe U For our example, we will study a data set of the salaries of employees at a large public institution, which we obtained from easily-accessed public records. Here we actually know U, the total population of the universe, because we know every individual salary. This ability to know the whole universe is very unusual. We are using it as a special test case here, but this will almost never be the case in reality, where the universe is typically unknown or unknowable. (And, of course, this is only "the universe" for this particular institution. If we were interested in all such institutions, then this data set becomes only a sample, and we would not know "the universe".)

Salary distributions, notoriously, are skewed to the right, so we do not want to use the mean as our measure; we will use the median instead. **In this universe U, the median salary is $58,041 and the mean salary is $81,073** (Figure 4.9). Notice the difference: the mean is much greater than the median; because the distribution is skewed to the right, the presence of those very high values pulls the average to the right. As we argued in Chapter 2 (*Describing and Presenting Data*), in skewed distributions, the mean value is a poor descriptor of the data set.

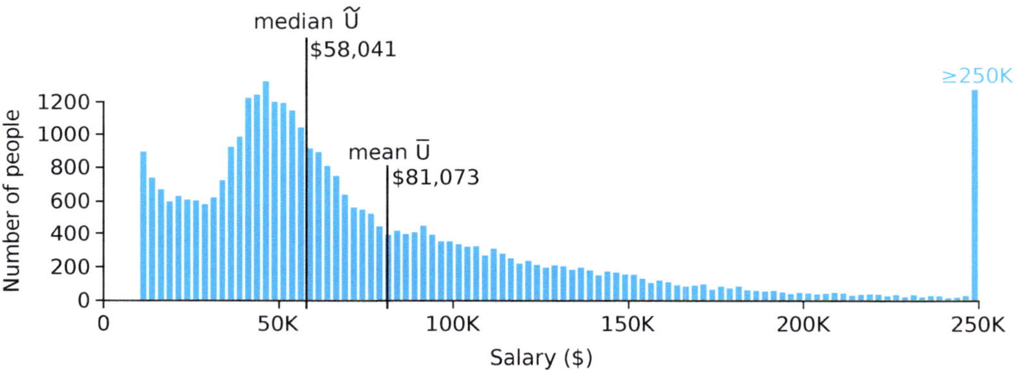

Figure 4.9 Histogram of salaries of employees at a large institution. The right-skewness makes the mean a poor descriptor of this population.

The observed data set In our experiment, we randomly pick $n = 10$ people from the universe U and record their salaries. Note that $n = 10$ is a very small sample; we will soon see the consequences of this in the very wide confidence intervals it produces. In any event, this is our observed data $X = \{x_1, x_2, \ldots, x_{10}\}$ (Figure 4.10),

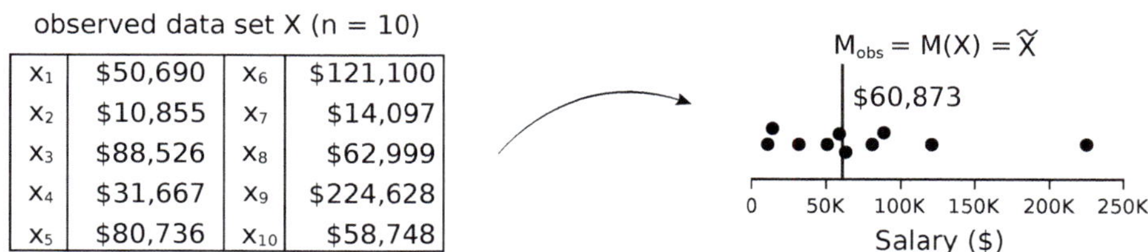

observed data set X (n = 10)

x_1	$50,690	x_6	$121,100
x_2	$10,855	x_7	$14,097
x_3	$88,526	x_8	$62,999
x_4	$31,667	x_9	$224,628
x_5	$80,736	x_{10}	$58,748

Figure 4.10 Observed data set X and the observed measure M_{obs}.

Recall that $\tilde{X}$, "X tilde", is the median of the data set X. Similarly, $\tilde{U}$ is the median of the data set U.

We then apply the measure M, in this case, the median, to the sample X and get its median

$$M_{obs} = M(X) = \tilde{X} = \$60{,}873$$

That is, the median salary of our observed sample is around $60,873.

The question is then: How well does this observed median $\tilde{X}$ estimate the "true median", $M_{true} = \tilde{U}$, which we know to be $58,041 (Figure 4.11)?

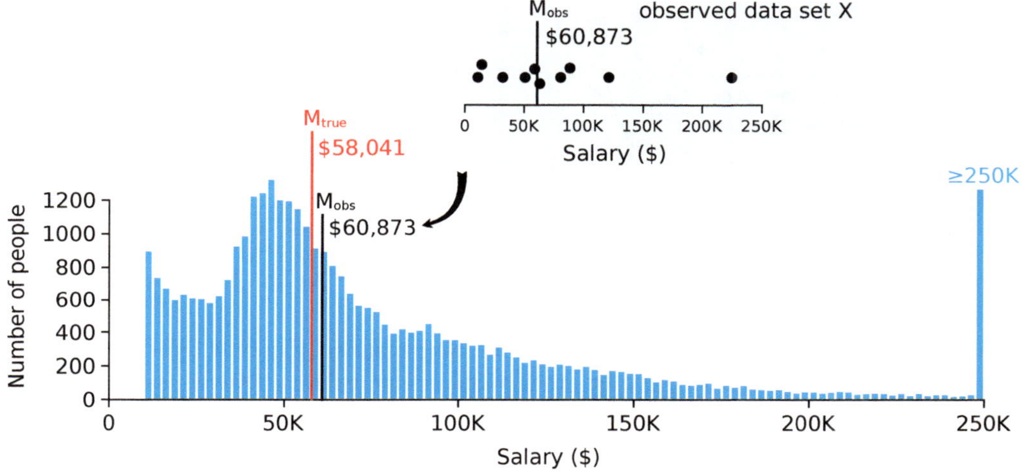

Figure 4.11 Comparing M_{obs} to M_{true}.

Resampling from X Now we will pretend for the moment that we don't know the universe U and all we have is the 10 data points in our sample X. We will then resample X to make inferences about U, knowing that in this artificial case, we can compare those inferences to the true value in U, M_{true}. (In section 4.5 we will use our knowledge of U to validate the resampling procedure.)

We've been talking about the theoretical but impossible ideal of 10,000 experiments in U, which we approximate by 10,000 resamplings of X. In this example, we can carry out this resampling procedure and compare it to the ideal of 10,000 experiments in U, which we can actually do.

The distribution of the 10,000 median values from the resamplings of X is called the **resampling distribution**. The resampling distribution will naturally cluster around M_{obs}, which is not the true value M_{true}. But, while the resampling distribution is not clustered around M_{true}, the resampling distribution allows us to understand something very important: the variability due to sampling.

Let's work this out for our sample of 10 salaries.

Our first resampling of X, which we will call X_1, is carried out by sampling with replacement 10 times from X (Figure 4.12). Notice that, because this is resampling with replacement, it is very possible that the same element in X will be selected multiple times. In X_1, the largest element in X was selected three times. The smallest two elements in X were selected twice each. And the 5th largest element in X, \$62,999 was selected three times. The first resample, X_1, has its own median, $\widetilde{X_1} = M_1 =$ \$62,999.

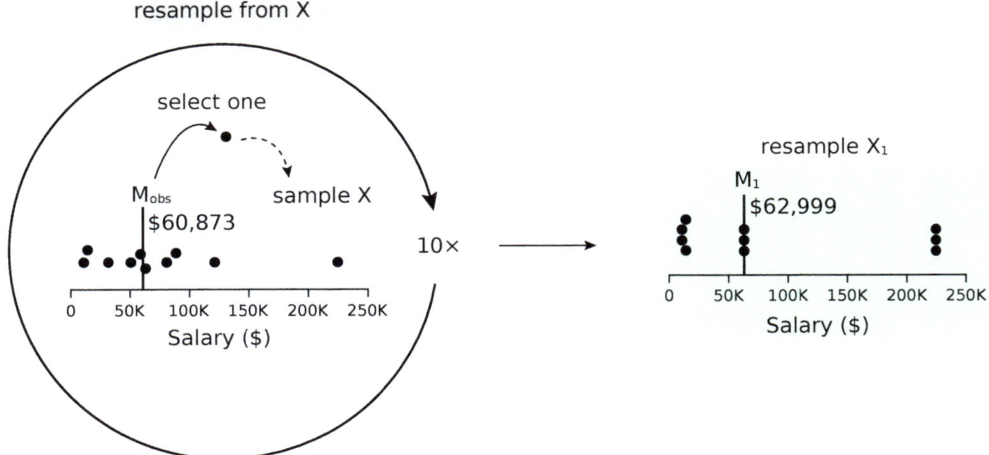

Figure 4.12 Resample from X to generate the first resample X_1 and the measure on it M_1.

The second resampling, X_2, is generated in the same manner (Figure 4.13). By chance, the higher values in X were not selected, but other middle range values in X were selected multiple times, resulting in a slightly higher estimate of the median: $\widetilde{X_2} = M_2 =$ \$69,742.

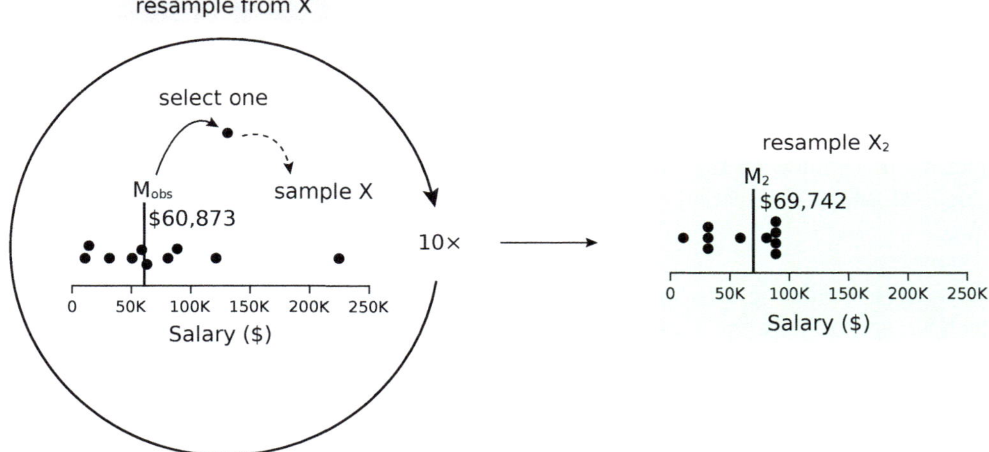

Figure 4.13 Resample from X to generate the second resample X_2 and the measure on it M_2.

We then repeat this procedure for a total of 10,000 times, resulting in 10,000 sample estimates $M_1, M_2, \ldots, M_{10,000}$, from which we generate the resampling distribution (Figure 4.14).

Confidence interval To construct the 99% confidence interval, we first rank the 10,000 sample estimates $M_1, M_2, \ldots, M_{10,000}$ from smallest to largest. Then from the ranked list, we count up from the smallest and down from the largest to get the 50th smallest and the

50th largest values, calling them M_{lower} and M_{upper}. When we did this, M_{lower} and M_{upper} were \$14,097 and \$121,100. Following the procedure of the previous section (*How to calculate the Confidence Interval* on page 149), the 99% confidence interval is

$$\begin{aligned} \text{99\% confidence interval} &= \left[2 \times M_{obs} - M_{upper},\ 2 \times M_{obs} - M_{lower}\right] \\ &= [2 \times 60{,}873 - 121{,}100,\ 2 \times 60{,}873 - 14{,}097] \\ &= [\$646,\ \$107{,}649] \end{aligned}$$

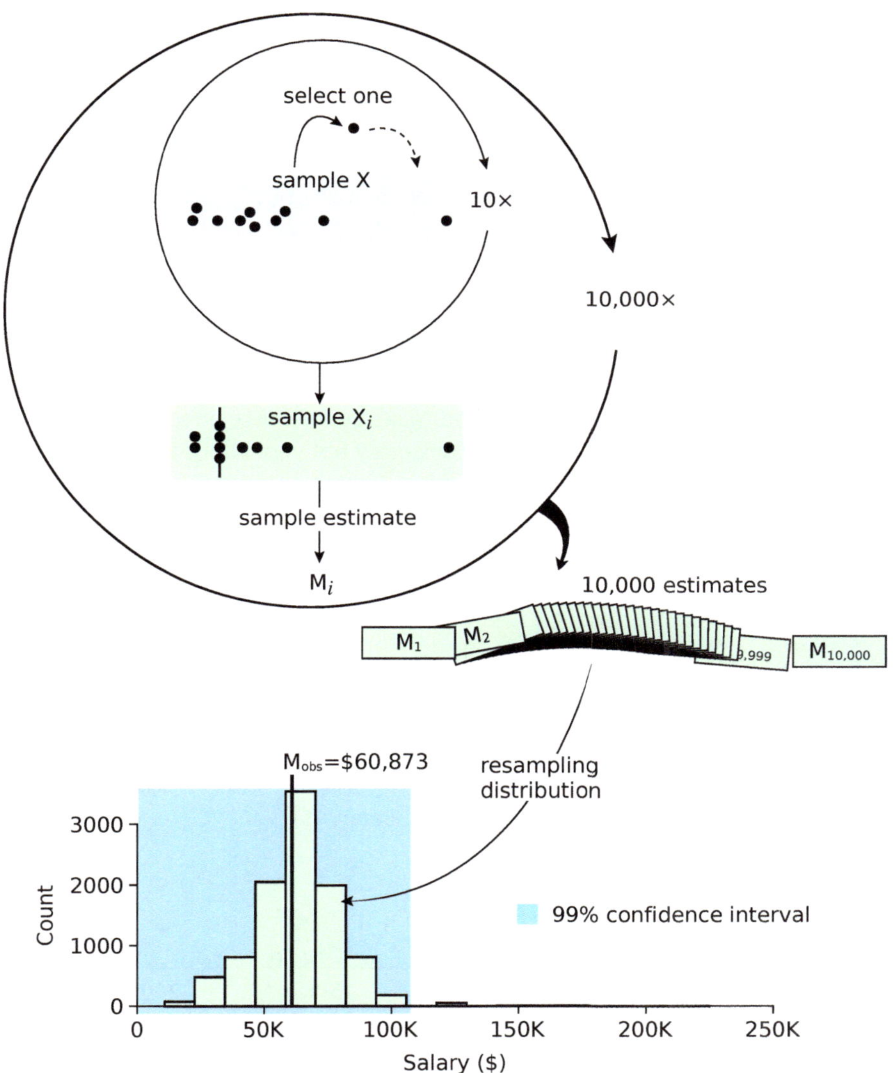

Figure 4.14 Generating 10,000 resamples from the observed data set X. We then collect the 10,000 values of sample estimates M_1, M_2, . . . , $M_{10,000}$ into a histogram, called the resampling distribution. The 99% confidence interval is then calculated and superimposed on the resampling distribution.

What have we learned about U? **Very little. The confidence interval is overwhelmingly wide. And so we have gained very little information about the true median, other than that it probably lies between $646 and $107,649 a year!**

Exercise 4.3.1 Your friend has collected data from an experiment and wants to understand the uncertainty around the median. Knowing you're taking a statistics course, they ask for your help. Explain the concept of a confidence interval and describe the resampling process used to construct a 99% confidence interval for a sample median.

Exercise 4.3.2 Suppose you randomly sampled 100 students and asked them how often they ate breakfast, where each student reports a percentage. From this sample, you calculate a 99% confidence interval of [75%, 92%]. If instead a 90% confidence interval was calculated, how would it differ from the 99% confidence interval?

a. The 90% confidence interval would be narrower than the 99% confidence interval.
b. The 90% confidence interval would have the same width as the 99% confidence interval.
c. The 90% confidence interval would be wider than the 99% confidence interval.
d. There is no way to tell if it's wider or narrower without actually calculating the 90% confidence interval and comparing them.

4.4 A bigger sample, a better estimate

The previous section used a sample of n = 10 to make an inference about the universe. We saw an extremely wide confidence interval, which is obviously not very informative. This precision will improve when we take a larger sample. We will now repeat the previous section and construct a confidence interval for the median of the salary data, with a sample size n = 100.

We begin by randomly drawing a new sample of 100 from our universe U (Figure 4.15). Note that the median of this larger sample, our new M_{obs}, is $55,094.

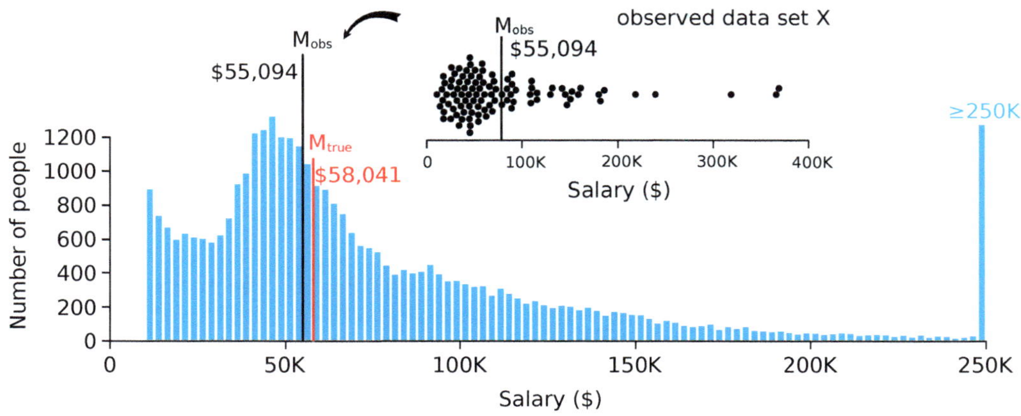

Figure 4.15 A new sample from U with n = 100.

Resampling from X We will construct the confidence interval by sampling from our new larger sample with replacement 100 times, to find our first resampling estimate, M_1, corresponding to our first resampled data set X_1 (Figure 4.16). The first resample, X_1, has its own median, $\widetilde{X_1} = M_1 = \$61,899$.

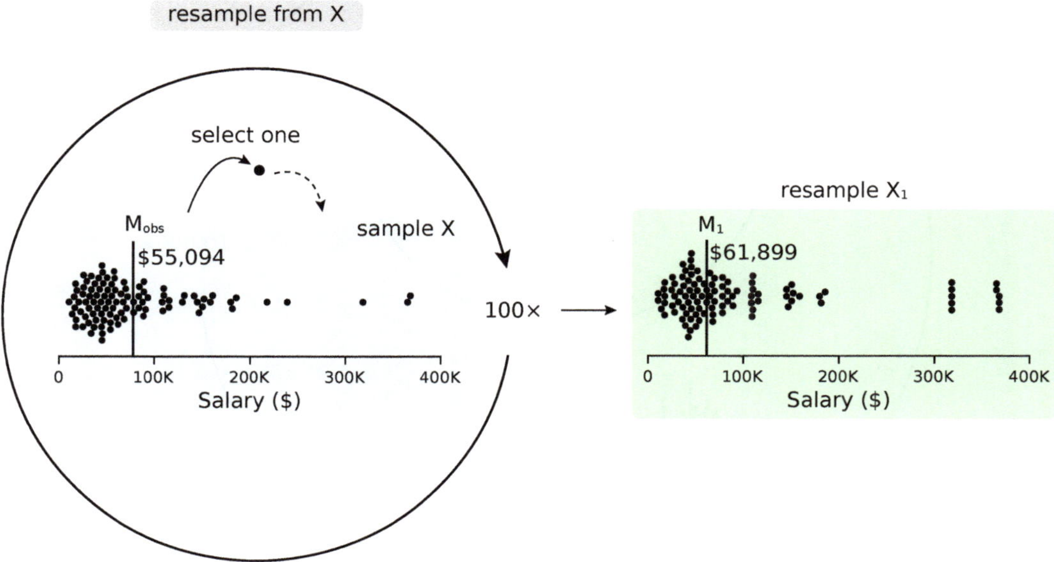

Figure 4.16 Resample from X to generate the first resample X_1 and the measure on it M_1.

A second resampling X_2 produces a second estimate M_2 (Figure 4.17).

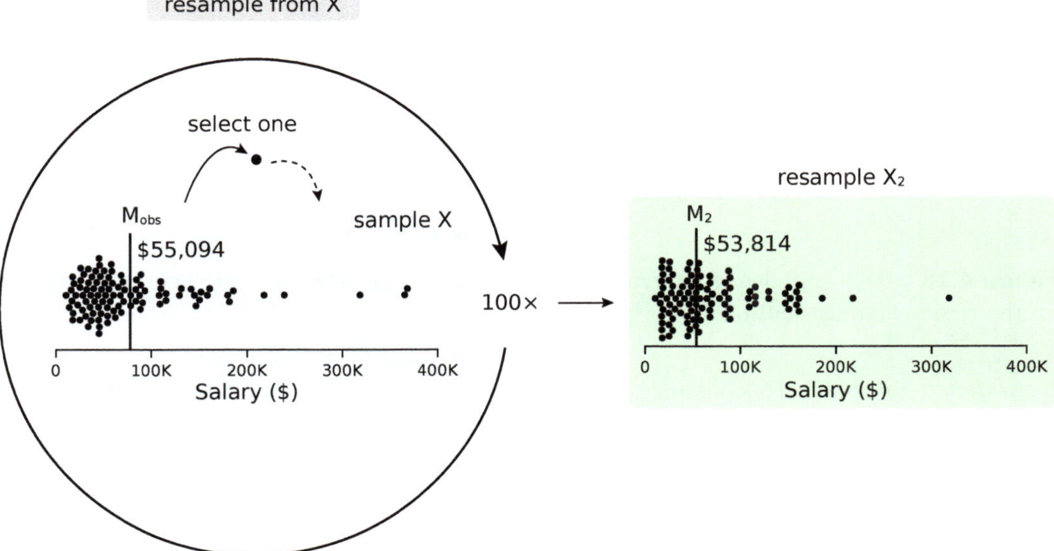

Figure 4.17 Resample from X to generate the second resample X_2 and the measure on it M_2.

Then we repeat this resampling procedure 10,000 times, to produce the resampling distribution from sample estimates M_1, M_2, . . . , $M_{10,000}$ (Figure 4.18).

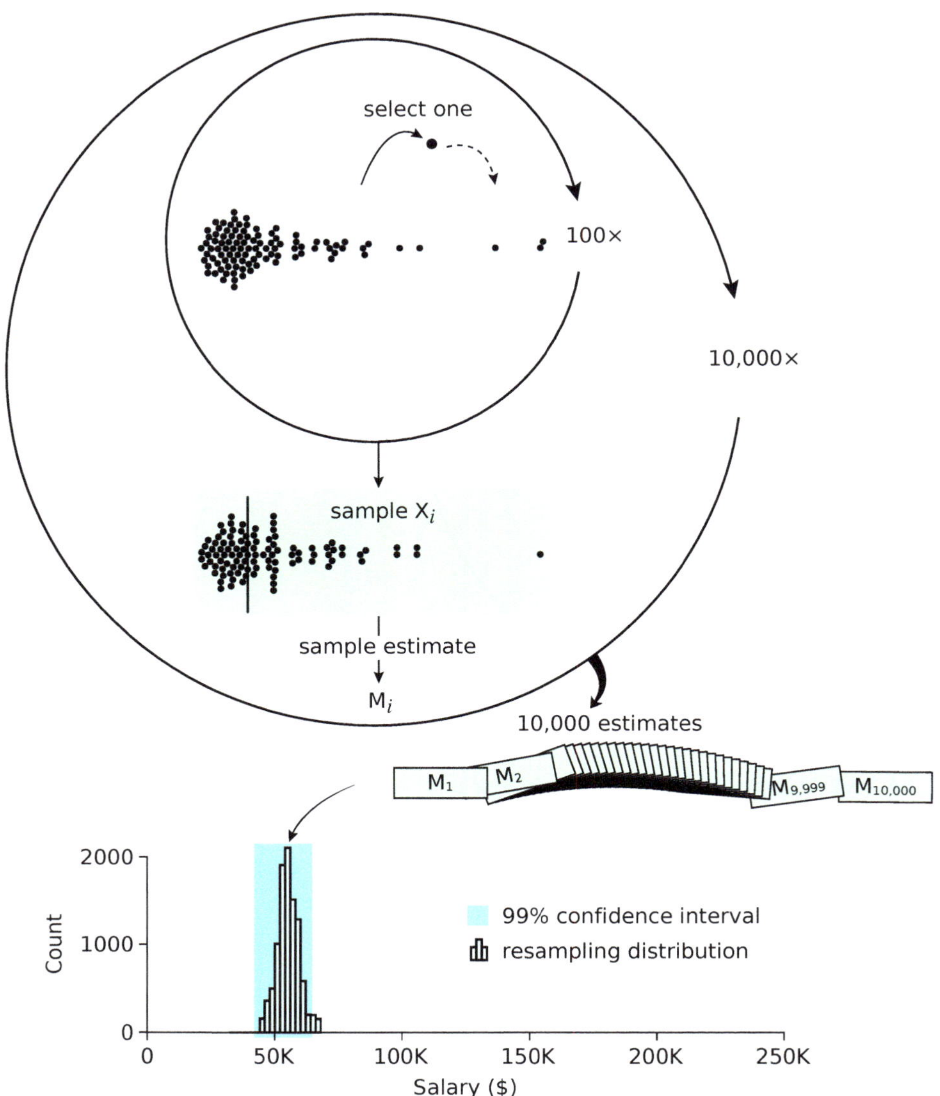

Figure 4.18 99% confidence interval for the median of the n = 100 data set, superimposed on the resampling distribution.

Confidence interval To construct the 99% confidence interval, we first rank the 10,000 sample estimates from smallest to largest. From the ranked list, count up from the smallest and down from the largest to get the 50th smallest and 50th largest values, calling them M_{lower} and M_{upper}. When we did this, M_{lower} and M_{upper} were $45,333 and $68,330. Following the procedure *How to calculate the Confidence Interval* on page 149, the 99% confidence interval is

$$\begin{aligned}\text{99\% confidence interval} &= \left[2 \times M_{obs} - M_{upper},\ 2 \times M_{obs} - M_{lower}\right] \\ &= [2 \times 55{,}094 - 68{,}330,\ 2 \times 55{,}094 - 45{,}333] \\ &= [\$41{,}858,\ \$64{,}855]\end{aligned}$$

What have we learned about U? A lot more. We now know that in our larger sample, the median salary is $55,094, with a 99% confidence interval of [$41,858, $64,855]. Compared to the smaller sample, this is a much tighter estimate, and our precision about M_{true} has improved greatly.

This is an important general principle: the bigger the data set, the tighter the confidence intervals, the more precision in our estimate, and the less the uncertainty.

Effect of sample size on confidence intervals As we saw, the confidence interval we calculated from the bigger study (n = 100) is a lot tighter than for the smaller study (n = 10).

In general, the larger the sample size, the narrower the confidence interval. As a further demonstration of this fact, we created a population (Figure 4.19). Then we randomly sampled from this population, and we compared the confidence intervals for samples of different sizes: n = 5 and n = 30 . Note how much wider the CIs are, and the less precision we have, when the sample sizes are small.

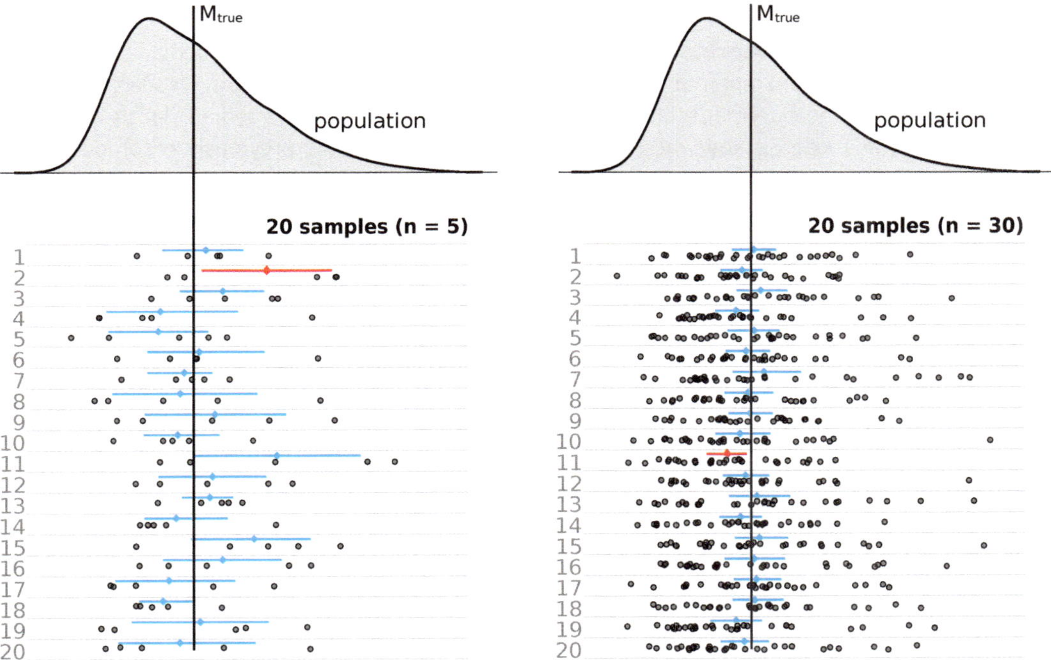

Figure 4.19 95% confidence intervals produced from samples of different sizes drawn from the same population. The gray dots show samples. The blue and red bars denote the confidence intervals. Note that, as to be expected for 95% confidence intervals, 1 out of the 20 confidence intervals, shown in red, has missed the true value.

Effect of confidence level on confidence intervals The size of a confidence interval reflects the level of confidence that we require. A 99% confidence interval would have to be larger than a 95% confidence interval, which would be larger than a 90% confidence interval. This may seem paradoxical, but a simple example might help. If we were asked to say, with 90% confidence where a given person is located, we can draw a smaller neighborhood, because we only need 90% confidence. If we now insist on 99% confidence, then we had better draw a bigger neighborhood.

FURTHER EXERCISES 4.4

1. Two confidence intervals for the same set of data are [30, 40] and [10, 70]. One is the 80% CI and one is the 99% CI. Which range is the 99% confidence interval?

   a) [30, 40] b) [10, 70] c) We have no way of knowing

2. For a 99% confidence interval, how would the width change if we increase the sample size from 100 to 1000? Explain your answer briefly.

3. A political poll finds that 305 out of 600 (50.8%) of a random sample of likely voters are in favor of the Blue party over the Gold party prior to an upcoming election. The p-value for whether the population proportion differs from 50% is 0.714. Which of the following is most likely the correct 95% confidence interval for the proportion of voters that will vote for the Blue party candidate? Explain your reasoning.

   a) [46.8%, 54.8%] b) [50.1%, 58.0%] c) [53.0%, 61.0%]

4. **Depth of the Tonga Trench.** A 1980 geological survey estimated the median depth of a section of the Tonga Trench at 6,000 m. Using modern sonar technology, researchers take 200 new depth measurements along the same section and find a median depth of 6,500 m. They want to determine whether this section is deeper than previously recorded.

   a. What is the Null Hypothesis?

      a) The median depth of this section of the Tonga Trench is 6,000 m.
      b) This section of the Tonga Trench is not deeper than the 1980 survey estimate.
      c) It was random chance that our sample found a median of 6,500 m, when the true median depth of this section is the same as the 1980 estimate (6,000 m).

   b. They want to calculate a 99% confidence interval for the median depth, using their 200 measurements. The observed median was 6,500 m. How many values should they randomly draw for each simulation?

      a) 200 b) 6,000 c) 6,500 d) 10,000

   c. In the procedure to construct a confidence interval, why do we draw 200 measurements for each simulation and calculate their median, repeating this 10,000 times? Could we just draw 2 million measurements and calculate their median?

      a) Because we want to "repeat the experiment many times" in order to calculate the frequency of an event.
      b) We need to replicate the original sampling via our resampling.
      c) It's ok to draw 2 million measurements from the box and we would get the same result.

   d. Researchers take 200 depth measurements, calculate the median value, and run 10,000 simulations to calculate a 99% confidence interval for the median. What is the sample size of this study?

a) 200 b) 10,000 c) 99

e. Researchers take 200 depth measurements and run 10,000 resampling simulations to calculate a 99% confidence interval for the median depth. How can they get a more accurate and precise estimate of the true median?

a) Run a million simulations
b) Take more depth measurements
c) Calculate a 95% CI
d) Calculate a 99.9% CI

f. The 99% confidence interval for median trench depth based on our sample is [6,406 m, 6,586 m]. What does this tell us?

a) We are 99% sure that the true median depth is in the interval [6,406 m, 6,586 m].
b) We are 99% sure that the interval [6,406 m, 6,586 m] contains the true value.
c) 99% of resampled medians are in the interval [6,406 m, 6,586 m].
d) If we went back and resurveyed this section many times, 99% of the confidence intervals we calculated would contain the true value.

4.5 Validating the resampling principle

In this artificial example of salary data, we really do know the universe U. So for this example, we can play the role of a deity, and actually do the experiment of going back into U 10,000 times to replicate the "10,000 (actual) friends".

Therefore, we can answer the question: how well did the resampling procedure compare with "reality"?

Remember the key assumption that resampling makes: suppose we were to go back into the universe U and reproduce our experiment many times, thereby producing thousands of M_{obs} values. These are the values $M(Y_i)$ in Figure 4.5. We assume that the sampling distribution, that is, the distribution of the values of $M(Y_i)$ around the true (unknown) measure M_{true} is approximately the same as the distribution of the resampled estimates $M(X_i)$ around M_{obs}. Recall that we abbreviate $M(X_i)$ as M_i.

resampling assumption distribution of $(M_{true} - M_{obs}) \approx$ distribution of $(M_{obs} - M_i)$

We can validate this assumption in this case, because we know U.

Distribution of (M_{true} − M_{obs}) To validate it, we went back into the universe U 10,000 times, drawing 10,000 fresh data sets, each with $n = 100$. We call these new data sets Y_1, Y_2, ..., $Y_{10,000}$. Each Y_i has an M-measure associated with it, $M(Y_i)$. We can think of these 10,000 $M(Y_i)$ as 10,000 M_{obs}-values from our "10,000 friends" (Figure 4.20 upper left).

Then, because we know M_{true} in this particular case, we can subtract each of the $M(Y_i)$ values from M_{true} to produce the distribution of sample estimates M_{obs} around M_{true} (Figure 4.20 upper right).

Distribution of (M_{obs} − M_i) From our $n = 100$ sample X, we generate 10,000 resamplings X_1, X_2, ..., $X_{10,000}$, calculate their M-values, $M(X_i)$, and plot them in a histogram (Figure 4.20

middle left).

We subtract each $M(X_i)$ from M_{obs} to produce the distribution of the resampled estimates $M(X_i)$ around M_{obs} (Figure 4.20 middle right).

Comparison When we overlay the sampling distribution and the resampling distribution, we see that the resampling assumption is a good approximation (Figure 4.20 lower right).

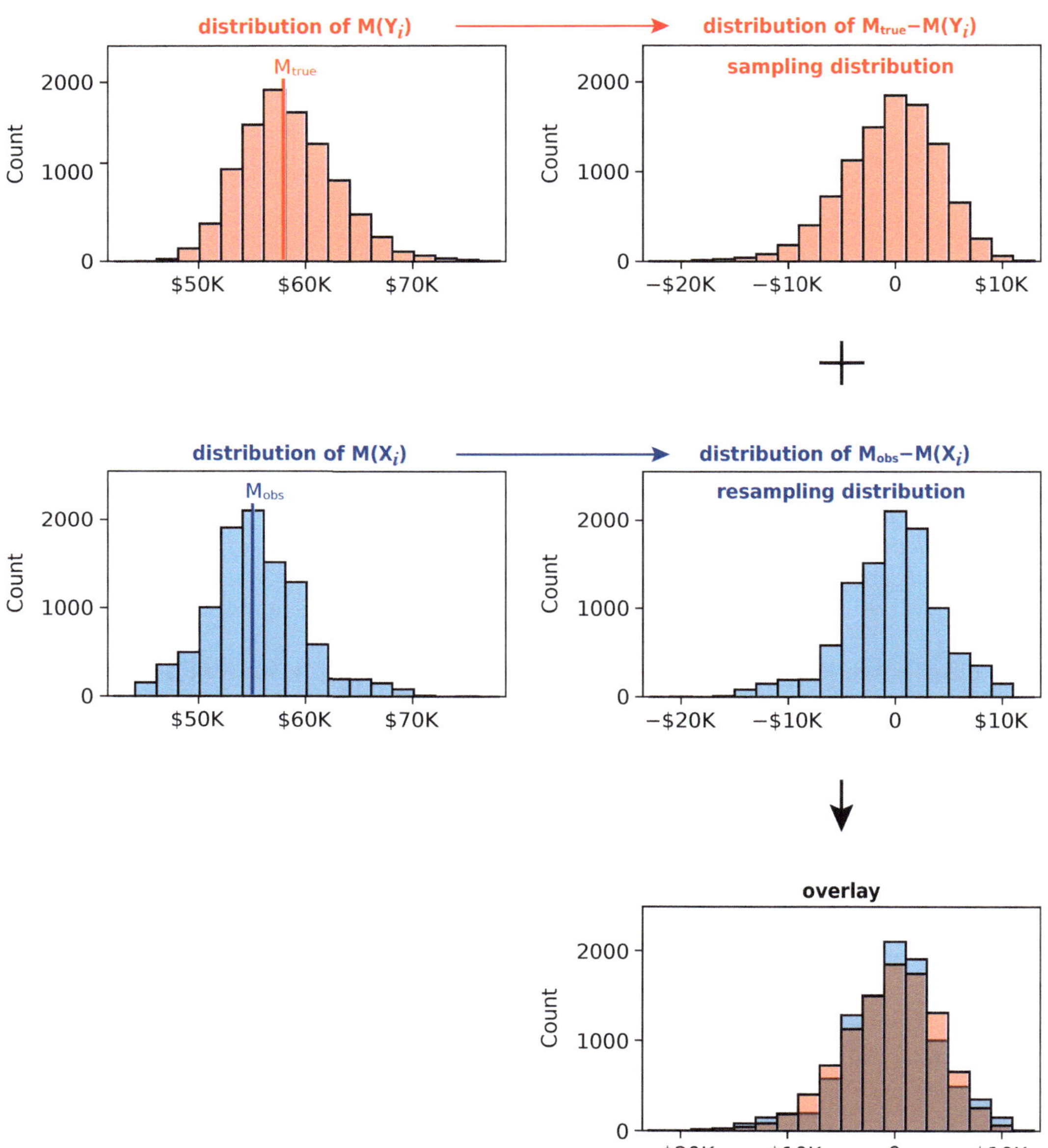

Figure 4.20 Validating the resampling procedure. Upper left: distribution of fresh experiments in U. Upper right: distribution of $M(Y_i)$ around M_{true}. Middle left: distribution of the M-values for 10,000 resamples of the data set X. Middle right: distribution of $M(X_i)$ around M_{obs}. Lower right: superposition of the two histograms above, showing their similarity.

Measures other than the mean

It is very important to understand that the resampling procedure we just sketched for a CI for the median works for absolutely any measure of any kind, whether that's the mean, median, or any other measure of central tendency. We have complete freedom of measure, knowing that regardless of what measure we choose as scientifically best for our data, we can calculate a confidence interval for it.

The same procedure can also be performed with descriptors other than measures of central tendency. Suppose for example, that we wanted to look at the variability of our data set, and used the MAD or the variance as our descriptor. Then we would find the 99% confidence interval for that measure in exactly the same way as we just did for the median.

FURTHER EXERCISES 4.5

1. **Pepto Bismol.** Diarrhea represents a significant public health challenge in underdeveloped nations, especially for infants. Investigators in Peru carried out a double-blind, randomized controlled trial, published in *The New England Journal of Medicine* ("A Controlled Trial of Bismuth Subsalicylate in Infants with Acute Watery Diarrheal Disease"), to assess whether bismuth subsalicylate (also known as Pepto Bismol) would improve the outcomes of infants suffering from diarrhea.

   According to the website of Professor Patrick Breheny (University of Iowa), "In their study, all infants received the standard therapy for diarrhea: oral rehydration. In addition to rehydration, 85 babies received bismuth subsalicylate, while 84 babies received a placebo. The total stool volumes for all infants over the course of their illness was measured. To adjust for body size, the researchers divided by body weight to obtain their outcome of interest: stool output per kilogram of body weight". (see `https://iowabiostat.github.io/data-sets/` for this and many other data sets.)

Total stool output (ml/kg)	
placebo (n = 84)	bismuth (n = 85)
131, 178, 127, 403, 413, 60, 859, 247, 278, 66, 210, 237, 105, 79, 307, 90, 244, 165, 59, 420, 159, 105, 667, 551, 61, 351, 59, 197, 173, 62, 386, 276, 85, 381, 567, 240, 539, 480, 157, 346, 69, 1101, 176, 193, 493, 113, 88, 55, 373, 73, 99, 234, 134, 57, 66, 84, 135, 371, 56, 199, 229, 384, 392, 490, 354, 162, 224, 62, 56, 210, 187, 141,114, 1677, 55, 615, 55, 193, 68, 590, 154, 130, 559, 75	237, 110, 106, 1099, 65, 74, 156, 60, 98, 234, 105, 79, 258, 61, 60, 196, 57, 55, 55, 62, 247, 121, 76, 365, 69, 245, 75, 236, 157, 100, 55, 185, 122, 126, 73, 133, 86, 430, 127, 56, 113, 55, 67, 116, 361, 168, 427, 570, 306, 107, 57, 231, 60, 398, 367, 137, 58, 66, 968, 78, 229, 69, 155, 115, 71, 55, 88, 1025, 108, 102, 371, 163, 349, 69, 208, 91, 84, 100, 61, 94, 175, 57, 356, 144, 99

   For the placebo group:

   a) Visualize the data using an appropriate plot. Explain why you chose that plot.

   b) Calculate a measure of central tendency and a measure of variability. Explain why you chose these measures.

c) Construct a 99% confidence interval for the chosen measure of central tendency using resampling. Show the resampling distribution and indicate the observed measure of central tendency and the calculated confidence interval.

d) What do you conclude from your analysis in parts (a – c)? Write a short paragraph summarizing your conclusions, including any relevant figures, numerical results, and their interpretation.

For the bismuth group, repeat steps (a – d) from above, using the same measures as in the placebo group. How do the two confidence intervals compare?

4.6 Other definitions of the confidence interval

The confidence interval we have defined is sometimes called the **pivotal confidence interval**. It's also referred to as the "reverse percentile confidence interval", and the "basic bootstrap confidence interval". These terms all describe the same method.

There are other definitions of the confidence interval. The most widely used alternative is called the **percentile confidence interval**.

To calculate a percentile confidence interval, we carry out exactly the procedures outlined in the two upper panels in Figure 4.8: we do 10,000 resamples, and choose the 0.5 and 99.5 percentiles as M_{lower} and M_{upper}. Then we use M_{lower} and M_{upper} directly as the lower and upper limits for the 99% confidence interval, without further transformation. We don't use the percentile method here because it does not give us information about $M(U) = M_{true}$, as we said.

Another approach to the confidence interval is called the **Bias Corrected accelerated** method, or BCa for short. The BCa method corrects for the fact that the median of the bootstrap estimates $M(X_i)$ may not be centered around $M(X)$, also called M_{obs}, by using the "fraction of bootstrap samples $< M_{obs}$" as the bias correction factor. (If that fraction is 1/2 then the distribution is symmetric around M_{obs}, and there is no bias.) The BCa correction also employs an "acceleration factor", which reflects how the variability of M_{obs} changes with respect to $M(U)$. The reader may want to experiment with both the percentile method and the BCa method and compare the results. BCa calculators are available in many software packages.

There is substantial debate in the literature about the relative strengths and weaknesses of each approach. The true test of a confidence interval method is its "coverage probability", the percentage of confidence intervals generated by that method that actually contain the true value in the underlying population. Ideally, the coverage probability of a 95% confidence interval should be 95%. However, different confidence interval methods have different coverage probabilities, especially under deviations from symmetry (such as skewness), sample size, even the type of estimator. The reader is advised to study the advanced literature if this becomes an issue.

4.7 Are there formulas for confidence intervals?

In some special cases, there are formulas for Confidence Intervals. As is often the case, if the data set 1) has very special properties, and 2) the measure we choose is a very simple one, then there may be a formula for the confidence interval for the observed measure.

In the best-known case, we must assume that the data set X is large and Normally distributed, and we *must* use *only* the mean value as our measure M(X). These are strong limitations, but they are absolutely required in order to derive a formula.

If those assumptions are met, then there is a theorem. The theorem tells us that the 99% confidence interval for the mean of a Normally distributed data set is

$$99\%\ \text{CI} = \text{mean} \pm (2.58 \times \text{s.e.m.})$$

Recall from Chapter 2 (section 2.4 *Summarizing a data set: measures of variability*) the definition of the standard error of the mean

$$\text{s.e.m.} = \frac{\text{s.d.}}{\sqrt{n}} \qquad \text{where s.d. = the standard deviation}$$

Note that this formula relies on mean and variance, concepts that are applicable to distributions that are unimodal, thin-tailed and symmetric, what we call "bell-shaped" or Normally distributed.

Therefore, the formula for the confidence interval for a mean should be taken with a large grain of salt, and its assumptions must be checked before the formula is applied. This formula-based confidence interval

- assumes that the data set is large and bell-shaped
- only works to estimate a confidence interval for the *mean* value of the data set

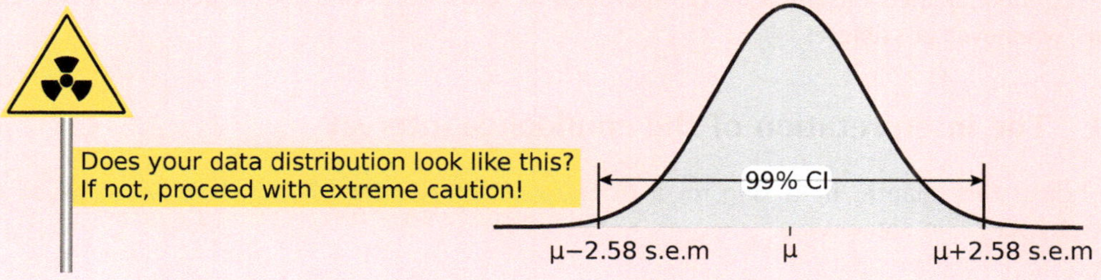

For almost all measures, and almost all distributions, there is no formula for confidence intervals. However, using a resampling-based approach, we can calculate a confidence interval for *any* measure and *any* distribution. For example, the salary data in Figure 4.9 is definitely *not* bell-shaped, but we were able to construct a confidence interval for the median, which was the better measure of our distribution.

Rather than trying to fit our problem into a form suitable for a formula, we use resampling methods, which work in every case and make no assumptions about the distribution of the data.

4.8 How to present confidence intervals

When it's time to present our result, we should use the form "result = effect size (99% CI = [CI_{lower}, CI_{upper}])". For example, "the median age was 28 years (99% CI = [25, 32])."

And if we are presenting the result graphically, we can show the confidence interval as a colored region and/or with arrows as shown in Figure 4.21.

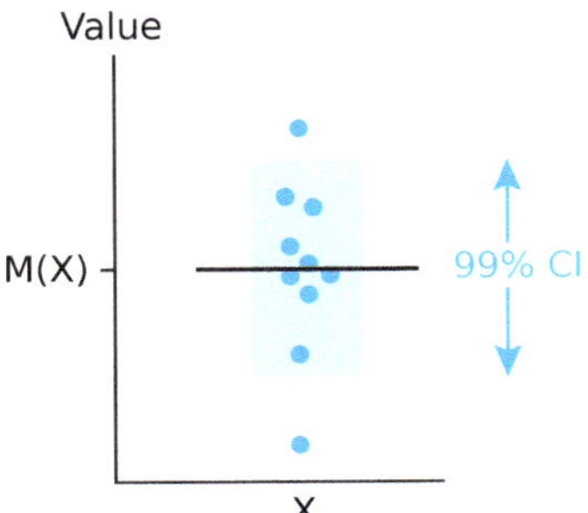

Figure 4.21 Graphical presentation of a 99% confidence interval together with data.

Remember that, back in Chapter 2, we were discussing "error bars" as the plunger in the Dynamite Plunger Plot. We noted that it was usually unclear what the "error bar" stood for, and, especially, why the standard deviation (s.d.) or standard error of the mean (s.e.m.) should be the "error". Now we can make a partial answer in that discussion. A better "error bar" is the confidence interval, which answers the question: we made an estimate from the data. How reliable is that estimate?

Why does the literature still use the s.d. or the s.e.m. as error bars? Because the s.e.m. is used in the formula for the confidence interval for the mean of a Normal distribution. It has become an unfortunate practice to show the s.d. or s.e.m. as "error bars" even when the distributions and/or measures make this inappropriate.

Confidence intervals are a better approach to "error bars" and should be used as our "error bar" whenever possible.

4.9 The interpretation of the confidence interval

We began this chapter by asking for a Confidence Interval for M_{true}. Have we achieved this? The answer is: in some ways.

Frequentism: CIs and p-values Both CIs and p-values are concepts that are defined in the framework that is called "Frequentism". There is a strong parallel in their interpretation. Both require us to imagine what would happen in hypothetical repetitions of the Null Hypothesis or our experiment.

As we mentioned in section 3.1 (*Modeling and simulating probabilities*), in the framework of Frequentism, probability statements must be interpreted as frequencies of occurrences. "X is probable" means "in many trials, X happens frequently". If we say that getting at least one Head in 10 flips of a coin is highly probable, we mean "if we do a 10-flip set many times, the frequency of 10-flip sets that have at least one Head will be very high.

In Chapter 3, we developed the concept of p-values. As we saw, the most fundamental problem with the p-value approach is simply that it is seeking an answer to the wrong question. We had some Hypothesis H (for example the Null Hypothesis H_0), and some Data D, and what we really wanted to know was $p(H \mid D)$: how likely is our Hypothesis given this Data? But the p-value is not the answer to this question. Instead, it's the answer to the inverted question, what is $p(D \mid H)$? or, how likely is our Data given the Hypothesis?

This inversion of the real question produces a frequentist concept, which is the p-value.

Frequentist approach to p-values

Frequentist: "If the Null Hypothesis is true and we were to repeatedly implement the Null Hypothesis, many times, how likely would we be to see data this extreme?"

- We assume that H_0 is true
- We treat the data as random
- We get $p(\text{Data} \mid H_0)$, which is the p-value.

In this chapter, we developed a resampling-based approach to understand and calculate Confidence Intervals. However, just like p-values, confidence intervals must also be interpreted in the backwards-logic framework of frequentism.

In this frequentist interpretation of confidence intervals, the statement "we are 99% confident that the 99% CI contains M_{true}" means "99% of the time, an interval produced by this method will contain M_{true}." So "99% CI = $[CI_{lower}, CI_{upper}]$" does not mean that *this* confidence interval probably contains M_{true}. What could that mean? M_{true} is some number with a fixed value, say, 37.5. What could it mean to ask "what is the probability that 37.5 is in *this* interval?"

The probability statement "what is the probability that M_{true} is in this interval?" can only mean that confidence intervals *like these* tend to contain the true value 99% of the time. The statement is about the long-run behavior of the confidence interval generating process. It's the likelihood of these CIs to contain M_{true}, not about the value of M_{true} lying in *this* CI.

Frequentist approach to Confidence Intervals

Frequentist: "If we were to repeat this experiment, many times, 99% of the resulting confidence intervals would contain the true value."

- We assume that the unknown true value, M_{true}, is fixed
- We treat our confidence interval as random (across many hypothetical replications of the experiment)
- We get $p(\text{CI containing } M_{true} \mid \text{experimental procedure})$, where the "experimental procedure" includes the data collection and CI construction.

Relationships between confidence Intervals and p-values There is a close, but not perfect relationship between confidence intervals as we have defined them, and p-values.

- If the null value, the value of the measure under the Null Hypothesis, lies outside the 99% confidence interval, then the Null Hypothesis can be rejected at the $\alpha = 0.01$ level, and we can declare a statistically significant finding at the 0.01 level.
- However, the converse is not always true. If the null value *is* contained in the 99% confidence interval, the observed result may still be significant at the $p < 0.01$ level when we calculate p-values. But such disagreement is rare, and happens only in borderline cases, where the null value is near the edge of the CI. In these cases, the conflict means that the

evidence for statistical significance found by the p-value is weak, and extra care should be taken to make sure that we are not p-hacking.

Let's look at an example where we calculate both p-values and confidence intervals.

In Chapter 3, we looked at several hypothetical scenarios for the sex ratio of births around power plants and did NHST to determine statistical significance for each case.

In the original study, we observed a 52% girl birth rate in a sample of 300 births. We wanted to know if this discrepancy from 50% is due to random chance or not. We found a p-value for our observed result: $p \approx 0.54$. If we calculate the confidence interval for our study, we see the 95% confidence interval is [46.3%, 57.7%] (Figure 4.22).

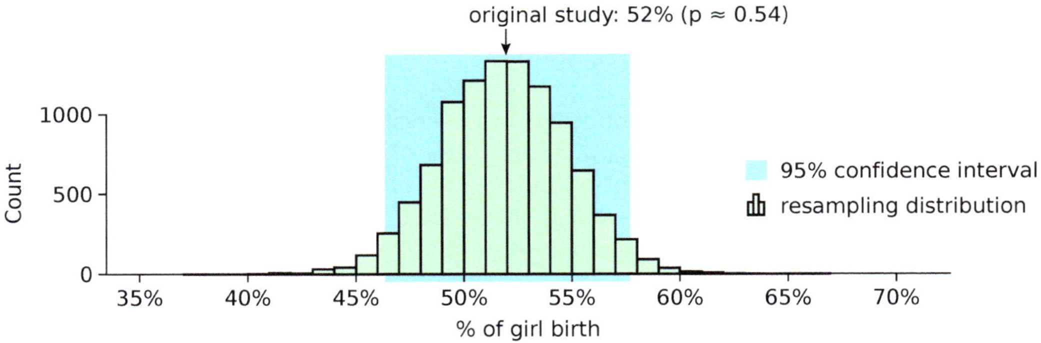

Figure 4.22 The 95% confidence interval for the study that observed 52% girl births in a sample of 300 births near power plants was [46.3%, 57.7%].

This interval includes the null value 50% and therefore we cannot reject the Null Hypothesis. The conclusion from the confidence interval therefore agrees with the NHST conclusion that the result is not statistically significant ($p \approx 0.54$) (Figure 3.12).

Then we presented a new study where we observed a 56% girl birth rate in a sample of 300. We calculated the 95% CI to be [50.3%, 61.7%]. We see that the 95% confidence interval is now roughly centered around the new observed value of 56%, and that the null value (50%) is just outside the lower limit of the interval (Figure 4.23). This is consistent with a p-value very close to 0.05 (Figure 3.13).

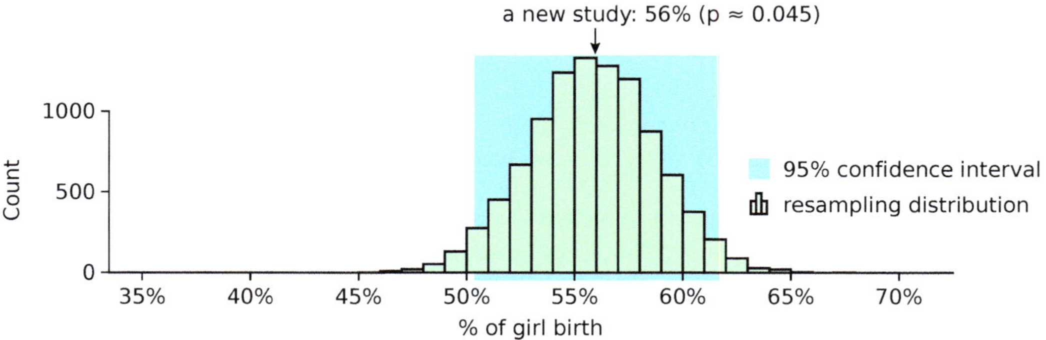

Figure 4.23 The 95% confidence interval for a new study that observed 56% girl births in a sample of 300 births near power plants.

Exercise 4.9.1 In Figure 4.23, the 95% CI barely excludes the null value of 50%. If we considered instead a 99% CI, what would be the conclusion?

To highlight the relation between CIs and NHST, we can compare the null distribution and the resampling distribution with confidence intervals (Figure 4.24). The null distribution centers around the null value while the resampling distribution centers roughly around the observed value. Note that with a bigger effect size, the CI is shifted to the right, away from the null value.

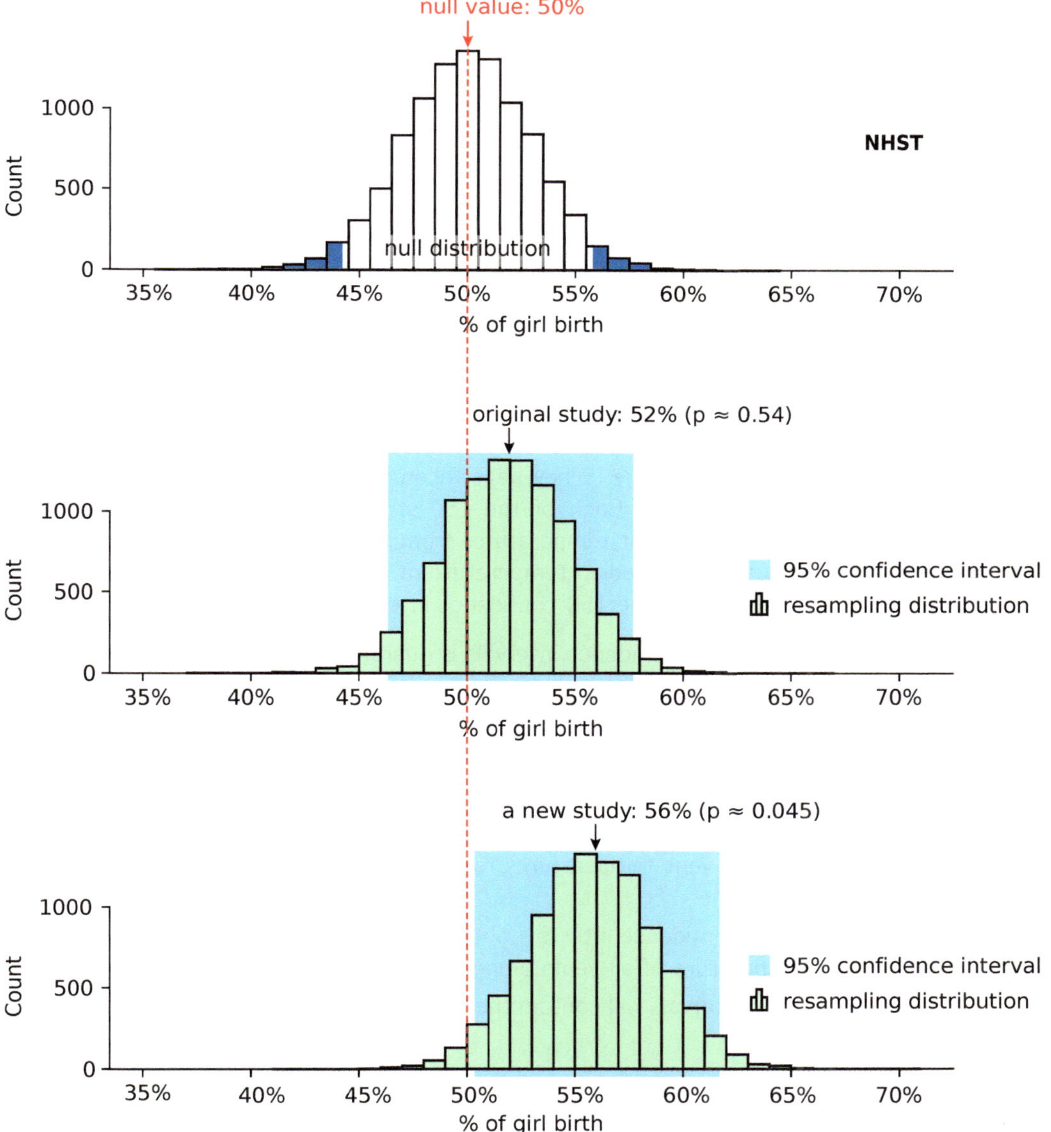

Figure 4.24 Relation between NHST and CIs. Upper: null distribution for the study of sex ratios near power plants. Regions for $p < 0.05$ are shown in blue. Middle: 95% CI for the study that observed 52% girl births in a sample of 300 births near power plants. Lower: 95% CI for a different study that observed 56% girl births in a sample of 300 births.

FURTHER EXERCISES 4.9

1. **Sleep deprivation and reaction time.** A neuroscience lab recruits 80 healthy adult volunteers and measures their reaction time (in milliseconds) on a standardized task before and after 24 hours of sleep deprivation. The study finds a median increase in reaction time of 30 ms, with a 99% confidence interval of [18, 42].

   a. Which of the following true median increases in reaction time are NOT reasonably compatible with your data?

      a) 10 ms b) 25 ms c) 35 ms
      d) They're all reasonably compatible.

   b. Is the observed increase in reaction time statistically significant at $\alpha = 0.01$, and how can you tell?

   c. A previous study found a median increase of 20 ms after 24 hours of sleep deprivation. Is this consistent with your data, and why?

   d. The researchers claim that sleep deprivation has a practically meaningful effect on reaction time if the true median increase exceeds 15 ms. Based on your confidence interval, is this claim supported, and why?

   e. If the researchers had constructed a 95% confidence interval instead of a 99% CI, would it be wider or narrower? How would this affect your answer to part c)?

2. **Hydrothermal vents of the Mid-Atlantic Ridge.** Researchers are studying hydrothermal vents along the Mid-Atlantic Ridge. They have recorded the temperature (in °C) of 500 vent openings across the ridge. Each of the 150 students in a marine science course draws a random sample of 40 vent temperatures from this dataset and constructs a 99% confidence interval for the true median temperature of all vents along the ridge. Which of the following statements about the 150 confidence intervals is most accurate?

   a. At least one of the 150 confidence intervals is guaranteed to contain the true median temperature of all vents along the ridge.

   b. About 99% of the time, the median temperature of a student's sample will equal the true median temperature of all vents along the ridge.

   c. About 99% of the time, the median temperature of a student's sample will be contained in their constructed confidence interval.

   d. At least 99% of the vent temperatures recorded will fall within any one of the 150 confidence intervals.

   e. About 99% of the confidence intervals obtained by the 150 students will contain the true median temperature of all vents along the ridge.

   f. All 150 confidence intervals will contain the true median temperature of all vents along the ridge.

3. **Truffle detection in pigs.** A researcher is studying whether pigs can detect truffles buried underground. Under the null hypothesis, the pig is guessing at random and will correctly identify the truffle location 20% of the time across 5 marked plots. After repeated trials, the researcher obtains a p-value of 0.03. They also construct 90%, 95%, and 99% confidence intervals for the true proportion of correct identifications. Which of these confidence intervals likely contain 0.2, and why?

Chapter 5

Normal Distributions

LEARNING OBJECTIVES

In this chapter, you will learn

- what a Normal distribution is.
- how the Normal distribution arises.
- what the special properties of the Normal distribution are.
- how we can tell whether a given data set is Normally distributed.

There is one particular distribution that has played a huge role in statistics, which we will now study.

We've been talking, ever since Chapter 1, about distributions that are "bell-shaped". What we meant by "bell-shaped" was

- symmetric
- unimodal (one-humped)
- thin-tailed

We said that when a data set met those strong assumptions, we could use various formula-based techniques that were applicable to "bell-shaped" distributions.

We now want to be more precise, and talk about the standard mathematical "bell-shaped" distribution, which is called the "Normal" distribution (Figure 5.1). (It was studied by Gauss, and is often called the Gaussian distribution.)

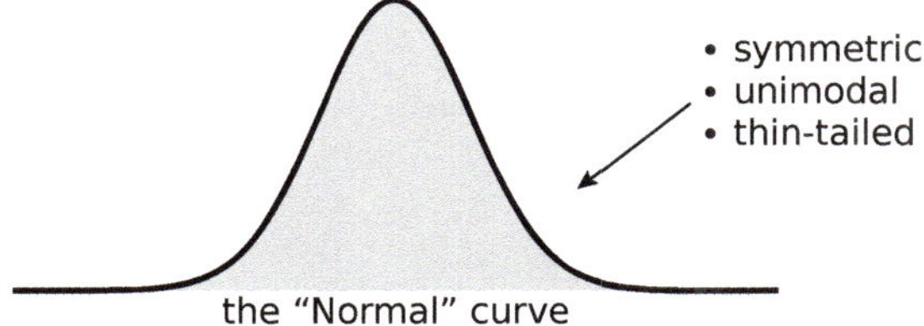

Figure 5.1 The "Normal" curve, also known as a Gaussian distribution.

A. Garfinkel and Y. Guo, *Understanding Data*,
https://doi.org/10.1007/978-3-032-18600-3_5

5.1 Mechanism of the Normal distribution

Flipping coins

flipping 1 coin Suppose we flip a fair coin once, and write down whether it is Heads (H) or Tails (T). Then suppose we were to repeat that experiment 10,000 times, and made a histogram of the outcomes. The resulting histogram is going to be fairly flat, with roughly 50% Heads and 50% Tails. As the number of trials goes up, the histogram will approach a perfect 50/50 distribution (Figure 5.2 left).

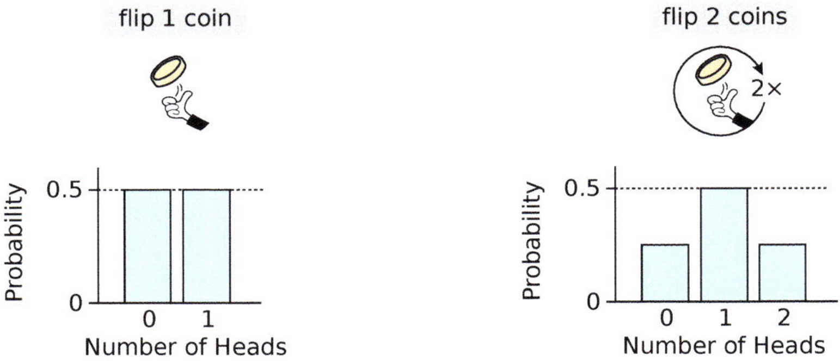

Figure 5.2

flip of 2 coins Now let's imagine an experiment in which we flip two fair coins, and record the outcomes. Now there are four possible outcomes: HH, HT, TH, and TT. We are not going to count the order of the heads and tails, so an HT is equal to a TH, and both count as "1 H, 1 T". Let's say we have performed thousands of flips of the 2 coins, and recorded the outcomes in a histogram. This histogram will have 3 possible outcomes: 2H, 1H1T, and 2T. The distribution will be approximately 25% (or $\frac{1}{4}$) , 50% and 25%, approaching those percentages exactly as the number of flips gets large (Figure 5.2 right).

flip of 3 coins, 4 coins Next let's try flipping 3 coins. The possible outcomes are 3H, 2H1T, 1H2T and 3T, and, as the number of flips gets large, they will occur in fractions of 1/8th, 1/4th, 1/4th and 1/8th respectively (Figure 5.3 left).

With 4 coins, the outcomes are 4H, 3H1T, 2H2T, 1H3T, and 4T, and the corresponding fractions are 1/16th, 4/16th , 6/16th, 4/16th and 1/16th, respectively (Figure 5.3 right).

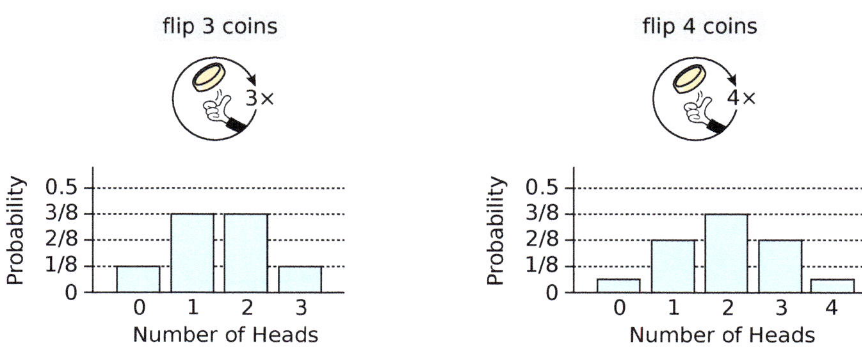

Figure 5.3

10 coins, 100 coins As we keep increasing the number of coins, the histogram gets finer and finer, and approaches the orange curve (Figure 5.4). The orange curve is the Gaussian distribution.

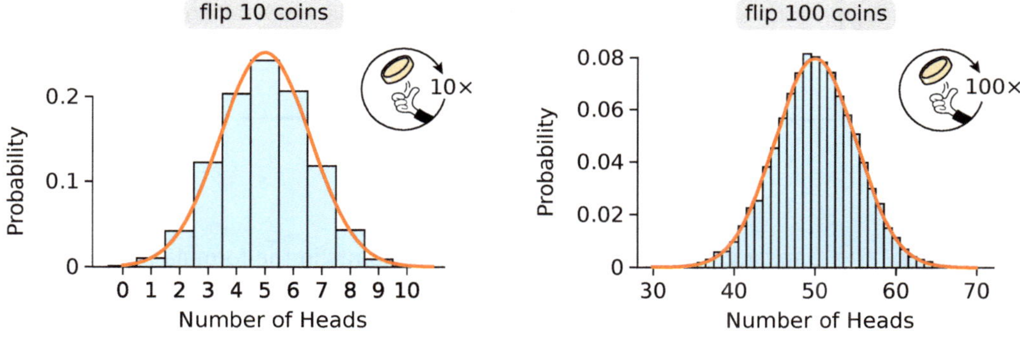

Figure 5.4

As the number of coins approaches infinity

In each histogram, the number of bins represents the set of all possible outcomes. The height of each bin is the probability for that outcome to occur. Because all possibilities are represented, the total sum of all bins must be equal to 1.

As the number of coins approaches infinity, the number of bins approaches infinity, and the histogram approaches a continuous curve, the orange curve.

The formula for the Gaussian distribution

Gauss (and later Laplace) was able to prove a theorem. For our purposes, we will summarize the theorem in non-technical terms: **the limit of a large number of small independent causes is a Gaussian distribution**, whose formula is

$$Y = \frac{1}{\sqrt{2\pi\sigma^2}} e^{-\frac{(X-\mu)^2}{2\sigma^2}}$$

where

$$\mu = \text{mean}$$
$$\sigma = \text{standard deviation}$$

Probability density function Let's talk about what this formula means. First of all, notice that Y is a continuous function of the continuous variable X. So X can be thought of as the continuous approximation to the measure we are studying, whether it be height, or the number of Heads, or blood pressure. In the coin flipping case, X = the number of Heads in one coin flip experiment.

The quantity Y is what is called a **probability density function**. We have to be careful about its interpretation. We cannot pick a single value of X and ask what is the probability of X? That's because there are infinitely many values of X (= infinitely many bins), so the vertical axis is no longer a probability, it becomes a probability *density*. That means that we can assign a probability to X lying between X = a and X = b. The probability of X falling between a and b is equal to *the area under the curve* between X = a and X = b (Figure 5.5).

Because all possibilities are covered, the total area under the curve must be 1.

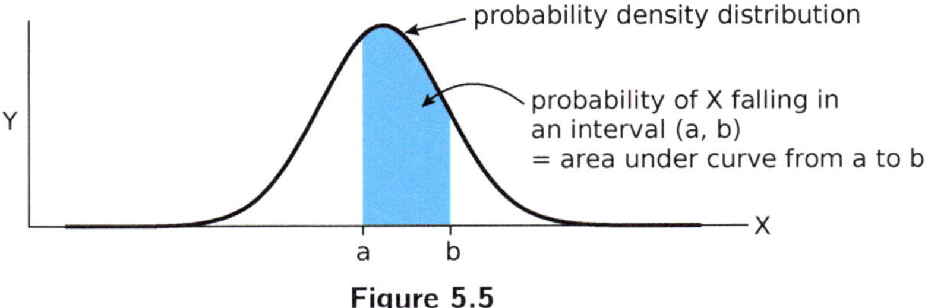

Figure 5.5

The shape of the Gaussian curve There are many constants in the formula for the Gaussian distribution. If we abstract away from them, the basic form of the formula is

$$Y = e^{-X^2}$$

This says that as X gets large (positively or negatively), Y dies off as e^{-X^2}. We know about "exponential decay" (e^{-X}), and how quickly exponential decay dies off. But the Gaussian curve declines much faster, not as e^{-X}, but as e^{-X^2}, which is much more drastic (Figure 5.6).

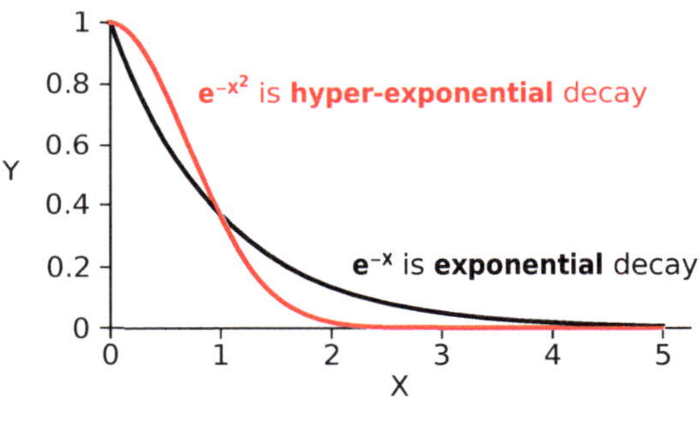

Figure 5.6

The drastic die-off of e^{-x^2} means that the tails are extremely thin. As we'll see, the thin tail is going to cause problems (section 5.3 *Are real-world distributions really Normal?*).

The coin flipping example gives us an insight into what is meant by the phrase "a large number of small independent causes". Let's take one particular flip experiment, that has landed on a certain value of X, that is a certain number of Heads. How did it get there? It got there by getting an H in coin #1, a T in coin #2, a T in coin # 3, ... **Each coin in this particular flip played a small role in where this flip ended up on the axis. Moreover, each coin is independent of every other coin, so the final location of this flip is indeed the sum of a large number of small independent causes.**

The Quincunx

There is a device with the odd name of "Quincunx" that also illustrates the mechanism of the Gaussian distribution (Figure 5.7).

The Quincunx is a board mounted vertically, with pins in the board at small regularly spaced intervals, making up a triangle. A ball is then dropped on the top pin. It can fall either to the

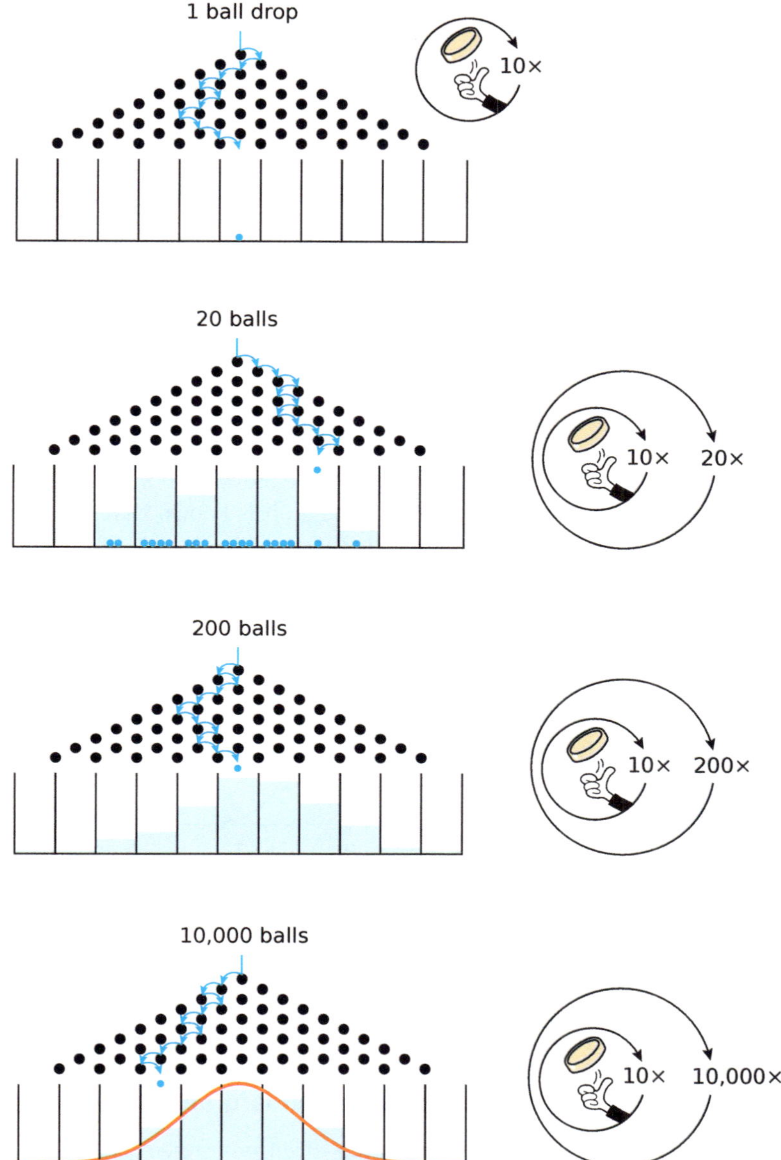

Figure 5.7 The Quincunx. With 20 balls, the histogram has no obvious shape, but with 200 balls, the overall shape of the Normal distribution is beginning to emerge. With 10,000 balls, the distribution is closely approximated by a Normal distribution curve (orange curve).

left or the right of the pin, causing it to drop down to the second level, where it falls on a pin and again goes either to the right or to the left, and falls to the third level. The ball keeps dropping down through what we assume is a large number of levels, until it falls out the bottom and is collected into a bin immediately below.

Clearly, **the final position of a ball, which bin it ends up in, is the result of a large number of small, independent causes.** The number of rows is assumed to be large, each row results in a small deviation to the left or right, and the outcomes of the rows are independent of each other.

5.2 Properties of the Normal distribution

The center and the spread

There are two parameters in the formula for the Normal distribution, μ and σ. They are the mean and the standard deviation. μ is the center of the distribution and σ gives the spread of the distribution. The traditional notation for Normal distributions is to write them as $\mathcal{N}(\mu, \sigma^2)$, where σ^2 is the variance.

For example, $\mathcal{N}(0, 1)$ denotes the Normal distribution with $\mu = 0$ and $\sigma^2 = 1$, the Normal distribution centered at 0 and having a variance of 1.

It's a very special property of the Normal distribution that the mean and standard deviation completely characterize the distribution. Once we specify the mean and the standard deviation, we have completely specified the distribution.

The distributional properties of the Gaussian curve are well known. Like all probability density functions, the total area under the curve has to be 1. It is known that in a Gaussian curve, 68% of the total probability lies within 1 standard deviation of the mean, 95% of the total probability lies within 2 standard deviations of the mean, and 99.7% of the total probability lies within 3 standard deviations of the mean (Figure 5.8).

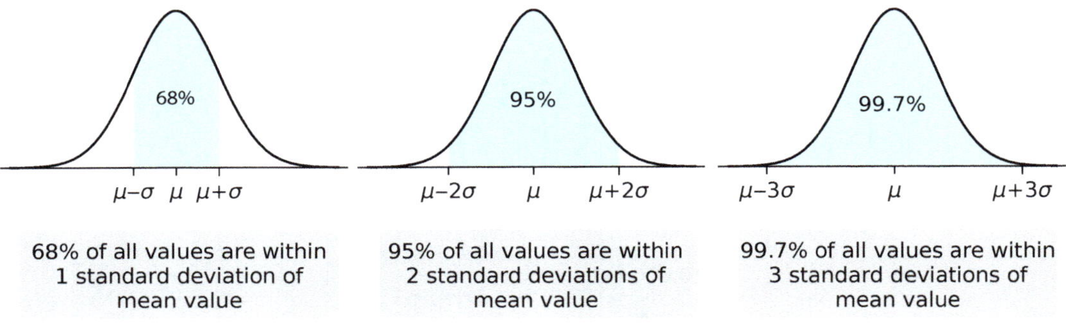

Figure 5.8

Where did the (bad) idea of "means and standard deviations" come from?

In the scientific literature, data distributions are often not shown, but are only summarized as "means and standard deviations".

As we saw in Chapter 2, this distorts our understanding of the data. Remember when we posed a question: "the average tax return is up $40 for this year. What can *you* expect?", 78% of the respondents gave a completely wrong answer, saying that the average, $40, would be a typical or at least frequent outcome.

Based on the news headline, which of the following is true?

A. I can expect around $40 more this year.

B. I can expect to get a bigger refund, amount uncertain.

C. Many people will get approximately $40 more, but, as they say, "Individual results may vary".

D. A typical refund will be $40 more.

E. None of the above.

The responses were

Why did so many people get that wrong? But this is not true! The false assumption that generated the wrong answers is the almost subconscious assumption that the amounts of tax returns must be distributed according to a Normal distribution. If they were, then the average would indeed be the most frequent outcome. But tax return amounts are *not* distributed Normally. As with so many other cases involving personal incomes, the distribution of tax returns is significantly skewed to the right: most people get very little, and some people get a lot.

The bad practice of reporting only "means and standard deviations" stems from the same subconscious or conscious error, that the distribution in question is Normal. The mean and the standard deviation characterize the distribution **if it's Gaussian (Normally distributed), but not if it's not!**

Unfortunately, many investigators, having conducted an experiment or a survey, do not take the trouble to look at the data by plotting it. This is statistical malpractice!

Fitting a Normal distribution to data

The procedure for fitting a Normal distribution to any data set is straightforward. First, calculate the mean of the data.

$$\mu = \overline{X} = \frac{1}{n}\sum x_i$$

Then calculate the standard deviation of the data.

$$\sigma = \sqrt{\frac{1}{n}\sum(x_i - \overline{X})^2}$$

It is a theorem that the Normal curve that best fits the data is the Normal curve with that mean and that standard deviation.

z-score When data is distributed Normally, there's a particularly simple description of any data point that measures how far away it is from the mean. We simply ask: how many standard deviations is this data point from the mean?

This measure is called the **z-score**.

$$z = \frac{x - \mu}{\sigma}$$

where as usual μ is the mean and σ is the standard deviation.

Using z-scores enables us to compare individual values from two different data sets, A and B, provided that both are Normally distributed. The two data sets might have different means, and different standard deviations. When this is the case, how can we say that one point in A is larger or smaller than another point in B? The answer is that we can compare them by transforming each of them into a z-score. Then we can say, for example, that the point in A is 2 standard deviations above the mean of A (z = 2), while the point in B is 1.5 standard deviations above the mean of B (z = 1.5), therefore, the point in A is "larger" than the point in B.

Exercise 5.2.1 You scored 80 on a test where the mean was 70 and the standard deviation was 8. Your friend scored 70 on a test where the mean was 65 and the standard deviation was 2. Assuming the grade distributions are roughly Gaussian (which is almost certainly false!), who performed better relative to their class?

Exercise 5.2.2 Assume that the fasting plasma glucose levels in the population is approximately Normally distributed (which is almost certainly false!) with a mean of 90 ml/dL and a standard deviation of 9 ml/dL. Tim went in to get his glucose tested and got the result of 126 ml/dL. What is his z-score?

Exercise 5.2.3 Before going to Medical school in the US, students need to take an exam called the MCAT. Before applying for graduate school, students need to take an exam called GRE. Here are some summary statistics for each exam:

Exam	Mean	Standard deviation
MCAT	500	10.6
GRE	302	9.6

Clarice took both exams. She scored 520 on the MCAT and 330 on GRE. On which exam did she perform relatively better?

The z-score is used by universities, for example, to compare the scores of people who took the SAT exam to the scores of people who took the ACT exam. The two exams have different means and different standard deviations, so a person's score on the SAT can be compared to another person's score on the ACT by directly comparing their z-scores.

Engineered Normal distributions

This comparison assumes that the grades in each exam are Normally distributed. Is this true? Yes. But this is because each exam is **artificially engineered** to produce a Normal distribution: first, test questions are chosen from a large bank of questions, in order to produce a Normal distribution. And then, raw scores are mathematically transformed into a

Normal distribution (see *Phony Normal distributions: the case of IQ* on page 184).

Testing whether a given distribution is Normal

Make a picture The most important step is a visual inspection: make a picture of the distribution (beeswarm, histogram, KDE, etc.) and ask is it:

• symmetric • unimodal • thin-tailed

Fit a Normal curve As a further step, we can find the best fit Normal distribution, given the mean and standard deviation of the data set, and compare the Normal curve to the histogram of the data itself (see for example the 1,000 adult weights data in Figure 5.9).

Numerical tests While we can fit a Normal distribution to any data, there are a number of numerical and graphical tests that can be used to test quantitatively how good this fit is. They have names like Shapiro-Wilk, QQ plots, χ^2 tests, and others.

We will not develop any of these in this book. If you need to prove numerically that some distribution is not Normal, then descriptions of these techniques can be easily found by searching. For example, the Python call for a Shapiro-Wilk test of a set "data" is `Shapiro(data)` and the Python call for a QQ plot is `sm.qqplot(data, line ='45')`, then use the Python call `py.show()` to show the two plots together.

Rather than worrying about whether a data set is Normal, so that we can apply concepts that are only valid for Normal distributions, **we recommend using resampling-based techniques, which are valid for all distributions**.

FURTHER EXERCISES 5.2

1. Based on this news headline, which of the following is true?

Forbes

PERSONAL FINANCE • EDITORS' PICK

Average Tax Refund Up 11% In 2021

a. A typical refund will be 11% more.

b. I can expect around 11% more this year.

c. Many people will get approximately 11% more, but, as they say, "individual results may vary".

d. I can expect to get a bigger refund, amount uncertain.

e. None of the above.

5.3 Are real-world distributions really Normal?

As we said in Chapter 2, "most distributions are not Normal" (see page 46). While it is widely believed that most distributions in nature are Normal, in fact, **almost no quantities found in biology and in society conform to the Normal distribution.**

There are three major reasons why biological quantities fail to be Normal.

- Almost all biological quantities can take only positive (or zero) values: population numbers, heights, weights, distances, etc. Their distributions all have a hard wall at zero.

  By contrast, the Normal distribution is perfectly symmetric and extends to infinity in both directions.

  Advocates of the Normal curve can argue that if the distribution mean is very far from 0, with highly tapered tails, then the distribution can be approximately Normal. But this is a question that must be addressed case-by-case.

The other two reasons why real-world distributions fail to be Normal are

- Most biological populations are "mixtures".
- Most biological populations have "Long Tails".

We will now discuss these two in detail.

Mixed populations

If you were to ask someone to give an example of a quantity that is Normally distributed, many people would say that *weights* are Normally distributed, as are *heights*. Let's examine these claims.

Weights We downloaded 1,000 adult weights from the Bureau of Labor Statistics. The results are shown in the histogram in Figure 5.9. We calculated the mean and standard deviation of this data set, and constructed the best fit Normal distribution from these numbers. The result is the Normal curve shown in purple superimposed on the histogram of the data.

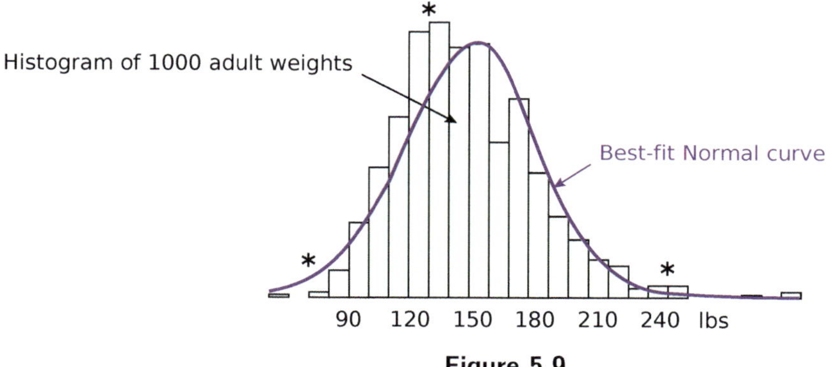

Figure 5.9

There are 3 places where the data significantly differs from the Normal curve (shown by asterisks).

1. On the left side of the curve, we see a significant deficit of severely underweight people (< 80 lbs). This is because this degree of underweight is not compatible with life. People in this category tend to be sick, such as cachexic cancer patients.[1]

2. On the right hand side of the curve, we see an excess of overweight people (> 240 lbs). These people are football players or, at the extreme, medically obese people.

3. But the most puzzling deviation from the Normal curve is just to the left of the peak. There is a very large, highly significant excess of slightly underweight Americans, including a large number of people who weigh 110–130 lbs. **Who are these slightly underweight Americans? They're women!**

The problem here is that weights do not form a Normal distribution because what we have here is a heterogenous population with two major subgroups: men and women, and the sum of two different Normal distributions is *never* a Normal distribution.

Heights The same problem applies to the distribution of adult heights (Figure 5.10). The mean and standard deviation of female heights and the mean and standard deviation of male heights are known. We plotted Normal distributions with those means and standard deviations, representing the female and male populations. Then we added those two populations to produce an estimate of what the overall distribution of heights should look like (black curve). The black curve is obviously not a Normal distribution, again for the reason that the sum of two different Normal distributions is *never* a Normal distribution!

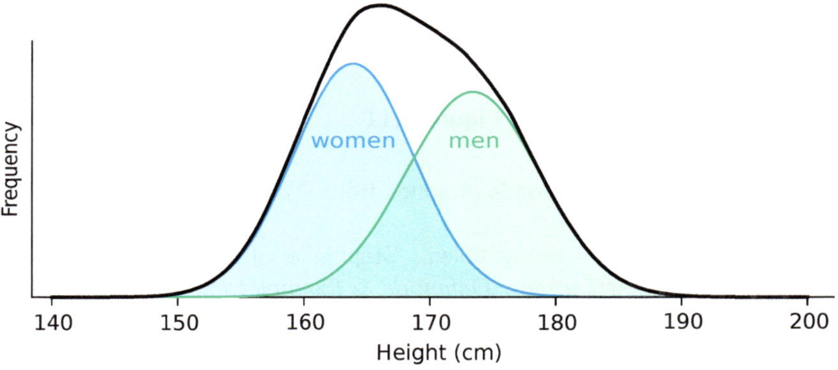

Figure 5.10

The problem here, as before, is that we have a "mixed" population. But what is a pure population? Are women a pure population? Certainly not! "Women" includes many different races, many different geographic backgrounds, and many different living standards. Therefore, "women" are a mixture!

[1] A similar point can be made about the distribution of metrics in many illnesses. For example, if we consider blood pressures in a population, there will be very few people with systolic blood pressures over 240 mmHg, because people with these blood pressures do not live very long.

Long tails

Remember that the Normal distribution declines very fast, giving it very very thin tails (Figure 5.6). Recent research has shown that many, if not most, distributions in nature have much thicker tails than a Normal distribution. **The Normal distribution is not normal!**

A recent best seller by Chris Anderson, *The Long Tail* documents a number of interesting examples.

e-commerce For example, if we graph book sales at Amazon.com, plotting sales against rank, the curve obviously declines, because lower rank means fewer sales, but the decline is very slow compared to the Normal distribution.

Even the book ranked #1,000,000 is generating sales, and the sales are nowhere near zero. So the decline is very slow. Anderson explains that the reason for this long tail is that Amazon is an online seller and consequently has very little additional cost to carry a large inventory, as opposed to the bricks-and-mortar bookstores, who have to focus on the higher-selling items. The tail becomes bigger and longer in new markets (depicted in red) (Figure 5.11). In other words, whereas traditional retailers have focused on the area to the left of the chart, online bookstores derive more sales from the area to the right.

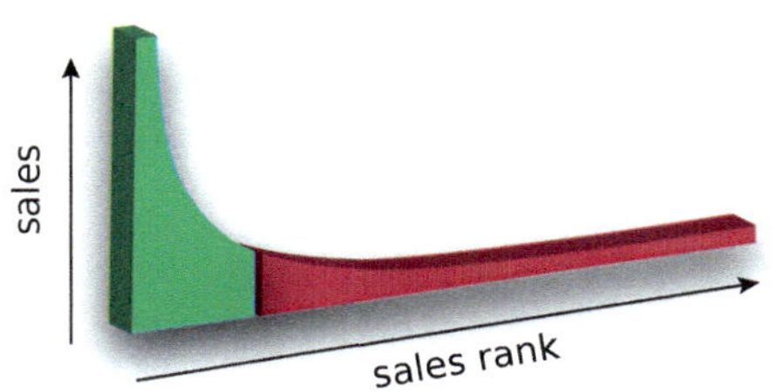

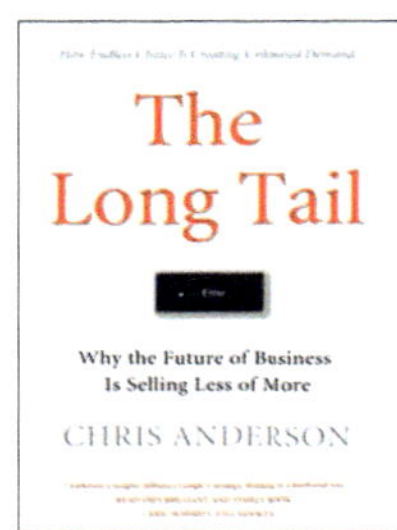

Figure 5.11

Similar facts are true about downloads of songs from iTunes, or followers of influencers on social media, etc.

Many real world situations are heavy tailed. Stephanie Glen, in her article *Heavy Tailed Distribution & Light Tailed Distributions: Definition & Examples (StatisticsHowTo.com)*, gives a number of additional examples of long tailed distributions:

- The top 0.1% of the population in the USA owns as much as the bottom 90%.
- File sizes in computers tend to be small, with a few very large files thrown into the mix.
- Web page sizes and computer systems' workloads tend to be heavy-tailed.
- Insurance Payouts and Financial Returns follow a similar pattern.

Tall women Assume that women's heights are Normally distributed with a mean given by μ = 62.4 in, and a standard deviation given by σ = 2.2 in. If a woman is randomly selected, find the probability that she might be 6' 6" or taller.

Let's first calculate how far 6' 6" is from the mean.

$$6'\,6'' = 78'' = 62.4'' + 15.6''$$

and

$$15.6 \approx 7 \times 2.2 = 7\sigma$$

So 6'6" is 7 standard deviations away from the mean (Figure 5.12).

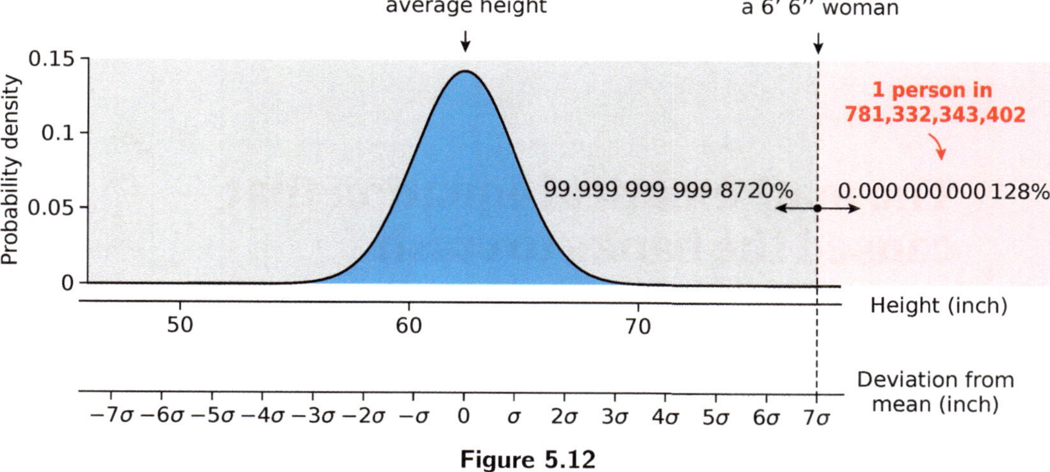

Figure 5.12

In other words, there should be *one* 6'6" woman in 800 billion. Probabilistically, there should not exist even one on the planet, which has a mere 4 billion women. But there are plenty of examples; eight are currently playing pro basketball in the US alone.

Cases like these tall women, who probably should not exist under the Normal distribution, have been called "Black Swans".

The Black-Scholes equation: the financial meltdown of 2008 The false assumption of a Normal distribution was the cause of the financial meltdown of 2008 (Figure 5.13).

Speculators in the financial markets often trade in what are called "options" or "derivatives". These are contracts to buy or sell a given amount of X company stock for a given price by a given date. They enable speculators to leverage their bets.

Obviously, the price of one of these options depends in part on the price of X company stock at that time. If the stock is going up, then the price of an option to buy will also go up.

Financial mathematicians came up with a differential equation to describe how the option price changes with the price of the underlying stock. It's called the Black-Scholes equation.

$$\frac{1}{2}\sigma^2 S^2 \frac{\partial^2 V}{\partial S^2} + rS\frac{\partial V}{\partial S} + \frac{\partial V}{\partial t} - rV = 0$$

where

V = price of option

S = stock price

r = interest rate

σ = "volatility" (= standard deviation of price fluctuations)

The impending disaster is hidden in the symbol σ. It is called "volatility", but what it really is is the standard deviation of expected fluctuations in the price of the stock. The use of the standard deviation signals that the authors are assuming that the expected fluctuations form a Normal distribution.

Normal distributions, as we have seen, have very thin tails. In this context, that simply means that large fluctuations are extremely rare. But in reality, the distributions of price fluctuations have much thicker tails. Therefore, the assumption of a Normal distribution resulted in an overvaluing of the options, because large fluctuations are assumed to be rare. But what was believed to be a once-every-200-years fluctuation was in fact once-every-7-years, and when it happened, prices crashed.

The mathematical equation that caused the banks to crash

The Black-Scholes equation was the mathematical justification for the trading that plunged the world's banks into catastrophe

In the Black-Scholes equation, the symbols represent these variables: σ = volatility of returns of the underlying asset/commodity; S = its spot (current) price; δ = rate of change; V = price of financial derivative; r = risk-free interest rate; t = time. Photograph: Asif Hassan/AFP/Getty Images

It was the holy grail of investors. The Black-Scholes equation, brainchild of economists Fischer Black and Myron Scholes, provided a rational way to price a financial contract when it still had time to run. It was like buying or selling a bet on a horse, halfway through the race. It opened up a new world of ever more complex investments, blossoming into a gigantic global industry. But when the sub-prime mortgage market turned sour, the darling of the financial markets became the Black Hole equation, sucking money out of the universe in an unending stream.

Figure 5.13 Article by Ian Stewart in *The Guardian* in 2012.

Phony Normal distributions: the case of IQ

When challenged to produce an example of a Normally distributed quantity, many people give the example of IQ scores, which are Normally distributed with mean = 100, and standard deviation = 15. However, this belief is misinformed, because the Normal distribution of IQ scores is intentionally and artificially manufactured, using a mathematical trick.

In fact, IQ is *defined* using this mathematical trick. There is a theorem that says that, given *any* ranked set (that could have *any* shape), you can define a Normally distributed quantity with *any* desired mean and standard deviation, that has the same rankings as the original set (Figure 5.14).

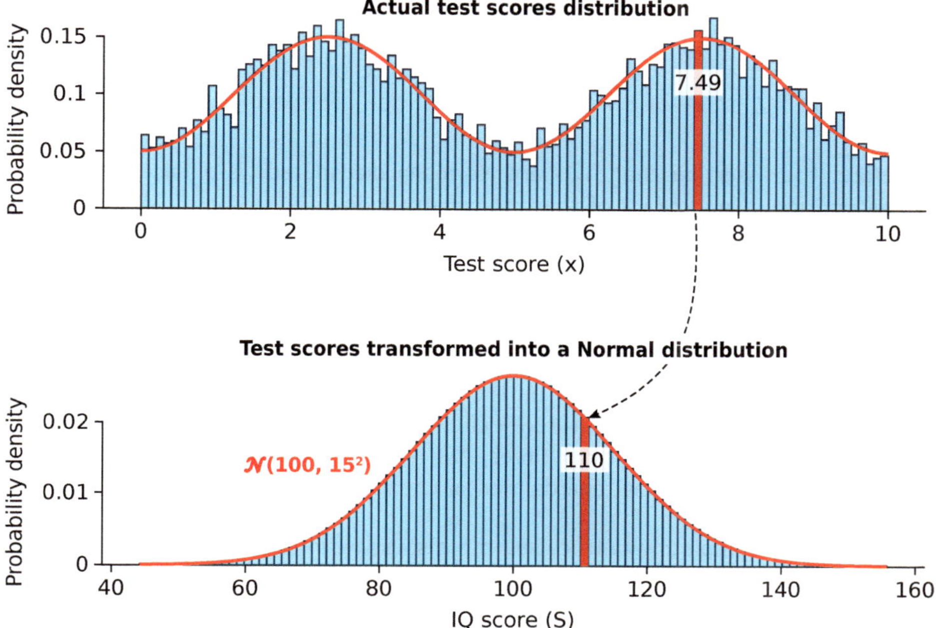

Figure 5.14 Transform an arbitrary distribution into a Normal distribution with a mean of 100 and the standard deviation of 15.

For an absurd example, suppose we have a ranked list of the top 500 College Football teams. Then we can plug those rankings into the theorem, and get a Normally distributed quantity, let's call it the CFQ (for College Football Quotient), with mean = 100, and standard deviation = 15, just like IQ.

"*You can always map the scores so that any distribution, whatever it is, becomes Normal.*"[2]

This is possible for any test with a distribution $f(x)$ of scores x. Let f be any continuous distribution whatever. f could be a bimodal distribution, as in Figure 5.14, or the distribution of rankings of college football teams, or any other continuous distribution.

We can then convert f into a Normal distribution with an arbitrary mean μ and an arbitrary standard deviation σ, using a formula. For example, suppose we have a bimodal function f describing a grade distribution, and a grade value x in this distribution. If we want to convert f into a Normal distribution $\mathcal{N}(100, 15^2)$, and thereby convert x into an "IQ"-like score S, we would use the formula:

$$\text{IQ}(S) = 100 + 15\sqrt{2}\ \text{erfc}^{-1}\left(2 - 2\int_{-\infty}^{S} f(x)\,dx\right)$$

The basic logic of this transform function is as follows: suppose a person receives a raw test score of x on a test, which places them in the kth percentile. Then, we decide, absolutely

[2] https://math.stackexchange.com/questions/1311149/why-are-iq-test-results-normally-distributed

arbitrarily, on what we want the mean and standard deviation of the Normal distribution to be. For this example, we chose mean = 100 and standard deviation = 15, corresponding to the distribution of IQ scores. Then the transform function calculates what the kth percentile of this Normal distribution would be and then assigns that number to the actual test score x.

The StackExchange post[3] provides Python code to carry this out. We ran this code on a hypothetical bimodal distribution of a test score x on a 0 to 10 scale (Figure 5.14). The code, as promised, converted the bimodal distribution into a Normal distribution with mean = 100 and standard deviation = 15. For example, a raw score of 7.49 in the original distribution is transformed to an "IQ" score of $S = 110$ in this artificial, pre-specified, Normal distribution.

Therefore, the assertion that "IQs are Normally distributed" is a false "discovery": the Normal distribution is an artifact of the test construction.

Why did people *ever* think that there was something called "intelligence" that was distributed Normally? The eugenicist Francis Galton simply pontificated it in a wild and false extrapolation. He said (1869):

> "Now, if this [normal distribution] be the case with stature, then it will be true as regards every other physical feature — circumference of head, size of brain, weight of grey matter, number of brain fibres, &c.; and thence, by a step on which no physiologist will hesitate, as regards mental capacity."

This is of course completely untrue, see the discussion in section 10.5 *The ugly history of correlation, regression, p-values and NHST*.

Why aren't real quantities Normal?

1. They are "mixtures" of distributions, for example, combining males and females.
2. Most biological quantities are "bounded" (e.g., they cannot be negative).
3. Most distributions have naturally "long" or "thick" tails.

[3] https://math.stackexchange.com/questions/1311149/why-are-iq-test-results-normally-distributed

Chapter 6

Comparing Two Groups

LEARNING OBJECTIVES

In this chapter, you will discover

- how to visualize the differences between two groups.
- how to use resampling techniques to determine if two groups are statistically different.
- how to calculate confidence intervals for the difference between two groups.
- when formulaic approaches (such as the t-test) are valid, and when not, and what the relation is between formulaic approaches and the resampling approach.

The most common statistical comparison is the case where we have two groups, A and B. For each individual in each of the two groups, we measure some quantity. Now we have a set of numbers for A and a set of numbers for B.

We want to ask: is group A "significantly different" from group B with regard to this measure?

This is probably the most frequently asked question in the literature. We can think of A and B, for example, as a Treated group and a Control group. We measure their blood pressures, or their survival times, or whatever quantity we are interested in, and then we ask if there is a difference between the two groups, and whether the difference is "statistically significant". The groups A and B can be any two groups: men and women, treatment A and treatment B, old and young, etc. In each case, we want to ask "do the two groups differ"?

Our overall strategy for approaching this problem is a resampling-based approach, which can be divided into 4 steps:

1. **Data presentation and choice of group measure.** First we need to visualize the two groups and decide on what measure to use to describe each of the groups.
2. **Choice of effect size.** Then we need to decide how we are going to compare the two group measures to get an effect size, that is, the observed "difference", Δ_{obs}, which could be an algebraic difference or some other comparison measure, like a ratio, etc.
3. **Calculating p-values by resampling.** Having calculated Δ_{obs}, we next need to find the p-value for this observed difference by resampling, either Big Box or Re-centered Two-Box.

A. Garfinkel and Y. Guo, *Understanding Data*,
https://doi.org/10.1007/978-3-032-18600-3_6

4. **Constructing confidence intervals for the effect size.** As we've said many times, an approach that is superior to the p-value calculation is to construct a confidence interval for the observed difference, Δ_{obs}.

You might have heard of something called the "t-test", and you may be thinking that it might apply here. Certainly, anyone who has taken a typical statistics course is thinking that this is the situation that calls for a "t-test".

The t-test is a formula-based approach; we plug numbers into a formula, and the formula spits out a p-value. However, applying the formula requires a large number of stringent assumptions to be valid, assumptions which are not met in most real-life applications. On the contrary, the resampling-based approach we'll be developing *applies to all situations.* For the sake of completeness and your own self-defense, we'll discuss the t-test and compare it to the resampling-based approach in section 6.9. But now let's dive into the resampling-based approach for two-group comparisons!

6.1 Data presentation and the choice of group measure

As we saw in Chapter 2 (*Describing and Presenting Data*), the decision to use a summary measure at all, and which one to use if we do use one, is not automatic, but depends on the shape of the data distribution and the scientific question we are addressing.

So how are we to approach the problem of choosing a summary measure? We begin with our standard approach: **look at the data distribution; make a visualization of the data.** That's the prerequisite for everything we do, and it is equally if not more important here.

What kind of picture should we make? The same types of visualizations as we did for one group. In Chapter 2, we learned to make dot plots, histograms, and Kernel Density Estimates for one group. Now we are going to make two visualizations, side by side, one for each group, and visually compare them.

Data presentation

Dot plot + box plot side-by-side First, let's take a case in which the data set is small enough to use the dot (or beeswarm) plot method (with an adjacent box plot showing, say, the median (when appropriate), the 25th and 75th percentiles, and the max and min). For two groups, we simply put the two dot plots next to each other (Figure 6.1).

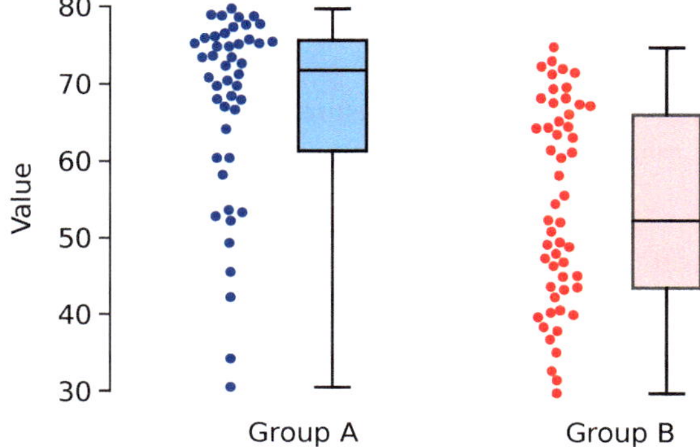

Figure 6.1 Dot plot and box plot for each of the two exemplary data sets A and B.

Let's stare at the picture and see what we can see in it:

- group A is markedly bunched up to the high side, that is, skewed "left" or to the lower values.
- group B looks fairly uniform in its distribution, with a hint of bimodality.

These are important qualitative differences between the two groups, and should be noted by the researcher in describing this data.

Histograms side-by-side To look at the distributions, we will use histograms. When we are using histograms, we should look at the two histograms, side by side or even superimposed on each other.

Here, the histograms of the same data sets A and B suggest two distributions that are very much *not* bell-shaped. Group A is heavily skewed to the left, and group B has a fairly flat distribution, possibly even bimodal (Figure 6.2).

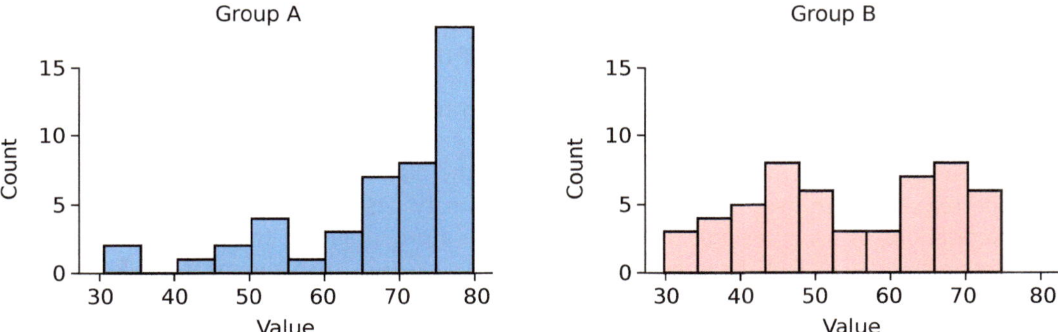

Figure 6.2 Histograms for the two groups in Figure 6.1.

KDEs or violin plots side-by-side Here are the violin plots of the same data sets A and B (Figure 6.3). Notice the qualitative shapes of the two violins, and how they differ. The violin plots confirm the impression of the histograms: Group A is heavy to the right (skewed left), and group B is bimodal.

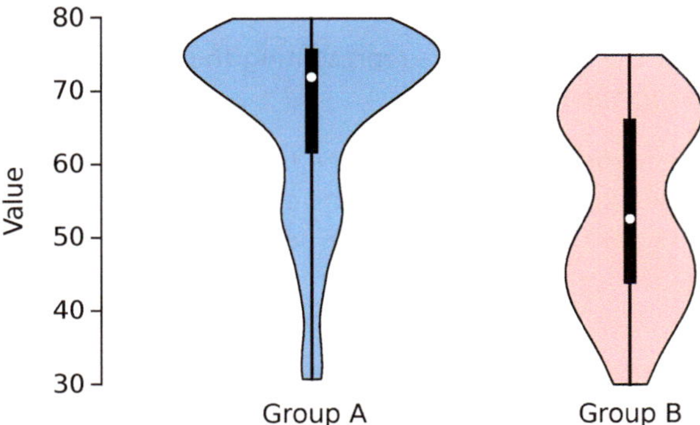

Figure 6.3 Violin plots for the same two groups in Figure 6.1. Also shown are the medians and 25th and 75th percentiles for the data sets.

It is absolutely critical to look at the two distributions side-by-side to make a qualitative assessment of the differences. **This qualitative evaluation can be the most important "analysis" of the data, and often gives us much more information than just using a summary measure.**

For example, consider the data sets above. Our first reaction to the data would be to comment on the change of shape between A and B. We should note that data set A is heavily skewed, and has high representation in the high values. Group B, on the other hand, has a much flatter distribution, with a suggestion of bimodality.

Choice of summary measure for the two groups

Once we have the distributions in sight, we can then decide whether to use a summary measure or not, just as we did for one group in Chapter 2.

It's important to stress that in comparing two groups, we have to make three big decisions:

- **Is it a good idea to use a single summary number to describe each of these distributions?** This is a scientific question, not a mathematical one. We have to ask whether our scientific purpose is expressible as a change in a summary number, or, for example, whether what is more important is the change in the shape of the distribution.

- **If we do want a summary measure, then which number should we use? (mean, median, variance, etc.)** Sometimes, the shape of the distribution will push us towards or away from using certain measures. For example, if the distribution is bell-shaped, then it may be reasonable to use the mean value as a summary descriptor if we keep in mind 1) the limitations of using a single number to describe a group and 2) the further limitations of using the mean. But if the distribution is heavily skewed, then the mean could be a poor choice of descriptor, because it will be distorted in the direction of the skew. And if the two distributions show markedly different spreads, that may itself be a valuable observation, and we would want a measure that reflects this difference in spread, such as the variance or the MAD.

  But the choice of group measure is not only dependent on the shape of the data; it should also reflect our scientific purpose. In Chapter 2, we saw the example of survival times with a particular form of cancer. In that case, the average survival time told the patient very little. What if, out of 10 people, 9 died within weeks, and the 10th lived for 10 years. Should we say that "the average survival time is 1 year?" Surely the *median* survival time is more informative for the patient.

- **How should we compare the two numbers representing the two groups? (difference, ratio, etc.)** This we'll discuss in the next section.

If it is appropriate to use a summary measure for the two groups, we will call them M(A) and M(B). In the case of the above data sets, based on the skewed (Group A) and bimodal-looking (Group B) distributions, we choose the median as our descriptor.

$$M(A) = \tilde{A} \qquad M(B) = \tilde{B}$$

Example 1 Evolution in family planning practices

Here's an interesting example, from a *New York Times* article (2018), comparing two histograms of "age at first pregnancy" of the US female population in 1980 and 2016.

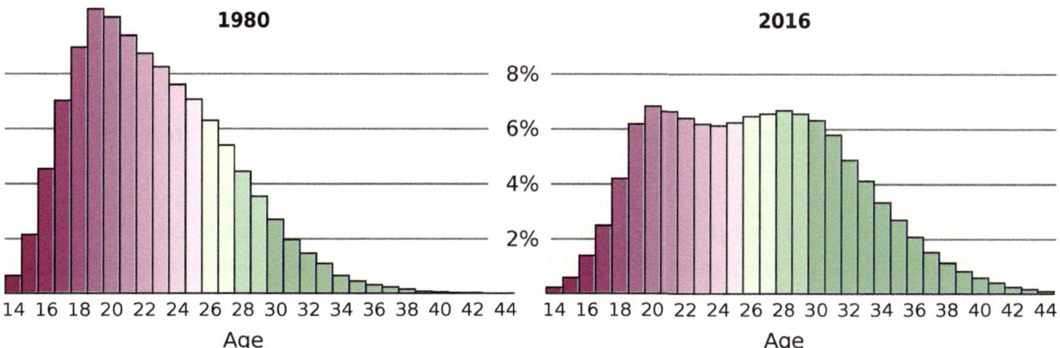

Question Comment on the shape of the two histograms and compare the two.

Answer **Here, the most striking difference is in the shapes of the two histograms.** The 1980 histogram peaks at age 18, with a gradual decline into the 30's that tapers off by 40. Now consider the 2016 histogram. The early peak has shifted to age 19, but now the curve hardly declines at all through the 20s, and indeed, has another peak at 29, declining more slowly into the mid-40s.

Question What are the advantages of presenting the histograms?

Answer **These observations of the histograms indicate a profound change in US family-planning practice.** What is important for us is that the insight here is gained through comparing the shapes of the histogram; almost all of these insights would be lost if we had immediately adopted some kind of summary descriptions of the two data sets. The most valuable insights into the data are the qualitative observations above. We should not rush to stick numbers on things before looking at the distribution.

FURTHER EXERCISES 6.1

1. **Sea turtle egg hatching rates.** Researchers recorded the hatching rate (%) of sea turtle eggs at two nesting sites. Site A experienced normal temperatures while Site B was affected by a prolonged heatwave. Using the histograms below, what summary measure should we use to compare the central tendency between the two sites? What if we are interested in comparing the variability between the two sites?

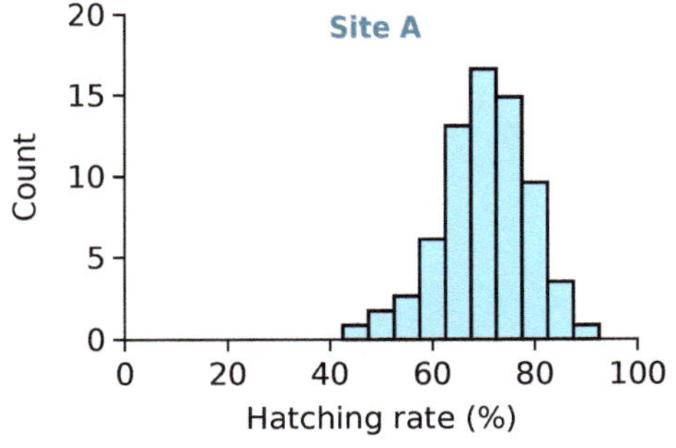

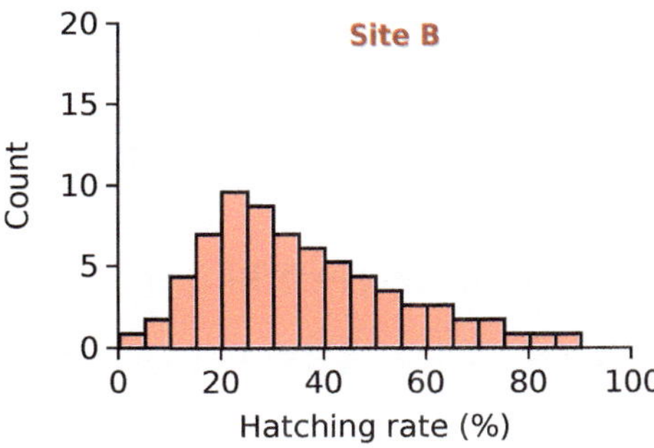

2. When comparing the central tendency between two groups, select all the reasons that should steer you toward choosing the median as the summary measure?

   a) The distributions are skewed.
   b) The distributions are bell-shaped.
   c) The distributions are uniform.
   d) The median is more interpretable.

3. In Example 1 *Evolution in family planning practices* on page 190, you were given two groups of data on the age at first pregnancy of US females in the year 1980 and in 2016. Formulate the scientific questions you are interested in and decide what summary measure(s) you should use. Discuss the pros and cons of various measures.

6.2 Choice of group comparison (effect size): Δ_{obs}

After choosing the summary measure for the two groups, we need to compare them to get an effect size. The effect size is also called the observed group difference. We use Δ_{obs} as the notation. This comparison could be an algebraic difference, it could be a ratio, or whatever serves our scientific purpose.

Algebraic difference We often talk about the "difference" between the two groups, and there is a temptation to think that this must mean the *algebraic* difference, written with the minus sign. If M is the measure we use (mean, median, etc.), so that M(A) and M(B) are the measures of the two groups, then we might think that the "difference" between M(A) and M(B) must be expressed as M(A) − M(B), the algebraic difference.

If the difference of the two groups is measured as the algebraic difference, it could be

$$\Delta_{\text{obs}} = M(A) - M(B) \qquad \text{or} \qquad \Delta_{\text{obs}} = M(B) - M(A)$$

Ratio But while we often use the algebraic difference, it is not automatic, and sometimes, other comparators can or should be used. For example, it often makes sense to use the *ratio* $\frac{M(A)}{M(B)}$ as our comparison. If we were comparing incomes, expressing one group measure as a *multiple* of the other may make more sense, and so the "difference" between M(A) and M(B) is the ratio $\frac{M(A)}{M(B)}$.

And if we were interested in a percentage change, or a "fold" change, we would take the ratio of the measures of the two groups, as the "difference". We'll discuss ratios as effect sizes in section 6.8 (*When the effect size is a ratio*).

If the difference of the two groups is measured as the ratio, it could be

$$\Delta_{\text{obs}} = \frac{M(B)}{M(A)} \qquad \text{or} \qquad \Delta_{\text{obs}} = \frac{M(A)}{M(B)}$$

6.3 Is the difference large?

We have two groups, A and B, and so far we have chosen the group measure and calculated the effect size Δ_{obs} (the observed group difference). Now we want to know if the two groups "differ". But notice that the answer to the question: "do the two groups differ?" is almost always going to be "yes"; there is almost *always* going to be *some* difference.

We could ask: yes, but is the difference "large"? is it "significant"?

But what can we possibly mean by "large"? What can we mean by "significant"?

There are two very different answers to this question.

1. **Scientifically or medically significant.** This means that the "difference" is large enough to make a difference to science or medicine or public policy. For example, if we are looking at the effect on survival of a drug vs. control, and the difference in the median survival times is 1 day, that would be an example of a difference which is *not* medically significant. No one is interested in a drug that prolongs survival by 1 day, regardless of the statistical significance of the result!

   Notice that this is not a statistical question. We are comparing the data to an external standard of real-world importance.

2. **Statistically significant.** This is a different question, one that can be answered by purely appealing to the data and nothing else. This can be a strength or a weakness. This is the concept we will now discuss.

It is extremely important to remember that "real world significant" is not equivalent to "statistically significant". They are very different concepts.

"Statistically significantly different" The question is "is Δ_{obs} large?" In the statistical interpretation, this means **large compared to what we might expect to occur just from random sampling.** After all, *any* two samples from the same population will have *some* difference. If we could say that the observed difference Δ_{obs} is much bigger than the differences that would arise just due to random sampling, that would be an interpretation of "statistically significantly large".

But how can we estimate how large a difference might have been seen simply due to random sampling? The best, most intuitive and natural answer would be to just *do it* and see how large the differences are. We could make a model of the underlying population U from our data and then do the experiment of randomly sampling n_A and n_B elements from U, where n_A is the number of data points in group A and n_B the number of data points in group B, and calculating their Δ-value: *if we did that many times, how large would those Δs be?*

There is a close parallel between this question and the questions we were asking in Chapter 3, when we were asking whether an observed distribution was "significantly different" from a theoretical prediction. Let's say we flipped a coin 100 times and got 60 Heads. In order to answer the question: "is 60 Heads out of 100 flips statistically significantly different from the theoretical prediction of 50?", we would formulate a Null Hypothesis, which is that the coin was fair, and then we would simulate 1000's of times by sampling 100 under the Null Hypothesis, and counting how many times 60 or more Heads or 60 or more Tails occurred (for a two-sided test). We carried out this simulation and found that "60 or more Heads or 60 or more Tails" occurred 5.98% of the time, so the p-value of our finding is 0.0598.

So the key idea was: simulate the Null Hypothesis, 1000s of times, and form a set of expectations of what you would be likely to see under the Null Hypothesis (that is, if the Null Hypothesis were true). In that case, the Null Hypothesis was given by a theoretical model.

Here we are not comparing data to a model; we're comparing two data sets. And so, the Null Hypothesis is different. **Now, the Null Hypothesis is: there is no real difference between A and B and the differences observed are merely the random effects of sampling 2 groups from the same underlying population.**

That tells us what we need to do to judge the statistical significance of Δ_{obs}. First, assume that A and B are two samples from the same underlying population, and then sample thousands of new pairs of groups, pseudo-As and pseudo-Bs, the same sizes as A and B, from this underlying population. We make a histogram of the Δ-values for these new pairs. This is the null distribution to which we will compare Δ_{obs}. From the histogram, we calculate the probability of a difference as extreme as Δ_{obs} happening if there were no difference between the groups (Figure 6.4).

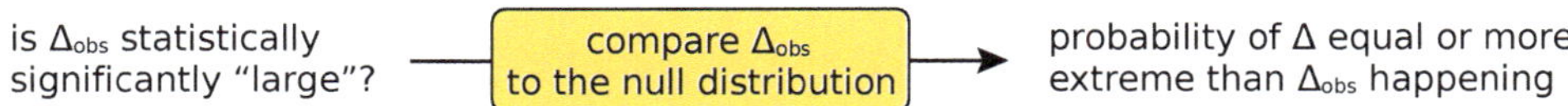

Figure 6.4 Definition of statistical significance for comparing two groups.

The only problem is that we have no idea what the underlying distribution is from which A and B are drawn!

Any "theoretical" assumptions about what the underlying population *must* look like are highly debatable and very possibly wrong. In particular, the idea that "if you don't know the distribution, assume that it's Normal" is a terrible idea and often wrong.

The Big Box method

The Null Hypothesis In fact, the only thing we can say about reality, about the underlying population, is: **it must look like our data, since our data are drawn from it.** We are, as always, assuming that our samples A and B are not biased, and therefore, **the best guess about what the underlying population looks like is to put A and B into a big box: that's our best model of the underlying population** (Figure 6.5).

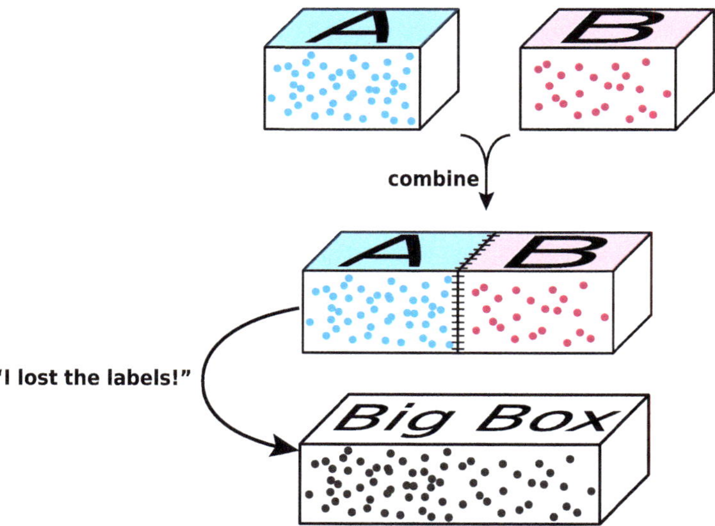

Figure 6.5 How to form a Big Box model.

Criteria for using the Big Box When we throw A and B into a Big Box, we are embodying the Null Hypothesis of "no difference between the groups". For us to do that, there would have to be no obvious differences between the two distributions of A and B.

For example, if A and B have markedly different variabilities, markedly different spreads, then we couldn't assume that they were drawn from the same underlying distribution. We could reject the Null Hypothesis that A and B are drawn from the same underlying distribution by showing that it is highly unlikely for two draws from the same underlying distribution to have such different variabilities.

When the two groups have markedly different variabilities, we can't use a Big Box model. Rather, a different procedure, the Re-centered Two-Box model, must be used (see *The Re-centered Two-Box method* on page 202).

Calculating the observed group difference Δ_{obs} Choose the group measure and method of comparison (Figure 6.6 part **1**).

Resampling from the null world Assuming that the two groups have roughly equal variabilities, and we've formed the Big Box as the null world, we then repeatedly resample from this null world to generate estimates of how large a difference between the two groups we might expect to see under the Null Hypothesis.

Suppose the number of data points in group A is n_A and the number of data points in group B is n_B. Then the number of data points in the Big Box is $n_A + n_B$.

Now we randomly select, with replacement, n_A data points from the Big Box to form a new sample A_1, and n_B data points from the Big Box to form a new sample B_1. For this first resample, A_1 and B_1, we'll calculate the group measure $M(A_1)$ and $M(B_1)$. Then we calculate the group difference between A_1 and B_1, and call that Δ_1. This gives our first estimate of how large a difference we would expect to see under the Null Hypothesis.

We then repeat this resampling procedure 10,000 times to generate 10,000 estimates Δ_1, $\Delta_2, \ldots, \Delta_{10,000}$ under the Null Hypothesis (Figure 6.6 part **2**).

Basic idea behind this course: resampling

If you want to know how often something will happen due purely to chance, find out (experimentally) how often it happens due purely to chance. Don't make a bunch of extreme assumptions about idealized populations in order to derive a *formula* for how often it occurs, just see how often it occurs with your actual data.

Calculating the p-value Given the 10,000 estimates from the null world, we now compare the simulated differences Δ_i ($i = 1, 2, \ldots, 10{,}000$) to the observed difference, Δ_{obs}, in order to estimate how often a result as extreme as or more extreme than Δ_{obs} happened. Here, as in Chapter 3 (section 3.2 *Null Hypothesis Significance Testing*), we use a 2-sided test, and ask how many simulations had a Δ_i equally or more extreme than Δ_{obs} *and* how many had a Δ_i equally or more extreme than $-\Delta_{obs}$.[1]

The number of times an equally or more extreme result was seen in the random resamplings, divided by the total number of resamples (here 10,000), is the p-value of the observed group difference Δ_{obs} (Figure 6.6 part **3**).

[1] Here we are assuming that the group difference is the algebraic difference. If our group comparison was a ratio, we would be asking how many simulations produced differences as or more extreme than either Δ_{obs} or $1/\Delta_{obs}$, see section 6.8 *When the effect size is a ratio*.

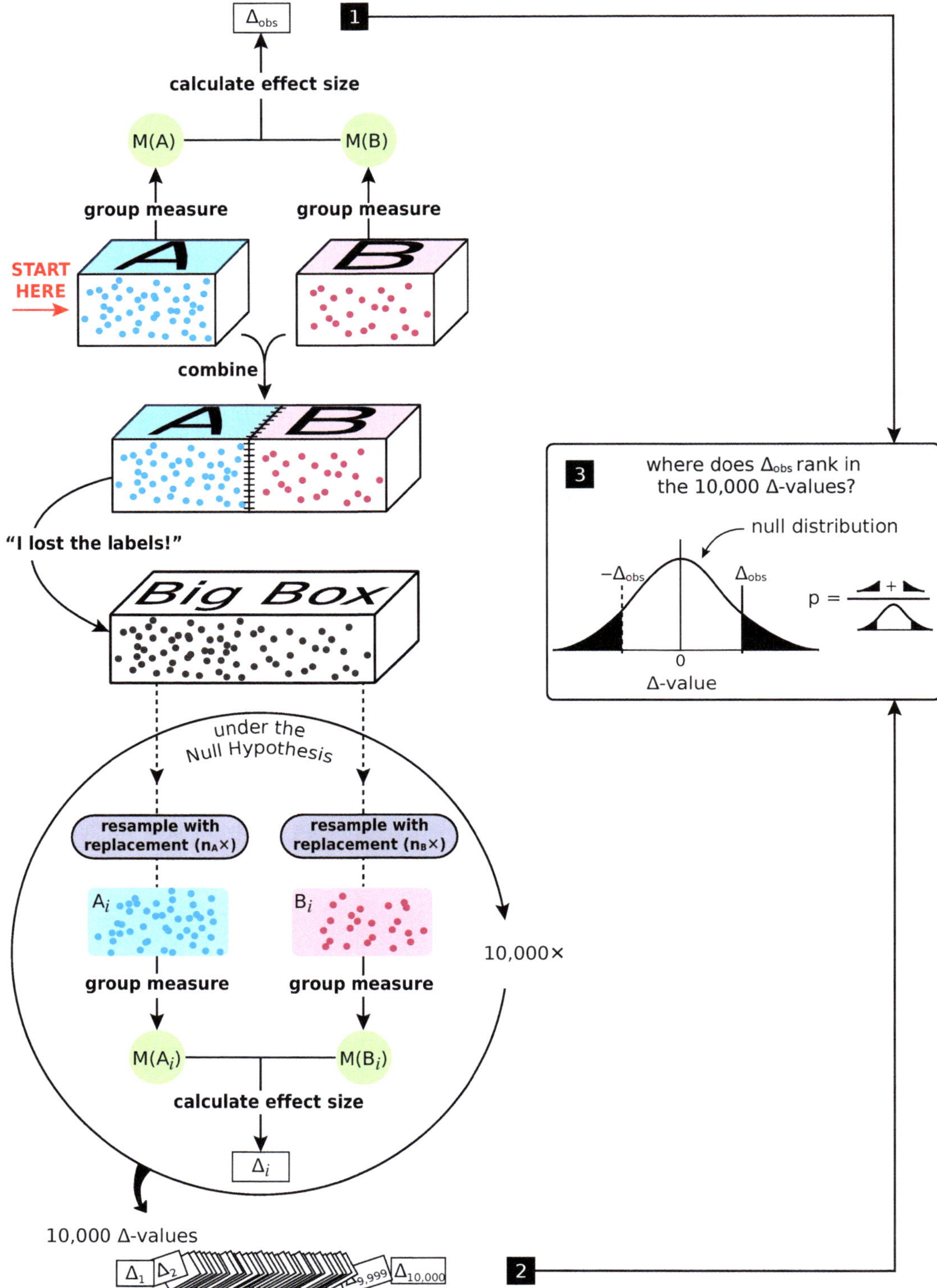

Figure 6.6 The Big Box method to determine the statistical significance of the observed effect size between two groups A and B.

The Big Box procedure

We can summarize the Big Box procedure in three parts (Figure 6.6).

1 **Analyzing the observed data.** Choose a summary measure (mean, median, MAD, ...) and a way of comparing two summary measures (the algebraic difference, ratio, ...) to get an effect size. These choices are based on the shape of the data distributions of the two groups and on our scientific questions.

2 **Resampling the data.** Form the Big Box model that embodies the Null Hypothesis. Then resample two new groups from the Big Box 10,000 times and keep track of the calculated effect size Δ_i for each resample. This gives us 10,000 estimates of the effect size we would anticipate under the Null Hypothesis.

3 **Interpreting and presenting the results.** Collect the 10,000 estimates Δ_1, Δ_2, ..., $\Delta_{10,000}$ and rank them from smallest to largest. Compare the actually observed group difference Δ_{obs} to this list. Where would Δ_{obs} rank in the list of 10,000 Δ_i's?

Let's say we use the algebraic difference to compare the two groups, and Δ_{obs} is positive. Suppose that 41 Δ_i-values are $\geq \Delta_{obs}$. To do a 2-sided test, we also need to ask how $-\Delta_{obs}$ ranks in the list. Let's say there are 53 Δ_i-values $\leq -\Delta_{obs}$. Then the number of resamples that produced Δ_is as extreme or more extreme than Δ_{obs} is 41 + 53 = 94. The p-value we should report is therefore $\frac{94}{10,000} = 0.0094$.

The Big Box procedure gives out a p-value: how often results as extreme or more extreme than the observed difference Δ_{obs} would occur if groups A and B come from the same population. As in all significance testing, we need to choose a threshold α to determine whether the p-value is small enough to declare "statistical significance" (Figure 6.7).

Figure 6.7 The Big Box method produces a p-value for the question "is Δ_{obs} statistically significantly large?" We recommend giving the observed result Δ_{obs} and the actual p-value.

Null Hypothesis Significance Testing

NHST and p-value. We have now calculated a p-value for the group difference Δ_{obs}. But, strictly speaking, that is not Null Hypothesis Significance Testing. NHST, which, unfortunately, is often used in the medical and scientific literature, results in a yes-or-no answer, "significant" or "not significant".

In Chapter 2, we saw how Dynamite Plunger Plots are often used inappropriately to present data and how they actually conceal data. When we see Dynamite Plunger Plots with little stars above the bars, the little stars mean "significant".

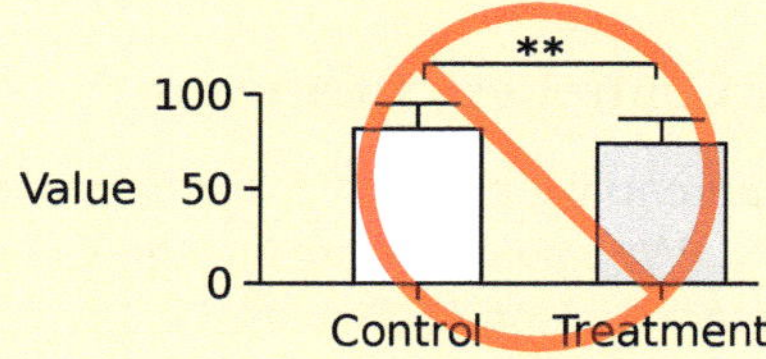

Often, one star means $p < 0.05$, two stars mean $p < 0.01$, etc. Here we show the standard NHST using two stars to denote a significant result at $p < 0.01$ level between the two groups. **We reiterate that Dynamite Plunger Plots are a poor practice and should be avoided!**

How to do an NHST using resampling? In order to do NHST using resampling, we would have to decide on an α, the threshold for declaring "statistically significant". That corresponds to taking the null distribution (the histogram of the Δ_i's) and drawing two cutoff lines on it, so that the areas outside the cutoffs, the number of simulations, are exactly α of the total number of simulations. The two cutoffs are

$$\frac{\alpha}{2}\% \quad \text{and} \quad (1-\frac{\alpha}{2})\%$$

For example, if our α is 0.01, the two cutoffs are the 0.5^{th} percentile and the 99.5^{th} percentile.

Then, if Δ_{obs} is outside the boundaries, we say Δ_{obs} is "statistically significant", otherwise, we say Δ_{obs} is "not statistically significant". This is simply a yes-or-no answer.

We show this here to illustrate what NHST is doing. But this is not recommended, as we are obviously throwing away information by just reporting whether the p-value was inside the boundaries or not, as opposed to giving its actual value.

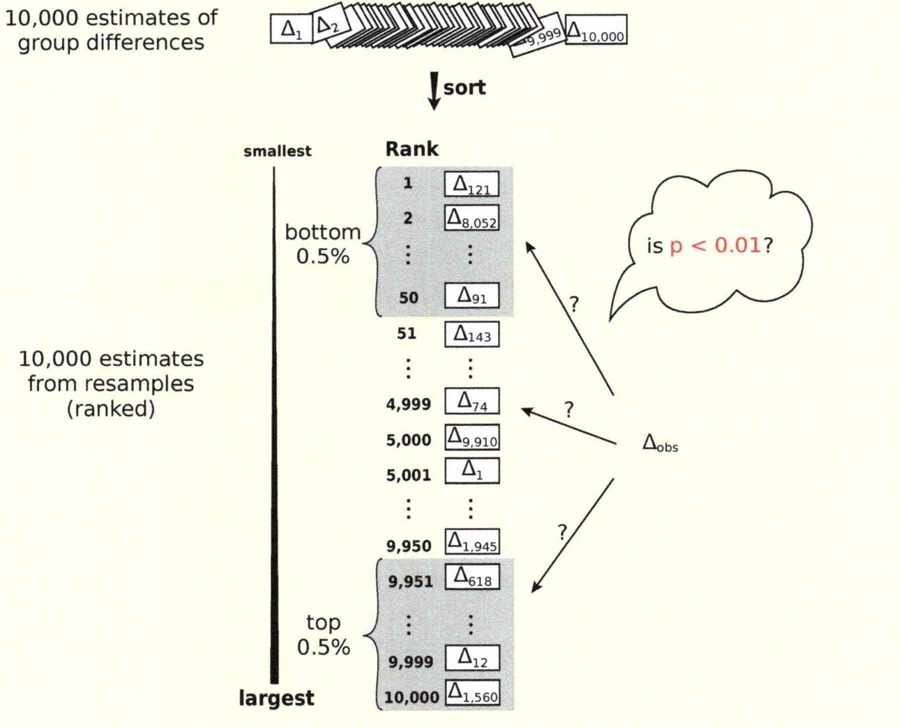

A case study: T-cell counts in Control vs. Treatment

As an example, let's say we have a potential treatment for an immune disease, and so we tested it by applying it to T-cells in culture. We have 49 plates in the Control group, and another 45 plates in the Treatment group. We are looking to increase T-cell counts, so the more T-cells, the better the treatment.

In this example, we have two groups, the Control group A and the Treatment group B, with $n_A = 49$ and $n_B = 45$ measurements in them, respectively, where each measurement is a T-cell count per unit area for one plate. The data for these two groups would look like Table 6.1.

T-cell count (per unit area)	
Control group (A) $n_A = 49$	Treatment group (B) $n_B = 45$
53.2, 54.6, 55.1, 55.4, 55.8, 56.2, 55.7, 56.5, 56.6, 56.9, 57.2, 56.7, 57.9, 57.6, 57.8, 58.5, 58.9, 59.0, 59.1, 58.7, 58.6, 58.8, 59.7, 59.4, 59.8, 60.1, 60.8, 60.5, 61.0, 61.7, 61.3, 61.9, 61.5, 62.8, 62.5, 62.6, 62.9, 63.5, 63.8, 65.5, 65.1, 64.9, 65.2, 66.0 ,67.9, 67.8, 68.1, 69.5, 70.0	57.9, 59.6, 61.0, 61.6, 62.4, 62.8, 63.1, 64.0, 64.7, 65.2, 65.9, 66.2, 66.4, 66.5, 68.4, 68.9, 68.3, 67.8, 67.7, 68.0, 69.0, 68.8, 69.5, 70.0, 70.1, 70.8, 71.2, 71.3, 72.0, 72.1, 71.1, 70.5, 71.1, 71.9, 71.5, 72.5, 73.1, 73.3, 72.8, 73.8, 73.5, 74.9, 75.2, 76.2, 75.8

Table 6.1 Data set for the Control group and the Treatment group in a hypothetical T-cell study. Note that the sizes of each group do not have to be the same.

Data visualization

Beeswarm plot First let's visualize the two data sets (Figure 6.8).

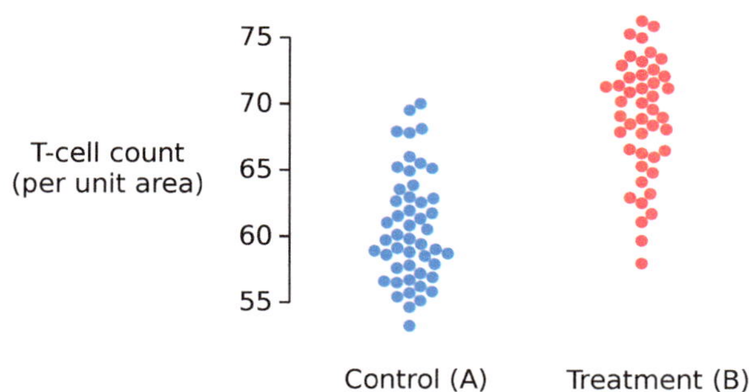

Figure 6.8 Beeswarm plots for the data set in Table 6.1.

We can form a visual impression of the shape of the distributions as well as their locations on the axis.

1. **overall distribution shape.** Neither data set is bell-shaped. The Control group is asymmetric, with many more values bunched up on the low side (that is, skewed to the right). The Treatment group is oppositely asymmetric, with many more data points on the high side (that is, skewed to the left).

2. **location.** It looks like the Treatment group has a lot more high values than Control. Since higher values are good, this means that the treatment could be having a good effect.

Histograms If we superimpose the two histograms (Figure 6.9), it confirms the impression that the Treatment group has many values to the high side, and the Control group has many values to the low side.

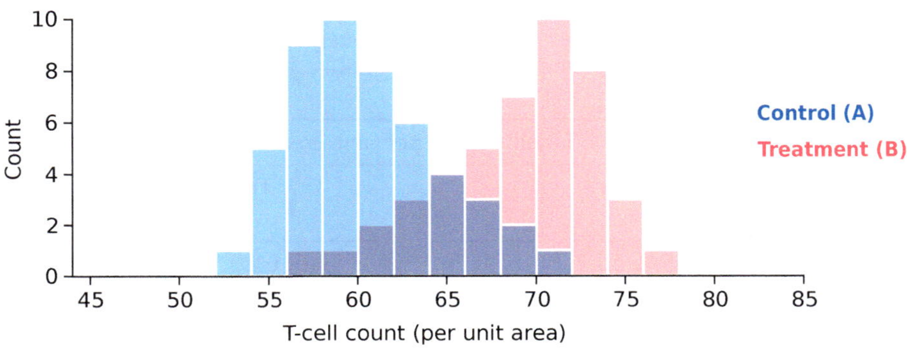

Figure 6.9 Histogram presentations of the two data sets in Table 6.1.

Histogram + KDE The KDE gives an especially clear view of the difference between the two groups, and confirms our overall impression that the treatment has had some good effect: the curve has been shifted toward the right (Figure 6.10).

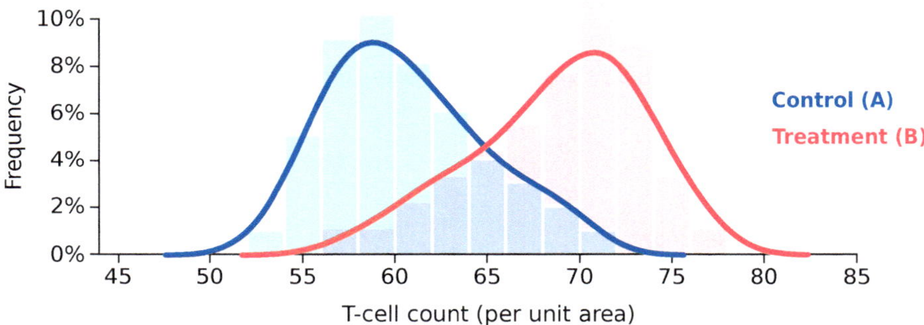

Figure 6.10 Histogram and KDE presentations of the two data sets in Table 6.1.

Choosing a summary measure for the two groups

Now that we have our overall impression, it is time to think about putting numbers on this effect, which means turning to summary measures.

What summary measures should we use to describe the groups? Here are two considerations:

1. The data are not really supportive of the "bell-shaped" assumption, since both groups are asymmetric, in opposite ways.

2. Let's think about the scientific meaning of this experiment and this data: what measure best captures what we really want to know about this drug? Is it the mean T-cell count? The mean suffers from all the defects discussed in Chapter 2: if the drug had one very bad plate, it would pull the average down substantially, and if it had one very good plate, it would pull the average up substantially. But we don't want that. It would give us misleading information about the efficacy of the drug.
   Moreover, it may well be the case that the drug, in order to be effective, has to have substantial T-cell production. The median value would give us a piece of valuable information, which is what kinds of T-cell counts we might typically expect.

Therefore, we will choose the *median* as our summary measure for the two groups: it is more appropriate to the distribution *and* gives us more useful information (Figure 6.11).

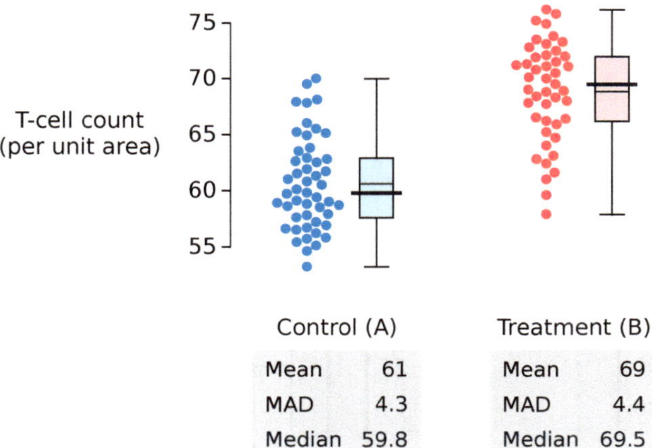

Figure 6.11 Choosing the summary measure for the two groups.

The two medians are:

control $M(A) = \text{median}(A) = \widetilde{A} = 59.8$ T-cells per unit area

treatment $M(B) = \text{median}(B) = \widetilde{B} = 69.5$ T-cells per unit area

Comparing the two groups: what's the observed difference?

We now have two numbers describing the two data sets. Next, we need to make another decision, about how to compare the two numbers to come up with a "difference". In other words, what is the effect size?

The idea that ratios often are more interpretable than algebraic differences could apply to our T-cell drug case. We might be more interested in the "percentage change" or the "fold change". Then our effect size would be stated as "the drug resulted in a 30% increase", or "the drug produced a 2-fold increase", that is, a ratio.

However, for this example, we will agree to use the *algebraic difference* of the two medians as our effect size. Our scientific justification could be that what we need here are absolute *numbers* of T-cells, and in this case, knowing that the drug produced, say, a 20% increase might not be helpful.

Later in this chapter we will revisit this case and treat it differently, where our effect size is a ratio (see section 6.8 *When the effect size is a ratio*).

$$\begin{aligned}\Delta_{\text{obs}} &= M(B) - M(A) \\ &= 69.5 \text{ T-cells/unit area} - 59.8 \text{ T-cells/unit area} \\ &= 9.7 \text{ T-cells/unit area}\end{aligned}$$

So, on these measures, the two groups are "different". How different? We have a number: a difference of 9.7 T-cells/unit area.

effect size $\Delta_{\text{obs}} = 9.7$ T-cells/unit area

Calculating the p-value

In order to answer this question, we compare Δ_{obs} to the null distribution generated by resampling from the Big Box. The use of a Big Box is justified by the fact that A and B have similar spreads (as measured by the MAD in Figure 6.11).

When we put the elements of A and B into a Big Box and resampled new A_i and B_i from the Big Box 10,000 times, we found that results equally or more extreme than Δ_{obs} or $-\Delta_{obs}$ never occurred. Hence the p-value is less than 0.0001. Therefore, we can conclude that the drug treatment produced an absolute increase in T-cell numbers that was statistically significant (Figure 6.12).

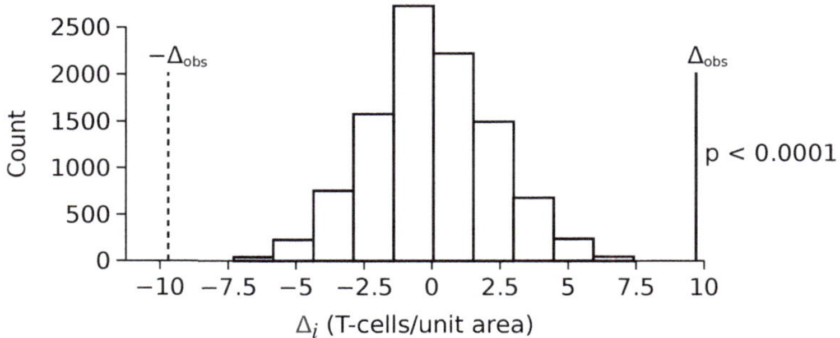

Figure 6.12 p-value calculation for comparing the two groups in Table 6.1, with the difference in median T-cell count between Treatment and Control (effect size Δ_{obs}) indicated on the null distribution.

Exercise 6.3.1 What are the advantages of reporting the effect size together with the actual p-value compared to the standard Null Hypothesis Significance Testing?

The Re-centered Two-Box method

The Big Box procedure gives us a way of testing whether two data sets might have come from the same underlying population.

But there are many cases in which we want to know "whether there's a difference" between two sets of data that clearly did *not* come from the same underlying population.

A weaker question Consider the two data sets in Figure 6.13, Group A is the wild type and Group B is the mutated strain. They clearly do not come from the same underlying distribution: A and B have markedly different spreads or variabilities, so they are not likely to have come from the same underlying distribution. **As a rule of thumb, we will define "markedly different" as "having standard deviations (or MADs) varying by more than 2×".**

We can no longer ask: do these two data sets come from the same underlying distribution (Big Box)? They obviously don't!

Let's say that we have decided to use the medians $\widetilde{A}$ and $\widetilde{B}$ as our group descriptive measure (thick black bars), and let's also agree that we will use the algebraic difference $\widetilde{A}-\widetilde{B}$ as our effect size, Δ_{obs}.

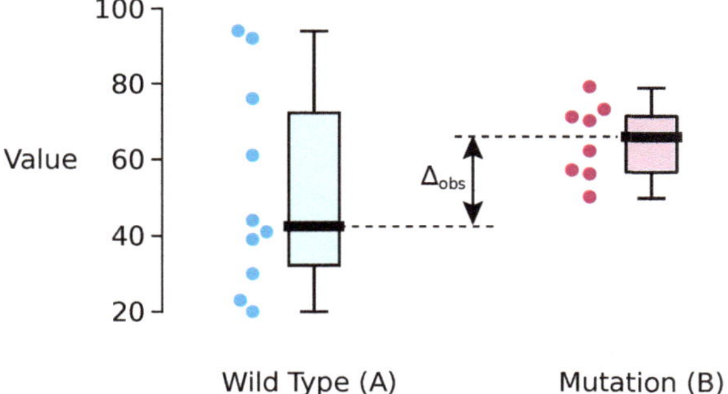

Figure 6.13 Dot plots and box plots for the two groups, Wild Type and Mutation. Note the different spreads.

We now want to ask if Δ_{obs} is "large" relative to what we would see "under the Null Hypothesis", but now we have a problem.

The Null Hypothesis in the Big Box case was that the two samples A and B came from the same underlying distribution, and we made a model of this underlying distribution by combining A and B into the Big Box.

Here, we have to ask a weaker question, namely, **is it possible that A and B were drawn from two different underlying distributions, but *with the same mean or median values?***

Creating the world of the Null Hypothesis by re-centering In order to test this weaker question with a resampling procedure, we first create the world of the Null Hypothesis. In this case, that means that A and B are drawn from two different distributions with the same mean or median values, but with different variabilities, corresponding to the variabilities of A and B. We can create that null world by taking the original distributions A and B, and simply sliding them with respect to each other so that their group measures (means, medians, etc.) coincide. Without loss of generality, we can set both group measures to be 0 (Figure 6.14).

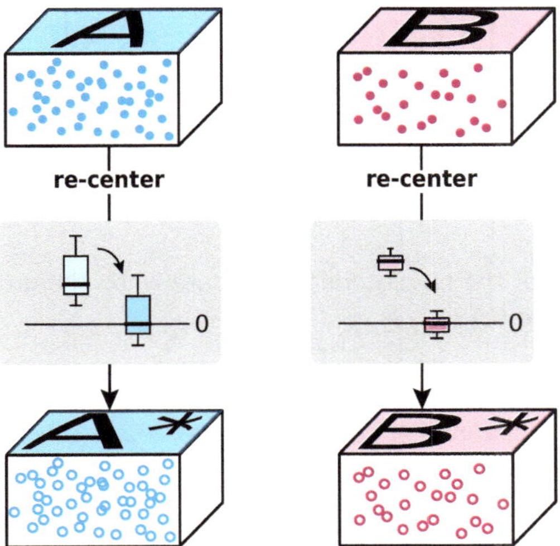

Figure 6.14 Forming the world of the Null Hypothesis by re-centering each data set by its respective group measure to 0.

Here we will illustrate this re-centering procedure when the group measure is the median value. We take as our null world the original data sets, but with the respective group median subtracted from every data point, so that the medians of the re-centered data sets, A^* and B^*, are both 0 (Figure 6.15).

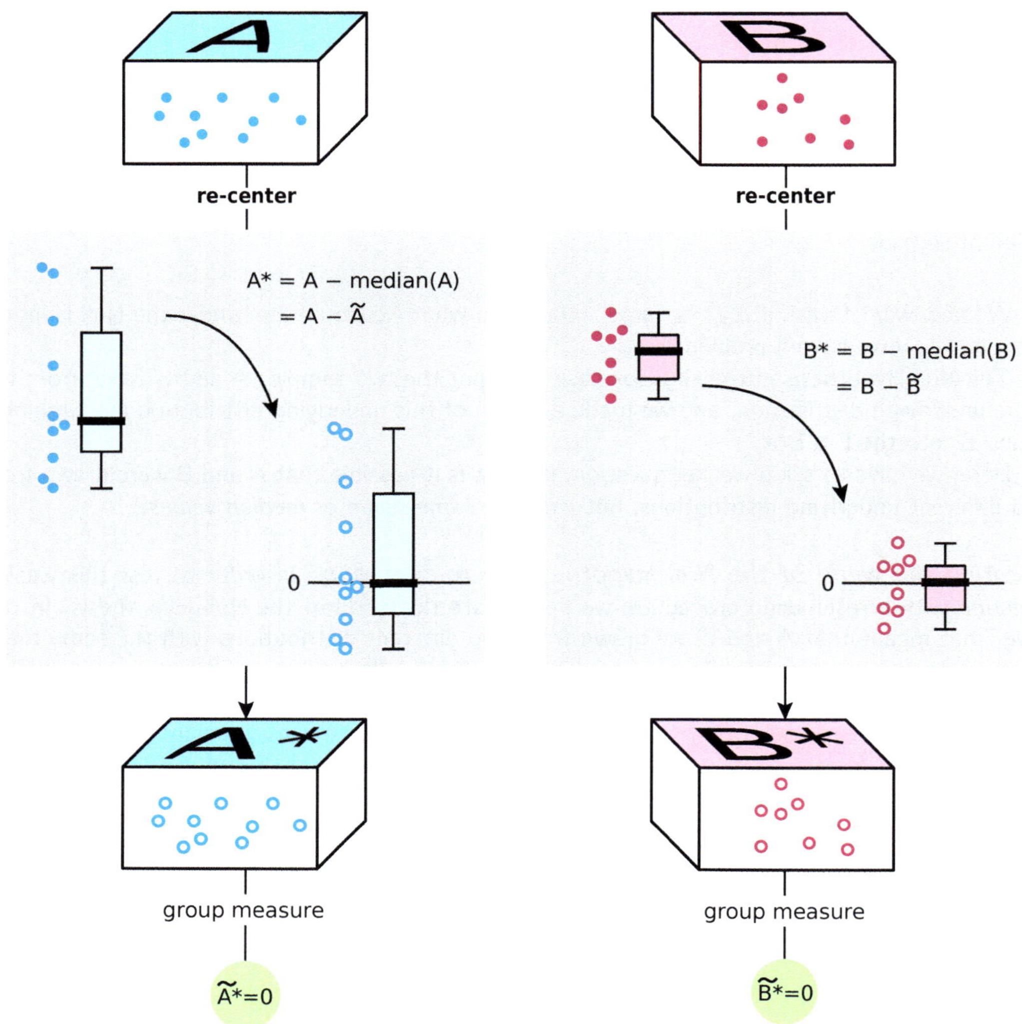

Figure 6.15 Illustration of the re-centering procedure when the group measure is the median, using the data set in Figure 6.13.

Exercise 6.3.2 What is the result of re-centering this data set, using the median?

$$A = \{2, 4, 5, 4, 6\} \quad \Longrightarrow \quad A^* = ?$$

a) {2, 4, 4, 5, 6}

b) {−2, 0, 0, 1, 2}

c) {−2, 0, 1, 0, 2}

d) {−2.5, −0.5, 0.5, −0.5, 1.5}

e) {−2.2, −0.2, 0.8, −0.2, 1.8}

f) None of the above

Resampling from the null world After we create the world of the Null Hypothesis, we can begin to answer the question: under the Null Hypothesis, how often do we observe a difference as extreme as or more extreme than the observed effect size Δ_{obs}?

To do this, we will resample a new A_1 out of the re-centered set A* and a new B_1 out of the re-centered set B*, take their median values $\widetilde{A_1}$ and $\widetilde{B_1}$, and calculate the difference between them ($\Delta_1 = \widetilde{A_1} - \widetilde{B_1}$) as our effect size for this resample, call it Δ_1, the first resampling estimate. Essentially, **this difference Δ_1 is an estimate of how large a difference we should see between the medians of two samples chosen from different underlying distributions with the same median value, but with different variabilities, equal to the variabilities of A and B**.

We repeat this resampling 10,000 times to obtain 10,000 estimates of the differences we would see under the Null Hypothesis, then compare our Δ_{obs} to the null distribution, the histogram of differences from the 10,000 resamplings, to determine what percent of them are equal to or more extreme than the observed effect size Δ_{obs}.

Calculating the p-value Just as in the case of the Big Box method, we use a two-sided test to determine the p-value: we count the number of simulations with a group difference Δ_i equally or more extreme than Δ_{obs} *and* how many had a Δ_i equally or more extreme than $-\Delta_{obs}$.

Since we are now resampling new As from the A* box and new Bs from the B* box, we call this the **Re-centered Two-Box method**. It also outputs a p-value to determine whether Δ_{obs} is "large" (Figure 6.16). In contrast to the Big Box method, the Re-centered Two-Box method answers a weaker question.

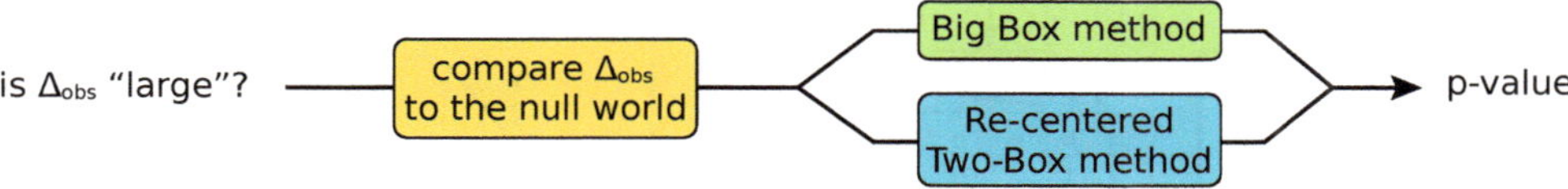

Figure 6.16 For a two-group comparison, which method should be used to calculate a p-value depends on the shape of the data.

The Re-centered Two-Box procedure

We can summarize the Re-centered Two-Box procedure in three parts (Figure 6.17).

1. **Analyzing the observed data.** This is the same as the first step in the Big Box model. Choose the group measure M. It could be mean or median, or MAD, etc. Then choose the method of group comparison. This could be −, or ÷, etc. Calculate the observed effect size Δ_{obs}.

2. **Resampling the re-centered data.** Re-center the two groups, calling them A* and B*. Then, resample a new A_1 from the re-centered set A*, and a new B_1 from the re-centered set B*, take their group measure $M(A_1)$, $M(B_1)$, and calculate the difference between them, Δ_1, as our first estimate of the effect size under the Null Hypothesis.

   Repeat this resampling process 10,000 times, generating the differences Δ_1, Δ_2, ..., $\Delta_{10,000}$. This gives us 10,000 estimates of the effect size we would anticipate under this Null Hypothesis.

3. **Interpreting and presenting the results.** Collect the 10,000 Δ_i's and compare Δ_{obs} to this list. For a two-sided test, the number of simulations that had a Δ_i equally or more extreme than Δ_{obs}, divided by 10,000, is the p-value of Δ_{obs}.

Because this is a two-sided test, when we say "more extreme than", we mean more extreme than Δ_{obs} or its opposite, whether that's $-\Delta_{\text{obs}}$ for algebraic differences or $1/\Delta_{\text{obs}}$ for ratios (see section 6.8 *When the effect size is a ratio*).

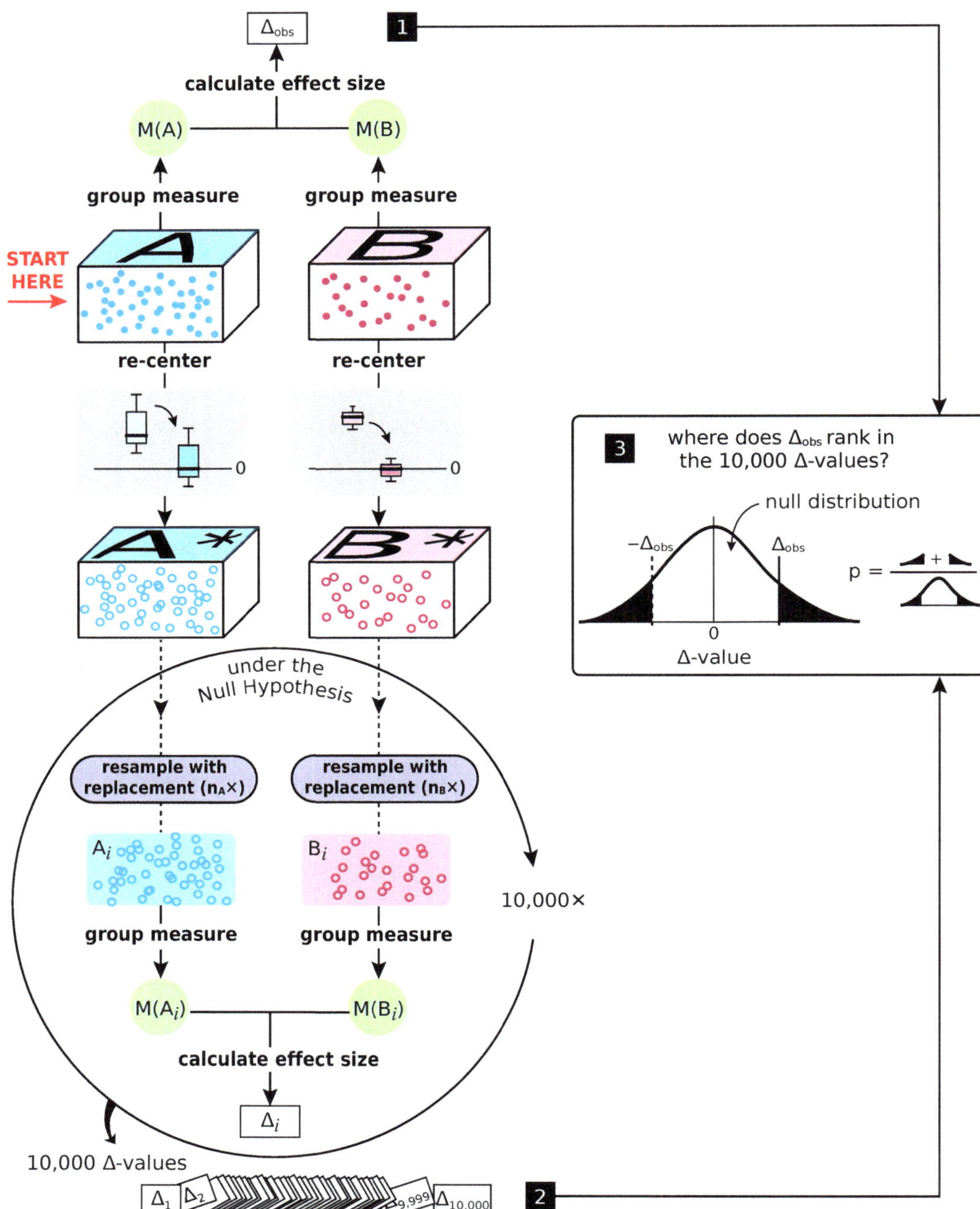

Figure 6.17 The Re-centered Two-Box method to determine the statistical significance of the observed effect size between two groups A and B.

A case study: a clinical trial of a cancer drug

In this example, we are evaluating a drug for its efficacy against a certain type of deadly cancer. We have chosen as our measure the survival times of patients who did or did not get the drug.

Survival times after diagnosis (years)	
Control group (A) $n_A = 55$	Treatment group (B) $n_B = 58$
2.0, 0.3, 2.9, 1.8, 0.8, 1.1, 2.7, 1.4, 1.0, 0.4,	2.3, 3.5, 7.5, 6.9, 3.0, 4.5, 6.1, 3.4, 3.9, 1.8,
1.5, 4.9, 2.6, 2.6, 1.7, 1.6, 2.9, 2.2, 1.1, 2.2,	2.2, 4.6, 10.3, 8.0, 2.8, 2.1, 0.6, 12.7, 5.5,
1.9, 0.6, 0.5, 0.9, 0.8, 0.6, 1.3, 2.7, 0.5, 0.8,	1.7, 1.1, 1.5, 4.7, 1.0, 2.2, 0.9, 4.9, 1.8, 2.8,
0.8, 1.0, 1.0, 0.7, 1.4, 2.4, 1.4, 0.5, 1.5, 2.2,	3.0, 5.9, 7.1, 4.0, 3.8, 6.0, 0.3, 9.1, 4.3, 9.9,
0.3, 2.4, 2.6, 2.3, 0.9, 1.3, 0.6, 0.7, 0.6, 3.9,	1.1, 8.8, 2.7, 3.2, 5.8, 14.3, 0.8, 8.3, 1.6,
1.2, 0.5, 0.8, 0.7, 2.0	9.0, 1.6, 1.7, 8.7, 3.0, 5.8, 5.5, 6.6, 3.9, 0.5

Table 6.2 Data sets for survival times (years) of the Control and Treatment groups.

Data visualization

Beeswarm plot Presented with this data, our first question is, as always: what are the distributions of the two data sets? To answer this, we make dot/beeswarm plots of the data (Figure 6.18).

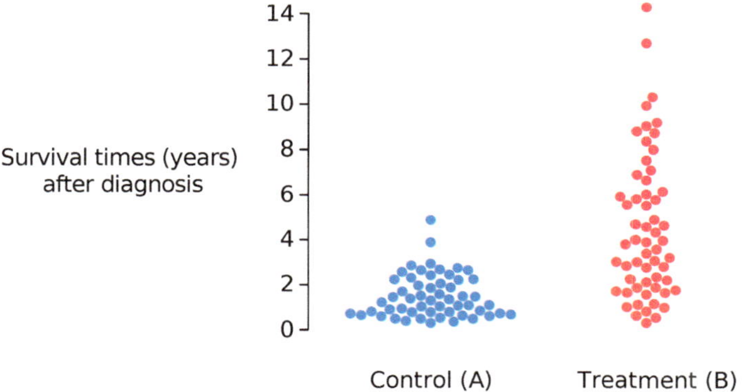

Figure 6.18 Beeswarm plots of the data sets in Table 6.2.

Both data sets are skewed to the right. And clearly, there are many more long-survivors in the Treatment group.

Histogram As a further guide, we decide to make histograms of the two data sets, and compare them visually by superimposing them (Figure 6.19).

The histograms seem to confirm our visual impression that the drug does have an effect.

We also see here that the two distributions are markedly *not* bell-shaped: they are both highly asymmetric and have long tails. So standard formulas that assume bell-shaped distributions can't be applied here.

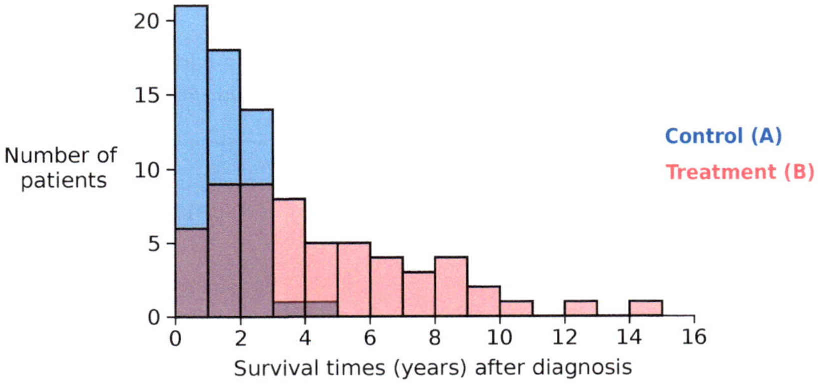

Figure 6.19 Histograms of the cancer drug data sets in Table 6.2.

Choosing a group measure for the two groups

Our next step is to choose a summary measure: we have two sets of survival times, and we have to decide whether we want to use a single number as "typical" and if so, what number that should be. We're going to decide that yes, we are going to use a summary measure (while keeping in mind the large amount of variability obvious in the dot plots). We are going to choose the median survival time as our summary measure, for two reasons:

1. the shapes of the distributions are very much not bell-shaped and have long tails, so the use of the average is a poor choice here, since outliers will pull the average up disproportionately.
2. the median is more interpretable: if I know the median survival, I know that there's a 50% chance of living that long or longer.

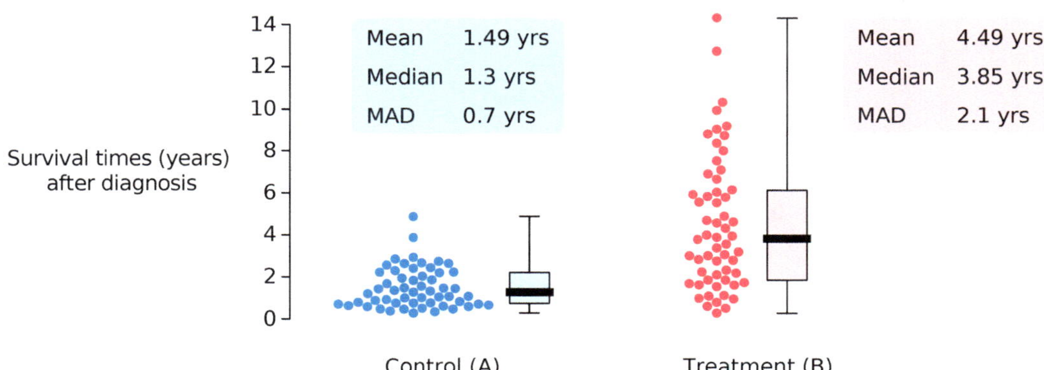

Figure 6.20 Choosing the summary measure for the two groups.

The two medians are:

$$\textbf{control} \quad M(A) = \text{median}(A) = \widetilde{A} = 1.3 \text{ years}$$
$$\textbf{treatment} \quad M(B) = \text{median}(B) = \widetilde{B} = 3.85 \text{ years}$$

Comparing the two groups: what's the difference?

Now that we have M(A) and M(B), we need to decide how to measure the effect size, and we choose the algebraic difference M(B) − M(A), meaning how much longer the median survival time in the Treatment group is relative to the Control group. And so our observed difference between the two groups is

$$\textbf{effect size} \qquad \Delta_{\text{obs}} = \text{M(B)} - \text{M(A)} = 2.55 \text{ years}$$

Re-centered Two-Box resampling: is the observed difference large?

Now we ask whether Δ_{obs} is large, compared to our expectations under the Null Hypothesis.

But what is the Null Hypothesis here? It can no longer be that the two data sets A and B come from the same underlying distribution, because they obviously don't: A and B have markedly different variability, so they are not likely to have come from the same distribution.

Instead, we have to ask a weaker question, namely, is it possible that A and B were drawn from two underlying distributions, with their own different variabilities, *but with the same median values?*

In order to test this weaker question with a resampling procedure, we have to create the world of the Null Hypothesis. In this case, that means that A and B are drawn from two different distributions with the same median values. We can create that dual universe by taking the original distributions A and B, and simply *sliding* them over so that their medians coincide. This is the Re-centered Two-Box model we just studied (Figure 6.14) and we have just re-centered each data set to have the same median.

Re-centered Two-Box resampling Thus we take as our dual universes the original data sets, but with each of their medians subtracted, so that both group medians = 0.

Then we follow the procedure for Re-centered Two-Box resampling. We let A* and B* be the two data sets with their medians subtracted. Then we sample n_A from A*, calling that A_1, and we sample n_B from B*, calling that B_1. We take the median of A_1 and the median of B_1 and calculate their difference, calling that Δ_1.

When we repeated this resampling process 10,000 times, we found that results equally or more extreme than Δ_{obs} or $-\Delta_{\text{obs}}$ never occurred (Figure 6.21). Hence the p-value is less than 0.0001.

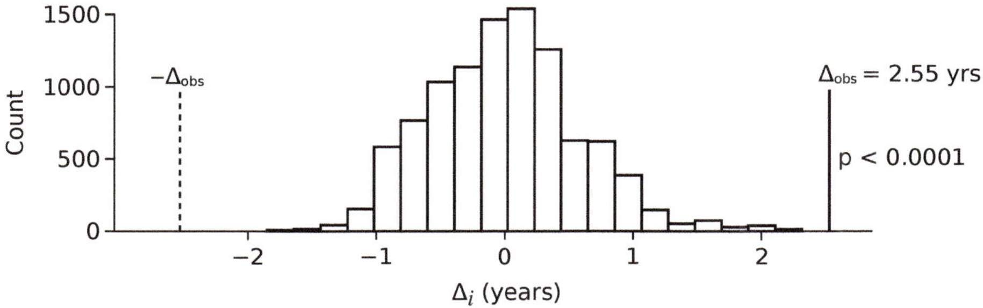

Figure 6.21 p-value calculation for comparing the two groups in Table 6.2, with the difference in survival years between the Treatment and Control (the effect size Δ_{obs}) indicated on the null distribution.

Survival curves

Another useful presentation of the data is to make "survival curves". This form of presentation of survival data is very widely used in medicine, and we include it here because you will be seeing a lot of these!

In Chapter 2, we introduced the idea of cumulative histograms. Given a histogram, its cumulative histogram shows the number of data points less than a given value (page 27). Thus cumulative histograms can never decrease.

A survival curve is just the opposite of a cumulative histogram: it shows the number of survival times *greater than* a given value, whereas a cumulative histogram shows the number of data points *less* than a given value. It is therefore (1 – the cumulative histogram) and can therefore be read as "the fraction of people still alive after time t".

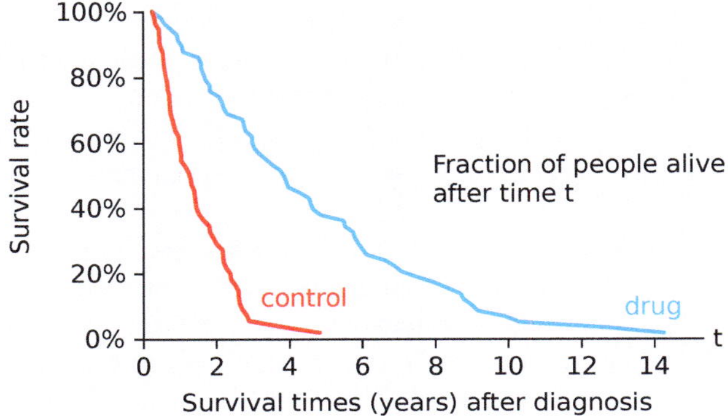

There are a number of special techniques for comparing survival curves. The details are beyond the scope of this text; our purpose here is that you be able to recognize this important type of curve, and know how to look it up for further details.

Understanding resampling and the p-value

In this section, we learned how to compare two groups by resampling from the null world. Each time, we calculated the observed difference Δ_{obs}, and obtained a p-value through a resampling technique. It's important to be able to interpret and visualize the meaning of the p-value in the context of two-group comparisons.

Let's say we are given two samples and want to know whether they are different in their median (Figure 6.22).

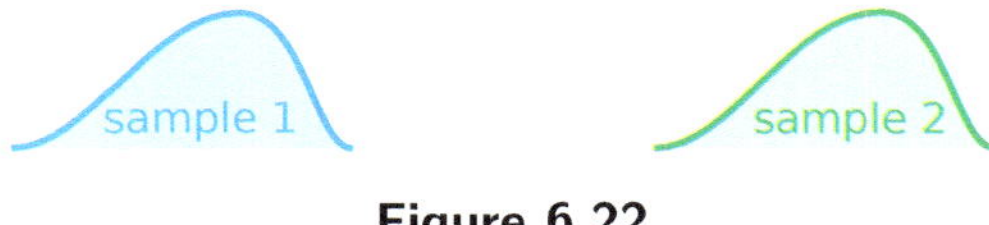

Figure 6.22

If random samples are drawn from populations with the same median, then it is more likely that the two sample medians will be close together. It is less likely, but always possible, that the sample medians will be far apart! (Figure 6.23).

In other words, the more different the sample medians are, the less likely it is that they are drawn from populations with the same median.

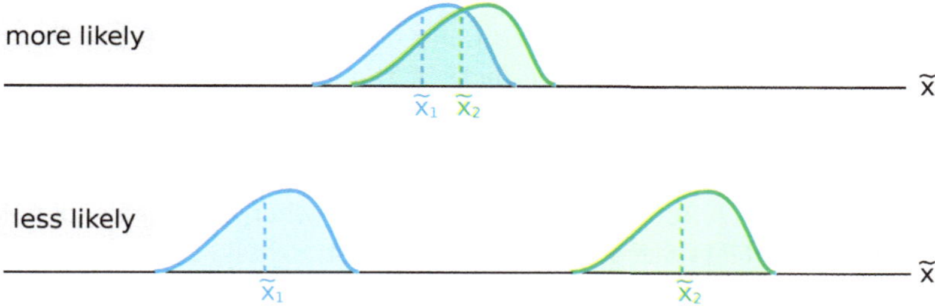

Figure 6.23 As the samples grow farther apart, it becomes less likely for them to have been drawn from populations with the same median.

So for our two samples, we ask, what is the *probability* of getting two sample medians that are at least this different if they were actually drawn from populations with the same median?

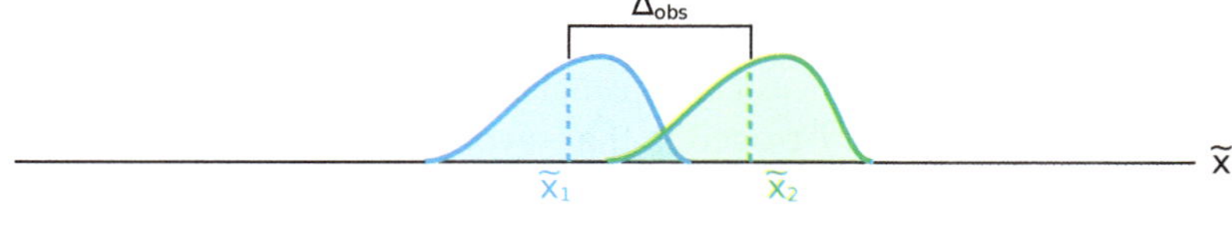

Figure 6.24

And the calculation of the p-value answers that probability question. For example, if a test for the two samples yields $p = 0.02$, that means there is a 2% chance of getting medians that are at least this different, if they were drawn from populations with the same median (Figure 6.25).

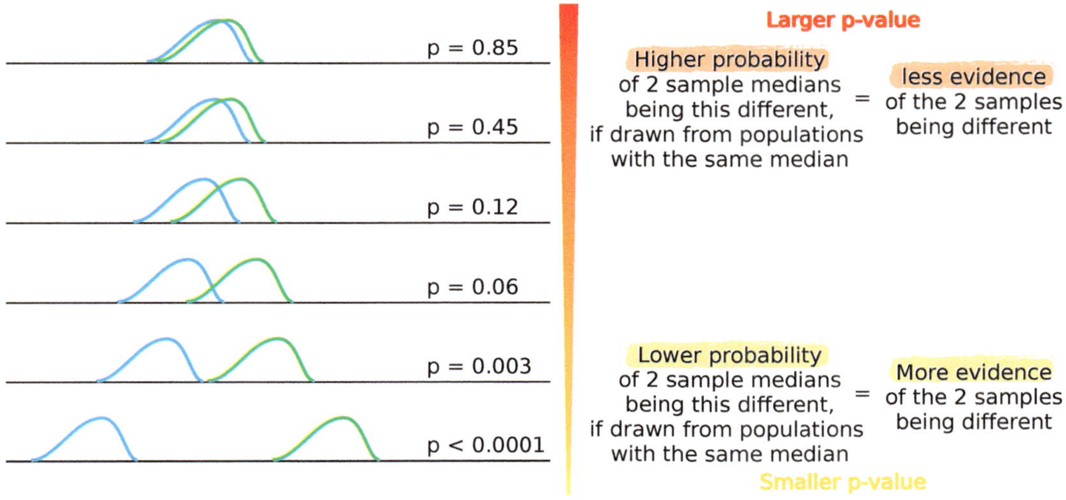

Figure 6.25

FURTHER EXERCISES 6.3

1. "We compared two groups and calculated $p = 0.2$, so there is no difference between the two groups." Is this interpretation correct?

2. "We compared two groups and calculated p = 0.005, so there's a 99.5% chance the groups are different." Is this interpretation correct? If the interpretation is incorrect, state what the interpretation should be. If we were doing a Null Hypothesis Significance Test with an $\alpha = 0.01$, what would the conclusion be?

3. **Blue light and sleep.** A researcher wants to study whether wearing blue-light-blocking glasses before bed affects sleep quality in college students. She recruits two groups of students: one group uses their phones as usual before sleeping, while the other wears blue-light-blocking glasses for the hour before bed. She measures each student's sleep quality using a standardized wristband tracker, and wants to know whether the difference between the median sleep quality scores between the two groups is statistically significant. She learns you're taking a statistics course and asks for your advice.

   a. She has heard that it's better to calculate p-values using simulations (resampling) rather than formulas, but isn't entirely sure why. Walk her through the reasoning.
   b. Using a resampling-based approach, you find a p-value below 0.05. How would you explain to the researcher when it might make sense to use a stricter significance level such as α of 0.01 and what tradeoffs that involves?
   c. The study yields a p-value of 0.026, which seems to support the effectiveness of blue-light-blocking glasses. If you rejected the null hypothesis on this basis, what is the probability that the glasses don't actually improve sleep?
   d. Because the study could contribute to healthy sleep guidance for college students, which is generally more harmful: claiming blue-light-blocking glasses help when they actually don't, or concluding they're ineffective when they actually do help?
   e. The researcher wants more conclusive results. Describe two changes she could make to the study design to strengthen the p-value, and for each, identify a practical tradeoff or challenge.

4. **Limits of resampling.** Below are two data sets.

$$A = \{8, 14, 14, 10, 16, 11, 3, 11, 7, 6\}$$
$$B = \{17, 6, 9, 11, 17, 8, 19, 16, 14, 16\}$$

   a. Calculate the observed difference in the medians of groups A and B.
   b. Formulate an appropriate Null Hypothesis.
   c. Throughout this book, we have calculated p-values by resampling data a "large" number (typically 10,000) of times, and comparing an observed test statistic (e.g., the difference in two medians) to a set of simulated differences. You may have wondered: why 10,000 times?

      To answer this question, we are going to resample the null world at increasing numbers of times, each time calculating a p-value, to explore how the number of resamples affects our calculation of the p-value.

      To start, state a Null Hypothesis that would be appropriate for evaluating the observed test statistic calculated above in (a).
   d. Based on the Null Hypothesis, use an appropriate resampling technique to calculate a p-value for the observed group difference. However, when conducting your simulation, only conduct 10 simulations (instead of the usual 10,000). Store your p-value into a new list.

e. Repeat part (d) an additional 10 times. Each time, increase the number of simulations, as follows: 50, 100, 200, 500, 1K, 5K, 10K, 20K, 50K, 100K. You will now have calculated 11 p-values in total. Store each p-value as the next element in the list created in (d), so that when you are complete, your list of p-values has 11 elements.

f. Create a simple line plot, with the number of simulations on the X-axis, and the p-value on the Y-axis. The plot will demonstrate how the p-value changes as a function of the number of simulations.

g. Comment on the above graph. What is the effect on the p-values as the number of simulations is increased? What do you think would be the calculated p-value with this data, if you had conducted a billion simulations?

h. Now, imagine that instead of increasing the number of simulations, we vastly increased the number of data points. Instead of 10 data points from each group, what if we had 50? We can simulate this by just replicating our data 5 times, such as:

$$A2 = 5 \text{ copies of } A$$
$$B2 = 5 \text{ copies of } B$$

We now have two groups, A2 and B2, each with 50 data points. Calculate the p-value for the difference between these two new groups, using 10,000 simulations.

i. How does the p-value calculated with 50 data points differ from the p-value calculated with 10 data points? What would happen if we used 1000 data points? Describe what is going on here. (This will relate to the concept of statistical power in Chapter 11.)

6.4 Confidence Intervals

As we have been saying, when describing results, giving an effect size and providing an estimate of the precision of that effect size is greatly superior to simply giving p-values, or, even worse, only doing NHST. This is especially true when we are reporting on the "difference" between two groups (Figure 6.26).

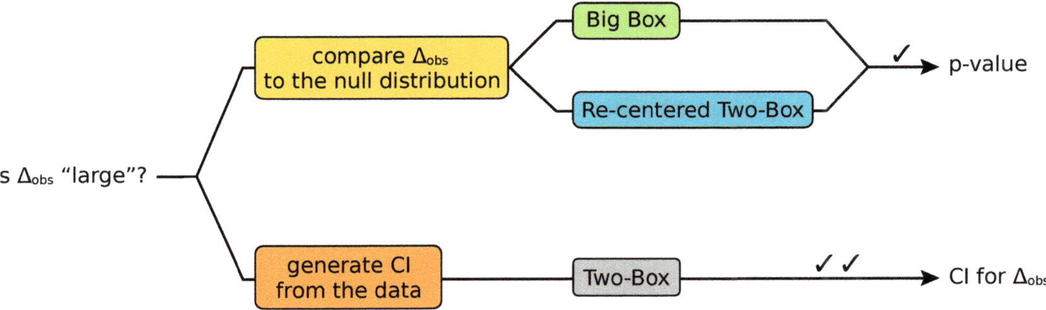

Figure 6.26 Two different approaches for answering the question "is Δ_{obs} large?": p-values and confidence intervals.

We do not want to fall into the trap of just saying "group A values were larger than group B values (p < whatever)." We want to be able to say "The observed difference between group A and group B was Δ_{obs}, with a 99% confidence interval $[\Delta_{low}, \Delta_{high}]$."

The procedure for obtaining a confidence interval for the effect size is straightforward (Figure 6.27).

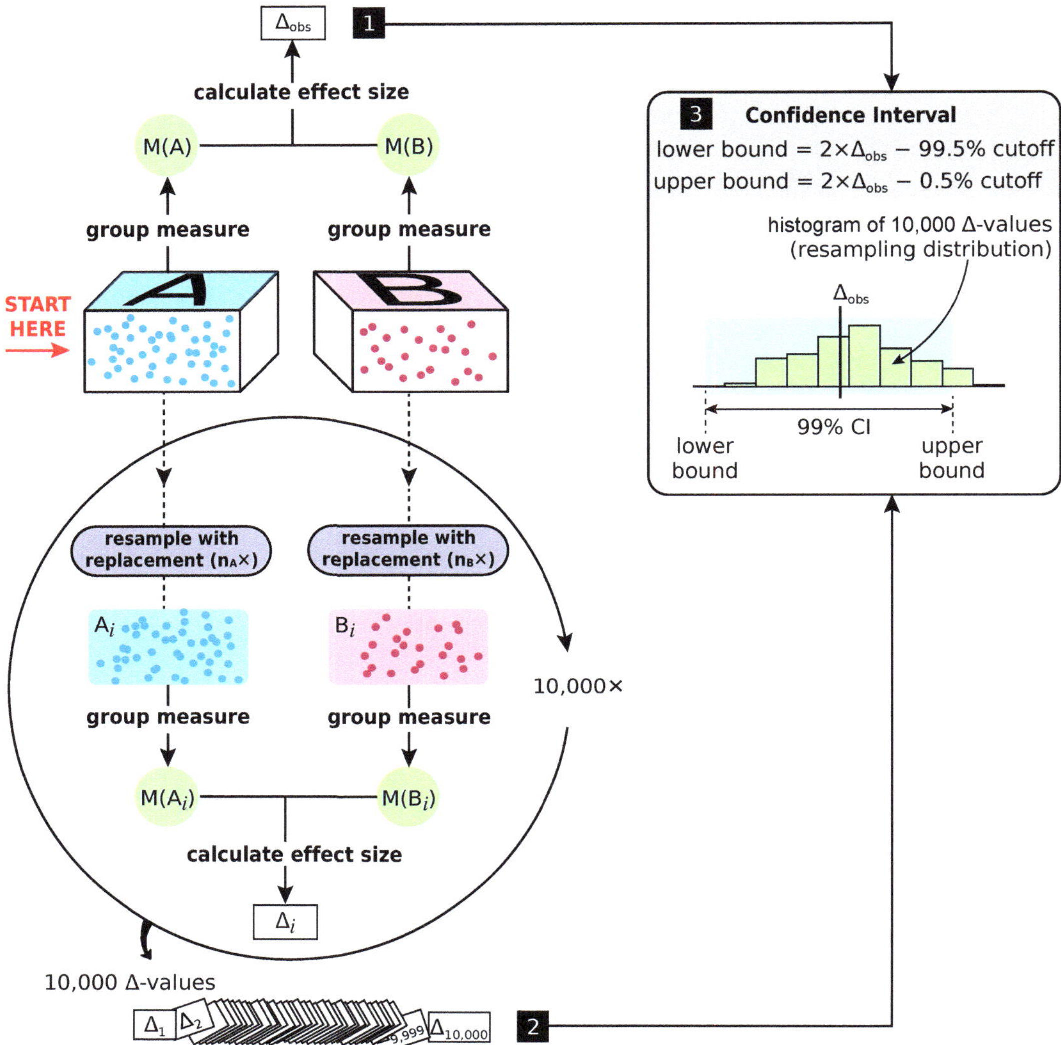

Figure 6.27 Constructing a confidence interval for a group difference.

The Two-Box procedure

We can summarize the Two-Box procedure in three parts.

1. **Analyzing the observed data.** Choose the group measure M. It could be mean or median, or MAD, etc. Then choose the method of group comparison. This could be −, or ÷, etc. Calculate the observed effect size Δ_{obs}.

2. **Resampling each observed group.** If the two group sizes are n_A and n_B, we draw n_A from A with replacement, calling that A_1, and n_B from B with replacement, calling that B_1. We take the group measures of this first resample: $M(A_1)$ and $M(B_1)$. Then we

calculate their difference, and call that Δ_1, as our first estimate of the observed effect size if we were to do this experiment again.

Repeat this process many times, resampling with replacement from the two boxes A and B, generating the differences $\Delta_1, \Delta_2, \ldots, \Delta_{10,000}$. This forms the **resampling distribution**.

3 **Constructing the confidence interval.** After resampling 10,000 times from the original data, collect the 10,000 Δ_i's and rank them from smallest to largest.

We then calculate the confidence interval from the list of Δ_i's following the procedure as outlined in section 4.2 *Constructing the confidence interval* on page 149.

For example, if we are looking for a 99% confidence interval, we rank the 10,000 Δ_i's from smallest to largest. Then we count up 0.5% of the ranked list, or 50 Δ_i's from the smallest, to get $\Delta_{0.5\%}$, and count down 50 from the largest to get $\Delta_{99.5\%}$.

$$\begin{aligned}\text{99\% confidence interval} &= \left[\text{lower bound}_{99\%},\quad \text{upper bound}_{99\%}\right] \\ &= \left[2\cdot\Delta_{\text{obs}} - \Delta_{99.5\%},\quad 2\cdot\Delta_{\text{obs}} - \Delta_{0.5\%}\right]\end{aligned}$$

Confidence Interval for Δ_{obs}

Step 1 ► Generate 10,000 Δ_i values by resampling from each of the two groups.

Step 2 ► Rank the 10,000 Δ_i from smallest to largest.

Step 3 ► Find the cutoff for the smallest 0.5%, call it $\Delta_{0.5\%}$.

Step 4 ► Find the cutoff for the largest 0.5%, call it $\Delta_{99.5\%}$.

Step 5 ► Construct the confidence interval as $[2 \times \Delta_{\text{obs}} - \Delta_{99.5\%},\ 2 \times \Delta_{\text{obs}} - \Delta_{0.5\%}]$.

Confidence interval for the T-cell experiment

We calculated the confidence interval for the treatment effect in the T-cell experiment (data presented in Table 6.1). The presentation of the result in words is "The Treatment group had higher median T-cell counts/unit area; the difference in median counts was 9.7 (99% confidence interval [7, 13])." The graphical presentation of the resulting CI is shown in Figure 6.28.

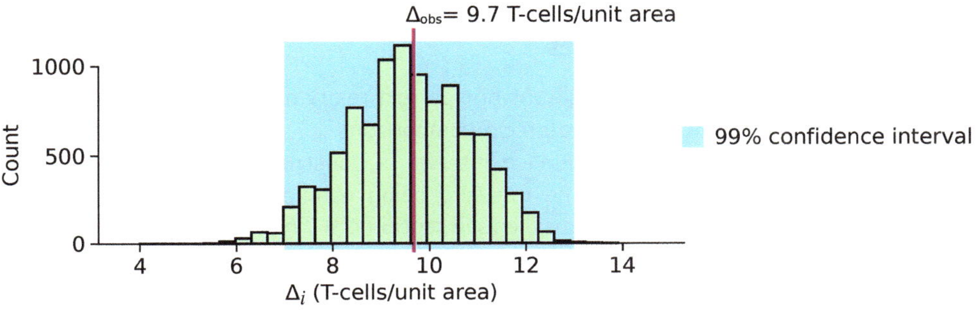

Figure 6.28 99% confidence interval for the difference in median T-cell counts of the Treatment relative to the Control group.

Exercise 6.4.1 We used the Big Box method to calculate the p-value for the observed difference in the median T-cell counts between the Control and Treatment groups (Figure 6.12). Which method should we use to calculate the 99% confidence interval for the effect size?

a) 1 Big Box
b) 2 boxes, randomly assigned
c) 2 boxes, re-centered
d) 2 boxes of original data

Confidence interval for the cancer drug data

We also calculated the confidence interval for the cancer drug data (data presented in Table 6.2). The presentation of the result in words is "The Treatment group had longer median survival times, with a difference in median survival time of 2.55 years (99% confidence interval [0.74, 3.8])". The graphical presentation is shown in Figure 6.29.

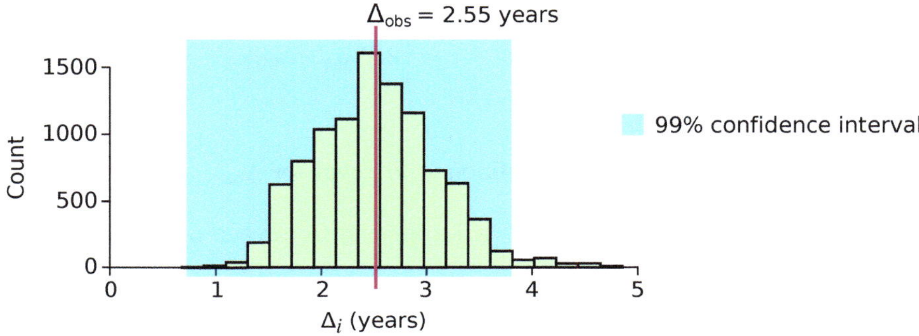

Figure 6.29 99% confidence interval for the improvement in survival times of the Treatment group relative to the Control group.

Exercise 6.4.2 In the calculation of the p-value for the difference observed in the median survival time of patients in the two groups, we used the Re-centered Two-Box method (Figure 6.21). Which method should we use to calculate the 99% confidence interval for the effect size?

a) 1 Big Box
b) 2 boxes, randomly assigned
c) 2 boxes, re-centered
d) 2 boxes of original data

Confidence intervals and p-values

In Chapter 4, we talked about the close-but-not-perfect relationship between confidence intervals and p-values. This is also true for two-group comparisons.

We said that when a 99% confidence interval is constructed, all values in that interval are considered plausible values for the measure being estimated, M_{true}. Values outside the interval are rejected as relatively implausible (see section 4.9 *The interpretation of the confidence interval*).

Similarly, when we have two groups and we make a confidence interval for the group difference, then values outside that confidence interval can be rejected as relatively implausible. **In particular, if the null value is zero and the confidence interval does not contain zero, then zero is not a plausible value, that is, the Null Hypothesis can be rejected. If the confidence interval** *does* contain zero, this in itself gives no evidence to reject the Null Hypothesis. However, a p-value calculation may still reject the Null Hypothesis.

Confidence intervals and p-values

If we calculate confidence intervals for the difference between two groups, we can generally predict statistical significance.

For example, below are 99% confidence intervals generated from three different studies. All studies have a null value of 0. The confidence interval in study 1 (top) *clearly* includes zero, therefore, we can conclude that the result is not statistically significant. The confidence interval in the second study (middle) does not contain zero, therefore the result is statistically significant at the $p < 0.01$ level. The third study (bottom) *barely* contains zero, and it therefore is *probably* not statistically significant, but independent p-value testing must be done in this borderline case, and additional care must be taken to avoid p-hacking.

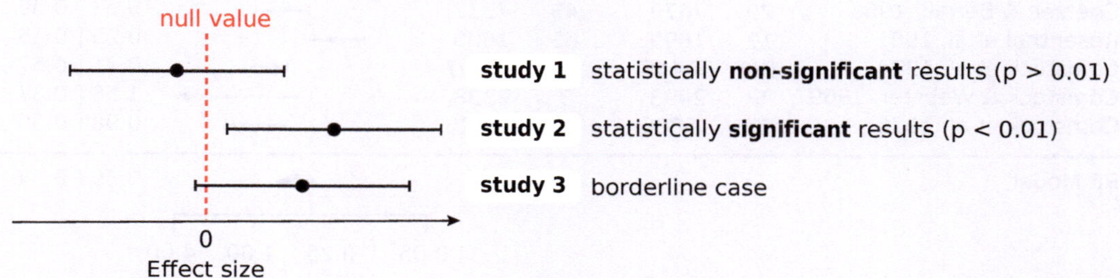

Overall,

- Given a confidence interval, we can often read off the p-value within the precision of the confidence interval, when the CI clearly excludes the null value.
- The other advantage of confidence intervals is that we can see how far the confidence interval is from the null value. Is this a borderline case or an open-and-shut case?
- Another feature we can read off confidence intervals is the quality of the study. Is this a good study or a poor study? We can judge it from the tightness of confidence intervals. When we have a narrow confidence interval, that means a small margin of error. If that margin of error is wide, it is a low quality study, and the precision of the study's estimate is poor.
- On the other hand, knowing the p-value tells us nothing about the effect size, and little about the confidence interval.

Exercise 6.4.3 We calculated the 99% CI for the difference in median survival time (years) under two different treatments to be [0.74, 3.8]. What can we say about the relevant p-value?

a) The p-value will be larger than 0.99.
b) The p-value will be larger than 0.05.
c) The p-value will be larger than 0.01.
d) The p-value will be smaller than 0.01.
e) We can't say anything about the p-value until we run NHST.

Forest plots There is an important subject called **meta-analysis**. In meta-analysis, a research group reviews the literature on a given subject, and presents a summary of all the findings. So these are important papers, and are usually highly cited.

For example, here is a survey of 13 published articles on the effectiveness of Tuberculosis vaccine (Figure 6.30).

Author(s) and Year	Vaccinated TB+	Vaccinated TB−	Control TB+	Control TB−	Relative Risk [95% CI]
Aronson, 1948	4	119	11	128	0.41 [0.13 , 1.26]
Ferguson & Simes, 1949	6	300	29	274	0.20 [0.09 , 0.49]
Rosenthal et al, 1960	3	228	11	209	0.26 [0.07 , 0.92]
Hart & Sutherland, 1977	62	13536	248	12619	0.24 [0.18 , 0.31]
Frimodt-Moller et al, 1973	33	5036	47	5761	0.80 [0.52 , 1.25]
Stein & Aronson, 1953	180	1361	372	1079	0.46 [0.39 , 0.54]
Vandiviere et al, 1973	8	2537	10	619	0.20 [0.08 , 0.50]
TPT Madras, 1980	505	87886	499	87892	1.01 [0.89 , 1.14]
Coetzee & Berjak, 1968	29	7470	45	7232	0.63 [0.39 , 1.00]
Rosenthal et al, 1961	17	1699	65	1600	0.25 [0.15 , 0.43]
Comstock et al, 1974	186	50448	141	27197	0.71 [0.57 , 0.89]
Comstock & Webster, 1969	5	2493	3	2338	1.56 [0.37 , 6.53]
Comstock et al, 1976	27	16886	29	17825	0.98 [0.58 , 1.66]
RE Model					0.49 [0.34 , 0.70]

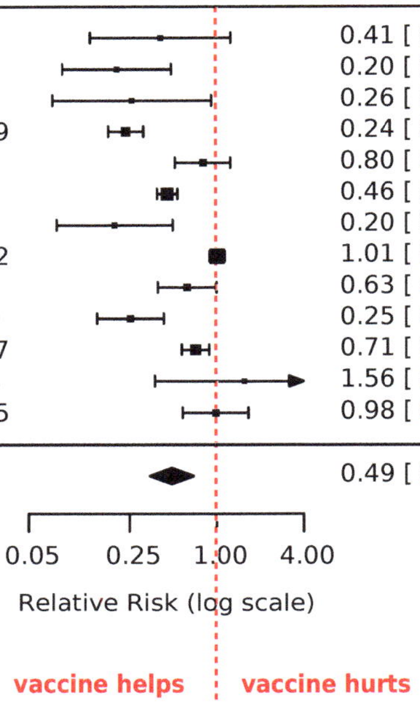

0.05 0.25 1.00 4.00

Relative Risk (log scale)

vaccine helps | vaccine hurts

Figure 6.30 A forest plot summarizes 13 studies of the effectiveness of Tuberculosis vaccine in the Vaccinated versus the Control group, with corresponding 95% confidence intervals for the individual studies (Viechtbauer, 2010).

Here, the effect size is the relative risk for the vaccine, which is the disease rate in the vaccinated divided by the disease rate in the unvaccinated. It is a ratio. (We'll spend more time and dive deeper on the concept of relative risk in section 6.8 *When the effect size is a ratio* and section 8.8 *Relative Risk: an effect size.*)

$$\text{Relative Risk} = \frac{\text{disease rate in vaccinated}}{\text{disease rate in unvaccinated}}$$

When we use relative risk as the effect size, the Null Hypothesis value is 1. If the vaccine has no effect, then the disease rate in the vaccinated will be the same as the disease rate in the unvaccinated. So RR = 1 means "no effect".

If we plot the relative risk for the vaccine that each of these studies found, we see that almost all of them found a number strictly lower than 1. This suggests that the vaccine has some positive effect, because lower than 1 means the vaccine helps, and greater than 1 means the vaccine hurts. This presentation of confidence intervals is called a **forest plot**. The forest plot lists all the results that these studies found together with their confidence intervals.

If we examine the results, we see several well-done studies (tight confidence intervals) indicating that the vaccine helps, such as Hart & Sutherland, Stein & Aronson and Comstock *et al.*, 1974.

If we look at the two studies that found RR-values greater than 1, we see that Comstock & Webster, which found RR = 1.56, had a very wide 95% CI including many values below 1. Wide confidence intervals are indicative of poor studies, typically caused by small sample sizes. TPT Madras, on the other hand, had a very tight 95% CI, indicating a large sample size (note n ≈ 176,000). Their observed RR was 1.01, which is not different from the null value.

There are also several studies that found a positive effect (RR < 1), but can't rule out the Null.

Overall, this is an effective way of summarizing all of these studies. This graphic tells it all, it tells us everything we want to know: here are all the studies, their findings, their confidence intervals, and the quality of each study.

In addition to meta-analysis, forest plots can also be used to summarize a series of similar findings in a single study. Prof. Morgan Tingley (UCLA) was investigating microbiota, the gut microbiome, in birds, by taking fecal samples and looking at them using molecular biology techniques. His question was whether the microbiome differs between high elevation and low elevation birds. Here is his answer (Figure 6.31).

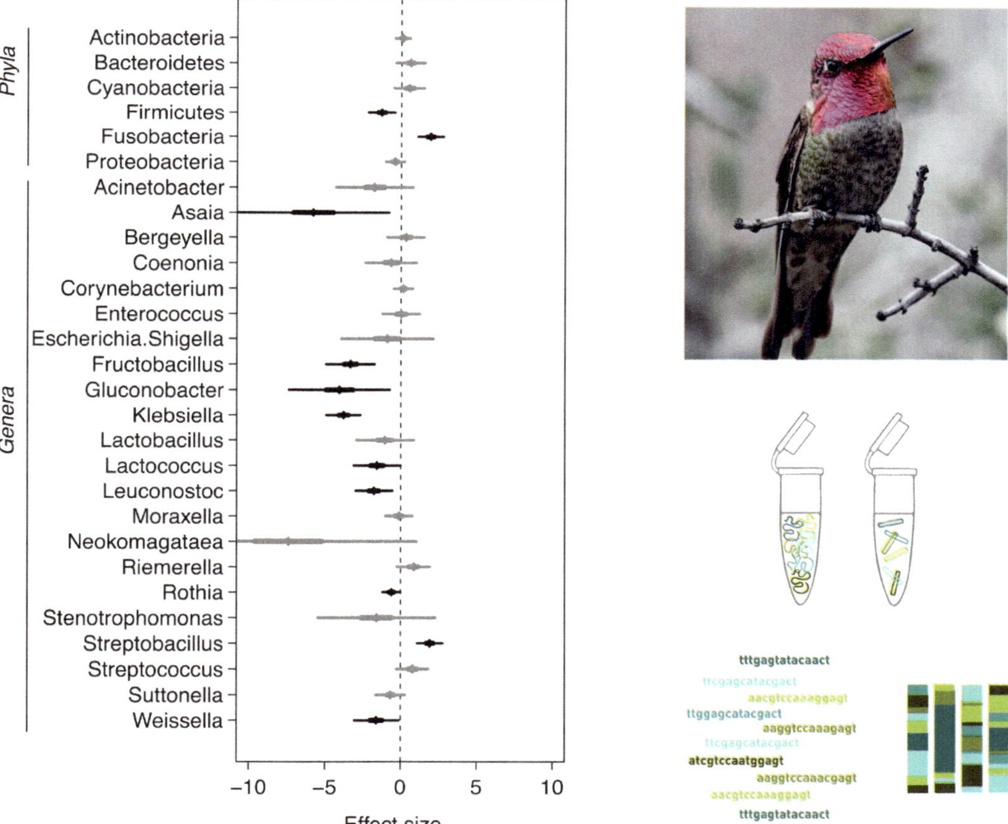

Figure 6.31 A forest plot summarizes the differences in bacterial species between high and low elevation birds. The central diamond is the mean estimate (high – low), the thicker line shows the 50% confidence interval, and the thinner line shows the 99% confidence interval. Black indicates a significant difference, while gray denotes non-significant differences (Herder *et al.*, 2021).

Note that this is not a meta-analysis, this is a review of every bacterial species that is found in the gut of the birds. The dashed line indicates a difference of zero (the null value). The grayed-out cases are bacteria for which the confidence interval includes zero, and the black are bacteria for which the confidence interval does not include zero, therefore indicating a significant difference between high elevation and low elevation.

Looking at the forest plot, we can conclude that many species of microbiota do not differ in their abundance between high and low elevations. But for some, there are significant decreases in the high elevation birds (black confidence intervals to the left of the dotted line), while for 2 others, there are significant increases.

Comparing two groups using confidence intervals for group measures

Another strategy for comparing two groups is to look at confidence intervals for the two group measures, to see if they overlap. The method for constructing the confidence interval for one group on one measure was laid out in section 4.2 *Constructing the confidence interval*.

If the two 99% confidence intervals do not overlap, we can conclude that the two groups are statistically significantly different at the $p < 0.01$ level, but the converse is not true: if the two confidence intervals overlap substantially, it is unlikely that the groups are significantly different; if the two confidence intervals overlap slightly, it is possible that the test for the difference between the two groups may still produce $p < 0.01$.

FURTHER EXERCISES 6.4

1. **Coral reefs.** Marine biologists are investigating whether establishing protected zones helps coral reef recovery. They measure coral coverage (%) at 40 randomly selected sites: 20 within a marine protected area and 20 in nearby unprotected waters off the coast of Australia's Great Barrier Reef. Below are the 99% and 95% confidence intervals for the difference in mean coral coverage between the two groups (H_0: there is no difference in coral coverage by protection status).

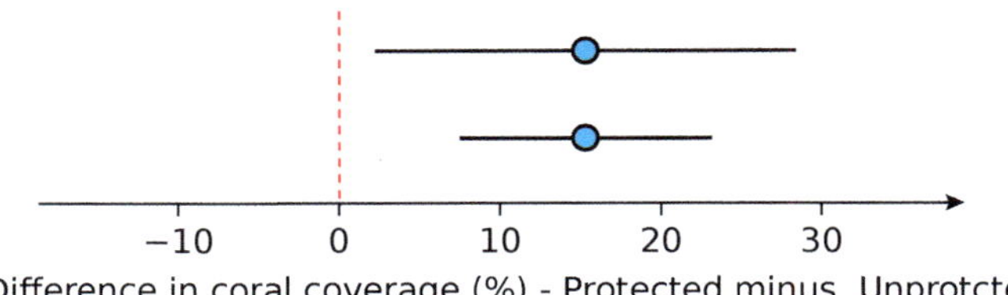

   a. Which of the two confidence intervals is the 99% CI, and how can you tell just by looking at the plot?

      a) Top b) Bottom

   b. Based on the confidence intervals alone, what can you conclude about whether marine protected areas make a difference for coral coverage?

   c. What is your best estimate of the p-value based on this information?

      a) $p > 0.05$ b) $0.01 < p < 0.05$
      c) $p < 0.01$ d) We can't tell

2. In order to use resampling to calculate confidence intervals for an observed effect size, which of the following statements are correct?

    a) Both groups need to be Normally distributed (bell-shaped) with same variance.
    b) The two groups need to have the same sample size.
    c) The parameter of interest should be the difference between the means of the two groups.
    d) Both groups need to be representative samples of their population.
    e) The resampling-based approach to calculate confidence intervals, unlike the formula-based approach, is applicable to all measures and all distributions.

3. For two-group comparisons (A vs. B), we always sample from the two boxes (A and B) without re-centering, to compute confidence intervals for effect sizes, such as the difference between group medians. Why can't we use Big Box sampling for confidence intervals as we sometimes do for NHST?

4. **Frogs.** In a study of frogs from Northern and Southern California, a 99% confidence interval for the difference in the median jump distance is [−0.4 m, 0.2 m].

    a. From this, what do we know about Northern and Southern California frogs?
    b. State an appropriate Null Hypothesis.
    c. Based on this confidence interval, should we reject the Null Hypothesis?

5. **Glucose levels.** A researcher wants to determine the effect of a drug on blood glucose levels in diabetes patients, so she measures the blood glucose levels of two groups of patients, one taking the drug and the other taking a placebo. The two data sets are somewhat skewed. In lieu of "doing a t-test", you convince her that it would be better to construct a 95% confidence interval on an effect size, namely, the difference in the median blood glucose levels of the two groups.

    a. Give three facts you would use to convince her.
    b. She's heard that it's better to calculate confidence intervals with simulations instead of using formulas, but isn't quite sure why. State two reasons why.
    c. Succinctly describe the resampling process for constructing a 95% confidence interval for the difference between medians. Include the box model(s), sample size(s), whether to sample with or without replacement, and how to calculate the confidence interval.
    d. The 99% confidence interval for the effect size of the drug is [−30, 5] mg/dL. What does this tell you about the true value of the effect size? Is there likely to be a statistically significant difference between the two groups?
    e. Another member of the lab did their own analysis and summarized their result by saying "We compared the two groups and calculated p = 0.005, so there's a 99.5% chance the groups are different." Is this interpretation of the p-value correct? If the interpretation is incorrect, state what the interpretation of the p-value should instead be. If we were doing a Null Hypothesis significance test with $\alpha = 0.01$, what would the conclusion be?
    f. Compare the 99% confidence interval and the p-value from NHST. What's the conclusion?

6. **Evaluating sentencing reform.** A state reforms its criminal sentencing law, replacing life without parole (LWOP) for certain offenses with parole eligibility after 25 years. Researchers analyze sentences before vs. after reform using the same data, summarized two ways:

   - 95% CI for difference in means (After – Before): [−18, −10] years
   - 95% CI for difference in medians (After – Before): [−1, 1] years

   a. Which CI suggests the reform clearly reduced sentences?

   b. What does the median CI say about the typical case?

   c. How can the mean sentence drop by many years while the median barely changes?

   d. If your priority is reducing prison overcrowding, which CI is more relevant?

   e. If your priority is to affect the sentences of most offenders, which CI is more relevant?

   f. Is it misleading to summarize the reform using only the mean CI?

7. **Cloud seeding.** In an experiment to determine whether seeding clouds with silver iodide increases rainfall, 52 isolated clouds were randomly assigned to be seeded or not. The amount of rain they generated was then measured.

Amount of rain	
Seeded	Unseeded
39.8, 27.7, 10.6, 37.1, 83.5, 19.6, 30.9, 93, 25.2, 94.1, 88.4, 54.5, 42.1, 194, 896.5, 143.3, 13.5, 4.2, 196.6, 445.3, 1626.9, 7.1, 441.2, 500.3, 0.7, 20.2	353.7, 84.3, 1634.2, 223.7, 201.7, 2770.7, 526.1, 283.3, 157, 725.2, 103.5, 30.4, 342.3, 92.8, 37.8, 164.5, 4.4, 78.4, 16.8, 316.3, 21.3, 1265.4, 2163.7, 188.6, 281.7, 6.3

   a. Visualize the data in both groups using an appropriate plot. Explain why you chose that plot.

   b. After data visualization, what measure would you choose to describe these two groups? Why?

   c. What type of two-group comparison (box model) will you use? List your reason(s) for choosing this method.

   d. Compute the p-value. Does silver iodide make a statistically significant difference in the amount of rain?

   e. Calculate the 95% confidence interval for the effect size of silver iodide seeding. What's your interpretation of the confidence interval?

   f. Based on the 99% confidence interval and the p-value from NHST, what's the conclusion?

6.5 Presenting the result

We can present the result by briefly stating, in the Methods section, how the p-value and the confidence interval were calculated. In the Results section, state the observed effect size, its p-value and confidence interval.

When it comes to writing up the result, here's how to say it:

The T-cell experiment

T-cell stimulation properties of an experimental treatment

Methods

Statistical comparisons Because the two groups were not Normally distributed, we used resampling methods to test the Null Hypothesis that there is no difference between the treated and untreated groups in their median T-cell counts. Briefly, we conjoined the two data sets into a single set, and resampled (with replacement) "Control" and "Treatment" groups from this set, and calculated the difference between their median T-cell counts. This was repeated 10,000 times, and the p-value of the observed result was calculated as the number of simulations producing results as extreme or more extreme than the observed result, divided by 10,000.

We also calculated a 99% confidence interval for the observed group difference, Δ_{obs}, by resampling a new "Control" group from the Control group, a new "Treatment" group from the Treatment group, and taking the difference in their medians. This was repeated 10,000 times, and the 99% confidence interval was calculated as $[2{\cdot}\Delta_{obs} - \Delta_U, 2{\cdot}\Delta_{obs} - \Delta_L]$, where Δ_U and Δ_L are the cutoffs for the 0.5^{th} and 99.5^{th} percentile of the resampled effect sizes.

Results

The treatment increased median T-cell counts by 9.7 T-cells/unit area, $p < 0.0001$, 99% confidence interval [7, 13].

The cancer drug study

Survival times under an experimental drug

Methods

Statistical comparisons Because the Control and Treatment groups were not Normally distributed, and had markedly different MADs, we used resampling methods to test the Null Hypothesis that the two groups have the same median values.

By subtracting the median values from each of the two groups, we created two data sets with the same median values and the same variabilities as the actual control and treated groups. We then resampled separately from these two data sets to estimate how large a difference we would expect under the Null Hypothesis. This was repeated 10,000 times, and the p-value of the result was calculated as the number of simulations producing results as extreme or more than ours, divided by 10,000.

We also calculated a 99% confidence interval for the observed group difference Δ_{obs}, by resampling a new "Control" group from the Control group, a new "Treatment" group from the Treatment group, and taking the difference in their medians. This was repeated 10,000 times, and the 99% confidence interval was calculated as $[2{\cdot}\Delta_{obs} - \Delta_U, 2{\cdot}\Delta_{obs} - \Delta_L]$, where Δ_U and Δ_L are the cutoffs for the 0.5^{th} and 99.5^{th} percentile of the resampled effect sizes.

Results

The treatment increased median survival times by 2.55 years (99% confidence interval [0.74, 3.8]). This increase was statistically significant at the $p < 0.0001$ level.

FURTHER EXERCISES 6.5

1. Given two sets of data to compare, what is the first thing to do?

   a) Perform a *t*-test
   b) Calculate the means of the two groups
   c) Visualize the two groups
   d) Calculate the variance of the two groups

2. **Metabolic rate of Northern Fulmars.** The metabolic rate for 6 female and 8 male breeding Northern fulmars was measured.

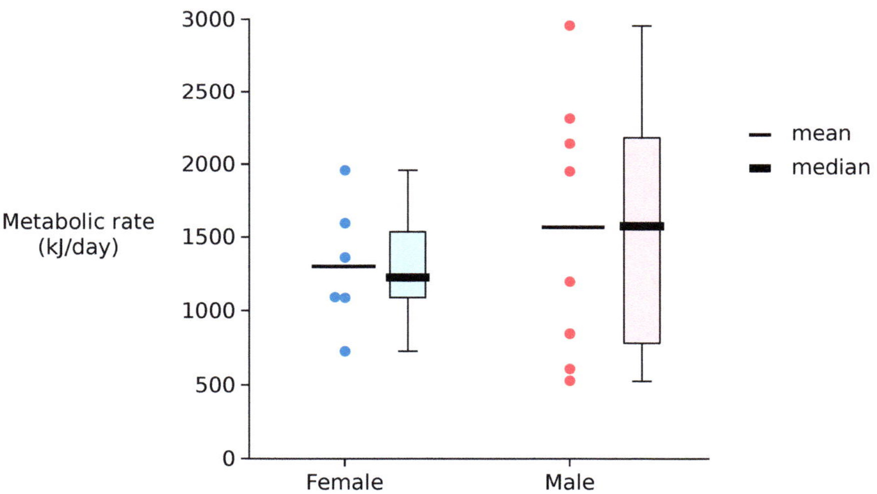

   a. Are these two data sets likely to have come from the same population?
   b. What is an appropriate Null Hypothesis?
   c. What type of two-group comparison should be used? List your reasons(s) for choosing this method.
   d. Using the two-group comparison method you chose above, write out the pseudo-code for how to conduct the resampling procedure.

3. **New medicine for the common cold.** You are a researcher hoping to develop a medication that will shorten the length of time that patients are symptomatic after contracting the common cold. In testing your medication, you have a Control group who take a placebo pill and a Treatment group who take your new medication. Below is the data from your experiment.

Days with symptoms after contracting the common cold	
Control group (A) $n_A = 20$	Treatment group (B) $n_B = 20$
7, 6, 4, 5, 6, 4, 3, 7, 10, 8, 5, 6, 7, 6, 10, 7, 6, 5, 4, 5	4, 5, 4, 7, 2, 7, 2, 5, 5, 5, 4, 3, 7, 8, 2, 2, 3, 2, 3, 1

   a. How would you present this data?
      a) Histograms
      b) Dot plots
      c) Box plots
      d) Beeswarm plots
      e) KDE plots
   b. After data visualization, what measure would you choose to describe these two groups? Why?
   c. What type of two-group comparison should be used in this situation? List your reason(s) for choosing this method.
   d. Using the two-group comparison method you chose above, write out the pseudo-code for how you will conduct the resampling procedure.
   e. Compute a p-value. Does your medication make a statistically significant difference?

4. **Wetlands and plant species richness.** Freshwater wetlands are important ecosystems for many plants, invertebrates, and vertebrate animals. They can either be permanent or ephemeral (i.e., wet for part of the year and dry for part of the year). Scientists collected data for randomly selected wetlands in a preserve in Wisconsin. Determine whether plant species richness (the number of species) differs between permanent and ephemeral wetlands in this preserve. *(Study and data from Amanda Little, University of Wisconsin - Stout.)*

Plant species richness	
Permanent	Ephemeral
42, 31, 34, 37, 18, 14, 34, 45, 47, 44, 37, 46, 41, 40, 15, 31, 43, 37, 25, 28, 29, 23, 14, 53	18, 24, 35, 37, 51, 32, 22, 36, 34, 32, 25, 47, 25, 15, 47, 17, 45, 33, 42, 49, 13, 51, 28, 24, 13, 22, 20, 33, 19, 30, 18, 24, 44

   a. Visualize the data in both groups using an appropriate plot. Explain why you chose that plot.
   b. Calculate the appropriate measures of central tendency and variability for each group. Explain why you chose those measures.

c. For each data set, use simulation to calculate the 99% confidence interval for your chosen measure of central tendency. For each group (permanent and ephemeral), include the resampling distribution with the observed effect size and confidence interval indicated.

d. Before you start coding to determine whether richness differs between the two groups, plan your analysis.
   - Summarize your goal/output.
   - Determine what measure and metric of comparison to use for this study.
   - State the Null Hypothesis of the study.
   - Describe your box model(s) for Null Hypothesis Significance Testing and confidence interval for the effect size.
   - Define the sample size for this study.

e. Compare the two groups and report the p-value. Use comments to explain briefly what each line of your code does. Include the null distribution and indicate the observed effect size.

f. Construct a 99% confidence interval for the observed effect size. Use comments to explain briefly what each line of your code does. Include the resampling distribution and indicate the observed effect size and the limits of the confidence interval.

g. What do you conclude from your analysis? Write a paragraph summarizing your conclusions, including any numbers and references to figures that should be reported and their interpretation.

5. **Pepto-Bismol and infant diarrhea.** Determine whether administering bismuth subsalicylate to infants with diarrhea affects stool output, according to the data set from FE 4.5.1 on page 163.

   (Hint: Previously, you already did some steps, including data visualization, choosing an appropriate measure and calculating 99% confidence intervals for each group. Now review those steps and carry out the necessary steps to determine whether the stool output from the bismuth subsalicylate group is statistically significantly different from the placebo group.)

6.6 Ranks

We know that the type of test we can use depends on the distribution of the data. But if the data sets are very small (say, $n \leq 10$), it is impossible to guess what the underlying distribution might have been.

When we don't know the underlying distribution, we don't know the variability or the potential range of the data. So we can say that 31 is 10 times 3.1, but if we don't know the range of the data, we don't know if 10 times is a big difference. For example, if the range of the data is $\sim 10^{12}$, then 31 isn't much different from 3.1 (Figure 6.32).

The alternative approach in this low-n situation is to consider not the data values, but only their relative *ranks*.

The idea is that when we have a very small n, we can't really trust the raw numbers; all we can trust is their relative position to one another, that is, their relative ranks. **If the data set is**

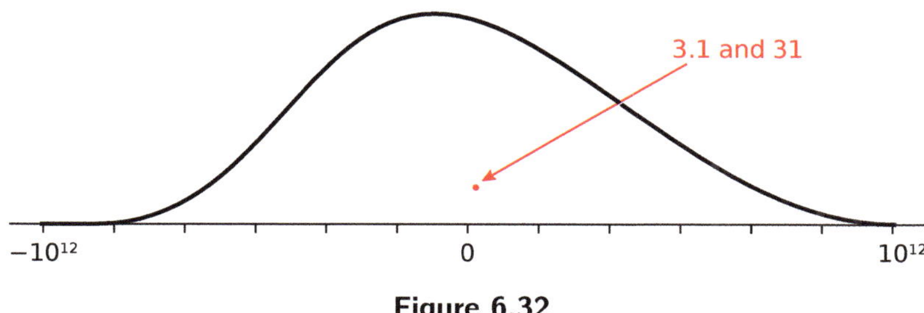

Figure 6.32

too small to say whether or not it is Normally distributed, we can't use methods that depend on the assumption of a Normal distribution.

In these cases, in which we can only trust the relative positions of one value to another, we replace the actual data values by their relative ranks.

Exercise 6.6.1 You have two data sets: {3, 10, 7, 4, 16} and {31, 3, 9, 40}. How would you describe the distributions of the populations these data sets represent?

a) Similar shape and similar variability
b) Similar shape, but different variability
c) Similar variability, but different shape
d) Different shape and different variability
e) I have no idea

Comparing two groups: the effect size

We can then compare two small groups A and B by putting all of their elements into one list, ranking them, and asking: does one of these groups have more highly ranked elements than the other? **If one of the two sets has more higher-ranked elements than the other, that is some reason to say that the two sets are different.**

To compare the two groups using ranks, our data takes the form of two sets of ranks, the ranks in group A and the ranks in group B. Our group measure will be the average rank for each group, and the overall effect size is $\overline{B_{rank}} - \overline{A_{rank}}$.

Figure 6.33 shows an example of calculating the effect size between two groups using ranks. In this example, the two groups we want to compare are

$$A = \{16, 10, 3, 7, 4\} \qquad B = \{9, 31, 40, 3\}$$

To calculate the effect size of the two groups using ranks, we put A and B into a single list, and sort and rank the combined list. Then we replace each element of each group by its corresponding rank in the combined ranked list.

$$A_{rank} = \{1.5, 3, 4, 6, 7\} \qquad B_{rank} = \{1.5, 5, 8, 9\}$$

We then use the average rank as the summary measure for each group, which gives us

$$\overline{A_{rank}} = \frac{1.5+3+4+6+7}{5} = 4.3 \qquad \overline{B_{rank}} = \frac{1.5+5+8+9}{4} = 5.875$$

The effect size is the difference between the two group measures,

$$\Delta_{obs} = \overline{B_{rank}} - \overline{A_{rank}} = 5.875 - 4.3 = 1.575$$

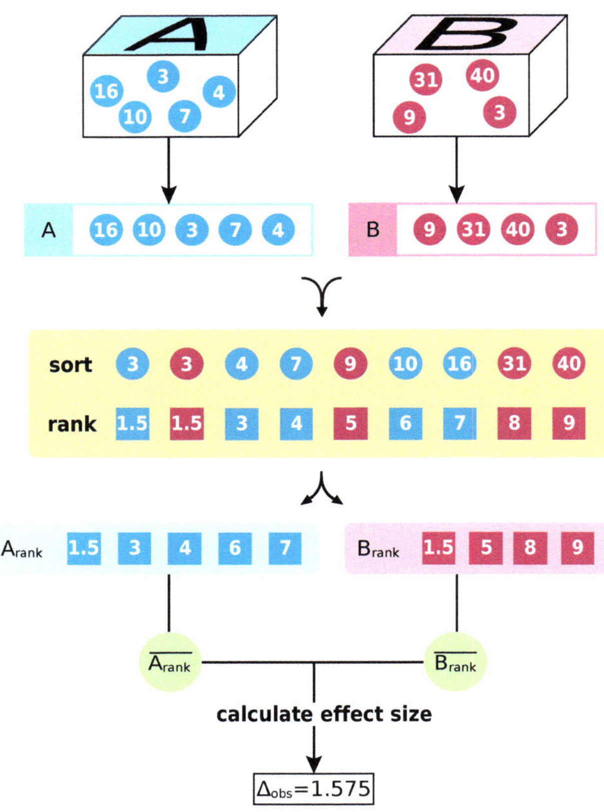

Figure 6.33 An example illustrating the procedure of calculating the effect size between two groups using ranks.

effect size (ranks)

Step 1 ► Let the two data sets be $A = \{a_1, a_2, \ldots a_{n_A}\}$ and $B = \{b_1, b_2, \ldots b_{n_B}\}$.

Step 2 ► Combine both data sets into a single list. Then rank the list from smallest to largest. In the case of ties, assign the average rank, so, for example, if there are two values tied for rank #10, they are both assigned rank #10.5.

Step 3 ► We now replace each element of both data sets by their corresponding ranks in the combined ranked list, giving us two new data sets, A_{rank} and B_{rank}.

Step 4 ► We can then use the average rank as our summary measure for each data set, which gives us $\overline{A_{rank}}$ and $\overline{B_{rank}}$.

Step 5 ► Calculate the effect size as the difference between the average ranks of the two groups A and B, call it the observed difference Δ_{obs}.

$$\Delta_{obs} = \overline{B_{rank}} - \overline{A_{rank}}$$

Determining statistical significance

We now know how to calculate Δ_{obs}, the observed effect size for two-group comparisons using ranks (part **1** in Figure 6.34). Then the next question is: is Δ_{obs} statistically significantly large, by which we mean: is Δ_{obs} more extreme than we would expect under the Null Hypothesis of no difference between the groups?

To form the distribution of expectations for the effect size Δ under the Null Hypothesis, we use resampling.

Null Hypothesis We put the two groups A and B into a Big Box. This is the world of the Null Hypothesis, that there is no difference between the two groups.

Resample We resample from the Big Box to generate many estimates of Δ under the Null Hypothesis. Here is the procedure to generate one estimate of the group difference using ranks:

Generating one estimate from one resample under the Null Hypothesis (using ranks)

Step 1 ► Resample n_A and n_B elements (with replacement) from the Big Box to get the new sets A_1 and B_1.

Step 2 ► Combine the elements of A_1 and B_1 into a single list, and rank the list from smallest to largest.

Step 3 ► Replace each element of A_1 and B_1 by its corresponding rank to get $A_{1\text{-rank}}$ and $B_{1\text{-rank}}$.

Step 4 ► Calculate $(\overline{B_{1\text{-rank}}} - \overline{A_{1\text{-rank}}})$ and call that Δ_1.

That gives us one estimate of how large a difference between ranks might have arisen simply due to sampling variability. Of course that is only one estimate. Now we need to repeat the resampling procedure 10,000 times, and use the 10,000 Δ_i's to form our "expectations under the Null Hypothesis" (part **2** in Figure 6.34).

Calculate the p-value The p-value of the observed effect size Δ_{obs} is the number of estimates Δ_i that are as extreme or more extreme than either Δ_{obs} or $-\Delta_{obs}$, divided by 10,000 (part **3** in Figure 6.34).

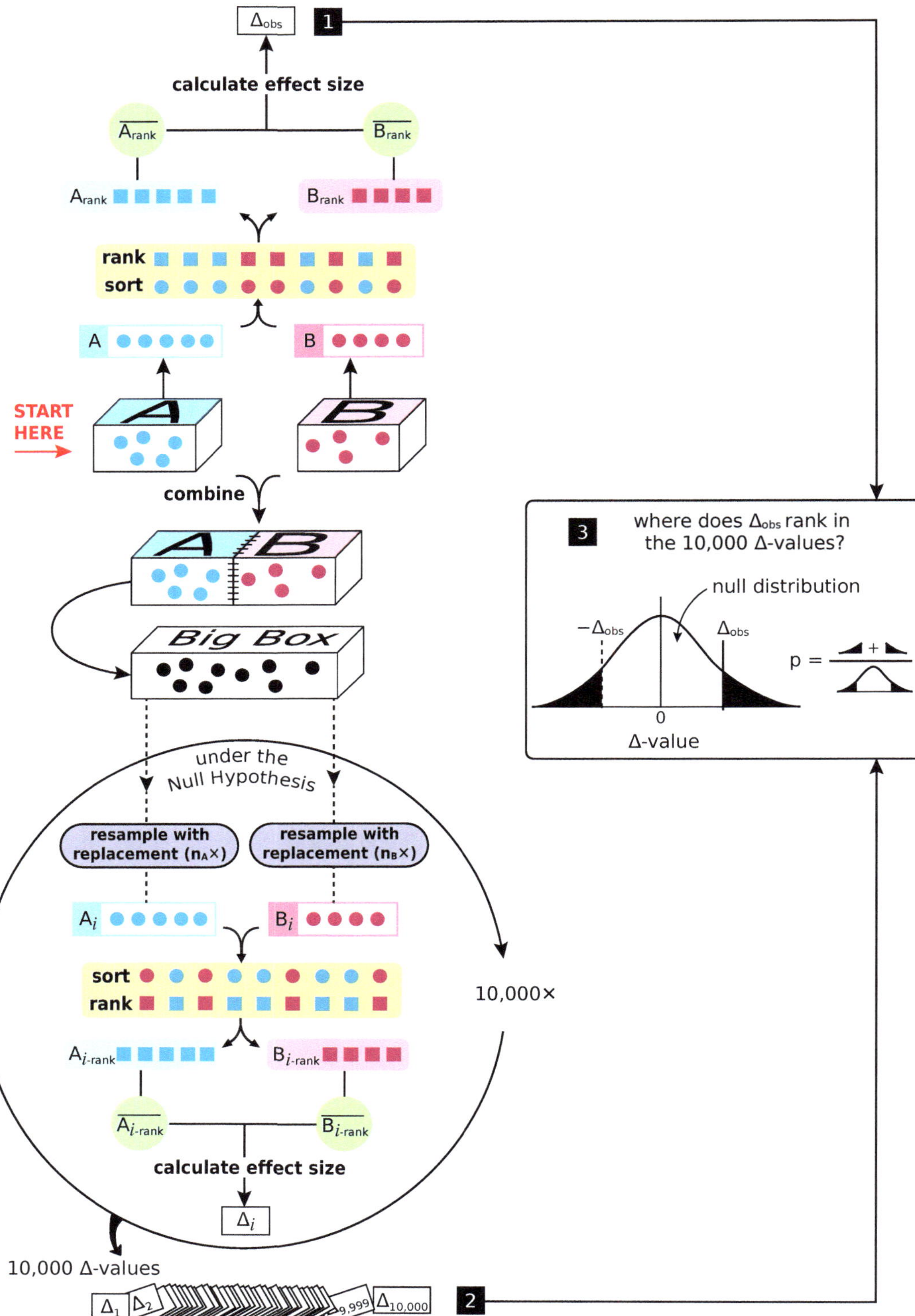

Figure 6.34 Big Box procedure for two-group comparisons using ranks.

To continue with our example in Figure 6.33, we calculated the observed group difference in ranks to be $\Delta_{obs} = 1.575$. We will now answer the question "is Δ_{obs} large?", by comparing Δ_{obs} to the null distribution (Figure 6.35); we see that there are 2,157 simulations with $\Delta_i \geq \Delta_{obs}$ and 2,091 simulations with $\Delta_i \leq -\Delta_{obs}$, this gives us a p-value of $(2{,}157 + 2{,}091)/10{,}000 = 0.4248$. This Δ_{obs} is clearly not statistically significant.

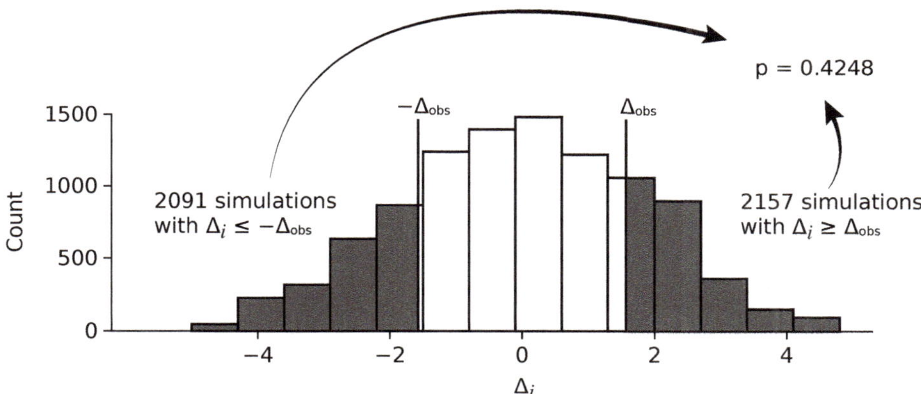

Figure 6.35 Calculating a p-value for the observed group difference in Figure 6.33. The histogram is the null distribution with the observed group difference indicated. A p-value is calculated by comparing the observed group difference to the null distribution.

Calculating a p-value using ranks

1. **Analyzing the observed data.** Combine the raw data from the two groups A and B into a list and rank the entries from smallest to largest. Then put the ranks back in their original groups in place of the raw data. Calculate the average rank for each group as the group measure, and subtract the two averages as the group difference. This gives the observed effect size, Δ_{obs}.

2. **Resampling from the Big Box.** Combine the two groups A and B into a Big Box. Resample a new A_1 and a new B_1 from the Big Box. Analyze the new data sets as in Step 1 to generate an estimate of the effect size under the Null Hypothesis. Repeat 10,000 times.

3. **Interpreting and presenting the results.** Using the 10,000 estimates under the Null Hypothesis, for a two-sided test, the p-value is the number of resamples that had a group difference equally or more extreme than Δ_{obs} or $-\Delta_{obs}$, divided by 10,000.

Example 2 Small sample size

Question When you are given samples with small sizes, and you don't have enough context to determine what the numbers mean, what should you do?

Answer The best thing is to get more data. The second best is to use ranks.

FURTHER EXERCISES 6.6

1. **Grandma's home remedy.** Your grandmother gives you a home remedy that she swears by to shorten the duration of colds. You want to determine if this actually works, so you recruit 5 sick friends to ingest your grandmother's remedy and 5 sick friends to ingest a similar-tasting placebo. Symptom durations for your friends taking the remedy are 3.5, 2.3, 4.7, 1.5, and 3.7 days. Symptom durations for your friends taking the placebo are 5.3, 3.6, 4.3, 5.7, and 6.7 days.

   a. Visualize the data.

   b. Describe the appropriate Null Hypothesis, the measure to describe the groups and the method for comparing the group measures.

   c. Calculate the observed effect size, that is, the difference between the two groups. Describe the resampling procedure to determine whether the observed effect size is statistically significant.

   d. Assuming you don't worry too much about false positives, does your grandmother's home remedy work?

2. **Mercury in fish.** Upon local reports that unregulated mining operations might be causing elevated mercury levels in a nearby recreational lake, a county environmental agency set out to investigate. Five striped bass in that lake were caught and their tissues were tested for mercury. For comparison, four striped bass in a lake far away from the mining operations were also caught and tested. The fish tissue mercury levels (in mg/kg) are 0.645, 0.591, 0.567, 0.598, 0.741 from the nearby recreational lake, and 0.371, 0.582, 0.249, 0.365 from the faraway lake.

   a. Visualize the data.

   b. Describe the appropriate Null Hypothesis, the measure to describe the groups and the method for comparing the group measures.

   c. Calculate the observed effect size, that is, the difference between the two groups. Describe the resampling procedure to determine whether the observed effect size is statistically significant.

   d. Does the data provide sufficient evidence to conclude that fish in the nearby lake have elevated levels of mercury?

6.7 "Paired" study design

So far, we've been talking about comparing two groups, that is, measurements of the same variable in two different groups of subjects, such as the survival times of cancer patients taking a drug vs. survival times of patients taking a placebo. The treatment and placebo groups contain different individuals, and they are *independent* groups, for example, they don't need to have the same number of data points.

However, there's another study design that is often classified under "two groups": **paired study design**, in which a variable is measured twice in the same subject, under two different conditions. Since two measurements of the same variable are made on the same i^{th} subject, this generates two *dependent* groups, in which each data point in group A, a_i, is uniquely *paired* to a data point in group B, b_i. The a_i-b_i linkage is what gives it the name "paired" (Figure 6.36).

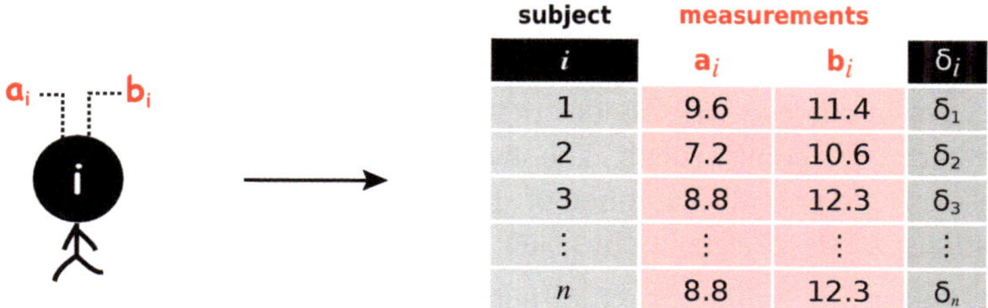

Figure 6.36 An illustration of a paired study design.

Paired study design is essentially one group of subjects with two measurements of the same variable on each subject. The paired constraint guarantees that the two groups A and B will always have the same number of data points. Similar to all cases of two-group comparisons, the two measurements must be of the same variable, meaning two heights, or two weights, or two speeds, or .... Since the purpose of paired study design is to ask whether there is a difference δ between the A measurement and the B measurement in each subject, the A's and B's must be subtracted from each other. You can't subtract heights from weights.

The more general case, where the A's and B's are measurements of two *different* quantities in the same individual, like their height and their age, is treated in Chapter 9 *Bivariate Data: Correlation and Beyond.*

Scenarios that call for a paired study design include:

1. **before and after** considers subjects pre- and post-treatment, in which the same factor is measured before and after the treatment.
2. **matched subjects** uses separate groups, but relies on matching every subject in one group with an equivalent in the other.
3. **cross-over trials** where individuals are randomized to two treatments and then the same individuals are crossed-over to the alternative treatment.
4. **repeated measures** on the same subject.
5. any circumstance in which each data point in one sample is uniquely matched to a data point in the second sample.

The first thing that needs to be observed is that this is not really "two groups"; it's one *group, with two measurements for each subject.* Thus, we could really have moved this discussion to Chapter 9 (Bivariate Data). We leave it here under "two groups" for historical reasons.

Taking the example of before and after study design, *it is clear that what we really have is one group of δs,* where the δs are the differences between the "before" value and the "after" value in the same subject.

The advantage of using a paired study design is that individuals can be compared to themselves under a different condition, thereby removing individual variability from obscuring the true effect of the different condition.

Exercise 6.7.1 An investigator hypothesizes that patients ask female physicians more questions than they ask male physicians. She designs a study in which patients who come to an urgent care clinic for the first time are randomly assigned to a female or male doctor, and an observer records the number of questions the patient asks. The next time the patient comes in, they see a doctor of the opposite gender than they had seen before, and again an observer records the number of questions asked. The investigator of this study will have two numbers associated with each patient: the number of questions asked of a female physician and the number of questions asked of a male physician.

We could simply forget the paired structure and ask the two-group question: do patients ask more questions of female doctors than they do of male doctors? That would reduce this data set to two piles of numbers: the numbers of questions asked of male doctors and the number of questions asked of female doctors.

But that would be to throw away a key part of this data: that we have two numbers *for each patient*, and therefore we can ask whether the gender of the doctor made a difference to each patient, by looking at the *difference* between the number of questions that each patient asks of the two genders.

Patient ID	**Number of questions asked by patients**		δ
	Female doctor	Male doctor	
1	8	5	3
2	6	2	4
3	5	7	−2
⋮	⋮	⋮	⋮

Which of the following is this study?

a) Before and after
b) Matched subjects
c) Cross-over trials
d) Repeated measures
e) None of the above

Exercise 6.7.2 For each of the following research questions, indicate whether you could design a study that will provide paired data samples. If you think this question can be addressed with a paired study design, describe what it would be and which kind of study it is (before and after, matched subjects, etc). If a research question cannot be addressed using a paired design, explain why.

a. Does a 6 month stay in the International Space Station impact astronauts' white blood cell count?

b. Do books cost more on average at the local bookstore than through Amazon.com?

c. Do freshmen students use the library to study more often than senior students?

d. Will taking summer school improve reading ability for kindergarteners going into first grade?

e. Does replacing butter with margarine reduce the lipoprotein levels of subjects with hypercholesterolemia.

Swimsuit vs. wetsuit study

The 2008 Olympics saw controversy about new wetsuits possibly providing unfair advantages to swimmers, leading to new international rules that came into effect January 1, 2010, regarding swimsuit coverage and material.

Can a wetsuit really make someone swim faster? How much faster?

Here's a classic example, a study of performance speeds of swimmers who wore conventional swimsuits versus the same swimmers wearing wetsuits. Twelve swimmers swam 1500 m at maximum speed twice each, once wearing a conventional swimsuit and once wearing a wetsuit. The order of the trials was randomized. Each time, maximum speed (m/sec) was recorded. The table shows their maximum swimming speeds (Table 6.3).

Swimmer	**Swimsuit** (m/sec)	**Wetsuit** (m/sec)
1	1.49	1.57
2	1.37	1.47
3	1.35	1.42
4	1.27	1.35
5	1.12	1.22
6	1.64	1.75
7	1.59	1.64
8	1.52	1.57
9	1.50	1.56
10	1.45	1.53
11	1.44	1.49
12	1.41	1.51
mean	1.429	1.507
stdev	0.135	0.130

Table 6.3 Swimming speeds of 12 swimmers using conventional swimsuits and wetsuits (De Lucas *et al.*, 2000).

Exercise 6.7.3 Which of the following is this swimsuit vs. wetsuit study?

a) Before and after
b) Matched subjects
c) Cross-over trials
d) Repeated measures
e) None of the above

How should we approach this data?

Independent sample analysis

First, let's ask: *what if we ignored the paired structure, and just treated it as two independent groups, as it were, two different groups of people?* (Figure 6.37).

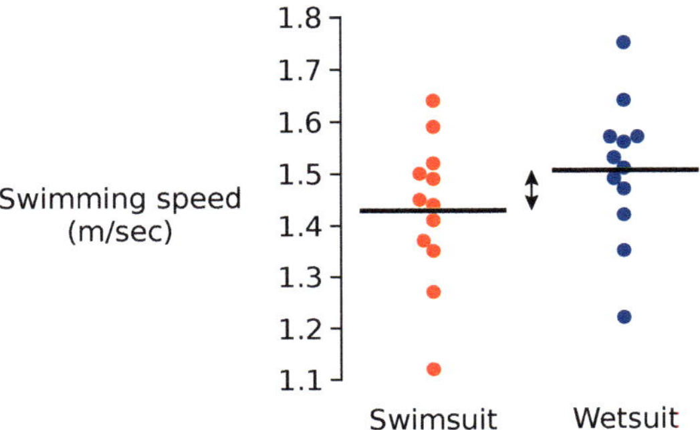

Figure 6.37 Swimsuit and wetsuit speeds, as in Table 6.3, represented as dot plots, except that we have forgotten which swimsuit speed is paired with which wetsuit speed. The mean of each group is represented as a horizontal line.

If we use the mean speed as our group measure, then the effect size is

effect size (independent)
$$\Delta_{\text{obs}} = \begin{matrix}\text{average speed}\\ \text{for wetsuits}\end{matrix} - \begin{matrix}\text{average speed}\\ \text{for swimsuits}\end{matrix}$$
$$= 1.507 \text{ m/sec} - 1.429 \text{ m/sec}$$
$$= 0.078 \text{ m/sec}$$

The difference 0.078 m/sec is much less than the standard deviations, which are both around 0.14. There are substantial within-group differences, due to the large diversity in the subjects (men and women, trained athletes and amateurs, etc.). So, from a statistical point of view, the across-group difference is much smaller than the within-group differences. This means that the across-group difference is not going to be statistically significant.

And indeed, all forms of two-group testing, whether by formula or resampling, return a conclusion of "not significant". For example, when we ran the two-group testing using a Big Box method, we saw no evidence to reject the null, with $p \approx 0.18$ (Figure 6.38).

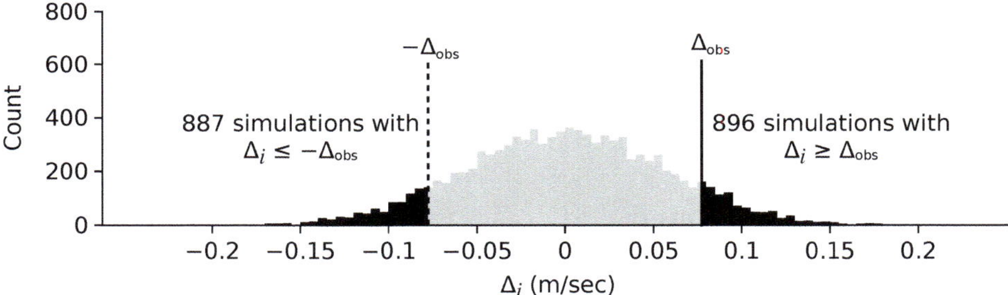

Figure 6.38 When we ignore the paired structure, and treat the two groups as independent samples, we do not find any statistically significant association between wearing a wetsuit and swim speed.

With this analysis, treating the two groups as independent samples, we would conclude that the wetsuit had little effect.

But this would be a mistake!

Paired analysis

The error can be seen easily if we make a new kind of plot, a **paired differences plot**, in which we draw lines connecting each swimsuit speed with the wetsuit speed for the same swimmer (Figure 6.39).

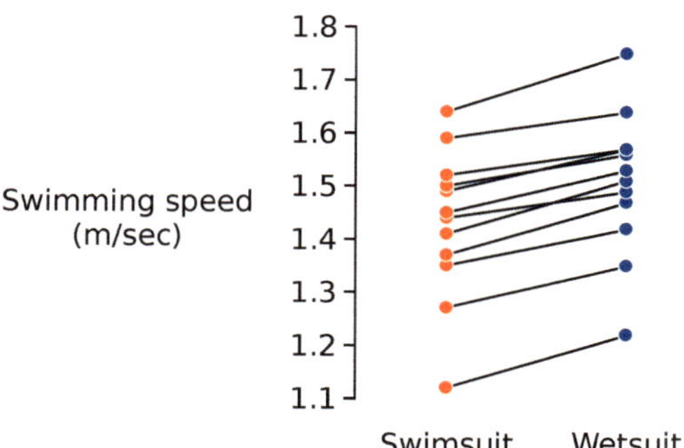

Figure 6.39 Paired differences plot for the swimsuit-wetsuit data in Table 6.3.

The pattern, hidden before, now becomes obvious: in every case, the wetsuit improved the speed of the swimmer, whether amateur or pro, etc.

What the graphic is doing is focusing our attention, not on the data values, but on the differences, the δs, between each person's swimsuit performance and their wetsuit performance. The statistical analysis should also focus on these δs (δ = wetsuit − swimsuit). Let's take a look at them: notice that *all* of them are positive (Table 6.4).

Swimmer	**Swimsuit** (m/sec)	**Wetsuit** (m/sec)	δ (m/sec)
1	1.49	1.57	0.08
2	1.37	1.47	0.10
3	1.35	1.42	0.07
4	1.27	1.35	0.08
5	1.12	1.22	0.10
6	1.64	1.75	0.11
7	1.59	1.64	0.05
8	1.52	1.57	0.05
9	1.50	1.56	0.06
10	1.45	1.53	0.08
11	1.44	1.49	0.05
12	1.41	1.51	0.10
mean	1.429	1.507	($\Delta_{obs} = \overline{\delta} =$) 0.078
stdev	0.135	0.130	

Table 6.4 Paired differences δs for each swimmer are shown in green.

Our data is the set of δs. Let's agree to use the mean $\overline{\delta}$ as our measure.[2] In this case $\overline{\delta}$ is 0.078 m/sec.

[2] It might be objected that in this example, the set of δs is markedly non-Normal, raising the question, why are

effect size (paired) $\Delta_{obs} = \overline{\delta} = 0.078$ m/sec

Our next question is "is Δ_{obs} 'large'?", and, as usual, the reply is "large relative to what?", and the answer is "large with respect to what we would see if the Null Hypothesis were true".

Creating the world of the Null Hypothesis But what is the Null Hypothesis here? It clearly has to be the assumption that the treatment, the wetsuit, makes no difference relative to the swimsuit. To put it mathematically, the Null Hypothesis is that the average of the δs will be zero. Some of the δs will be positive and some negative, so their averages will cluster around zero. We implement this Null Hypothesis by assuming that the swimsuit speed might just as well have been the wetsuit speed, and vice versa. Therefore, the Null Hypothesis and the Alternative Hypothesis are:

H_0 The wetsuit makes no difference in swim speed relative to the swimsuit, $\Delta = 0$

H_a Wetsuit performance is significantly different from swimsuit performance, $\Delta < 0$ or $\Delta > 0$

The implementation of the Null Hypothesis would be to take the set of δs, and then multiply each one randomly by a 1 or a −1, representing the idea that the difference could have been what it was (δ), or it could have been the other way around ($-\delta$). This gives us the first resample, we call it δ_1. Taking the mean of the set δ_1 gives us the first estimate of the speed difference between swimsuit and wetsuit groups under the Null Hypothesis, $\Delta_1 = \overline{\delta_1}$ (Figure 6.40).

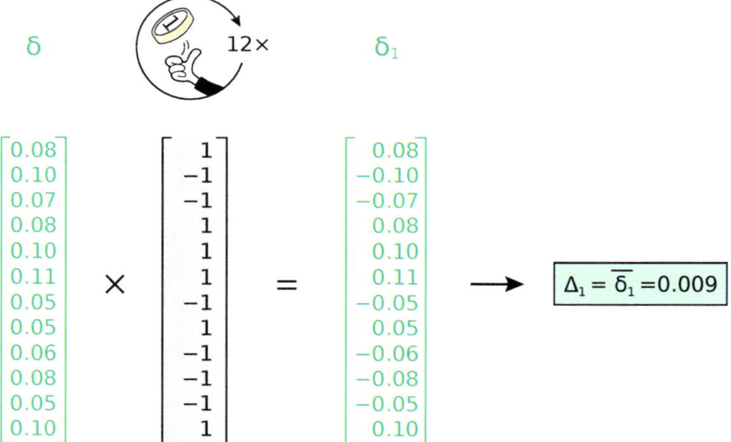

Figure 6.40 Generating the first resample under the Null Hypothesis and calculating its corresponding effect size Δ_1.

We then repeat the resampling procedure 10,000 times, which gives us 10,000 estimates, $\Delta_1, \Delta_2, \ldots, \Delta_{10,000}$, of the difference between swimsuit and wetsuit groups, under the Null Hypothesis (Figure 6.41).

we using the mean and not the median? The answer is that in small data sets like this one, the median is highly unstable with respect to changing even a single data point. Therefore, a resampling procedure using medians of small data sets would be highly unstable, and the results would be highly clustered. If, on the other hand, the data set is sufficiently large, then it might well be appropriate to use a median value. Thanks to Gaston Pfluegl (UCLA) for pointing this out to us.

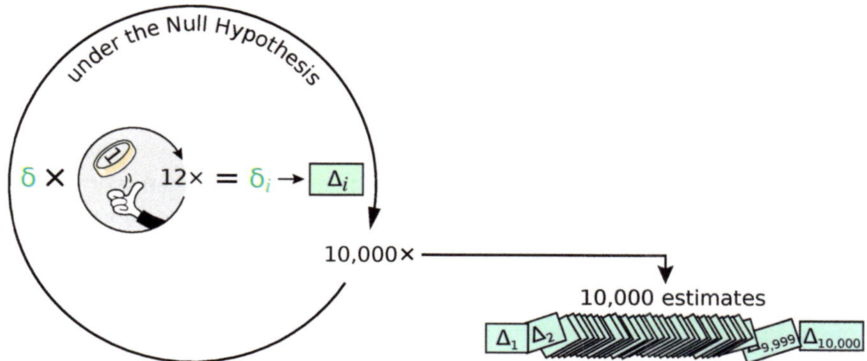

Figure 6.41 Generating 10,000 expectations of group differences, under the Null Hypothesis, for the swimsuit vs. wetsuit data set.

Calculating a p-value for a paired study design

Step 1 ► Take the list of actually observed paired differences, δ. Calculate its mean and call it Δ_{obs}. This is our observed group difference.

Step 2 ► For each element in the set δ, choose a 1 or a -1 randomly and independently to multiply it. Call that new set δ_1. Calculate its mean and call it Δ_1. This is the first estimate of the group difference under the Null Hypothesis.

Step 3 ► Repeat Step 2 for a total of 10,000 times to generate 10,000 estimates: Δ_1, Δ_2, ..., $\Delta_{10,000}$.

Step 4 ► The p-value for the observed difference Δ_{obs} is then the number of Δ_i's that were more extreme than either Δ_{obs} or $-\Delta_{\text{obs}}$, divided by 10,000.

One-sided or two-sided p-value calculation? Note that, in Step 4 above, we said that the total number of simulations that count against the Null Hypothesis are all cases where Δ_i has a value that's more extreme than either Δ_{obs} or $-\Delta_{\text{obs}}$. That is, we are carrying out a 2-sided test.

We discussed 1-sided vs 2-sided testing in section 3.2 *Null Hypothesis Significance Testing* on page 103. We said that 1-sided testing is never valid unless the quantity in question can mathematically only take positive values (such as distances). Here the Δs could be fall on either side the null value, and so a 2-sided test must be used.

This is true even if your *scientific* hypothesis is that the wetsuits are faster than the swimsuits, not the other way around. It doesn't matter if you are "uninterested" in the opposite outcome; occurrences where this happens count against the surprise value of your finding.

So, if the conclusion of the two-sided test is that your result is statistically significant, and if your observed result is that wetsuits are faster than swimsuits, then you have confirmed that wetsuits are statistically significantly faster than swimsuits. If the observed result was the other way and was statistically significant by a 2-sided test, then you would have confirmed that wetsuits are inferior to swimsuits.

For two-group comparisons, including paired study design, regardless of your *scientific* hypothesis, you must use a two-sided test to determine the statistical significance.

Using a paired analysis, we carried out 10,000 simulations, under the Null Hypothesis, for the swimsuit-wetsuit data set. The observed difference was more extreme than *all* of the 10,000 simulated values, so the p-value of this observation is less than 1 in 10,000, or $p < 0.0001$ (Figure 6.42).

With a p-value < 0.0001, we have very strong evidence against the Null Hypothesis and can conclude that wearing a wetsuit does increase, on average, your maximum swim speed. We can draw a cause-and-effect conclusion since the researchers used random assignment for the order of the two conditions (conventional swimsuit and wetsuit) for each swimmer. If all the swimmers had started with the wetsuit first, then the conventional swimsuit, someone could argue that the slower speeds recorded on the second attempt were due to fatigue.

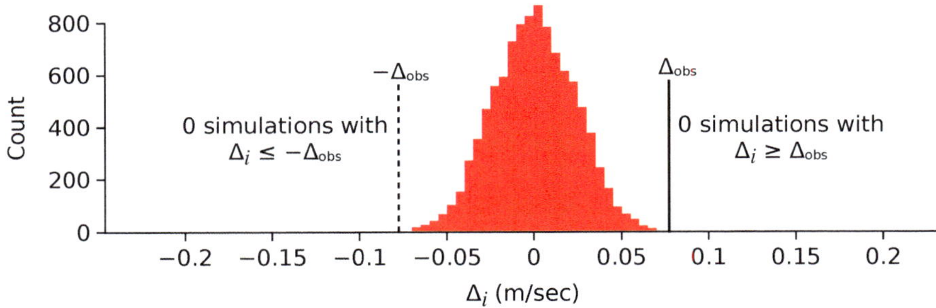

Figure 6.42 Distribution of the speed differences between wetsuits and swimsuits under the Null Hypothesis when using paired analysis.

Why use a paired analysis? The two analyses, independent and paired, that we just carried out, and the opposite conclusions we got from them, give us an important insight into the core idea of statistical significance.

When we ignored the pairing, and looked at the data as two separate piles, we calculated the difference between the wetsuit group average speed and the swimsuit group average speed to be 0.078 m/sec. This was not even close to being statistically significant, however. Then we *did* include the pairing, and calculated the mean difference between the pairs, and it was 0.078 m/sec, exactly the same as the unpaired difference!

Taking the pairing into account does not change the effect size. What it does change is the comparison for statistical significance. In the unpaired case, we have to consider the within-group variation, since "statistically significant" means "effect size much larger than within-group variation". When we do that, 0.078 looks very small. It's one half of one standard deviation.

The big effect of the pairing is to remove this within-group variation

The big effect of the pairing is to remove this within-group variation by comparing each wetsuit value with the matched swimsuit value. Thus, the same effect size has now become extremely statistically significant, by basically eliminating the within-group differences. This is why study designers love paired data designs: they eliminate a great deal of within-group variability.

As an interesting visual comparison, when we overlay the null distributions of the speed differences when the data is considered paired versus independent, we see that the paired analysis produced a null distribution that is a lot narrower, therefore, it can better detect differences between conditions, since it has removed individual variability among the individual swimmers (Figure 6.43).

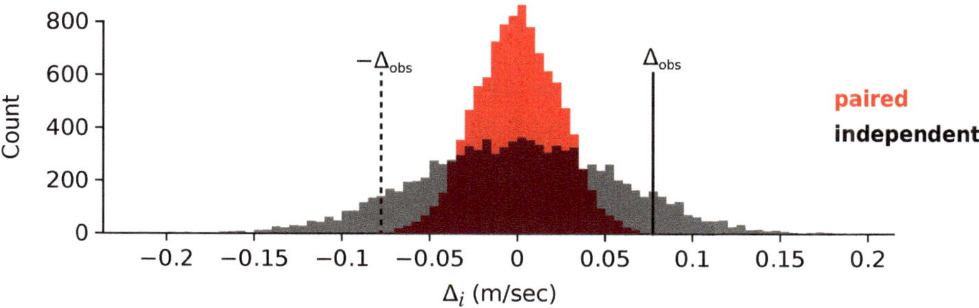

Figure 6.43 Null distributions of the speed differences between wetsuits and swimsuits when the data is considered paired versus independent.

Confidence intervals

We can also calculate a confidence interval for Δ_{obs}. It is really nothing new: we have a column of numbers and an average value (the δ column and $\overline{\delta}$ in Table 6.4), and we calculate the confidence interval for the average as in Chapter 4. We resample (with replacement) the column of δs 10,000 times and calculate the 10,000 averages. Then we list them from smallest to largest, calling the 50th smallest Δ_L and the 50th largest Δ_U . Then we calculate the 99% confidence interval as

$$99\% \text{ confidence interval} = [2{\cdot}\Delta_{obs} - \Delta_U, 2{\cdot}\Delta_{obs} - \Delta_L]$$

We present the result of the 99% confidence interval together with the resampling distribution (Figure 6.44). We would state this result as "wetsuits improved swimming speeds over traditional swimsuits (Δ_{obs} = 0.078 m/sec, 99% CI [0.062, 0.092])".

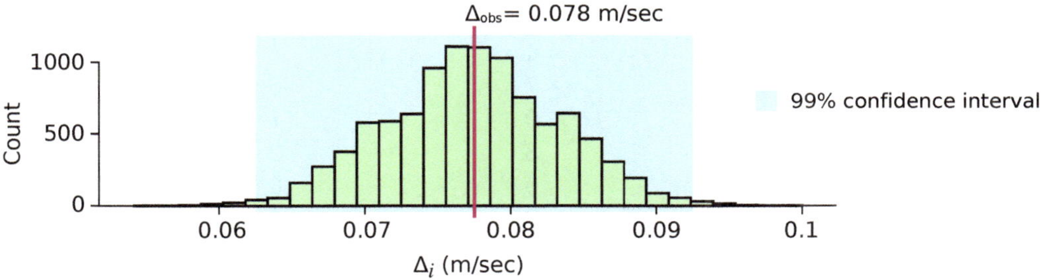

Figure 6.44 99% confidence interval for the observed group difference Δ_{obs} of swim speed between swimsuit and wetsuit.

Exercise 6.7.4 Simulating H_0 for paired design. How does multiplying each paired difference by −1 or +1 (chosen randomly) simulate the Null Hypothesis (for example as seen in Figure 6.40)?

Exercise 6.7.5 Confidence interval for paired design. How should we calculate a 95% confidence interval for the effect size in a study with paired design?

FURTHER EXERCISES 6.7

1. Why might researchers want to use a paired study design rather than just independent samples?

2. To construct the confidence interval for a parameter, we form the resampling distribution of the parameter by repeatedly drawing samples from the original data set. This simulation assumes which hypotheses to be true?

   a) The Null Hypothesis
   b) The Alternative Hypothesis
   c) Both hypotheses
   d) Neither hypothesis

3. All 110 people enrolled in a clinical trial received both the placebo and the treatment, delivered for a month each, in random order. Which paired design does this describe?

   a) Before and after
   b) Matched subjects
   c) Cross-over trials
   d) Repeated measures
   e) None of the above

4. **Questions for doctors.** A researcher hypothesizes that patients ask female physicians more questions than they ask male physicians. The researcher designs a study where patients who come to a clinic for the first time are randomly assigned to a female or male doctor. The next time the patient comes in, they see a doctor of the opposite sex than they had seen before. For both instances, an observer out of sight records the number of questions the patient asked their respective physician. Here is the data from the study:

Patient ID	Number of questions asked by patients	
	Female doctor (Q_F)	Male doctor (Q_M)
1	8	5
2	6	4
3	5	4
4	4	4
5	7	6
6	9	7
7	3	2
8	6	8
9	8	4
10	7	5
11	2	1
12	10	8
13	5	4
14	4	4
15	5	7

   a. What type of study design should you use to analyze this data set?
   b. What is the Null Hypothesis H_0, and the Alternative Hypothesis H_a for this study?
   c. Assume you define δ to be $Q_F - Q_M$. Calculate the difference for each patient.

d. As always, before we start performing statistical analysis, we should first visualize the data. In this case, the appropriate visualization is a paired differences plot (see below). What is your first impression?

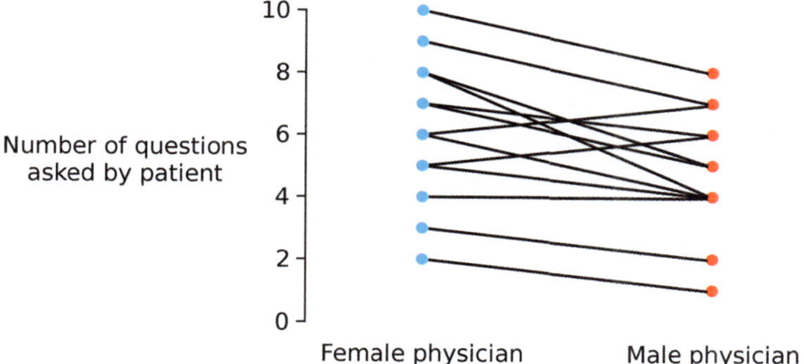

e. Define a statistic of interest that you would use to compare the difference between numbers of questions that patients ask female physicians vs. male physicians. Calculate the effect size according to your definition, and write the calculation that you used to get the answer.

f. Describe the procedure to simulate the Null Hypothesis.

g. A resampling-based method was used to develop the null distribution of your statistic of interest. The null distribution is shown below. Circle the bars that would represent the numerator of the p-value. Calculate your p-value. Does your calculation confirm your visual inspection? What's your conclusion?

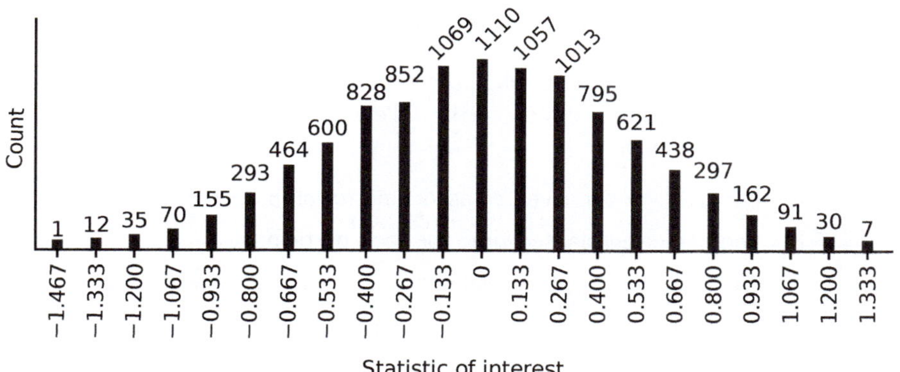

h. Looking at those results, if you were to compute the 95% confidence interval of your statistic of interest, do you think this interval would contain the value 0?

5. **Revisit the study on wetlands and plant species richness.** In the study of wetlands and plant species richness in FE 6.5.4 on page 225, scientists collected a variety of data for randomly selected wetlands in a preserve in Wisconsin and tried to determine if plant species richness (number of species) differs between permanent and ephemeral wetlands in this preserve.

   Can we analyze the data in that study as paired data to study the same research question? If yes, describe what it would be and what type of study it is (before and after, matched subjects, etc.).

6. **Quit smoking.** When people quit smoking, some might have a hard time concentrating or sleeping, have strong urges to smoke, or just feel generally uncomfortable. These feelings are called withdrawal. How does withdrawal affect the perception of time?

   20 daily cigarette smokers (12 male, 8 female) were asked to estimate the duration of a 45-second period of time in a laboratory setting. Smokers participated in two sessions: once after smoking ad-lib and once after objectively confirmed 24-hour smoking abstinence. The sequence of the two sessions was randomly assigned to each participant.

Perceived elapsed time (in seconds)					
participant ID	ad-lib	withdrawal	participant ID	ad-lib	withdrawal
1	46	58	11	53	63
2	42	79	12	40	125
3	44	180	13	35	39
4	51	65	14	44	120
5	44	45	15	47	47
6	50	90	16	51	51
7	40	66	17	39	150
8	44	54	18	47	71
9	42	86	19	53	90
10	52	52	20	45	68

a. Which of the following is this study?

   a) Before and after b) Matched subjects c) Cross-over trials

   d) Repeated measures e) None of the above

b. An independent variable is what the experimenter determines and a dependent variable is the outcome. Which of the following is a dependent variable? Which is an independent variable?

   a) The average perceived elapsed time for all participants.
   b) The perceived elapsed time for each participant.
   c) Whether or not withdrawal changes the perception of time.
   d) Whether or not the person was suffering from withdrawal.

c. Which of the following is the Null Hypothesis, and which is the Alternative Hypothesis?

   a) Withdrawal increases smokers' perceived length of time.
   b) Withdrawal decreases smokers' perceived length of time.
   c) Average perceived time for ad-lib $\neq$ average perceived time for withdrawal.
   d) Average perceived time for ad-lib = average perceived time for withdrawal.
   e) Withdrawal has no effect on smokers' perceived time.
   f) Withdrawal changes smokers' perceived time.

d. Calculate a p-value and include the null distribution, with the observed difference indicated. Based on your p-value, write the conclusion for this study.

7. **Amiloride and lung function.** Researchers want to determine whether a drug, amiloride, improves lung function in patients with cystic fibrosis. They measure reduction in lung function over 25 weeks in 14 patients, each with the drug and a placebo separately, with order of treatment randomly assigned. Determine if amiloride improves lung function. Note: a smaller loss of lung function is an improvement (see "A pilot study of aerosolized amiloride for the treatment of lung disease in cystic fibrosis", *New England Journal of Medicine*, 1990).

Patient ID	**Reduction in lung function**	
	Drug	Placebo
1	213	224
2	95	80
3	33	75
4	440	541
5	−32	74
6	−28	85
7	445	293
8	−178	−23
9	367	525
10	140	−38
11	323	508
12	10	255
13	65	525
14	343	1023

a. For this study, should you use a two-group comparison or a paired approach? Why?

b. If it is a paired study design, which of the following is it?

a) Before and after b) Matched subjects c) Cross-over trials
d) Repeated measures e) None of the above

c. Visualize the data by group.

d. Describe the appropriate Null Hypothesis, the measure to describe the groups, the method of comparing the group measures, and the resampling procedure to calculate a p-value.

e. Calculate a p-value for this result. Include the null distribution with the observed result indicated.

f. Calculate a 99% confidence interval for the effect size(s).

g. Interpret your results.

8. **Alpha-wave and physical exercise.** An experiment was performed to see whether physical exercise has any effect on the alpha-wave patterns produced by the brain. 10 healthy adults were instructed to complete four weeks of exercise training. The alpha-wave frequencies were measured for all subjects before the training and repeated again after the training.

Subject ID	Alpha-wave frequencies (Hz)	
	Initial measure	Four weeks later
1	9.6	9.9
2	9.5	11.1
3	10.1	10.4
4	10.9	10.4
5	10.6	11.2
6	9.6	10.7
7	10.4	10.7
8	10.3	10.9
9	9.7	10.3
10	9.8	10.5

a. Which of the following is this experiment?

a) Before and after b) Matched subjects c) Cross-over trials
d) Repeated measures e) None of the above

b. Visualize the data.

c. Describe the appropriate Null Hypothesis, the measure to describe the groups, the method of comparing the group measures, and the resampling procedure to calculate a p-value.

d. Calculate a p-value. Include the null distribution with the observed effect size indicated.

e. Calculate a 99% confidence interval for the effect size.

f. Interpret the result.

6.8 When the effect size is a ratio

Algebraic differences vs. ratios

So far, we've been focusing on examples where we have measures M_A of set A and M_B of set B, and we compare them by subtracting one from the other. The effect size, in that case, is an algebraic difference, $M_B - M_A$. But, while it's true that we talk about "the *difference* between group A and group B", that does not necessarily imply the algebraic minus sign. Sometimes, the best way to compare M_A and M_B is to take their ratio $\frac{M_B}{M_A}$.

The choice of ratio versus algebraic difference is dictated by the practical meaning of the "difference" between A and B. We considered the case where the data sets in question are the survival times (in days) of groups on and off some drug. There, the only "difference" that makes sense is the algebraic difference. That is the quantity that gives you useful information: how much longer will I live (on average or median) on drug A? It wouldn't make sense to ask for a ratio, that is, an answer like "group A lived 1.5 times as long as group B", if in fact group A lived for an average of 1 day and group B lived for an average of 1.5 days. The fact that these are days of survival means that we want to know "how many more days?" and that is asking for an algebraic difference, not a ratio.

But it is equally true that sometimes, the quantities in question make better scientific sense if they are compared by a ratio. The comparison of incomes is one obvious case. As we mentioned above, if we are comparing incomes of, say, workers and management, it makes more sense to ask for the ratio, which automatically adjusts for inflation. Other cases include all examples where we are interested in the "fold difference", that is, is the measure M_A two-fold the measure M_B? three-fold? As we saw, in the example of a drug being tested on dishes of cells, where the measures are "T-cells/unit area" in group A and "T-cells/unit area" in group B, then it may make more sense to ask for the "percentage change", that is, the ratio or fold change. It doesn't help that much to be told that "group A had 1000 more T-cells/unit area" if that is only a 1% improvement!

Comparing ratios[3]

When we decide, on scientific or practical grounds, that a ratio is the best way to compare M_A and M_B, we have to adapt our procedures for doing Null Hypothesis Significance Testing and for obtaining confidence intervals.

The problem is this: suppose our result, our effect size, is $\frac{M_B}{M_A}$, rather than the algebraic difference $M_B - M_A$. Let's call $\frac{M_B}{M_A}$ by a new name, r_{obs}, to emphasize that it is a ratio as opposed to Δ_{obs}, which is the difference $M_B - M_A$.

ratio $$r_{obs} = \frac{M_B}{M_A}$$

algebraic difference $$\Delta_{obs} = M_B - M_A$$

Let's say that in a given case $r_{obs} = 2$, so M_B is twice as big as M_A. Let's plot this r_{obs} on an axis, and let's also plot the Null Hypothesis value, which is $r = 1$ (that is, $M_A = M_B$) (Figure 6.45). We can see a problem right away. We see the two markings on the axis, and our eye wants to turn that into a judgment of the distance between the two; that's how our eyes see differences.

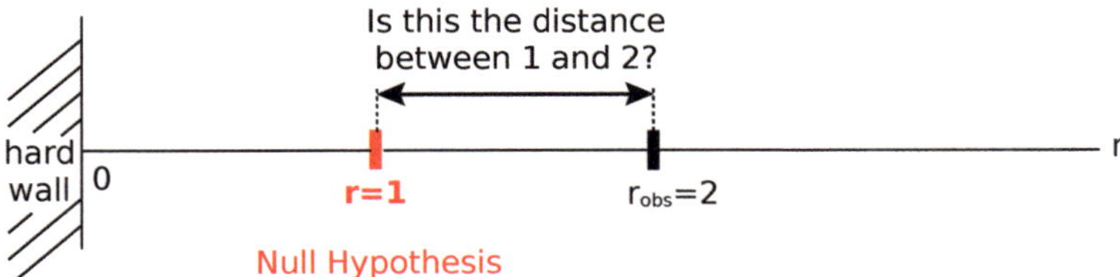

Figure 6.45 There is a hard wall at $r = 0$, because negative numbers don't make sense when we use ratios.

But r_{obs} needs to be compared to the Null Hypothesis ($r = 1$), not by its algebraic difference from it, but by its ratio.

Looking at Figure 6.46, we see that there is a deep asymmetry in this situation. In theory, there are two possibilities: either M_A is bigger than M_B or M_B is bigger than M_A. These two possibilities should obviously be treated symmetrically. But on the graph, "$M_A > M_B$" means that $r < 1$. We have only the interval of length 1, from the hard wall at $r = 0$ to $r = 1$ to

[3] With thanks to Morgan Tingley (UCLA).

represent that possibility. But the other possibility, $M_B > M_A$, gets the whole infinite line from 1 to infinity to represent its possible values!

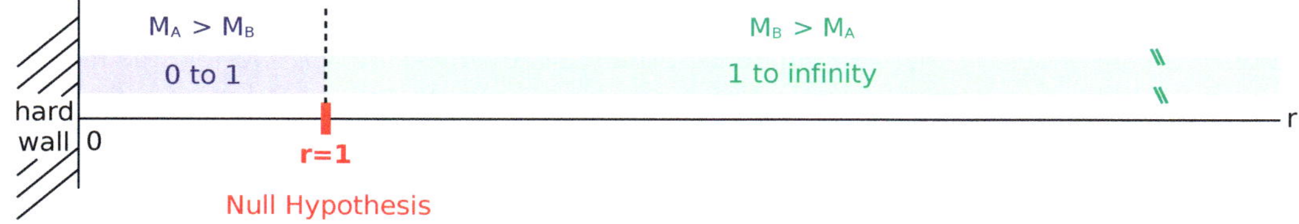

Figure 6.46 The asymmetry in comparing ratios.

Let's consider two symmetric possibilities: M_A is twice M_B and M_B is twice M_A. Those are obviously symmetrical conditions. If we try to put those on the ratio axis, we see that they are not equally "distant" from 1! As we see in Figure 6.47, $M_A = 2 \times M_B$ is at $r = \frac{1}{2}$, while $M_B = 2 \times M_A$ is at $r = 2$. **These are not equally distant from 1 because "distant" means distant in the additive sense.**

1. M_A is twice M_B $\Longrightarrow$ $r = \dfrac{1}{2}$
2. M_B is twice M_A $\Longrightarrow$ $r = 2$

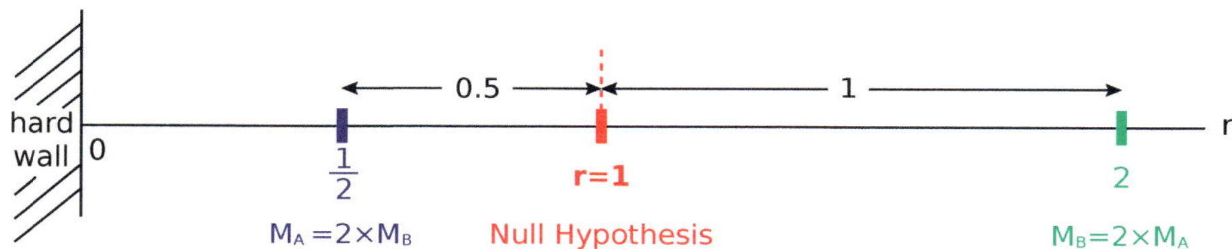

Figure 6.47 Two symmetric cases but the distances are not symmetric on the ratio axis (r-axis).

The remedy for this asymmetry is to turn multiplication into addition by the use of logarithms.

If we take the log (base e, usually written as ln) of r, then the two symmetric possibilities become

1. M_A is twice M_B $\Longrightarrow$ $r = \dfrac{1}{2} \xrightarrow{\text{take the log}} \ln(\dfrac{1}{2}) = \ln 1 - \ln 2 = -\ln 2$
2. M_B is twice M_A $\Longrightarrow$ $r = 2 \xrightarrow{\text{take the log}} \ln(2) = \ln 2$

The Null Hypothesis, which is $M_A = M_B$, will be transformed to 0 by the logarithm operation, $\ln(1) = 0$. Thus the two possibilities are equally spaced on the $\ln(r)$ axis, being exactly $\ln(2)$ to the left and $\ln(2)$ to the right of the Null Hypothesis (Figure 6.48).

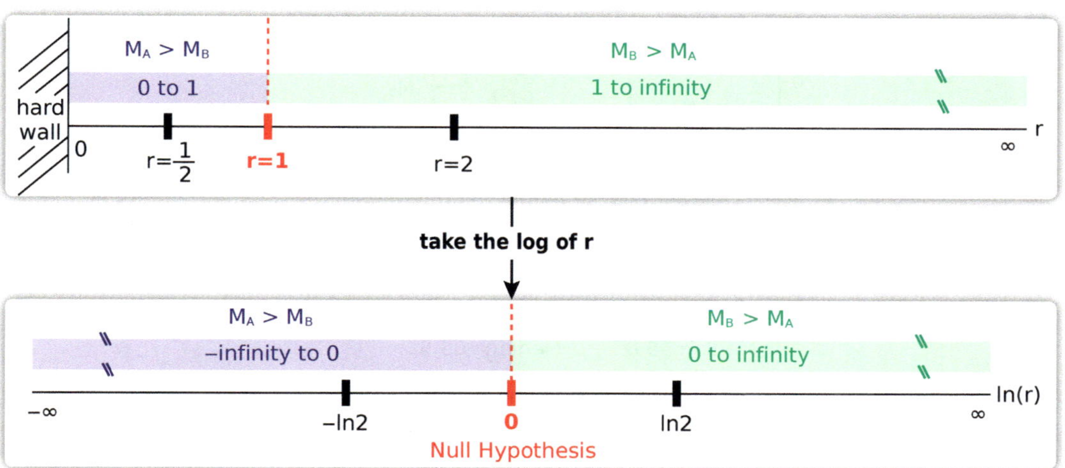

Figure 6.48 Taking the log of r has transformed the asymmetric axis to be symmetric.

Now let's consider *any* two symmetric possibilities: M_A is k times M_B, and M_B is k times M_A. Taking the logarithm of each group difference r will transform any two symmetric possibilities to be equally spaced on the ln(r) axis, being exactly ln(k) to the left and ln(k) to the right of the Null Hypothesis, which is $\ln(1) = 0$.

Two symmetric possibilities

- M_A is k times M_B $\implies$ $\ln(r) = \ln(\frac{M_B}{M_A}) = \ln(\frac{1}{k}) = \ln(1) - \ln(k) = -\ln(k)$
- M_B is k times M_A $\implies$ $\ln(r) = \ln(\frac{M_B}{M_A}) = \ln(k)$

Statistical significance for ratios

Now that we see how to use logs to compare ratios, the approach to testing the Null Hypothesis is straightforward.

Sample data sets Let's revisit the T-cell count example (*A case study: T-cell counts in Control vs. Treatment* on page 198). This time, we will use the ratio to compare the two groups (Figure 6.49).

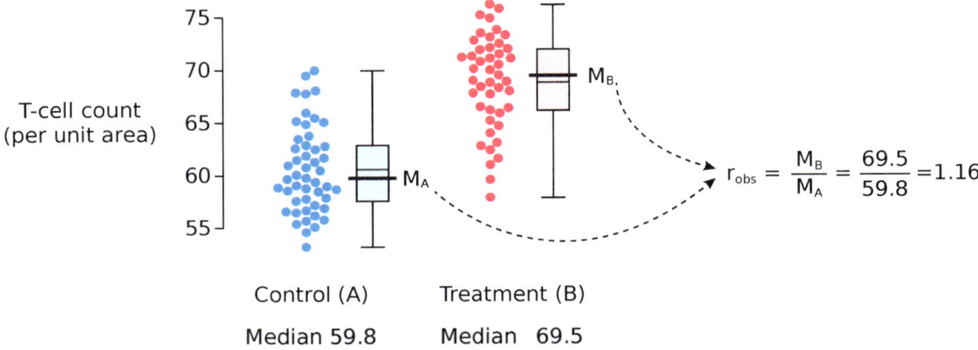

Figure 6.49 Beeswarm plots for the T-cell counts of the control and treatment groups (data set shown in Table 6.1). The group measure is the median and the ratio is used to measure the group difference. The observed ratio is $r_{obs} = 1.16$.

Calculating r_{obs} We choose the median as our summary measure for the two groups, and the observed group difference is:

$$\textbf{effect size (ratio)} \qquad r_{obs} = \frac{M_B}{M_A} = \frac{69.5}{59.8} = 1.16$$

Is r_{obs} significant? Now let's ask if that r_{obs} is statistically significant.

We will perform a Big Box version of Null Hypothesis Significance Testing (consult the workflow in Figure 6.6), *except that here we are using division to compare the two measures, not subtraction*.

First we put A and B into the Big Box, then we resample with replacement 49 from the Big Box to give us A_1, and another 45 to give us B_1. Then we calculate the medians of the two samples, $M(A_1)$, $M(B_1)$ and compute the ratio $\frac{M(B_1)}{M(A_1)}$, calling it r_1. This gives us the first estimate of the effect size under the Null Hypothesis. We repeat this procedure 10,000 times, and form the null distribution from the 10,000 estimates $r_1, r_2, \ldots, r_{10,000}$.

We did this for the T-cell count study, and obtained the null distribution in Figure 6.50. To calculate a 2-sided p-value, we mark our observed result r_{obs} and the equivalent opposite case $1/r_{obs}$, and count the number of simulations that have a group "difference" r_i equally or more extreme than r_{obs} or $1/r_{obs}$. There are none, so $p < 0.0001$, that is, less than 1 in 10,000,

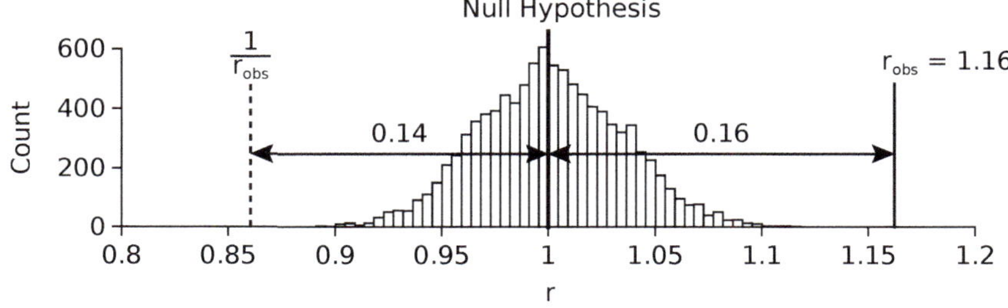

Figure 6.50 **Null distribution of the T-cell count study.** Histogram of the 10,000 r_i values generated by Big Box resampling. Also shown are the observed result r_{obs} and the symmetric opposite case $1/r_{obs}$. Note the different distances of those two symmetric cases to the null value, which is 1.

because a result this extreme didn't happen even once in the 10,000 resampling trials. Therefore, the observed ratio $r_{obs} = 1.16$ is statistically significant at the $p < 0.0001$ level.

Notice an asymmetry of those two symmetric cases on the r-axis. This asymmetry is inherent in comparing ratios using algebraic differences (Figure 6.48). If we make the null distribution symmetric by using logs, it doesn't change the result of the p-value calculation. That is because, essentially, a p-value is the fraction of the 10,000 simulations that are more extreme than our observed result, and logging doesn't change the rank order of each simulation.

When we use logs, we will take the list of 10,000 r_i's and log them to produce a new list: $\ln(r_1), \ln(r_2), \ldots, \ln(r_{10,000})$. This forms a new distribution centered at the null value 0 (= ln(1)) (Figure 6.51).

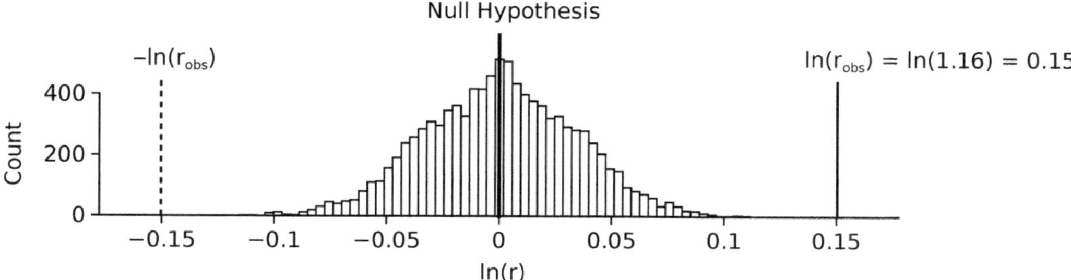

Figure 6.51 Null distribution of the T-cell count study when using logs. Histogram of the logs of the 10,000 r_i values. Also shown are the two symmetric cases of the observed result in the logged space: $\ln(r_{obs})$ and $-\ln(r_{obs})$. Note how these two cases, which ought to be symmetric, now symmetrically flank the null value ($\ln(1) = 0$).

The conclusion is the same as before, the observed result of $\ln(r_{obs}) = \ln(1.16) = 0.15$ is highly statistically significant ($p < 0.0001$).

Confidence Intervals for ratios

Finally, we will sketch the procedure for obtaining confidence intervals for an observed difference r_{obs}.

As usual for confidence intervals, we use a Two-Box resampling procedure. We generate the first pseudo-sample by drawing 49 times with replacement from A, calling it A_1, and 45 times with replacement from B, calling that B_1. Then we take the median as the group measure, $M(A_1) = \widetilde{A_1}$ and $M(B_1) = \widetilde{B_1}$. Now we calculate the effect size in our first resample by taking the ratio of two group measures, $r_1 = \frac{M(B_1)}{M(A_1)}$.

For confidence interval calculations, logged space is mandatory to preserve the symmetry. Since we are operating in the logged space, we take the log of the effect size, calling it $\ln(r_1)$.

We repeat the procedure 10,000 times, generating 10,000 resamples, each with its own log transformed effect size,

$$\ln(r_1), \ln(r_2), \ldots, \ln(r_{10,000})$$

We rank the list of 10,000 log transformed effect sizes from lowest to highest, and find the cutoff for the lowest 50 values as the 0.5% cutoff, calling it $\ln(r)_{lower}$, and for the highest 50 values as the 99.5% cutoff, calling it $\ln(r)_{upper}$.

On the logged axis, $\ln(r)$, the lower and higher cutoffs for the 99% CI are calculated using the same procedure that we introduced in Chapter 4 (see *How to calculate the confidence interval* on page 149). And the resulting 99% CI on the logged axis, $\ln(r)$, is shown in Figure 6.52.

ln(r) axis 99% CI = [CI lower bound, CI upper bound] = [0.11, 0.20]

After obtaining the confidence interval in the logged space, we need to transform it back into the original unlogged space to make sense of it.

r axis 99% CI = $[e^{\text{CI lower bound}}, e^{\text{CI upper bound}}] = [e^{0.11}, e^{0.20}] = [1.12, 1.22]$

Now we can say that the observed "difference" is 1.16, or a 16% increase in T-cell count in the treatment relative to the control, and that the 99% confidence interval is [1.12, 1.22].

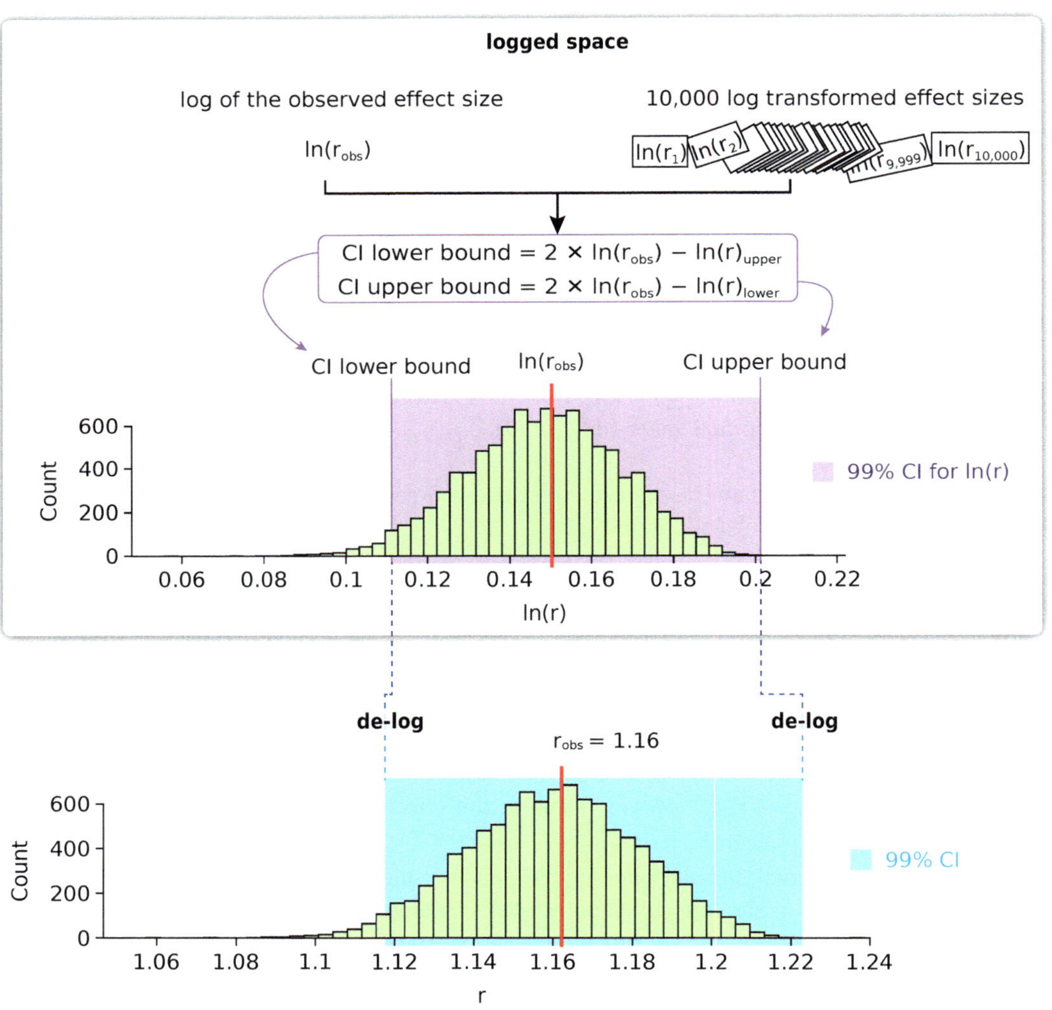

Figure 6.52 99% confidence interval for the T-cell count study when using ratio as the effect size to compare the control and treatment groups. In the logged space, the histogram shows the resampling distribution, the observed effect size is indicated in red, and the 99% confidence interval is in pink. After de-logging to return to the original space, the observed effect size is in red, the 99% confidence interval is in blue.

6.9 The "*t*-test"

This section is highly technical and is not necessary for understanding anything in this book! It's a response to some defenses of the *t*-test formula approach. Do not read this section unless you absolutely have to!

What is a *t*-test?

Most scientists are familiar with the term "*t*-test". It is the most widely used form of statistical comparison in the literature. As one source remarked:

> "it is used in the literature because it's taught in courses, and it's taught in courses because it's used in the literature."

So what is a "*t*-test"? The *t*-test is a formula. What most people do in practice is plug their data into a calculator, which "does a *t*-test", the nature of which is hidden from the user. Behind the screen, and invisible to the user, it does two calculations. The first is to calculate the value of the "*t*-statistic" for the present data set. The second is to compare that *t*-statistic to a theoretical distribution for the *t*-statistic, and return a p-value that reflects how extreme the calculated *t*-statistic is in the theoretical distribution. If that p-value is less than the cutoff value α (alpha), the program then declares that the two groups are "statistically significantly different".

At a slightly higher level, many traditional statistics courses will give the formula that the computer is using, and let you calculate the *t*-statistic by hand. The most common formula is called "Student's *t*-test".[4] It applies to groups A and B, uses the formula

$$t = \frac{\sqrt{n}\,(\overline{A} - \overline{B})}{\sigma}$$

where

1. n = the sample size of both A and B (assumed to be equal)
2. $\overline{A}$ and $\overline{B}$ are the mean (average) values of the data sets A and B
3. σ = the standard deviation of both A and B, assumed to be equal

You calculate t from your data, and then you look up that value in a printed table of values (old school) or you let a computer program do it for you (more recent). Either way, if the calculated *t*-statistic from the data is greater than the cutoff *t*-value given in the table, then you can say "$p < 0.01$" or whatever.

There is something deeply unsettling about this procedure. Leave aside for the moment the assumptions about $\sigma_A = \sigma_B$ and $n_A = n_B$, and let's look at the overall procedure: you use a formula whose origin is mysterious, in order to calculate a number, and then you compare that number to a table, which is based on another formula of mysterious origin, to see if it is bigger than a certain cutoff value.

[4] We shouldn't think that the word "Student" means that this is for students! In fact, "Student" was the pen name chosen to protect the anonymity of its author, W. S. Gossett, an employee of the Carlsberg Brewing Company.

This is, obviously, extremely unsatisfying and a recipe for Bad Practice: we shouldn't be doing things we don't understand and are merely following blindly, what students sometimes describe as "pluginski". What we are emphasizing in this book is: you should understand what you are doing! If it's mumbo-jumbo to you, don't do it!

If we look under the hood (or behind the screen), what we find is truly problematic.

The formula that describes the distribution of t-values is based on a theorem. As we know, any theorem has a set of assumptions that must be met for the theorem to be valid. In the case of the t-test, these assumptions are very demanding.

Student's t-test The first version of the theorem is one that can be found on many websites and in many articles.

Student's t-test (theorem)

IF

groups A and B are Normally distributed
and we use the mean ($\overline{A}$ and $\overline{B}$) to describe the groups
and we use subtraction to compare the two group means
and A and B have the same size n
and A and B have the same variance σ^2

THEN

under the Null Hypothesis that A and B are drawn from the same Normal distribution, the quantity $t = \frac{\sqrt{n}(\overline{A}-\overline{B})}{\sigma}$ will be distributed like a theoretical distribution called "the t-distribution". If an observed t-statistic is highly unlikely according to this distribution, then the Null Hypothesis is disconfirmed.

The t-distribution is a probability density distribution for the values of t, given that A and B are drawn from the same Normal distribution, which is the Null Hypothesis. Given the sample size n, $p(t \mid v)$ is the probability of observing a given value of t. It is defined as follows

$$p(t \mid v) = \frac{\Gamma\left(\frac{v+1}{2}\right)}{\sqrt{v\pi}\,\Gamma\left(\frac{v}{2}\right)} \left(1 + \frac{t^2}{v}\right)^{-\frac{v+1}{2}}$$

$$\text{where} \quad v = n-1, \quad \Gamma(z) = \int_0^{\infty} x^{z-1} e^{-x} dx$$

The t-test procedure is to take the t_{obs} that was calculated from your data, and look up that value in the theoretical t-distribution. If t_{obs} is in the top 0.5%ile or the bottom 0.5%ile, you can then say that the difference is statistically significant at the $p < 0.01$ level.

Our first problem with the t-test is the total lack of transparency of the formula for the probability density distribution, the t-distribution. Where does it come from? What does it mean? What is the Gamma function, and what is it doing in this formula? Why a $\sqrt{\pi}$ in the denominator? None of these questions can be answered by the typical user, even the typical professor, so we are left in the unenviable position of blindly applying a formula that we do not understand.

Regardless of that misgiving, we might think that at least, the theorem holds when its assumptions are met. But what is it really doing?

It is important to see that the distribution $p(t\,|\,\nu)$ is the probability density distribution of the t-statistic of two samples of size $n=\nu+1$ drawn from the same Normal distribution; its extreme values mean that so large a difference is very unlikely.

But we've already seen that Big Box resampling is an simulation-based answer to the question: how big a difference could we expect from two groups sampled at random from the same underlying distribution? Moreover, Big Box resampling is valid for *any* distribution, not just Normal distributions, and is valid for *any* group measure, not just the mean, and for *any* difference measure, not just subtraction. Student's t-test applies only to a very special case of the much broader question that Big Box resampling answers.

Student's t-test vs. Big Box resampling

The theoretical t-distribution $p(t\,|\,\nu)$ is what we would expect to see if we drew two samples of the same size $n=\nu+1$ from the same Normal distribution, calculated their means and standard deviations, and plugged them into the formula to calculate the t-statistic.

But the t-statistic is just the difference of the means of the two groups, with corrections for n and σ. Therefore, given n and σ, the p-value as determined by the t-distribution ought to be identical to the p-value of the differences in means $(\overline{A}-\overline{B})$ as determined by Big Box resampling. In other words,

IF both groups are bell-shaped AND have the same size AND the same variance AND we use the mean AND we use subtraction to compare.
THEN Big Box resampling gives the same p-value as Student's t-test.

This can be verified by choosing an example, and showing that the Big Box procedure gives the same answer as the t-test when the data meet the assumptions of the t-test.

To set up our test example, we chose two n = 40 samples at random from the same Normal distribution, $\mathcal{N}(30,3^2)$.

The data sets A and B are shown below.

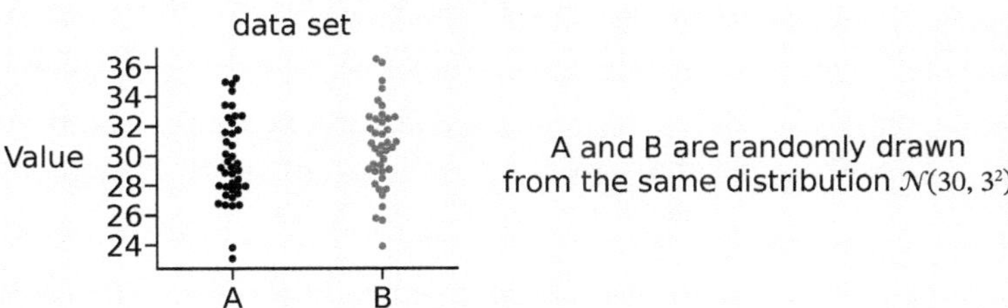

We calculated Δ_{obs} for A and B, and used Big Box resampling to obtain a p-value for Δ_{obs}. As can be seen in the figure below, the resampling procedure returned a p-value of p = 0.190. When we subjected the same data to Student's t-test, using the Python package `scipy`, we obtained a p-value p = 0.198, very close to the resampling result.

Therefore, Student's t-test is the formulaic approximation to the Big Box resampling procedure, in the special case in which the data sets are Normally distributed, and we use the difference of means as our effect size. Big Box resampling, of course, is valid much more generally.

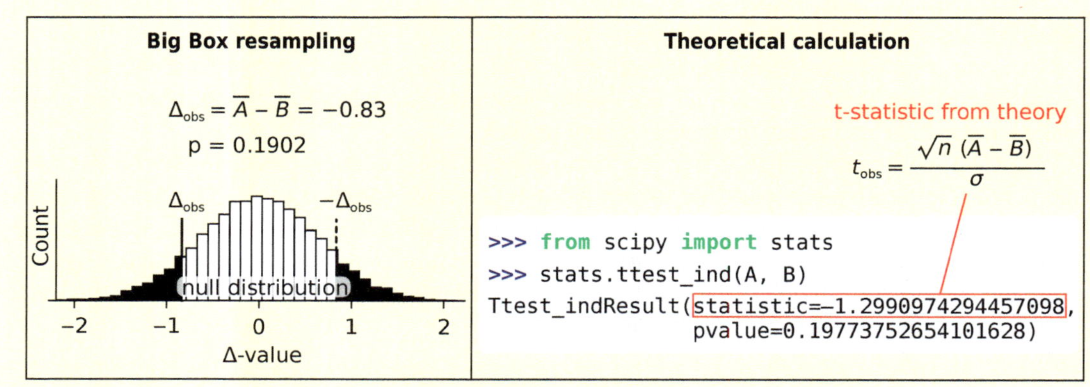

It is crucial to look carefully at the assumptions for the applicability of Student's t-test. When we do, we find that they are frequently NOT met by data sets, and so the test is being widely misused.

Welch's t-test Statistical theoreticians have been able to relax the conditions that $n_A = n_B$ and that $\text{var}_A \approx \text{var}_B$.

Remember that for the resampling approach, it was perfectly OK to use the Big Box approach to simulation even if n_A did not equal n_B. We simply mirrored the actual data, and compared apples to apples: if n_A does not equal n_B, then we selected n_A from the Big Box for our pseudo-A group and n_B for our pseudo-B group. Different sample sizes is not a problem for Big Box resampling.

The other requirement, that the variances are approximately equal, is more problematic. As we saw, if the variabilities are quite different (more than 2:1 was our rule of thumb) then we had to use the Re-centered Two-Box method.

For the formula-based approach, the possibility that n_A is not equal to n_B, and the possibility of different variabilities, requires a different theoretical formula, called Welch's t-test.

Welch's t-test (theorem)

IF

groups A and B are Normally distributed

and we use the mean ($\overline{A}$ and $\overline{B}$) to describe the groups

and we use subtraction to compare the two group means

THEN

there is a new quantity, also called t, defined as

$$t = \left(\sqrt{\frac{1}{n_A} + \frac{1}{n_B}}\right)^{-1} \left(\frac{\overline{x}_A - \overline{x}_B}{\sqrt{\dfrac{(n_A - 1)\widehat{\sigma_A}^2 + (n_B - 1)\widehat{\sigma_B}^2}{n_A + n_B - 2}}} \right)$$

This new quantity t then has a new probability density distribution, which is called Welch's t-distribution, defined as follows

$$p(t \mid v) = \frac{\Gamma\left(\frac{v+1}{2}\right)}{\sqrt{v\pi}\,\Gamma\left(\frac{v}{2}\right)} \left(1+\frac{t^2}{v}\right)^{-\frac{v+1}{2}}$$

$$\text{where} \quad v = \frac{\left(\frac{\widehat{\sigma_A}^2}{n_A} + \frac{\widehat{\sigma_B}^2}{n_B}\right)^2}{\left(\frac{\widehat{\sigma_A}^4}{n_A^2\, v_A} + \frac{\widehat{\sigma_B}^4}{n_B^2\, v_B}\right)^2}, \quad \Gamma(z) = \int_0^\infty x^{z-1} e^{-x} dx$$

Welch's t-test relaxes the equal size and equal variance requirements. The two groups could have different sample sizes, $n_A \neq n_B$ and/or different standard deviations, $\sigma_A \neq \sigma_B$.

Welch's t-test vs. Re-centered Two-Box resampling

Welch's t-distribution is the theoretical probability density distribution of this new t-statistic when A and B, with distinct sample sizes n_A and n_B and distinct variances σ_A^2 and σ_B^2, are drawn from two different Normal distributions with the same mean and differing variances σ_A^2 and σ_B^2. It answers the question: how large a difference in means would we be likely to see from drawing randomly from two distinct Normal distributions with the same mean and these variances? With the exception of the Normal distribution requirement, *this exactly describes the requirements for the Re-centered Two-Box resampling test.*

Therefore, Welch's t-test is the formulaic approximation to the Re-centered Two-Box resampling procedure, in the special case in which both data sets are Normally distributed, and we use the difference of means as our effect size. Re-centered Two-Box resampling, of course, is valid much more generally.

In other words,

IF	both groups are bell-shaped ~~AND the same size AND the same variance~~ AND we use the mean AND we use subtraction to compare.
THEN	Re-centered Two-Box resampling gives same result as Welch's t-test.

Similarly, this can be easily verified by choosing two samples from two different Normal distributions and comparing the results of the t-test formula and Re-centered Two-Box resampling.

Thus, compared to Student's t-test, Welch's t-test applies to a wider class of cases (unequal n and unequal σ), *but it still assumes that the two groups are Normally distributed and that we are using the algebraic difference of the mean values as the group difference.*

Problems with the t-test

The assumption of Normally distributed data sets We've now addressed the problem of unequal n and unequal σ. But one of the biggest limitations of the t-test is another one: the requirement that A and B both be Normally distributed.

This requirement is prominently displayed on many websites. It is common to see the requirement stated for both the Student's version and Welch's version.

For example, a credible website, at UC Berkeley,[5] says

[5] https://statistics.berkeley.edu/computing/r-t-tests

"The assumption for the test is that both groups are sampled from normal distributions with equal variances."

This is obviously a very strong requirement. **It means that a great number of applications of Student's and Welch's t-tests are invalid. It is simply wrong to ignore the fact that the data are not Normal and apply the t-test anyway.**

Unfortunately, this is done a lot of the time in the scientific and medical literature. Many researchers apply the test in spite of their distributions being markedly non-Normal. This is a significant source of statistical error in the literature.

Do we really require Normal distributions? As we stressed this requirement to our colleagues, who were not always happy to hear it, we got some pushback. No, we were told, the t-test does *not* require that A and B be Normally distributed. Rather, we were told, the requirement is much weaker. They said "no, the theorem doesn't require that the data sets be Normally distributed, it only requires that the *averages* be drawn from a Normal distribution, not the data itself."

The question then is: how can we tell if two averages are drawn from a Normal distribution? The answer we are given is that something called "The Central Limit Theorem" applies, which says that the averages of large samples from almost any distribution will follow a Normal curve.

For example, Wikipedia takes this more general point of view. According the posting under "Student's t-test", the data doesn't have to be Normally distributed, rather, "the means of the two populations being compared should follow Normal distributions."

But how are we to know whether the two means we observed come from a Normal distribution? Here, we are told, there is a theorem that comes to our aid. Wikipedia continues immediately:

1. Under weak assumptions, this follows in large samples from the Central Limit Theorem, even when the distribution of observations in each group is non-Normal.

2. To cover both cases we use the Lindeberg-Feller Central Limit Theorem. An approximate translation of this result is that if $Y_1, Y_2, \ldots, Y_n$ are a large collection of independent random variables with variances $s_1^2, s_2^2, \ldots, s_n^2$, then the average

$$\overline{Y} = \frac{1}{n}\sum_{i=1}^{n} Y_i$$

is approximately Normally distributed, with mean equal to the average of the means of the Ys and variance equal to the average of their variances, under two conditions:

a) The variance of any single observation is small compared to the sum of the variances.

b) The number of outcomes that are extreme outliers, more than $\sqrt{n}$ standard deviations away from their mean, is small.

So we are told that this is true "under weak assumptions". But when we examine these "weak assumptions", we see the problem. Notice the clause that we highlighted above: each of the possible distributions *must have a finite variance!* That's what is meant by "has a variance s_1^2": it has to be a finite number.

But don't all distributions have finite variances? No, they do not! First of all, notice that they are not talking about *finite* distributions. Of course every finite set has a variance. They are referring to an infinite distribution, and here it is not true that every infinite distribution has a finite variance.

One of the most important kinds of distribution, that has received intense attention in recent decades, are the so-called "long tail" distributions. As detailed in many articles and in Chris Anderson's book *The Long Tail* as well as other books, long-tailed distributions, like those following a power law, $f(x) = \frac{1}{1+x^2}$, describe important distributions much better than do Gaussian (Normal) distributions, which die off <u>much</u> (!) faster, as e^{-x^2}. Examples of long-tailed distributions include the size distribution of earthquakes, book sales at Amazon.com, movie rentals at Netflix, etc.

The key point is: **power law distributions do not have finite variances! And neither do many other distributions.** So we would have to assume that they are not the underlying distributions from which we are sampling.

But so many distributions in reality are power laws, so how could we make this assumption of finite variances?[6]

And then, even if we ignore all these warnings, and swallow the pill despite all the specific labels warning us not to use it in this case, there is still a limitation that cannot be overcome.

What about freedom of measure? We see that the t-test (even Welch's version) is only of the *algebraic difference* of the *means* of two groups. But what if those two choices are not appropriate for our data?? We are out of luck: the t-test is strictly of the algebraic difference of two means. If we want freedom of measure, say, to use ratios instead of differences, or medians[7] instead of means, we have to look elsewhere.

There's an old proverb that says "To someone with only a hammer, everything looks like a nail", and that definitely applies here. Many texts and lectures that push us to use the t-test based on these very doubtful criteria, also have chapters extolling the mean as having nice mathematical properties. We are not interested here in nice mathematical properties, but rather whether procedures accurately describe the data and answer our question. As we have seen, the mean is often is a poor choice of measure from the scientific point of view.

In summary, freedom of measure is out of the question for formula-based approaches, such as the t-test. Resampling-based approaches, on the contrary, give us freedom of measure, saying to us "make your best scientific choice of measure, knowing that you will be able to test its statistical significance".

---

[6] It might be objected that if we were to truncate the power-law distribution to a finite interval, say between -10^{100} and 10^{100}, then the variance would have to be finite, and so the Central Limit Theorem would apply. But in that case, the variance in a typical sample will be huge, and the difference between two typical sample means will be much smaller than those variances, so the t-test will need much larger effect sizes and/or sample sizes to achieve a power (ability to find true associations) that is comparable to that of more general tests, such as rank-based tests. This makes it much harder to find statistical significance.

[7] There is a mathematical approach to the difference of medians. Suppose our Null Hypothesis is that the medians of the two groups are the same. Then Mood's Median Test will test this Null Hypothesis by asking how often a randomized (Big Box) set would produce the same relative distributions above and below the medians of groups A and B. This can be done by formula or by simulation methods. However, the test relies on the strict assumption that "the distributions of the populations the samples were drawn from all have the same shape."

Advantages of resampling-based approaches

The "t-test" is a theoretically derived formula. The necessary theoretical assumptions result in three major limitations. Resampling-based approaches are not subject to these limitations.

t-test	resampling-based approach
Two Normally distributed groups. The t-test requires that the measurements in each of the two groups be Normally distributed.	**No distributional assumption.** The two groups do not need to be Normally distributed.
Means. The t-test requires you to choose the mean as the summary measure for each of the two groups.	**Freedom of measure.** We can choose any measure to describe the two groups: mean, median, MAD, ...
Algebraic difference. The t-test only allows you to compare the algebraic difference between the two group means.	**Freedom of comparison.** We can compare the two group measures by subtraction, or division, or absolute differences ....

FURTHER EXERCISES 6.9

1. Select all requirements that are necessary in order to use Student's t-test to compare two groups.
   a) Both groups are Normally distributed (bell-shaped) with same variance.
   b) The two groups have the same sample size.
   c) The parameter of interest is the difference between the means of the two groups.
   d) Both groups are representative samples of their population.

2. Select all requirements that are necessary in order to use Welch's t-test to compare two groups.
   a) Both groups are Normally distributed (bell-shaped) with same variance.
   b) The two groups have the same sample size.
   c) The parameter of interest is the difference between the means of the two groups.
   d) Both groups are representative samples of their population.

3. Select all requirements that are necessary in order to use the resampling-based approach to compare two groups.
   a) Both groups are Normally distributed (bell-shaped) with same variance.
   b) The two groups have the same sample size.
   c) The parameter of interest is the difference between the means of the two groups.
   d) Both groups are representative samples of their population.

6.10 What test should we use?

In this chapter, we have learned the resampling-based approach for two-group comparisons, including NHST and confidence intervals.

We use the Two-Box method to construct confidence intervals, whereas for the p-value calculation, we take care to choose an appropriate resampling technique to simulate the Null Hypothesis (Figure 6.53).

For completeness, we have included the two formula-based tests (Student's and Welch's

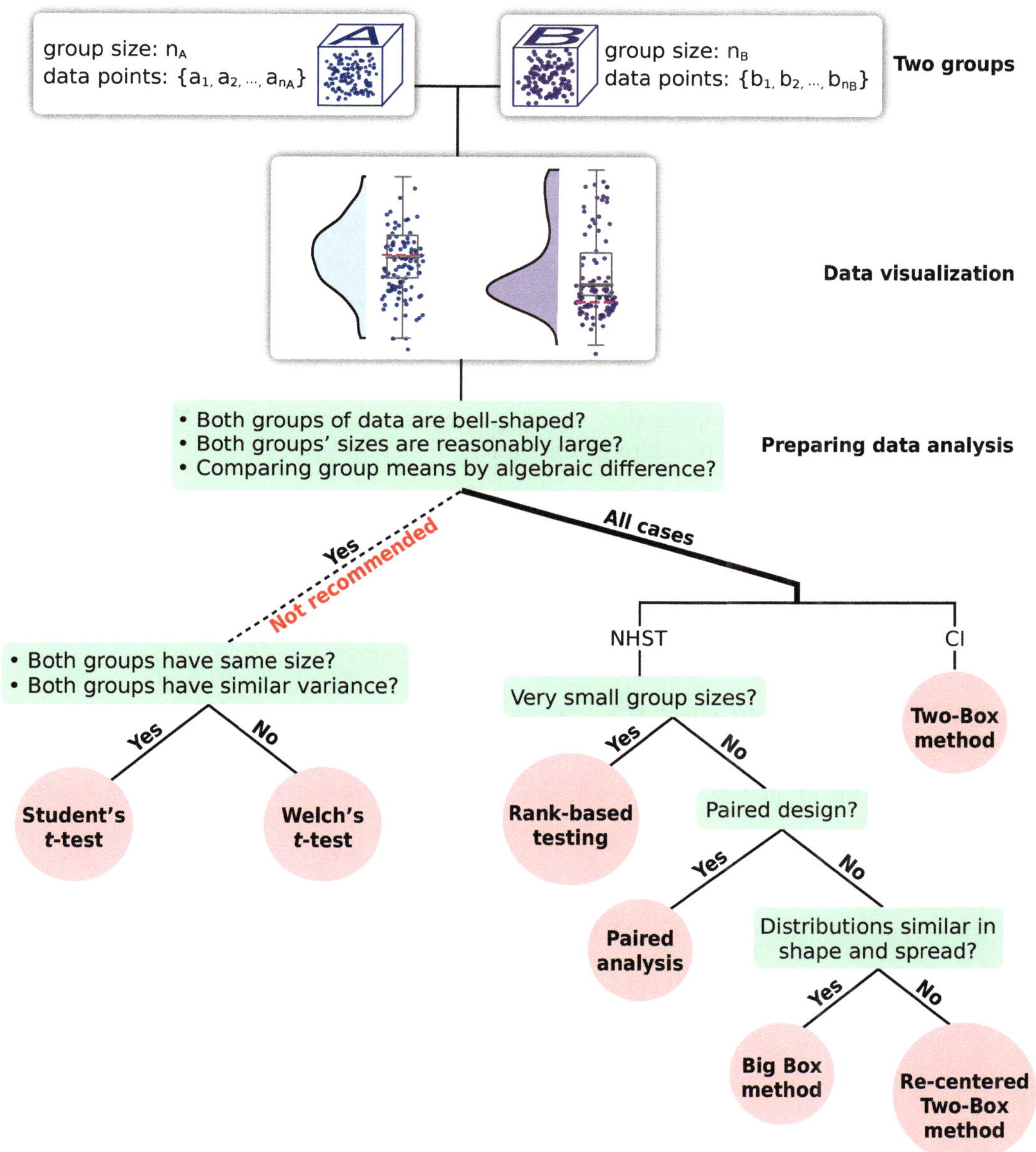

Figure 6.53 Decision tree for two-group comparisons.

t-tests). However, we do not recommend using them, ever, even on the rare occasion when the technical assumptions are met. The would-be user has to face the burden of first demonstrating that the distributions are in fact Normal. This must be shown, and can't simply be assumed. For example, the *Nature* family of journals has the following in their "Statistical Guidelines":

> **Normal distribution:** Many statistical tests require that the data be approximately normally distributed; when using these tests, authors should explain how they tested their data for normality. If the data do not meet the assumptions of the test, then a nonparametric alternative should be used instead.

As we've said, the resampling approach, which is nonparametric, is *always* applicable and should be our method of choice.

FURTHER EXERCISES 6.10

1. Why is it important to visualize raw data before doing any statistical significance testing?
2. Give and explain three reasons why comparing two groups using simulations is superior to using formulas.
3. You are studying the effect of a new diet on cholesterol levels. Each patient's cholesterol is measured before they begin the diet and again after they have followed it for three months.
   a. What method would you choose to conduct an NHST on the data?
   b. What could you compute that would give more information than NHST? Describe how this would help to interpret the data.
4. **Sunscreen pilot study.** A researcher conducted a pilot study to compare two types of sunscreen: one that is environmentally safe for coral reefs (reef-safe, e.g., zinc-based) and one that is not (standard). Five beachgoers were recruited for each sunscreen type and asked to rate their satisfaction on a scale from 0 (hated) to 10 (loved).

   Reef-safe scores: $\{5, 6, 4, 7, 5\}$ Standard scores: $\{6, 8, 7, 6, 8\}$

   Identify the appropriate effect size statistic for this study, calculate its value from the sample data, conduct the relevant statistical analysis, and interpret your findings.
5. Select all forms of analysis we could use for the data set below. Describe each analysis.

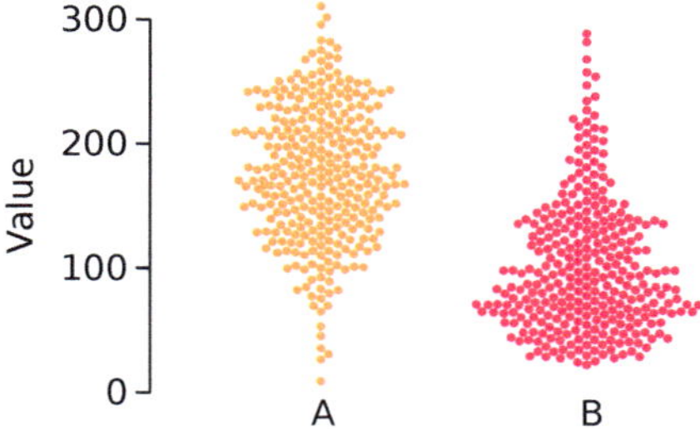

   a) Formula-based t-test
   b) Big Box NHST for two groups
   c) Re-centered Two-Box NHST
   d) Paired NHST
   e) Rank-based NHST
   f) CI on effect size

6. **Revisit grandma's home remedy.** In the FE 6.6.1 on page 232, your grandmother gives you a home remedy that she swears by to shorten the duration of colds. You set out to determine if this actually works.

   Your friend who works at a drug company is interested in testing this home remedy for possible distribution. What should they do differently from you in designing their study and analyzing their data? Why?

7. Here are four sets of two-group sample data sets. Based on the following graphical presentations, use the decision tree to pick a suitable type of two-group comparison for each pair of data.

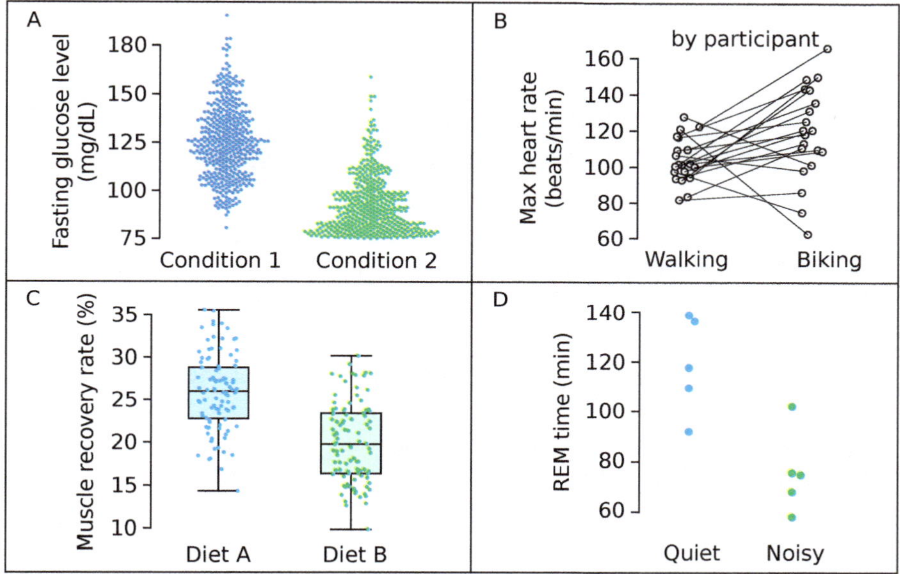

8. **Myocardial infarction.** The principal investigator (PI) of your research lab asks you to compare the number of years patients lived after a myocardial infarction (MI, a.k.a. heart attack) after using the standard treatment (control) and a new drug (treatment). The clinical trial randomly assigned 30 people to the control group and 30 people to the treatment group. You visualize the distribution and calculate the summary statistics:

   - For the control group, the mean is 16.87, the median is 15, and the MAD is 5.
   - For the treatment group, the mean is 30.13, the median is 23.7, and the MAD is 6.7.

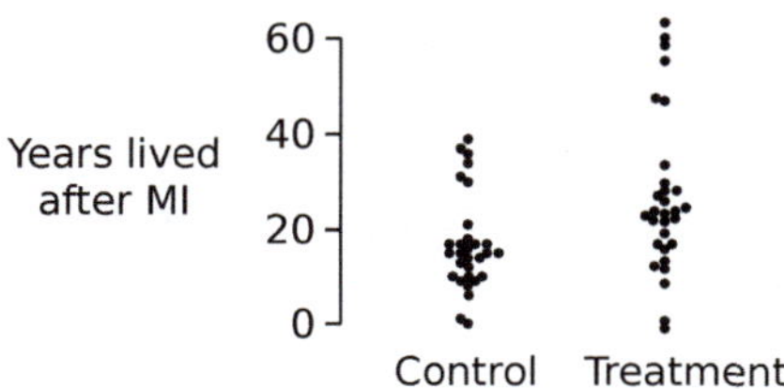

   a. A graduate student in your lab recommends using a t-test for this analysis. How should you answer?

b. What is the statistic of interest (effect size) in this study? What is its value for these samples?
c. State the Null Hypothesis in the context of this study and as a mathematical statement using the statistic of interest.
d. You want to determine if the treatment has a statistically significant effect. How should you conduct resampling?
e. While calculating a p-value for the statistic of interest using resampling, you run 10,000 simulations. Based on the information you have, which of the four plots below is most likely to be the histogram of your simulations?

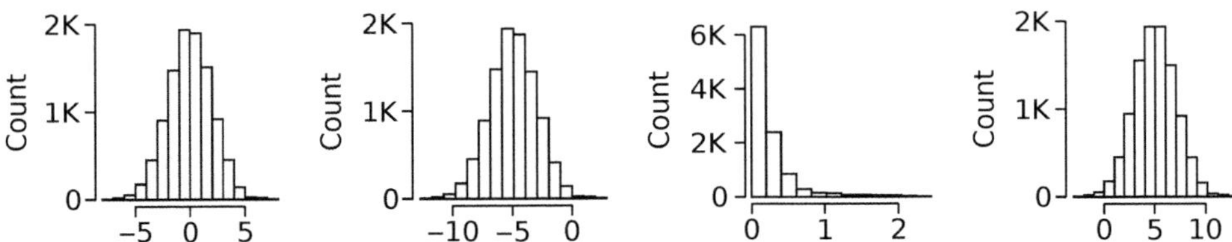

f. Suppose 90 of your 10,000 simulations have an effect size greater than or equal to the observed effect size, and 93 others have an effect size less than or equal to minus the observed effect size. What is your p-value?
g. If you set $\alpha = 0.01$, what should you conclude from the significance testing?
h. You also want to calculate a 99% confidence interval for the effect size. How should you conduct the resampling?
i. You find that the 99% confidence interval for the statistic of interest is [−1, 17]. Based on the information above, do you think you did your analysis correctly?
j. Your lab's PI realizes they made a mistake! They forgot to tell you that actually the clinical trial was set up such that each person in the control group had a counterpart in the drug group who had the same race, gender, age, and cholesterol level. What is the type of the study design?
k. Based on the new study design, what kind of plot should you use to visualize the data?
l. You want to incorporate this new information about the study design into your analysis of statistical significance. How should you conduct resampling to determine if the treatment has a statistically significant effect?
m. How do you expect the p-value and confidence interval for the effect size to change once you analyze the data with this new information?

9. **Length of salmon: similar variability.** Researchers want to determine whether the median length of salmon differs between two rivers: the North River and the South River. They collect a sample of 100 salmon from each river. Suppose the MADs of the two samples are similar ($MAD_{North} \approx MAD_{South}$), suggesting both samples have comparable spread.

   a. What is the Null Hypothesis?

      a) Median salmon lengths in the two rivers are the same.
      b) Mean salmon lengths in the two rivers are the same.
      c) Sample salmon from the two rivers are actually from the same population.

b. What does the box model look like for NHST, and why?

a) Two boxes (one for sample data from each river, de-medianized), because the spreads are too different to combine.

b) Two boxes (one for sample data from each river), because the medians are different.

c) One big box (sample data from both rivers combined), because the spreads are similar enough to treat both samples as from the same population.

d) One big box (sample data from both rivers combined), because the sample sizes are equal.

c. To conduct an NHST, each time you resample, what do you draw from the box model?

a) 200 salmon for a combined sample
b) 100 salmon in total
c) 50 salmon + 50 salmon
d) 100 salmon + 100 salmon

10. **Length of salmon: very different variability.** Researchers want to determine whether the median length of salmon differs between two rivers: the North River and the South River. They collect a sample of 100 salmon from each river. Suppose the salmon lengths in the two rivers show very different variability, $\text{MAD}_{\text{North River}} > 3 \times \text{MAD}_{\text{South River}}$, meaning the North River salmon vary much more in length than those in the South River.

a. What is the Null Hypothesis?

a) Median salmon lengths in the two rivers are the same.

b) Mean salmon lengths in the two rivers are the same.

c) Sample salmon from the two rivers are actually from the same population.

b. What does the box model look like for NHST, and why?

a) Two boxes (one for sample data from each river, de-medianized), because the spreads are too different to combine.

b) Two boxes (one for sample data from each river), because the medians are different.

c) One big box (sample data from both rivers combined), because the spreads are similar enough to treat both samples as from the same population.

d) One big box (sample data from both rivers combined), because the sample sizes are equal.

c. To conduct an NHST, each time you resample, what do you draw from the box model?

a) 200 salmon for a combined sample
b) 100 salmon in total
c) 50 salmon + 50 salmon
d) 100 salmon + 100 salmon

11. **Two very small samples.** You have two data sets: {2, 4, 5, 4, 6} and {3, 8, 4, 5, 4}.

a. How would you describe the data sets and how would you go about comparing the two groups? What's the effect size?

b. Which of the following is an appropriate box model to calculate a p-value?

a) {2, 4, 5, 4, 6, 3, 8, 4, 5, 4}

b) {2, 4, 5, 4, 6} and {3, 8, 4, 5, 4} (separately)

c) {2, 3, 4, 5, 6, 8}

d) {−2, 0, 1, 0, 2} and {−1, 4, 0, 1, 0}

c. Describe the resampling procedure to obtain a p-value.

12. **Huntington's disease.** A researcher conducted an experiment on survival times of patients with Huntington's disease under treatments A and B. Below are the results: descriptive statistics and dot plots for each group (upper), the null distribution with observed difference in medians (middle), and the confidence interval for the effect size (lower).

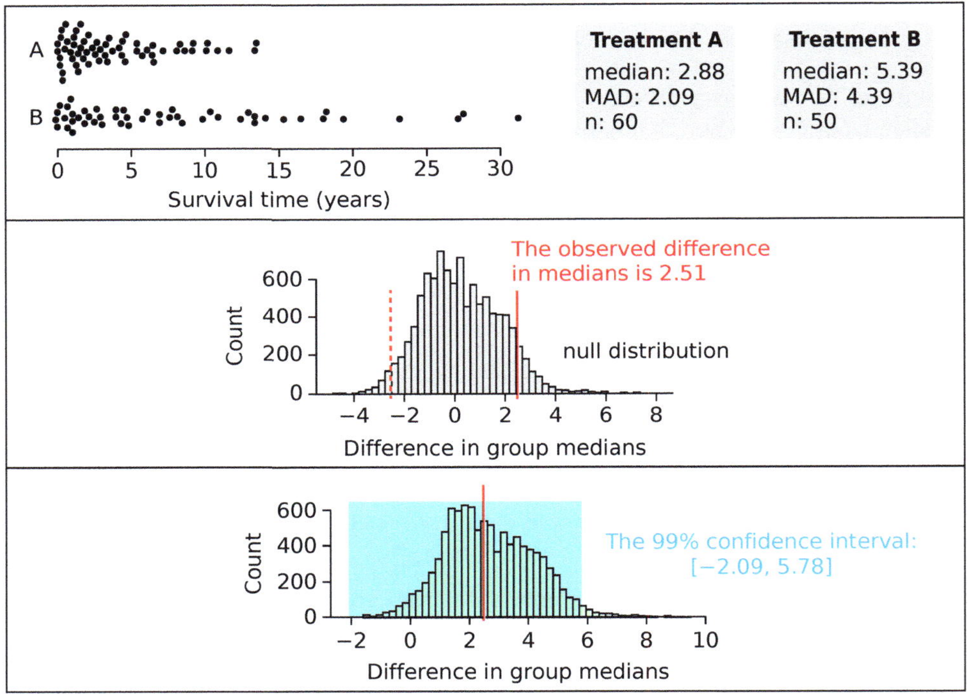

Answer the following questions:

a. What is the researcher's Null Hypothesis?

b. What method should the researcher use to calculate a p-value? Briefly describe it.

c. Write a sentence interpreting the p-value.

d. Describe how the researcher should calculate a confidence interval for the effect size.

e. Write a sentence interpreting the confidence interval.

13. **CI and p-value.** We found that a 95% confidence interval for the observed difference Δ_{obs} between group A and group B is [3, 8]. Assuming the null value is 0, what does this imply about the p-value? What if the 95% confidence interval was [−2, 6]?

a) $p > 0.01$ b) $p < 0.99$ c) $p < 0.01$ d) $p > 0.05$
e) $p < 0.05$ f) It doesn't tell me anything about the p-value

14. **Stopping distance.** A tire manufacturer tested the braking performance of one of its tire models. The company tried the tires on 10 different cars, recording the stopping distance for each car on both wet and dry pavement.

Car ID	Stopping distance (m)	
	Dry	Wet
1	160	208
2	156	212
3	143	192
4	147	159
5	130	178
6	134	172
7	134	161
8	128	180
9	137	196
10	156	207

a. Visualize the data.
b. Describe the data distributions for both groups.
c. Define an appropriate measure for each group, and a measure to compare the two groups.
d. What is the Null Hypothesis?
e. What is the observed test statistic? Interpret this observed test statistic.
f. Using an appropriate method, simulate the Null Hypothesis and calculate a p-value.
g. Is your p-value one-sided or two-sided?
h. Given the result of the p-value calculation, state the conclusions from the study.
i. Calculate a 95% confidence interval for the observed effect size. Interpret the result.

15. **No pressure.** A friend who has not taken any statistics courses shows you an online article that claims that new research shows that organic vegetables lower blood pressure. The article says there is only a 5% chance the finding is false.

a. Explain to your friend where that 5% number likely comes from and what it actually means. Your explanation should be understandable to someone with no statistics background.
b. Moreover, should you believe this finding? What other information would you like to know?

Inspired by your friend's claim, you teamed up with researchers in your lab to try to replicate the finding. The scientists enlisted 40 participants, 20 of which ate only organic vegetables for two months, and 20 of which ate conventional vegetables. At the end of the two months, they measured the changes in their blood pressure.

	Change in blood pressure in mmHg (change = after − before)
Organic	−8, −10, −23, +6, 0, −8, −11, +4, −4, −16, −5, −1, +6, 0, +12, −10, −2, −9, −14, +8
Non-organic	+6, 0, −4, +11, +7, −2, +13, +3, −5, +6, +3, +12, +8, +6, −2, +16, +8, +11, +10, −5

c. Visualize the data.

d. Describe the data distributions.

e. Define an appropriate measure to compare the two groups.

f. You prepare to conduct a statistical test comparing the groups. What is the Null Hypothesis?

g. What is the observed test statistic? Interpret this observed test statistic.

h. Using an appropriate method, simulate the Null Hypothesis and calculate a p-value.

i. Is your p-value one-sided or two-sided?

j. Given the result of the Null Hypothesis significance test, state your conclusions.

k. Finally, your friend wants to know, from your study, just how much it might truly help his blood pressure if he starts eating organic vegetables. Calculate an estimate with 95% confidence.

l. Provide an interpretation of the 95% confidence interval in terms your friend might comprehend.

Chapter 7

Comparing Three or More Groups

LEARNING OBJECTIVES

After reading this chapter, you will be able to:

- perform a resampling-based analysis in answer to the question "are there significant differences among these groups?"
- if there are significant differences, identify which groups are significantly different.
- discuss good and bad ways to present a data analysis comparing three or more groups.
- discuss the perils of multiple comparisons, and be able to employ techniques to control the rate of false positives.

We now know how to compare two groups on a single outcome, like males and females in terms of blood pressure, or Diet A vs. Diet B on hours of sleep. But we often want to compare an outcome across *many* groups. We might want to look at the 50 US states in terms of their income levels: can we say which states have statistically significantly higher or lower incomes? Or we might want to look at the effects of 6 different types of exercise (biking, hiking, tennis, weightlifting, walking and swimming) on glucose levels, and ask if any of the forms of exercise have statistically significantly better outcomes.

Suppose, for example, we considered 6 different types of exercise. For each type, we asked a distinct set of subjects to work out using that type of exercise, and then measured their change in glucose levels over the next week (Table 7.1).

Biking	Hiking	Tennis	Weightlifting	Walking	Swimming
−3	+1	−4	−2	0	−2
+2	−3	−5	−4	−1	−3
⋮	⋮	⋮	⋮	⋮	⋮
+1	−1	+2	−3	+5	−6

Table 7.1 Hypothetical example of a 6-group comparison of the effects of different types of exercise on subject's glucose levels (mg/dL).

A. Garfinkel and Y. Guo, *Understanding Data*,
https://doi.org/10.1007/978-3-032-18600-3_7

Similarly, if we were looking for statistically significant differences among the states in terms of their income levels, we would sample the incomes of people from the 50 states, and would then have a 50-group comparison (Table 7.2).

Alaska	Alabama	Arizona	Arkansas	California	...	Wyoming
50.1	60.2	82.3	79.0	50.5	...	78.4
85.5	90.3	77.6	43.1	120.4	...	90.3
⋮	⋮	⋮	⋮	⋮	⋱	67.5
70.8	40.3	110.6	45.5	88.8	...	100.1

Table 7.2 Hypothetical example of a 50-state comparison of annual incomes (dollars in thousands).

How can we carry out a multi-group comparison? We know how to make comparisons of *two* groups. If we were just comparing California to Oregon, we would use a two-group comparison as in the previous chapter. We would sample people from California and people from Oregon, measure their income levels, and use two-group methods to ask if the median (if that is the appropriate measure) income is significantly higher in one state vs. the other.

However, we can't just start making multiple comparisons of group A against group B, group A against group C, etc. As we'll see, there's a problem with that: multiple comparisons generate large numbers of false positives. We will spend significant time developing methods for controlling the occurrence of false positives in multi-group comparisons, what is called "*post hoc* analysis". We will show how to make useful comparisons of one group with another without running afoul of the false positives problem.

In order to see the structure of multi-group comparisons, we will introduce a new concept.

- A **categorical variable** is a variable that has a discrete set of values that are not numbers but properties ("categories"). So "Gender" is a categorical variable, and its values could be {male, female} in one study, or {male, female, nonbinary} in another, or even more nuanced categories, as required by science and the intended application. "Day of the week" is a categorical variable that has seven values {Monday, Tuesday, Wednesday, . . . , Sunday}.

  In the examples above, "US State" is a categorical variable, with 50 values {Alabama, Alaska, Arizona, . . . , Wyoming}, and "Exercise Regimen" is a categorical variable that has, in this case, 6 values {biking, hiking, tennis, weightlifting, walking, swimming}.

- A **numerical variable** (also called a **quantitative variable**) is a variable that takes numbers as values, such as height, weight, blood pressure, etc.

We want to know whether there is some kind of association between the categorical variable (group membership) and the numerical outcomes (survival times, glucose levels, etc.). Do some exercise plans produce better results than others? Are there significant differences across the days of the week in the frequency of traffic accidents?

A typical multi-group comparison would have a data structure like Figure 7.1.

In this chapter, we will develop a general, resampling-based method for making multi-group comparisons. As usual, the resampling-based method makes no assumptions about the shape of the data sets, or choice of measure, and are applicable to a much wider class of cases.

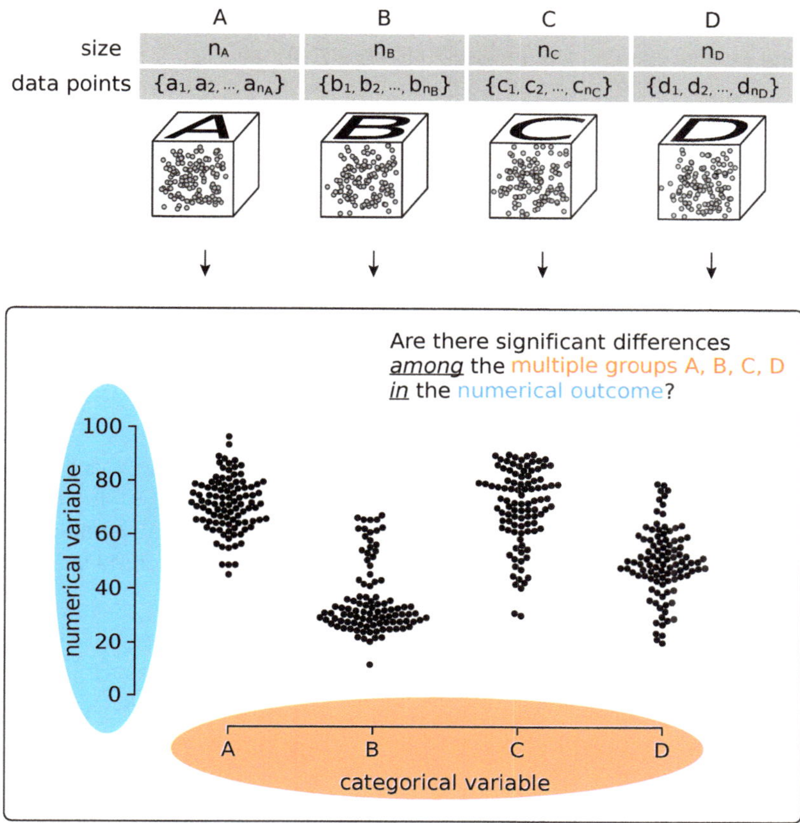

Figure 7.1 Typical set-up and presentation of raw data for a multi-group comparison of groups A, B, C and D (categorical) on an outcome variable (numerical).

At the end of the chapter, we will describe a purely mathematical theory for the comparison of many groups, the theory that is called Analysis of Variance (ANOVA). It says that if all the groups in question are Normally distributed, *and* they meet several other strong requirements, *and* we agree to use the mean values as descriptors, then there is a mathematical theory of the distribution of the algebraic differences of the means (only!) of such groups.

ANOVA is a formula-based approach. We'll compare it to resampling-based approaches. This will be another example of resampling-based approaches vs. formula-based approaches.

Due to the stringent requirements of formula-based ANOVA, we do not recommend it. Instead, we recommend a resampling-based approach.

7.1 Visualizing the data and choosing the measures

Visualizing the data

First, as always, we visualize our data and decide on our measures. For our example, we will consider a case where we have a control group and two drug treatments. We will call the groups A, B and C. We will be considering the control and the two drugs in terms of how they affect T-cell counts in an experiment (higher counts are better). Let's suppose we took some measurements, and we have measurements a_1 through a_{n_A} from group A, b_1 through b_{n_B} from group B, and c_1 through c_{n_C} from group C (Table 7.3).

groups	A	B	C
size	n_A	n_B	n_C
data points	a_1	b_1	c_1
	a_2	b_2	c_2
	$\vdots$	$\vdots$	$\vdots$
	a_{n_A}	b_{n_B}	c_{n_C}

Table 7.3 A typical data set with three groups.

We make a beeswarm plot of the data (Figure 7.2). Here, the categorical variable is the "drug type" (none, drug B or drug C), and the outcome variable is "T-cell count". We want to know if there is a significant association between the categorical variable and the numerical variable, that is, if any of the drugs significantly improves T-cell counts vs. control, and if one drug is significantly better than the other.

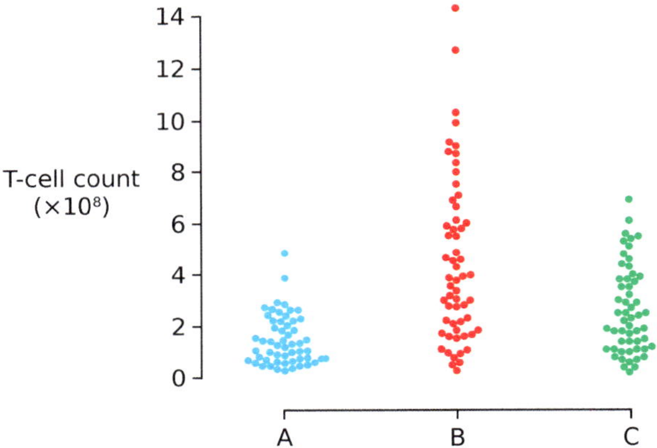

Figure 7.2 Beeswarm plots of the three groups A, B and C.

The visual impression that these three groups are asymmetric, skewed to the right, is confirmed by violin plots (Figure 7.3).

Another useful visual presentation of the data is a box plot (Figure 7.4). Recall from Chapter 2 that the box plot shows the mean or median (in this case we are showing both), the 25th and 75th percentiles of the data, and the max and min of the data.

Notice, in the box plots, how the fourth quartile of the data, the highest quartile of each plot, is much larger than any of the other quartiles, indicating the skew to the "right" (that is, toward the higher values).

Another confirmation of the rightward skew of the data can be seen in the fact that the mean of each data set (the dashed lines) is to the right (larger) than the median (solid line).

We then consider the shape of the data, and let that guide our choice of measures. It is clear, in this case, that the three data sets are all asymmetric, skewed to the right.

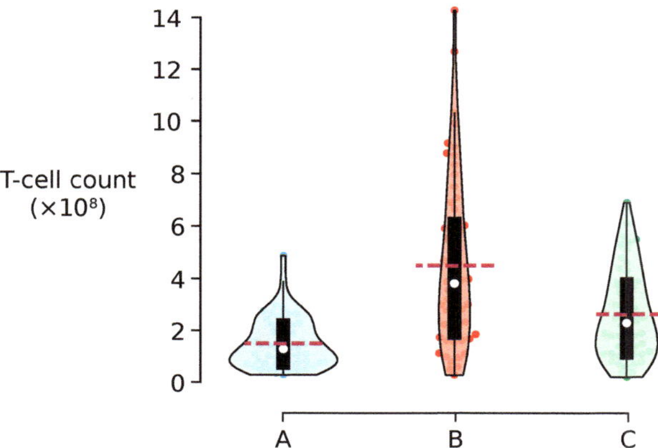

Figure 7.3 Violin plots of the groups A, B and C. Medians are shown in white circles, means in dashed lines. Thick black bars indicate the interquartile range.

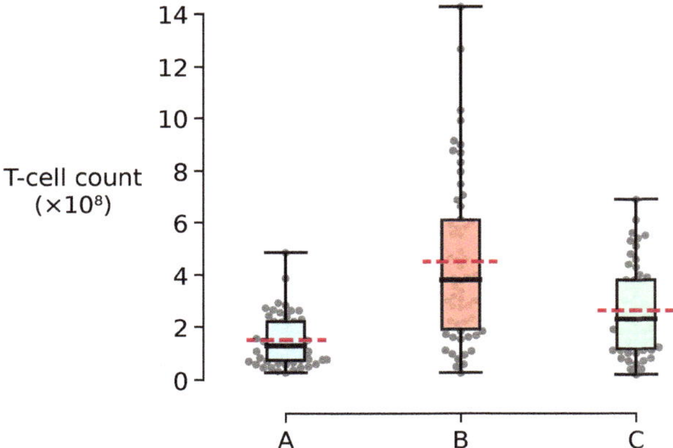

Figure 7.4 Box plots of the data sets A, B and C. Medians are shown in solid lines, means in dashed lines.

Choosing the measures

After consulting our visualizations of the data, we decide that because the data sets are skewed, the mean (average) value is not a good representation of the data. It's also true that a median T-cell count gives us a better estimate, scientifically speaking, of the efficacy of the drug. The median is a better descriptor, so we use it.

Then we need to decide how we want to compare group medians; we decide to use, as our comparison tool, the absolute value of the differences between group medians.

We now want to ask:

- Is there a significant difference among these three groups?
- How do we make sense of the idea of a "difference among three groups"?

Data preparation

When it comes to comparing three or more groups, here is the overall procedure for data preparation (Figure 7.5).

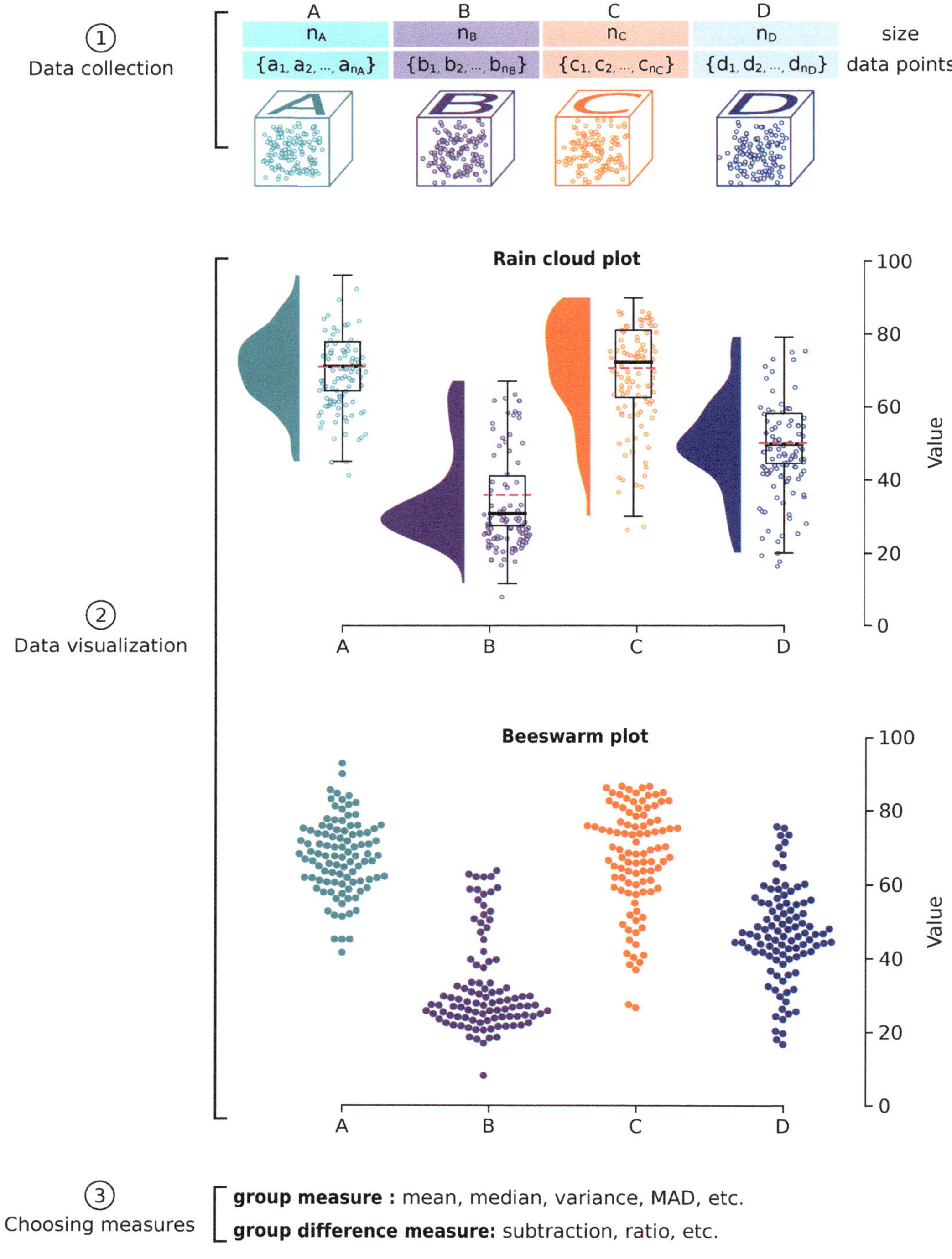

Figure 7.5 Data preparation for multi-group comparisons.

① **Collect all data.** From your experiment, collect all the data into a digital format that can be analyzed.

② **Visualize data.** Plot the data and comment on the shape and distribution.

③ **Choose measures.** Decide what measure to use to describe the groups, and what measure to use to compare the group measures.

FURTHER EXERCISES 7.1

1. The General Social Survey (GSS) is a yearly survey of adults in the United States conducted since 1972, on a variety of topics. Here are a few regularly asked questions from year to year. In each case, identify the type of variable of the answer for each question. Is it quantitative or categorical?

    a. In what state are you currently living?
    b. What is your age?
    c. Are you currently employed?
    d. What is your annual income?
    e. Are you currently: married, widowed, divorced, separated, or never married?
    f. About how much time per week do you spend sending and answering e-mail?
    g. Besides e-mail, do you ever use the Internet?
    h. How many hours did you work last week, at all jobs?

2. For the following scenarios, identify the group membership variable and the outcome variable. Are they categorical or quantitative?

    a. A study that tests the effects of sleep duration on newborn babies' growth.
    b. A clinical trial that looks into a new drug for the optimal dosage that will reduce tumor growth in lung cancer.
    c. What are the best lights to grow tomato plants? Fluorescent, incandescent, or natural light?
    d. A biofuel researcher is studying how a culture medium's pH affects the growth rate of algae.
    e. To test the visual range of different red light wavelengths in humans, each subject was asked to identify if they observed a flash.
    f. A study was conducted to test the effects of meditation duration on cortisol (stress hormone) levels.
    g. Researchers investigate whether different teaching methods (lecture-based, flipped classroom, or project-based learning) affect student test scores in mathematics.
    h. A study examines whether commute type (driving, public transit, or cycling) is associated with self-reported stress levels among urban workers.
    i. A clinical trial compares three pain management approaches (medication only, physical therapy only, or combined treatment) on patient-reported pain levels using a 0-10 scale.

3. **Commute times.** The histograms below show one-way commute times (in minutes) for six different communities, where the horizontal axis ranges from 0 to 90 minutes.

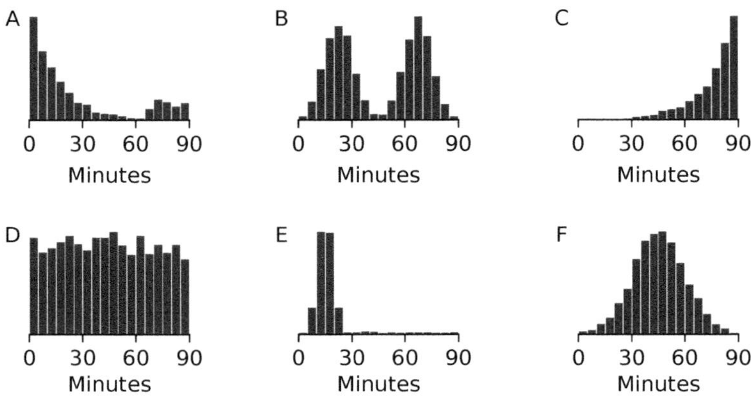

a. Two of these communities would produce nearly identical box plots. Which ones are they, and why? What does this tell you about the limitations of box plots?
b. For which communities would the mean be a misleading measure of the typical commute time? Explain your reasoning for each.
c. Which of the following sets of statistics most likely describes community E?

   a) Mean 30, Median 45, MAD 6
   b) Mean 30, Median 30, MAD 20
   c) Mean 30, Median 30, MAD 6
   d) Mean 45, Median 30, MAD 6

d. For each community, briefly describe the shape of the commute time distribution. Match the descriptions below to the histogram that best fits it.

   a) A sprawling city where most residents drive long distances to work
   b) A compact city where most residents walk or bike to work
   c) A city split between short local commutes and long suburban ones
   d) A large city where most residents rely on public transit.

7.2 The perils of multi-group testing

Can't we just do multiple two-group comparisons?

We might think: "well, I know how to do a two-group analysis, that was the previous chapter, so I'm just going to beat on this three-group problem by doing three separate two-group analyses. I'm going to do A against B, then B against C, then A against C."

So can we just do that? The answer is: not exactly. In order to see what's wrong with just doing three separate two-group analyses, think about, say, *seven* groups.

Suppose we were doing a study of the traffic on a particular highway at 9 am, and each day, around 9 am, we go out and sample 10 cars and record their speed. Now our data has 7 groups (the days of the week), and for each group, we have 10 numbers, the speed (in m/h) of the 10 cars we sampled on that day (Table 7.4).

Monday	Tuesday	Wednesday	Thursday	Friday	Saturday	Sunday
a_1	b_1	c_1	d_1	e_1	f_1	g_1
a_2	b_2	c_2	d_2	e_2	f_2	g_2
$\vdots$	$\vdots$	$\vdots$	$\vdots$	$\vdots$	$\vdots$	$\vdots$
a_{10}	b_{10}	c_{10}	d_{10}	e_{10}	f_{10}	g_{10}

Table 7.4 Hypothetical example of a 7-group comparison. Each entry represents the speed of a randomly sampled car on that day of the week.

If we just wanted to do two-group comparisons, how many two-group comparisons are there? There are 7 days, and each day has to be paired with 6 others, then 7 times 6 is 42 comparisons. But Tuesday vs. Wednesday is the same as Wednesday vs. Tuesday, so there are only 42 divided by 2, that is, 21 unique comparisons: Sunday to Monday, Sunday to Tuesday, Monday to Tuesday etc.

And what's wrong with doing 21 two-group analyses? **The most important problem is the tendency of multiple comparisons to generate false positives.** We saw, in *Problem of multiple testing* (section 3.3), that if we carry out many comparisons of completely random numbers, and if we choose "$p < 0.05$" as the threshold for statistical significance, then 5% of the studies on random numbers will yield "statistically significant" findings (at the $p < 0.05$ level).

That's just the definition of "$p < 0.05$": to say "happens by chance less than 5% of the time" means "happens by chance as much as 5% of the time". Those two sentences mean one and the same thing. Therefore, as we saw, the more comparisons you make, the more the chance of getting a false positive.

Green jelly beans

There's a great cartoon illustrating this, from *XKCD*. The cartoon is so important, **we would rather have you understand this cartoon than memorize any formula in any statistics book.**

The cartoon begins with a lab chief directing lab members to investigate whether there is a link between eating jelly beans and acne. The lab members do a study, and they come back and report "no, we found no significant association between jelly beans and acne".

But the lab chief does not take "no" for an answer. Instead, he hypothesizes, "maybe it's a certain *flavor* of jelly beans". So one group is asked to test red jelly beans, another group to test turquoise jelly beans, another magenta jelly beans, another yellow jelly beans, and another salmon jelly beans. Each of those five groups came up empty-handed.

Then the boss orders even more tests: grey jelly beans, tan jelly beans, chocolate jelly beans, etc. A total of 20 different studies were ordered, and, in the course of doing them, the team that was assigned green jelly beans found a statistically significant result at the $p < 0.05$ level.

The boss then gets very excited, and announces to the news media "Green jelly beans linked to acne!"

The problem with this, of course, is that if you do 20 tests on **absolutely random numbers**, and use a $p < 0.05$ standard, the odds are that one of them will turn out to be statistically significant. And if these are all random numbers, **that positive result is necessarily a false positive** (Figure 7.6)

Figure 7.6 The green jelly beans fallacy.

p-hacking

This is exactly the situation for which the term "p-hacking" has been coined (Austin *et al.*, 2006; Head *et al.*, 2015): carrying out lots of tests in order to find a few that turn out to be positive, and which are therefore likely to be false positives.

Recreating a green jelly beans scenario

As a demonstration of this fallacy, we created a 7-group comparison (Figure 7.7).

For each of the 7 groups, we chose 20 data points at random from the same Normal distribution, with a mean of 50 and a standard deviation of 10. We then carried out the 21 two-group comparisons, each using the two-groups approach in Chapter 6.

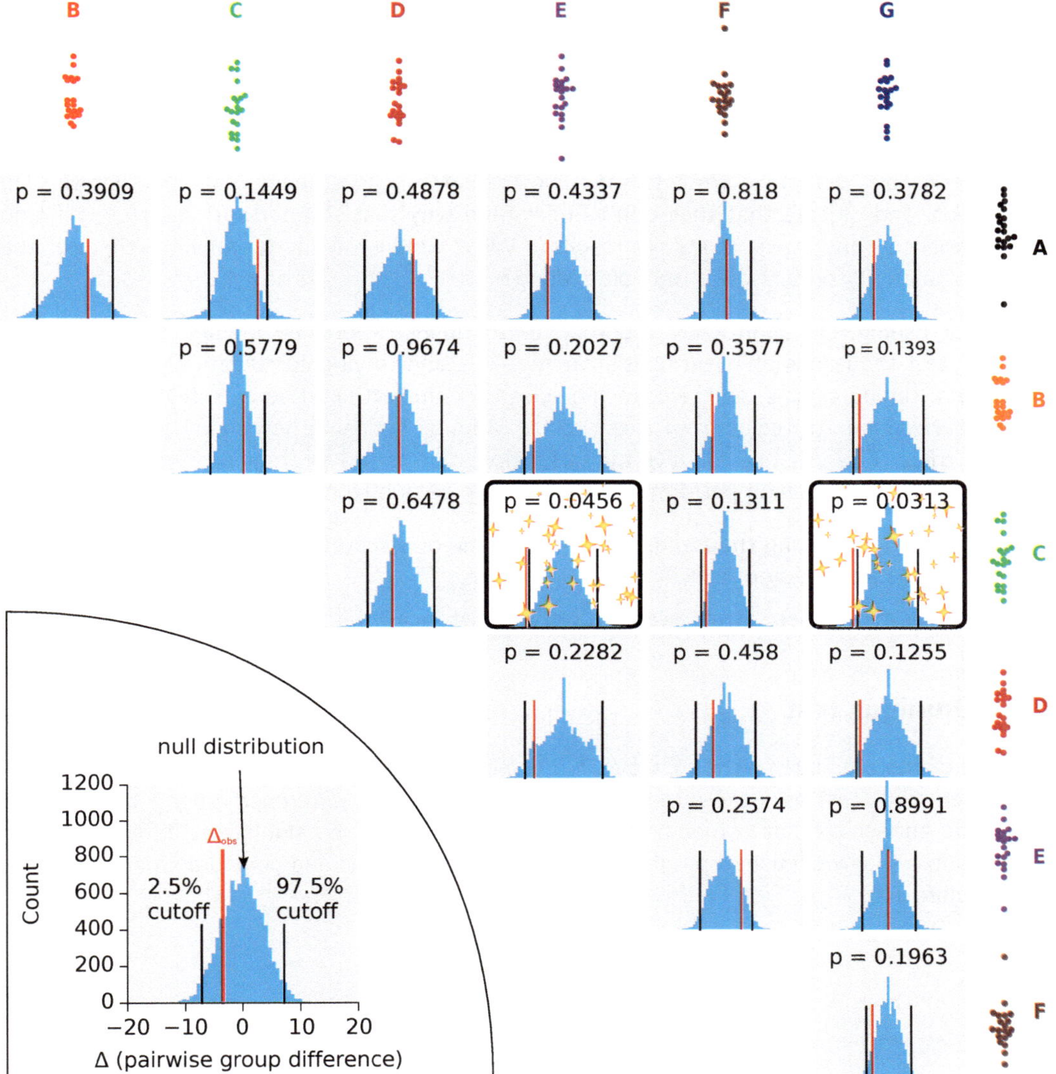

Figure 7.7 **Recreating the "green jelly beans" fallacy.** 7 groups (A through G) of 20 data points, each randomly selected from the same Normal distribution, give rise to $\frac{7\times6}{2} = 21$ two-group comparisons. In this case, 2 of the 21 two-group comparisons yielded statistically significant differences at the $p < 0.05$ level, shown sprinkled with stars. The inset shows the scales.

We would have expected at least one of those comparisons to come out statistically significant at the $p < 0.05$ level. Here, by chance, there were 2.

These two significant findings are, necessarily, spurious findings.

In order to prevent these false positives, when we deal with many groups, we first have to do a preliminary screening test called an "omnibus test".

FURTHER EXERCISES 7.2

1. **The diet dilemma.** A student tells you that their lab has studied 10 different popular "fad" diets. In head-to-head comparisons, they found that two comparisons showed statistically significant differences in weight loss after 1 week, each with a p-value < 0.05. Incredulous, you ask, "but did you correct for multiple testing?" The student tells you that the lab did not, and objects that they don't understand why that's important, retorting, "if I do the science right, then results are results." What would you say to convince them of the importance of correcting for multiple testing?

2. **Factor fishing.** In quantitative approaches to finance, portfolio managers and analysts often test thousands of investment strategies — using varying combinations of stocks, risk factors, timing signals, and rebalancing rules. A common practice is to highlight only the top-performing strategies from these tests, without disclosing how many total strategies were tried. This practice can give the false impression that a strategy has predictive power, when in reality, its "high performance" might well be due to random chance.

   a. How does testing thousands of strategies and only reporting the best-performing ones create an illusion of skill?

   b. Why might these strategies perform well historically but fail in the future?

7.3 Omnibus test

The role of the omnibus test is to act as a preliminary screening to prevent false positives.

The omnibus test asks the question: *is there a significant difference among these groups?* And if the answer to that comes back negative, "No, there is no significant difference among these groups" then we must stop right there. We cannot proceed and go and do many two-group tests (Figure 7.8).

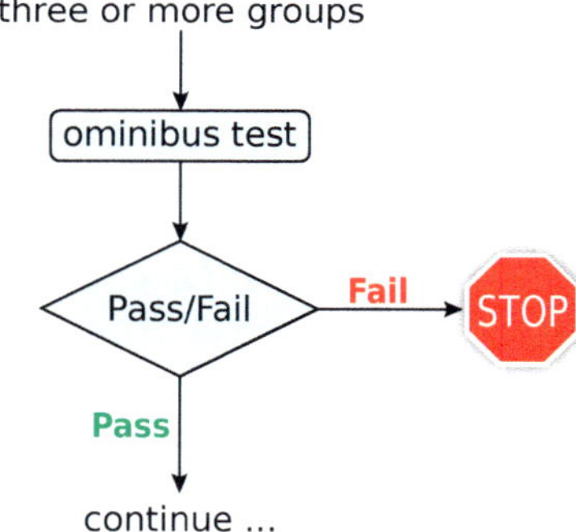

Figure 7.8 The omnibus test is the first step in a multi-group comparison.

The purpose of the omnibus test is to look at the total, aggregate differences among the many groups, and ask whether they, in total, amount to a statistically significant aggregate difference.

Therefore, in order to create an omnibus test, we first have to ask how we can express the idea of the **aggregate "difference" among many groups**.

The F-statistic

Suppose we have three groups, A (control), B and C, and we have chosen the median as our group descriptor. If we were to consider only the medians (Figure 7.9) and ask: are the groups very different?, we might give an answer in terms of the real-world difference, medical significance, and say: "well, group B median is 50% bigger than control (group A), so that might be a difference worth noting."

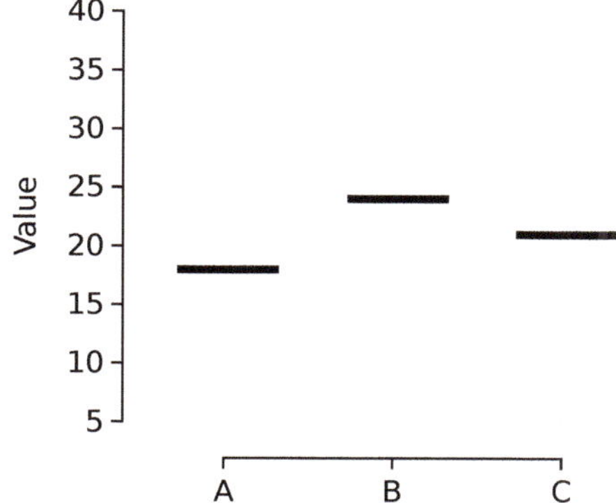

Figure 7.9 The medians of the three groups.

But if we are asking: are the groups *statistically significantly* different, then it is obvious that we are missing a critical piece of information; we also need to know how much variation there is *within* each group.

Small within-group variation If the within-group differences are small, the data might look like Figure 7.10. In this kind of case, we might suspect that these differences among the group medians are statistically significant, because **the across-group differences are fairly large with respect to the scatter within the groups**. The medians of the three groups are further apart from each other than any pair of points within a group. In other words, the differences among group medians could not easily have arisen by chance sampling from within a group.

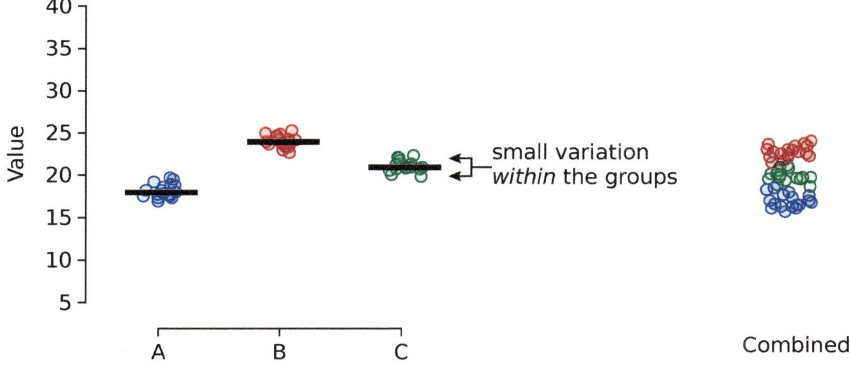

Figure 7.10 Small variation within each group.

This analysis is further supported by the last column in Figure 7.10. Here we plotted the data from all three groups, color coded by group, with group A in blue, group B in red and group

C in green. The separation of the three colors along the axis is clear: the reds are on the top, the green points are in the middle, and the blue points are near the bottom, although there are a few blue points in the green area. The overall pattern is that there are three distinct clusters of data points.

Large within-group variation By contrast, let's consider a different data set (Figure 7.11).

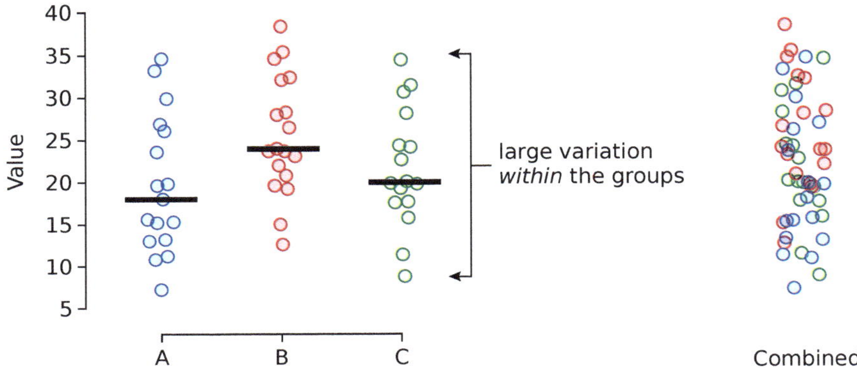

Figure 7.11 Group medians unchanged from the data sets in Figure 7.10, but with large variation within each group.

Here, the groups A, B and C have exactly the same medians as previously, but now there is a lot more scatter of points within each group, that is, there is more within-group variation. Visual inspection tells us that there is going to be a much weaker overall effect here. Consistent with this, in the combined plots, the colored dots are much more interspersed with each other.

In this case, the given differences among the group medians could easily have arisen by chance from sampling within a group, so we can suspect that this difference is not going to be statistically significant.

Statistics can be seen as the battle between across-group differences (effect sizes) and within-group differences (variability). Statistical significance means that across-group differences are much larger than within-group differences.

The concept of the F-statistic **We want a measure that will reflect this fundamental fact, and give us a way of describing the battle between across-group differences and within-group differences.**

We will proceed by, first, defining a measure of the aggregate (total) *across*-group difference, and then we will define a measure of the aggregate *within*-group difference. The "aggregate difference" among a set of groups is then defined as the total across-group difference *divided by* the total within-group difference.

If that ratio is large, that is, if across-group differences are much larger than within-group differences, then the aggregate difference will be large. But if the within-group differences are large, and/or the across-group differences are small, then the aggregate difference will be small.

Therefore, we have just specified a test statistic that reflects this tension between across- and within-group differences. We will call this test statistic the **F-statistic**.

$$\text{F-statistic} = \frac{\text{total across-group difference}}{\text{total within-group difference}} = \frac{\text{effect size}}{\text{variability}}$$

Clearly, if the F-statistic is large, then across-group differences are bigger than within-group differences, and we might have a significant result.

Exercise 7.3.1 Which of the two data sets (Figures 7.10 and 7.11) has the larger F-statistic?

Now we need to think about how to define these two quantities.

Across-group difference The definition of across-group difference is straightforward. We choose a group measure, and then we define the total across-group difference as the sum of the distances from the group measures to the measure of the whole data set.

For example, if we are using means as our measures, then the mean of all the data is called the Grand Mean, and the group measures are the group means. We would then take as our definition of across-group differences the sum of the distances of the group means to the Grand Mean, with each distance weighted by how many data points the group has.

The same logic applies to other measures.

Let's say, for the data set in Figure 7.11, we decided to use medians to describe the groups, so we define the Grand Median ($\widetilde{G}$) as the median of all the data. Then our definition of the across-group differences will be to sum up the distances from each group median ($\widetilde{A}$, $\widetilde{B}$ and $\widetilde{C}$) to the Grand Median, incorporating weights that reflect how many subjects there are in each group.

In Figure 7.12, the horizontal dotted line is the Grand Median. The three pink arrows show the distances from the group medians to the Grand Median.

Then we need to define how we are going to measure "distance". We will use the sum of the **absolute values** of the differences between the group medians and the Grand Median,

As we will see later in this chapter, in classical ANOVA, it is traditional to use the **squares** of the distances. We see no advantages, and several disadvantages, to squaring everything, so here we will use the absolute values.

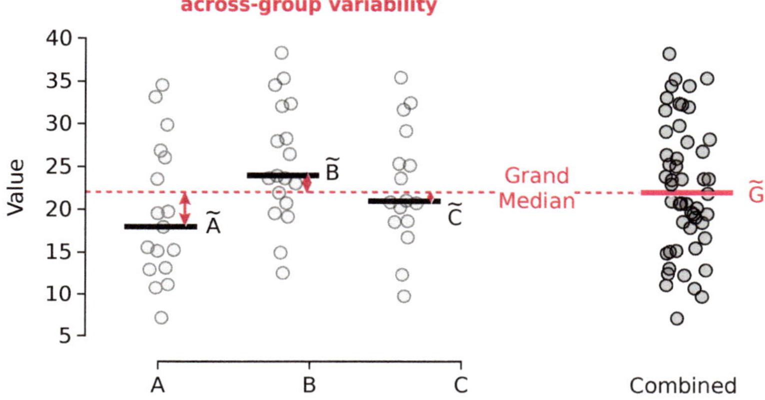

Figure 7.12 Across-group differences with the median as the group measure.

Using the median as the group measure, and absolute values as our "distance" measure, we define the aggregate across-group difference of groups A, B and C as:

$$\text{total across-group difference} = n_A \cdot |\widetilde{A} - \widetilde{G}| + n_B \cdot |\widetilde{B} - \widetilde{G}| + n_C \cdot |\widetilde{C} - \widetilde{G}|$$

where n_A, n_B, n_C are the number of data points in groups A, B and C.

Within-group difference For our definition of within-group differences, we will use the sum of the distances from each data point to the median of its group (Figure 7.13).

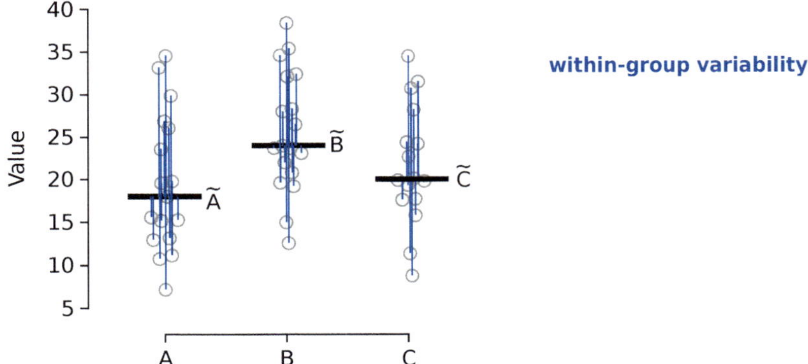

Figure 7.13 Within-group differences with median as the group measure.

Similar to our definition of across-group differences above, we will use absolute values as our distance measure. Again, classical ANOVA would use squares, and is committed to means only, whereas the resampling-based approach can use absolute values and has total freedom of measure (mean, medians, MAD, etc.).

$$\text{total within-group difference} = \sum_{i=1}^{n_A} |a_i - \widetilde{A}| + \sum_{i=1}^{n_B} |b_i - \widetilde{B}| + \sum_{i=1}^{n_C} |c_i - \widetilde{C}|$$

Defining the F-statistic Having defined the across-group difference and the within-group difference, their ratio is the quantity we call F. This is the test statistic for the aggregate difference among groups.

$$F = \frac{\text{total across-group difference}}{\text{total within-group difference}} = \frac{n_A \cdot |\widetilde{A} - \widetilde{G}| + n_B \cdot |\widetilde{B} - \widetilde{G}| + n_C \cdot |\widetilde{C} - \widetilde{G}|}{\sum\limits_{i=1}^{n_A} |a_i - \widetilde{A}| + \sum\limits_{i=1}^{n_B} |b_i - \widetilde{B}| + \sum\limits_{i=1}^{n_C} |c_i - \widetilde{C}|}$$

If this ratio is large, then across-group differences are bigger than within-group differences, which would indicate the possibility of a significant difference among the groups, that is, a positive omnibus test.

F_{obs} We have now defined a measure for the aggregate difference among groups, the F-statistic. The F-statistic calculated from the observed data is a quantity we will call F_{obs}.

In classical ANOVA, F is defined only for group means and only for squared distances. On the contrary, in the resampling-based approach, we have freedom of measure, and we do not have to distort distances by squaring everything.

Possible ways of defining F are illustrated in Figure 7.14. When we have freedom of measure, we can use any group descriptor and methods for measuring distances that makes sense for this data and our scientific purposes.

Group measure	Distance measure	Definition of F	
mean	absolute distance	$\text{across-group diff} = n_A \cdot \lvert\bar{A}-\bar{G}\rvert + n_B \cdot \lvert\bar{B}-\bar{G}\rvert + n_C \cdot \lvert\bar{C}-\bar{G}\rvert$ $\text{within-group diff} = \sum_{i=1}^{n_A}\lvert a_i-\bar{A}\rvert + \sum_{i=1}^{n_B}\lvert b_i-\bar{B}\rvert + \sum_{i=1}^{n_C}\lvert c_i-\bar{C}\rvert$ $F = \frac{\text{across-group diff}}{\text{within-group diff}} = \frac{n_A \cdot \lvert\bar{A}-\bar{G}\rvert + n_B \cdot \lvert\bar{B}-\bar{G}\rvert + n_C \cdot \lvert\bar{C}-\bar{G}\rvert}{\sum_{i=1}^{n_A}\lvert a_i-\bar{A}\rvert + \sum_{i=1}^{n_B}\lvert b_i-\bar{B}\rvert + \sum_{i=1}^{n_C}\lvert c_i-\bar{C}\rvert}$	These are the methods we use.
median	absolute distance	$\text{across-group diff} = n_A \cdot \lvert\tilde{A}-\tilde{G}\rvert + n_B \cdot \lvert\tilde{B}-\tilde{G}\rvert + n_C \cdot \lvert\tilde{C}-\tilde{G}\rvert$ $\text{within-group diff} = \sum_{i=1}^{n_A}\lvert a_i-\tilde{A}\rvert + \sum_{i=1}^{n_B}\lvert b_i-\tilde{B}\rvert + \sum_{i=1}^{n_C}\lvert c_i-\tilde{C}\rvert$ $F = \frac{\text{across-group diff}}{\text{within-group diff}} = \frac{n_A \cdot \lvert\tilde{A}-\tilde{G}\rvert + n_B \cdot \lvert\tilde{B}-\tilde{G}\rvert + n_C \cdot \lvert\tilde{C}-\tilde{G}\rvert}{\sum_{i=1}^{n_A}\lvert a_i-\tilde{A}\rvert + \sum_{i=1}^{n_B}\lvert b_i-\tilde{B}\rvert + \sum_{i=1}^{n_C}\lvert c_i-\tilde{C}\rvert}$	
mean	squared distance	$\text{across-group diff} = n_A \cdot (\bar{A}-\bar{G})^2 + n_B \cdot (\bar{B}-\bar{G})^2 + n_C \cdot (\bar{C}-\bar{G})^2$ $\text{within-group diff} = \sum_{i=1}^{n_A}(a_i-\bar{A})^2 + \sum_{i=1}^{n_B}(b_i-\bar{B})^2 + \sum_{i=1}^{n_C}(c_i-\bar{C})^2$ $F = \frac{\text{across-group diff}}{\text{within-group diff}} = \frac{n_A \cdot (\bar{A}-\bar{G})^2 + n_B \cdot (\bar{B}-\bar{G})^2 + n_C \cdot (\bar{C}-\bar{G})^2}{\sum_{i=1}^{n_A}(a_i-\bar{A})^2 + \sum_{i=1}^{n_B}(b_i-\bar{B})^2 + \sum_{i=1}^{n_C}(c_i-\bar{C})^2}$	This is the method used in traditional statistics courses.

Figure 7.14 Different ways of defining the F-statistic. $\bar{A}$, $\bar{B}$, and $\bar{C}$ are the group means, and $\bar{G}$ is the Grand Mean. $\tilde{A}$, $\tilde{B}$, and $\tilde{C}$ are the group medians, and $\tilde{G}$ is the Grand Median.

And then the next question becomes: is my F_{obs} large? And, just as before, when we ask: large relative to what?, our answer is: large relative to the same measure when calculated on randomized versions of the observed data set.

The randomization procedure and calculating the p-value

After calculating F_{obs} we now need to create **the world of the Null Hypothesis, the world in which all across-group differences have been eliminated**. Creating this world will enable us to answer the question: how big an F would we expect simply due to chance sampling from this null world?

To answer this question, we sample a new data set, A_1, B_1, and C_1 from the world of the Null Hypothesis, and calculate a new F-value, which we will call F_1, for that sample. We can view F_1 as a sample answer to the question: how big an F might we see under the Null Hypothesis?

By repeating this procedure, we generate 10,000 F_i values, for 10,000 different (re)samplings of a new A_i, B_i, and C_i, **all generated under the Null Hypothesis of no difference among the groups** (Figure 7.15).

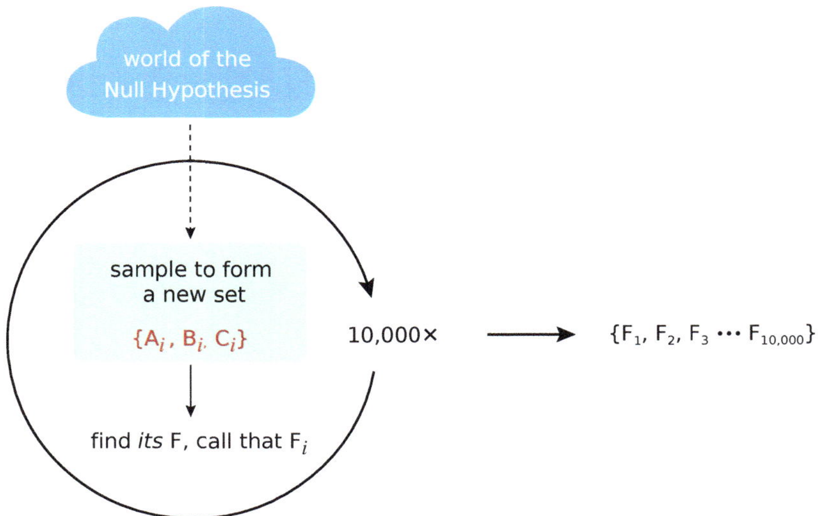

Figure 7.15 Generating 10,000 expected values of F under the Null Hypothesis by repeatedly sampling from the null world.

These 10,000 F_i values under the Null Hypothesis then give us a set of expectations for how large an F-statistic we might see due to random sampling.

The question is: how to randomize the data set in order to implement the Null Hypothesis in our statistical simulations? And the answer is exactly what it was in the two-group comparison case. In Chapter 6, we saw that the choice of randomization procedure depended on the *relative variabilities* of the two groups: if the two groups had roughly the same variability, we used a Big Box model, but if the variabilities of the two groups were different by more than a factor of 2, we used a Re-centered Two-Box model.

Here, we use the same principle: if the variabilities among the groups differ by more than a factor of 2, we use a Re-centered multi-box model for the Null Hypothesis, else we use a Big Box.

Variability and shape Note that we are using the word "variability" here, which is not a technical term. It's a little vague, and it can have several precise definitions.

And how do we define variability? One way is to use the definition of *variance*, the average squared distance of data points to the group mean. But if we don't want to square things unnecessarily, and/or don't wish to be committed to the average, we will use measures other than variance. For example, we could use the MAD (the median or mean absolute distance), that is, we will take the median (or mean, as appropriate) of the absolute values of the distances of each data point to its group mean (or median). In all cases, we will use the "more than a factor of 2" rule of thumb for when to use a Re-centered multi-box method vs. a Big Box method.

Another factor we have to take into consideration is the shape of the distribution. Remember that our original question was: could these groups have come from the same underlying distribution? But if their variabilities and/or shapes are different, say, A is skewed left and B is skewed right while C is bimodal, or A, B and C have widely different variabilities, then that is evidence in and of itself that A, B and C were not drawn from the same box. Three different draws from the same box ought to have roughly the same shape, namely the shape of the data set in the box, and roughly the same variability, namely the variability of the data set in the box.

When these requirements are not met, we can no longer use a Big Box model because we have already disconfirmed the Big Box hypothesis. When the groups *obviously* did not come from the same Big Box, we need to retreat to a weaker Null Hypothesis, namely, that A, B and C were drawn from three different boxes, but with each having the same summary statistic. That is the Re-centered multi-box approach.

How do we randomize?

- If the distributions A, B and C have roughly the same shape, and the variabilities in A, B and C are roughly equal, then Big Box with replacement.
- If the shapes and/or the variabilities in A, B and C are NOT roughly equal, then Re-centered n-box with replacement.

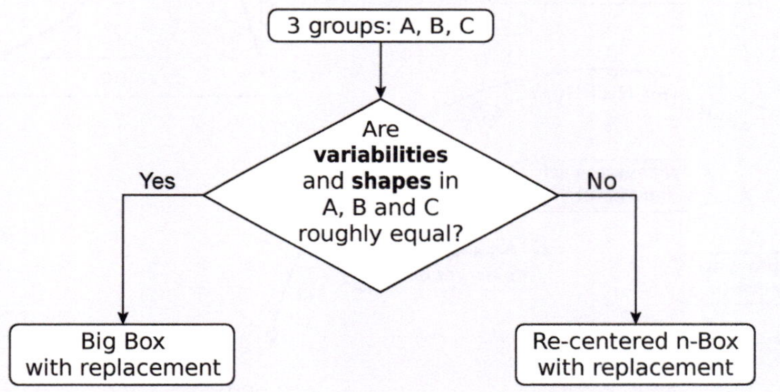

Big Box: roughly equal shapes and variabilities When the distribution shapes and variabilities in the three groups are roughly equal, we throw all the data into a big box to form the world of the Null Hypothesis (Figure 7.16).

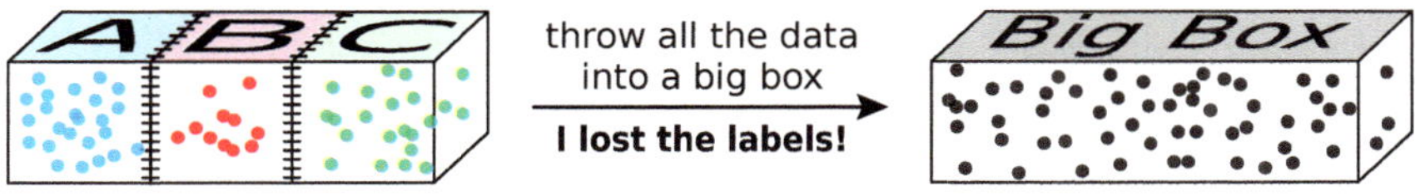

Figure 7.16 Forming the Big Box model.

The procedure to conduct the omnibus test using the Big Box model is shown in Figure 7.17.

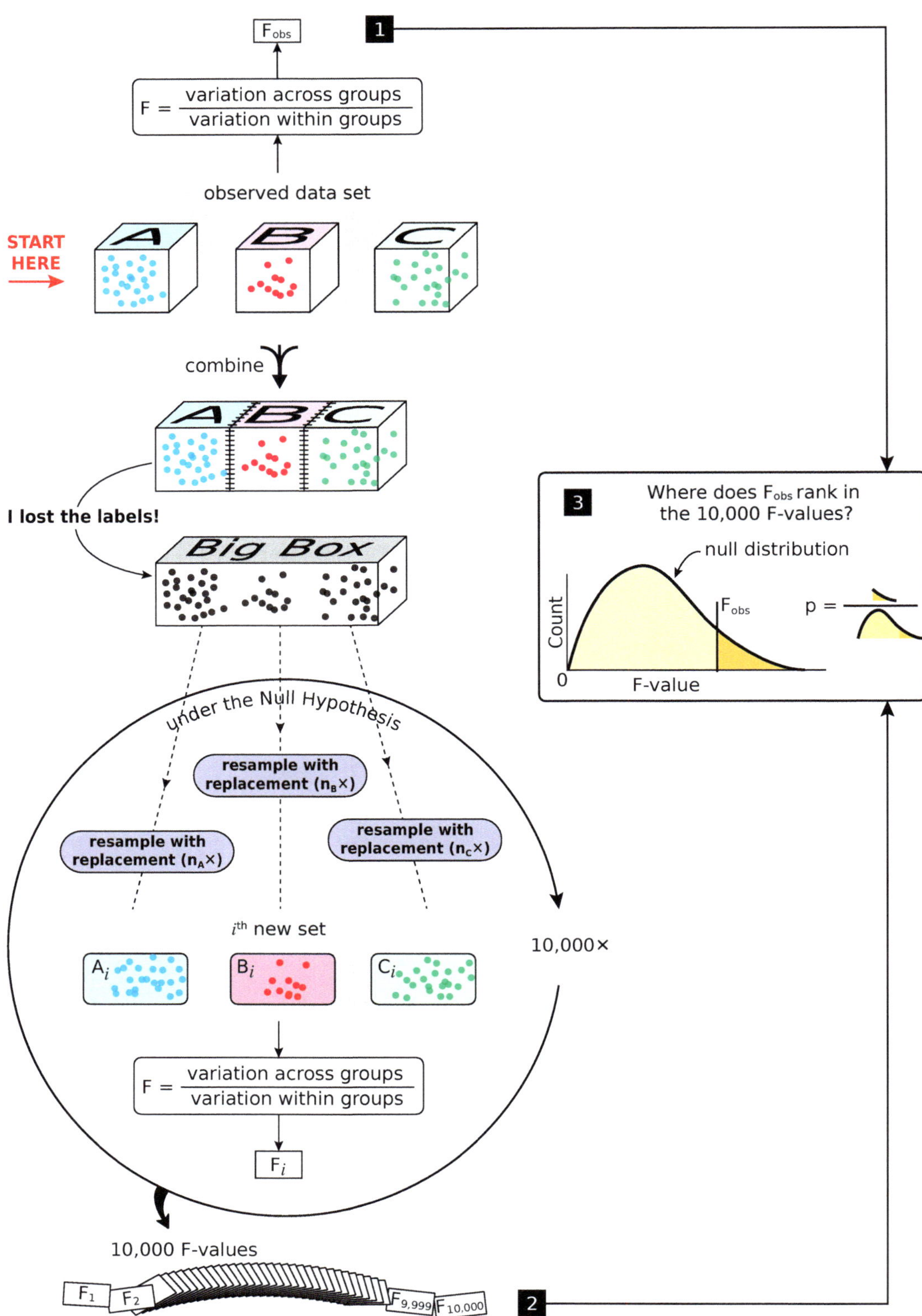

Figure 7.17 Omnibus test using the Big Box model.

We can summarize the Big Box omnibus procedure in three parts:

1 **Analyze the observed data.** Calculate the F-statistic for the observed data, call it F_{obs}.

2 **Simulate the Null Hypothesis by resampling.** Resample with replacement a new A_1, B_1, and C_1 out of the Big Box, and calculate the F statistic for A_1, B_1, and C_1. Write down that value of F, calling it F_1.

Repeat the resampling procedure 10,000 times, generating 10,000 F_i values, estimates of F in a world governed solely by chance.

3 **Calculate a p-value.** Now that we have the 10,000 F_i's, we can compare our F_{obs} to them to determine its p-value and statistical significance.

If F_{obs} is greater than the 99th percentile of the F_i's, then F_{obs} is statistically significant at the $p < 0.01$ level. This means that the omnibus test has rejected the Null Hypothesis of "no difference among the groups", which we are calling a positive omnibus test. Otherwise, the omnibus test has failed to find a statistically significant aggregate difference among the groups, and the multi-group analysis stops.

Re-centered n-Box: unequal shapes and variabilities When we have different distribution shapes and variabilities in the three groups, remember our procedure when we had those differences in the case of two groups: we used a Re-centered Two-Box model. We re-centered A by subtracting the mean (or median or ...) of A from every element of A, and then resampled from this re-centered box to get a pseudo-A. Then we did the same for B. Now we will do the same for C (Figure 7.18).

Note that, for variety, here we are using the mean as the group measure. So for the illustration of the Re-centered n-Box model, we are re-centering by subtracting the means. If we were using the median as the group measure, we would re-center by subtracting the medians.

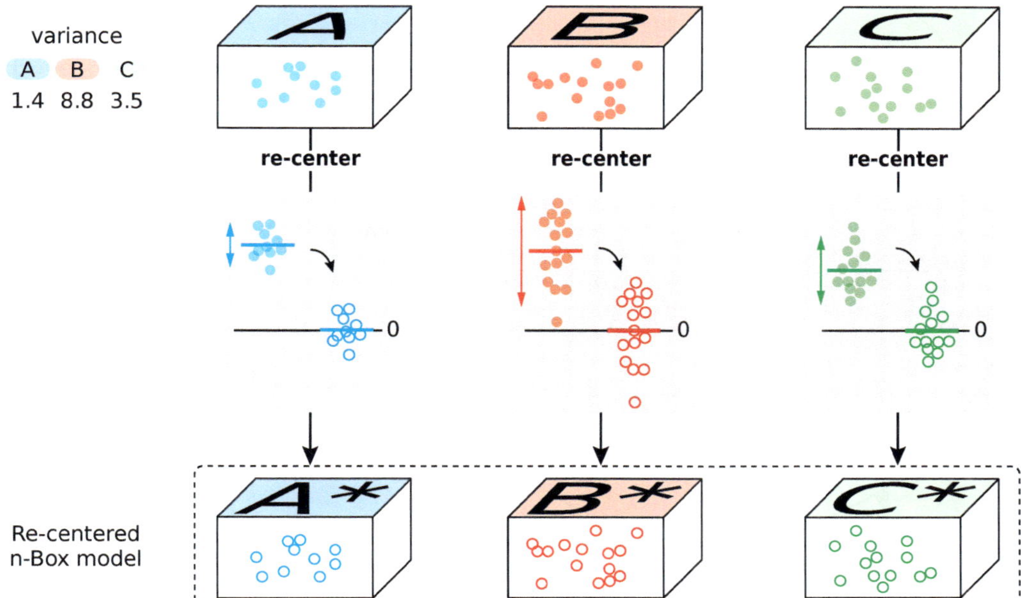

Figure 7.18 Forming the Re-centered n-box model. Note that the variances differ by more than a factor of 2.

The procedure to conduct the omnibus test using the Re-centered n-Box model is shown in Figure 7.19.

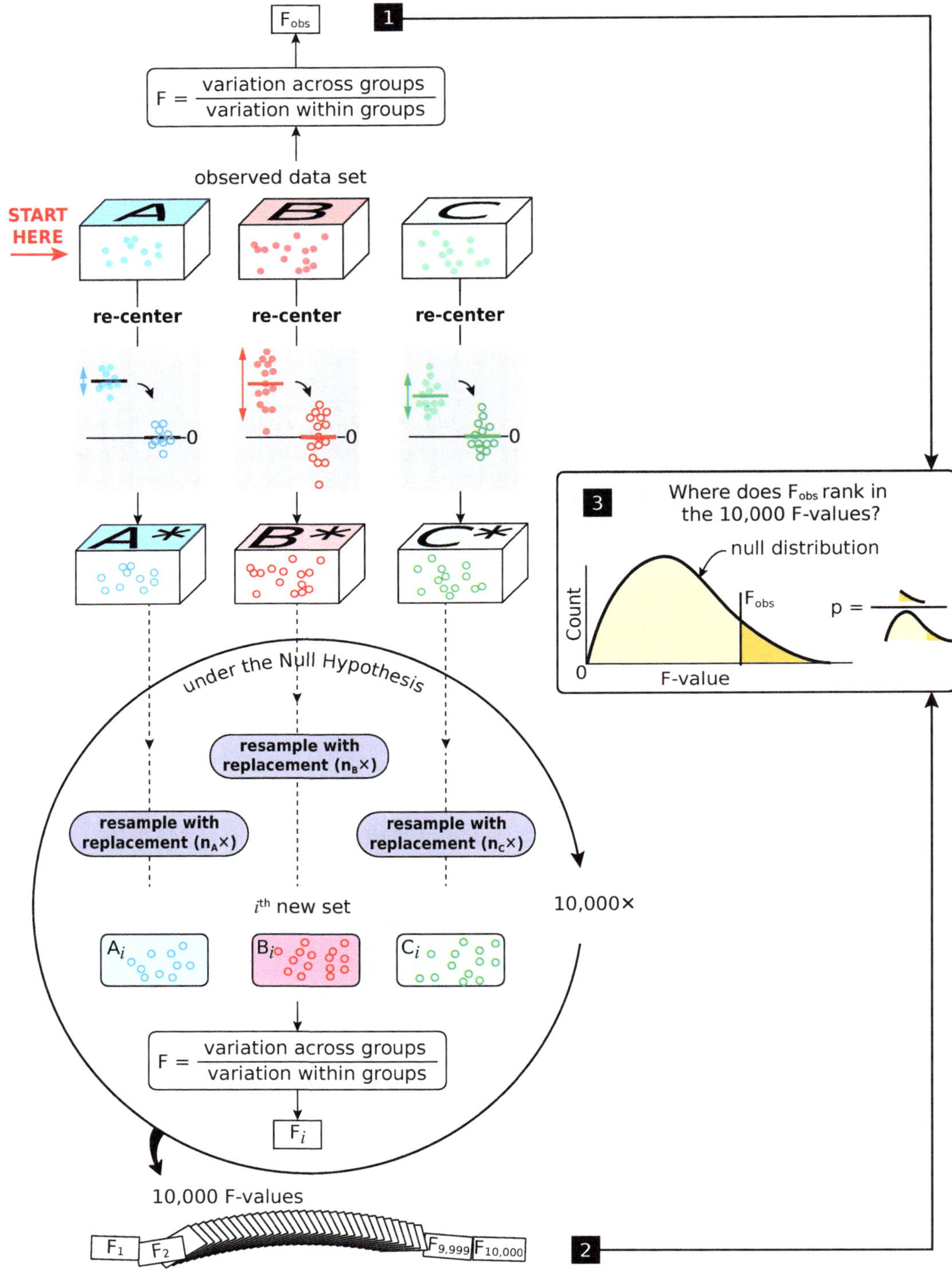

Figure 7.19 Omnibus test using the Re-centered n-Box model.

Similar to the Big Box omnibus procedure, we can summarize the Re-centered n-Box procedure in three parts.

1 **Analyze the observed data.** Calculate the F-statistic for the observed data, call it F_{obs}.

2 **Simulate the Null Hypothesis by resampling.** We form the Re-centered n-Box model, and resample A^* to get A_1, resample B^* to get B_1 and resample C^* to get C_1. Calculate the F-value for this new data set $\{A_1, B_1, C_1\}$, and call it F_1.

We then repeat this procedure 10,000 times, generating 10,000 F_i's, from the world of the Null Hypothesis. Any variation in these F_i's has arisen solely by chance.

3 **Calculate a p-value.** Just as for the Big Box model, now that we have the 10,000 F_i's, we can compare our F_{obs} to them to calculate a p-value and determine statistical significance.

If F_{obs} is greater than the 99th percentile of the F_i's, then F_{obs} is statistically significant at the $p < 0.01$ level, and we have a positive omnibus test.

Reporting the result We report the result of the omnibus test by citing the value of F_{obs} together with its p-value.

Note that we are using a one-sided test here. Only a one-sided test makes sense, because F can never be negative, and small values of F are non-significant by definition.

Omnibus test for the T-cell counts study

We carried out an omnibus test on our original data set of T-cell counts in a control group and two drug groups (Figure 7.20).

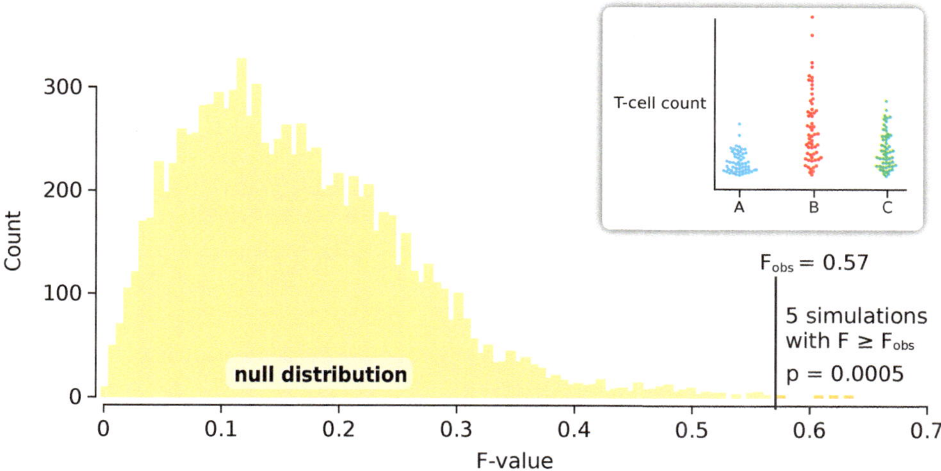

Figure 7.20 Omnibus test for the data set of T-cell counts in three groups (a control group and two drug groups, original data presented in Figure 7.2).

The results were highly significant. The results of the 10,000 randomized simulations form the null distribution. The 99th percentile of the results is much less than the observed value F_{obs}. Thus our omnibus test is positive: we have found an aggregate group difference that is significant at the $p < 0.01$ level, in fact, $p = 0.0005$.

Omnibus test cures the "green jelly beans" fallacy

As an illustration of the power of the omnibus test to correct for false positives, we will apply it to a simulation model of the "green jelly beans" fallacy.

Recall that we created a world of seven groups A through G, with each of the groups being a random draw of 20 data points from the same Normal distribution, $\mathcal{N}(50, 10^2)$. In this setting, *any* finding of a significant difference must be a false positive, because the seven groups exactly satisfy the Null Hypothesis that they are random samples from the same population (Figure 7.7).

When we did multiple pairwise two-group comparisons, 21 in all, we got the results shown in Figure 7.7. Notice that, in true "green jelly beans" style, 2 of the tests came out statistically significant!

But, unlike the lab chief in the cartoon, we are not going to alert the news media. We will first apply the omnibus test (Figure 7.21).

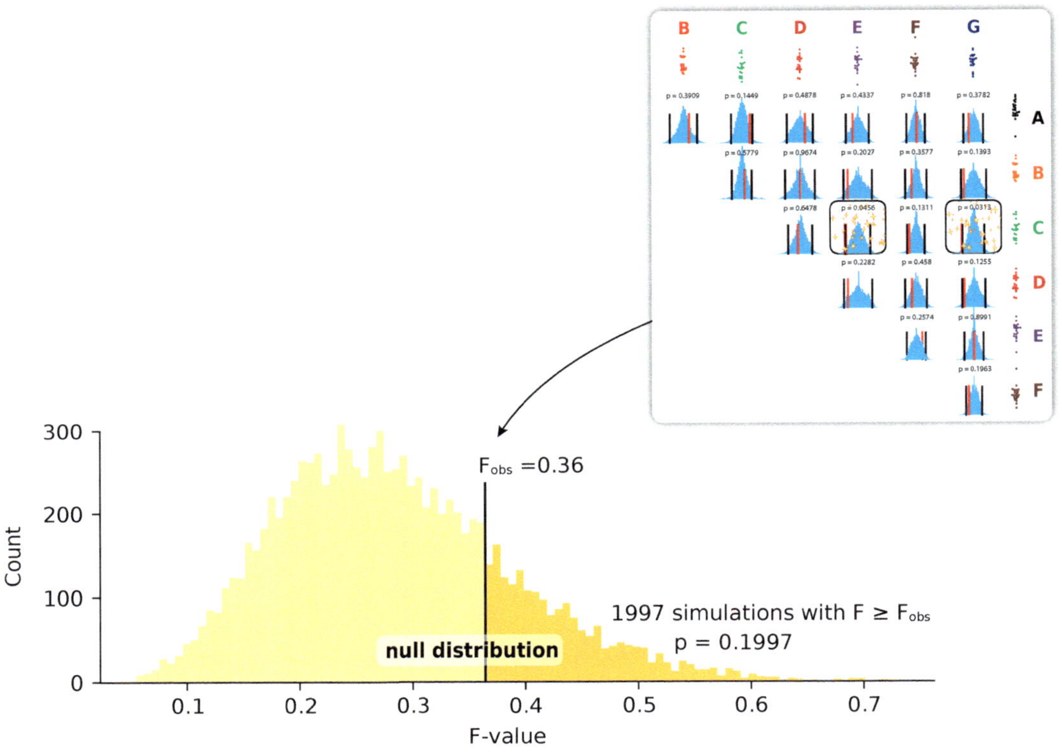

Figure 7.21 Omnibus test cures the "green jelly beans" fallacy.

The results were not significant at all. The results of the 10,000 randomized simulations form the null distribution. Note that almost 20 percent of the results are equally or more extreme than the observed value F_{obs}. In other words, what we see in the "green jelly beans" result is not rare at all. Thus our omnibus test is negative: we have found no significant differences among the seven groups. Since the omnibus test is negative, we must stop here and conclude that there are no significant group differences!

If an omnibus test turns out to be positive, we can then proceed to find out *where* the differences are. This is *post hoc* analysis, which will be treated in the next section.

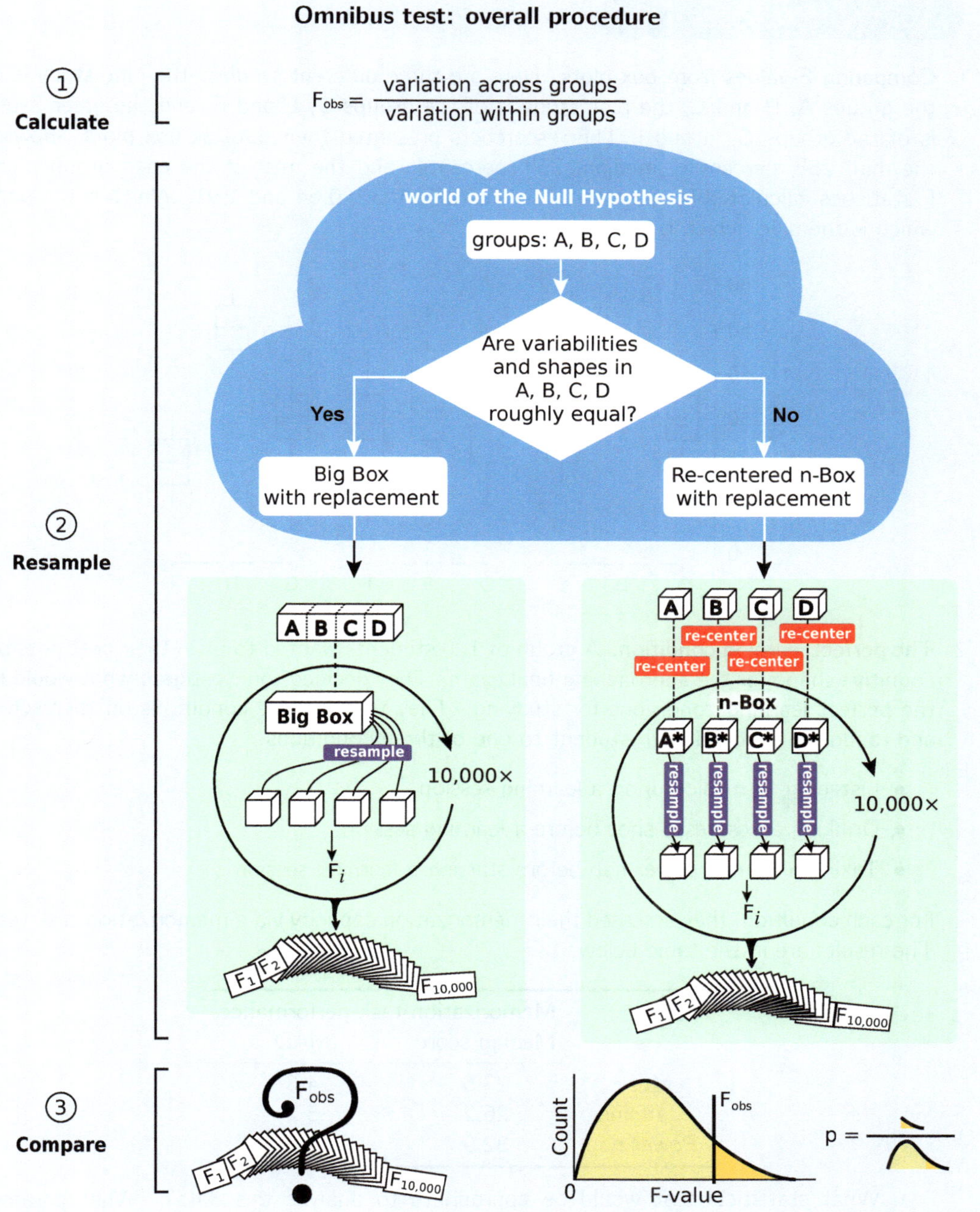

① **Calculate F_{obs}.** This gives the observed aggregate difference among all the groups.

② **Sample from the world of Null Hypothesis.** We simulate the null world many times to form the null distribution, simulating the situation where there is no aggregate difference among the groups.

③ **Compare F_{obs} to the null distribution.** How rare is the observed F_{obs} compared to the null distribution of F?

FURTHER EXERCISES 7.3

1. **Comparing F-values from box plots.** Here are three different studies: the blue study is of the groups A, B and C, the pink study is of the groups D, E and F, and the green study is of the groups G, H and I. The researchers presented their data as box plots, showing the min, 25th percentile, median, 75th percentile and the max of the each group. The F-statistics calculated in these three studies were 0.32, 0.64 and 1.91. Which is the pink, which is the blue, which is the green?

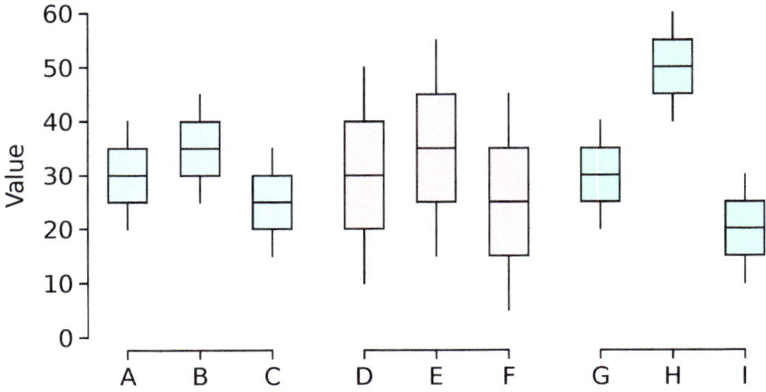

2. **The perfect learning condition.** A group of 120 students wanted to be in their best possible cognitive shape for the approaching final exam. They decided to investigate what would be the perfect learning conditions for studying. They tested three conditions on themselves and randomly assigned each student to one of three conditions.

   - Listening to music during a learning session
   - Drinking an espresso shot before a learning session
   - Taking a 30 min power nap before starting a learning session

   For each condition, they assessed their memorization capacity via a memorization task test. The results are in the table below.

	Memorization task performance	
	Median score	MAD
Music	23.5	4.6
Caffeine	26.2	5.2
Power nap	32.9	4.8

   a. What statistical test would be appropriate to analyze this data? What parameter/statistic would you use?

   b. You quickly wrote a piece of code to calculate the statistic of interest for this test and found −2.6. Should you worry or do you carry on your analysis?

   c. You were able to run the appropriate test of significance and find a p-value of 0.004. What does this p-value tell you? What should you do next?

7.4 post hoc analysis part I: pairwise group comparisons

Suppose, then, that we have conducted an omnibus test, and found a positive result: a statistically significant aggregate difference among the groups.

Now we are licensed to go and look at the pairwise group differences, because the omnibus test came back positive and said, there's something in there. Therefore, we now have the right to go look and see where it is. And that is the subject that is called *post hoc* analysis (Figure 7.22).

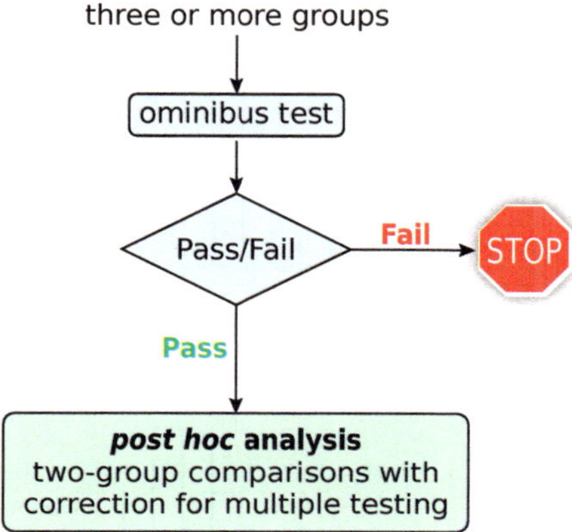

Figure 7.22

"*post hoc*" is a Latin phrase, meaning "after this", that is, after the omnibus test. If the omnibus test is positive, we can go on to look at which groups are significantly different from which others. However, a positive omnibus test does not guarantee that at least one two-group comparison will be statistically significant; it is possible to have a positive omnibus test and have no two groups statistically significantly different from each other.

post hoc analysis therefore has two parts.

- **Pairwise group comparisons.** The first part is simply to make the pairwise group comparisons we talked about at the beginning of this chapter: A against B, B against C, etc. We use whatever two-group method (Big Box resampling? Re-centered Two-Box resampling?) is appropriate for this particular data set. The purpose of this is to get p-values and confidence intervals for these two-group comparisons.

- **Correction for multiple comparisons.** The second part of *post hoc* analysis is to correct for the tendency of multiple comparisons to produce false positives.

p-values

We carried out the first part of the *post hoc* analysis for the T-cell count study of two drugs B and C versus control A (data presented in Figure 7.2).

For this study, there are three pairwise comparisons we need to do: A vs. B, A vs. C, and B vs. C (Figure 7.23). T-cell counts in Group B were highly significantly greater than in Group A ($p = 0.0001$), and T-cell counts in Group C were significantly greater than in group A ($p = 0.008$). The plates for drug B did better than those for drug C, but the difference was not significant ($p = 0.0219$).

Each pairwise group comparison is essentially the same as the two-group p-value calculation we discussed in Chapter 6. In a two-group p-value calculation, we simulate the Null Hypothesis many times to form the null distribution, we then compare the observed difference Δ_{obs} against that null distribution to see how rare our observation is under the Null Hypothesis, which yields the p-value.

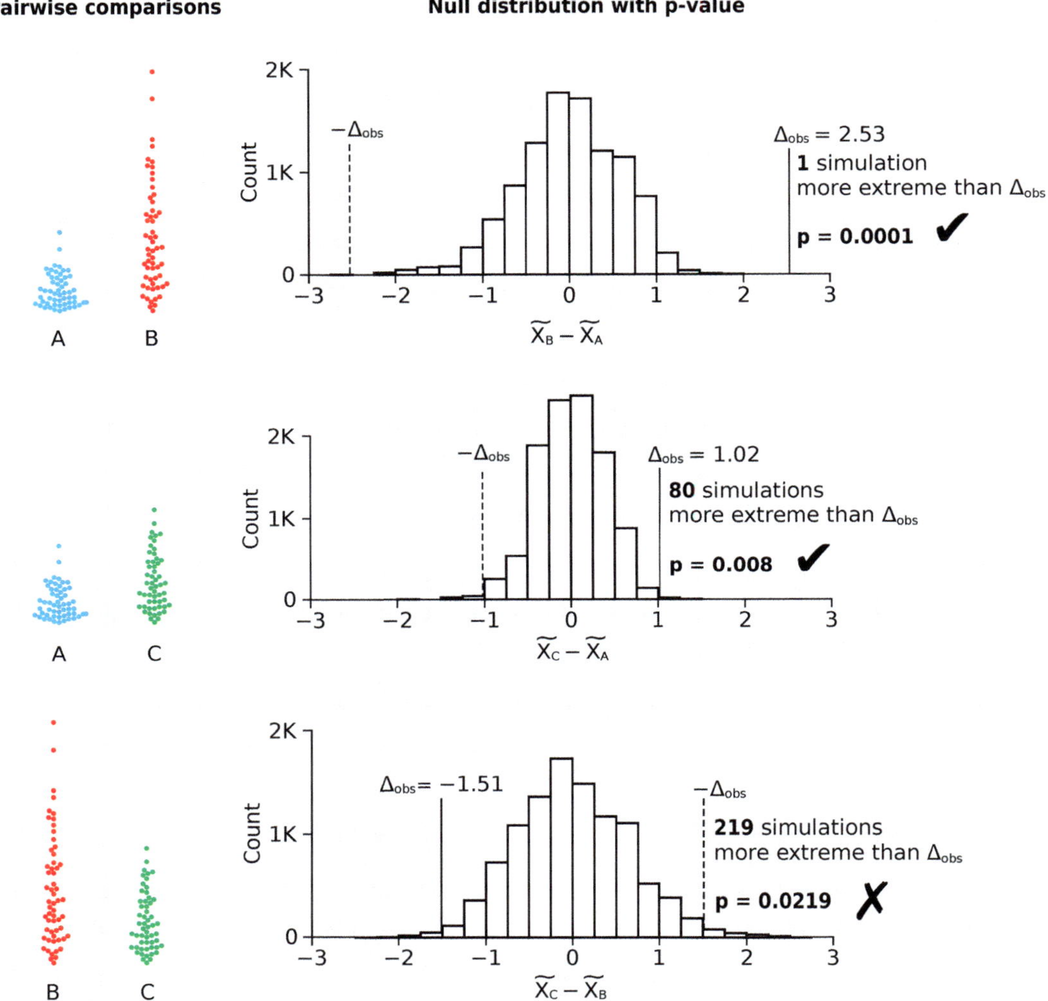

Figure 7.23 Pairwise comparisons for groups A, B and C, each yielding a p-value.

Confidence intervals

But there is an even more informative and useful method: *making confidence intervals for the differences between the groups.* We've already developed this two-group comparison method, in the previous chapter (section 6.4 *Confidence Intervals*). Now we are going to apply this method three times, once for each pair of groups.

Remember our procedure for constructing a confidence interval in the two-group case. We used a Two-Box model and resampled from it many times. We had a Group A and a Group B.

We resampled a set A_1 from A, and resampled a set B_1 from B. Then we took our measures of A_1 and B_1, and calculated the group difference (or ratio, or ...) to compare those two groups. We wrote that down, calling it Δ_1. Then we repeated the resampling procedure 10,000 times and constructed a *confidence interval on the A vs. B difference.*

In this case, we're going to do that three times, once for the A vs. B difference, once for the A vs. C difference and once for the B vs. C difference. We will do a 99% confidence interval for each comparison. The result for the T-cell drugs is shown in Figure 7.24.

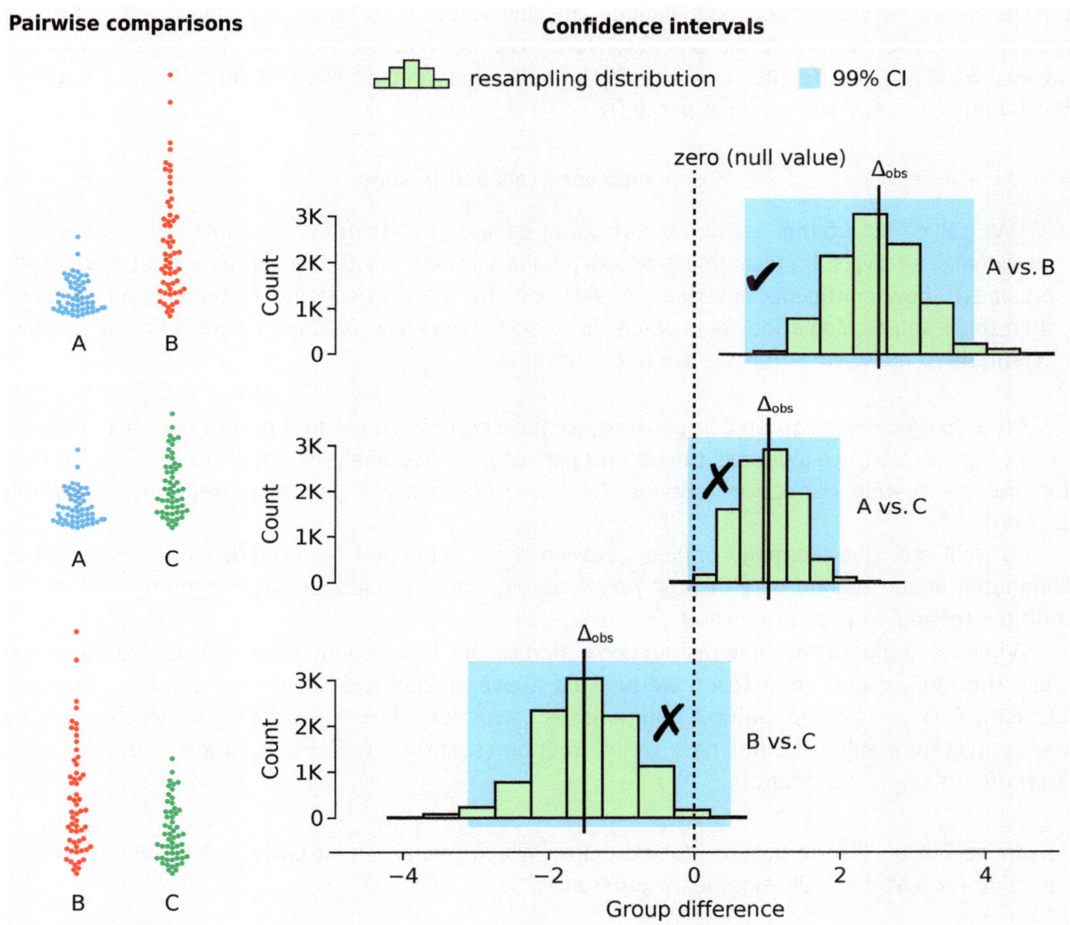

Figure 7.24 Pairwise comparisons for groups A, B and C, each with a 99% confidence interval for the group difference.

Now we can draw a conclusion. Notice that the null value zero is outside the confidence interval of the A-B difference, implying that A and B are significantly different. But zero is not outside the confidence interval of the A-C difference and it is not outside the confidence interval of the B-C difference. Therefore we can conclude that A is statistically significantly different from B, but the other group differences are probably not statistically significant.

Resolving conflicts Note an interesting phenomenon that happened in this data set.

We did confidence intervals on the group differences, and found that the 99% confidence interval on the A-C difference included 0. But when we calculated the p-value of the A-C difference, it came out as $p = 0.008$, which would indicate a significant difference ($p < 0.01$). These two results might appear to be in conflict, one saying the difference is not significant, and the other saying that it is.

This difference can be explained. First of all, while it is true that if a 99% confidence interval for the group difference does *not* include the null value, then there is a significant difference between the two groups at the $p < 0.01$ level, the converse is not true: the 99% confidence interval for a group difference can include the null value, and yet the test for difference between the two groups may still produce $p < 0.01$.

Confidence intervals and p-values

We talked about this asymmetry of going back and forth from resampling-based pivotal confidence intervals to resampling-based p-values in section 6.4 (*Confidence intervals and p-values*): how confidence intervals provide a higher bar in declaring statistical significance than the p-value calculation. And when those borderline cases happen, we need to pay special attention to make sure that we are not p-hacking.

After doing pairwise group comparisons, we have completed the first part of *post hoc* analysis. In the next section, we'll discuss the second part of *post hoc* analysis, and introduce the method for multiple testing correction (section 7.5: *post hoc analysis part II: correction for multiple testing*).

As we'll see, this seeming conflict, between the p-value and the confidence interval, will be eliminated in the second part of *post hoc* analysis, where p-value penalties are imposed by the multiple testing correction method.

When we apply the multiple testing correction to the T-cell count study, we will see there that while the multi-group comparison we just did above yielded two significant p-values, only one of them, A-B, passed the multiple comparisons correction. The A-C difference, which produced the conflicting results, did not pass the correction (section 7.6 *BH application to the study of drug effects on T-cell counts*).

Exercise 7.4.1 Of the groups plotted below, which two are most likely to have the smallest p-value for a statistically significant difference?

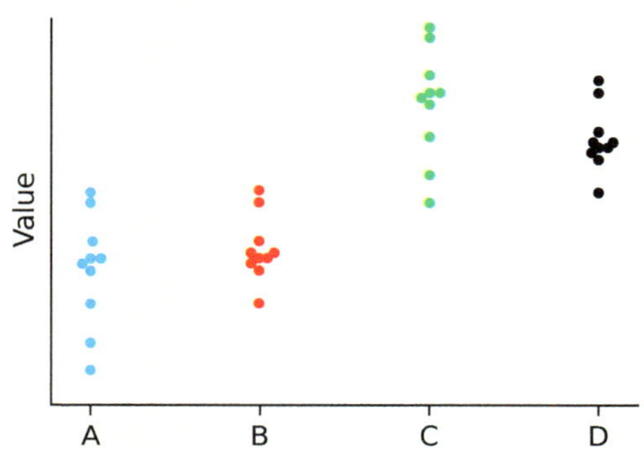

Pairwise group comparisons

After a positive omnibus test confirming that there is a significant difference among the groups, we are then licensed to carry out the *post hoc* analysis to identify where exactly the differences are. To do so, we have the options of either using p-values or confidence intervals. The confidence intervals can be for group differences or for group measures.

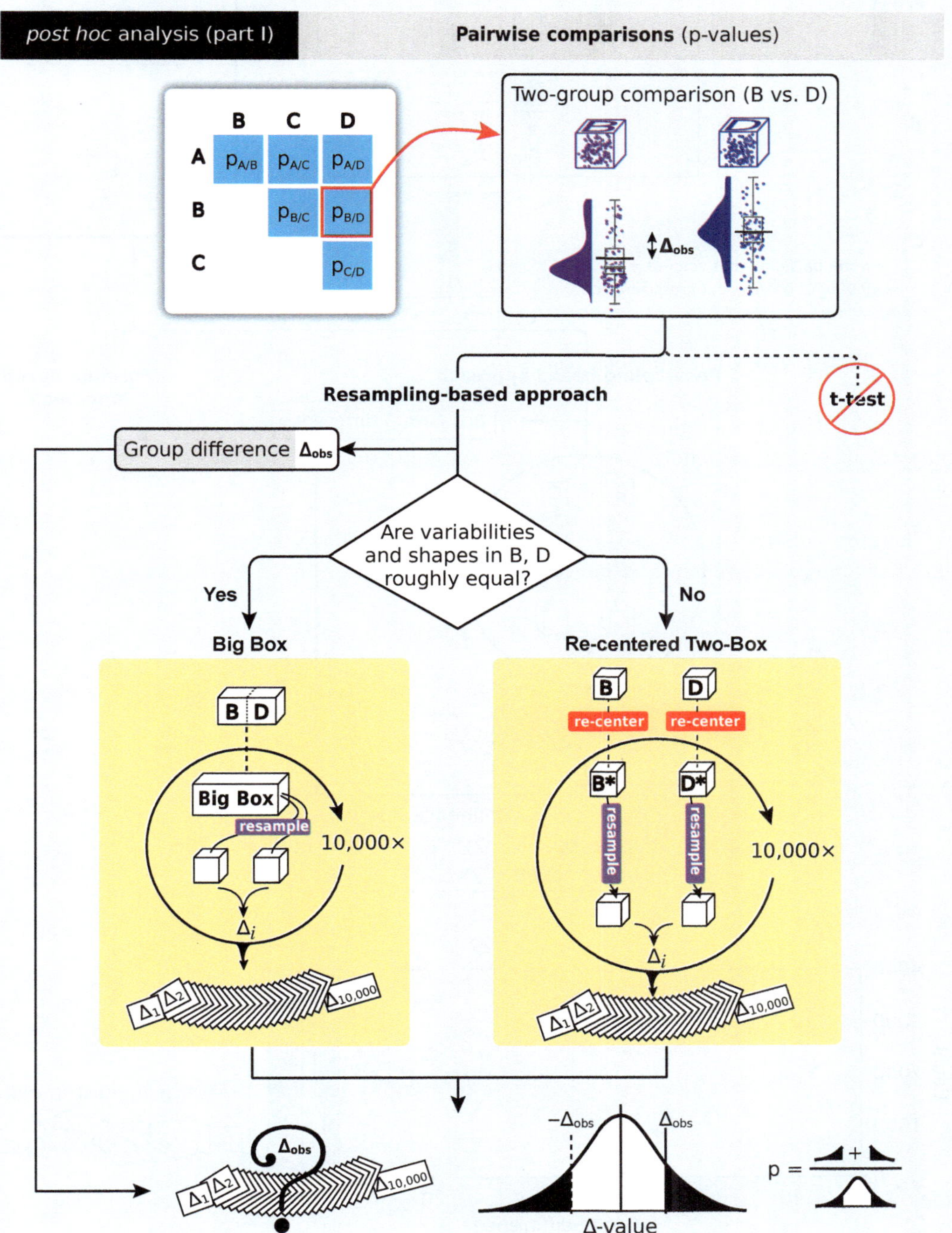

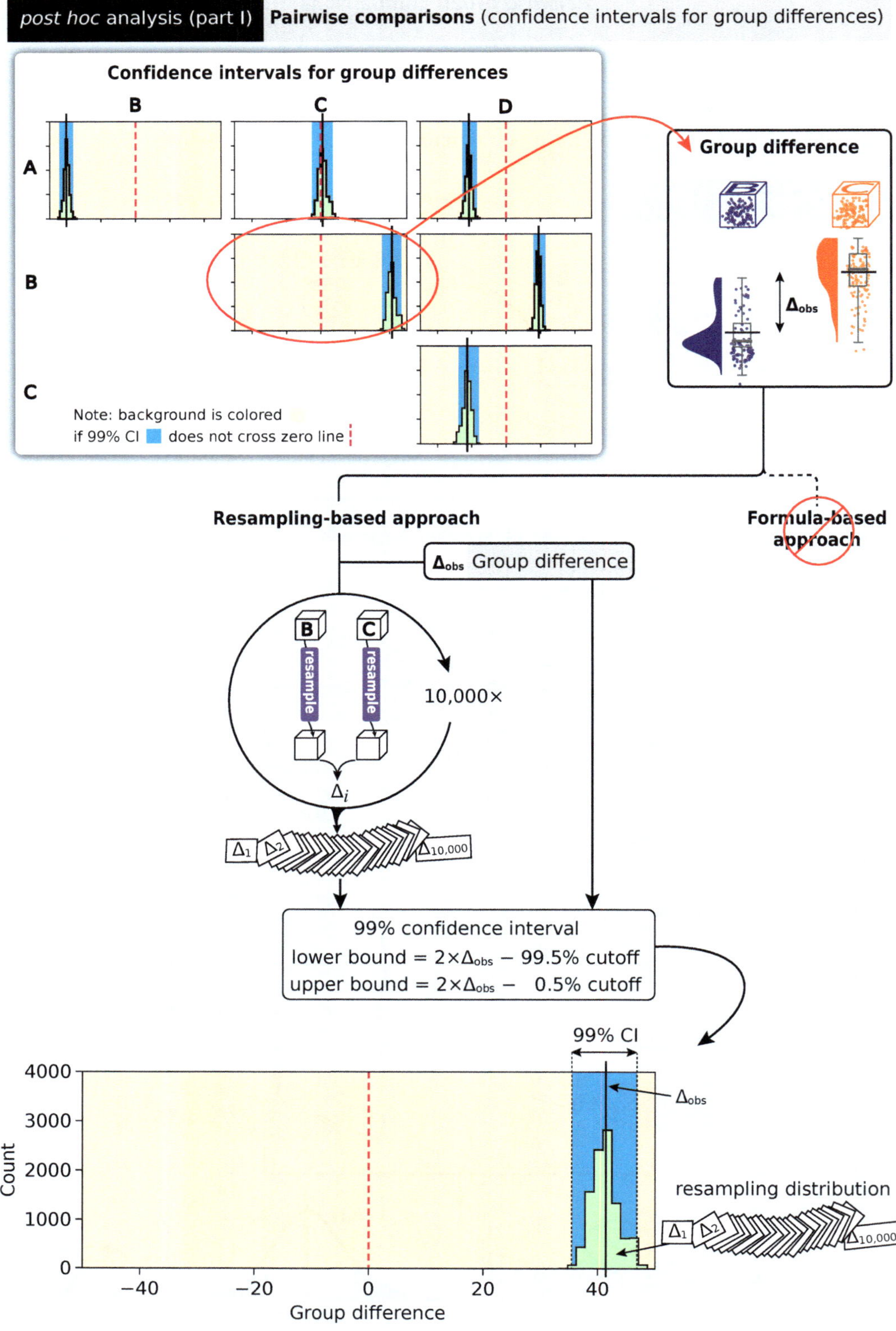

post hoc analysis (part I)
Pairwise comparisons (confidence intervals for group differences)
Confidence intervals for group differences
B
C
D
A
B
C
Note: background is colored
if 99% CI does not cross zero line
Group difference
Δobs
Resampling-based approach
Formula-based approach
Δobs Group difference
B
C
resample
resample
10,000×
Δi
Δ1
Δ2
Δ10,000
99% confidence interval
lower bound = 2×Δobs − 99.5% cutoff
upper bound = 2×Δobs − 0.5% cutoff
99% CI
Δobs
resampling distribution
Count
Group difference
4000
3000
2000
1000
0
−40
−20
0
20
40

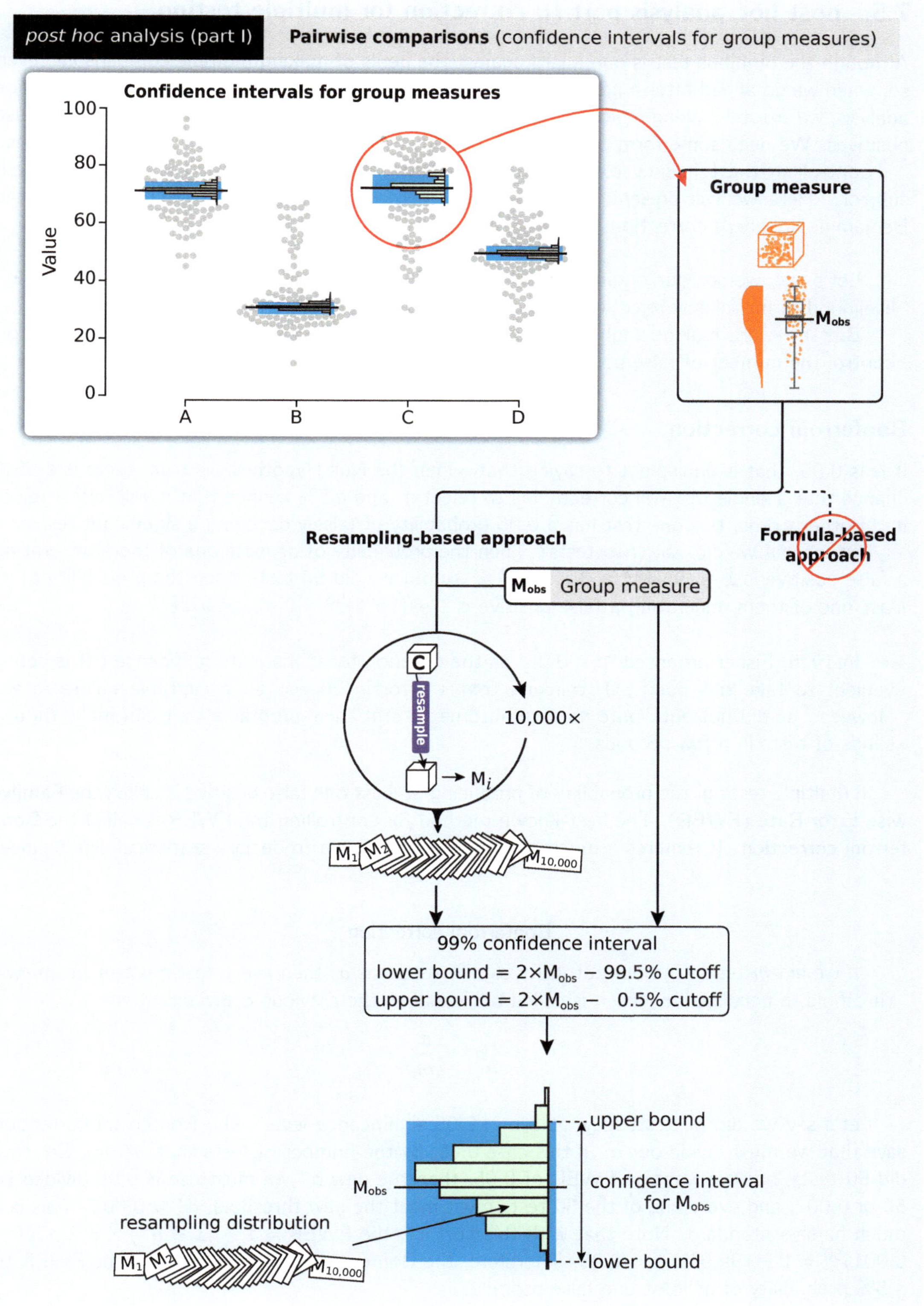

post hoc analysis (part I)
Pairwise comparisons (confidence intervals for group measures)
Confidence intervals for group measures
Value
100
80
60
40
20
0
A
B
C
D
Group measure
Mobs
Resampling-based approach
Formula-based approach
Mobs Group measure
C
resample
10,000×
Mi
M1 M2 M10,000
99% confidence interval
lower bound = 2×Mobs − 99.5% cutoff
upper bound = 2×Mobs − 0.5% cutoff
upper bound
confidence interval for Mobs
lower bound
Mobs
resampling distribution
M1 M2 M10,000

7.5 post hoc analysis part II: correction for multiple testing

Although the omnibus test is one guard against the perils of making multiple comparisons, even so, when we go ahead after a positive omnibus test and make our group-against-group *post hoc* analysis, we are still making multiple comparisons, and therefore we are still risking increased false positives. **We need some form of correction to reduce the tendency to produce false positives.**

Controlling the tendency to produce false positives in multiple testing is a large research subject. Here we will present two well-known methods: the Bonferroni correction and the Benjamini-Hochberg correction.

Let's say, we set our original α to be 0.05. If we were only making one test, then the likelihood of mistakenly rejecting a Null Hypothesis and declaring a significant result is 0.05.

But if we are making multiple tests, we have to use a lower value of α, called α^*, to control the number of false positives.

Bonferroni correction

If α is 0.05, that is equivalent to saying that, when the Null Hypothesis is true, there is a 95% chance that a single test will correctly fail to reject it, and a 5% chance that it will falsely reject it. In other words, this one test has a 0.05 probability of falsely declaring a significant result.

But what if we did, say, *two* tests? Then the probability of *at least* one of them generating a false positive is $1 - (95\%)^2 \approx 0.1$, or 10%. And if we did 50 tests, then the probability of *at least* one of them generating a false positive is $1 - (95\%)^{50} \approx 0.92$, or 92%.

In 1925, Fisher proposed "$p < 0.05$" as the criterion for statistical significance ("It is convenient to take this point..."), thinking that 1 error in 20 was an acceptable failure rate. However, he did not anticipate modern automated statistical programs that can make thousands of tests in a few seconds.

In multiple testing, the probability of producing at least one false positive is called the **Family-wise Error Rate (FWER)**. The best-known method for controlling the FWER is called the **Bonferroni correction**. It requires a much more stringent α value to declare statistical significance.

Bonferroni correction

If we did m tests, and our cutoff for significance was α, then *every* test must meet a new threshold, a penalized α, which is called α^*, equal to the previous α divided by m.

$$\alpha^* = \frac{\alpha}{m}$$

Let's say we did 50 tests, say, at a $p < 0.05$ significance level. The Bonferroni correction says that we must divide our α, in this case 0.05, by the number of tests that we did. So if we did 50 tests, and we want an FWER of 0.05, then the new α^* we must use is 0.05 divided by 50 or 0.001, and every one of the 50 tests must meet the new threshold, $\alpha^* = 0.001$. This is a much harsher standard. Note that with this new α^*, the FWER $= 1 - (1 - \alpha^*)^{50} = 1 - (1 - 0.001)^{50} = 1 - (99.9\%)^{50} = 0.05$. Therefore, the Bonferroni correction restores the FWER to a 5% probability of at least one false positive.

Astrological signs and medical conditions

Recall, from Chapter 3, the study of zodiac signs and medical diagnoses carried out by researchers at the University of Toronto (Austin *et al.*, 2006). They randomly divided the residents 50/50 into a derivation cohort and a validation cohort. After identifying 24 statistically significant associations in the derivation cohort, these 24 hypotheses were then tested in the validation cohort.

> "We conducted a study of all 10,674,945 residents of Ontario aged between 18 and 100 years in 2000. We searched through 223 of the most common diagnoses for hospitalization until we identified two for which subjects born under one astrological sign had a significantly higher probability of hospitalization compared to subjects born under the remaining signs combined ($p < 0.05$). Residents born under Leo had a higher probability of gastrointestinal hemorrhage ($p = 0.0447$), while Sagittarians had a higher probability of humerus fracture ($p = 0.0123$) compared to all other signs combined."

This study was published, not seriously, but as an illustration of the importance of multiple-testing corrections and the dangers of data mining.

First, this study is doing many tests: 2 most frequent diagnoses × 12 astrological signs, etc, so they definitely should use a multiple testing correction. The authors, correctly, did this. As they stated,

> "by not making appropriate adjustments for the testing of multiple hypotheses, we greatly increased our risk of falsely "uncovering" an association between astrological sign and illness. Had we instead endeavored to preserve an overall type I error rate of 0.05, we would have had to use a significance level of 0.00213 for each of the 24 individual hypothesis tests (this is marginally less conservative than a Bonferroni correction, which would have used a significance level of 0.05/24 = 0.00208. . .). Using this significance level, none of the 24 hypothesized associations would have been significant."

Benjamini-Hochberg correction

The fact that the Bonferroni correction is overly harsh can be seen in an extreme case: suppose our α was 0.01. Say we did 10 tests, and got p-values that were all around 0.005. So we have 10 tests, all statistically significant, from which it follows that none are statistically significant!

A well-known and widely-used approach to reduce false positives in multiple testing focuses not on the FWER, but on the likelihood of a positive finding being a false positive. This probability is the **False Discovery Rate (FDR)**, which is defined as the fraction of declared positives that are false positives.

From an investigator's perspective, facing multiple positive findings, FDR is more relevant, since it is telling us how reliable these positive results are.

$$\mathrm{FDR} = \frac{\mathrm{FP}}{\mathrm{FP} + \mathrm{TP}}$$

A sliding threshold The best-known method for controlling the FDR is called the **Benjamini-Hochberg correction** (BH). In the Benjamini-Hochberg procedure, we are still penalized by the total number of tests we did, but we are also rewarded by the fraction of those tests that came out significant.

Basic thinking behind BH correction

If there is a lone significant p-value out of, say, 20 tests, that p-value is suspicious, and it had better be very highly significant, because it needs to pass a Bonferroni-like harsh penalty. But, if there are many significant p-values, the penalty will get lighter and lighter as we go through those many significant p-values.

So let's say we did a total of m tests, resulting in m p-values, p_1 through p_m. Let's assume that these are ordered from smallest to largest: p_1 is the smallest p-value, p_2 is the next smallest p-value, p_3 is the third smallest p-value, etc.

Let's say that our m in this case is 25; we have done 25 tests. Let's assume that our original α was 0.01.

Now we ask the question: is that first p-value, p_1, less than α divided by m? (here, $\frac{\alpha}{m} = \frac{0.01}{25}$ or 0.0004) (For this smallest p-value, BH correction is equivalent to the Bonferroni correction). This is a yes/no question; let's write down the answer.

Now let's look at p_2, the second smallest. Now the question is: is p_2 less than α times 2 over m, $\alpha \cdot \frac{2}{m} = 0.01 \times \frac{2}{25}$ or 0.0008? which is a less stringent requirement; we're taking the first criterion and doubling it. Then write down the answer: yes or no.

Next we ask: is p_3 less than α times 3 over m, $\alpha \cdot \frac{3}{m} = 0.01 \times \frac{3}{25}$ or 0.0012, which is an even more relaxed criterion, and write down that answer.

We ask that question of every one of the m p-values, p_k, $k = 1$ through $k = m$. Let $\widehat{k}$ be the largest number for which the answer is yes, that is, $\widehat{k}$ is the largest k for which p_k is less than α times k over m, $\alpha \cdot \frac{k}{m}$. Let's say that in this case $\widehat{k} = 15$.

The Benjamini-Hochberg procedure is to declare as statistically significant every p_k for every k up to and including $\widehat{k}$ (Figure 7.25). If $\widehat{k} = 15$, then $p_1, p_2, \ldots, p_{15}$ will be considered to have survived the BH correction and are declared as statistically significant.

	rank		p-value		threshold		
smallest	1	is	p_1	<	$\alpha \times \frac{1}{m}$	?	✓
	2	is	p_2	<	$\alpha \times \frac{2}{m}$	?	✓
	3	is	p_3	<	$\alpha \times \frac{3}{m}$	?	✓
	⋮	is	⋮	<	⋮	?	✓
	$\widehat{k}$	is	$p_{\widehat{k}}$	<	$\alpha \times \frac{\widehat{k}}{m}$	?	✓
	$\widehat{k}+1$	is	$p_{\widehat{k}+1}$	<	$\alpha \times \frac{\widehat{k}+1}{m}$	?	✗
	⋮	is	⋮	<	⋮	?	✗
largest	m	is	p_m	<	$\alpha \times \frac{m}{m}$	?	✗

Figure 7.25 The Benjamini-Hochberg procedure uses a sliding threshold to find $\widehat{k}$.

Benjamini-Hochberg procedure

Step 1 ► Let $p_1 \leq p_2 \leq \ldots \leq p_m$ be the ordered, observed p-values.

Step 2 ► Calculate $\widehat{k} = \max_{1 \leq k \leq m} \left\{ p_k \leq \alpha \cdot \frac{k}{m} \right\}$

Step 3 ► If $\widehat{k}$ exists, then reject Null Hypotheses corresponding to $p_1 \leq p_2 \leq \ldots \leq p_{\widehat{k}}$. Otherwise, reject nothing.

As $\widehat{k}$ is the largest number that satisfies the condition for statistical significance, this means *not every p-value up to $\widehat{k}$ needs to meet the statistical significance criterion against the sliding threshold.* As long as a k is less than or equal to $\widehat{k}$, p_k is declared significant.

There is a useful graphical presentation of the BH procedure that will help us visualize what the algorithm is doing (Figure 7.26).

We use the ranking index k as the horizontal axis. For each k, we show the value of p_k (solid

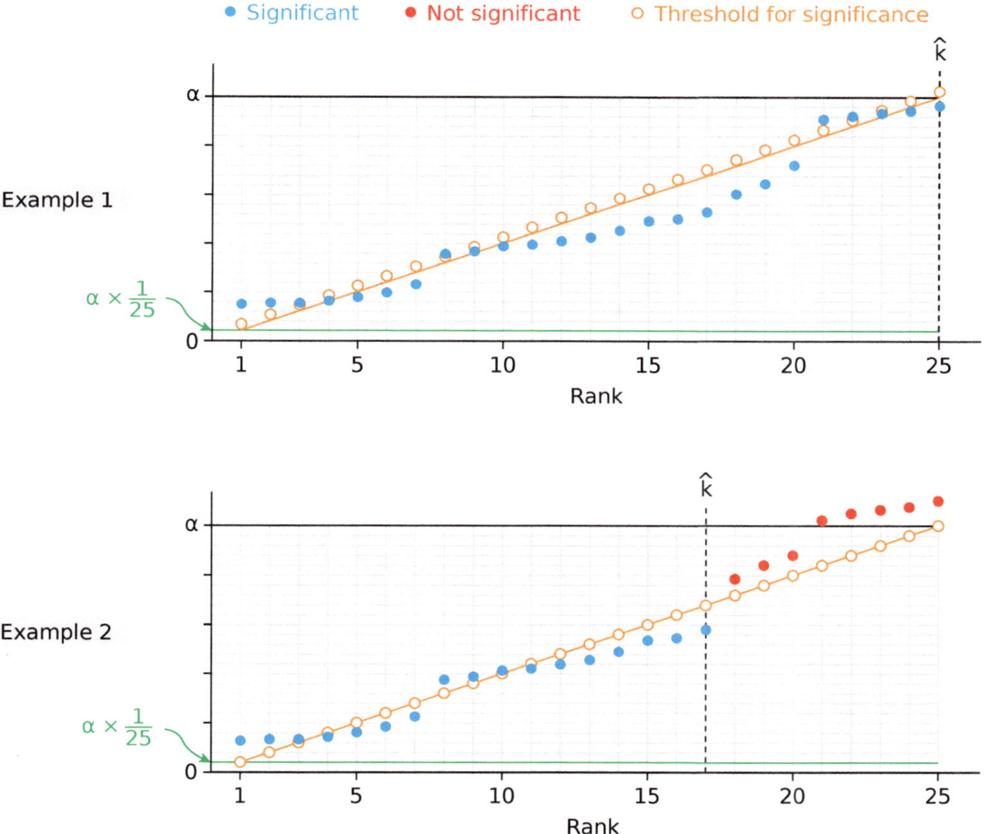

Figure 7.26 Two hypothetical examples of the Benjamini-Hochberg correction. Note that all p-values less than or equal to $p_{\widehat{k}}$ are declared significant. Solid dots are p-values, blue is significant and red is not significant. The orange hollow dots are the sliding thresholds for statistical significance. The black horizontal line is the original α value. The green horizontal line is the Bonferroni criterion.

dots). Then we show the value of the BH-corrected threshold for significance (orange hollow circles), connected by the orange line that passes through (0, 0) and has slope $\frac{\alpha}{m}$. We also plot the original α value, 0.01 (horizontal black line), and the Bonferroni criterion (horizontal green line).

Example 1 shows a hypothetical case in which we did 25 tests, and for which $\hat{k}$ is equal to 25. Note that in the BH correction, all 25 tests are considered significant, even those that failed to meet their individual thresholds. Example 2 shows another hypothetical case, in which 25 tests were done and $\hat{k}$ is equal to 17. Note that all of the smallest 17 p-values are therefore declared significant.

In the BH correction, the lowest p-value receives a harsh penalty, equal to Bonferroni, but the penalty gets lighter and lighter as the number of initially significant results increases.

Using the Benjamini-Hochberg correction will guarantee that the overall False Discovery Rate will be approximately α, thereby correcting for multiple testing.

Benjamini-Hochberg corrected confidence intervals

Confidence intervals are often used in contexts where they are *not* hypothesis testing. For example, we might be using the CI as a measure of the precision of our estimate of the effect size. This is how we first used CIs back in Chapter 4: as the true "error bars" around the estimate of the effect size. In this context, we are not doing covert hypothesis testing, and so there is no need to penalize CIs used for this purpose. There is no need for a multiple testing correction because there is no "testing" at all.

But CIs are also used for backdoor hypothesis testing, when we say things like "the 95% confidence interval excluded the null value, therefore $p < 0.05$". In this case, we do have to make a multiple testing correction.

It makes sense that if a penalty is made on the p-values of multiple comparisons, we also have to penalize, and be more demanding of, confidence intervals when we make multiple comparisons.

Benjamini and Yekutieli (2001) show that if we really want our overall False Discovery Rate to be α, then we have to calculate, not a $(1-\alpha)$ CI, but rather a more stringent (hence wider) CI, given by

$$1 - \frac{\hat{k}}{m} \times \alpha$$

where

$\hat{k}$ is the maximum value of k for which p_k which was declared significant under BH
m = total number of tests

For example, if we did 3 tests ($m = 3$), of which 2 p-values passed the BH criterion ($\hat{k} = 2$), then, if we want an overall false discovery rate of 0.01 (α), we need to take confidence intervals not of 99%, but of $(1 - \frac{\hat{k}}{m} \times \alpha)$, that is

$$1 - \frac{2}{3} \times 0.01 = 99.333\%$$

Note that the BH correction for confidence intervals requires that we have already done a BH correction on the p-values. This is necessary because we need to know $\hat{k}$ to do the CI correction.

FURTHER EXERCISES 7.5

1. Which is more "conservative", meaning it will produce the results with the fewest significant findings: the Benjamini-Hochberg correction or the Bonferroni correction? Describe the advantages and disadvantages of using each.

2. You compare the number of mosquito bites volunteers receive with three different types of insect repellents. There are 30 volunteers in each of the three groups (total of 90). You calculate a p-value for the omnibus test that is greater than the α you pre-selected for this study. How do you interpret this? What further steps should you take to further investigate this issue?

3. Sara conducted a study that passed the omnibus test. She then proceeded to conduct 5 pairwise group comparisons and got the following p-values: 0.009, 0.2, 0.02, 0.005, 0.1.

   a. She wants to correct for multiple testing using the Benjamini-Hochberg method to achieve a 5% FDR. Which p-values are still significant after the correction?

   b. What if she corrects for multiple testing using the Bonferroni method to achieve a 5% FWER. Which p-values are still significant after the correction?

4. A pharmaceutical company is screening 40 candidate drugs for their ability to inhibit cancer cell growth. Each drug is tested independently, and they want to identify which drugs show promise for further development. Here are the p-values from 40 independent tests (sorted):

Drug	p-value	Drug	p-value	Drug	p-value	Drug	p-value
1	0.0001	11	0.0028	21	0.0065	31	0.0350
2	0.0003	12	0.0030	22	0.0070	32	0.0500
3	0.0005	13	0.0035	23	0.0080	33	0.0750
4	0.0008	14	0.0038	24	0.0090	34	0.1000
5	0.0010	15	0.0040	25	0.0100	35	0.1500
6	0.0012	16	0.0045	26	0.0120	36	0.2500
7	0.0015	17	0.0048	27	0.0140	37	0.3500
8	0.0018	18	0.0050	28	0.0160	38	0.5000
9	0.0020	19	0.0055	29	0.0200	39	0.7000
10	0.0025	20	0.0060	30	0.0250	40	0.9000

   a. **Bonferroni Correction.** Using $\alpha = 0.05$ with Bonferroni correction, how many drugs would be declared significant? What is the adjusted threshold?

   b. **Benjamini-Hochberg:** Using FDR = 0.05 with a B-H procedure, how many drugs would be declared significant?

   c. **Business Impact.** If each drug costs $10 million to advance to the next stage of development: What is the total investment under each method? How many additional drugs does B-H identify?

   d. **Trade-offs.** Now consider opportunity costs: if a successful cancer drug can generate $10 billion in revenue, what is the cost of missing one blockbuster drug by being too conservative? Which type of error is more costly in drug development: advancing a few drugs that will fail in the next stage (animal studies, Phase I/II/III trials), or missing promising drugs entirely? Research what pharmaceutical companies actually do: do they tend to use Bonferroni or FDR-based methods in early screening?

7.6 Benjamini-Hochberg in action

The BH procedure when there are real differences

It's useful to see how Benjamini-Hochberg works in a case where there *are* real differences. To test this, we created a real-difference world.

We chose two distributions that were substantially different from each other: a Normal distribution A with mean $\mu_A = 50$ and standard deviation $\sigma_A = 5$, and another Normal distribution B with mean $\mu_B = 60$ and standard deviation $\sigma_B = 8$.

The first experiment consisted of sampling 20 elements randomly from A and 20 elements randomly from B, then making a two-group comparison of the resulting data sets and writing down the p-value of that comparison. Then this experiment was repeated for a total of 20 times, resulting in 20 p-values. Now we will apply the BH procedure (Table 7.5).

BH procedure We ranked these 20 p-values from smallest to largest, calling the result p_1, $p_2, \ldots, p_{20}$.

When we apply the BH test, we see that while 17 of our experiments returned a p-value that was statistically significant at the $p < 0.01$ level, the BH procedure disqualifies the highest 2 of those, leaving us with 15 significant comparisons.

k	p_k	$\alpha \cdot \frac{k}{m}$	Is $p_k < \alpha$?	Is $p_k < \alpha \cdot \frac{k}{m}$?
1	<0.0001	0.0005	✔	✔
2	<0.0001	0.0010	✔	✔
3	<0.0001	0.0015	✔	✔
4	<0.0001	0.0020	✔	✔
5	0.0001	0.0025	✔	✔
6	0.0002	0.0030	✔	✔
7	0.0002	0.0035	✔	✔
8	0.0004	0.0040	✔	✔
9	0.0006	0.0045	✔	✔
10	0.0006	0.0050	✔	✔
11	0.0007	0.0055	✔	✔
12	0.0011	0.0060	✔	✔
13	0.0013	0.0065	✔	✔
14	0.0051	0.0070	✔	✔
15	0.0075	0.0075	✔	✔
				$\hat{k} = 15$
16	0.0089	0.0080	✔	✘
17	0.0090	0.0085	✔	✘
			$\alpha = 0.01$	
18	0.0125	0.0090	✘	✘
19	0.0125	0.0095	✘	✘
20	0.0177	0.01	✘	✘

Table 7.5 p-value calculations for the 20 two-group comparison experiments (see text). The two largest initially "significant" p-values have been disqualified.

The graphical presentation of the BH procedure is shown in Figure 7.27.

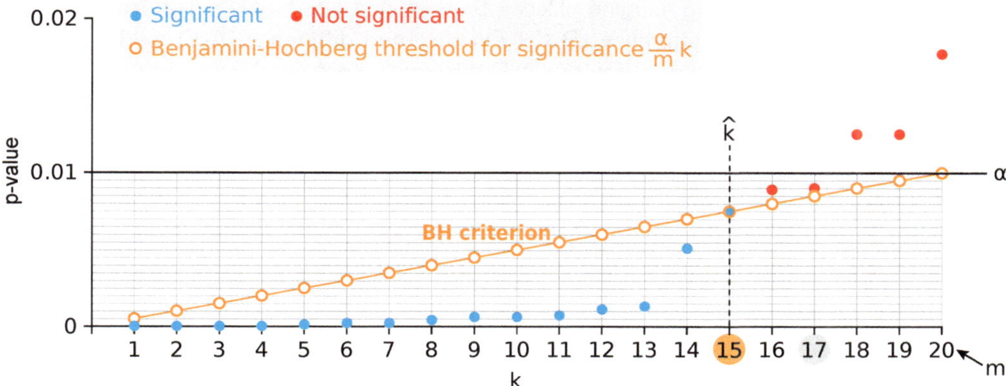

Figure 7.27 Graphical visualization of Table 7.5. $\widehat{k}$ is the last point for which the p-value is below the BH criterion line. All the p-values p_k with $k \le \widehat{k}$ are significant (blue dots). All the p-values p_k with $k > \widehat{k}$ are not significant (red dots).

BH, Bonferroni and uncorrected We can compare the BH correction to the Bonferroni and the uncorrected approaches (Figure 7.28). Using the data from our example, the uncorrected $\alpha = 0.01$ is the most lenient, and allows for 17 significant p-values, while the Bonferroni is the harshest, and allows only 8. The BH is in the middle, and its sliding scale allows for 15 significant p-values.

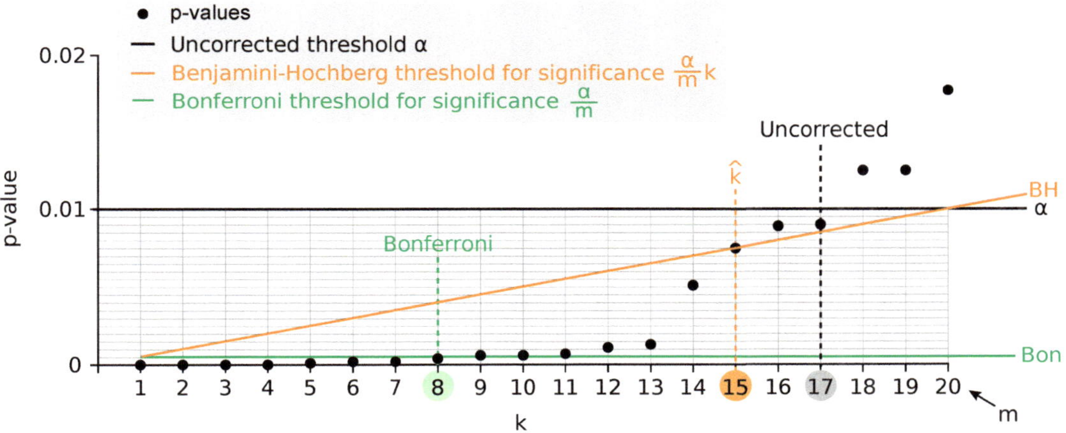

Figure 7.28 Comparison of the three methods for reducing false positives: uncorrected, Bonferroni and BH.

We recommend the BH correction. It's built into Python and almost all other programming languages. It's the most generally recommended method for controlling false positives.

BH application to the study of drug effects on T-cell counts

Let's return to our example of the T-cell counts study (data presented in Figure 7.2), where we introduced a three-group comparison: a control group A and two drugs B and C, in terms of their median T-cell counts. After a positive omnibus test (Figure 7.20), we carried out part I of a *post hoc* analysis: pairwise group comparisons.

At that time, we left an issue hanging. Recall that we had one highly significant difference (A-B) and one non-significant difference (B-C). On the third difference, A-C, we had a minor conflict: the 99% confidence interval for the A-C difference contained zero, and yet the two-group comparison of A and C yielded a p-value that was significant at the 0.01 level (Figures 7.23 and 7.24).

While this is not a mathematical contradiction, it is still conflicting advice.

We had promised that the conflict would be resolved when we apply a false positive correction to this analysis. The result is that the p-value in conflict, for the A-C comparison, is ruled non-significant by the BH correction, thus resolving the conflict (Table 7.6 and Figure 7.29).

Comparison	k	p_k	$\alpha \cdot \frac{k}{m}$	is $p_k < \alpha$?	is $p_k < \alpha \cdot \frac{k}{m}$?
A vs. B	1	0.0001	0.0033	✔	✔
A vs. C	2	0.0080	0.0067	✔	✘ ($\widehat{k} = 1$)
B vs. C	3	0.0219	0.01	✘ ($\alpha = 0.01$)	✘

Table 7.6 The BH correction applied to the drug test on T-cell counts data.

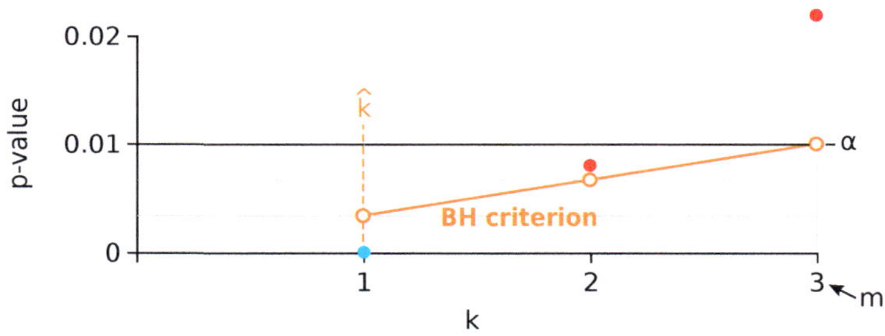

Figure 7.29 Graphical presentation of Table 7.6.

BH-corrected confidence intervals Suppose that we now wanted to construct confidence intervals on the group differences. If we are using the confidence intervals for their own sake, to illustrate the precisions of our estimates, then we do not correct for multiple testing because there is no testing at all.

But if we are constructing confidence intervals to do backdoor significance testing, that is, declaring a significant difference when the confidence interval excludes the null value, then we need to correct for multiple testing. Therefore, in this case, we must use BH-corrected confidence intervals.

In the T-cell count study, we have 3 groups, hence 3 tests, so $m = 3$. Only p_1 passed the BH criterion, so $\widehat{k} = 1$.

Benjamini and Yekutieli then tell us that if we wanted an overall 0.01 FDR, we should insist on $\left(1 - (\frac{1}{3} \times 0.01)\right) = 99.67\%$ confidence intervals instead of 99%.

FURTHER EXERCISES 7.6

1. **Regimen rivalry.** You're studying the effectiveness of four different fitness regimens for building lower-body strength. The data set passed the omnibus test. You then conduct pairwise group comparisons for each pair of treatments. Your p-values are

Pairwise group comparison	p-value
T1 vs. T2	p = 0.34
T1 vs. T3	p = 0.001
T1 vs. T4	p = 0.0084
T2 vs. T3	p = 0.003
T2 vs. T4	p = 0.013
T3 vs. T4	p = 0.006

Use the Benjamini-Hochberg procedure to determine which regimens are significantly different from each other to maintain a 1% FDR.

2. **Hand swelling.** Contrast baths are a common treatment to reduce hand swelling in patients with carpal tunnel syndrome. Researchers randomly assigned individuals to the treatments "Bath", "Bath+Exercise", or "Exercise" and measured change in hand volume.

Bath	Bath+Exercise	Exercise
5, 10, −4, 11, -3, 13, 0, 2, 10, 6, −1, 8, 10, −9	6, 10, 0, 14, 0, 15, 4, 5, 11, 7, 20, 9, 11, 21	−12, −10, −7, −1, −1, 0, 0, 0, 0, 0, 2, 4, 5, 5

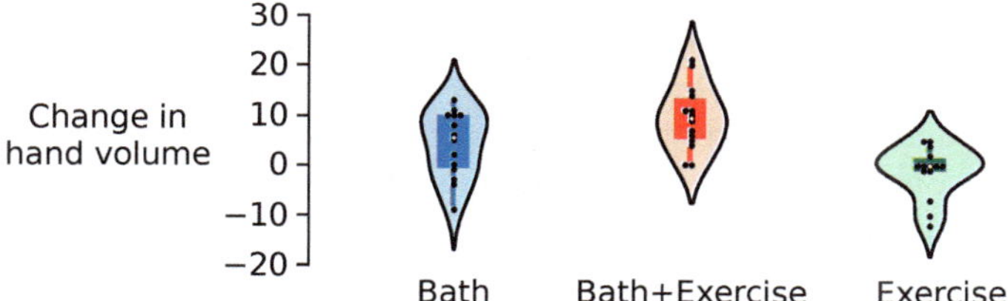

a) Comment on the shape of the data distributions.

b) Describe the appropriate Null Hypothesis.

c) Based on your observations, determine what measure to use to describe the groups, and what measure to use to compare the groups.

d) Conduct an omnibus test and interpret the result.

e) If the omnibus test is positive, calculate p-values for pairwise group comparisons and correct for multiple testing.

f) Calculate 95% confidence intervals for the group differences.

g) Interpret your results from steps (d) and (e) (use $\alpha = 0.05$).

h) According to this data, which treatment would you recommend to reduce hand volume the most?

7.7 Writing up the results

Let's say that we have now done our analyses, and it's time to show our results in an oral presentation or for publication.

The simplest form of presentation of the *post hoc* results of a multi-group comparison is to present the group measures (means or medians, or ...) together with, say, 95% confidence intervals for each of the group measures, constructed by the resampling methods in Chapter 4. In addition, we should always either show the data directly as dot plots/beeswarm plots, or, if the data set is large, as histograms/violin plots.

We now present the T-cell count study (data presented in Figure 7.2), a three-group comparison of median T-cell counts for drugs B and C and a control group A (Figure 7.30).

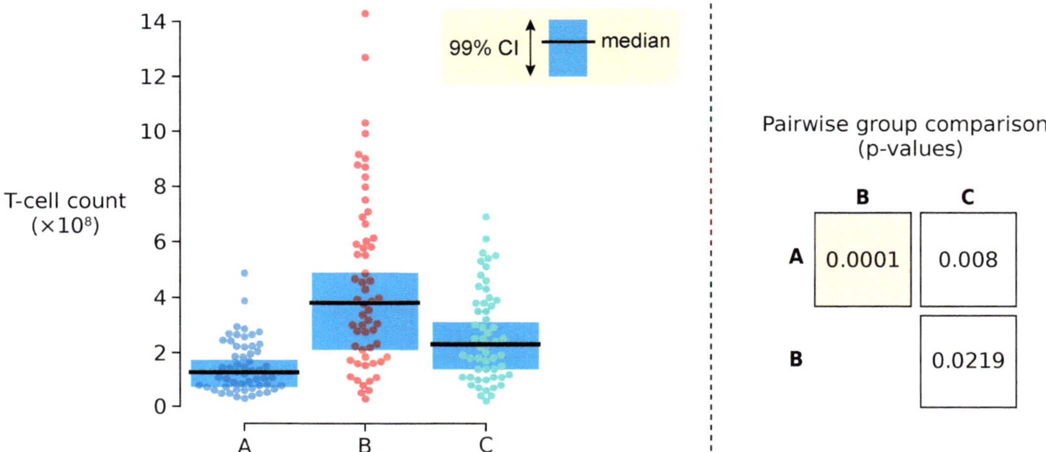

Figure 7.30 Presentation of the results of the T-cell count study. The raw data are shown as beeswarm plots. Superimposed on the beeswarm plots are the median values for the three groups (horizontal black lines) and the 99% confidence intervals for the medians (blue rectangles). The pairwise group comparison p-values are shown on the right. Recall that only the A-B comparison passed the Benjamini-Hochberg correction.

This is a useful presentation of the results of a three-group comparison, and allows us to draw some tentative conclusions about the data. Notice how much richer this is than the unfortunately frequent practice of showing means and standard deviations (or standard errors) as dynamite plunger plots (see Chapter 2).

Under no circumstances should results be presented simply as dynamite plunger plots, that is, bar graphs with "error bars".

Generally, published papers have a "Methods" section, where you state and defend your methods, and a "Results" section, where you set out the results that these methods produced.

In the Methods section, you will have to defend your choice to use resampling methods. Most readers and reviewers will look at a multi-group study and think "ANOVA" (see section 7.10 *The mathematical theory: traditional (formula-based) ANOVA*). Unfortunately, the reality is that most people simply aren't aware of the heavy limitations of ANOVA and the many stringent assumptions that are required for its application. So it's important to say up front why we aren't using formula-based methods like ANOVA. Otherwise, readers will have the lingering question: "Why aren't they using ANOVA?"

Here is a presentation, as if for publication, of the T-cell count study.

A study of the T-cell Stimulation Properties of Drugs B and C

Methods

Statistical Comparisons Because the data did not meet the requirements of Normality, multi-group comparisons were made using resampling methods (Efron and Tibshirani, Statistical data analysis in the computer age, *Science* 1991).

Briefly, for an omnibus test, we considered the ratio of across-group differences to within-group differences in the data, and then compared that ratio to similarly calculated ratios observed under the Null Hypothesis of no difference among the groups, as realized by 10,000 randomized versions of the same data. If the data set passed the omnibus test ($p < 0.01$), we proceeded to conduct *post hoc* analyses consisting of pairwise group comparisons using resampling-based methods. Differences were considered statistically significant at the $p < 0.01$ level, after Benjamini-Hochberg correction for multiple testing.

Results An omnibus test for the multi-group comparison proved highly significant ($p = 0.0005$, see Supplemental Methods), indicating that there were significant differences among the groups.

In *post hoc* analysis, Drug B was seen to improve median T-cell counts vs. Control A, by 2.53 ($\times 10^8$ cells per unit area), with a 99% CI [0.58, 3.9], $p = 0.0001$, significant after Benjamini-Hochberg correction. When we compared Drug C to Control A, the 99% confidence interval [−0.09, 2.02] for the C–A difference included the null value zero (suggesting no significant difference). While the p-value of the C–A difference was 0.008, the Benjamini-Hochberg correction disqualified this p-value as not significant. Drugs B and C were not significantly different from each other (99% CI [−3.6, 0.66]) ("Figure 1", corresponding to Figure 7.30).

Supplemental Methods

1. The median was chosen for the descriptors of the three groups, because each of the data sets showed significant skew.

2. We carried out an omnibus test of the three groups. For across-group differences, we summed the absolute values of the differences of the Group Medians from the Grand Median, and for within-group differences, we summed the absolute values of the differences between each data point and its group median. We defined F as the ratio of across-group to within-group differences, and compared F as calculated from the data to 10,000 values of F calculated from randomized versions of the same data ("Figure S1", corresponding to Figure 7.20).

7.8 Ranks

As we saw in two-group comparisons, if we don't know the true variability of the data set, we don't really know whether a numerical difference is "large" or not.

This applies in particular when the data sets are small, making it impossible to guess what the underlying distribution is or what the underlying variability might have been.

Suppose, for example, we have 4 subjects in each of three groups A, B and C. How can we analyze this data? The answer is that we use ranks, similar to how we use ranks to compare two small groups (section 6.6). Here we are making a multi-group comparison, and because there are more than two groups, we have to guard against false positives from multiple testing by first doing an omnibus test. If the omnibus test is positive, we then go on to do *post hoc* analysis to see where the differences are.

Omnibus First, we define our across-group and within-group differences.

Take the raw data numbers from the 3 groups A, B and C, and put them in a single list. Order that list from smallest to largest and then put rank numbers on it, #1, #2, #3, ..., #12. Then put those rank numbers back into the three groups, replacing each raw number by its rank.

We will use the average rank in each group as the group measure and the overall average rank as the Grand Mean. Define the across-group difference as the sum of the weighted distances from each group's average rank to the Grand Mean, where each of the 3 distances is weighted by its group size. For the within-group differences, we use the sum of the distances from each ranking to the average ranking for its group. We then calculate our F_{obs} from this data, the across-group differences divided by the within-group differences.

Next we compare F_{obs} to its randomized counterparts.

We carry out the randomization by throwing all the raw data into a Big Box and resampling a new set A_1, B_1 and C_1 out of the box. Then we calculate F_1 on this new set using ranks.

We repeat this randomization procedure 10,000 times, generating 10,000 estimates: F_1, F_2, ..., $F_{10,000}$. Then we compare F_{obs} to the 10,000 F-values to obtain a p-value, which is the fraction of simulations that have an F-value greater than or equal to F_{obs}.

post hoc analysis When the omnibus test is positive, we can go on to make multiple pairwise comparisons: A vs. B, A vs. C, and B vs. C. For each pairwise group comparison, we use the same procedure as in two-group comparisons using ranks, generating 3 p-values. The p-values must then be subjected to a multiple testing correction, such as Benjamini-Hochberg.

FURTHER EXERCISES 7.8

1. **Sunscreen.** In preparation for next summer, you decide to find the best sunscreen that will feel good and take care of your skin. In two separate experiments, you buy three kinds of sunscreen and give out samples to beach-goers. (Each person tries one kind of sunscreen.) At the end of the day, you record how much each person liked their sunscreen on a scale from 0 to 15 (hated = 0, loved = 15).

   a. Based on the beeswarm plots, predict which experiment has the larger F-statistic.
   b. When conducting significance testing for this study, is it reasonable to use $\alpha = 0.05$? Explain why.
   c. When we're comparing three or more groups, why do we look for statistically significant differences among all the groups before comparing each pair of groups?

d. Why do we recommend the Benjamini-Hochberg method instead of the more strict Bonferroni method for multiple test corrections?

e. Carry out the appropriate analysis for experiment 1 and interpret your result.

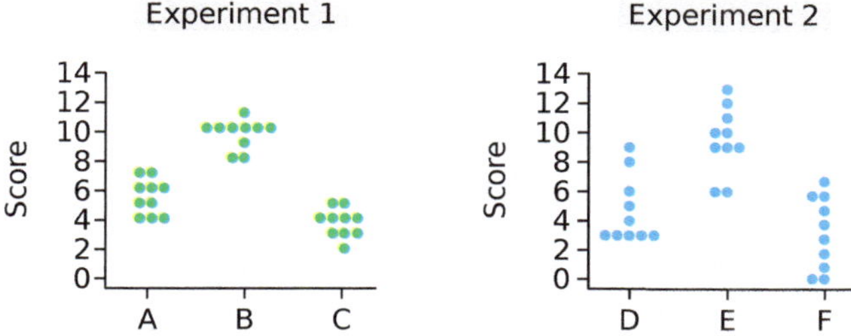

7.9 Two-way (or two-factor) multi-group comparisons

An example: the effect of sunlight and watering on plant height

Frequently, we have not just one type of cause for a given effect, but two or more types of causes. We might be looking at plant height as the outcome, and one causal factor is the amount of sunlight, and the second causal factor is the amount of food given to the plant. So now we have two causal factors operating, and we want to know what their effect is on the outcome variable, plant height.

A hypothetical data set might look like this (Figure 7.31):

		Sunlight Exposure			
		None	Low	Medium	High
Watering Frequency	Daily	3.8	4.4	6.4	6.5
		3.6	5.0	6.2	6.3
		3.2	5.2	5.1	5.8
		3.3	4.3	5.5	5.6
		4.2	4.6	5.8	5.9
	Weekly	4.2	5.6	5.8	5.4
		3.8	5.3	6.2	4.9
		4.4	4.8	6.3	5.2
		3.9	5.4	6.5	5.3
		3.7	4.7	5.5	5.1

Figure 7.31 Data set for the effect of sunlight and watering on plant height (m).

Down the left hand side, there are two values of watering frequency: daily and weekly. And then, along the top, we have exposure to sunlight, broken down into four categories: none, low, medium and high.

And now there are three questions we could ask.

Two-way (or two-factor) multi-group comparisons

Q1: is there an A (sunlight) effect (one-factor comparison)?

Q2: is there a B (watering) effect (one-factor comparison)?

Q3: is there an interaction effect? (Does the A effect depend on the B levels? Does the B effect depend on the A levels? Is there "synergy" between A and B? If there is, is it positive or negative?)

Three questions

Q1 The first question is: is there an A effect? In this case, is there a sunlight effect? We answer this question by simply forgetting that there are two different values of watering frequency, as if we erased the horizontal line that separates the two watering frequencies, thereby giving us four columns of 10 data points each for the four levels of sunlight exposure (Figure 7.32). This converts the problem to a one-factor multi-group comparison with four groups.

Sunlight Exposure			
None	Low	Medium	High
3.8	4.4	6.4	6.5
3.6	5.0	6.2	6.3
3.2	5.2	5.1	5.8
3.3	4.3	5.5	5.6
4.2	4.6	5.8	5.9
4.2	5.6	5.8	5.4
3.8	5.3	6.2	4.9
4.4	4.8	6.3	5.2
3.9	5.4	6.5	5.3
3.7	4.7	5.5	5.1

Figure 7.32

Q2 The second question is: is there a B effect? And that is the question: is there a watering frequency effect? To answer this, we just forget the distinction into different levels of sunlight exposure, erasing the vertical lines that separate the sunlight levels, and now we have 20 numbers for the daily watering frequency, and another 20 numbers for the weekly watering frequency (Figure 7.33). Then we do a one-way multi-group comparison on those two levels. (A multi-group comparison with two levels is equivalent to a two-group comparison, as in Chapter 6.)

Watering Frequency					
	Daily	3.8	4.4	6.4	6.5
		3.6	5.0	6.2	6.3
		3.2	5.2	5.1	5.8
		3.3	4.3	5.5	5.6
		4.2	4.6	5.8	5.9
	Weekly	4.2	5.6	5.8	5.4
		3.8	5.3	6.2	4.9
		4.4	4.8	6.3	5.2
		3.9	5.4	6.5	5.3
		3.7	4.7	5.5	5.1

Figure 7.33

Q3 The third question, and this is a critical question, is: **is there an interaction effect?** The interaction effect is something that is new in two-factor analysis. We interpret that question to mean: does the A effect depend upon the B levels? and does the B effect depend upon the A levels? This is a new kind of question, and it is very important, so we will spend some time developing it and showing examples.

The interaction plot

There's a new type of graphic that answers all three questions for a two-way multi-group comparison, including revealing the interaction effect. It's called an **interaction plot**.

We'll begin with the simplest possible abstract example, to illustrate the concept.

How to read an interaction plot

Let's consider a two-factor analysis where there are only 2 levels of factor A, called A_1 and A_2, and 2 levels of factor B, called B_1 and B_2. So the data looks like:

factor A

factor B		A_1	A_2
	B_1	$\{A_1, B_1\}$	$\{A_2, B_1\}$
	B_2	$\{A_1, B_2\}$	$\{A_2, B_2\}$

In each of the four cells, we have the summary measure for that cell, whether it's the mean or the median or ....

Then we make the interaction plot. Choose one of the two factors (it doesn't matter which) to be our X-axis. Here we chose the B factor. So the X-axis is going to have 2 values: B_1 and B_2. The Y-axis is going to record the summary measures for each cell.

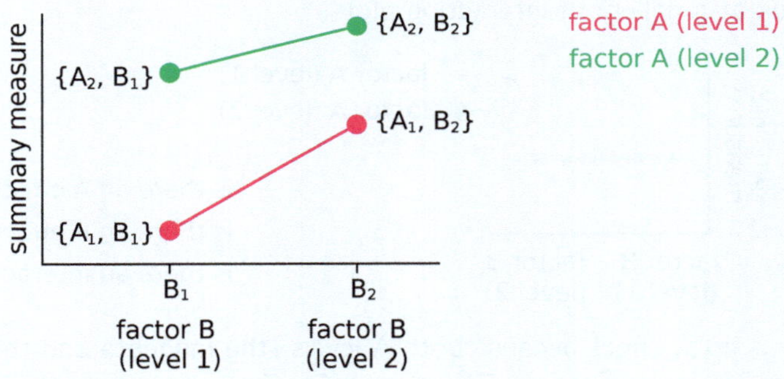

Then at the value $B = B_1$, we plot the 2 summary values for the 2 cells in B_1, that is, $\{A_1, B_1\}$ and $\{A_2, B_1\}$. At the value $B = B_2$, we plot the 2 summary values for $\{A_1, B_2\}$ and $\{A_2, B_2\}$. Then, and this is the key, *we draw a straight line connecting the two A_1 values, and another straight line connecting the two A_2 values.* This gives us the interaction plot.

The three questions of two-way multi-group analysis can then be translated into questions about the points and lines on an interaction plot. Of course, these visual impressions must then be verified by formal two-way calculations.

Q1. is there an A effect? This is the question: are the A_1 values *everywhere* either substantially higher or lower than the A_2 values? This is equivalent to asking if the A_1 line (magenta) is substantially higher or substantially lower than the A_2 line (green)? (Here, "higher" could mean that all points are higher, or that some points are *much* higher, making the average higher.) In this example, the answer is yes; all A_1 values are below all A_2 values, so it looks like there is an A effect: A-level 2 is higher than A-level 1.

Q2. is there a B effect? Here we are asking whether the B_1 values, $\{A_1, B_1\}$ and $\{A_2, B_1\}$, that is, the left hand side of the plot, are higher or lower than both B_2 values, $\{A_1, B_2\}$ and $\{A_2, B_2\}$, that is, the right hand side of the plot.

In this case, the answer appears to be 'yes', because, although $\{A_2, B_1\}$ is slightly above $\{A_1, B_2\}$, $\{A_1, B_1\}$ is substantially below $\{A_2, B_2\}$, so the average on the left side will be lower than the average on the right side.

Q3. is there an interaction effect? Is the question, *are the lines non-parallel?* If the lines are roughly parallel (that is, they maintain a roughly constant distance from each other), then that says that the B treatment is doing roughly the same thing to both levels of A, and A is doing the same thing to both levels of B. In our example, the lines are not parallel, so there is an interaction effect: going from B_1 to B_2 has a bigger effect on A_1 than it does on A_2.

We should emphasize that these interaction plots are giving us rough qualitative visual comparisons. The effects that we see in the plots may or may not be statistically significant; the diagrams have to be supplemented by quantitative analysis.

A gallery of interaction plots

Let's look at a gallery of interaction plots.

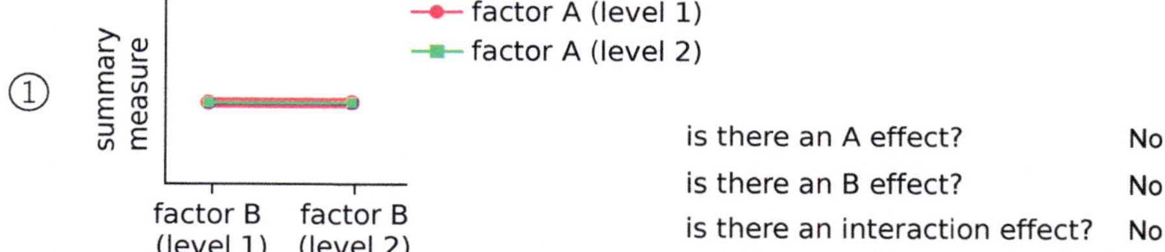

There is no A effect because both A levels (the magenta and the green lines) have the same summary measures. There is no B effect because both B levels (left and right) have the same summary measures. And there is no interaction effect because the two lines are parallel.

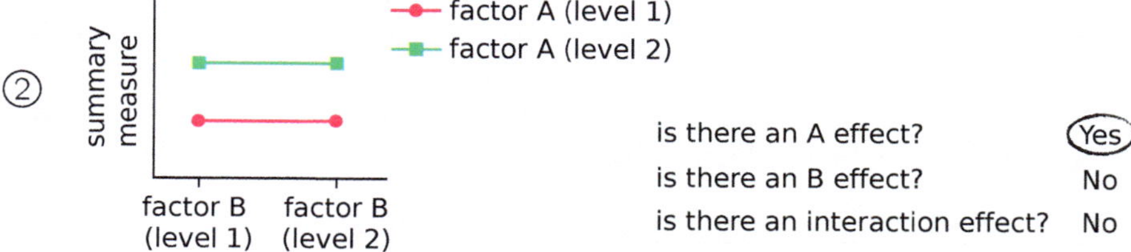

There is an A effect because the green line is everywhere higher than the magenta line. There is no B effect because both B levels (left and right) have equal summary measures. And there is no interaction effect because the two lines are parallel.

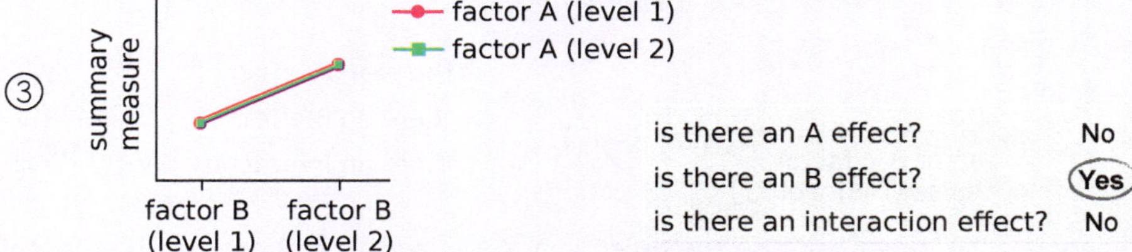

There is no A effect because both A levels (the magenta and the green lines) have equal summary measures. There is a B effect because B-level 2 is always higher than B-level 1, regardless of A. And there is no interaction effect because the two lines are parallel.

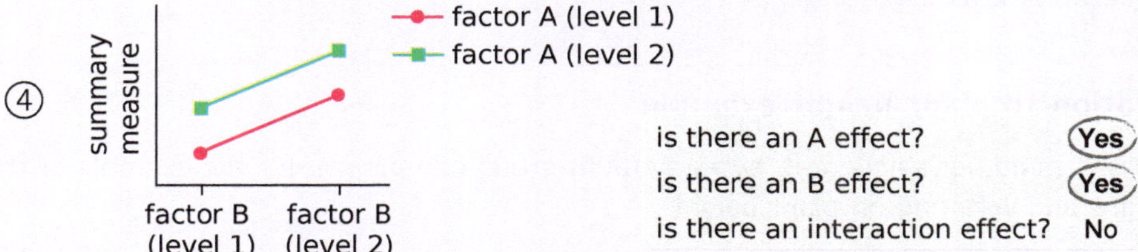

There is an A effect because the green line is everywhere higher than the magenta line. There is a B effect because B-level 2 is always higher than B-level 1, regardless of A. And there is no interaction effect because the two lines are parallel.

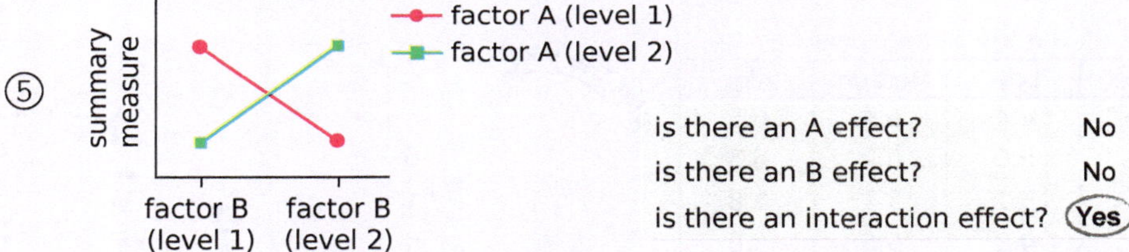

There is no A effect because the two lines are not consistently higher or lower than the other. There is no B effect because the two B levels on the left are equal to the two B levels on the right. There is a strong interaction effect because the effect of going from left to right (B-level 1 to B-level 2) depends on the A level. A-level 1 goes down, A-level 2 goes up.

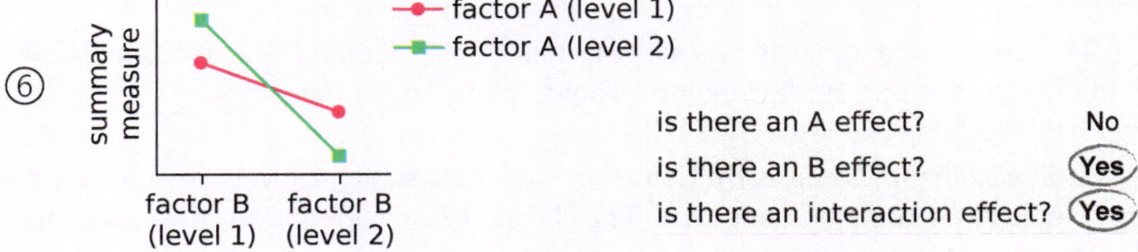

There is no A effect because the magenta dots (A-level 1) are not consistently higher or lower than the green dots (A-level 2). There is a B effect because the two dots on the left (B-level 1) are both higher than the two dots on the right (B-level 2). There is an interaction effect because going from left to right (B-level 1 to B-level 2) has a greater effect on the green line (A-level 2) than it does on the magenta line (A-level 1).

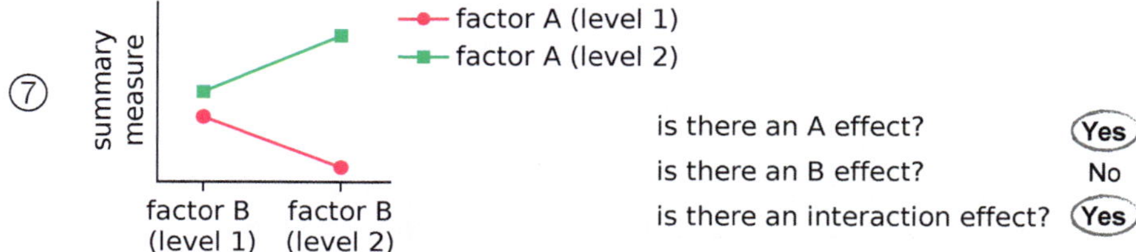

There is an A effect because the green line is everywhere higher than the magenta line. There is no B effect because the average of B-level 1 (left) is about the same as the average of B-level 2 (right). There is a strong interaction effect because the effect of going from left to right (B-level 1 to B-level 2) depends on the A level. A-level 1 goes down, A-level 2 goes up.

Application to plant height example

With this in mind, let's now do a two-way multi-group comparison for the example of the effect of sunlight and watering on plant height.

Q1: is there a Sunlight Exposure effect?

Plot the data Our first question is the one-factor question about the 4 levels of Sunlight Exposure (Figure 7.34 Left). First we plot the data (Figure 7.34 Right).

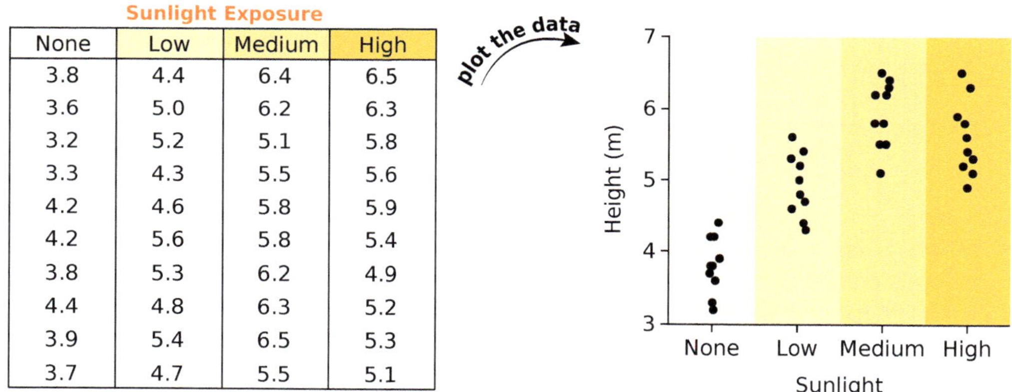

Sunlight Exposure			
None	Low	Medium	High
3.8	4.4	6.4	6.5
3.6	5.0	6.2	6.3
3.2	5.2	5.1	5.8
3.3	4.3	5.5	5.6
4.2	4.6	5.8	5.9
4.2	5.6	5.8	5.4
3.8	5.3	6.2	4.9
4.4	4.8	6.3	5.2
3.9	5.4	6.5	5.3
3.7	4.7	5.5	5.1

Figure 7.34 Left: data table for a one-way multi-group comparison for Sunlight Exposure. Each of the 4 columns has 10 data points. Right: plots of the data.

Our overall visual impression is that all three non-zero sunlight levels are superior to "None", and that "Medium" is superior to "Low". The "High" looks superior to "Low" but may or may not be superior to "Medium". Now we need to confirm (or disconfirm) this using quantitative analysis.

Calculating across-group and within-group differences The first step in the omnibus test is to define what we mean by across-group and within-group differences. Here we are using the group means and the Grand Mean as our measures, so the total across-group differences is the sum of the 4 weighted distances from the group means to the Grand Mean, with each distance weighted by its group size. The total within-group differences is calculated as the sum of the distances from each data point to its own group mean (Figure 7.35).

Group means and variances are shown in the table on the left, along with the differences of the means (column minus row) in the matrix on the right. Looking at the group differences, all are positive, except the last, which indicates that "High" is actually somewhat inferior, on average, to "Medium".

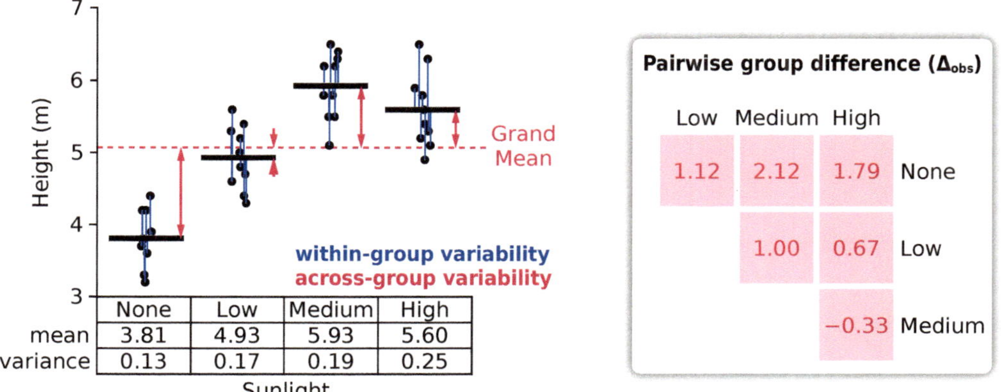

Figure 7.35 Left: across- and within-group differences for the one-factor analysis of the effect of sunlight on plant height (data presented in Figure 7.34). Right: pairwise group differences (column minus row).

Omnibus test for the sunlight effect For our data, we calculated F_{obs} to be 1.9.

$$F_{obs} = \frac{\textbf{total across-group difference}}{\textbf{total within-group difference}} = \frac{\sum|\updownarrow\times\text{group size}|}{\sum|\,\text{I}\,|} = 1.9$$

Note: here we are using absolute distance as the "difference" measure.

When we ran 10,000 simulations of the Null Hypothesis (Big Box model, see Figure 7.17), an F-value greater than 1.9 never occurred, so our p-value for the omnibus test is $p < 0.0001$ (Figure 7.36). **We conclude that there is a significant difference among the groups.**

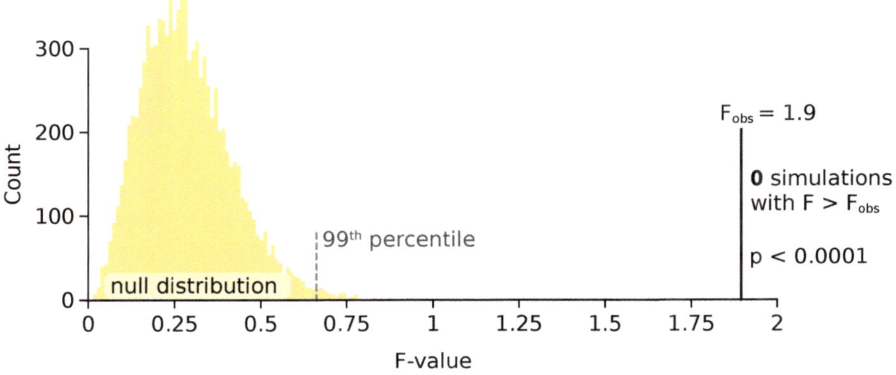

Figure 7.36 Omnibus test for the sunlight effect.

post hoc analysis Having passed the omnibus test, we can now proceed to make pairwise group comparisons (Figure 7.37). Since the variances of the 4 groups were not very different (< 2×), we used a Big Box resampling approach to compare each group to each other.

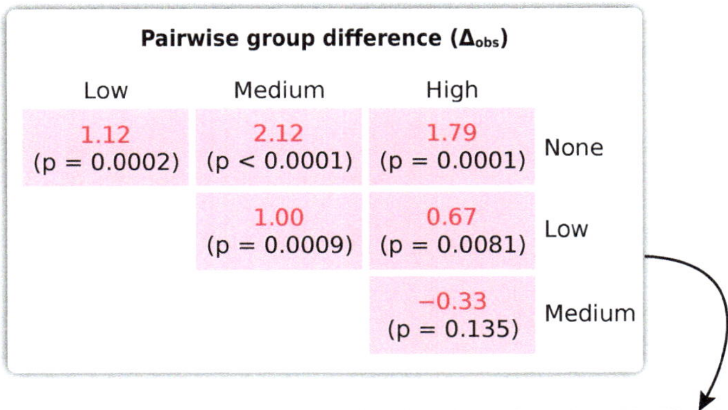

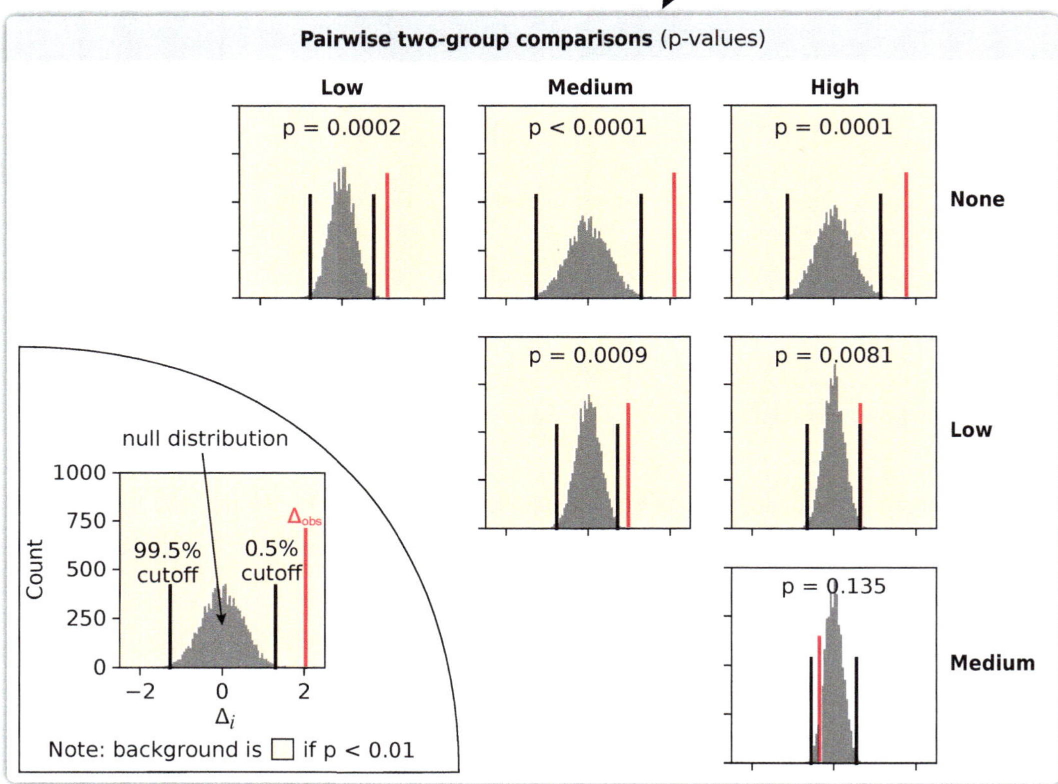

Figure 7.37 ***post hoc* analysis (part I) of the effect of levels of sunlight on plant height (p-values)**. Each sunlight level was compared to the other three, giving 6 pairwise group comparisons in all. Upper: the matrix shows the observed group differences and their p-values. Lower: for each pairwise two-group comparison, the observed difference in their means (column minus row) is shown as a red vertical line. The histograms are the null distributions, showing the simulated group differences in 10,000 simulations of the Null Hypothesis, using a Big Box resampling method. The black vertical lines are the cutoffs for the 99th percentile of the simulations. The insert shows the scale.

For each pairwise group comparison, we simulated the corresponding Null Hypothesis 10,000 times, generated the null distribution, compared the observed group difference to the null distribution, and obtained a p-value for the observed group difference.

As we see, of the $\frac{4\times3}{2}$ = 6 pairwise comparisons, 5 came out statistically significant at the $p < 0.01$ level. They are shaded in the figure.

However, as we must always do when we make multiple comparisons, we have to use a correction for false positives. Here we will use the Benjamini-Hochberg procedure (Table 7.7 and Figure 7.38).

Since all 5 significant p-values survived the BH correction, we can proceed to say that those 5 comparisons are statistically significant at the $p < 0.01$ level.

k	(grp_A, grp_B)	p_k	$\alpha \cdot \frac{k}{m}$	is $p_k < \alpha$?	is $p_k < \alpha \cdot \frac{k}{m}$?
1	(None, Medium)	<0.0001	0.0017	✔	✔
2	(None, High)	0.0001	0.0033	✔	✔
3	(None, Low)	0.0002	0.0050	✔	✔
4	(Low, Medium)	0.0009	0.0067	✔	✔
5	(Low, High)	0.0081	0.0083	✔	✔
				$\alpha = 0.01$	$\widehat{k} = 5$
6	(Medium, High)	0.1350	0.0100	✘	✘

Table 7.7 ***post hoc*** **analysis (part II) of the effect of levels of sunlight on plant height.** Applying the Benjamini-Hochberg correction to the p-values in the Sunlight Exposure analysis (p-values are shown in Figure 7.37).

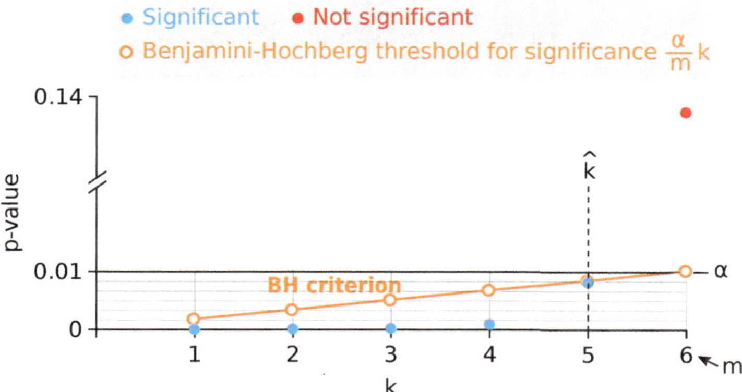

Figure 7.38 Graphical representation of the Benjamini-Hochberg correction for the Sunlight Exposure analysis. Note that all 5 significant p-values remained significant after BH correction.

Confidence intervals for pairwise group comparisons However, simple p-values are not the best answer, and should not be the final step. The last step in the one-way comparison of Sunlight levels is to do the confidence intervals for the differences between the groups (Figure 7.39).

The confidence intervals confirm our conclusions above about the pairwise group differences: all three non-zero sunlight levels are superior to "None", and both "Medium" and "High" are superior to "Low". However, we now have confidence intervals on the differences, giving us much stronger information (effect size and CI) than simply p-values.

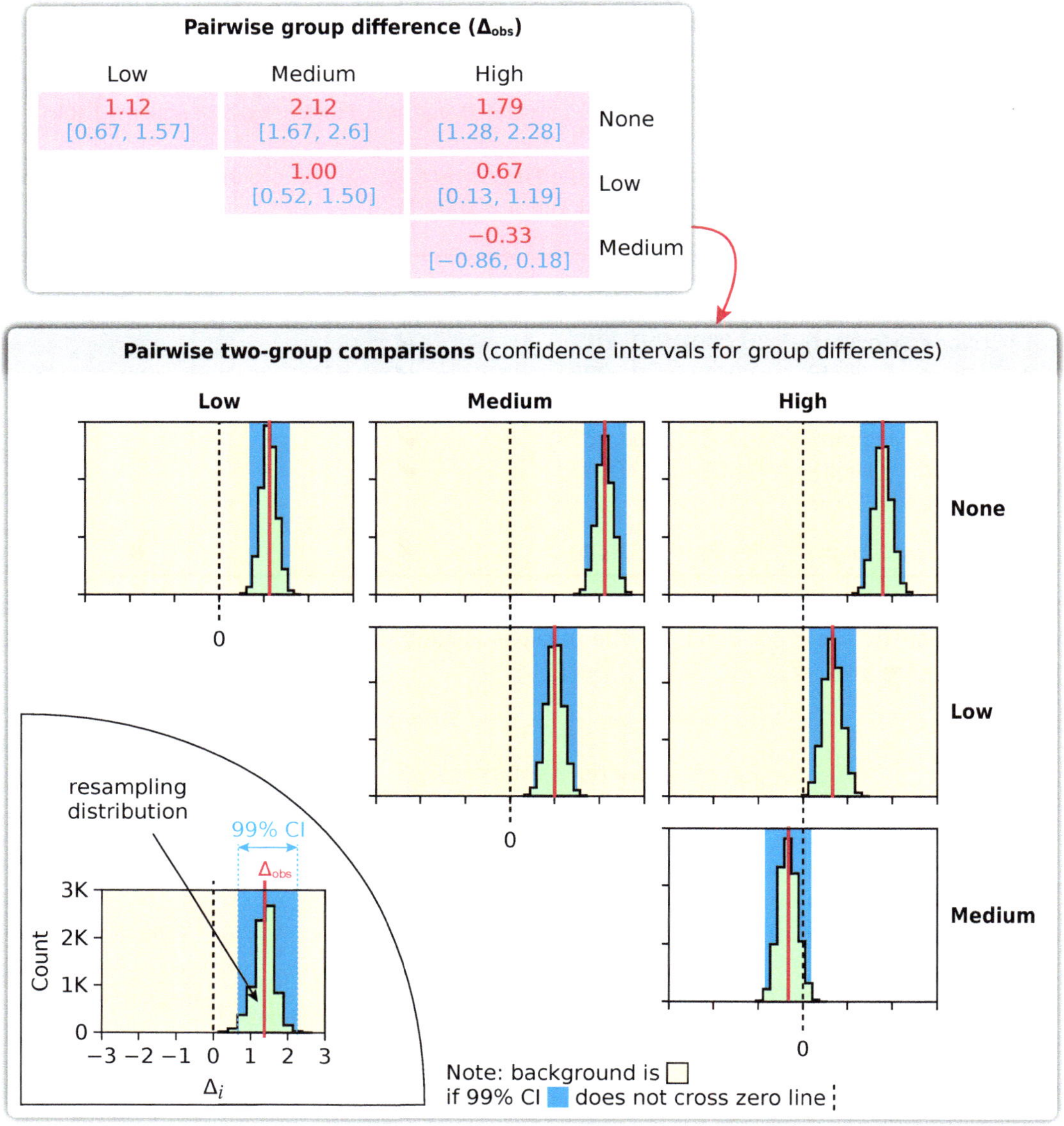

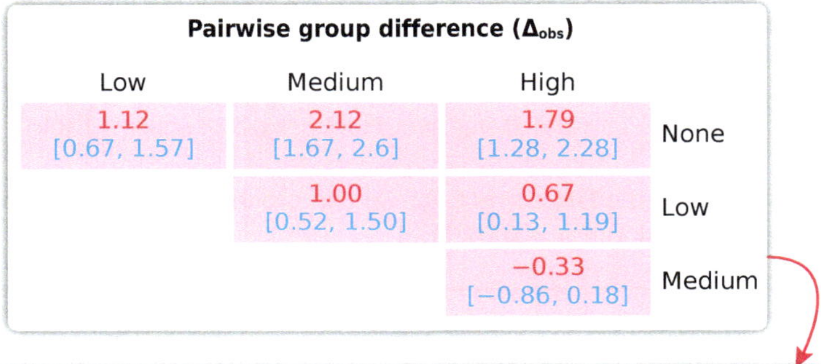

Figure 7.39 Confidence intervals for the effect of levels of sunlight on plant height. Each sunlight level was compared to the other three, giving 6 pairwise group comparisons in all. Upper: the matrix shows the observed group differences. Lower: for each pairwise two-group comparison, the observed difference in their means (column minus row) is shown as a red vertical line. The histograms are the resampling distributions, and the dashed black vertical lines are the null values, with 99% confidence intervals in blue rectangles. The insert shows the scale.

Sunlight effect A number of comparisons were statistically significant at the $p < 0.01$ level (Figures 7.38 and 7.39). The quantitative analysis confirms our visual impression of the raw data: all three non-zero sunlight levels are significantly superior to "None", and "Medium" is significantly superior to "Low", but "High" is significantly better than "Low" but not better than "Medium".

Q2: is there a Watering Frequency effect?

Q2 is the question: is there a Watering Frequency effect? When we plot the data for the two groups "Daily Watering" and "Weekly Watering", we see that they are quite similar. Indeed, the differences in the group means is only 0.1, and that is in sets whose variance is at least 6 times larger than the effect size! (Figure 7.40)

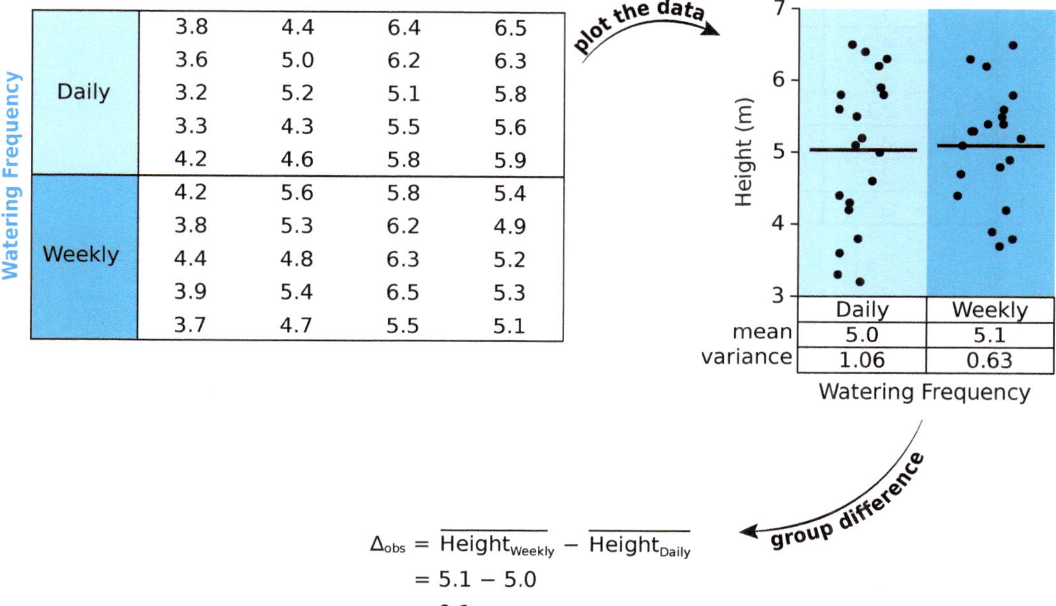

Figure 7.40 One-factor multi-group comparison for effect of Watering Frequency on plant height.

Watering effect This is hardly likely to be statistically significant, and we confirm this using a resampling-based approach. Since the two data sets have similar variances, we use a Big Box method to assess the difference in their means. As we see, differences as big as or bigger than the observed difference Δ_{obs} occurred 83 percent of the time. We conclude that there is no evidence for a Watering Frequency effect (Figure 7.41).

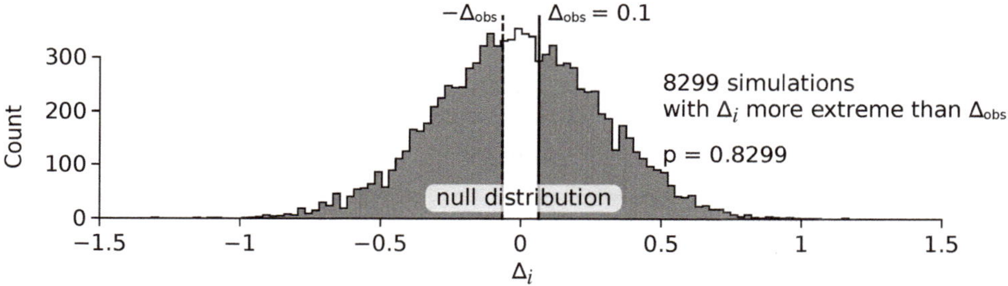

Figure 7.41 p-value calculation for the Watering Frequency analysis.

Q3: is there an interaction effect?

Now that we've done the two one-way comparisons for each factor, it's time to look at the interaction effects. The first step is to make the interaction plot (Figure 7.42).

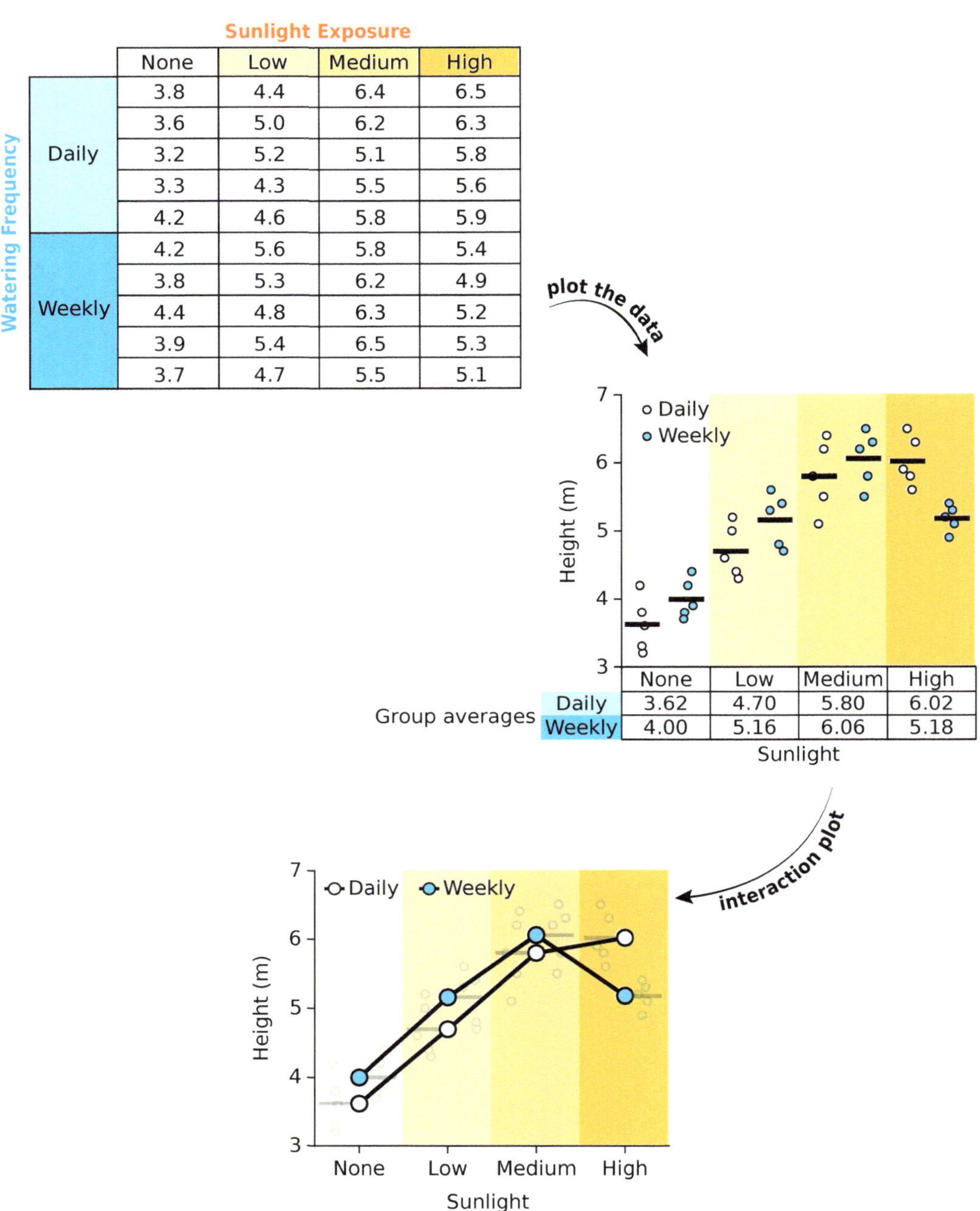

Watering Frequency	Sunlight Exposure: None	Low	Medium	High
Daily	3.8	4.4	6.4	6.5
Daily	3.6	5.0	6.2	6.3
Daily	3.2	5.2	5.1	5.8
Daily	3.3	4.3	5.5	5.6
Daily	4.2	4.6	5.8	5.9
Weekly	4.2	5.6	5.8	5.4
Weekly	3.8	5.3	6.2	4.9
Weekly	4.4	4.8	6.3	5.2
Weekly	3.9	5.4	6.5	5.3
Weekly	3.7	4.7	5.5	5.1

Group averages

	None	Low	Medium	High
Daily	3.62	4.70	5.80	6.02
Weekly	4.00	5.16	6.06	5.18

Figure 7.42 Generating the interaction plot. The upper figure shows the raw data for all 8 groups, and the lower graph shows the interaction plot, with the 4 means for daily watering connected by one broken line, and the 4 means for weekly watering connected by a second broken line.

Interaction plot Inspecting the interaction plot, we see some evidence for interaction: the line for weekly watering was above the line for daily watering, except for the "High" sunlight condition, where it went substantially below. This lack of parallelism suggests an interaction effect, since the superiority of weekly watering is reversed in the high sunlight condition. However, this is not a very large effect, and may well fail to be statistically significant.

We now want to confirm this with a mathematical analysis. The main thing we need is an F-statistic, F_{AB}, a measure of the ratio of across-group differences to within-group differences.

In order to do this, we think of the basic data structure as consisting of 8 "cells", one for each condition. So there is a (sunlight = None, watering frequency = Daily) cell, a (sunlight = None, watering frequency = Weekly) cell, etc.

Within-group difference The total within-group difference is defined exactly as before: we sum up the distances from every data point to its cell mean. We can use the squares of the differences, as in classical ANOVA, or we can choose to not square everything, and use instead the absolute values of the differences. Here we will use the absolute values, because we want to avoid the distortions imposed by squaring. That is, we will add up the lengths of the blue lines in Figure 7.43 as our total within-group difference.

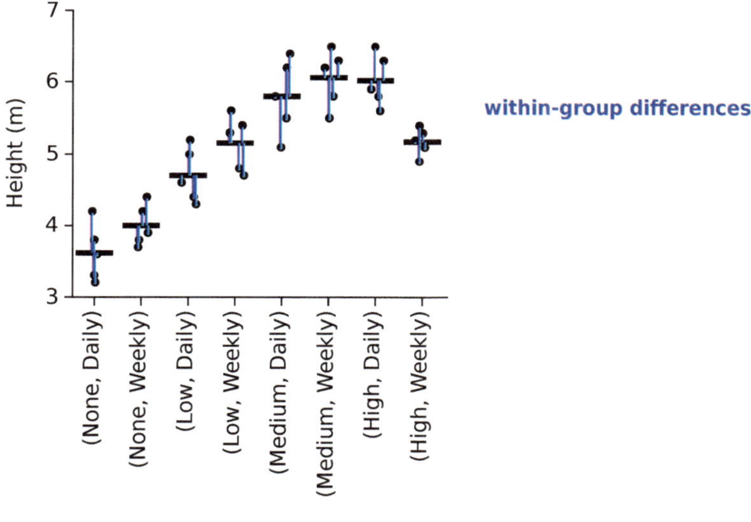

Figure 7.43

$$\text{total within-group difference} = \sum_{\text{all data points}} \left|\text{data point} - \text{cell mean}\right|$$

Across-group difference Then we need to define the total across-group difference. In one-factor comparisons, we summed up the weighted distances from each cell mean to the Grand Mean, with each distance weighted by its cell size, n_{cell}. Here we have two factors. Intuitively, for each cell, we would like to take the distance from its cell mean to the Grand Mean, and then subtract the row effect (that is, the watering frequency effect) and the column effect (that is, the sunlight effect). This yields one cell's contribution to the total across-group difference. We then add up each cell's contribution to get the total across-group difference.

So for each cell, we start with the computation of

$$(\text{cell mean} - \text{GM}) - \Big(\text{row effect} + \text{column effect}\Big)$$
$$= (\text{cell mean} - \text{GM}) - \Big((\text{row mean} - \text{GM}) + (\text{column mean} - \text{GM})\Big)$$
$$= (\text{cell mean} - \cancel{\text{GM}}) - \Big((\text{row mean} - \cancel{\text{GM}}) + (\text{column mean} - \text{GM})\Big)$$

Note that two terms are "−GM" and "+GM", which cancel out to give us

$$\text{cell mean} - (\text{row mean} + \text{column mean} - \text{GM})$$

Then we have to make a decision: to square these terms as in traditional theory, or to use the absolute value as we have been doing previously. We use the absolute value.

$$\Big|\text{cell mean} - (\text{row mean} + \text{column mean} - \text{GM})\Big|$$

This term then has to be weighted by the number of entries in that cell, which we call n_{cell}. This gives us our equation for the across-group differences:

$$\text{total across-group difference} = \sum_{\text{all cells}} n_{cell} \times \Big|\text{cell mean} - (\text{row mean} + \text{column mean} - \text{GM})\Big|$$

For example, let's calculate the contribution of the cell (sunlight = None, watering frequency = Daily) to the total across-group difference (Figure 7.44).

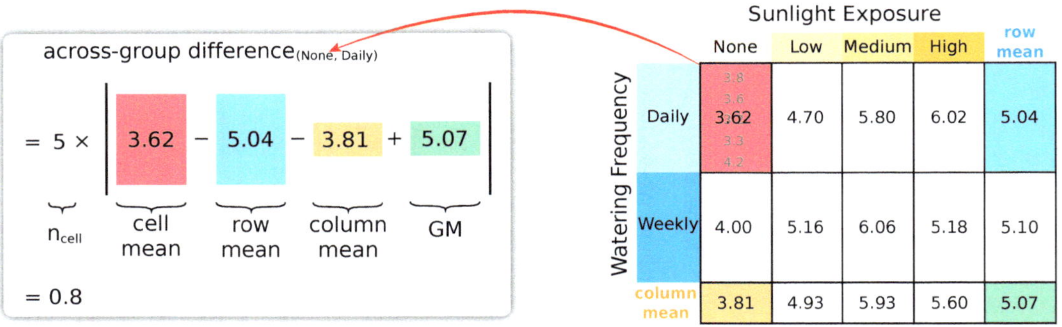

Figure 7.44 Calculation of one cell's contribution to the total across-group difference.

The F-statistic F_{AB} is then defined as usual, as the ratio of the total across-group difference to the total within-group difference.

$$F_{AB} = \frac{\text{total across-group difference}}{\text{total within-group difference}}$$

$$= \frac{\sum_{\text{all cells}} n_{cell} \times \Big|\text{cell mean} - (\text{row mean} + \text{column mean} - \text{GM})\Big|}{\sum_{\text{all data points}} \Big|\text{data point} - \text{cell mean}\Big|}$$

When we compute F_{AB} for this data set, it is 0.153.

p-value The next step is to compare this observed F_{AB} with the result of 10,000 simulations of the Null Hypothesis, that is, 10,000 recalculations of F, computed from a Big Box resampling into 8 new data sets.

We carried this out, and found that of the 10,000 simulations of the Null Hypothesis, 54 of them had F-values that were greater than or equal to our observed value. Therefore, we conclude that the interaction effect is statistically significant, with p = 0.0054 (Figure 7.45).

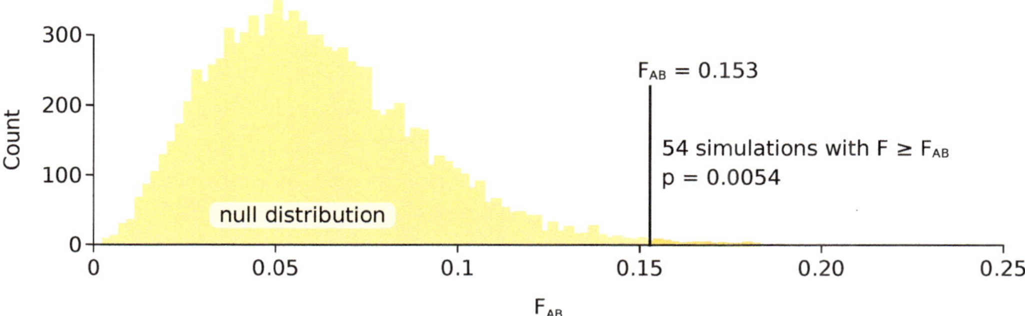

Figure 7.45 p-value calculation for the interaction effect F_{AB}. The null distribution is formed by 10,000 values of F_{AB} from randomized simulations of the Null Hypothesis.

Interaction effect We conclude that there is a statistically significant interaction effect. By looking at the interaction plot, we can see what it is: high levels of sunlight make weekly watering superior to daily watering whereas at the lower levels of sunlight, daily watering is superior to weekly watering.

This visual impression can be confirmed by statistical analysis. The "interaction effect" test serves as an omnibus test: if the omnibus test is positive, it tells us that there are significant differences, which we can then go on to look for in *post hoc* analysis. In two-way multi-group comparison, the *post hoc* analysis takes the form of pairwise group comparisons, comparing each cell to each other. For example, in the sunlight and watering example, we have 2 levels of watering frequency, and 4 levels of sunlight exposure, yielding 8 cells in all. If we want to quantify that daily watering is superior to weekly watering when sunlight is high, we would do a pairwise group comparison between the two cells: (sunlight = High, watering frequency = Weekly) vs. (sunlight = High, watering frequency = Daily). And, as with all multi-group comparisons, the *post hoc* analysis must include a correction for multiple testing.

We can summarize our findings for the plant height data as follows:

- is there a sunlight effect? Yes. All three non-zero sunlight levels are significantly superior to "None"; "Medium" is significantly superior to "Low", and "High" is significantly superior to "Low" but not significantly better than "Medium".
- is there a watering frequency effect? No.
- is there an interaction effect? Yes. "High" sunlight has the opposite effect on the daily watering than it does on the weekly watering; for daily watering, "None", "Low" and "Medium" levels of sunlight increases plant height but "High" level of sunlight decreases plant height.

Example 1 Are men paid more than women?

The Bureau of Labor Statistics collects data on earnings of workers in the United States, classified according to various characteristics. Here are the median weekly earnings (in dollars) of men and women who were working full-time in the first quarter of 2021, in two age groups (`https://www.bls.gov/cps/earnings.htm`).

Age	Women	Men	All
16–19	$467	$511	$494
20–24	607	667	628
All	593	634	611

For example, the table says that the median weekly income of women aged 16-19 was $467, and that the median weekly income of all women aged 16-24 was $593.

If we graph these numbers we get:

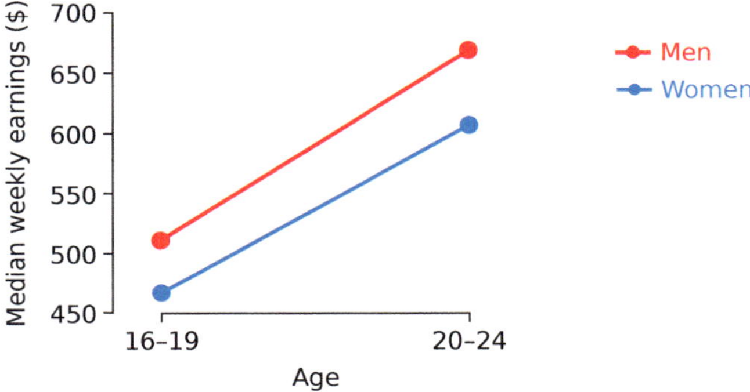

Question Is there a gender effect?

Answer Yes, since the red line (men) is everywhere above the blue line (women). If we take the medians of the genders over both age groups, we see that the median for males ($634) is higher than the median for females ($593). But whether this difference is statistically significant or not depends on the formal analysis.

Question Is there an age effect?

Answer Yes, since both genders are paid more in the higher age group (the right-hand side is strictly higher than the left-hand side). Whether that difference is statistically significant or not depends on the formal two-group comparison.

Question Is there an interaction effect?

Answer No, since the two lines are parallel. Age does not change the gender effect, and gender does not change the age effect.

A more complete picture

These 2 age groups are only the first 2 out of 7. The full data set is:

Age	Women	Men	All
16–19	$467	$511	$494
20–24	607	667	628
25–34	850	950	901
35–44	999	1232	1131
45–54	1002	1334	1161
55–64	964	1224	1108
65+	911	1102	997
All	900	1089	989

When we plot these data, we get a rather different-looking figure:

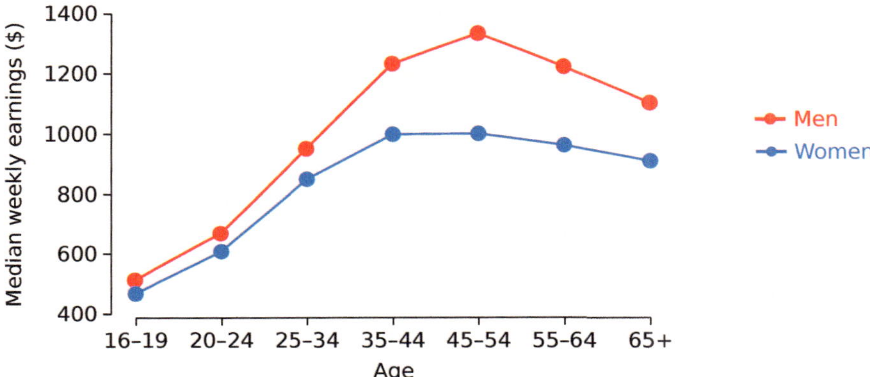

Question Is there a gender effect?

Answer Yes, because the red line (men) is always above the blue line (women), at all age groups.

Question Is there an age effect?

Answer Yes, because the older age groups are paid more than the younger ones.

Question Is there an interaction effect?

Answer Yes, because the lines are not parallel: they grow further apart as age increases, and then narrow slightly at the highest age groups.

Example 2 Battery life

Battery performance is affected by operating temperature. Different types of batteries respond to temperature differently. Here is an example of a two-way group comparison involving three different battery types and three different temperatures.

For this example, we'll use medians as group measures and absolute distances to compare groups. We use medians because the median battery life is more informative than the average, which can easily be pulled up or down by outlying values. We use median absolute deviation

(MAD) to measure the within-group variability. The outcome variable is hours of battery life, so more is better. For statistical significance, we use $\alpha = 0.05$.

Temperature	Material Type I		Material Type II		Material Type III	
15°F	130	155	150	188	138	110
	74	180	159	126	168	160
	135		155		125	
70°F	34	40	126	122	174	120
	82	58	106	115	150	139
	40		120		150	
125°F	100	70	25	70	96	104
	82	58	58	45	82	60
	75		40		55	

Question Is there a material type effect?

Answer No, there isn't a statistically significant material type effect.

To see whether there is a material type effect, we disregard temperature and form a new data set of the three material types. We then plot the data and visually inspect it. We see that there is relatively big variation within each group, and relatively small across-group differences.

This observation will need to be confirmed by a quantitative analysis.

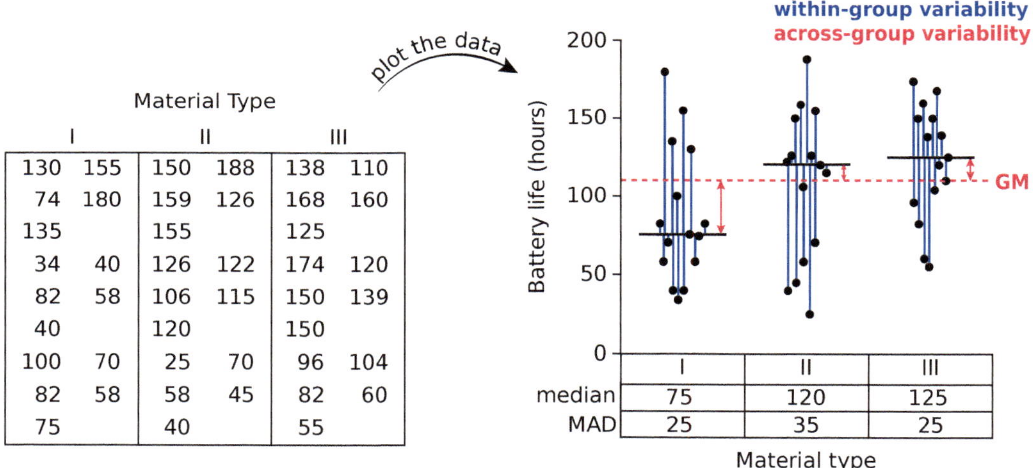

Material Type I		Material Type II		Material Type III	
130	155	150	188	138	110
74	180	159	126	168	160
135		155		125	
34	40	126	122	174	120
82	58	106	115	150	139
40		120		150	
100	70	25	70	96	104
82	58	58	45	82	60
75		40		55	

	I	II	III
median	75	120	125
MAD	25	35	25

We calculate the within-group differences and across-group differences and obtain an F-statistic of 0.6.

$$F_A = \frac{\text{total across-group difference}}{\text{total within-group difference}} = \frac{\sum|\updownarrow|\times\text{group size}}{\sum|\,|\,|} = \frac{900}{1509} = 0.6$$

Since the three battery type groups have similar variability, we use a Big Box model for the omnibus test. A p-value of 0.1123 confirms our visual impression that there isn't a statistically significant material type effect.

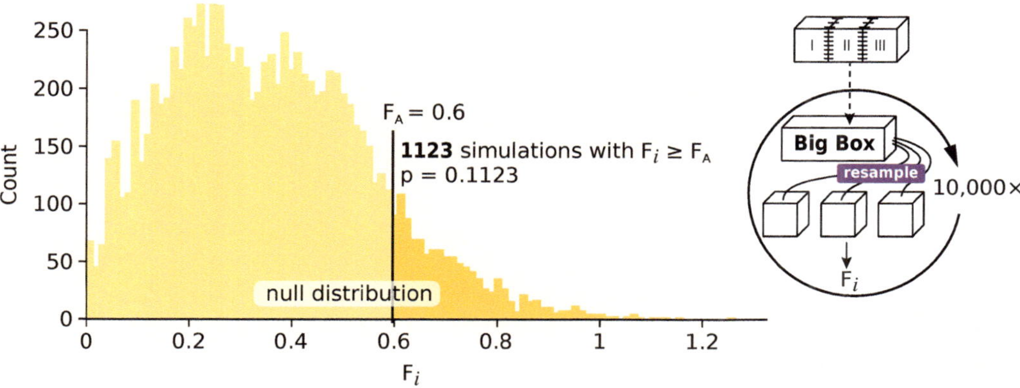

Question Is there a temperature effect?

Answer Yes, there is a statistically significant temperature effect. The higher the temperature, the lower the battery life.

We disregard material type and form a new data set of the three temperatures. We then plot the data and visually inspect it. We see relatively larger across-group differences, but there is substantial within-group variability, so a quantitative analysis is necessary.

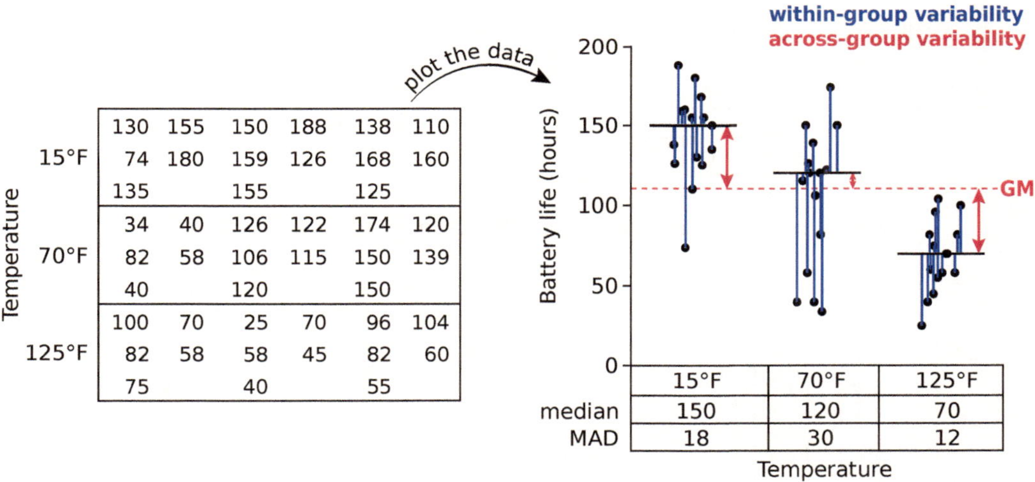

Temperature						
15°F	130	155	150	188	138	110
	74	180	159	126	168	160
	135		155		125	
70°F	34	40	126	122	174	120
	82	58	106	115	150	139
	40		120		150	
125°F	100	70	25	70	96	104
	82	58	58	45	82	60
	75		40		55	

	15°F	70°F	125°F
median	150	120	70
MAD	18	30	12

First, as an omnibus test for the temperature effect, we calculate the within-group differences and across-group differences and obtain an F-statistic of 1.23.

$$F_B = \frac{\text{total across-group difference}}{\text{total within-group difference}} = \frac{\sum |\updownarrow| \times \text{group size}}{\sum |\,|\,|} = \frac{1350}{1101} = 1.23$$

Since the three temperature groups have quite different variabilities (>2×), we use a Re-centered n-Box method for the omnibus test. We found that only 15 simulations out of 10,000 were as or more extreme than the observed F_B. p = 0.0015 confirms our visual impression that there is indeed a statistically significant temperature effect.

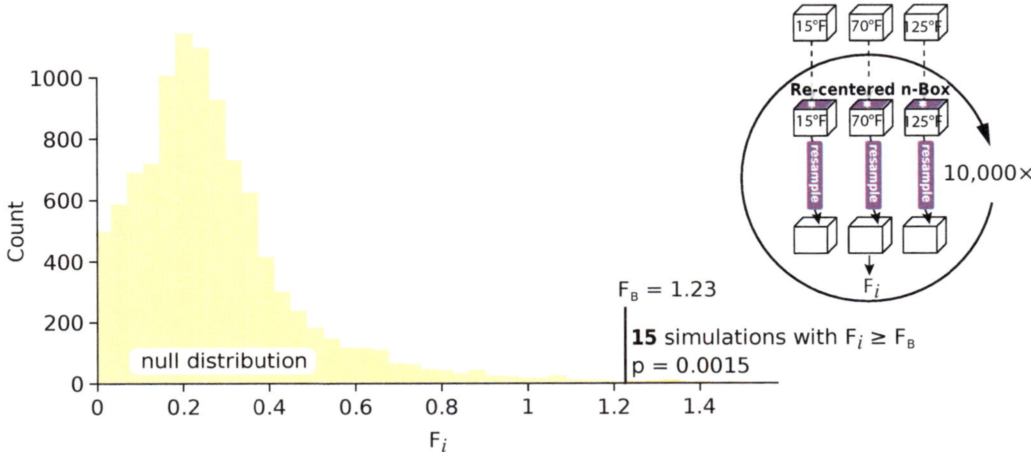

Having passed the omnibus test, we can now proceed to do a *post hoc* analysis of the temperature effect and see where the differences are. We conduct three pairwise group comparisons.

	70°F	125°F
15°F	Δ_{obs}=30 p = 0.0558	Δ_{obs}=80 p = 0.0046
70°F		Δ_{obs}=50 p = 0.0297

Note: background is ☐ if pairwise comparison is statistically significant after BH correction

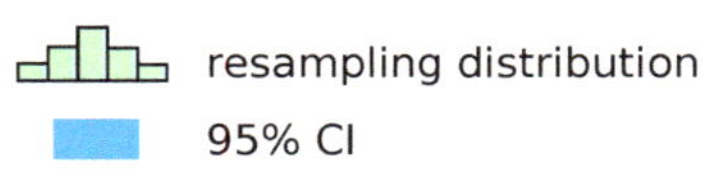

Pairwise two-group comparisons (confidence intervals for group differences)

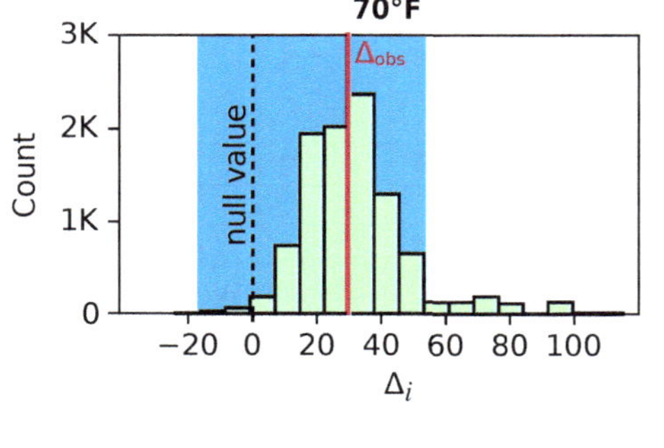

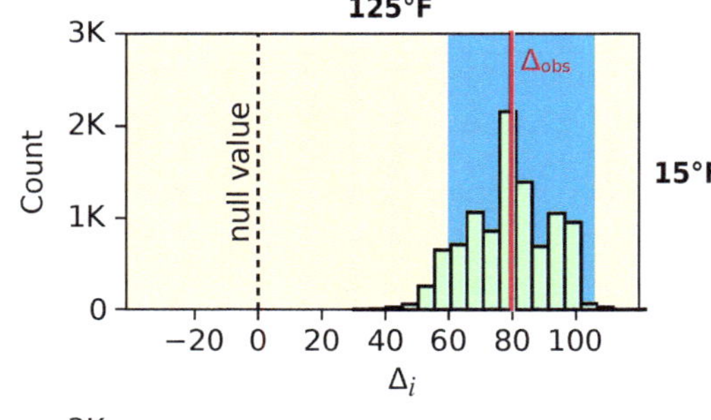

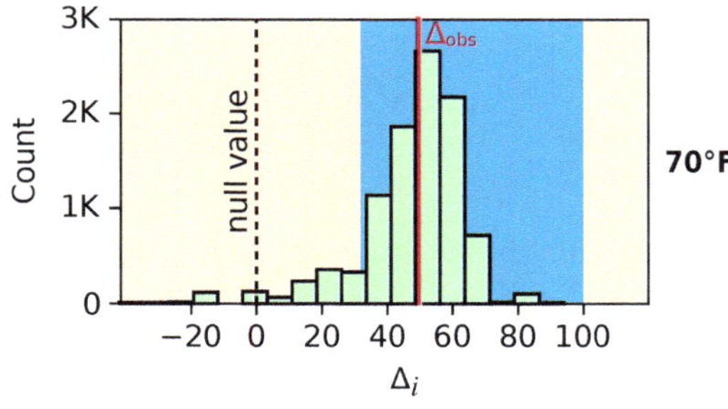

Question Is there an interaction effect between material type and temperature?

Answer Yes, there is a statistically significant interaction effect. First, we plot the data and make the interaction plot.

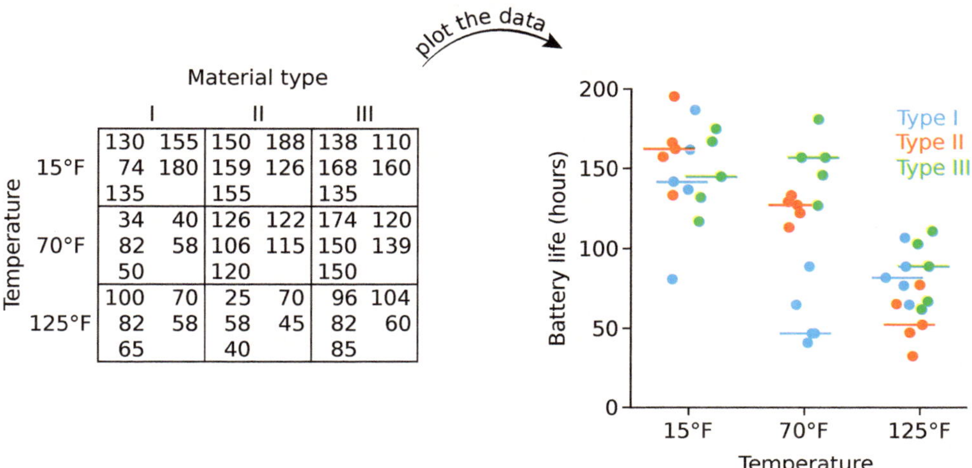

Temperature	Material type I	Material type II	Material type III
15°F	130 155 74 180 135	150 188 159 126 155	138 110 168 160 135
70°F	34 40 82 58 50	126 122 106 115 120	174 120 150 139 150
125°F	100 70 82 58 65	25 70 58 45 40	96 104 82 60 85

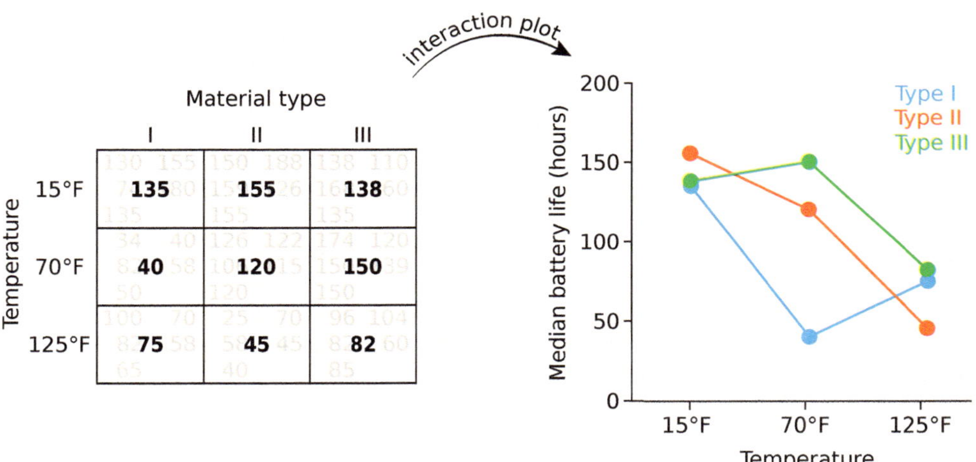

Temperature	Material type I	Material type II	Material type III
15°F	**135**	**155**	**138**
70°F	**40**	**120**	**150**
125°F	**75**	**45**	**82**

We see that at the low temperature, all material types perform well; at the medium temperature, material type I performs poorly; at the high temperature, material type II performs poorly.

Then we confirm our visual inspection by quantitative analysis. Total within-group differences are calculated as the sum of the distances from every data point to its cell median.

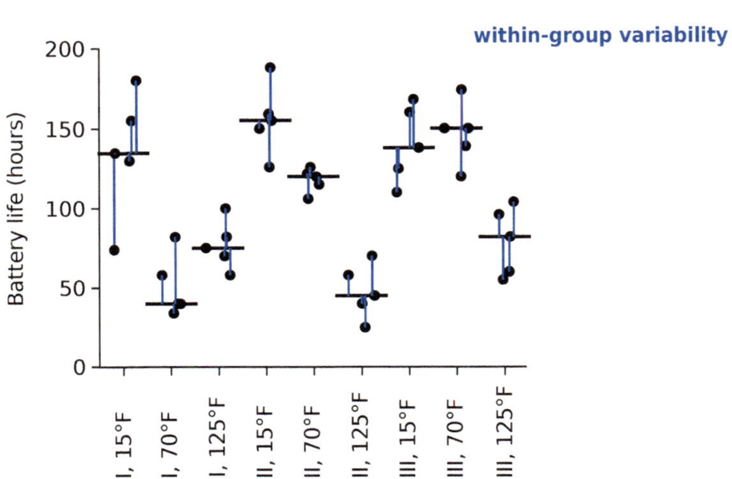

	I, 15°F	I, 70°F	I, 125°F	II, 15°F	II, 70°F	II, 125°F	III, 15°F	III, 70°F	III, 125°F
Median	135	40	75	155	120	45	138	150	82
MAD	20	6	7	5	5	13	22	11	22

The total across-group difference is the sum of each cell's contribution. We use medians and absolute distances. Here we show the cell (Type I, 15°F) to illustrate this calculation.

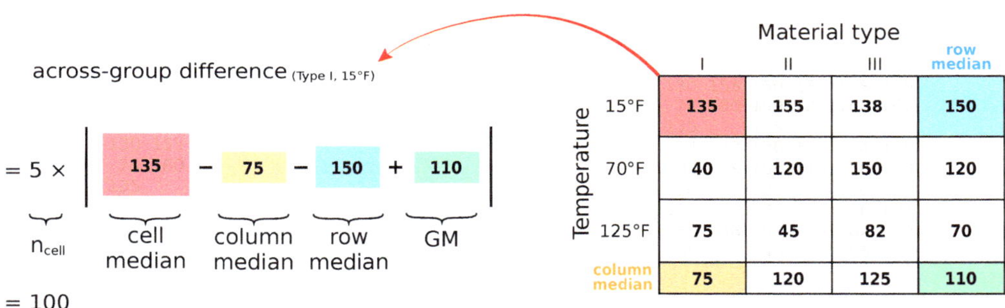

After summing up each cell's contribution, we calculated the across-group differences and within-group differences and obtained an F-statistic of 2.64.

$$F_{AB} = \frac{\text{total across-group difference}}{\text{total within-group difference}} = \frac{\sum_{\text{Type}}^{\text{I, II, III}} \sum_{\text{Temp}}^{15°\text{F}, 75°\text{F}, 125°\text{F}} \text{across-group difference}_{(\text{Type, Temp})}}{\sum |\ |} = \frac{1730}{655} = 2.64$$

Since the 9 cells have quite different variances, we use a Re-centered n-Box model for the interaction effect. The p-value for the observed interaction effect F_{AB}, $p < 0.0001$, confirms our visual impression that there is indeed a statistically significant interaction effect.

After this positive omnibus test, we can then do *post hoc* analysis to quantify our visual impressions. Here, we have 3 material types and 3 temperatures, giving us 9 cells in all. The *post hoc* analysis would then consist of pairwise group comparisons among the 9 cells, with correction for multiple testing.

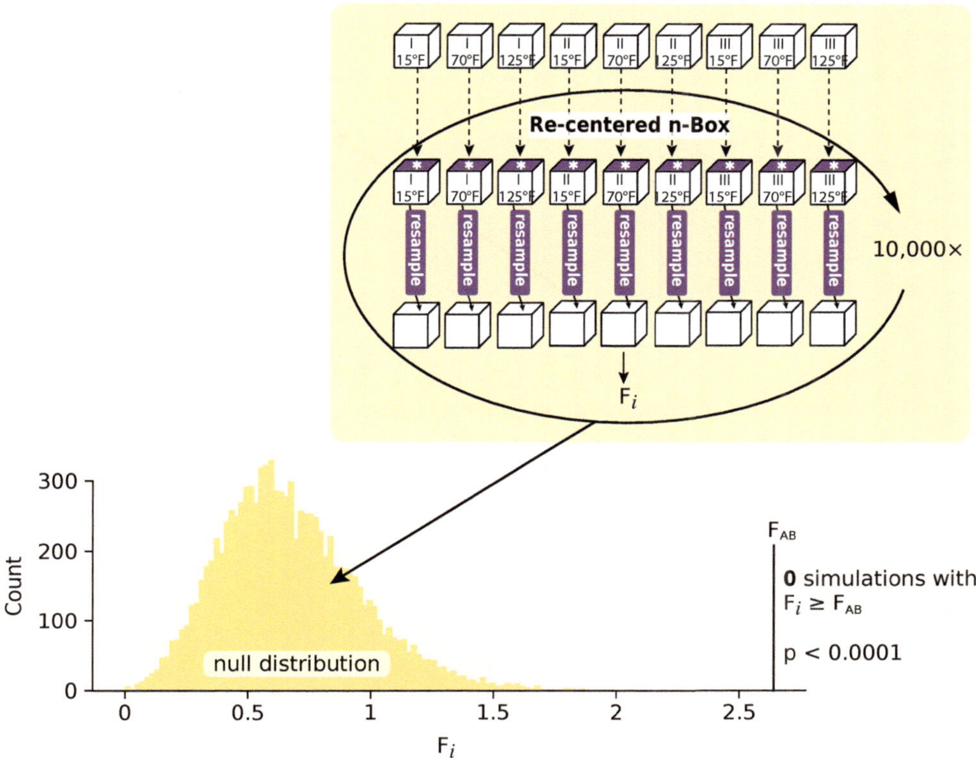

FURTHER EXERCISES 7.9

1. **Note-taking.** A study includes a two-factor analysis of the performance on factual and conceptual questions for students who took notes using longhand (handwriting) and laptops. The interaction plot is shown below.

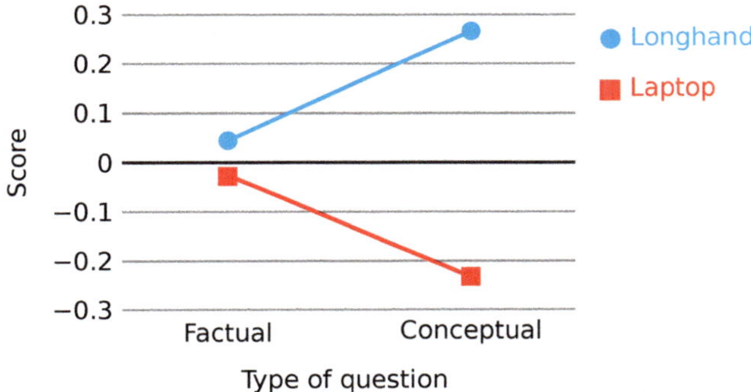

   a. Is there an effect of note-taking type?

   b. Is there an effect of type of question?

   c. Is there an interaction effect?

2. **Exercise schedule.** A study includes a two-factor analysis of hours of sleep for people who exercised in the morning or evening at light or heavy intensity. The interaction plot is shown below.

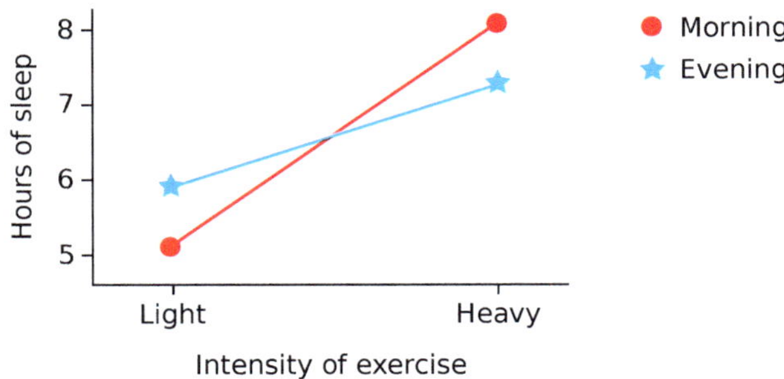

   a. Is there an effect of exercise intensity?
   b. Is there an effect of time of day of exercise?
   c. Is there an interaction effect?

3. **Race and death penalty.** A study looked into 326 murder cases by the victim's race, the defendant's race, and whether the defendant was sentenced to death (see article *Racial Characteristics and the Imposition of the Death Penalty*, *American Sociological Review*, 1981). The data looks like:

Did white defendant receive death penalty?		
	Yes	No
White victim	19	132
Black victim	0	9
Did black defendant receive death penalty?		
	Yes	No
White victim	11	52
Black victim	6	97

   Treat this data as a two-factor multi-group comparison, with "victim's race" and "defendant's race" as the factors, and "percentage receiving death penalty" as the outcome. Draw the interaction plot to answer the three questions:

   a. Is there a victim's race effect?
   b. Is there a defendant's race effect?
   c. Is there an interaction effect?

4. **Chocolate and happiness.** A group of life science students set out to study the effects of chocolates on happiness. They chose three of the most popular chocolate brands, Kokoloco, Chocaholic, and Dachoc, and tested their effects on the level of happiness on a scale of 1 to 10, in a group of people diagnosed with clinical depression. The score is based on a self-evaluation by the study subjects.

a. Describe a step-by-step procedure to conduct a resampling-based omnibus test to figure out whether different chocolate brands affect happiness differently (in this patient population).

b. Suppose the omnibus test resulted in a p-value $< \alpha$, and here is the result of 95% CIs for the pairwise group comparisons. Which chocolate was associated with the highest happiness? Which groups are likely to be statistically different?

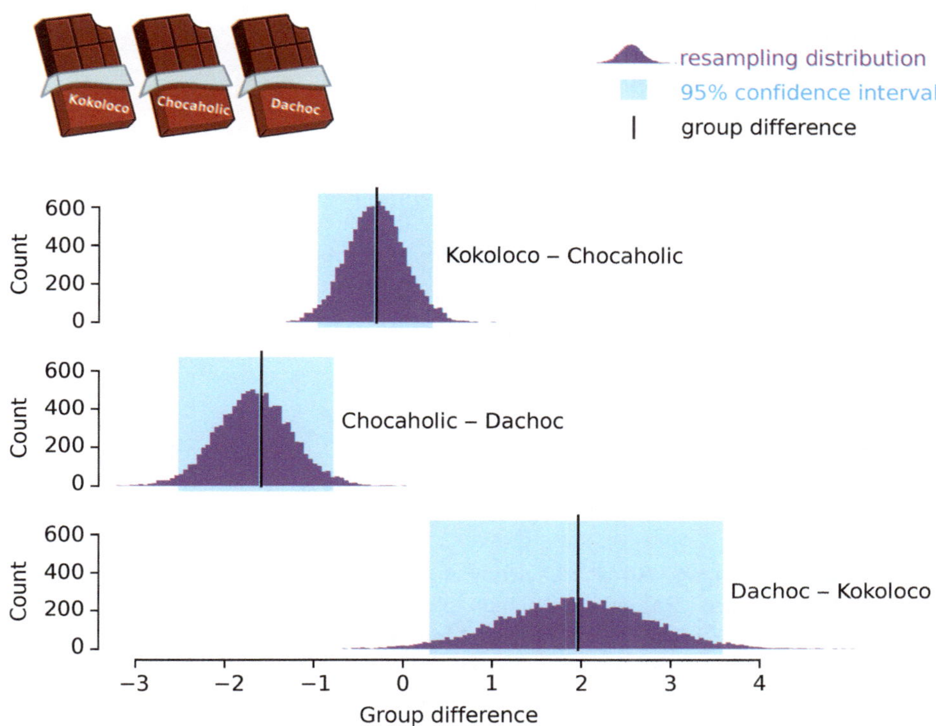

c. Suppose you change your study and add a control group of healthy people and now perform a two-way analysis measuring the interactions between different brands of chocolates and group membership (healthy vs. depressed). Based on the interaction plot below, answer the following. Is there a chocolate brand effect? Is there an effect of being healthy versus depressed? Is there a brand-health interaction effect?

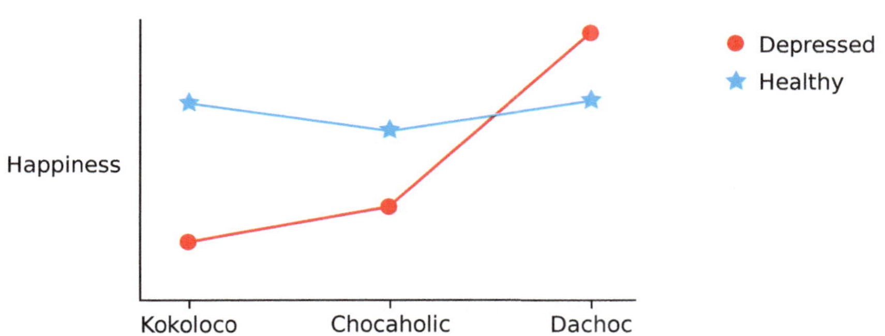

5. **Condiments on foods.** We'll look at the effect of 2 foods (hot dogs and ice cream) and 2 condiments (chocolate sauce and mustard) on "enjoyment scores". Most people like mustard on hot dogs and chocolate sauce on ice cream, but not mustard on ice cream or chocolate sauce on hot dogs.
   a. Sketch an interaction plot corresponding to this description. Make sure to provide all necessary labels.
   b. Is there an effect of type of food?
   c. Is there an effect of condiment?
   d. Is there an interaction effect?
6. **Cell type and temperature.** You measure gene expression in two types of cells (A1 and A2) at two temperatures (B1 and B2). Which of the interaction plots below shows an interaction between cell type and temperature? How would you describe that interaction?

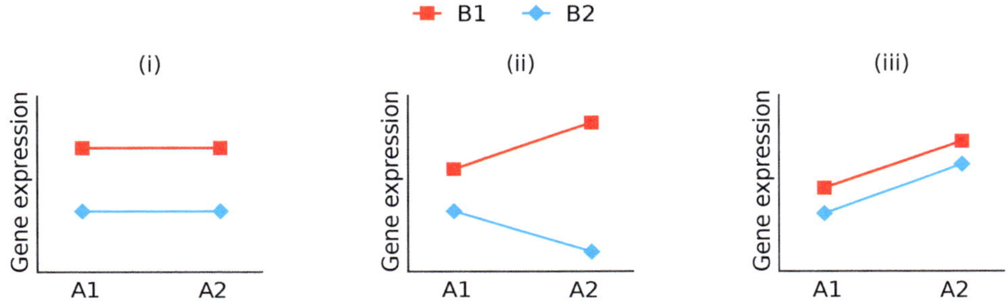

7. **Drug dose and timing.** A research study is examining the effects of drug dosage and timing on blood pressure. Patients with high blood pressure are randomly assigned to one of four groups: Low/High drug dose and Morning/Evening administration. The researchers measure blood pressure after six months with the assigned therapy.

   Each interaction plot below is a potential outcome of this study, with blood pressure on the Y-axis, drug dose on the X-axis, with morning in red and evening in blue. Describe each interaction plot (remember that here, lower blood pressure is considered to be better).

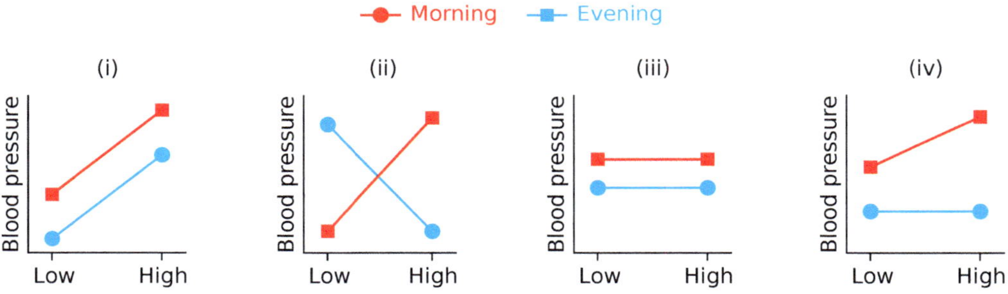

8. **Coffee? a nap? both?** A study is analyzing whether drinking coffee and/or taking a nap affect reaction time in driving. Suppose the study shows that people have the quickest reaction time when they drink coffee and take a nap, second most when they drink coffee without a nap, third most when they take a nap without drinking coffee, and have the longest reaction time when they don't drink coffee or take a nap.

a. Sketch an interaction plot corresponding to this description. Make sure to provide all necessary labels.

b. Is there an effect of drinking coffee?

c. Is there an effect of napping?

d. Is there an interaction effect?

7.10 The mathematical theory: traditional (formula-based) ANOVA

The t-test for two groups: review

In the previous chapter, we discussed the method for comparing two groups A and B on some measure M. If we use M(B) − M(A) = Δ_{obs} to denote the observed group difference, then the question was "how likely is it that a difference at least as extreme as Δ_{obs} could have happened purely due to chance?"

Formula-based approach We saw that if the two groups A and B satisfied a list of strong assumptions:

- A and B are both Normally distributed
- The two group sizes n_{A} and n_{B} are equal and large: $n_{\text{A}} = n_{\text{B}} = n$
- The two groups have the same standard deviation: $\sigma_A = \sigma_B = \sigma$
- The group measure is the mean, μ
- We use subtraction to compare the two groups

then we could define a quantity t

$$t = \frac{\mu_{\text{B}} - \mu_{\text{A}}}{\sqrt{n} \cdot \sigma}$$

and use a theoretical formula (Student's t-test) to calculate the probability (that is, the p-value) of seeing at least this extreme a t-value from two groups drawn at random from the same Normal distribution.

We also saw that the assumptions of equal group sizes and equal group variances could be relaxed, in a more elaborate procedure (Welch's t-test).

Resampling-based approach However, we focused on learning to use resampling methods to give us a p-value, that is, a probability that a group difference at least this extreme would be seen from **this data** under the Null Hypothesis of no difference between the groups. The resampling method is valid for all situations, and when the highly restrictive requirements for the t-test are violated, which is almost always, resampling methods are the only valid approach.

Moreover, these resampling methods work for any choice of measure, not just the mean. In addition, we saw that we can use resampling methods to calculate confidence intervals for the observed difference between the two groups.

Extending the two-group theory to multi-group comparisons

We now want to ask the question: what about *three* groups? Or seven groups?

For multi-group comparisons, there is also a theory based on assumptions like those above, and there is also a resampling-based approach for when the assumptions fail, which is often. We have just developed the general, resampling-based approach to multi-group comparisons. Now, just as in the discussion of two groups, we will discuss the mathematical theory that exists for the cases in which all quantities are Normally distributed, *and* we use mean values as our measure, *and* we square all distances. In these cases, there is a formula-based mathematical theory of multi-group comparisons, the subject that is called **Analysis of Variance** or **ANOVA**.

Requirements for ANOVA

1. All groups are Normally distributed.
2. The variances of each group are approximately equal.†
3. All groups have the same n.†
4. We use the mean as the group measure.
5. We use the algebraic difference of means as the distance measure.
6. We square all distances.
7. There at least 10 elements in each group.

† These two conditions can be relaxed in the more general procedure called Welch's ANOVA.

The mathematical theory says that if all those requirements are met, then there is a certain version of the F-statistic, namely, the version that uses mean values and squared distances, whose probability distribution, called the F-distribution, can be derived theoretically. We can then calculate that version of the F-statistic for our data, and compare it to the percentiles of the theoretical F-distribution to see how often we should expect to see at least that big an F.

Suppose, then, that we have three groups A, B and C, and we have data a_1 through a_n, b_1 through b_n and c_1 through c_n. Here we are assuming that the group sizes are equal, but see below for Welch's ANOVA.

If we were proceeding as we usually do, we would now make graphic presentations of the data and use them to choose a descriptive measure for the set. Is it the mean? the median? another measure? Is a measure of central tendency even a good idea at all?

But in traditional ANOVA, we do not have freedom of measure: the theorem that gives the validity of the ANOVA procedure is a theorem about mean values.[1]

So our summary measure in traditional ANOVA has to be the averages, $\overline{A}$, $\overline{B}$ and $\overline{C}$. The ANOVA question is: is there a significant difference among $\overline{A}$, $\overline{B}$, and $\overline{C}$?

[1] There is also a mathematical theory of multi-group comparisons using the median. It's not part of ANOVA and doesn't use variances, but it does gives us a procedure for testing the Null Hypothesis that there is no difference among the medians of several groups. It's called Mood's Median test, and it looks at how many points over and under the group medians we would expect to see from a randomized (Big Box) version of the data sets.

It is also possible to compare medians using the analytical test called Kruskal-Wallis, which operates not on the data values, but on their relative *ranks* with respect to each other. The test also assumes that the groups all have roughly the same variance and that their histograms have the same shape.

Traditional ANOVA uses the same logic that we have been developing, where

$$F = \frac{\text{total across-group difference}}{\text{total within-group difference}}$$

We will now define what traditional ANOVA means by the terms "across-group differences" and "within-group differences" (Figure 7.46).

Across-group differences We define the Grand Mean ($\overline{G}$) as the mean of all the data. (If the three data sets have the same size, then the grand mean is the mean of the group means.)

Then, as our definition of the total across-group difference, we use the sum of the *squares* of the distances from each of the group means to the Grand Mean, that is, we take the lengths of the red arrows in the lower part of Figure 7.46, square them, and then add up the resulting squares, with each squared distance being weighted by the number of data points in that group.

$$\text{total across-group difference} = n(\overline{A} - \overline{G})^2 + n(\overline{B} - \overline{G})^2 + n(\overline{C} - \overline{G})^2$$

Within-group differences As the definition of within-group variability, ANOVA uses the sum of all the squared distances from each data point to its group mean, that is, it takes the lengths of all the blue lines in the bottom part of Figure 7.46, squares them, and then adds up the resulting squares.

$$\text{total within-group difference} = \sum_{i=1}^{n}(a_i - \overline{A})^2 + \sum_{i=1}^{n}(b_i - \overline{B})^2 + \sum_{i=1}^{n}(c_i - \overline{C})^2$$

Recall that previously, in our resampling-based analyses, we were using just the absolute values of the distances, that is, what we normally think of a distance, as in "the distance from LA to Chicago is around 2000 miles". But using absolute values makes it challenging or impossible for mathematicians to prove theorems, which generally require smooth functions. For example, if we compare squaring to the absolute value function, we see the sharp corner at the bottom of the absolute value function, so we can't use calculus to find the minimum value of the function, by setting the derivative = 0 (see *Finding the best fit line in absolute deviation regression* in section 10.2 on page 511).

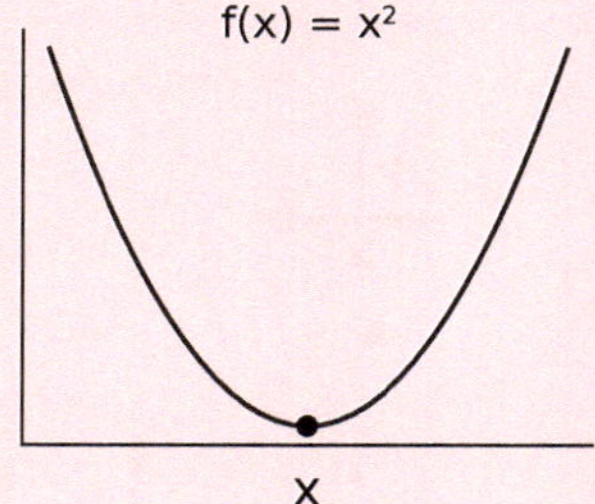

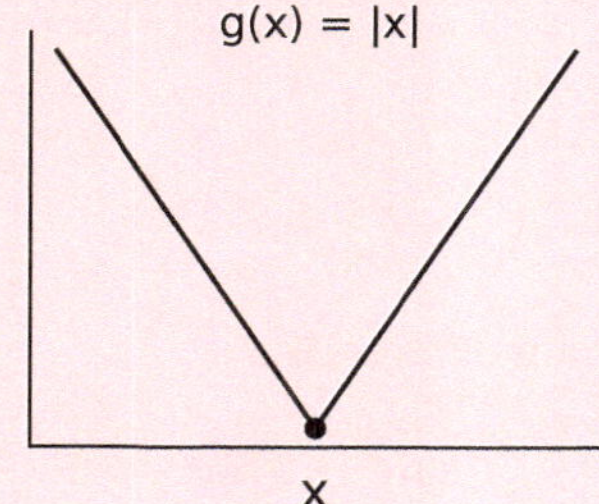

So instead, in order to achieve a provable formula, in traditional ANOVA, we agree to square everything, and the total across-group difference becomes the sum of the squares of the distances from each group mean to the Grand Mean, with each term weighted by the number of data points in that group.

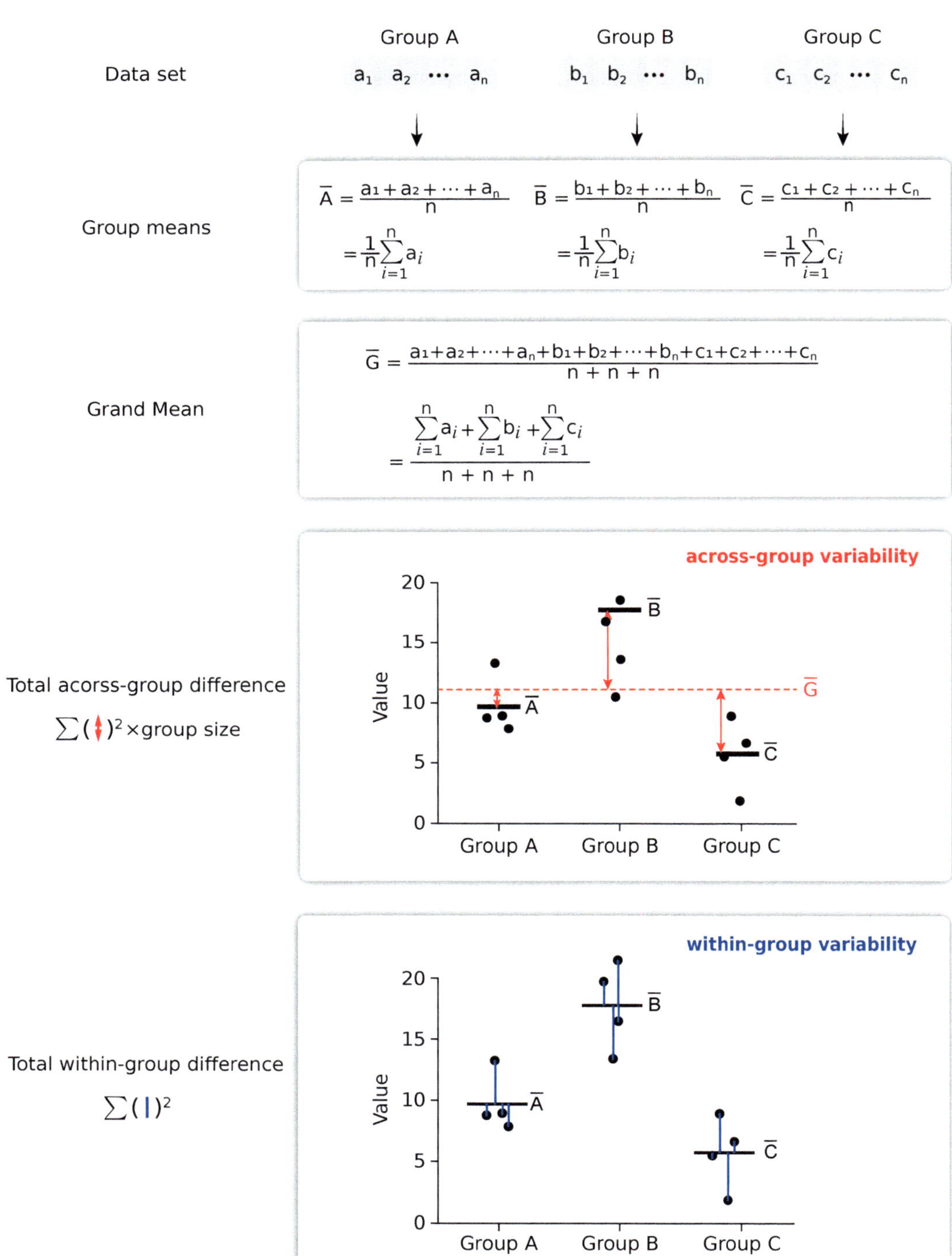

Figure 7.46 Definition of across-group and within-group differences in traditional ANOVA.

The F-statistic We can now define the F-statistic in traditional ANOVA.

For the numerator, the across-group variability, we take the sum of the squared distances from the group means to the Grand Mean, with the squared distance for each group weighted by the number of data points in that group.

The analytical theory then requires that this quantity must be divided by something called the "degrees of freedom". This is an absolute requirement for the theoretical calculation of the p-value for the observed F-statistic, because we will be comparing the observed F to a theoretically derived distribution called $\mathscr{F}$, (read "script F"), and there is a different theoretical $\mathscr{F}$-distribution for every k and every n, that is, the theoretical $\mathscr{F}$-distribution depends on the number of groups and the number of elements within the groups.

How do we calculate the "degrees of freedom"? If the groups all have the same size, the degrees of freedom for the numerator is $k-1$, where k is the number of groups. In our example, there are 3 groups, so the degrees of freedom for the numerator is 2.

For the denominator, the within-group variability, we use the sum of all the squared distances from each data point to its group mean. This must then be divided by the "degrees of freedom" for the denominator, which is the total number of data points minus the number of groups. Again, this is a requirement for statistical theory.

These definitions of degrees of freedom apply only to ANOVA with equal group sizes and variances. In the more general case described by Welch's ANOVA, the degrees of freedom are calculated using a complex formula that takes into account the group variances and sample sizes, resulting in a "pooled" degrees of freedom value that is generally *not* a whole number.

In the simple case where group sizes are equal and group variances are equal, we then define the test statistic F as the ratio of these two quantities:

$$\mathrm{F} = \frac{\text{across-group variability}}{\text{within-group variability}} = \frac{\dfrac{n(\overline{\mathrm{A}}-\overline{\mathrm{G}})^2 + n(\overline{\mathrm{B}}-\overline{\mathrm{G}})^2 + n(\overline{\mathrm{C}}-\overline{\mathrm{G}})^2}{k-1}}{\dfrac{\sum_{i=1}^{n}(\mathrm{a}_i-\overline{\mathrm{A}})^2 + \sum_{i=1}^{n}(\mathrm{b}_i-\overline{\mathrm{B}})^2 + \sum_{i=1}^{n}(\mathrm{c}_i-\overline{\mathrm{C}})^2}{n+n+n-k}}$$

That is the F-statistic as it is traditionally defined, and as found in most statistics texts, courses and software packages.

Let's suppose then that we have calculated some number for F for the data in our study.

The question is then: is my observed F "large"?

ANOVA answer this question by comparing the observed F to a theoretical distribution, $\mathscr{F}$. If the observed F is an extreme value in the $\mathscr{F}$-distribution, then that F may be considered to be statistically significant.

When a piece of statistics software "does an ANOVA", this is what it is doing: it calculates the F-statistic from your data, and then compares that to the percentiles of the $\mathscr{F}$-distribution from theory. If the F calculated from your data is in the 99th percentile of the $\mathscr{F}$-distribution, then the result will be called statistically significant at the $p < 0.01$ level.

ANOVA theorem

IF

all groups are Normally distributed
and we use the mean as the group measure
and the variances of each group are approximately equal†
and all groups have the same $n^{\dagger}$
and we use the algebraic difference of means as the distance measure
and we square all distances
and there are at least 10 elements in each group

THEN

the quantity F will be distributed like $\mathscr{F}$, a probability distribution.
The actual formula for the $\mathscr{F}$-distribution is a little dense:

$$\mathscr{F}(x; d_1, d_2) = \frac{\sqrt{\dfrac{(d_1 x)^{d_1} {d_2}^{d_2}}{(d_1 x + d_2 x)^{d_1+d_2}}}}{x B\left(\frac{d_1}{2}, \frac{d_2}{2}\right)}$$

$$= \frac{1}{B\left(\frac{d_1}{2}, \frac{d_2}{2}\right)} \left(\frac{d_1}{d_2}\right)^{\frac{d_1}{2}} x^{\frac{d_1}{2}-1} \left(1 + \frac{d_1}{d_2} x\right)^{-\frac{d_1+d_2}{2}}$$

where $B(x, y) = \int_0^1 t^{x-1}(1-t)^{y-1}\, dt$

d_1 and d_2 are the two degrees of freedom ($k-1$ and $n-k$).

† These assumptions can be circumvented by Welch's ANOVA, which has a more complicated formula for the $\mathscr{F}$-distribution. For example, the "degrees of freedom" will typically not be a whole number.

We might think: no worries, the computer program will calculate this! But it is essential, before pushing the ANOVA button, to verify that your data meets the list of requirements.

What we have, then, is a formulaic method for calculating the p-value for a multi-group comparison in which we agree to use the mean as our measure, and the squares of the differences from the means as the measure of variability, and in which all groups are Normally distributed. In addition, in order for the theory, which uses continuous functions, to be valid for discrete data sets, we have to assume that every group has ≥ 10 data points.

How does resampling compare to formula-based ANOVA, when the assumptions of traditional ANOVA are met?

In order to compare the formulaic method to our resampling method, we will create a universe that meets the requirements of formula-based ANOVA.

We created three Normally distributed groups that are substantially different from each other. Group A was 20 samples from $\mathscr{N}(6, 3)$, which is the Normal distribution with a mean of 6 and a variance of 3. Group B was 20 samples from $\mathscr{N}(10, 3)$ and group C was 20 samples from $\mathscr{N}(14, 6)$ (Figure 7.47 top).

Omnibus We carried out two resampling-based analyses, one using absolute distances and the other using squared distances. We then did a formula-based ANOVA in Python and compared it to the resampling using squared distances.

> **Absolute distances.** First we did an omnibus test in the manner recommended in this chapter: we used the absolute values of the distances for both the across-group and the within-group differences. The result is that the three groups were found to be highly statistically significantly different from each other, with $p < 0.0001$ (Figure 7.47 top row).
>
> **Squared distances.** We repeated the omnibus test, this time using using the squares of the distances, as in traditional ANOVA. Again, the resampling procedure said that the groups were highly significantly different, with $p < 0.0001$. Note that the F-statistic for the data was $F = 41.09$ (Figure 7.47 middle row).
>
> **Formula.** Lastly, we carried out a traditional formula-based ANOVA, using the `f_oneway` command from the `scipy.stats` package in Python (Figure 7.47 bottom row). Note that the Python program calculates an F-statistic of 41.09, exactly equal to our hand calculation. The formula-based calculation also finds the difference to be highly statistically significant.

post hoc analysis Since the omnibus test was highly significant, we proceeded to carry out a standard *post hoc* analysis, using resampling methods. For the analytical theory, we carried out multiple two-group comparisons using t-tests. The analytical t-tests in Python were in agreement with our resampling analysis (Figure 7.48). The three pairwise comparisons were all highly significant, with p-values well below 0.001, even after Benjamini-Hochberg correction.

Resampling-based methods and formula-based ANOVA produced similar results, so we can conclude that resampling-based multi-group comparisons are equivalent to results from ANOVA when the rigid requirements of ANOVA are met. And of course, resampling-based methods are the only options when the assumptions of traditional ANOVA are not met.

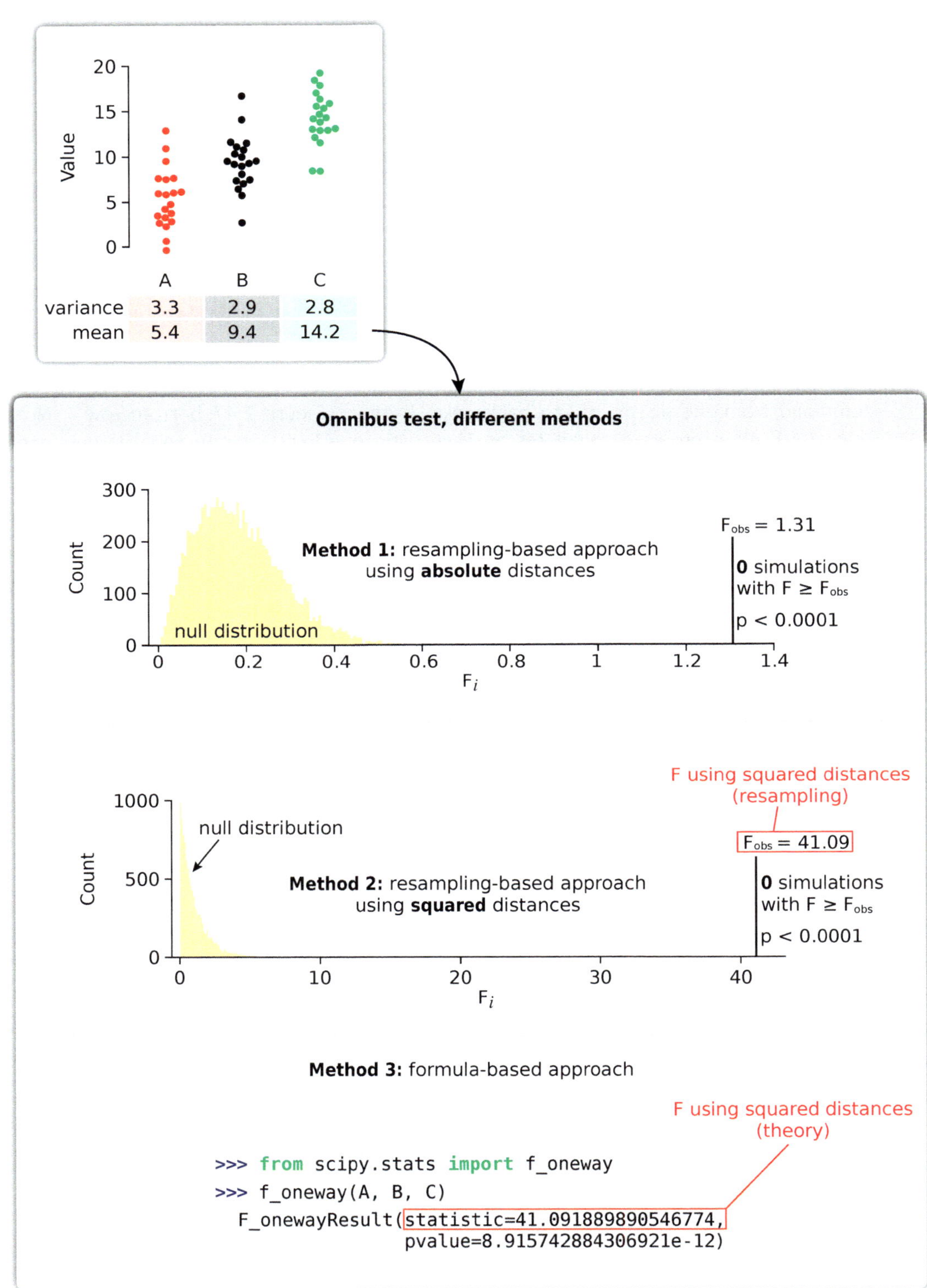

Figure 7.47 Three approaches to omnibus testing.

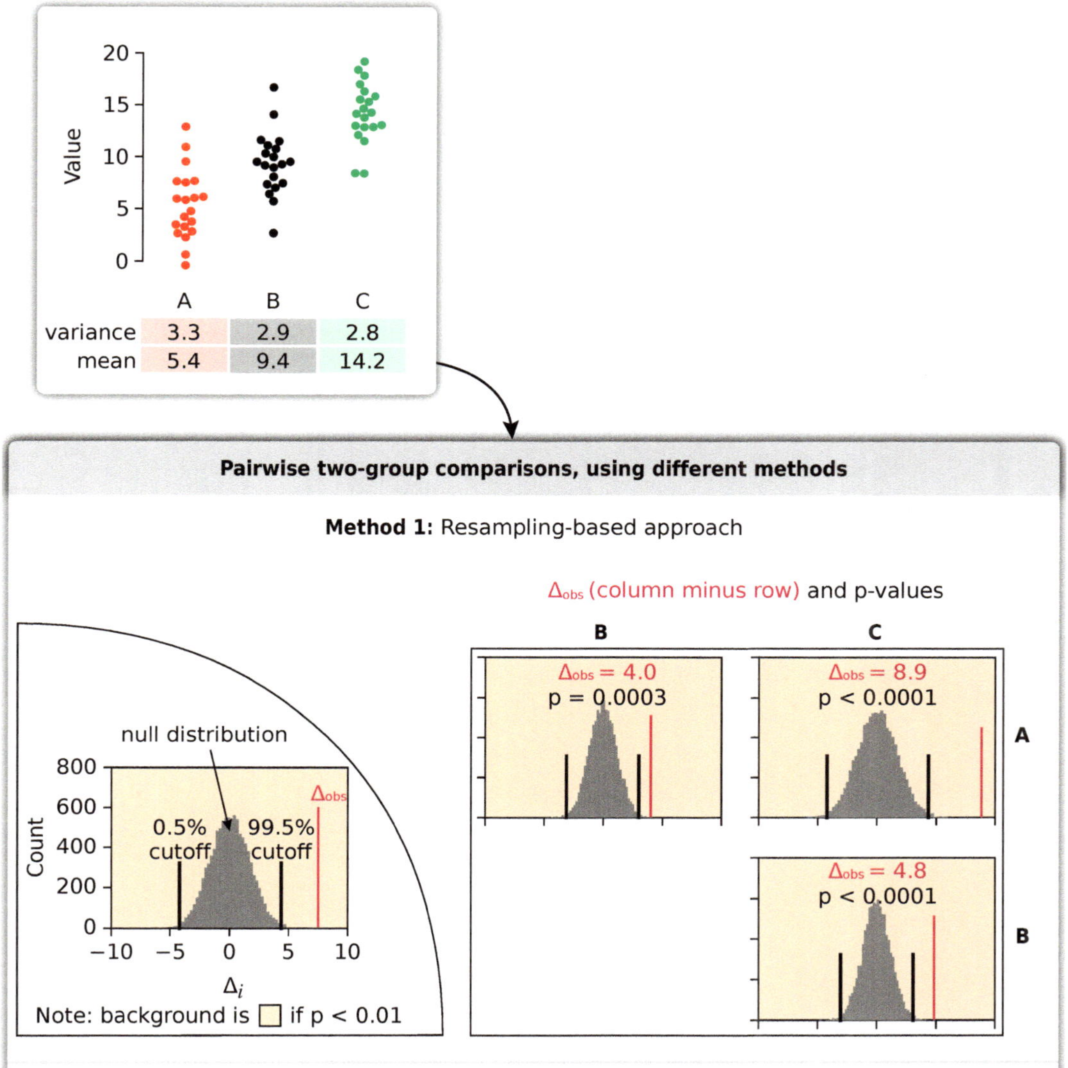

Value

20 15 10 5 0

A B C

	A	B	C
variance	3.3	2.9	2.8
mean	5.4	9.4	14.2

Pairwise two-group comparisons, using different methods

Method 1: Resampling-based approach

null distribution

Count

0.5% cutoff

99.5% cutoff

Δ_{obs}

Δ_i

Note: background is ☐ if $p < 0.01$

Δ_{obs} (column minus row) and p-values

	B	C	
	$\Delta_{obs} = 4.0$, $p = 0.0003$	$\Delta_{obs} = 8.9$, $p < 0.0001$	A
		$\Delta_{obs} = 4.8$, $p < 0.0001$	B

Method 2: Formula-based apporach (*t*-tests)

```
from scipy.stats import stats

>>> stats.ttest_ind(B, A)
Ttest_indResult(statistic=3.9722773314734705,
                pvalue=0.00030657358006771135)

>>> stats.ttest_ind(C, A)
 Ttest_indResult(statistic=8.94758867629546,
                pvalue=6.831257361061613e-11)

>>> stats.ttest_ind(C, B)
 Ttest_indResult(statistic=5.1844574292051355,
                pvalue=7.451822404310602e-06)
```

Figure 7.48 Resampling vs. formula-based *post hoc* testing.

Chapter 8

Independence, Proportions, and Relative Risk

LEARNING OBJECTIVES

After reading this chapter, you will be able to:

- decide when an observed histogram of outcomes is significantly different from expectations.
- decide whether two observed histograms are significantly different from each other.
- interpret claims that two factors are "independent" of each other.
- understand and calculate relative risks with respect to an outcome.

Suppose we are researching heart attacks, and we want to know whether a person's having diabetes "affects" their chances of having a heart attack. Another way this is frequently put is to say that we want to know whether diabetes is a "risk factor" for heart attacks. A third way is to ask if heart attacks are "independent" of diabetes status. All three questions are really asking the same thing.

Let's say we choose a sample population, and for each subject, we record their diabetes status (yes/no) and whether they have suffered a heart attack (yes/no). We now can make two distributions for heart attacks, one for the diabetic-yes population and one for the diabetic-no population. Each distribution has 2 outcomes, the number of 'heart-attack-yes' people and the number of 'heart-attack-no' people.

We can then interpret the questions

- does a person's having diabetes "affect" their chances of having a heart attack?
- is having a heart attack "independent" of diabetes status?
- is diabetes a "risk factor" for heart attacks?

as asking whether the distribution of heart attacks (yes and no) is *similar* in the diabetic-yes and the diabetic-no populations. Another way to put this question is to ask whether the "observed" distribution is similar to what we might "expect" to see if the outcomes (heart-attack-yes and heart-attack-no) were independent of diabetes status.

A. Garfinkel and Y. Guo, *Understanding Data*,
https://doi.org/10.1007/978-3-032-18600-3_8

In this chapter, we are going to develop a general method for taking an observed distribution of outcomes and comparing it to the distribution that would be expected under the Null Hypothesis.

Independence This will enable us to tackle questions like the question of independence: when are two factors independent of each other? We see this in questions like:

- In a trial of a drug, the question whether the drug had an effect on survival is the question whether a subject's survival was *independent* of whether or not the subject took the drug. We will interpret this as asking whether the distribution of 'survived/not survived' in the drug-yes population is *similar* to the distribution of 'survived/not survived' in the drug-no population.
- If we were studying educational outcomes, we might want to know whether success in a given course is *independent* of gender. We can interpret this as the question whether the grade distribution in males is *similar* to the grade distribution in females.

We will see that, in each case, we are asking a question about whether an observed distribution is similar to expectations.

- The question whether a drug had an effect on survival is the question whether the distribution of survival, in people who took the drug or didn't, is similar to what we would expect under the Null Hypothesis (that the two distributions are identical).
- If we were studying educational outcomes, we would want to know whether the distribution of grade outcomes by gender is similar to what we would expect under the Null Hypothesis (that the two distributions, for males and females, are identical).

Proportions Another way of formulating the Null Hypothesis that outcome X is independent of membership in group A vs. group B is to put it as a statement about *proportions*, namely, that the proportion of outcome X in group A is the same as the proportion of outcome X in group B.

Questions about independence are really questions about the equality of proportions.
Is the proportion of survivors the same in the drug-yes and drug-no groups?
Is the proportion of heart attacks the same in the diabetes-yes and the diabetes-no groups?
Are the proportions of the grades A, B and C the same in males as in females?

Relative Risk The question of independence and the question of equality of proportions are both yes/no questions. As we have repeatedly seen, yes/no questions are severely limited in the information that they give us, and need to be supplemented by numerical measures of effect size and confidence intervals for the observed effect size. This is supplied by the concept of Relative Risk.

The first step in this process is to understand how we compare a distribution of outcomes to a distribution of expectations.

8.1 Comparing histograms of observations to expectations

We will be comparing the outcome of an experiment, a histogram of observed counts, to a histogram of expected counts under the Null Hypothesis, and we want to know whether the observed histogram is "different" from what we expected. The expectations can come from many different sources.

The observed histogram

We imagine that we are presenting our data as a histogram of counts.

Observation: coin flipping Suppose we flip a coin 100 times and record the outcomes. We count the total number of Heads and the total number of Tails and compile them into a histogram. Let's say we got 60 Heads and 40 Tails (Figure 8.1).

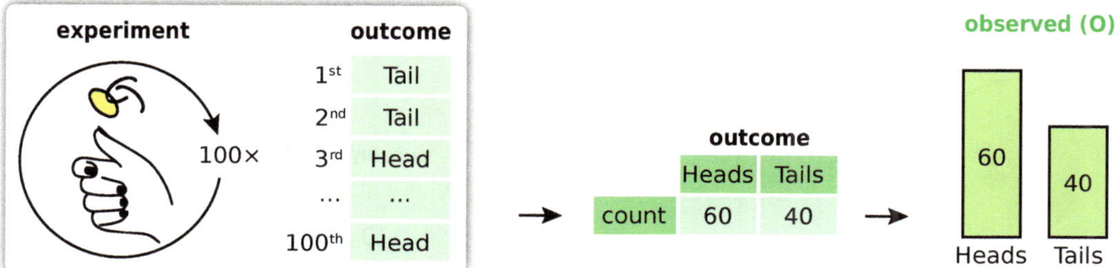

Figure 8.1 Observed histogram for an experiment of flipping a coin 100 times.

Observation: dice rolling The number of outcomes doesn't have to be binary (Heads/Tails, yes/no, lived/died, etc.); the number of outcomes can be more than 2. A single die can be rolled many times, and we can make a 6-element histogram of the observed outcomes. Suppose we did such an experiment, rolled the die 60 times, and recorded the outcome each time. Then we count the number of times each of the six faces occurred and put them into a summary table. Then we can construct the observed histogram for this experiment (Figure 8.2).

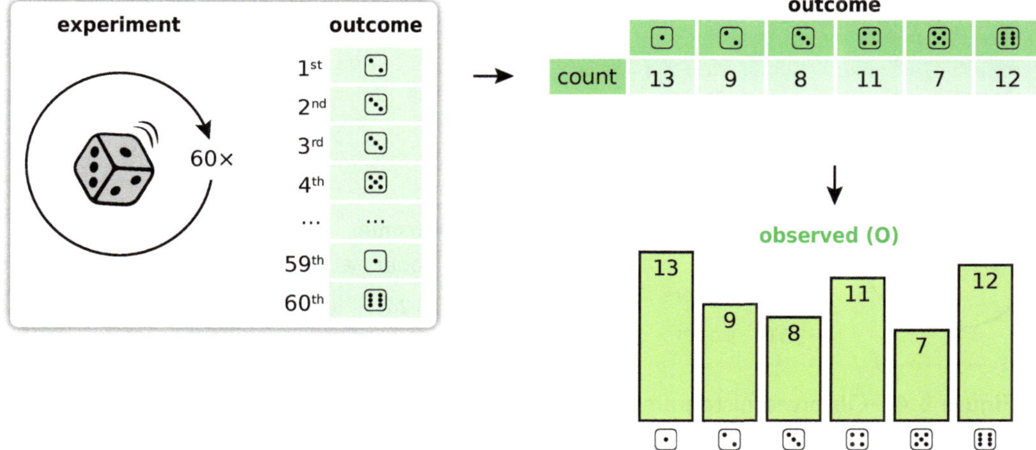

Figure 8.2 Observed histogram for an experiment of rolling a die 60 times.

Observation: genotype frequencies and the Hardy-Weinberg Law In population genetics, we learn the Hardy-Weinberg Law, which considers the frequencies of two genetic alleles, A and a, and tells us what frequencies to expect of the genotypes AA, Aa, and aa, *if* a number of assumptions are met, including the assumptions of a large population with random mating, no mutations, no natural selection and no migration.

Let's say we've observed 200 plants, and found that the occurrences of the genotypes AA, Aa, and aa were 58, 110 and 32 (Figure 8.3). We can then compare this to the expected frequencies under the Hardy-Weinberg law. The purpose of this comparison is to see whether the histogram is "different" from what is expected under the assumptions of a large population with random mating, no mutations, no natural selection and no migration. If they are sufficiently different, then that is a reason to think that the assumptions have been violated, and that, for example, evolution or non-random mating or migration has occurred.

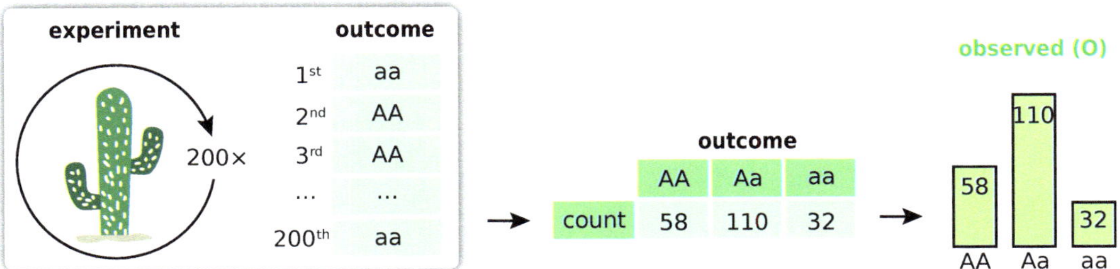

Figure 8.3 Observed histogram for the experiment of phenotyping 200 plants.

Observation: rock-paper-scissors There have been a number studies of how people play the 2-person game "rock/paper/scissors". The rules of the game are that each player, on a prompt, puts out a hand signal representing 'rock' or 'paper' or 'scissors'. Then the winner is decided by the ranking 'rock beats scissors', 'scissors beats paper' and 'paper beats rock'. This is then repeated many times, and a running score is kept.

Since the rules are perfectly symmetric among the 3 outcomes, we might expect that each sign is played an equal number of times. However, this is often not the case. One study for example found this result (Figure 8.4).

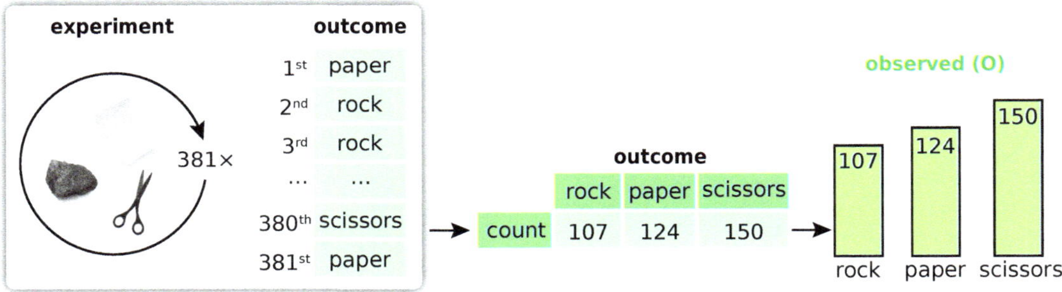

Figure 8.4 Observed histogram of the outcome of 381 games of rock-paper-scissors.

Observation: political affiliation of U.S. boards of directors A recent study did a survey of corporate boards of directors of public and private companies in the United States. The survey found that out of the 769 polled directors, 182 were Democrats, 384 were Republicans and 203 were Independents (Figure 8.5).

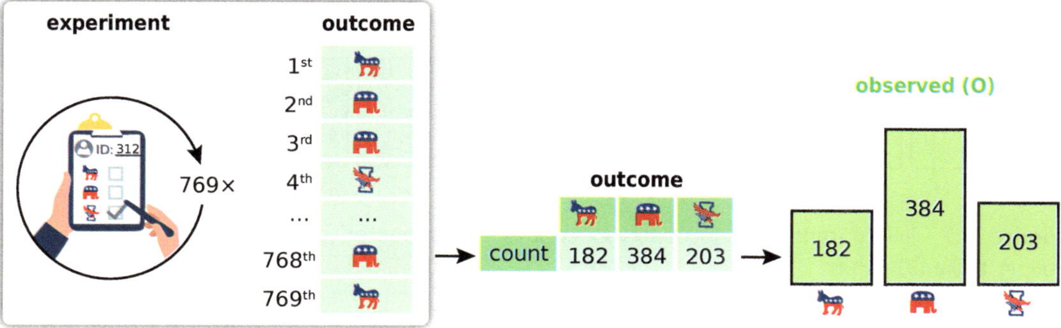

Figure 8.5 Observed histogram of political party affiliation in U.S. Boards of Directors. (source: Boris Groysberg and J. Yo-Jud Cheng, based on a 2015 survey.)

Obtaining the expectations

We then want to compare these observed histograms to "expectations", but where do we get these "expectations"? There are a number of ways:

Theory: In some cases, the expectations are given by a theory. For example, a "fair" coin has an equal probability of Heads and Tails, and a "fair" die has each of the 6 faces equally likely.

Background population: We may have some background population in mind, like "all US residents" to generate our expectations.

Theory: coin flipping We want to use the outcome from our experiment of 100 coin flips to address the question: is this a "fair" coin? To answer this question, of course we need to know what the expected counts would be. In this case, if it is a fair coin, the expected counts would be 50 Heads and 50 Tails. We obtain that number by first consulting the definition of "fair", which is 50% Heads and 50% Tails, and then applying that definition to the actual number of flips, which is 100. We now have two histograms, one for the observed counts and one for the expected counts (Figure 8.6).

Figure 8.6 Histograms of observed and expected outcomes for the 100 coin flips experiment.

Consequently, the question whether this is a fair coin becomes a question of the comparison of the two histograms, observed and expected.

Theory: dice rolling Is this die fair? If it were, out of the 60 rolls, we would expect to see 10 occurrences of ⚀, 10 occurrences of ⚁, ..., and 10 occurrences of ⚅.

The question of the fairness of the die then becomes the question of comparing two histograms, the histogram of observed counts and the histogram of expected counts (Figure 8.7).

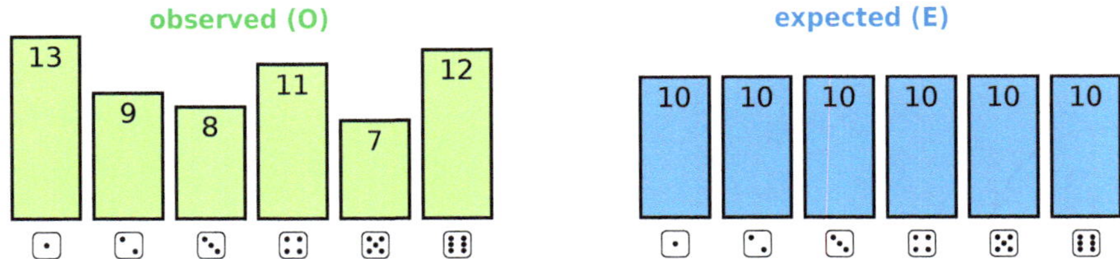

Figure 8.7 Histograms of the observed counts and expected counts for the experiment of rolling a die 60 times.

Similar to the case of the fairness of a coin in the coin flipping experiment, the question whether a die is fair becomes a question of the comparison of the two histograms, observed and expected.

Theory: Hardy-Weinberg Consider the frequencies of two genetic alleles, A and a. Let's say A is dominant and has probability p and a is recessive and has probability q. Since there are only two possibilities, $p+q=1$.

The Hardy-Weinberg law says that if we consider a large population with random mating, no mutations, no natural selection and no migration, the probabilities of the homozygous AA, heterozygous Aa, and homozygous aa genotypes will be p^2, $2pq$, and q^2, respectively.

In our experiment, we genotyped 200 subjects. Assuming alleles A and a have an equal chance of occurring, so $p=q=0.5$, the expected counts of the homozygous AA, heterozygous Aa, and homozygous aa will be 50, 100, and 50, respectively (Figure 8.8).

Figure 8.8 Histograms of the observed counts and expected counts for the experiment of surveying alleles in a population of 200.

Theory: hand signs in rock-paper-scissors games In the study of hand signs used by a group of students during 381 games, the number of occurrences of each hand sign was counted. In theory, we expect that each sign has an equal chance (Figure 8.9).

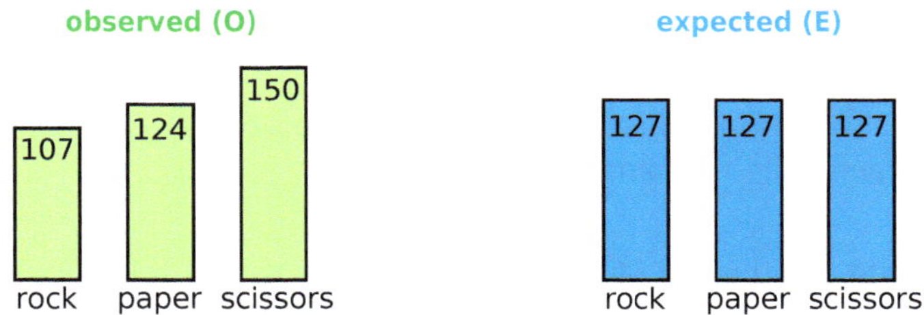

Figure 8.9 Histograms of the observed counts and expected counts for the experiment of 381 games of rock-paper-scissors.

Background population: political affiliation of U.S. boards of directors In the previous cases, the expected outcomes were given by a theoretical model: a fair coin has a 1/2 chance for each side to occur, a fair die has a 1/6 chance for each face to occur, the alleles are expected to occur in the ratio 1:2:1, the hand signs are expected to occur with equal probability.

However, expectations don't always come from a theoretical model. Sometimes we need to compare an observed histogram to expectations from a background population.

In the survey about the political makeup of U.S. boards of directors, if we were interested in the question whether U.S. boards of directors are politically representative of the general population, we would want to compare two histograms. The first would be the histogram of the observed counts in the survey, and the second would be the histogram of the counts that would be expected if the boards were representative of the general public.

Since the American public is 28% Republicans, 31% Democrats, and 39% Independents, in this case that would mean that, under the hypothesis that the boards are representative, we would expect to see $28\% \times 769 = 219.71$ Democrats, $31\% \times 769 = 243.26$ Republicans, and $39\% \times 769 = 306.03$ Independents (Figure 8.10).

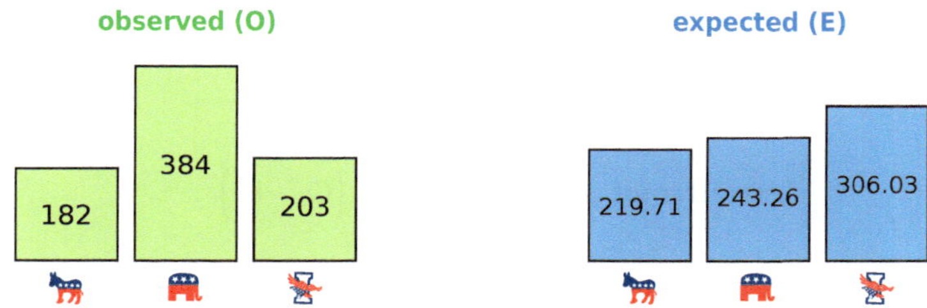

Figure 8.10 Histograms of the observed counts and the expected counts for the political party affiliation on U.S. Boards.

Similar to the case of the fairness of a coin or a die, the question whether a histogram of outcomes from an experiment is representative of an underlying population becomes a question of the comparison of the two histograms, observed and expected.

FURTHER EXERCISES 8.1

1. Assume the six colors (green, orange, blue, red, yellow, and brown) in an M&Ms pack are equally likely to occur. Given the following result from a bag, what is the expected count for each color?

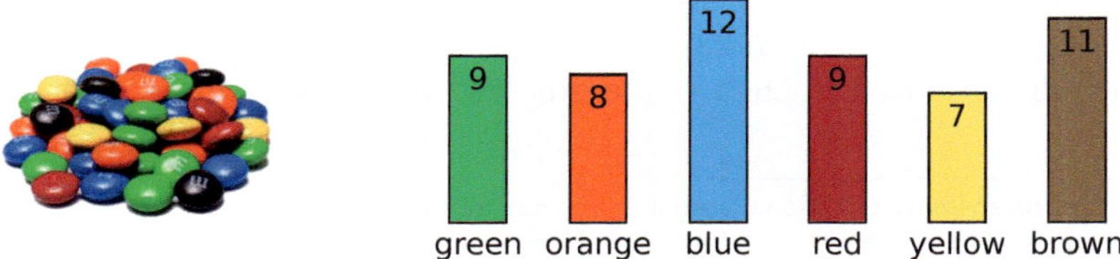

8.2 χ, the distance between observation and expectation

We have now seen how to turn the question of "fair" or "representative" into a question about the comparison of two histograms: observed and expected. "Fair" or "representative" means that the histogram of observations is "roughly equal to" or "similar to" the histogram of expectations. "Not fair" or "not representative" means that the histogram of observations is "different from" the histogram of expectations.

But of course, *any* two histograms will probably be at least slightly different, so what we really need is a measure of the *distance* between two histograms. If we had such a distance measure, we could ask whether the distances were "large" or, on the other hand, whether an observed distance at least this big could easily have arisen by chance.

The distance between two histograms: χ One obvious measure of the distance between two histograms would be to consider the differences between the counts in each bin, Observed – Expected, and add them up.

Let's say there are k bins in the histograms. If we denote the observed count in bin i as O_i and the expected count in bin i to be E_i, then our first attempt at a measure of distance is to add up the differences bin by bin:

$$\textbf{first attempt} \qquad \text{distance} = \sum_{i=1}^{k} \left(O_i - E_i \right)$$

One immediate drawback of this measure is that in this definition, large negative differences and large positive differences would be allowed to cancel each other, which is obviously undesirable.

There are two remedies for this.

The first is the one taken by traditional statistics, and that is the familiar one of just squaring everything to get rid of the negative values. This gives rise to a branch of statistical theory called the "Chi-Squared" statistic. It has a number of limitations, due to the squaring, as you might expect by now. We will discuss it in section 8.12 called *The theoretical approximation (χ^2)*.[1]

The second remedy is to use absolute values. Here, we will take that better approach, and use absolute values as we have before, to prevent positive and negative effects from cancelling each other out, without the distortions introduced by squaring.

$$\textbf{second attempt} \qquad \text{distance} = \sum_{i=1}^{k} \left| O_i - E_i \right|$$

This is an improvement, but it still suffers from a drawback: it is dependent on the sizes of the samples. If E_i is large we would expect to see bigger differences in absolute numbers. We control for this by normalizing each bin difference to its corresponding expected value, that is, expressing the difference as a fraction of the expected value. We call this measure χ.

---

[1] There is a second reason why traditional statistics squares all these quantifies. As we've said, it is much easier for mathematicians to prove theorems about squares of quantities than about absolute values of quantities. However this will not be a problem for us since we are not interested in deriving theoretical results, but rather, as usual, we will use simulations to make estimates.

Definition of χ

If there are k bins in each histogram, then

$$\chi = \sum_{i=1}^{k} \frac{|\mathrm{O}_i - \mathrm{E}_i|}{\mathrm{E}_i} = \sum_{i=1}^{k} \frac{|\mathrm{Observed}_i - \mathrm{Expected}_i|}{\mathrm{Expected}_i}$$

χ summarizes the **distance between observed and expected histograms**:

- The **smaller the overall distance** between the observed and the expected, the **smaller the value of χ**.
- Conversely, the **larger the distance** between the observed and the expected, the **larger the value of χ**.

Example 1 Calculating χ

Let's apply the χ measure to our examples of coin flipping, dice rolling, Hardy-Weinberg, rock-paper-scissors, and political party representation, and calculate the distance between the observed histogram and the expected histogram for each example.

1. **Coin flipping.** After constructing the observed histogram and deriving the expected histogram from theory, we calculate the distance χ between observed and expected histograms. In coin flipping, there are only two outcomes, Head or Tail, so in the $\sum$ term, the choice of i is the set {Head, Tail}, and there are 2 terms in total.

$$\chi = \sum_{i \text{ in } \{\text{Head, Tail}\}} \frac{|\mathrm{O}_i - \mathrm{E}_i|}{\mathrm{E}_i} = \frac{|60-50|}{50} + \frac{|40-50|}{50} = \frac{10}{50} + \frac{10}{50} = 0.2 + 0.2 = 0.4$$

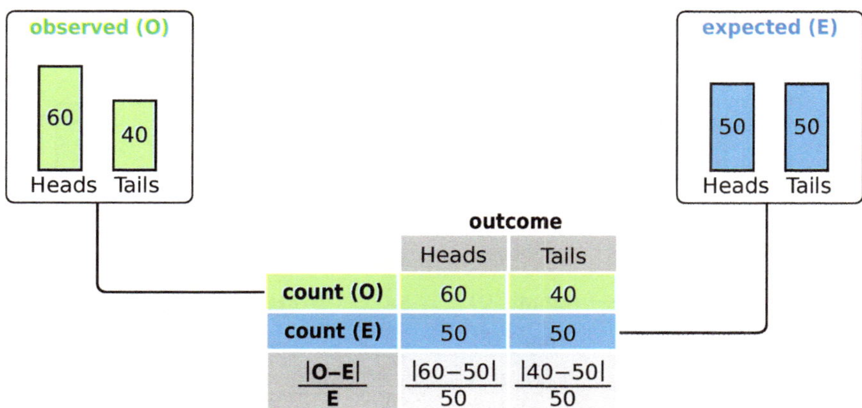

2. **Dice rolling.** Similarly, after constructing the observed histogram and deriving the expected histogram from theory, we calculate the distance χ between observed and expected histograms. In the case of dice rolling, there are 6 outcomes, so in the $\sum$ term, the choice of i is the set {⚀,⚁,⚂,⚃,⚄,⚅}, and there are 6 terms in total.

$$\chi = \sum_{i \text{ in } \{⚀,⚁,⚂,⚃,⚄,⚅\}} \frac{|O_i - E_i|}{E_i} = \frac{3}{10} + \frac{1}{10} + \frac{2}{10} + \frac{1}{10} + \frac{3}{10} + \frac{2}{10} = 1.2$$

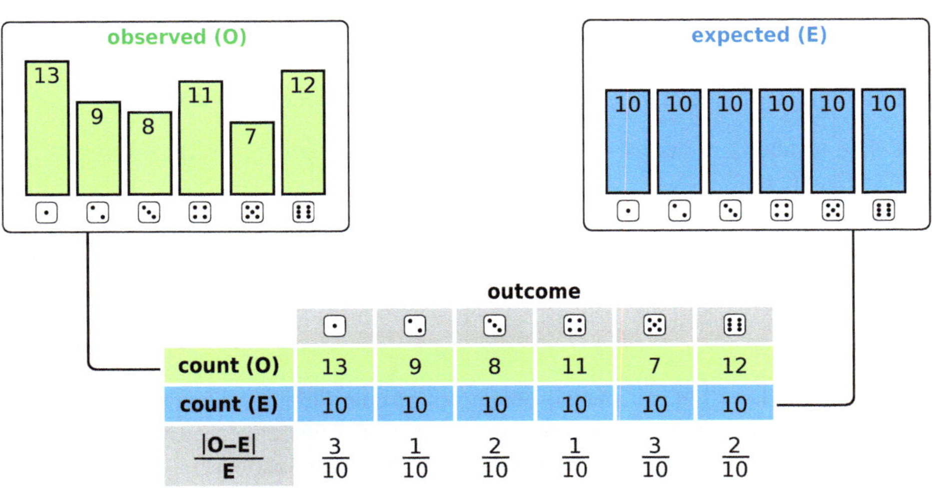

3. **Hardy/Weinberg.** There are 3 outcomes, so in the $\sum$ term, the choice of i is the set {AA, Aa, aa}, and there are 3 terms in the expression for χ.

$$\chi = \sum_{i \text{ in } \{AA,\ Aa,\ aa\}} \frac{|O_i - E_i|}{E_i} = \frac{|58-50|}{50} + \frac{|110-100|}{100} + \frac{|32-50|}{50} = \frac{8}{50} + \frac{10}{100} + \frac{18}{50} = 0.62$$

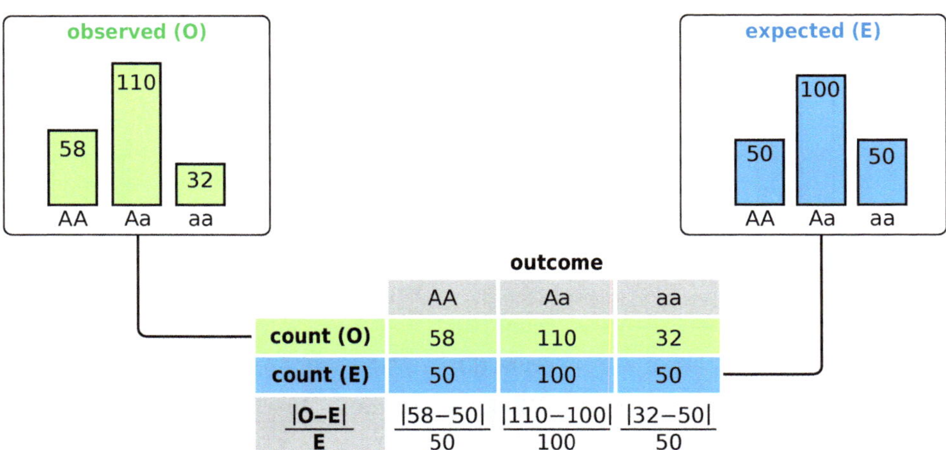

4. **rock-paper-scissors.** There are 3 outcomes, so in the $\sum$ term, the choice of i is the set {rock, paper, scissors}, and there are 3 terms in the $\sum$ expression for χ.

$$\chi = \sum_{i \text{ in } \{R,\ P,\ S\}} \frac{|O_i - E_i|}{E_i} = \frac{|107-127|}{127} + \frac{|124-127|}{127} + \frac{|150-127|}{127} = 0.362$$

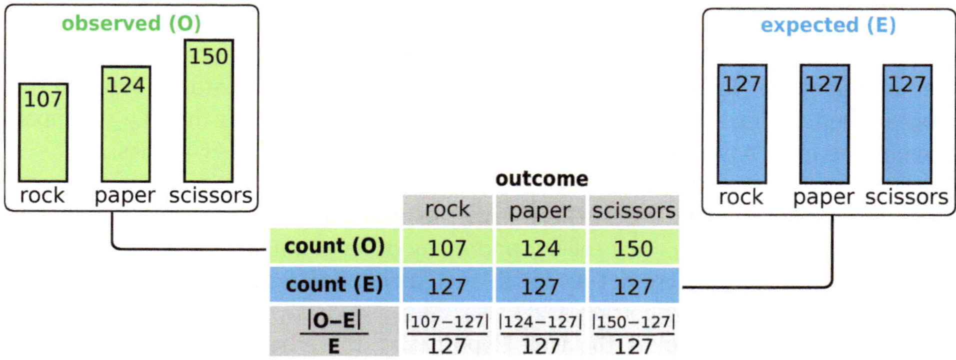

5. **Political affiliation of U.S. board directors.** There are 3 outcomes, so there are 3 terms in the $\sum$ expression for χ.

$$\chi = \sum_{i \text{ in } \{R,\, D,\, I\}} \frac{|O_i - E_i|}{E_i} = \frac{37.71}{219.71} + \frac{140.74}{243.26} + \frac{103.03}{306.03} = 1.09$$

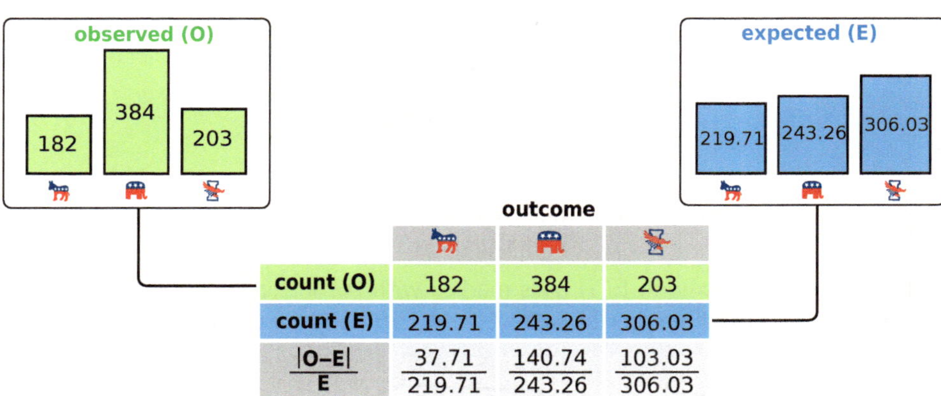

We have now calculated χ as the distance between observation and expectation. We will call the observed distance χ_{obs}.

The next question is, as usual, is χ_{obs} large?

FURTHER EXERCISES 8.2

1. χ calculates the distance between observed data and expected counts. The expected counts are generated from:

   a. The researcher's knowledge of this population
   b. A mathematical model
   c. Published data from other studies
   d. Assuming that the outcomes have equal probability
   e. The Null Hypothesis, which could be based on any of the above

8.3 Is χ_{obs} "large"? Null Hypothesis Significance Testing

The question "is χ_{obs} large?", as usual, means: is χ_{obs} large with respect to what we might see under a randomized Null Hypothesis? We know χ_{obs}, which is the distance between the observed histogram and the expected histogram. The question is: how does that χ_{obs} compare to what we might see from *any* random sample from the world of the Null Hypothesis?

And what is this world of the Null Hypothesis? It's the world in which randomness is the only factor in play. In many cases, this means the probabilities of each outcome are equal. This is the case for coin flipping, dice rolling, and many other examples. In other cases, probabilities under the Null Hypothesis are not all equal, such as in our Hardy/Weinberg example, where the three outcomes stand in proportion 1:2:1.

The expected histogram *is* the Null Hypothesis. If we need to make a world of the Null Hypothesis, we use the expected histogram, and sample randomly from it.

First, let's deal with the case in which the expected values are whole numbers. Then we will deal with the case where there are expected numbers that are fractions, like 2.6 people.

Sample from the world of the Null Hypothesis Let's say the expected histogram E has two outcomes, a and b, having bins with expected counts E(a) and E(b) (Figure 8.11).

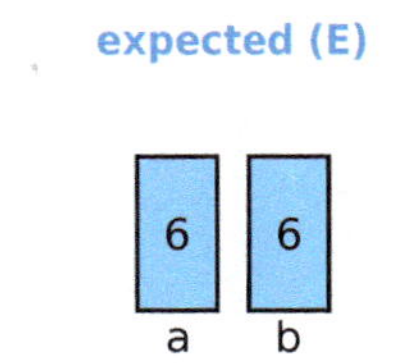

Figure 8.11 Exemplary histogram of expected counts.

Then we make a box with E(a) balls marked with the letter "a" and E(b) balls marked with the letter "b". The total number of balls, E(a) + E(b), we will call n (Figure 8.12).

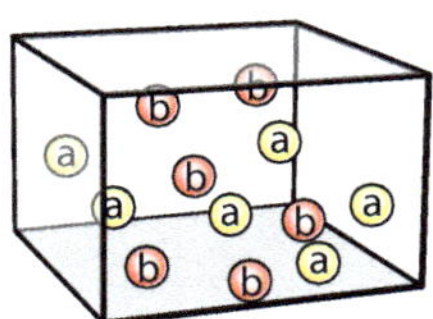

Figure 8.12 The world of the Null Hypothesis based on the expected histogram in Figure 8.11.

Then we sample (with replacement) n balls from the box.

Count the number of "a"-balls in this new sample of n. Call that number $E_1(a)$. Then count the number of "b"-balls in the sample and call that number $E_1(b)$. Combine the two values into a histogram, called pseudo-observed (O_1). This is our first simulation of what a random sample from the world of the Null Hypothesis might look like.

Calculate the distance χ from pseudo-observed O_1 to E and call that distance χ_1 (Figure 8.13).

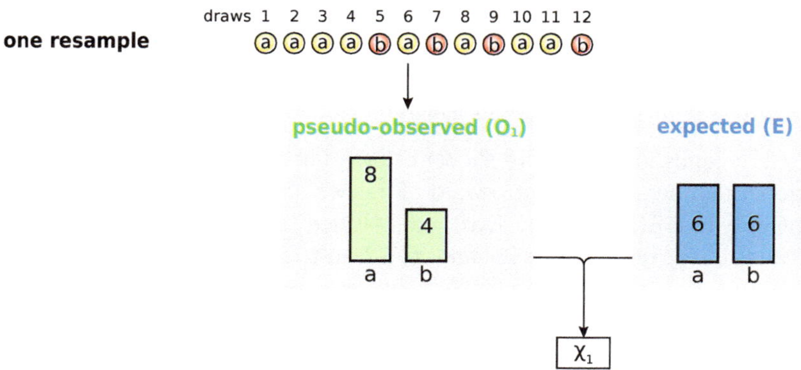

Figure 8.13 First resample from the world of the Null Hypothesis, the pseudo-observed O_1, and its distance to the expected histogram E, called χ_1.

Repeat the resampling procedure 10,000 times and collect the values χ_1, χ_2, ..., $\chi_{10,000}$ into a histogram, forming the null distribution. Then we ask: where does χ_{obs} rank in this histogram of 10,000 χ-values from samples from the world of the Null Hypothesis? The number of simulations in which χ_i was greater than or equal to χ_{obs}, divided by 10,000, is the p-value for χ_{obs} (Figure 8.14).

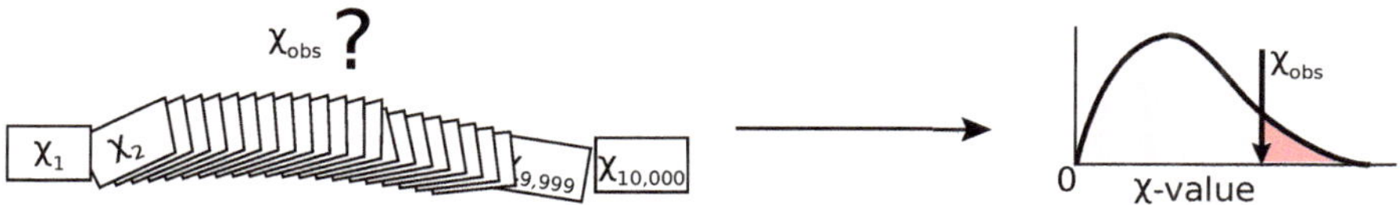

Figure 8.14 10,000 χ-values form the null distribution. The p-value of χ_{obs} is the fraction of the null distribution that has a χ-value greater than or equal to χ_{obs}.

We summarize the procedure of NHST for χ_{obs} in Figure 8.15.

Fractional expectations and probabilistic sampling Frequently, the entries in the expected histogram will not be whole numbers. We could very well have an expected number that is, say, 5.7. In these cases, we need to resort to a more sophisticated form of sampling.

Suppose that E(a), the expected value in the first bin of the histogram, is 5.7. The question is: how do we put 5.7 balls marked "a" into the box? We will give two answers to this question. One of them will be intuitive, using the balls-in-the-box metaphor. But the other is to look at the actual code that we use to produce this sampling.

- **Balls-in-a-box metaphor.** In terms of balls-in-a-box, when we have an expected value of 5.7 as E(a), we place 6 balls in the box. 5 of them are marked "a" and the other ball is marked "a(0.7)". Then, when we go to sample n balls from the box, if we draw an "a"-ball, we count it as an "a" and put it into our sample. But when we draw the "a(0.7)" ball, we consult a random number generator that gives us a number uniformly between 0 and 1. If that random number is ≤ 0.7, then we put an "a"-ball into the sample, and if it is > 0.7, we put the ball back into the box without selecting it.

  Similarly, for E(b), which is 6.3, we put 7 balls into the box. 6 of them are marked "b", and the seventh is marked "b(0.3)". When we sample a "b" ball from the box, we put it into the sample. But when we sample the "b(0.3)" from the box, we consult the random number generator, giving us a number from 0 to 1 with uniform probability. If it gives a number ≤ 0.3 we put a ball marked "b" into the sample, but if it gives us a number > 0.3, we put the ball back into the box without selecting it.

- **Actual code.** Commands in Python are shown below. The change from whole numbers of balls to fractional numbers is straightforward. First we show the Python code for whole number sampling where E(a) = 6 and E(b) = 6. Notice that the probabilities of the two outcomes 'a' and 'b' are given as '6/12' and '6/12' in the code line

```
p_outcomes=[6/12, 6/12]
```

sample with replacement (whole numbers)

```
1 import numpy as np
2 outcomes=['a', 'b']
3 n=12
4 p_outcomes=[6/12, 6/12]
5 pseudo_observed=np.random.choice(outcomes, size=n, replace=True,
   ↪ p=p_outcomes)
6 print(pseudo_observed)
7 # output of the "print" command
8 # a, a, b, b, a, a, a, b, a, b, b, b
```

If we need to do fractional sampling, say, 'a' has expected frequency E(a) = 5.7 and 'b' has expected frequency E(b) = 6.3, we simply change that code line to

```
p_outcomes=[5.7/12, 6.3/12]
```

sample with replacement (fractional numbers)

```
1 import numpy as np
2 outcomes=['a', 'b']
3 n=12
4 p_outcomes=[5.7/12, 6.3/12]
5 pseudo_observed=np.random.choice(outcomes, size=n, replace=True,
   ↪ p=p_outcomes)
6 print(pseudo_observed)
7 # output of the "print" command
8 # a, b, b, a, a, b, b, b, b, b, b, b
```

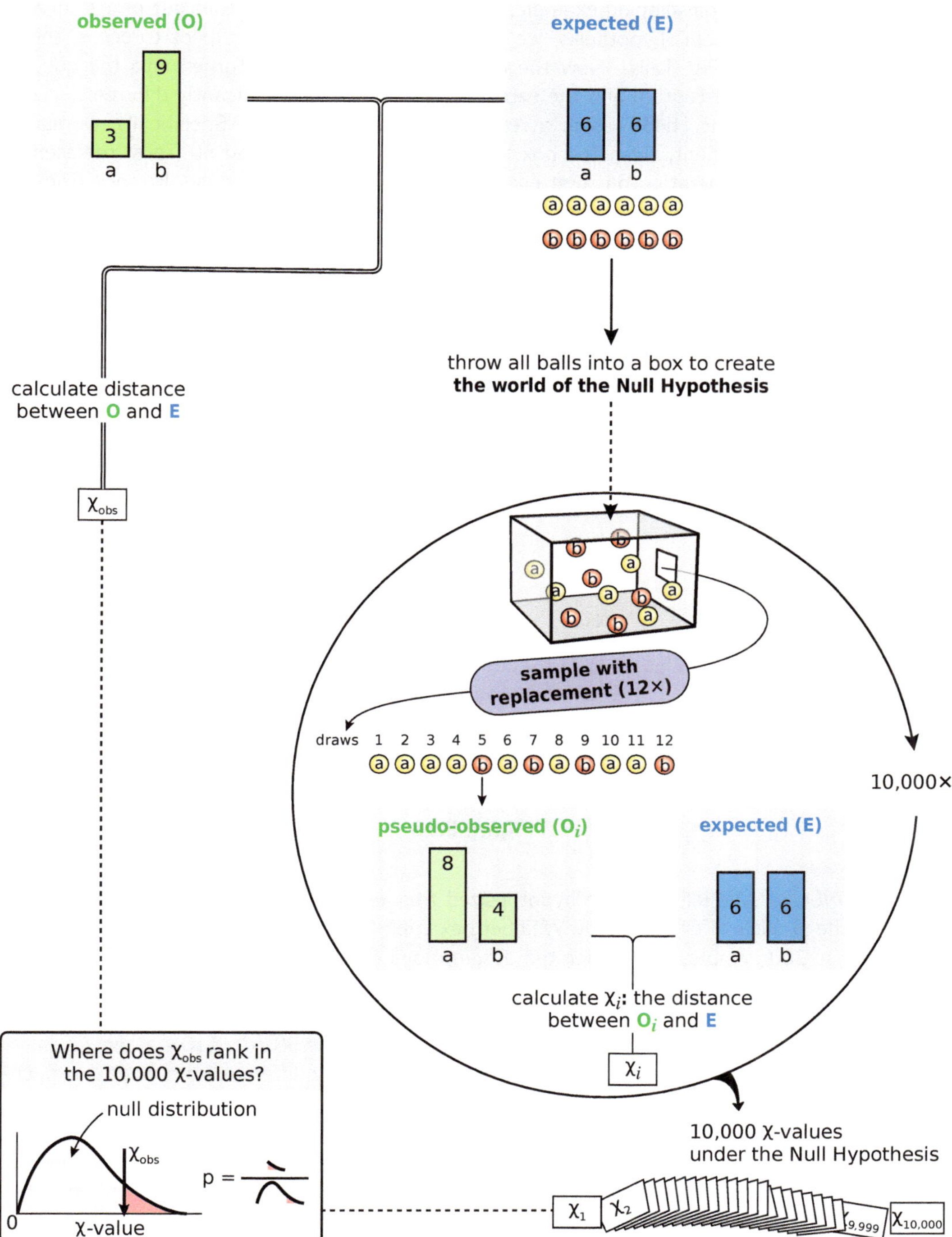

Figure 8.15 NHST for χ_{obs}. Sample with replacement from the world of the Null Hypothesis 10,000 times to construct the null distribution. A p-value is then calculated by comparing χ_{obs} against the null distribution, and asking what fraction of simulations are equally or more extreme than it. In this illustration, the expected histogram E has two bins, with expected counts E(a) = 6 and E(b) = 6. The total number of balls n is 12 (n = E(a) + E(b)).

Coin flipping In the coin flipping example, we had 60 Heads and 40 Tails out of 100 flips. In order to test the "fair coin" hypothesis, we construct what we would expect to see in theory, which is 50 Heads and 50 Tails. Now the question of "fair coin" is turned into the question whether the observed histogram and the expected histogram are significantly different.

We then conduct the NHST using a resampling-based approach. Specifically, we sample 100 balls, with replacement, from the box that contains 50 Heads and 50 Tails, representing the Null Hypothesis, generating our first pseudo-observed. We calculate the distance from this pseudo-observed to the expected, calling it χ_1.

We repeat this 10,000 times to generate 10,000 χ-values, forming the null distribution. Comparing the actual χ_{obs} to the null distribution, we see that the observation of 60 Heads and 40 Tails in 100 flips is not that extreme for a "fair" coin; in fact, cases at least this extreme occur approximately 5.74% of the times (Figure 8.16).

Note that we are using a one-sided p-value calculation. This is because the occurrence of small χ-values is expected under the Null Hypothesis, therefore small χ-values should not be counting against the Null Hypothesis.

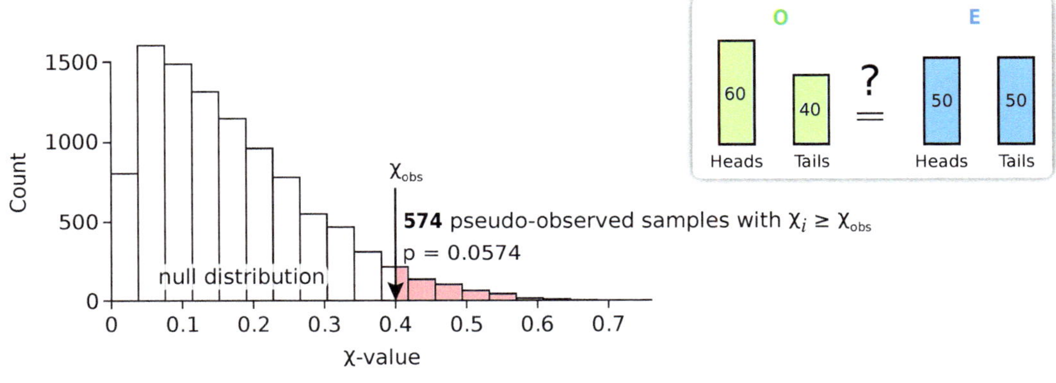

Figure 8.16

Hardy-Weinberg We let p be the hypothesized frequency of the A allele, and q be the frequency of the B allele. For our Hardy-Weinberg example, we had postulated that $p = q = 0.5$. We looked at 200 samples, so, under the assumptions of the Hardy-Weinberg law, we would expect the AA frequency to be $p^2 \times 200$, or 50, the Aa frequency to be $2 \times p \times q \times 200$, or 100, and the aa frequency to be $q^2 \times 200$ or 50.

In our observation, we found the genotype frequencies to be 58 AA, 110 Aa and 32 aa.

The question now is: is the observed histogram significantly different from what is expected under the Null Hypothesis, that is, the Hardy-Weinberg law?

To compare the two histograms, we conduct NHST using a resampling-based approach. First we create "balls in a box" under the Null Hypothesis which is the Hardy-Weinberg law, that is, a box with 50 balls marked "AA", 100 marked "Aa" and 50 marked "aa". Then we sample (with replacement) 200 balls from this box, record their markings, and calculate the distance from this pseudo-observed histogram to the expected, calling it χ_1.

Repeating this procedure 10,000 times, we form the null distribution from the 10,000 χ-values, and we see that the observed difference χ_{obs} is indeed quite extreme, with only 87 samples, out of 10,000, equal to or exceeding it, giving a p-value of 0.0087. Therefore, the NHST result says the observed distribution is unlikely to have occurred under the Hardy-Weinberg law (Figure 8.17).

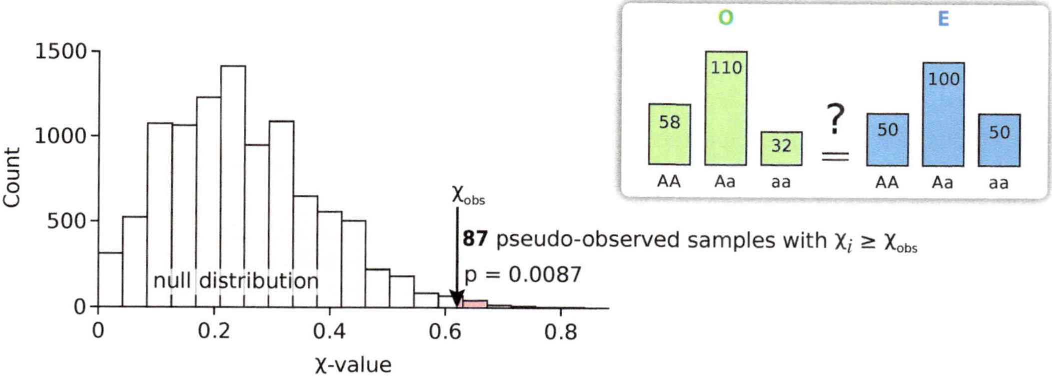

Figure 8.17

Rock-paper-scissors From the results of the 381 games of rock-paper-scissors, we construct the observed histogram. In theory, each hand sign should have an equal chance of occurring, giving us the expected histogram.

The question now is: is the observed histogram significantly different from the expected?

To compare the two histograms, we conduct NHST using a resampling-based approach. Specifically, we sample 381, with replacement, from the box that has 127 "rock" tokens, 127 "paper" tokens, and 127 "scissors" tokens. We then calculate the distance from this pseudo-observed sample to the expected histogram, calling it χ_1.

We repeat this procedure 10,000 times, generating 10,000 χ-values: $\chi_1, \chi_2, \ldots, \chi_{10,000}$, from which we form the null distribution.

We then compare the observed result χ_{obs} to the null distribution, obtaining a p-value of 0.0391 (Figure 8.18). We conclude there is some weak evidence that these players did not simply choose signs at random.

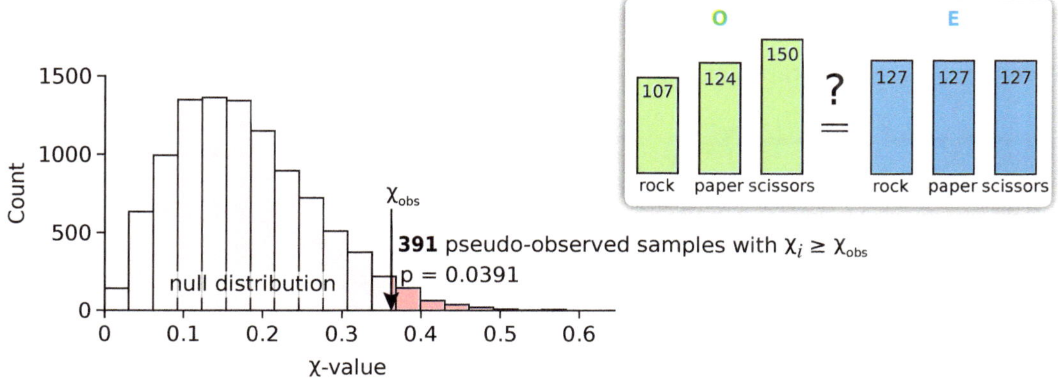

Figure 8.18

U.S. board political affiliation In looking at the political affiliation of board members in major U.S. companies, the survey result is the observed histogram, and the expected histogram is from the background population, that is, the Null Hypothesis is that the distribution of political affiliations of board members is the same as the distribution in the general public (equivalently, that board members were chosen randomly from the general population).

The question now is: is the observed histogram significantly different from the expected?

To compare the two histograms, we conduct NHST using a resampling-based approach. To simulate the Null Hypothesis, we sample 769, with replacement, from a box that has 219.71 D's, 243.26 R's, and 306.03 I's. We make a histogram of this pseudo-observed result, and calculate the distance from it to the expected.

We repeat this procedure 10,000 times and form the null distribution from the 10,000 χ-values.

Comparing χ_{obs} to the null distribution, we see that the observed difference χ_{obs} is indeed highly extreme; none of the 10,000 samples even came close to what was observed in the survey (Figure 8.19).

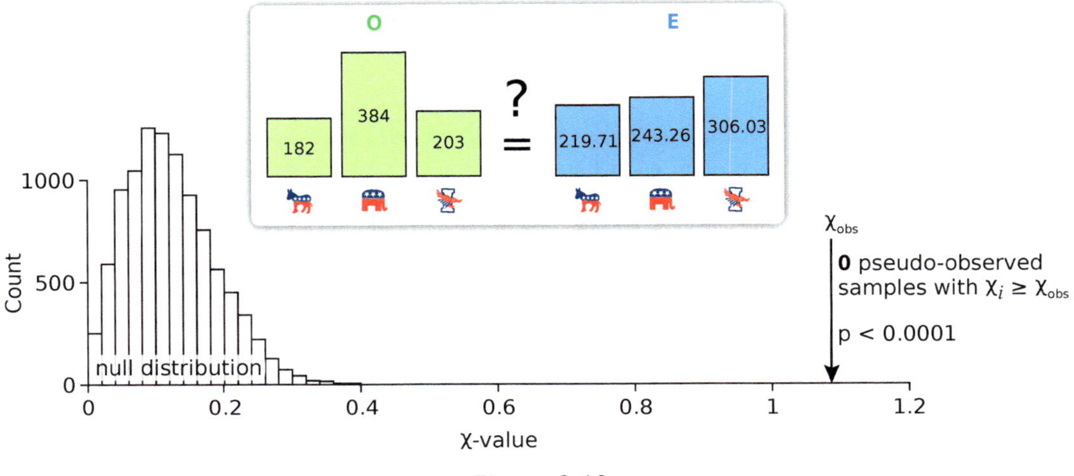

Figure 8.19

Therefore, the observed distribution is highly statistically significantly different from what we would expect under the Null Hypothesis, with $p < 0.0001$. In other words, we can reject the Null Hypothesis that the makeup of political affiliations of board members in major U.S. companies resembles or is representative of the general public.

FURTHER EXERCISES 8.3

1. ***Platorchestia platensis.*** In order to study the pressures of natural selection, a study examined variation at the *Mpi* locus in the amphipod crustacean *platorchestia platensis* collected from a single location on Long Island, New York (see article "Selection component analysis of the *Mpi* locus in the amphipod platorchestia platensis", *Heredity*, 1989).

   There were two alleles, which we'll call A and B. The observed genotype frequencies in the sample of 5,800 individuals were

   1203 AA, 2919 AB, 1678 BB.

   The estimate of the A allele proportion from the data is

$$\frac{2 \times \text{obs(AA)} + \text{obs(AB)}}{2 \times \left(\text{obs(AA)} + \text{obs(AB)} + \text{obs(BB)}\right)} = \frac{2 \times 1203 + 2919}{2 \times 5800} = 0.4591$$

Using the Hardy-Weinberg formula and this estimated allele proportion, the expected genotype proportions are

$$0.211 \text{ AA}, 0.496 \text{ AB}, 0.293 \text{ BB}.$$

Determine whether to reject the Null Hypothesis that the data fit the expected Hardy-Weinberg proportions.

a) Show the calculation that derived the expected genotype proportions.
b) Write the table for observed and expected counts. Calculate $\mathcal{X}_{obs}$.
c) Describe the appropriate Null Hypothesis.
d) Calculate a p-value for $\mathcal{X}_{obs}$. Include the null distribution with the observed result indicated.
e) Interpret the result.

2. **Cold symptoms.** A senior physician has noted that every year, around cold and flu season, the first symptoms that patients exhibit are in the following proportions: runny nose (25%), sore throat (25%), congestion (25%), fever (25%).

   However, after the recent cold and flu season, the physician went back to review their patients and wondered if they were missing a discrepancy. The distribution of first symptoms for the most recent cold and flu season is provided in the table below.

First presenting symptom	running nose	sore throat	congestion	fever
Number of patients	40	10	20	30

   a. State the Null Hypothesis, and create the table of expected counts.
   b. Calculate $\mathcal{X}_{obs}$, the difference between the observed and expected data.
   c. Having calculated $\mathcal{X}_{obs}$, determine whether it is statistically significant at the $\alpha = 0.01$ level.

3. **Mendel's pea experiment.** In 1866, Gregor Mendel, father of genetics, published the results of his experiments on peas. He found that his experimental distribution of peas closely matched the theoretical distribution predicted by his theory of genetics.

Phenotype	**Observed count**	**Theoretical proportion**
Round, Yellow	315	9/16
Round, Green	108	3/16
Wrinkled, Yellow	101	3/16
Wrinkled, Green	32	1/16

   a. What is the expected count for each pea phenotype according to Mendel's theory?
   b. What is the observed difference $\mathcal{X}_{obs}$?
   c. What is the Null Hypothesis?
   d. Calculate a p-value for $\mathcal{X}_{obs}$. Include the null distribution with the observed result indicated. Interpret the result.

8.4 post hoc testing: "Mt. Fuji plots"

The χ-statistic is like an omnibus test, meaning that if it is significant, that tells us that there is a difference among the groups, but it does not tell us *which* bins differ. In order to tell which bins differ, we need to do further "*post hoc* testing", which we will now address.

Suppose we are dealing with histograms that have multiple (more than 2) bins. Then, if the observed histogram is significantly different from the expected, as shown by a significant p-value in the χ-test, it makes sense to go further and ask *where* they differ, that is, in which bins?

Using the Hardy-Weinberg experiment as an example, we consider the 3 genotypes AA, Aa and aa in a population. Our sample of 200 shows that there is a significant difference among the observed histogram and the expected, with $p = 0.0087$ in the χ-test (Figure 8.17). Now we'd like to know *where* they differ, that is, in which genotype. This is the question *post hoc* testing answers (Figure 8.20).

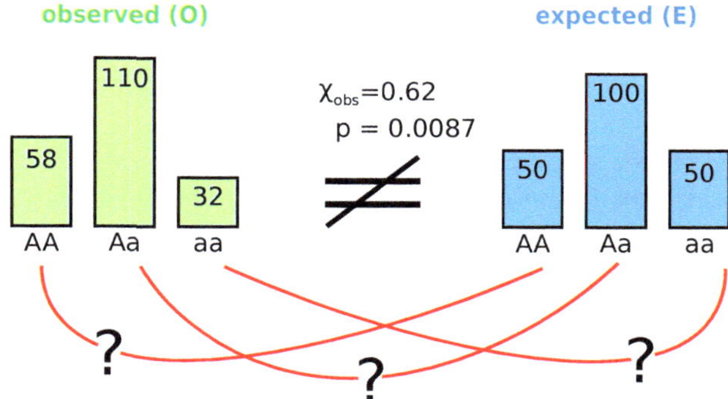

Figure 8.20 In the Hardy-Weinberg experiment, the χ-test confirmed that there is an aggregate difference between the observed histogram and the expected. However, in order to identify specifically in which genotype(s) the difference exists, we need to do *post hoc* testing.

How to construct a Mt. Fuji plot

There's a convenient graphical procedure (Figure 8.21) for this *post hoc* analysis, which we call a Mt. Fuji plot (we'll explain the terminology below).

Suppose we have a total of n data points in a study, and the observed histogram (O) has k bins, so $n = \sum_{i=1}^{k} O(i)$, where $O(i)$ is the number of data points in bin i. We then derive the expected histogram E from theory or a background population or some kind of Null Hypothesis.

Then, we sample n with replacement from the expected histogram (E), and construct our first histogram of expectations under the Null Hypothesis. We repeat this 10,000 times, and record and keep the 10,000 histograms.

For each bin i, there will be 10,000 values of $E(i)$.

We create an "insignificance band" (or "no-rejection zone") for each bin i by ranking the 10,000 values of $E(i)$ from the stored resamplings and finding the cutoffs for statistical significance. For example, if we were to construct a 95% insignificance band, we would count 250 up from the smallest $E(i)$ and down 250 from the largest as the 2.5 percentile and 97.5 percentile cutoffs. Any value that lies in between those cutoffs is by definition not statistically significant.

Then we use a shaded rectangle to mark each insignificance band on the observed histogram. This gives us the Mt. Fuji plot.

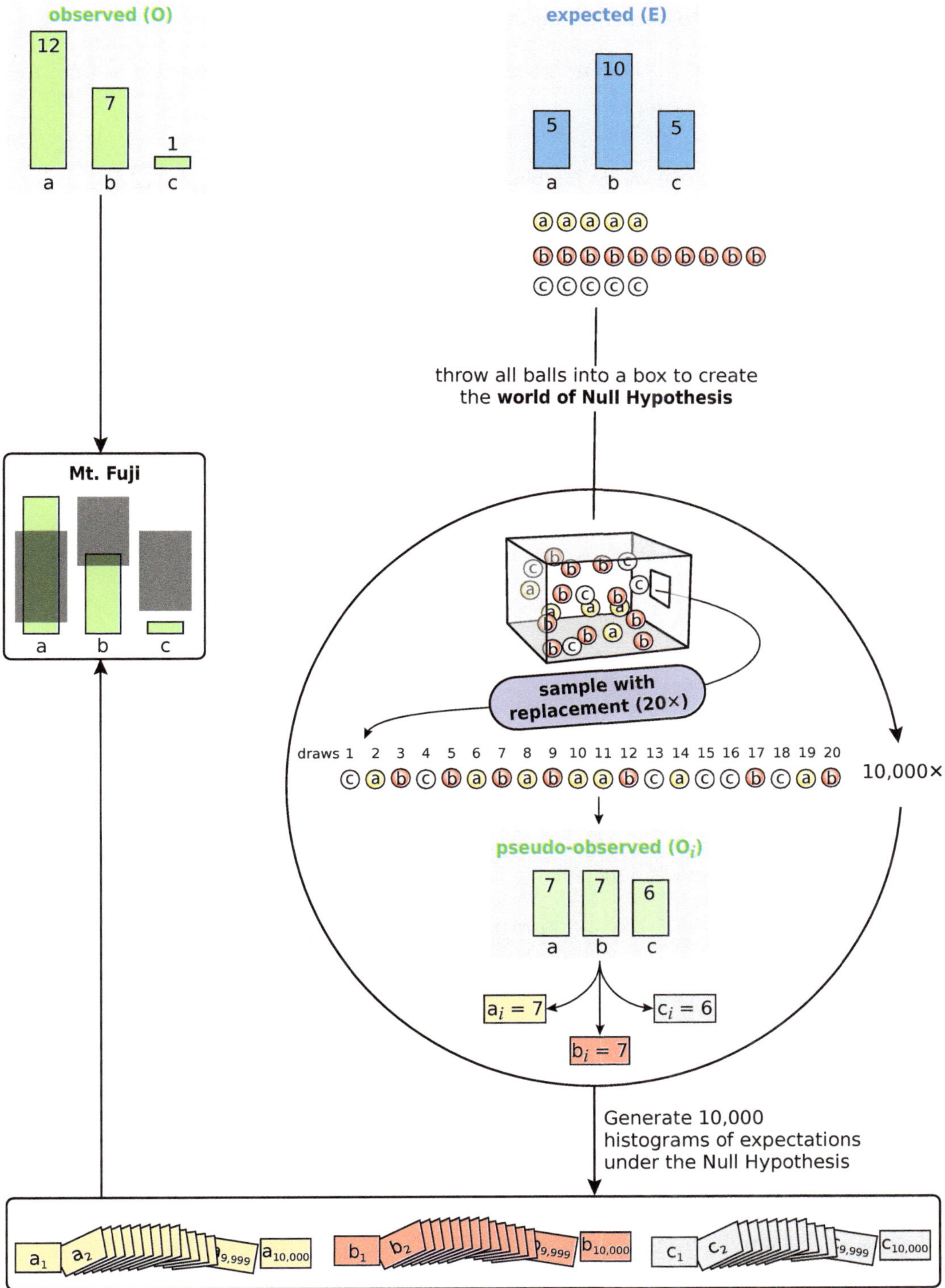

Figure 8.21 **Mt. Fuji plot.** Procedure to generate a Mt. Fuji plot using an artificial example that has three outcomes: a, b and c, where the observed histogram is (12, 7 ,1) and the expected histogram is (5, 10, 5).

How to construct an insignificance band for each bin

Step 1 ► Consider "bin a" of the histograms. We will have 10,000 values for "bin a" from the 10,000 samples.

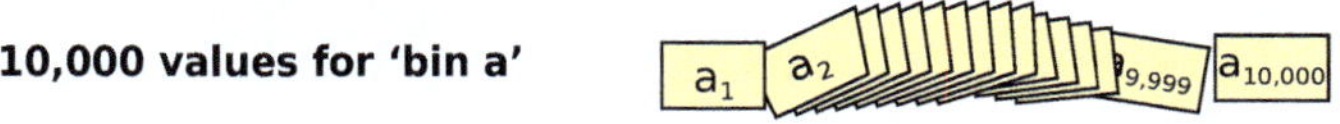

Step 2 ► For a $p < 0.01$ level of significance (recommended), rank the 10,000 "bin a" values from smallest to largest. Count up 50 from the smallest and call that value "bin-a bottom", and count down 50 from the largest and call that value "bin-a top".

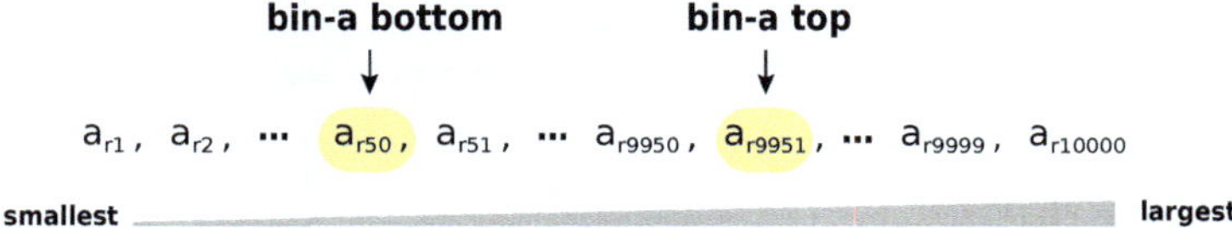

Step 3 ► Take the actually observed histogram (green histogram), and draw a rectangle (gray) on "bin a" whose bottom is "bin-a bottom" and whose top is "bin-a top". This is the insignificance band for bin a.

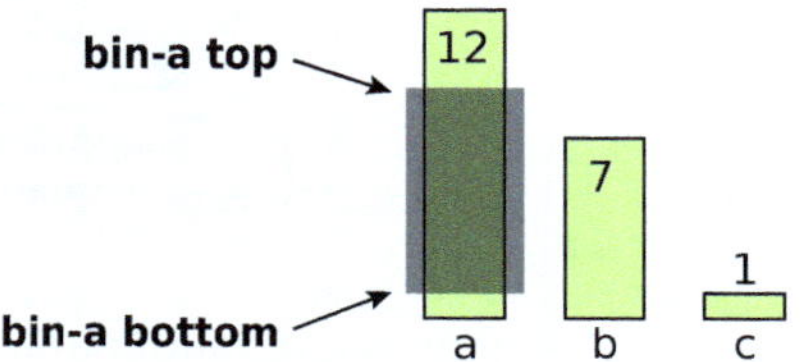

Repeat steps 1 through 3 for each bin and construct a 99% insignificance band for each bin, resulting in the following Mt. Fuji plot.

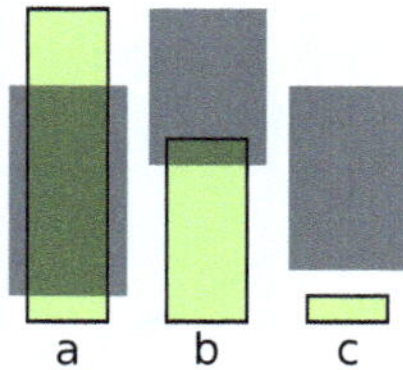

Interpreting the Mt. Fuji plot

Any bin in the actually observed histogram that sticks out above or below the shaded cloud is a candidate for being statistically significantly different from expectations, subject to multiple testing corrections. Note that the insignificance bands are not confidence intervals, they are thresholds for statistical significance at the chosen α level, in this case, 0.01.

In the artificial example in Figure 8.21, we see that

- "bin a" sticks above its cloud, so "bin a" is significantly larger than the expected.
- "bin b" is inside the cloud, so is not significantly different from the expected.
- "bin c" sticks below its cloud, and so is significantly smaller than the expected.

As another example, we constructed a Mt. Fuji plot for the Hardy-Weinberg example (Figure 8.22). Genotype AA and Aa are both *within* the clouds, therefore, are not significantly different from the expected, whereas genotype aa sticks below its cloud, therefore, is significantly lower than expected under the Hardy-Weinberg law, at the $p < 0.01$ level.

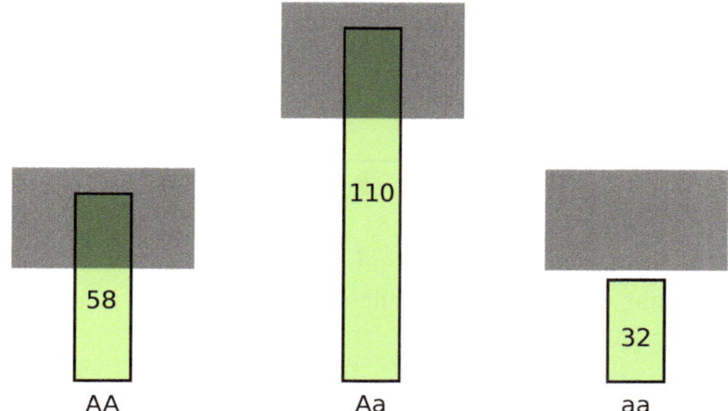

Figure 8.22 Mt. Fuji plot for the Hardy-Weinberg experiment in Figure 8.20. The insignificance bands are at the $p < 0.01$ level.

Origin of the term "Mt. Fuji plot" In trying to explain this graphical method to a colleague, we composed a haiku:

Statistical Significance

The data rise
like Mount Fuji
above the clouds of uncertainty

Some of our colleagues started calling these "Mount Fuji plots", and the name has stuck (Figure 8.23).

Figure 8.23 Left: Mount Fuji rising above the clouds. Right: here the peak lies within the clouds.

Exercise 8.4.1 After comparing the histogram of the political affiliation of U.S. board members to the general public, we concluded that the political affiliation makeup of the board members is significantly different from the general population at $p < 0.0001$ level (result of the χ-test presented in Figure 8.19). Now we want to pinpoint *where* the difference(s) are. Which party is significantly over/under represented? We constructed the Mt. Fuji plot at $\alpha = 0.01$ level. How would you describe the result?

Exercise 8.4.2 After comparing the histogram of the 381 games of rock-paper-scissors to theory, we concluded that the observed result does not deviate that far from the Null Hypothesis given by theory ($p = 0.0391$).

1. Which of the following statements are correct?
    a) Even though the overall histogram of the 381 games does not deviate that far from theory, there is still some chance for individual outcomes (rock, paper, scissors) to be significantly different from theory, so we should conduct a Mt. Fuji plot regardless just to rule out any possibilities.
    b) We cannot proceed to conduct a *post hoc* test since the χ-test is not significant.
    c) A Mt. Fuji plot should not be conducted when the omnibus test fails.
    d) *post hoc* testing is always necessary when we don't find any significant difference in the χ-test.
2. If you went ahead and did the Mt. Fuji plot, which one would it look like? (*Hint: remember each outcome is equally likely. What effect will that have on the location and the thickness of the clouds?*)

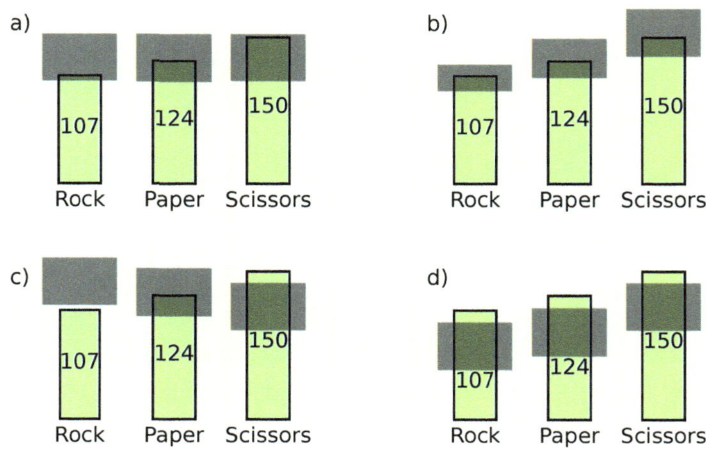

Is the roulette wheel biased?

A European roulette wheel has 37 slots, 0-36, distributed randomly around the wheel. (American wheels have an additional slot 00). The theoretical expectation, of course, is that all slots have an equal probability of landing the ball, giving us a flat histogram.

But real roulette wheels, even casino quality, don't always meet these expectations. To test this hypothesis, we downloaded a data set of 40,000 spins of a single roulette wheel.

The data are presented in a conventional histogram (Figure 8.24). From the histogram, we see that even with a fairly large n = 40,000, the histogram is far from flat.

We can then ask how many of these highs and lows are statistically significant. We carried out a Mt. Fuji procedure, sampling from the expected histogram, and constructing the 99% clouds of uncertainty. (The expected value is very close to 1100 in each bin.) The clouds are shown in blue.

Note the three gold arrows, which point to two adjacent slots that are above the cloud (slots 4 and 21), and one that is below (slot 14).

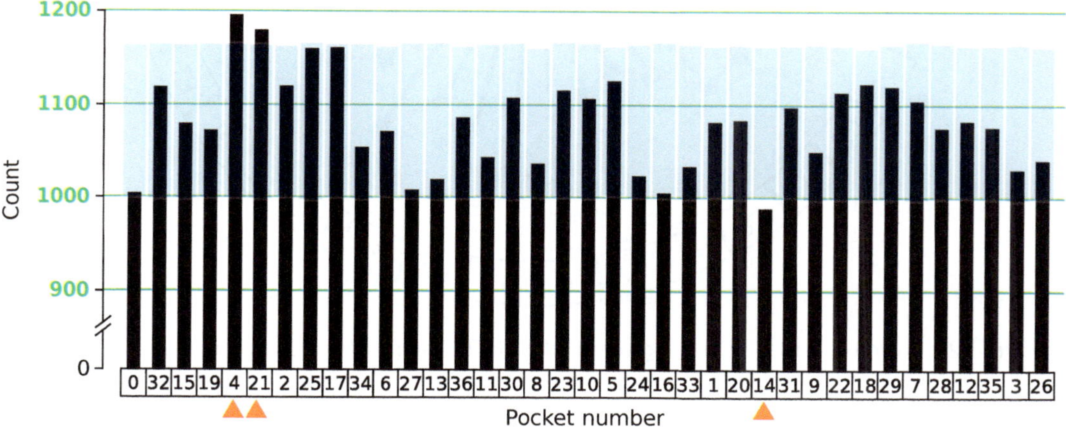

Figure 8.24 Histogram of the 40,000 spins of a roulette wheel, overlaid with its 99% clouds of uncertainty. Black bars are the observed results, showing how many times each slot landed the ball (the observed histogram). Light blue rectangles are clouds of uncertainty.

We then rolled up the 1-D histogram, arranging it in a circle (Figure 8.25). Another fact worth noting emerges: the significantly high counts are on the exactly opposite side of the wheel from the significantly low count. Keeping in mind that high counts in a slot mean a low point in the physical wheel, and low counts mean a high point in the physical wheel, this fact increases our confidence that this really is a feature of this particular wheel, due to either imperfect manufacture or wheel installation, and not just a statistical fluke.

The histograms of particular roulette wheels in particular Las Vegas casinos were the subject of intense practical study by a group from the UC Santa Cruz departments of math and physics, in the late 1970s and early 1980s. The story is told in Thomas Bass' *The Eudaemonic Pie* (Bass, 1985).

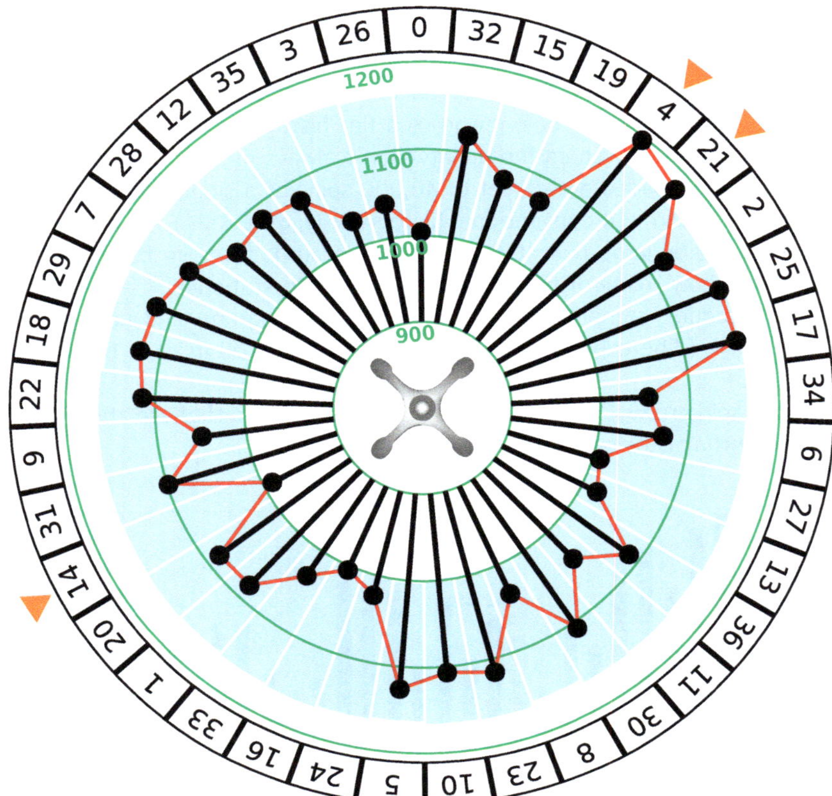

Figure 8.25 We took the 1-D histogram of the 40,000 spins of a roulette wheel in Figure 8.24, and rolled it up into a circle, to visualize it in the same arrangement as an actual roulette wheel. Black bars are the observed histogram, with the black dots indicating the bin tops. We have connected the bin tops by a red line to better visualize the histogram. Light blue rectangles are the 99% clouds of uncertainty.

Correction for multiple testing

When we make a Mt. Fuji plot, we are making multiple tests of significance, so we need to be concerned about correcting for multiple testing. We will use a Benjamini-Hochberg style correction (see section 7.5 *post hoc analysis part II: correction for multiple testing*).

First, we calculate the actual p-values for each bin. If the observed histogram has m bins, then there would be m p-values.

If the observed count is higher than that expected under the Null Hypothesis, the p-value of the observed count in a bin is the fraction of simulations in which the bin value was greater than or equal to the observed. If the observed count is lower than the expected, then its p-value is the fraction of simulations that had values *less* than or equal to the observed value.

Rank these p-values from smallest to largest as p_1, p_2, ..., p_m while keeping track of which bins they came from in the original histogram. Then apply a Benjamini-Hochberg correction to these p-values.

Benjamini-Hochberg correction

Step 1 ► Choose the cutoff for significance, say, $\alpha = 0.01$.

Step 2 ► For each of the m sorted p-values, ask the following question:

For p_1, ask whether p_1 is $< \alpha \times \frac{1}{m}$

For p_2, ask whether p_2 is $< \alpha \times \frac{2}{m}$

...

For p_m, ask whether p_m is $< \alpha \times \frac{m}{m}$

Step 3 ► Then find the largest p-value for which the answer is "yes" and declare statistical significance for it and all smaller p-values.

We applied the Benjamini-Hochberg procedure to the Hardy-Weinberg example (Figure 8.26).

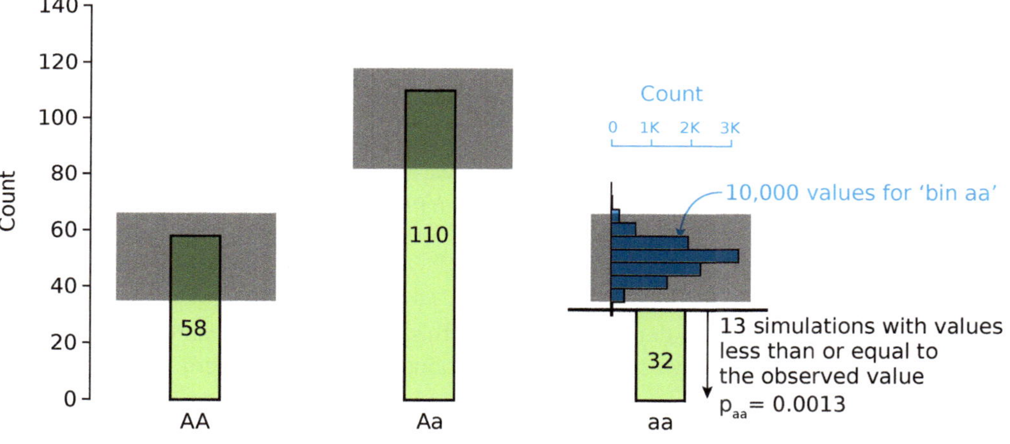

Figure 8.26

Here, only one of the comparisons was significant: the aa allele had a significantly low count. Its p-value was 0.0013. When we adjust the α value, 0.01, dividing it by 3 (the number of tests we made, which is m), and ask the question:

$$\text{is } p_1 = p_{aa} < \alpha \times \frac{1}{m} \text{ ?}$$

The answer is yes, and so we can declare that the aa allele has significantly low proportions in this sample.

Writing up the result

We have now carried out the full statistical analysis of the allele frequencies. The next step is to learn how to write up these results for a scientific paper. As a template, we wrote a little pseudo-paper for these (fictitious) results.

A study of genotypes in a population

Methods

Data Collection 200 plants of species X were collected from site Y, and their genotypes (AA, Aa and aa) were determined by method Z.

Statistical Comparisons The observed allele frequencies were compared to expectations derived from the Hardy-Weinberg law. We created a null model in which the allele frequencies were 25%, 50% and 25%, as predicted by Hardy-Weinberg. We then drew 200 samples, with replacement, from this model and determined allele frequencies.

This procedure was repeated 10,000 times, generating, for each genotype, 10,000 estimates of its frequency under the Null Hypothesis.

For each genotype, we then counted the top 0.5%ile and the bottom 0.5%ile of these 10,000 estimates (gray bars in Figure 8.26). Any observed value that was less than the bottom 0.5%ile or higher than the top 0.5%ile was considered a candidate for statistical significance, which would then be subject to a Benjamini-Hochberg correction to control false positives caused by the multiple comparisons.

Results

We found a deficit in the aa allele frequency that was statistically significant at $p = 0.0013$, which remained statistically significant after Benjamini-Hochberg correction.

Discussion

Further research is needed to determine whether this violation of the Hardy-Weinberg Law can be replicated, and, if so, which of the assumptions of Hardy-Weinberg has been violated to create this discrepancy.

Exercise 8.4.3 ***post hoc* analysis for the political affiliation of U.S. board members.** Recall the political affiliation survey (Example 1 on page 359). We calculated χ_{obs}, the distance from the observed histogram of the political makeup of U.S. board members to that of the general public, to be 1.09. Then we used a resampling approach to conduct Null Hypothesis Significance Testing, and found out that the political makeup of U.S. board members is significantly different from that of the general public, with $p < 0.0001$ (result shown in Figure 8.19).

We then showed the result of the Mt. Fuji plot (see Ex 8.4.1 on page 374) to identify where the differences are.

But we need to conduct a Benjamini-Hochberg correction to account for multiple comparisons. For each bin, we overlay the histogram of the 10,000 values generated in the resampling procedure on top of the 99% insignificance band of the Mt. Fuji plot as shown below:

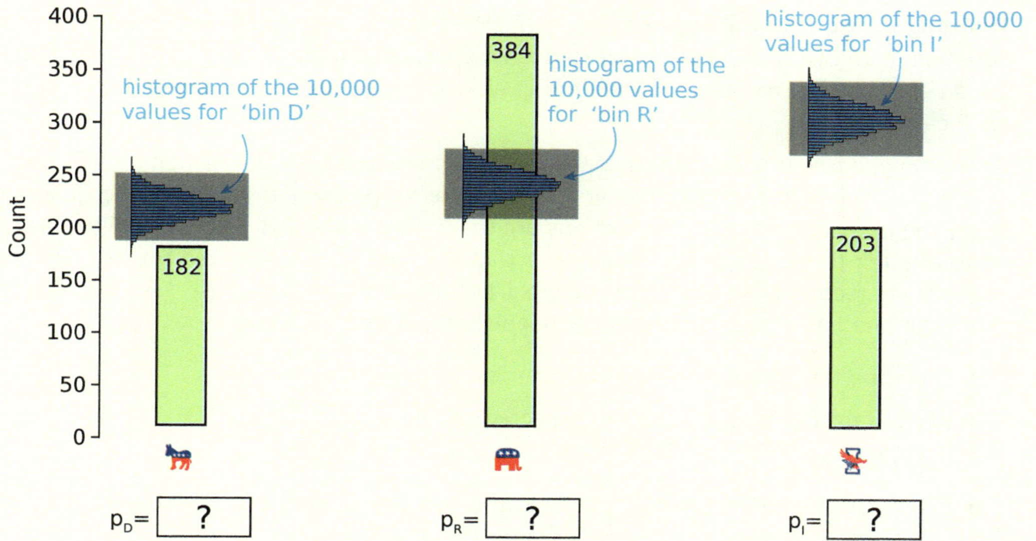

1. How many comparisons have we done in total?

   a) 3 b) 769 c) 10,000

2. Which of the following could be the p-value for each of the three political parties?

   a) $p_D < 0.0001$ $p_R < 0.0001$ $p_I < 0.0001$
   b) $p_D = 0.0016$ $p_R < 0.0001$ $p_I < 0.0001$
   c) $p_D = 0.0016$ $p_R = 1$ $p_I = 0$
   d) $p_D = 0.0016$ $p_R = 1$ $p_I = 0$
   e) $p_D = 0.9984$ $p_R = 0$ $p_I = 1$

3. Describe how you would do the Benjamini-Hochberg correction based on your answers above.

4. After the Benjamini-Hochberg correction, which of the following statements are correct?

   a) The proportion of Democrats in U.S. board members is significantly lower than the proportion of Republicans in U.S. board members.
   b) Independents are significantly underrepresented in U.S. board members compared to Republicans.
   c) Republicans are significantly overrepresented on U.S. boards relative to the general population.
   d) Democrats are significantly overrepresented on U.S. boards relative to the general population.
   e) Democrats are slightly overrepresented on U.S. boards relative to Independents.

5. Using the pseudo-paper "A study of genotypes in a population" on the facing page as a template, write up your results.

FURTHER EXERCISES 8.4

1. **HIV drug.** To study the likely effects of a certain HIV drug, a group of researchers examined variation at the UGT1A1 gene in HIV-infected adult patients in Portugal (this gene codes for an enzyme that metabolizes the drug). We'll focus on two of the alleles found, which we'll call A and B.

   The genotype frequencies in the sample of 62 adults were 24 AA, 33 AB, and 5 BB. Using the Hardy-Weinberg formula and the estimated allele proportion based on the observed data, the expected genotype proportions are 0.427 AA, 0.453 AB, and 0.121 BB. Determine whether to reject the Null Hypothesis that the data fit the expected Hardy-Weinberg proportions (see article "Variability in the UDP-glucuronosyltransferase 1A1 (UGT1A1) promotor in a HIV-infected Portuguese population", *Annals of Medicine*, 2019).

   a. Describe the appropriate Null Hypothesis.
   b. Write table(s) for observed and expected counts.
   c. Calculate χ_{obs}.
   d. Calculate a p-value for χ_{obs}. Include the null distribution with the observed result indicated.
   e. Interpret (d) in the context of this study.
   f. Briefly discuss how we can determine which groups are significantly high or low, violating the Hardy-Weinberg law.

2. An observer collected data on the X-values of a certain population. X could be their heights, weights, reproduction rates, etc. The black dots represent the observed counts. In this case, there was a theoretically expected value under the Null Hypothesis. This set of expected values was resampled, and the distribution of the number of points in each of the 14 bins was tabulated. This procedure was repeated 10,000 times, to give us 10,000 values of the expected outcome in each bin. For each bin, we found the 99.5% highest value and the 0.5% lowest value, and drew a red box between those two values. Given the information above, what is your interpretation of the plot below?

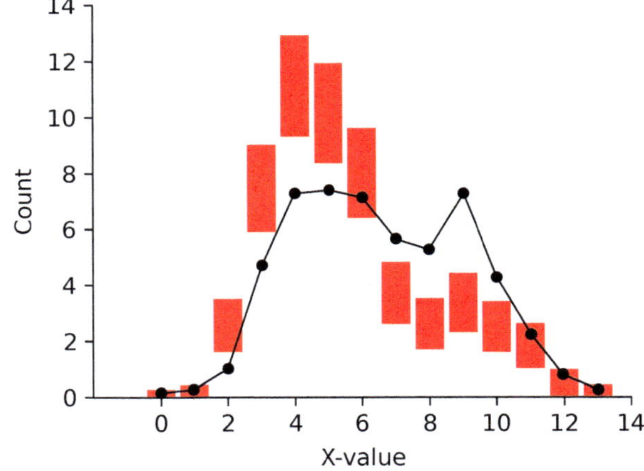

8.5 Comparing two observed histograms

We now know how to compare a histogram of observations to its corresponding histogram of expectations.

But in many cases, we want to compare two observed histograms to each other, and ask whether they are significantly different from each other, as opposed to comparing one histogram to expectations, as we have been doing up to now.

When are two factors "independent" of each other? The most important class of cases where we want to compare two observed histograms are situations where we want to ask whether one factor is *independent* of another. As we mentioned in the introduction to this chapter, this is an extremely frequent question.

1. **Heart disease and gender.** Is the likelihood of heart disease *independent of* gender?
2. **Clinical trial.** In a given clinical trial, was survival *independent of* whether the subject took the drug or not?
3. **Vaccination.** Was hospitalization *independent of* vaccination status?
4. **Titanic.** In a classic example, the survivors of the Titanic shipwreck in 1912 came from all 3 classes of ticket: First class, Second class, and Steerage (Third class). It was asked at the time whether survival on the Titanic was *independent* of ticket class.
5. **Political party preference.** Is political party preference *independent of* gender?

In each case, the question of independence is really a question about the similarity of two observed histograms.

1. **Heart disease and gender.** The question whether heart disease is *independent* of gender is asking whether the histogram of heart disease (yes/no) in the male population is *similar* to the histogram of heart disease (yes/no) in the female population
2. **Clinical trial.** We interpret the question whether the drug had an effect on survival as asking whether the histogram of 'survived/not survived' in the drug-yes population is *similar* to the histogram of 'survived/not survived' in the drug-no population.
3. **Vaccination.** The question whether hospitalization is independent of vaccination status is asking whether the histogram of 'hospitalized/not hospitalized' in the 'vaccination status = yes' group is *similar* to the histogram in the 'vaccination status = no' group.
4. **Titanic.** The question whether survival on the Titanic is independent of ticket class is asking whether the histograms of the 3 classes were *similar* in the survival-yes group compared to the survival-no group.
5. **Political party preference.** The question whether political party affiliation is *independent* of gender is asking whether the histogram of political party preference (D, R, I) is *similar* in the male and female populations.

Observed: a 2-D histogram

The key to comparing two 1-D observed histograms is to see that the two 1-D histograms are really *one* 2-D histogram. The problem of comparing two observed histograms to each other is represented by creating one 2-D histogram, made up of the two 1-dimensional histograms of the two observed populations, and then comparing that 2-D histogram to another 2-dimensional histogram of expectations under the Null Hypothesis of no difference between the two groups.

We convert the two 1-D histograms into one 2-D histogram by using the group identifier (male/female, drug/placebo, vaccinated/unvaccinated, Titanic ticket 1st/2nd/3rd class) as one axis, and the outcome (heart disease/not, lived/didn't, Democrat/Republican/Independent etc.) as the other axis.

Let's discuss this procedure in detail for the case of gender and heart disease.

Incidence of heart disease among males and females Is heart disease more prevalent in men than it is in women?

A study interviewed a number of subjects, and for each subject, recorded whether they were male or female, and whether or not they had heart disease (Janosi and Detrano, 1988).

In the male group, they found 114 males with heart disease and 92 males with no heart disease. In the female group, they found 25 females with heart disease, and 72 females with no heart disease.

The standard presentation of data like this is as a "2 × 2" table. We use the group identifier "male/female" as one axis, and the outcome "heart disease/not" as the other axis. Then we can turn the two observed 1-D histograms into one 2-D histogram. **Essentially, the 2-D histogram is the "2 × 2" table, with bars whose heights represent the count in each cell.**

Figure 8.27 illustrates this procedure. The new 2-D histogram becomes our observed.

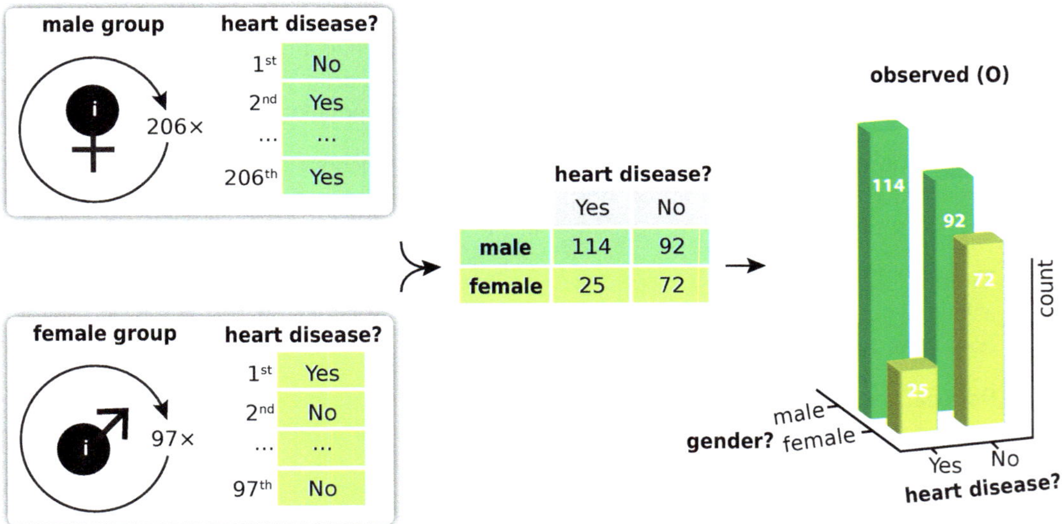

Figure 8.27 Assembling the 2-D observed histogram for the gender and heart disease study.

The 2 × 2 table is called a 2 × 2 **contingency table** in the literature. Contingency tables can have any number of dimensions m × n, where m is the number of groups and n is the number of outcomes.

The next question is: Now we have a 2-D histogram of observations. To what expectations do we compare this histogram?

Expected: a 2-D histogram

We construct the 2-D histogram of expectations by assuming that the Null Hypothesis (no difference) is true, and asking what we would expect to see if there were no difference between the groups.

In the gender/heart disease case, we are asking what we would expect to see if gender had no effect on heart disease. **Therefore, we generate our expectations by simply forgetting the gender of our subjects.** We have a total of 114 + 25 = 139 cases of heart disease, and 92 + 72 = 164 cases of no heart disease. **This means that the overall probability in our population, ignoring gender, of having heart disease, is $\frac{139}{139+164}$ or 46%** (Figure 8.28).

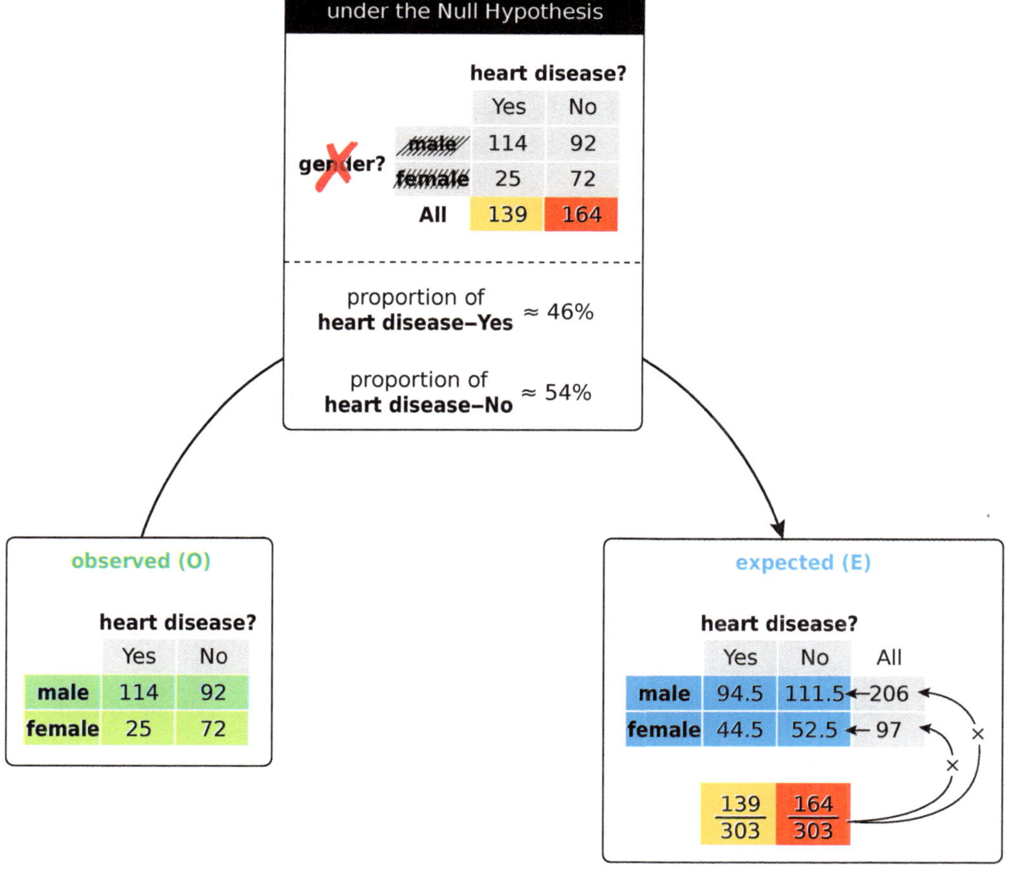

Figure 8.28 Generating the expected (E) histogram from the observed (O) histogram under the Null Hypothesis that gender does not play a role in the occurrence of heart disease. Note both observed and expected are 2-D histograms.

Now, **if heart disease were truly independent of gender, we would expect that 46% probability to fall equally on males and females.**

Therefore, the number of female cases of heart disease we would expect to see is 46% of the total number of females. 46% of 97 is 44.5, so we would expect, under the Null Hypothesis, to see 44.5 cases of heart disease.

For males, under the Null Hypothesis, we would also expect to see 46% of the total number of males to have heart disease. 46% of 206 is 94.5, so we would expect, under the Null Hypothesis, to see 94.5 cases of heart disease in the male population. Therefore, our expectations for the four categories are as in Figure 8.28.

χ_{obs}: comparing the observed and expected 2-D histograms

χ-statistic for 2-D histograms

If there are k cells in each 2-D histogram, then the distance χ between a 2-D observed histogram and its expected histogram is the sum of the normalized differences, for each cell, from the observed to its expected.

$$\chi = \sum_{i=1}^{k} \frac{|O_i - E_i|}{E_i} = \sum_{i=1}^{k} \frac{|\text{Observed}_i - \text{Expected}_i|}{\text{Expected}_i}$$

For example, if we denote each cell in the observed 2-D histogram as a, b, c and d, and corresponding frequency in the expected 2-D histogram as a_E, b_E, c_E, d_E.

observed (O)

	outcome?	
	Yes	No
group 1	a	b
group 2	c	d

expected (E)

	outcome?	
	Yes	No
group 1	a_E	b_E
group 2	c_E	d_E

Then applying the definition of χ, the distance between the observed and the expected is

$$\chi = \sum_{i=1}^{k} \frac{|O_i - E_i|}{E_i} = \frac{|a - a_E|}{a_E} + \frac{|b - b_E|}{b_E} + \frac{|c - c_E|}{c_E} + \frac{|d - d_E|}{d_E}$$

We can now compare the observed histogram to the expected, using the χ-measure. The comparison now has 4 terms, one for each cell in the histograms. The observed difference χ_{obs} is 1.2 (Figure 8.29).

But we do not know if $\chi = 1.2$ is statistically significantly large or not. We need to do NHST for χ_{obs}.

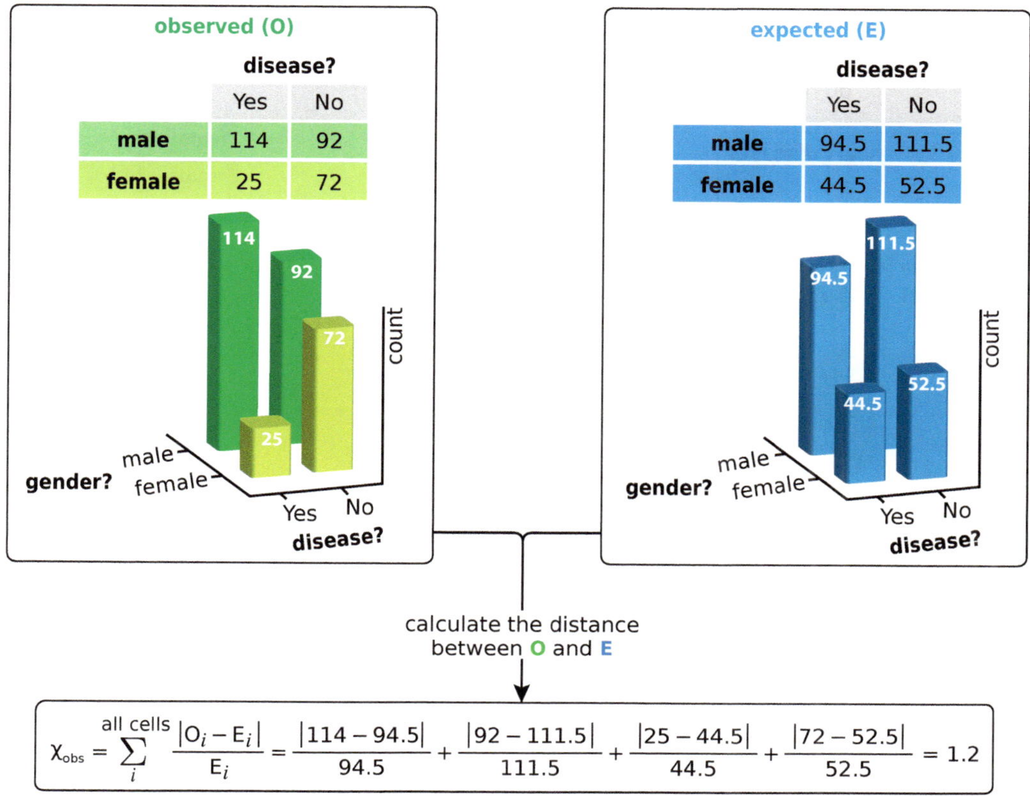

Figure 8.29 In the study of heart disease and gender, after constructing the observed 2-D histogram (left) and the expected 2-D histogram (right), we calculate χ_{obs}, the distance between the observed and the expected (bottom).

NHST for χ_{obs}

For the heart disease and gender study, we will use a somewhat different method of randomization, resampling *without* replacement, also called a permutation test, or in colloquial language "shuffle, deal, and count". We will explain why we are using this method a little later.

Big Deck We start with the observed histogram, and imagine that we have a deck with 139 "heart disease" cards and 164 "no heart disease" cards. This is the "Big Deck" that embodies the Null Hypothesis.

Shuffle, deal, and count We shuffle this Big Deck, and then deal out the cards into two piles to form two new groups, each with the same total number of cards as in the observed samples. This is our first estimate of what the male and female groups would look like under the Null Hypothesis. We assemble this estimate into a 2-D histogram, calling it the pseudo-observed (O_1). We then compute χ for this first randomized histogram with respect to the expected histogram, call it χ_1 (Figure 8.30).

The "shuffle, deal, count" method is equivalent to sampling without replacement. It is also called a permutation test.

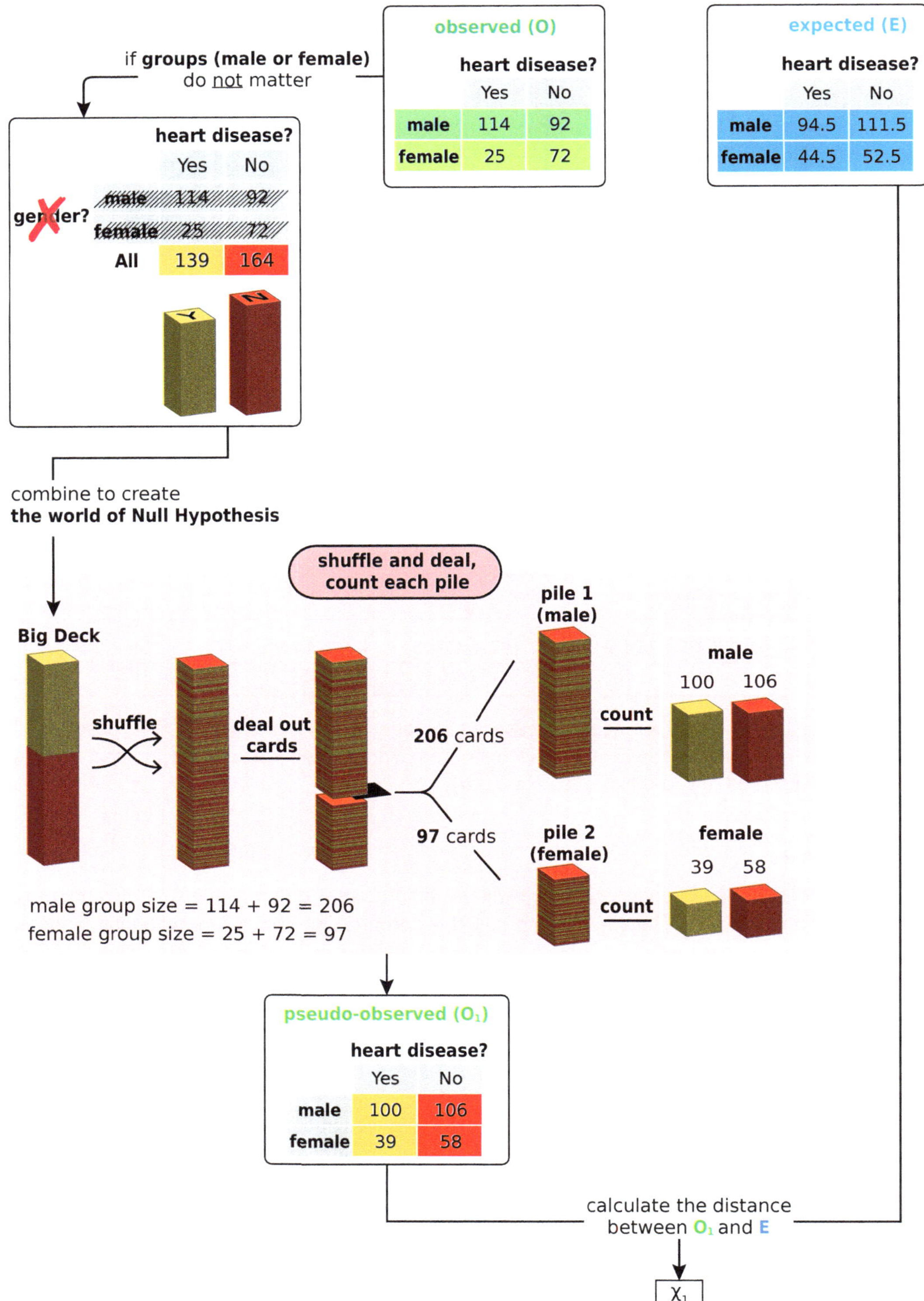

Figure 8.30 Generating the first estimate χ_1 by "shuffle, deal, and count" from the world of the Null Hypothesis. Here we illustrate this process using the heart disease and gender study.

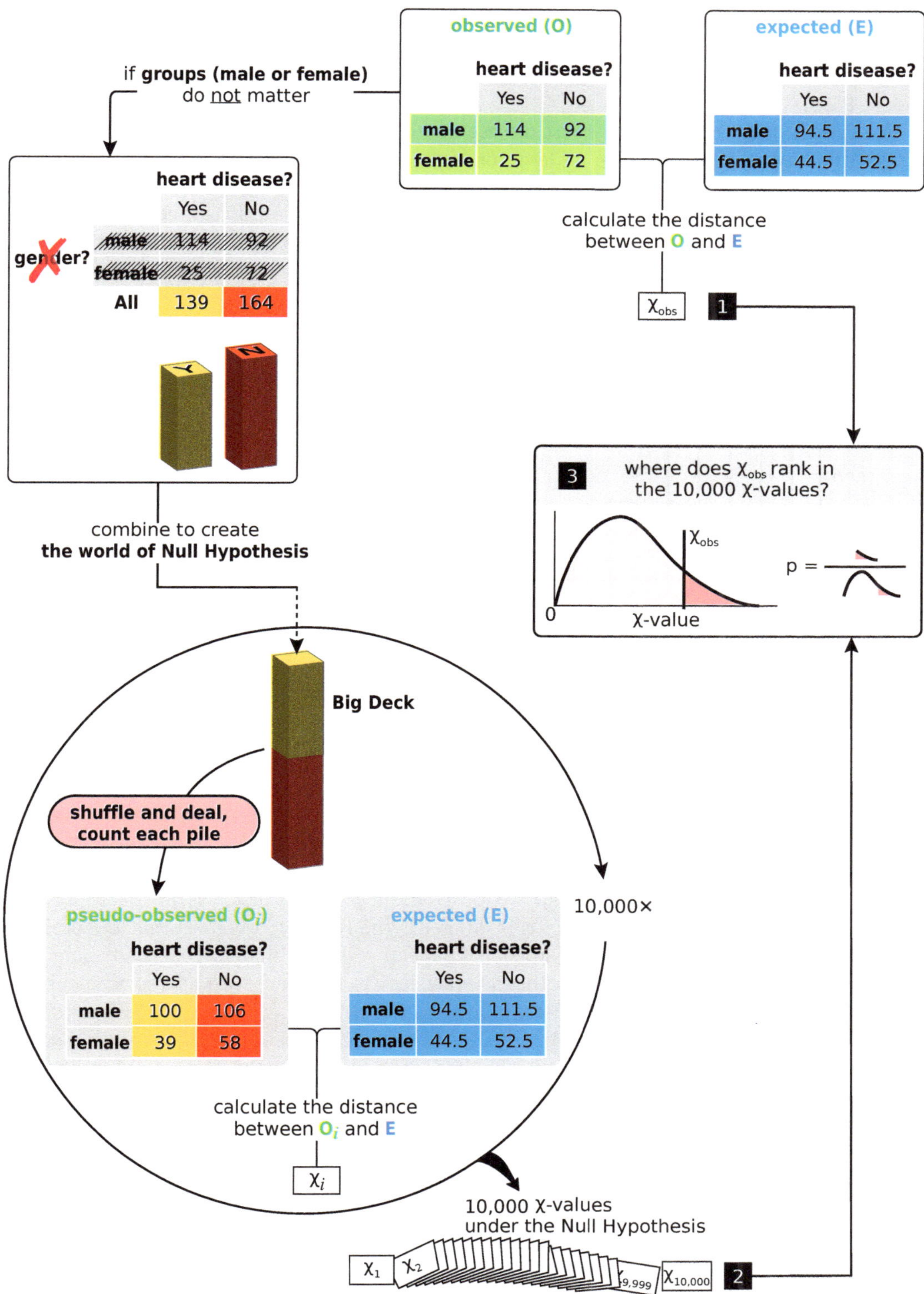

Figure 8.31 NHST for χ_{obs}. Shuffle the world of the Null Hypothesis 10,000 times to generate the null distribution and find the p-value for the observed difference χ_{obs}.

p-value This process is then repeated 10,000 times, and, as usual, we ask where did our actual χ_{obs} rank in the 10,000 χ-values generated under the Null Hypothesis. The number of χ-values greater than or equal to the observed, divided by 10,000, is the p-value of our finding (Figure 8.31).

We carried out NHST for the heart disease and gender study. We found that none of the 10,000 simulations had a χ-value as large or larger than our observed χ_{obs}, so $p < 0.0001$ (Figure 8.32).

Therefore, we can reject the Null Hypothesis, and state that, in this limited sample, gender is associated with heart disease, with men having a higher incidence ($p < 0.0001$).

Please note that this conclusion is far from conclusive. A number of other, larger studies argue for a heavier burden in females, and in fact, historically, women were often neglected in disease studies. Here we are only using this small study to illustrate the procedure of NHST.

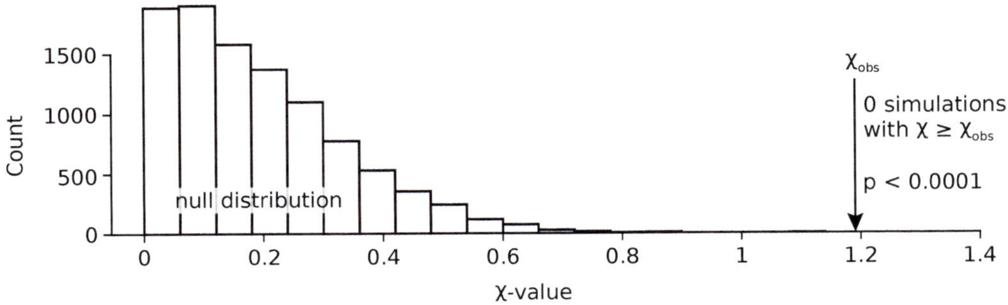

Figure 8.32 NHST for the heart disease and gender study. The Null Hypothesis that there is no gender difference in heart disease is rejected at $p < 0.0001$.

Shuffle, deal, and count (sampling without replacement)

Previously, when we were comparing one 1-D observed histogram to expectations, we did NHST for χ_{obs} by sampling *with* replacement from the world of the Null Hypothesis, that is, from the expected histogram (see section 8.3 and Figure 8.15).

However, when we are comparing two observed histograms (such as in the heart disease/gender study), the Null Hypothesis is different: the Null Hypothesis is that the proportion of heart disease is the same regardless of gender. We use "shuffle, deal, and count" to randomize the world of the Null Hypothesis, equivalent to sampling *without* replacement.

First, we ignore the group identifiers, producing two sets of cards: one marked "outcome = yes" and the other marked "outcome = no". We combine the two sets of cards into a deck. Then we shuffle this deck and deal out the cards into two piles (the same number of piles as the number of groups, with each pile having the same number of cards as the original group). We sort and count the cards in each pile into a 2-D histogram. This represents one implementation of the Null Hypothesis.

The "shuffle, deal, and count" randomization technique keeps the proportions of each outcome (in total, ignoring groups) the same throughout each resample. In other words, because the proportions of each outcome when group identifiers are ignored *is* the Null Hypothesis, the Null Hypothesis is respected in each resample. Thus, the "shuffle, deal, and count" is asking the question "given this *fixed* proportion of heart disease in the combined observed sample, what is the likelihood of a gender imbalance this large or larger occurring at random?"

On the contrary, if we had sampled *with* replacement, the Null Hypothesis becomes that the observed overall heart disease proportion is not fixed, but is a random sample from an underlying population whose disease proportions are unknown. As a consequence, the heart disease proportion will vary from one sample to another. We could have drawn more heart-disease cards in one sample and observed a higher overall disease proportion, but have drawn fewer heart-disease cards in another sample and resulted in a lower proportion. It is the direct result of sampling *with* replacement that we can draw the same card multiple times. Therefore, the "shuffle, deal, and count" Null Hypothesis of "fixed overall disease proportions" is violated, and the 10,000 samples will not have the same proportions as in the original study. As an extreme case, if we sampled with replacement, it would be possible to have a sample with a disease incidence of 100%!

In conclusion, sampling with replacement and without replacement test different Null Hypotheses (Fieberg *et al.* (2020)).

- "shuffle and deal" tests the Null Hypothesis that the overall proportion of diseased people is fixed in each resample, and the fixed number of "disease" and "no disease" cards is randomly dealt to each group.
- By contrast, resampling with replacement tests a different, broader Null Hypothesis, because we are allowing the proportion of diseased people in each resample to vary around the observed proportion of diseased people.

The Null Hypothesis of shuffle-and-deal is that the variability in outcome is strictly the result of the dealing process, whereas the Null Hypothesis of resampling is that both dealing and sample selection are allowed to vary from sample to sample.

The difference between the two methods is not very large, but it can be important. For the heart disease and gender example, the null distribution by using "shuffle and deal" method has many more very small χ-values (see figure below). This is because the variability in the samples is constrained by the requirement of fixed overall disease proportion, so the samples will resemble the observed sample more closely than if the proportion had been allowed to vary, resulting in more small χ-values.

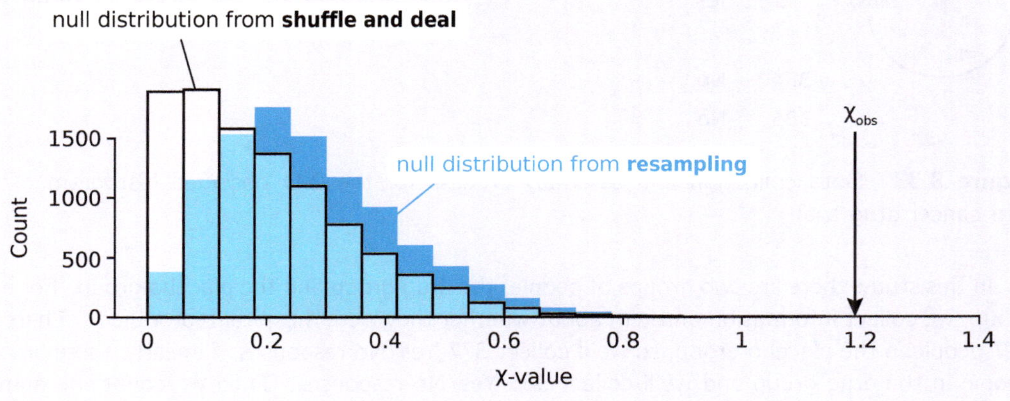

Clinical trial

Does our anti-cancer drug work? A study of the effectiveness of an anti-cancer drug found that in the placebo group (n = 377), 53 patients, or 14.1% were hospitalized or died. For those who received the drug (n = 385), 28, or 7.3% were hospitalized or died. The company claimed that their drug "significantly reduced the risk of hospitalization or death".

In the beginning of this section, we talked about how the question of whether the drug reduced hospitalization/death can be interpreted as the comparison of two observed 1-D histograms, so the question whether the drug group is statistically significantly different from the placebo group can be turned into the comparison of the 2-D observed histogram to the 2-D expected histogram.

Observation The data collection and assembly process is straightforward (Figure 8.33).

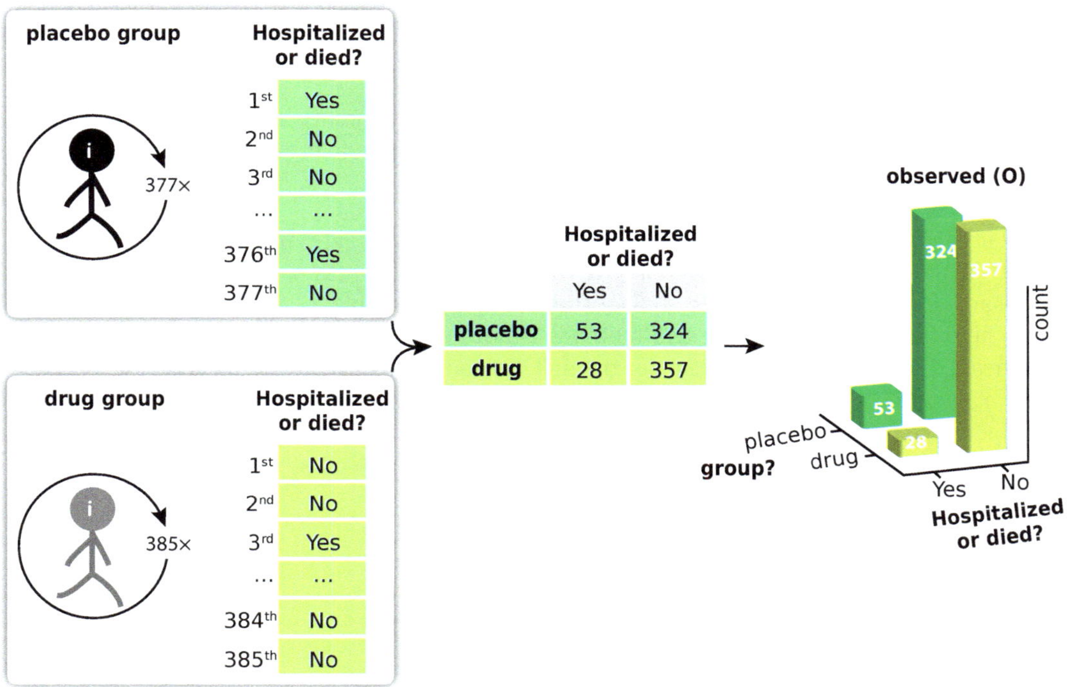

Figure 8.33 Data collection and assembly process for the 2-D observed histogram of the anti-cancer drug trial.

In this study, there are two groups of people, the drug group and the placebo group. For each group, we collect information on them about whether they were "hospitalized or died". There are 377 people in the placebo group, so we'll collect 377 Yes/No responses. Similarly, there are 385 people in the drug group and we'll collect 385 Yes/No responses. Then we count the number of Yes and No in each group and assemble them into a 2 by 2 table, or a 2-D histogram. In this 2-D histogram, we use the group identifier, placebo/drug, as one axis, and the information we collect on each person, Yes/No with regard to "hospitalized or died", as the other axis.

Expectation We generate our expectations by simply forgetting the group labels (drug or placebo) of our subjects (Figure 8.34).

Therefore, we can say that the total number of hospitalized or died is 53 + 28 = 81. Divided by the 762 people in the survey, that means an overall risk of hospitalization or death of $\frac{81}{762} \approx$ 11%.

Applying that 11% risk to the 377 in the placebo group, we would expect to see 40.1 deaths in the placebo group. Applying that 11% risk to the drug group, we would expect to see $385 \times \frac{81}{762} = 40.9$ deaths. We can now construct the expected counts under the Null Hypothesis, and form the expected histogram.

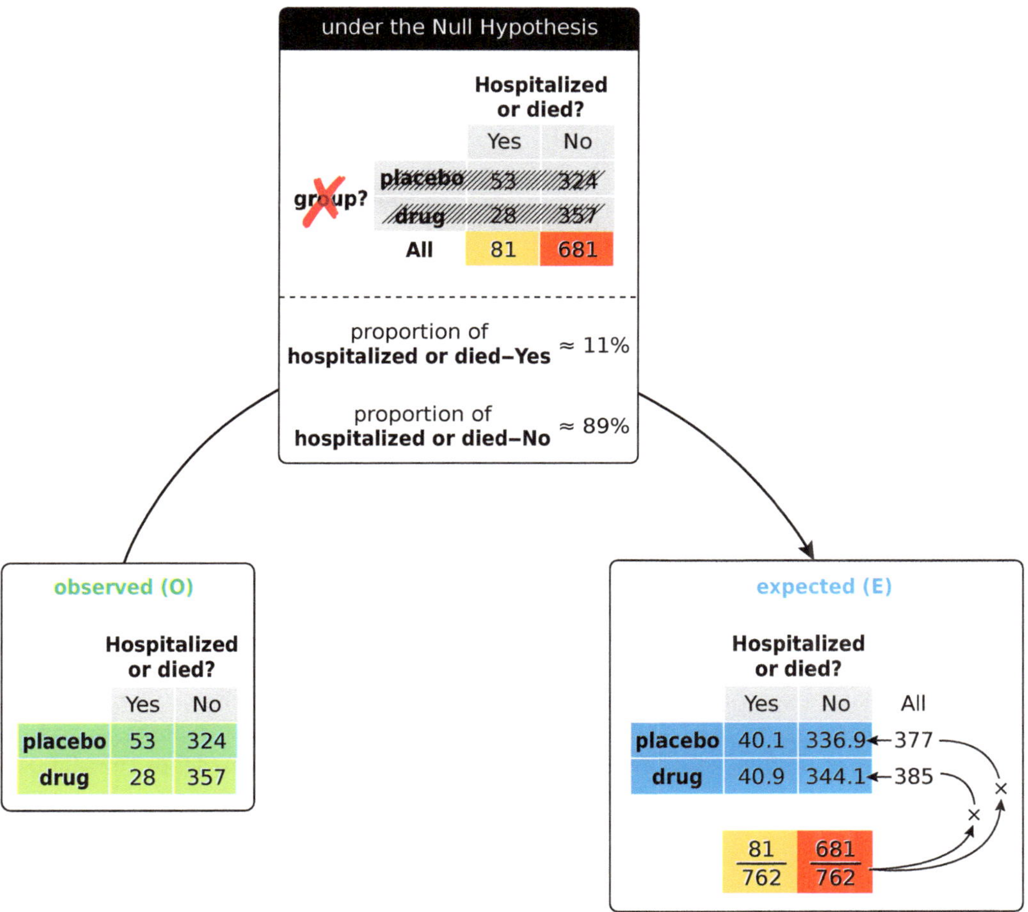

Figure 8.34 Deriving the expected (E) histogram from the observed (O) histogram under the Null Hypothesis, that the group a person belongs to (drug or placebo) does not play a role in the outcome (hospitalized/died or not).

χ_{obs} After constructing the observed 2-D histogram and the expected 2-D histogram, we compare the observed to the expected histogram by calculating χ_{obs}, the distance between the observed and the expected (Figure 8.35).

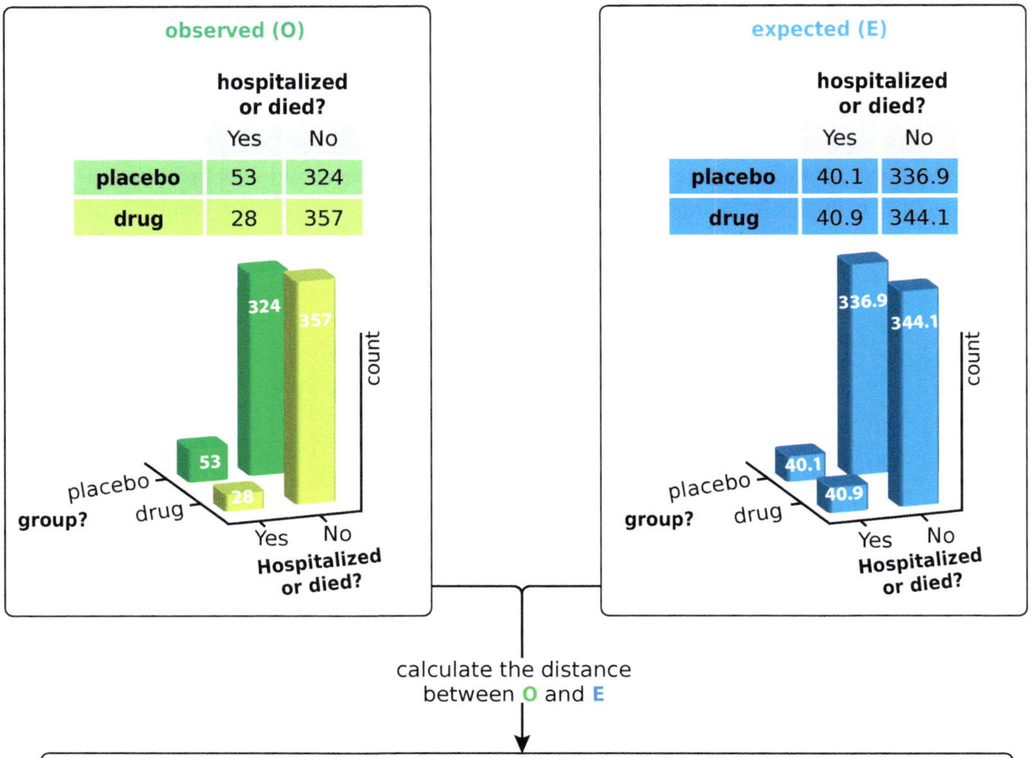

observed (O)

	hospitalized or died? Yes	No
placebo	53	324
drug	28	357

expected (E)

	hospitalized or died? Yes	No
placebo	40.1	336.9
drug	40.9	344.1

calculate the distance between O and E

$$\chi_{obs} = \sum_{i}^{\text{all cells}} \frac{|O_i - E_i|}{E_i} = \frac{|53-40.1|}{40.1} + \frac{|324-336.9|}{336.9} + \frac{|28-40.9|}{40.9} + \frac{|357-344.1|}{344.1} = 0.71$$

Figure 8.35 Calculating the observed distance χ_{obs} for the anti-cancer drug trial.

NHST for χ_{obs} To see if the observed difference $\chi_{obs} = 0.71$ is statistically significantly large, we carry out resampling-based NHST (Figure 8.36).

We found that 14 out of the 10,000 simulations had a χ-value as large or larger than the observed difference χ_{obs}, so $p = 0.0014$. Therefore, we can reject the Null Hypothesis, and state that the anti-cancer drug is associated with lower rates of hospitalization/death ($p = 0.0014$).

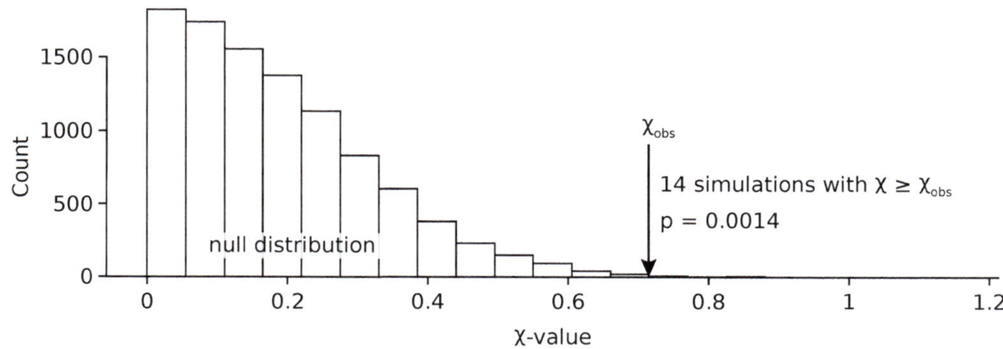

Figure 8.36 NHST for the anti-cancer drug trial.

8.6 Survival on the Titanic

When the Titanic collided with an iceberg and sank in 1912, many lives were lost. The survivors came from all 3 classes of ticket: First class, Second class, and Steerage (Third class). It was asked at the time whether survival on the Titanic was *independent* of ticket class, and therefore, independent of socioeconomic class.

Observation The raw statistics of survival by class on the Titanic are:

	survived? Yes	No	
1st class	201	123	324
2nd class	118	166	284
3rd class	181	528	709
	500	817	1317

Figure 8.37

Expectation The Null Hypothesis is that the likelihood of survival is the same for all classes.

In total, we see that there are 500 people who survived and 817 people who did not. This gives us the overall proportions of people who survived and people who did not.

Therefore, under the Null Hypothesis "same likelihood of survival in all classes", we can multiply the number of people in each class by the overall proportions of people who survived ($\frac{500}{1317}$) and people who did not ($\frac{817}{1317}$), to calculate the expected counts of people in each class.

For example, there are 709 people in the 3rd class, and under the Null Hypothesis, we would expect to see $709 \times \frac{500}{1317}$, or 269.2 people to have survived, and $709 \times \frac{817}{1317}$, or 439.8 people to have died.

We carry out the same calculation for each class and construct the expected histogram (Figure 8.38).

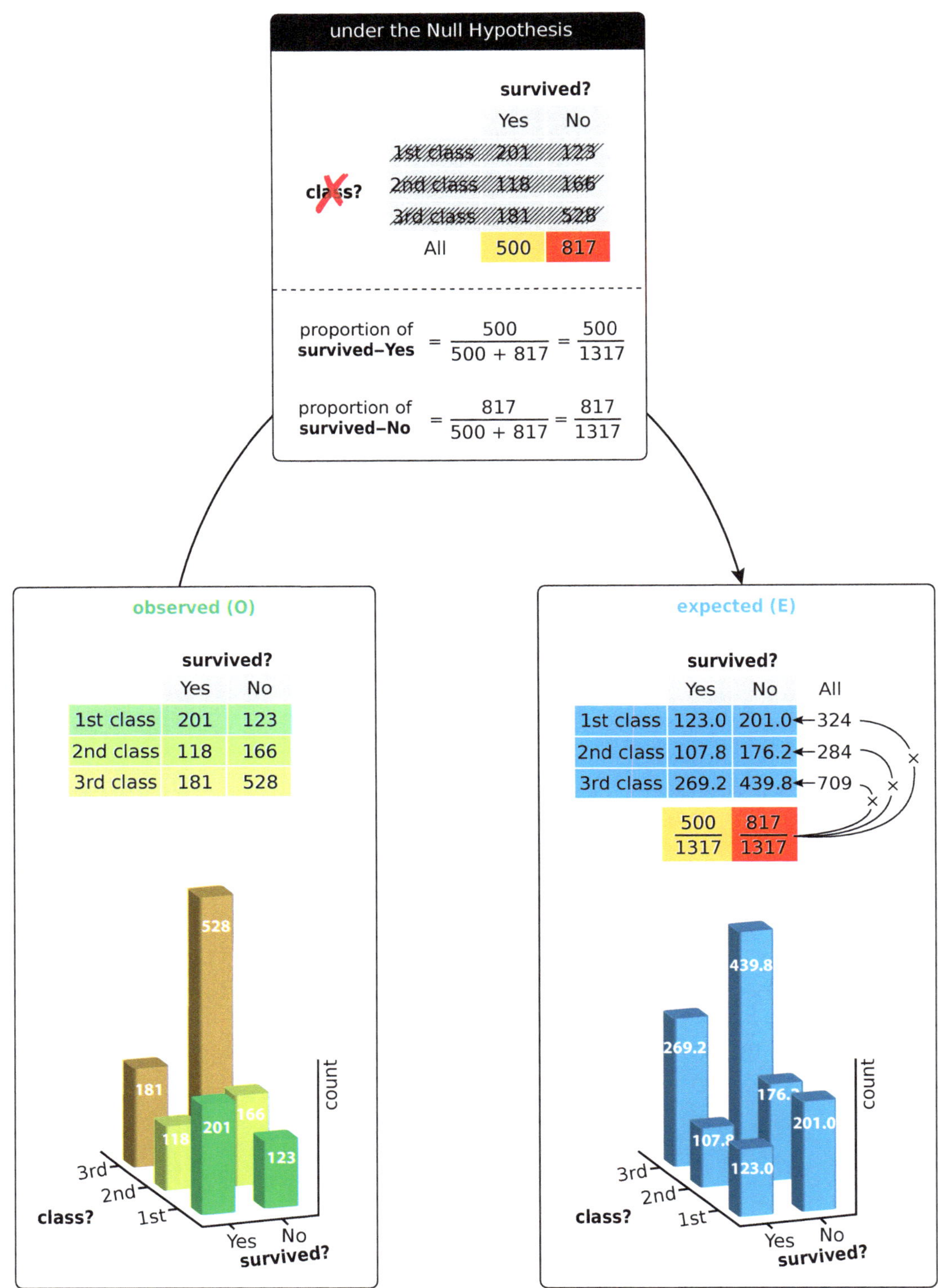

Figure 8.38 Generating the expected histogram from the observed for the Titanic data.

χ_{obs} After generating the observed 2-D histogram and the expected 2-D histogram, we calculate the distance from the observed histogram to the expected.

In the Titanic data, we have 6 cells in total,

$$\begin{aligned}
\chi_{\text{obs}} &= \left|\frac{O_{\text{1st class, survived}} - E_{\text{1st class, survived}}}{E_{\text{1st class, survived}}}\right| + \left|\frac{O_{\text{1st class, died}} - E_{\text{1st class, died}}}{E_{\text{1st class, died}}}\right| \\
&+ \left|\frac{O_{\text{2nd class, survived}} - E_{\text{2nd class, survived}}}{E_{\text{2nd class, survived}}}\right| + \left|\frac{O_{\text{2nd class, died}} - E_{\text{2nd class, died}}}{E_{\text{2nd class, died}}}\right| \\
&+ \left|\frac{O_{\text{3rd class, survived}} - E_{\text{3rd class, survived}}}{E_{\text{3rd class, survived}}}\right| + \left|\frac{O_{\text{3rd class, died}} - E_{\text{3rd class, died}}}{E_{\text{3rd class, died}}}\right| \\
&= \left|\frac{201-123.0}{123.0}\right| + \left|\frac{123-201.0}{201.0}\right| + \left|\frac{118-107.8}{107.8}\right| + \left|\frac{166-176.2}{176.2}\right| + \left|\frac{181-269.2}{269.2}\right| + \left|\frac{528-439.8}{439.8}\right| \\
&= 1.7
\end{aligned}$$

NHST for χ_{obs} We create the world of the Null Hypothesis and generate the first sample O_1 from it.

The world of the Null Hypothesis consists of 500 'life' cards and 817 'death' cards (the Big Deck). After shuffling the Big Deck, we cut it into 3 piles:

Pile 1 has the same number of cards as the number of people in 1st class: 324

Pile 2 has the same number of cards as the number of people in 2nd class: 284

Pile 3 has the same number of cards as the number of people in 3rd class: 709

Then we sort and count the cards in each of the three piles, and assemble the counts into a 2-D histogram, calling it O_1. This is the first implementation of the Null Hypothesis. We then calculate the distance from O_1 to the expected, calling it χ_1 (Figure 8.39).

We repeat this "shuffle and deal, then count cards in each pile" procedure 10,000 times to generate 10,000 histograms of the shuffled data. For each of these, we calculate its distance to the expected, generating 10,000 χ-values: $\chi_1, \chi_2, \ldots, \chi_{10,000}$. These 10,000 χ-values form the null distribution (Figure 8.40).

From the null distribution, we see that the observed difference χ_{obs} is far outside the range of the 10,000 χ-values from the 10,000 shufflings. In fact, 0 simulations out of the 10,000 had a χ-value as large as or greater than χ_{obs}. This tells us that the observed histogram is statistically significantly different from the expected under the Null Hypothesis. We can therefore reject the Null Hypothesis and state that survival on the Titanic was *not* independent of ticket class, and therefore, *not* independent of socioeconomic class.

This is a result, but it only tells us *that* there is an overall difference between the observation and expectation, but is silent on *which* group(s) differ, if any. It is, therefore, a kind of omnibus test. For the question of *which* group(s) differ, we need to make a *post hoc* test, which we'll do next, following the procedure in section 8.4 (*post hoc* testing: "Mt. Fuji plots").

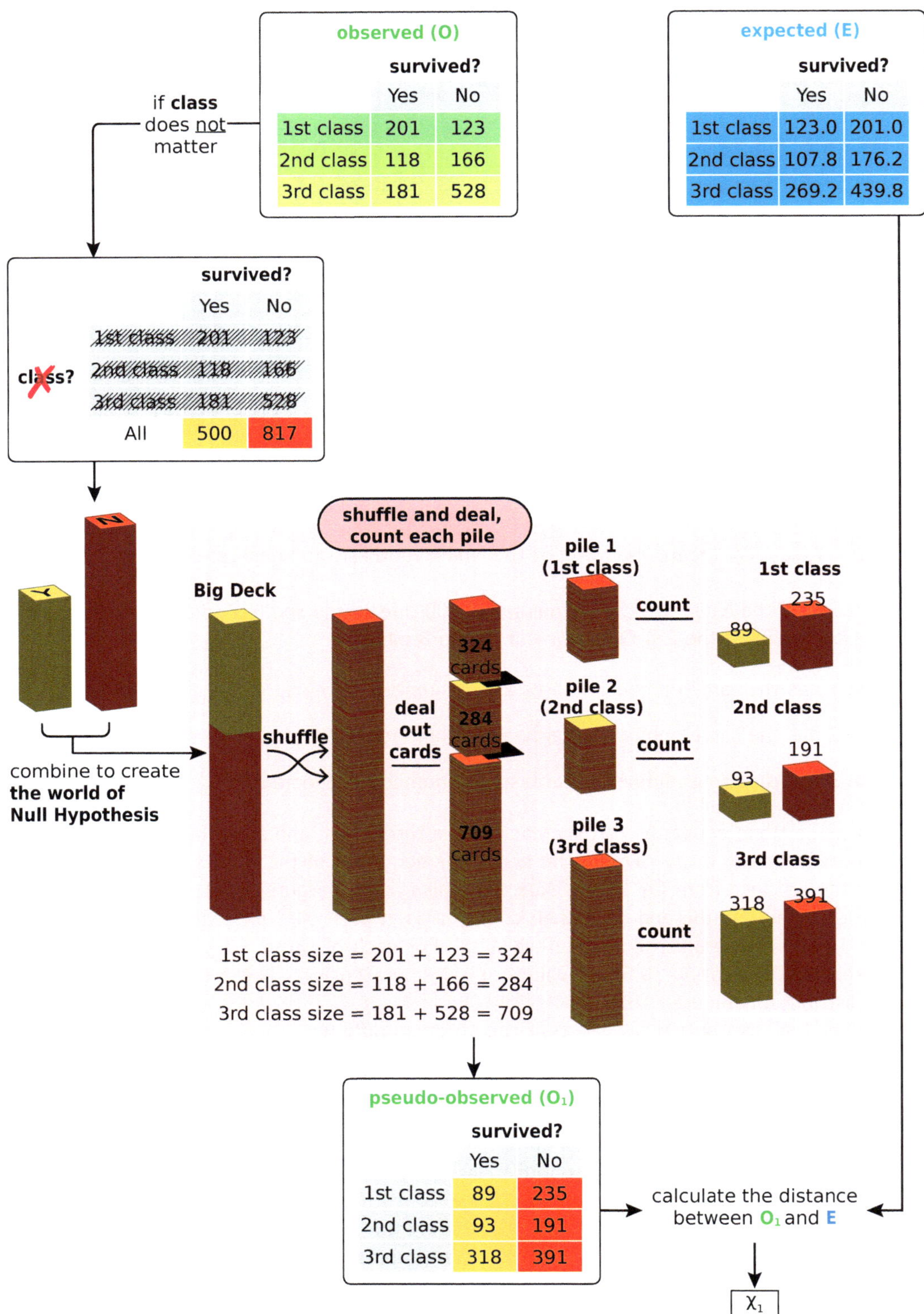

Figure 8.39 Generating the first estimate, χ_1, under the Null Hypothesis for the Titanic data.

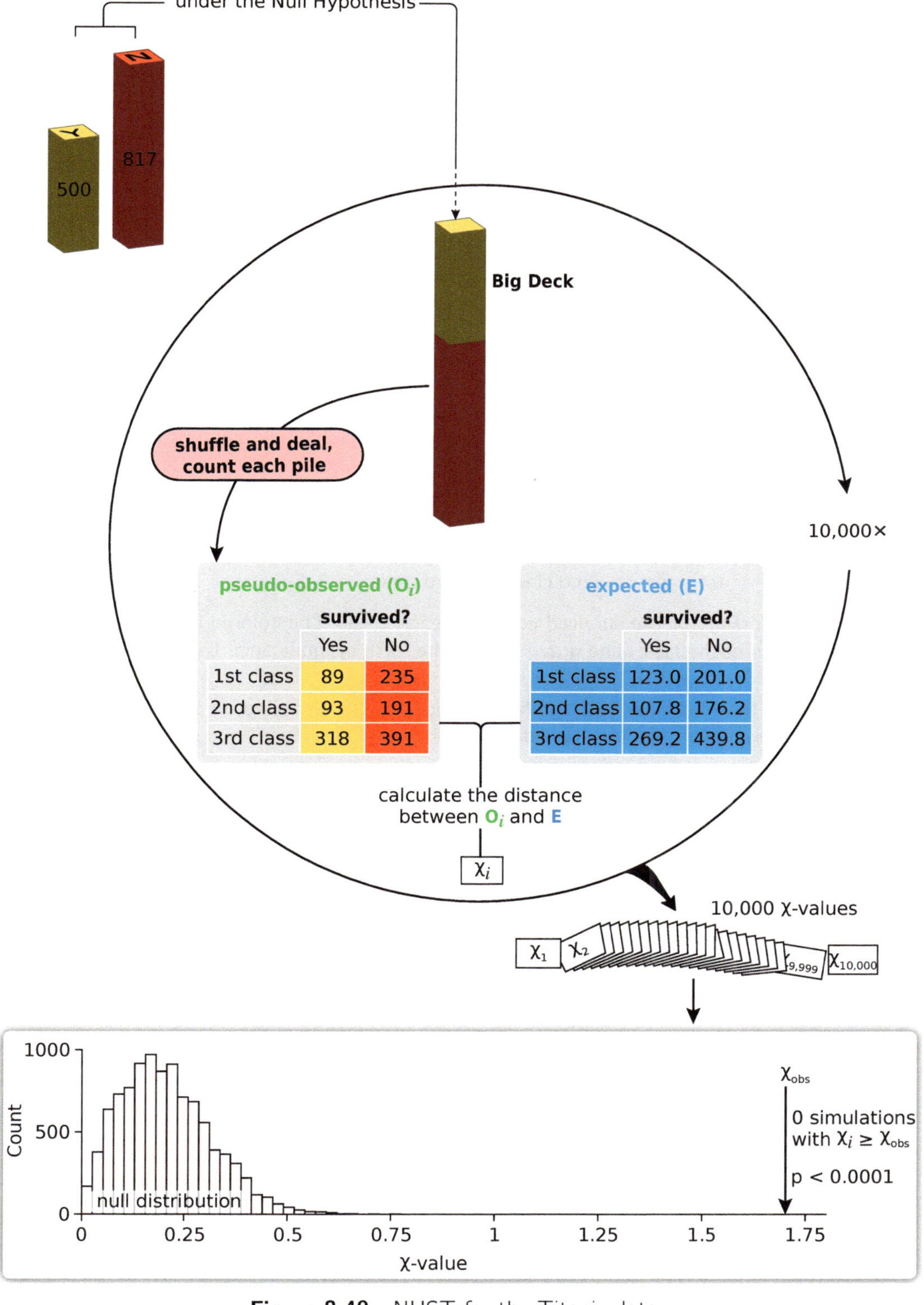

Figure 8.40 NHST for the Titanic data.

post hoc test: Mt. Fuji plot For the Mt. Fuji plot, we will sample with replacement from the expected histogram many times to generate a "band of insignificance" or "no-rejection zone" for each cell, representing the range of values which are not significantly different from the observed. We will then compare the observed histogram to these bands, to see where the observed histogram "sticks out" above or below the clouds of uncertainty.

We begin with the 10,000 shufflings of the world of the Null Hypothesis, which we carried out to do NHST. This gives us 10,000 values for each cell of the observed histogram. Now, we are going to form a band of insignificance for each cell by counting 50 up from the smallest and 50 down from the largest as the cutoffs for statistical significance at the 0.01 level. We constructed the Mt. Fuji plot for the survival counts (Figure 8.41).

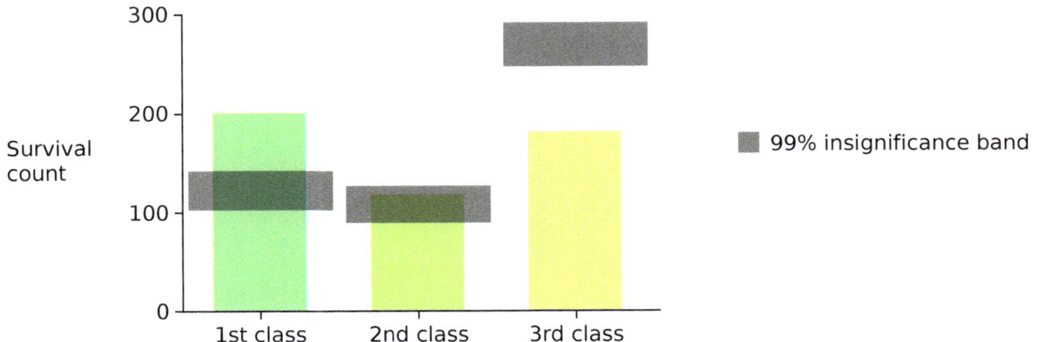

Figure 8.41 Mt. Fuji for the survived counts in each class. The colored bars are the observed survived counts in each class. The gray bars are the 99% insignificance band.

We conclude that 1st class had significantly more people surviving, in contrast to 3rd class, where significantly fewer people survived. Deaths in 2nd class were not significantly different from expectations.

Of course, we are making multiple comparisons, and so must correct for multiple testing. For 1st class and 3rd class, both have $p < 0.0001$. So they will survive not only the Benjamini-Hochberg correction but even the more stringent Bonferroni correction, and our conclusion stands.

Note that, even with the Mt. Fuji plot, this result is still only a Yes/No question. The survival rates are either significantly different (either higher or lower) from expectations or they are not. We have been saying all along that NHST needs to be supplemented by a notion of effect size, together with confidence intervals. We will soon discuss an important effect size, called "relative risk" (see section 8.8 *Relative Risk: an effect size*).

FURTHER EXERCISES 8.6

1. **Fecal transplant and *Clostridium difficile*.** Infection with the bacterium *Clostridium difficile* causes persistent diarrhea, and is difficult to treat. Failure rates of standard antibiotic therapy are high. To test a new therapy using fecal transplants, a group of researchers set out to run a clinical trial. 42 patients were recruited. They were randomly assigned to receive one of three therapies: infusion of a solution of feces from a healthy donor,

the standard antibiotic vancomycin, or the standard vancomycin plus a bowel lavage. Of 16 patients in the infusion group, 15 were cured. 4 out of 13 patients who received the standard vancomycin alone were cured. 3 of 13 patients receiving vancomycin with bowel lavage were cured (see the article "Duodenal Infusion of Donor Feces for Recurrent Clostridium difficile", *The New England Journal of Medicine*, 2013).

a. Write the observed count for each category.
b. State an appropriate Null Hypothesis.
c. Write the expected count for each category under the Null Hypothesis.
d. Calculate the distance between observed and expected.
e. Describe the procedure to conduct NHST for this study, and calculate a p-value. What's your interpretation of the p-value?
f. In order to find out which of the three therapies works, what test do you need to further conduct? Present and interpret your result.

2. **The lady smelling Parkinson's.** Joy Milne claimed she could detect the scent of Parkinson's. To test Milne's sense of smell, she was asked to classify 12 shirts worn by either healthy individuals or Parkinson's patients. The results were:

	Diseased	Healthy
Smell positive	6	1
Smell negative	0	5

a. If Joy Milne can't detect Parkinson's by smell, and is merely guessing randomly, what would be the expected counts in the 2 × 2 table?
b. Calculate the distance between the observed and expected.
c. How would you test her claim?

3. **Race and police searches.** Here is the racial makeup of motorist searches recorded in Maryland from January 1995 through June 2000. Out of the total 8,027 searches, 2,146 were on the major north-south interstate highway, the I-95 Corridor. Can you conclude that the police engaged racial profiling in choosing who to search in the I-95 Corridor? If yes, explain why. If not, what further information would you need? (see article "Road work, racial profiling and drug interdiction on the highway", *Michigan Law Review*, 2002.)

	I-95 Corridor (2,146)	**Elsewhere** (5,881)
White	33.3%	62.6%
Black	59.7%	32.2%
Hispanic	5.9%	3.8%
Other	1.1%	1.4%
Total	100%	100%

8.7 χ in action: comparing US and UK populations

In Chapter 2, we talked about how to describe and present data. One of our examples was looking at the distribution of generations in the US and the UK and trying to answer some simple questions, such as comparing the relative size of a particular age cohort in the two countries (Example 5 on page 25).

We presented the data as two histograms, which we reprint here.

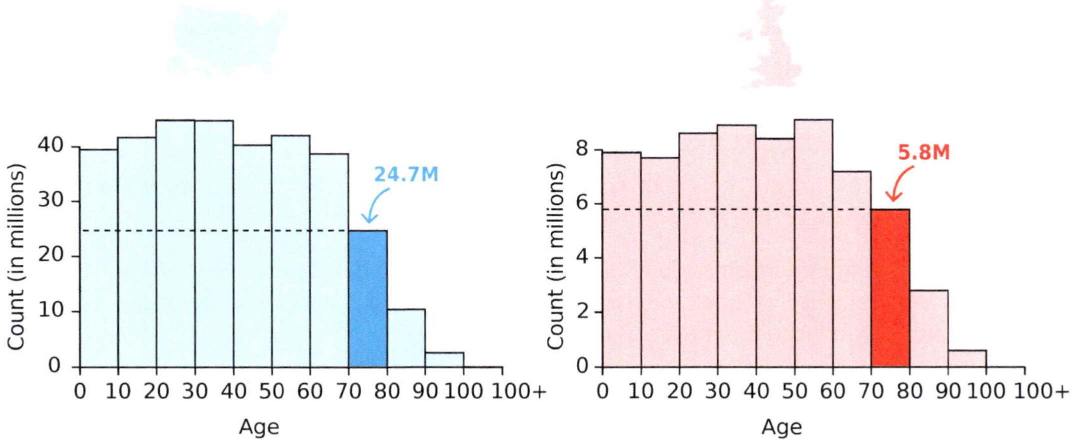

Figure 8.42 Distribution of generations in the US and the UK.

At the time, we didn't have the tools we have now, so we didn't know how to compare two observed histograms to answer questions like: is the distribution of generations in the US different from the UK? If so, *where* are the differences?

Instead, we normalized those two histograms into frequencies, and constructed two normalized histograms, read out the relative percentages of a particular age cohort in the two populations. We said that "From the normalized histogram, we easily read off that people in their 70s constitute 7.5% of the US population and 8.6% of the population in the UK." However, this is only describing what we see after presenting the data; we're not answering the question whether the two countries' histograms are significantly different.

We now want to ask: is the difference between 7.5% and 8.6% "statistically significant"? That is, is this difference "large" compared to what we would expect due to random sampling? What do we mean by "random" here? And what about the other age cohorts?

We now know how to compare two histograms, and we are going to apply what we've learned.

Observation First, let's recognize that we are comparing two observed histograms (section 8.5), and recall that **the key to comparing two observed histograms is to see that the two 1-D histograms are really *one* 2-D histogram**. We are now going to do that, and convert the two 1-D histograms into one 2-D histogram (Figure 8.43).

Expectation We now have our (2-D) histogram of observed counts. The next step is to construct a 2-D histogram of expected values, so we can compare it to the observed histogram. But what does "expectation" mean? Expectation means "Expectation under the Null Hypothesis" and here the Null Hypothesis is: the distribution of generations is independent of country. In

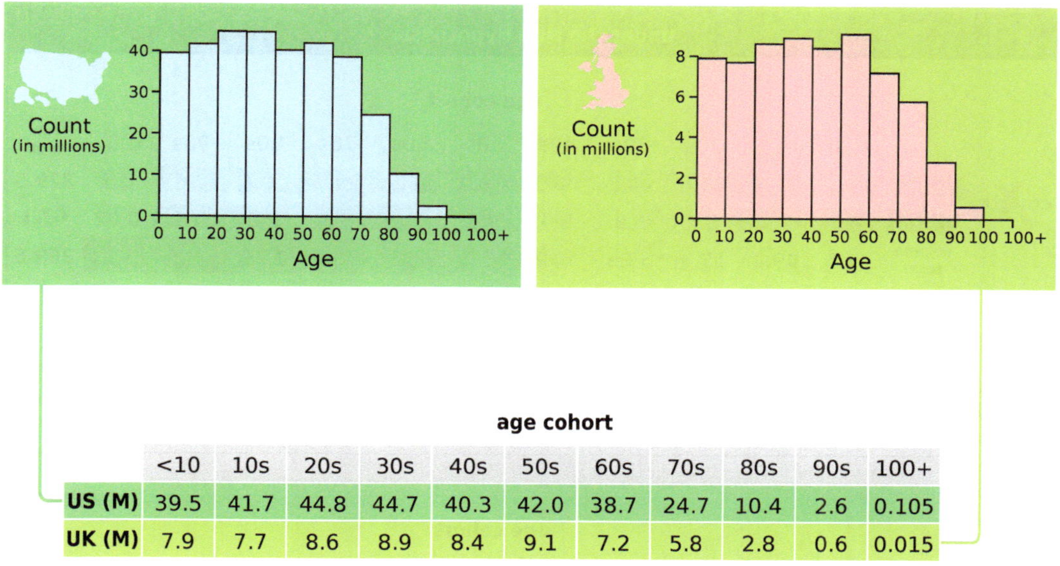

	<10	10s	20s	30s	40s	50s	60s	70s	80s	90s	100+
US (M)	39.5	41.7	44.8	44.7	40.3	42.0	38.7	24.7	10.4	2.6	0.105
UK (M)	7.9	7.7	8.6	8.9	8.4	9.1	7.2	5.8	2.8	0.6	0.015

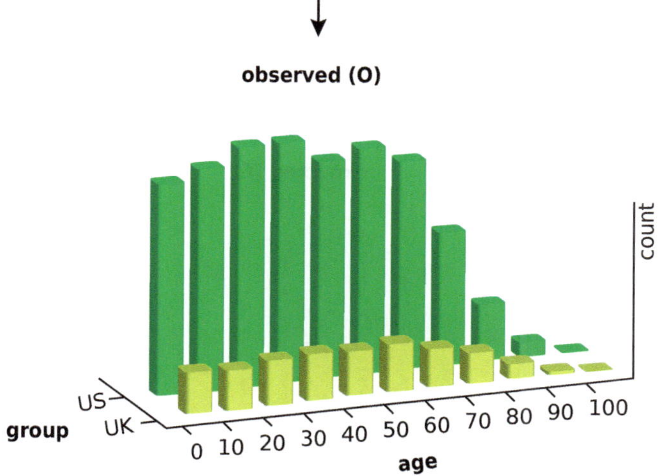

Figure 8.43 Distribution of generations in the US and the UK. Top: Two 1-D histograms representing the age distributions of the US and UK populations. Bottom: The two 1-D histograms are used to construct one 2-D histogram.

other words, the proportion of people that fall into each age cohort is the same for the two countries.

We construct the expected histogram by assuming that country does not make a difference. Therefore, we have one combined histogram, with each column being the sum of the US column and the UK column. This combined count is then divided by the total population of the two countries to get the proportion of that age cohort in the combined population. This percentage is then applied to the total US population to get the expected number for that age cohort in the US, and applied to the total UK population to get the expected number for the UK (Figure 8.44).

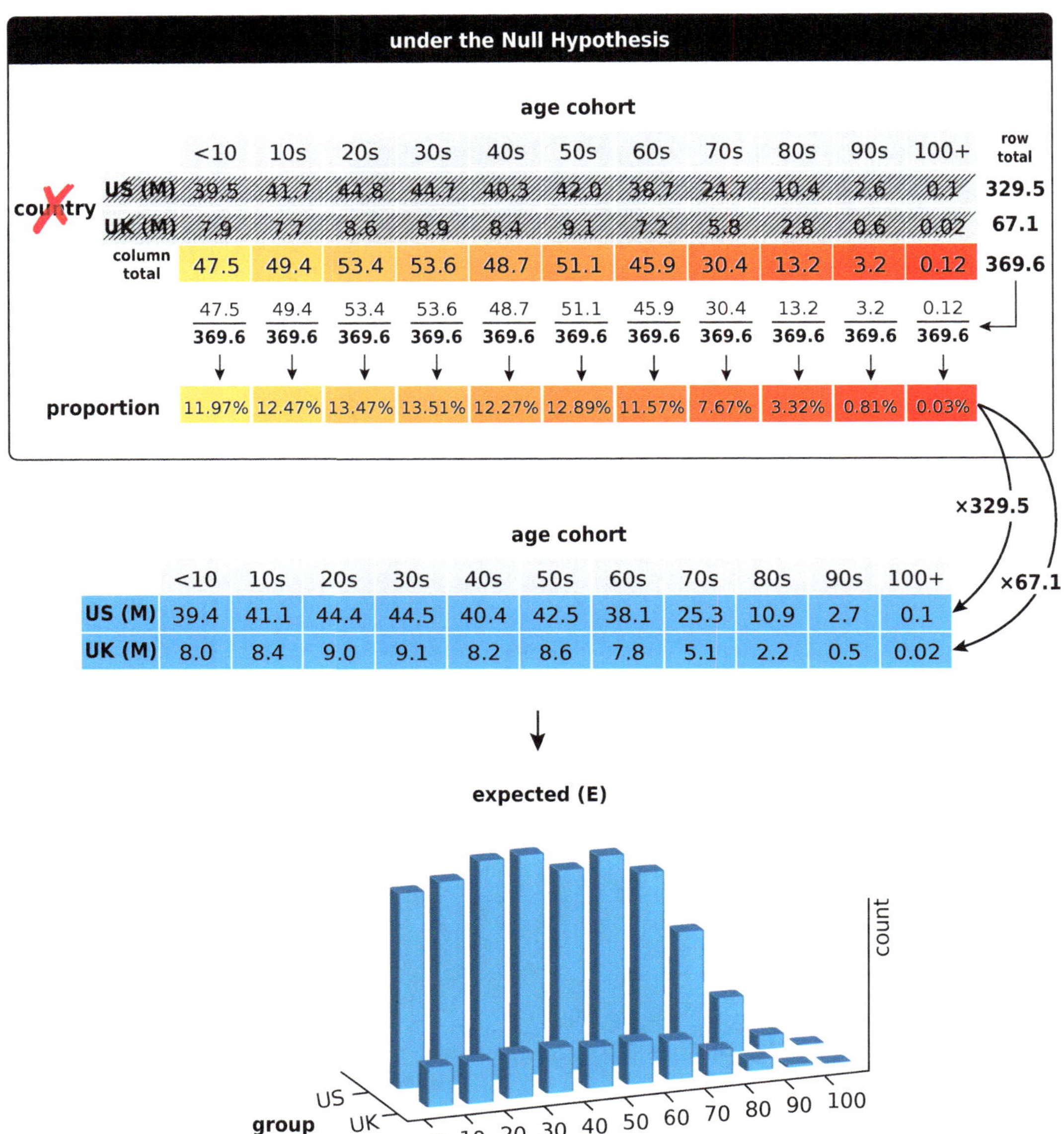

Figure 8.44 Expected histogram of the age distribution of the US and the UK populations, under the Null Hypothesis that "country doesn't matter".

χ_{obs} We then calculate the distance from the observed 2-D histogram to the expected 2-D histogram to give us our χ_{obs} (Figure 8.45).

NHST Now that we have a value for χ_{obs}, we want to ask if that value is statistically significant.

As usual, we interpret that to mean: **is χ_{obs} larger than 99% of what we would see for χ-values for samples randomly drawn from the world of the Null Hypothesis that country doesn't matter to the age distribution?**

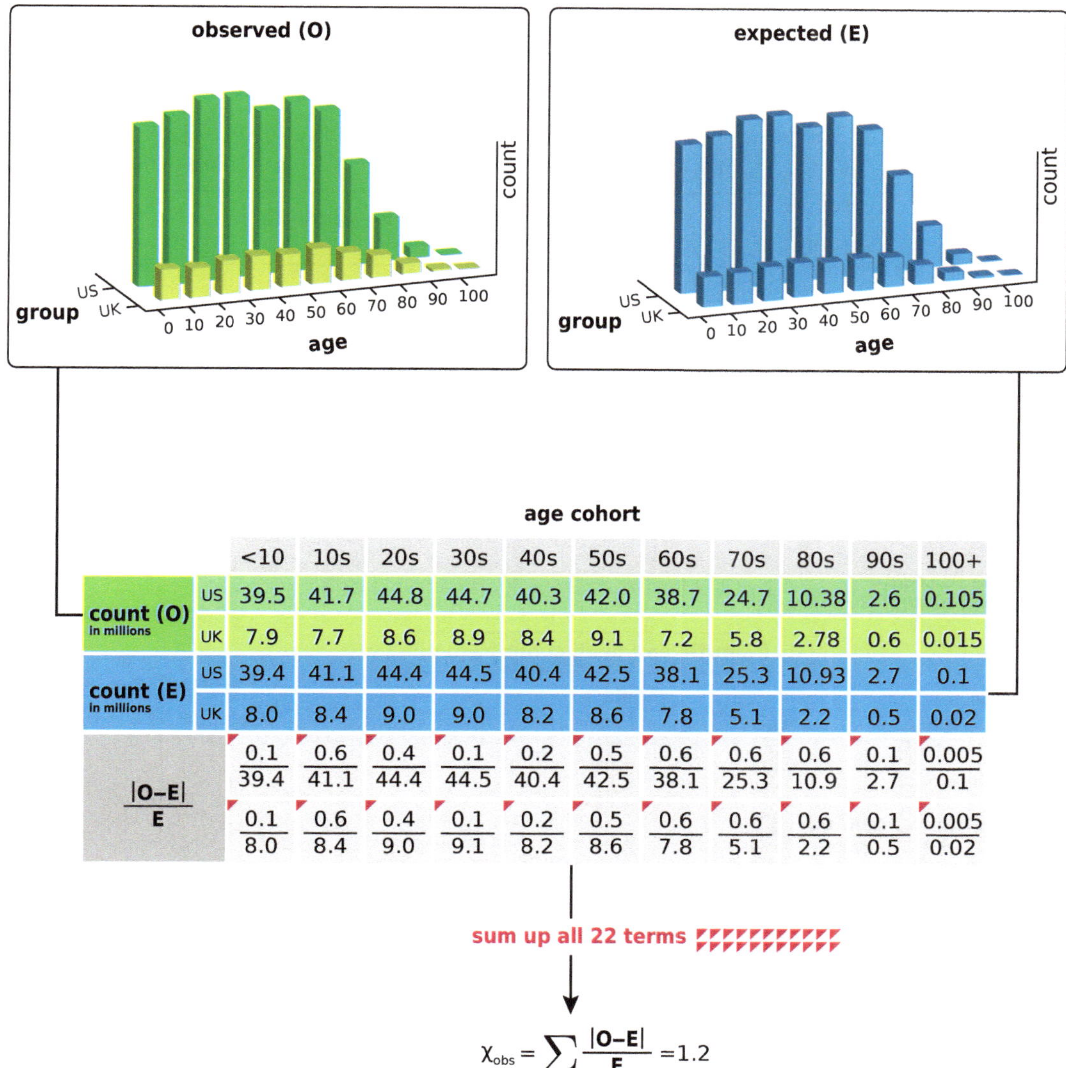

		<10	10s	20s	30s	40s	50s	60s	70s	80s	90s	100+
count (O) in millions	US	39.5	41.7	44.8	44.7	40.3	42.0	38.7	24.7	10.38	2.6	0.105
	UK	7.9	7.7	8.6	8.9	8.4	9.1	7.2	5.8	2.78	0.6	0.015
count (E) in millions	US	39.4	41.1	44.4	44.5	40.4	42.5	38.1	25.3	10.93	2.7	0.1
	UK	8.0	8.4	9.0	9.0	8.2	8.6	7.8	5.1	2.2	0.5	0.02
$\frac{\lvert O-E \rvert}{E}$		$\frac{0.1}{39.4}$	$\frac{0.6}{41.1}$	$\frac{0.4}{44.4}$	$\frac{0.1}{44.5}$	$\frac{0.2}{40.4}$	$\frac{0.5}{42.5}$	$\frac{0.6}{38.1}$	$\frac{0.6}{25.3}$	$\frac{0.6}{10.9}$	$\frac{0.1}{2.7}$	$\frac{0.005}{0.1}$
		$\frac{0.1}{8.0}$	$\frac{0.6}{8.4}$	$\frac{0.4}{9.0}$	$\frac{0.1}{9.1}$	$\frac{0.2}{8.2}$	$\frac{0.5}{8.6}$	$\frac{0.6}{7.8}$	$\frac{0.6}{5.1}$	$\frac{0.6}{2.2}$	$\frac{0.1}{0.5}$	$\frac{0.005}{0.02}$

$$\chi_{obs} = \sum \frac{|O-E|}{E} = 1.2$$

Figure 8.45 Calculating the difference from the observed (O) to the expected (E) using the χ-statistic. This is the observed χ, which we call χ_{obs}.

In order to answer this, we create many samples under the Null Hypothesis, and calculate their χ-values. We create these null worlds just as in the Titanic example: for each of the 11 columns (age cohorts), we use a color to denote each age cohort, and then create as many cards of that color as there are people in that combined column. The totality of these colored cards is the Big Deck.

Then we shuffle the Big Deck, and deal it out into 2 piles, of sizes (population of US) and (population of UK), and then count how many cards of each color we found in the newly-dealt piles. That gives us our first sample of what the age distributions would look like under the Null Hypothesis. We will call this the first pseudo-observed distribution O_1 (Figure 8.46).

Given this pseudo-observed histogram, we calculate its χ-value, the distance from it to the expected, and call it χ_1.

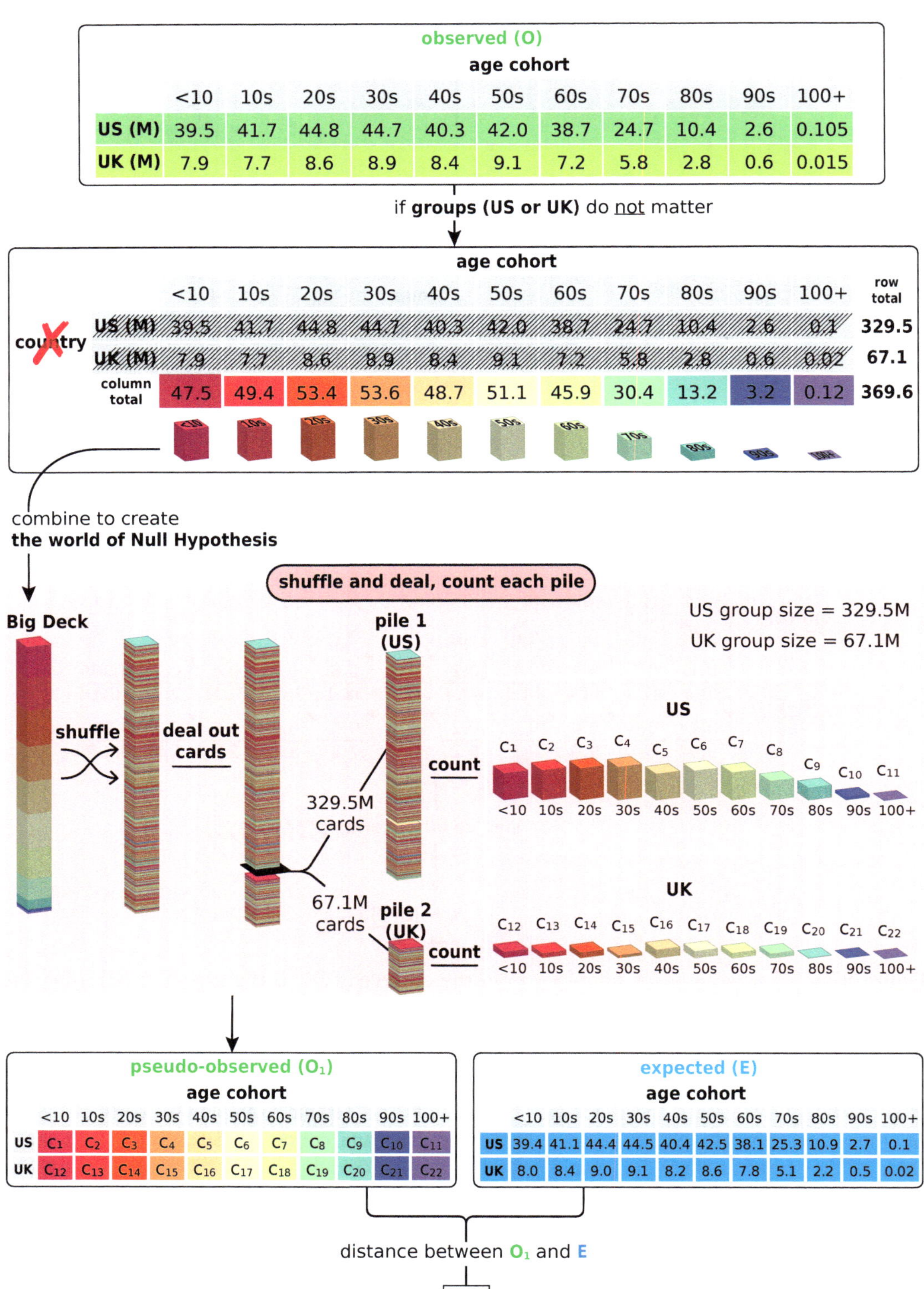

Figure 8.46 Generating one randomized sample of US and UK population distributions, the pseudo-observed (O_1). Then we calculate the distance between the pseudo-observed and the expected histograms, calling that χ_1.

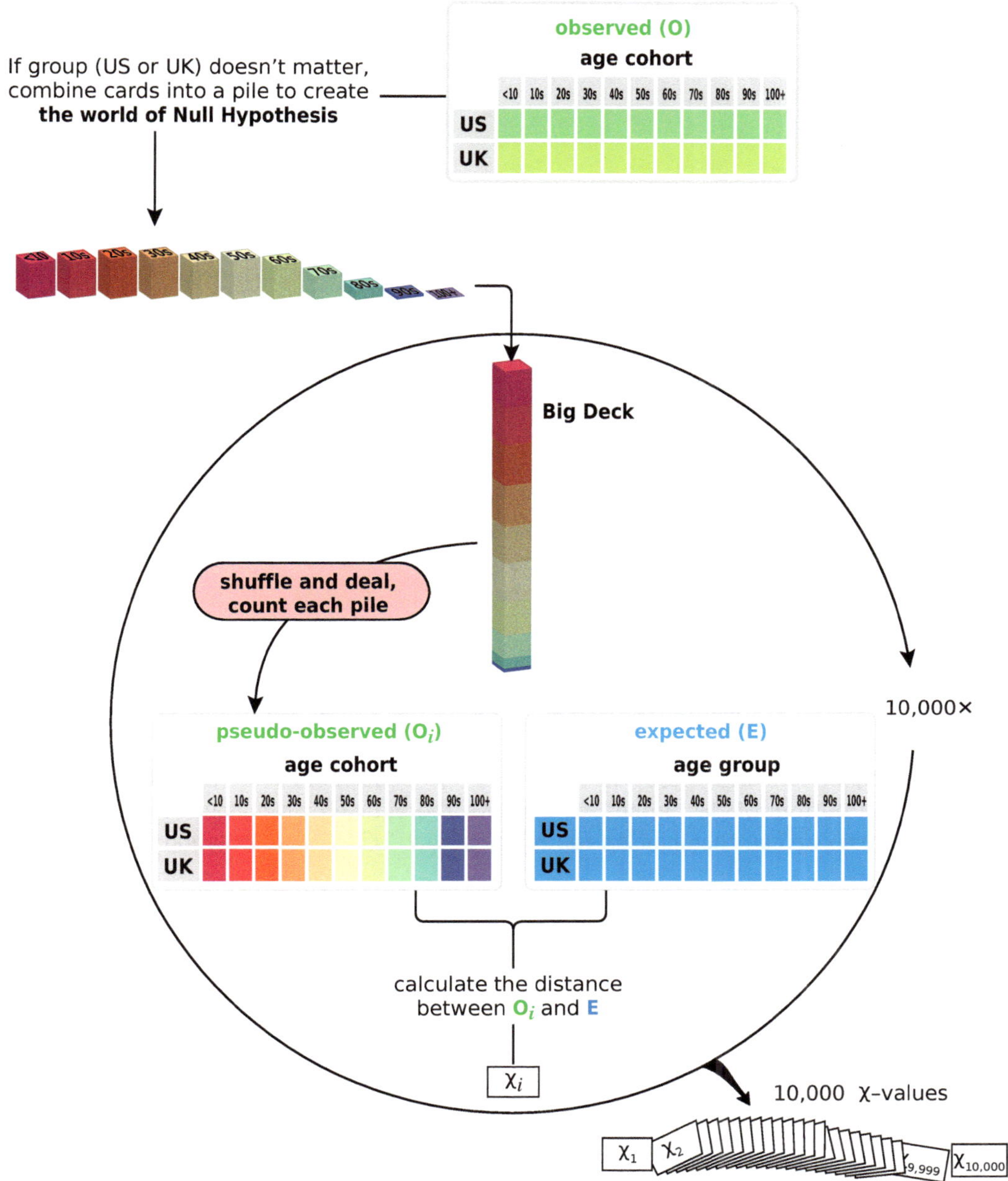

Figure 8.47 NHST for χ_{obs}. Under the Null Hypothesis, we generate the Big Deck of cards from the observed histogram, and then randomize by "shuffle and deal, count each pile" 10,000 times. For each randomization, we generate a pseudo-observed O_i for the US and the UK population, and we calculate the distance from that pseudo-observed to the expected, calling that χ_i. The 10,000 χ_i values will be used to construct the null distribution of the χ-statistic.

Then we repeat this randomization 10,000 times (Figure 8.47), plot these 10,000 values of χ_i, and ask where does our χ_{obs} rank in the 10,000 χ_i values? (Figure 8.48).

As can be seen from the relative position of χ_{obs} to the 10,000 randomizations, we found that in *none* of the 10,000 simulations did we see a χ_i as big as χ_{obs}, so we can declare that p < 0.0001.

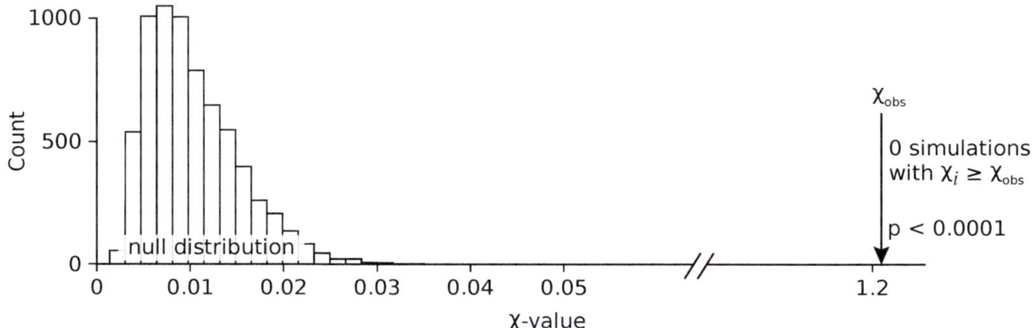

Figure 8.48 Result of NHST comparing the US/UK age distributions.

Mt. Fuji plot Since the two histograms, the US and the UK populations, were found to be highly significantly different, it makes sense to go on and ask *where* they differ, in *which* age cohort(s). The χ test, we saw, is like an omnibus test, and now we can go and ask for the *post hoc* differences.

We will do our *post hoc* test in the form of a Mt. Fuji plot. To construct the clouds of uncertainty, we go back to our 10,000 simulations under the Null Hypothesis. But now, instead of just recording the χ-value, we keep track of how many people showed up in each column in the shuffled populations (Figure 8.49).

For the US, we will have 10,000 values for the <10 population. If we take the 0.5% cutoffs at the top and bottom of this group, we will have the outline of the cloud for the "US < 10" population, which we can then plot on top of the actual "US < 10" population.

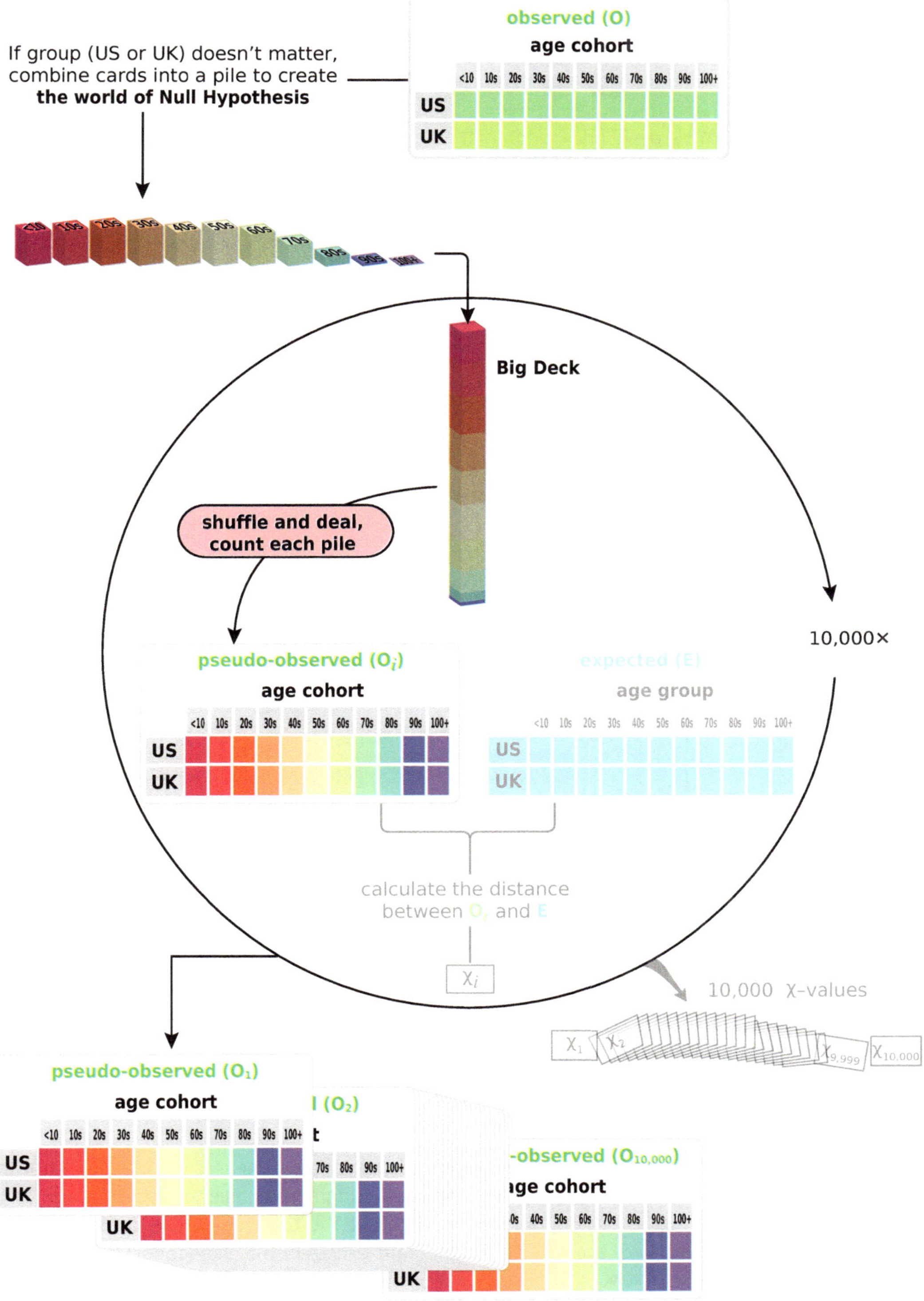

Figure 8.49 Procedure to generate the data for a Mt. Fuji plot for the comparison of the US and the UK populations.

By repeating this for the total 22 age cohorts, we construct the Mt. Fuji plot (Figure 8.50).

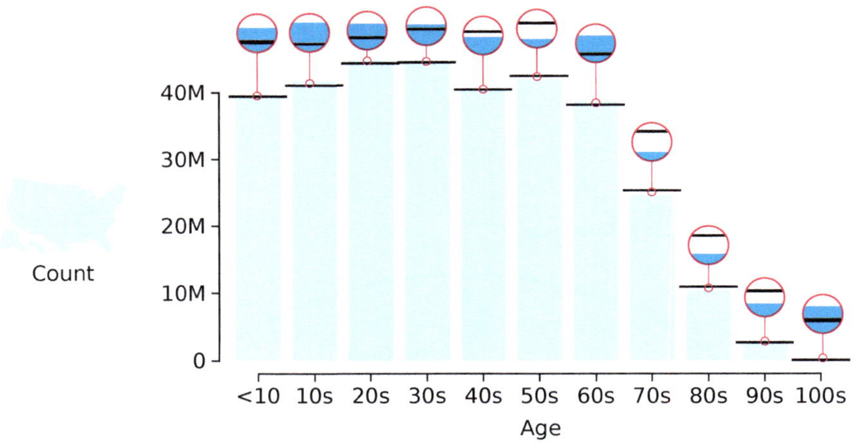

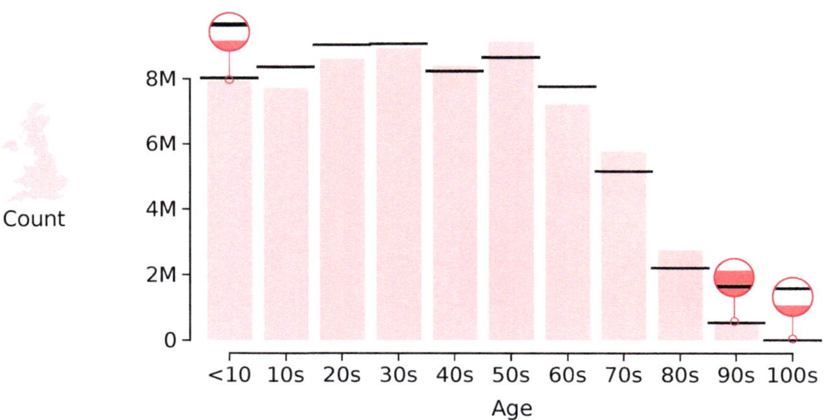

Figure 8.50 A Mt. Fuji plot shows that the clouds of uncertainty are extremely thin. This is due to the very large sample size.

There are several features that immediately strike us. In this Mt. Fuji plot, the clouds are very thin; they look like lines (black), and all the lines are very close to the tops of the bins, making the visual judgment of significance (does the peak stick above the clouds? does the peak stick below the clouds?) difficult. This is especially true in the US histogram due to the larger sample size.

Therefore, we provided insets for columns where the visual judgement is challenging, showing the very top of the bin and the cloud, making the visual judgments easier.

For example, for the "US < 10" population, we see that the top of the blue bar sticks above the (thin) black cloud, therefore, the US has a significant excess in that age cohort. By contrast, in the UK histogram, for the "UK < 10" age cohort, the top of the orange bar sticks below the cloud, indicating that the UK has a significant deficit in that age cohort.

Notice also that in this analysis *all* of the differences were statistically significant. The bins all stick either above or below the clouds, and none are within the clouds.

Of course, we are making multiple comparisons, and must correct for multiple testing. For each of the 11 age cohorts, we calculated the p-value, they are all $p < 0.0001$. So they will survive not only the Benjamini-Hochberg correction but even the more stringent Bonferroni correction.

This case presents some unusual features. First of all, the clouds of uncertainty are so thin that they look like lines, and they all tend to be very close to the tops of the histogram bars. The reason for this is straightforward: sample size.

The sample size here ($n \approx 350$ million) guarantees that virtually *any* difference will be statistically significant, no matter how small. In these very large samples, the relative variability will be very small, making the clouds very thin.

Therefore, to illustrate the impact of sample size on statistical significance, we will consider the same analysis done on two much smaller populations. We will take the US and UK populations scaled down by a factor of 1,000,000.

What if we scale down each age cohort by a factor of a million? Here is the resulting observed 2D histogram (Figure 8.51).

scaled down by 1,000,000 ⟶ **observed (O)**

age cohort

	<10	10s	20s	30s	40s	50s	60s	70s	80s	90s	100+
US	40	42	45	45	40	42	39	25	10	3	0
UK	8	8	9	9	8	9	7	6	3	1	0

Figure 8.51 The observed 2-D histogram after scaling down the US and UK populations by a factor of 1,000,000, rounded off to the nearest whole numbers.

NHST for the smaller sample We carried out the same NHST procedure, namely, "shuffle and deal, count the cards", repeating 10,000 times as before. The result is shown in Figure 8.52.

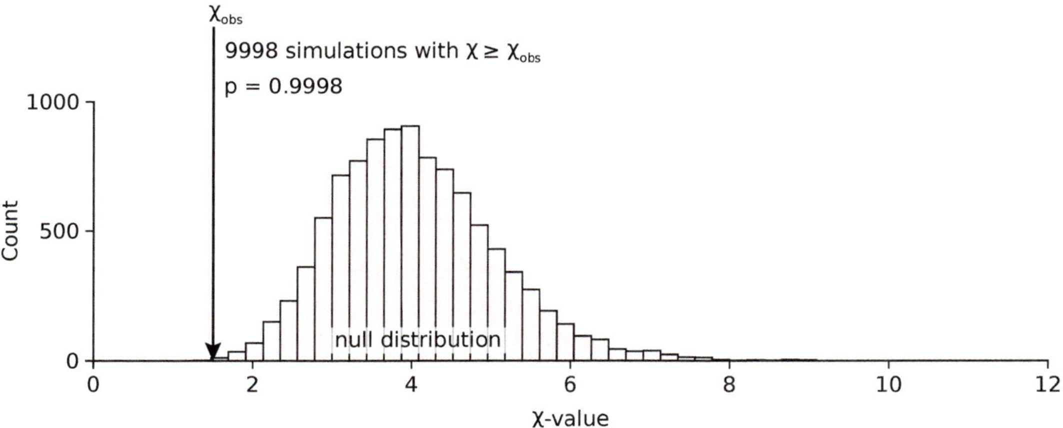

Figure 8.52 NHST for the two smaller populations (data presented in Figure 8.51).

First, we see that that observed distance χ_{obs} hasn't really changed, it's around 1.5 instead of 1.2 (the slight change is due to rounding numbers when scaling down). Comparing Figure 8.48 and Figure 8.52, we see that what has changed is the null distribution: it shifted well to the right of χ_{obs}, giving us a very *not* significant χ_{obs}, with a p-value of 0.9998.

Note how much the horizontal axis has expanded. The null distribution is *much* wider in Figure 8.52 compared to the null distribution with the large sample size in Figure 8.48.

Usually, when we obtain a highly non-significant omnibus result, we would not proceed to *post hoc* testing. However, for the purpose of illustration of the impact of sample size, we'll create a Mt. Fuji plot for this much small sample.

Mt. Fuji plot for the smaller sample After collecting the two populations into one Big Deck, and shuffling and dealing 10,000 times, we can construct the clouds for the Mt. Fuji plots for these smaller populations (Figure 8.53).

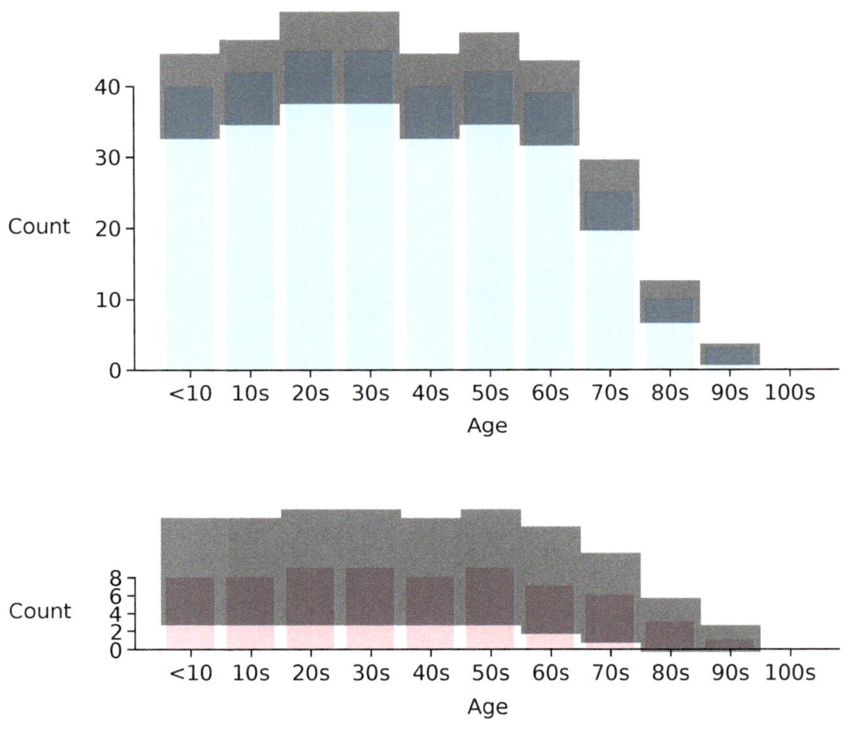

Figure 8.53 Mt. Fuji plot for the smaller sample, obtained by scaling down the US and UK population by a factor of a million (data presented in Figure 8.51).

As we see, now the clouds completely obscure and cover the peaks, so none of these differences can be considered statistically significant. This change in statistical significance is due to the smaller sample size and therefore larger variability from sample to sample. How sample size impacts statistical significance will be discussed in Chapter 11 (*Power*).

FURTHER EXERCISES 8.7

1. For each of the following studies, give the type of NHST analysis you would use and your reasons for that choice, the statistic you would calculate, and the Null Hypothesis for this particular study. If you think that more than one type of analysis may be performed, describe the one you think is best.

   a. A study seeks to find out if the new antiviral vaccination significantly reduces the risk of seasonal flu in children.

   b. A study seeks to identify if physical activity before lectures will improve student performance in math. The investigator of this study will have two sets of data associated with each student: the score with exercise and the score without exercise for the same student.

   c. A study would like to determine if the age of people experiencing COVID-19 disease is dependent on the country of residence.

   d. A clinical trial that tests how different drug delivery schedules (morning/evening) impact the length of recovery time.

   e. Three hundred registered voters were selected at random to participate in a study on attitudes about how well the governor is performing their job. They were each asked to answer a questionnaire, watched a video that presented information about the job description of governor, and answered the questionnaire again. The investigator of this study will have two sets of data associated with each person: the initial questionnaire scores and the follow-up questionnaire scores.

   f. A researcher is testing the success rate of a new treatment and visualizes her data as in the figure below. Blue is the new treatment and orange is a control.

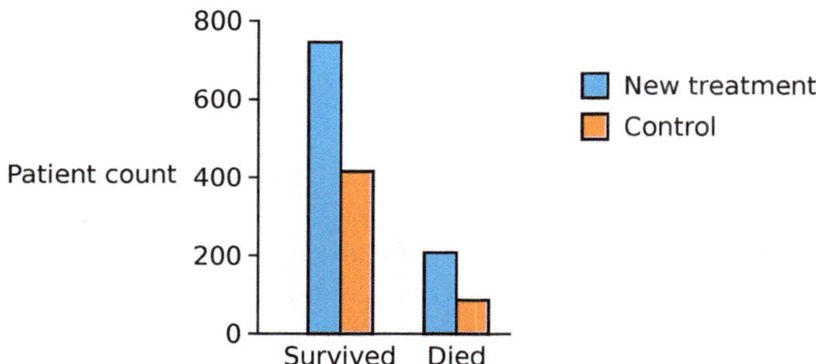

   g. Thirty dogs were selected at random from those residing at the humane society last month. The 30 dogs were split at random into two groups. The first group of 15 dogs was trained to perform a certain task using a food-reward method. The second group of 15 dogs was trained to perform the same task using an affection-reward method. The investigator of this study will have two sets of data: the learning times for the dogs trained with the food-reward method and the learning times for the dogs trained with the affection-reward method.

8.8 Relative Risk: an effect size

So far in this chapter, we've been trying to reject a Null Hypothesis that two properties, say gender and heart disease, were "independent" or "unrelated". And if we did reject the Null Hypothesis in a particular case, we said that the two properties were "associated". We said that gender is "associated with" heart disease, or that survival on the Titanic was "related to" (or "not independent of") class of ticket.

All statements of this kind suffer from the defect that they are only Yes/No questions. We can say that the two factors are "associated" (or not), but we can't yet say how large the effect is: How *strong* is the effect of ticket class on survival on the Titanic? How *strong* is the effect of gender on heart disease? To answer this question, we need an effect size.

We do not want to make the mistake of confusing statistical significance, which is a question about the **strength of the evidence**, with real-world significance, which is a question about the **strength of the effect**. We can have very strong evidence for a very small effect, and *vice versa*.

This is clearly a major drawback. We need to develop an effect size for these comparisons.

If the problem can be put in the form of a 2 × 2 design, then the most important, and the most frequently used measure of effect size is called **relative risk**.

Let's consider a 2 × 2 contingency table, where the two groups can be thought of as "treatment" and "control". For each group, we have a certain number of "outcome-yes" cases, and a certain number of "outcome-no" cases (Figure 8.54).

		outcome	
		Yes	No
group	treatment	a	b
	control	c	d

Figure 8.54 The structure of a 2 × 2 design.

Then the relative risk associated with the treatment is defined as the fraction of "outcome-yes" cases in the treatment group, divided by the fraction of "outcome-yes" cases in the control group. So the relative risk can be thought of as the change in risk that is associated with being in the "treatment" group relative to the "control" group. That relative change in risk can be either be an increase (relative risk > 1) or a decrease (relative risk < 1). If the relative risk is 1, then there is no change in risk.

$$\text{relative risk associated with the treatment}$$
$$= \frac{\text{fraction of ``outcome-yes'' cases in the treatment group}}{\text{fraction of ``outcome-yes'' cases in the control group}}$$
$$= \frac{a/(a+b)}{c/(c+d)}$$

Relative risk is truly an effect size: if the two groups are "disease" and "control", and the "outcome" is a bad one (such as death), then the *bigger* the relative risk, the worse the disease is in terms of that outcome. And if the treatment really is a treatment for a disease, and the control group is untreated people with that disease, then the *lower* the relative risk, the better the treatment is at preventing the bad outcome.

Relative risk in the heart disease/gender study The heart disease and gender study can be put in the form of a 2 × 2 study design. Therefore, in addition to calculating a χ-statistic and doing NHST, we can ask a different question: what is the relative risk of being a male for the outcome of heart disease?

We reprint the 2 × 2 design here with labels a, b, c and d for each cell (Figure 8.55).

		heart disease	
		Yes	No
gender	male	a 114	b 92
	female	c 25	d 72

Figure 8.55

The treatment here is "being a male", the control is "being a female", and the outcome is heart disease. Therefore, the relative risk of being a male for the outcome of heart disease is

$$
\begin{aligned}
\text{relative risk of being a male} &= \frac{\text{fraction of ``heart disease-yes'' cases in the male group}}{\text{fraction of ``heart disease-yes'' cases in the female group}} \\
&= \frac{a/(a+b)}{c/(c+d)} \\
&= \frac{114/(114+92)}{25/(25+72)} \\
&= \frac{114/206}{25/97} \\
&= 2.15
\end{aligned}
$$

Therefore, we can say that in this sample, males were more than twice as likely as females to have heart disease.

Example 2 Diabetes and gender: relative risk

Assume you are given the result of a study on diabetes and gender. In the study, there are 21,605 males and 24,008 females. In the male group, 2105 are diabetic while in the female group, 2335 are diabetic.

Question What is the relative risk of being a male for diabetes?

Answer We assemble the data into a 2 × 2 design.

		diabetes	
		Yes	No
gender	male	a 2,105	b 19,500
	female	c 2,335	d 21,673

Here, if we refer to Figure 8.54, "being male" is the "treatment", "being female" is the "control", and diabetic and non-diabetic are the outcomes. We calculate the relative risk of being a male for diabetes as:

$$\begin{aligned}
\text{relative risk of being a male} &= \frac{\text{fraction of "diabetes-yes" cases in the treatment group}}{\text{fraction of "diabetes-yes" cases in the control group}} \\
&= \frac{a/(a+b)}{c/(c+d)} \\
&= \frac{2105/(2105+19500)}{2335/(2335+21673)} \\
&= \frac{2105/21605}{2335/24008} \\
&= 1.0018
\end{aligned}$$

This relative risk is very close to 1, which is the "no effect" value, so we would say that, in this sample, the relative risk of being a male for diabetes is negligible.

If there are more than two groups in a particular study, we can still apply the concept of relative risk, but only to each pair of groups.

For example, in the Titanic case, there are three groups (First, Second and Third class), each group having two outcomes (survived or died). First, we choose one of the groups as the "control". Let's say we choose First class as the control. Then we calculate the relative risk of Second class for the death outcome as the fraction the people dead in Second class divided by the fraction of people dead in First class. Similarly, we can also calculate the relative risk of Third class for the death outcome as the fraction of the people dead in Third class divided by the fraction of people dead in First class.

FURTHER EXERCISES 8.8

1. If the relative risk of cancer for eating a diet item is 3, what would be the relative risk of not eating that item?

8.9 Absolute Risk Reduction: another effect size

In section 6.8, Algebraic differences vs. ratios, we mentioned that sometimes the effect size is best measured by the ratio of the group measures, but that, at other times, the algebraic difference gives a more useful measure. For example, if we are looking at survival on a drug vs. control, it would not be helpful to say that the drug had a Relative Risk of 0.5, if that represented a risk in the control group of 0.001 (one tenth of a percent) and the risk in the drug group being 0.0005. The halving of risk does not amount to much, since the background risk is so low.

A better measure here would be to use the algebraic difference of the risk ratios, which is the quantity called **Absolute Risk Reduction**, or **ARR**. The ARR tells us *how many* cases the drug would improve. If we were to ask how many people would we have to treat to prevent one outcome, that number, called the **NNT (Number Needed to Treat)**, can be defined as

$$\text{NNT} = \frac{1}{\text{ARR}}$$

The NNT gives us a very important real-world fact: in the above hypothetical example, the Relative Risk is 0.5, suggesting that the drug has a good effect, but the ARR tells us that we would have to treat 1/0.0005, or 2000 people, to avoid 1 bad outcome. These two facts must be balanced in any decision-making.

Example 3 Potential pitfall of Relative Risk

Question What is a greater concern?

- an additional 1% increase added to the rate of heart attacks
- a doubling in the rate of brain-eating amoeba infections

Answer Heart attacks are common, so even a small percentage increase in risk means many more people will have them. Infections by brain-eating amoebae are extremely rare, so even a doubling of rates leaves them extremely rare. When comparing relative risks, we need to think about the absolute risk of various conditions too!

Example 4 Aspirin in Heart Attack Prevention - ASPREE Trial

The ASPREE trial studied aspirin use in healthy elderly adults: "Of the 19,114 persons who were enrolled in the trial, 9525 were assigned to receive aspirin and 9589 to receive placebo. After a median of 4.7 years of follow-up, the rate of cardiovascular disease was 10.7 events per 1000 person-years in the aspirin group and 11.3 events per 1000 person-years in the placebo group (hazard ratio, 0.95; 95% confidence interval [CI], 0.83 to 1.08). The rate of major hemorrhage was 8.6 events per 1000 person-years and 6.2 events per 1000 person-years, respectively (hazard ratio, 1.38; 95% CI, 1.18 to 1.62; P<0.001)."

Question Calculate absolute risk and absolute risk difference over the 4.7-year study period.

Answer

- For **cardiovascular disease**, the absolute risks are:
  - Aspirin: 10.7 per 1,000 person-years × 4.7 years = 50.3 per 1,000 = 5.03%
  - Placebo: 11.3 per 1,000 person-years × 4.7 years = 53.1 per 1,000 = 5.31%

  ↓ *ARR (Absolute Risk Reduction)* = 5.31% − 5.03% = 0.28% over 4.7 years

  However, this small apparent benefit is not statistically significant. We calculated the 95% CI for ARR to be [−0.35%, 0.91%]. Since the CI crosses zero, we cannot confidently say aspirin has any effect on cardiovascular disease. This is consistent with what the study found concerning the hazard ratio "(hazard ratio, 0.95; 95% confidence interval [CI], 0.83 to 1.08)". Note that hazard ratio = 1 represents "no difference".

- For **major bleeding**, the absolute risks are:
  - Aspirin: 8.6 per 1,000 person-years × 4.7 years = 40.4 per 1,000 = 4.04%
  - Placebo: 6.2 per 1,000 person-years × 4.7 years = 29.1 per 1,000 = 2.91%

  ↑ *ARI (Absolute Risk Increase)* = 4.04% − 2.91% = 1.13%

  This increase in bleeding risk is statistically significant. We calculated the 95% CI for

ARI to be [0.62%, 1.65%]. The CI doesn't cross zero, confirming that aspirin significantly increases bleeding risk. This is consistent with what the study found concerning the hazard ratio "(hazard ratio 1.38, 95% CI: 1.18 to 1.62, P<0.001)".

Question Does benefits outweigh harms?

Answer

- For **cardiovascular disease** (benefit):

  NNT = 1/0.0028 = 357 people need to take aspirin for 4.7 years to prevent one cardiovascular event (not statistically significant)

- For **major bleeding** (harm):

  NNH = 1/0.0113 = 89 people taking aspirin for 4.7 years will cause one major bleeding event (statistically significant)

The harm-to-benefit ratio is roughly 4:1 ($357/89 \approx 4$) — you cause 4 major bleeding events for every cardiovascular event you prevent. The harms outweigh any potential benefits.

8.10 NHST for relative risk

Once we have calculated a relative risk in a given situation, the next question is: is this observed relative risk statistically significant, that is, statistically significantly different from 1, which is the Null Hypothesis?

We answer this question by a "shuffle, deal, and count" methodology. Our Null Hypothesis is that the outcome is unrelated to group membership, therefore, we implement the Null Hypothesis by gathering together all the outcome cards (outcome-yes and outcome-no) into a Big Deck. The Big Deck is then shuffled, and dealt out to the two groups, and the relative risk is calculated for this instance of the world of the Null Hypothesis.

The "shuffle, deal, and count" procedure is then repeated 10,000 times to form the null distribution. We calculate a p-value by comparing the observed relative risk RR_{obs} to the null distribution. When we are dealing with ratios, to do a 2-sided p-value calculation, the symmetric case of RR_{obs} is $1/RR_{obs}$, and the p-value is the fraction of the 10,000 simulations that are more extreme than either RR_{obs} or $1/RR_{obs}$ (see section 6.8 *Statistical significance for ratios*).

Here, we use the heart disease and gender study as our example to illustrate the procedure.

NHST for the relative risk in the heart disease/gender study The relative risk associated with being a male for heart disease was $RR_{obs} = 2.15$.

To implement the Null Hypothesis, we ignore the group identifiers male/female, and form a Big Deck of "heart disease-yes" and "heart disease-no" cards, shuffle the Big Deck, and deal out cards into two piles to form a new "male" group and a new "female" group, to get the first pseudo-observed (O_1). We then calculate RR_1, the relative risk of being male for heart disease in this first randomized sample (Figure 8.56).

We then implement the Null Hypothesis 10,000 times. Each time, the relative risk is calculated for a new randomized sample, generating RR_1, RR_2, ..., $RR_{10,000}$. Out of the 10,000 simulations, none are more extreme than either RR_{obs} or $1/RR_{obs}$, therefore, $p < 0.0001$ (Figure 8.57). The relative risk associated with being male for heart disease, $RR_{obs} = 2.15$, is statistically significant at the $p < 0.0001$ level.

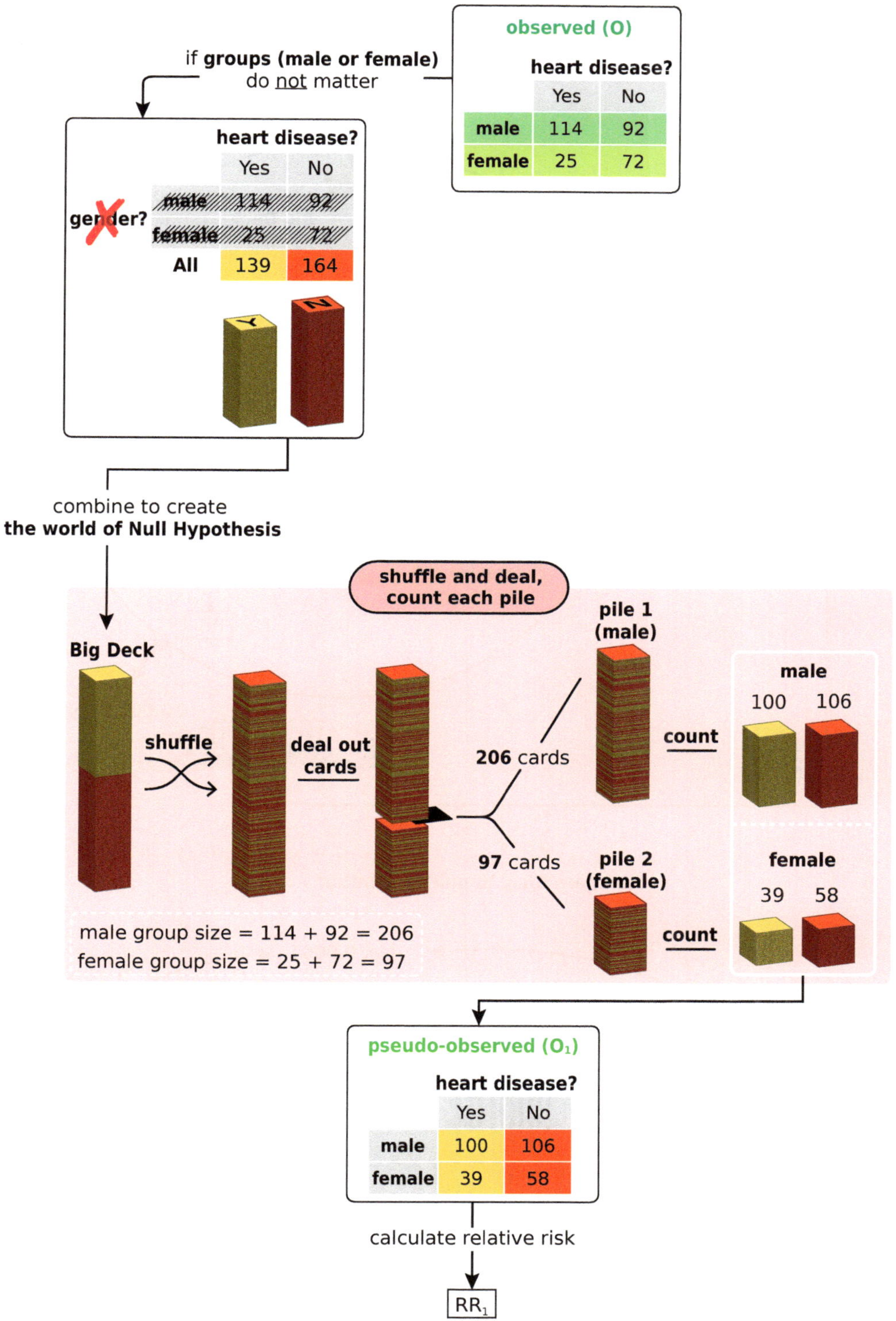

Figure 8.56 Generating RR_1, the first estimate for relative risk under the Null Hypothesis. In this illustration, we use the risk of being a male for heart disease as an example.

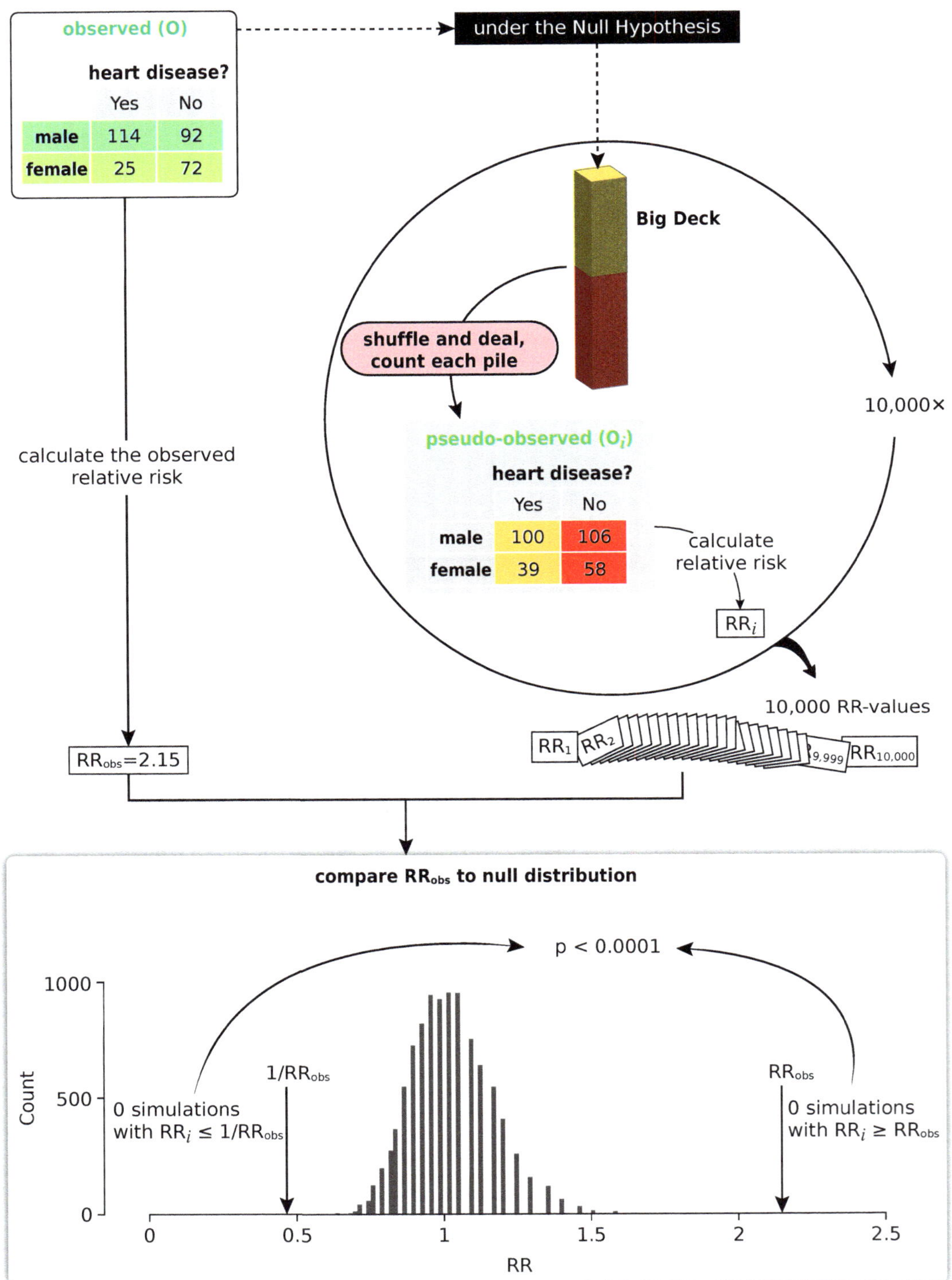

Figure 8.57 **NHST for relative risk in the heart disease/gender study.** We shuffled the world of the Null Hypothesis 10,000 times to generate the null distribution and found the p-value for the observed relative risk RR_{obs}. In this illustration, we used the relative risk of being a male for heart disease as an example.

8.11 Confidence intervals for relative risk

Since relative risk (RR) is an effect size, it makes sense to ask for a confidence interval. In keeping with our overall strategy to go beyond NHST to "effect size + confidence interval", we want to be able to give some precision to the estimate of relative risk. For example, we want to be able to say something like: "the relative risk of catching flu for someone who is vaccinated is x, with a 99% confidence interval [y, z]".

We construct confidence intervals for relative risk exactly as we have done before for other measures: we resample the original data and calculate the relative risk in that resampled data set. As always in this book, resampling means resampling with replacement. Each resampled data set gives a slightly different value for the relative risk; we are estimating the variability of relative risk that we might see simply due to sampling.

Repeating that procedure 10,000 times gives us 10,000 estimates of what the RR could have been if we were to redo this experiment many times.

As an example, let's construct the 99% confidence interval for the relative risk of being a male for the outcome of heart disease in a small sample data set (Figure 8.55). The observed RR was calculated to be 2.15.

To construct the confidence interval, we'll follow the workflow in Figure 8.58. We resample the original data set by putting the 206 males, heart disease and not, into a box, and then resampling a new set of 206 males from this set, and counting how many have heart disease and how many do not.

Next, we do the same thing for the 97 females, drawing 97 with replacement from the set of (25 heart disease, 72 no heart disease).

In this first resampled data set, we calculate the relative risk, calling it RR_1.

We repeat this resampling procedure to generate 10,000 estimates RR_1, RR_2, . . . , $RR_{10,000}$.

Using the 10,000 RR-values from the 10,000 resamples, we calculate a confidence interval. To do this, we have to remember that relative risk is a *ratio*, and therefore, we have to deal with that fact in constructing confidence intervals. Recall, from section 6.8 (*When the effect size is a ratio*), that effect sizes that are ratios have to be treated differently from the case where the effect sizes are algebraic differences.

The crux of the problem is that if RR = 1 is the Null Hypothesis, that means that the risk in group A is equal to the risk in group B. We can imagine two scenarios: the risk in group A is twice the risk in group B, and the risk in group B is twice that in group A. These are clearly symmetric scenarios, which would be written as RR = 0.5 and RR = 2.

But the "distance" from RR = 0.5 to RR = 1 is only 0.5, while the distance from RR = 1 to RR = 2 is 1. In section 6.8 *When the effect size is a ratio*, we corrected this asymmetry by taking as our effect size, not the ratio itself, but rather, the log of the ratio. When we take the logs, the distance between ln(0.5) and ln(1) is the same as the distance from ln(1) to ln(2).

$$\ln(1) - \ln(0.5) = 0.69$$
$$\ln(2) - \ln(1) = 0.69$$

We follow the same treatment by first logging the 10,000 estimates, and calculating the confidence interval by the method described in Chapter 4 (*Confidence Intervals: one measure, one group*): we rank the 10,000 logged RR-values, count 50 down from the largest to get the 99.5% cutoff ($\ln(RR)_{upper}$), count 50 up from the smallest to get the 0.5% cutoff ($\ln(RR)_{lower}$),

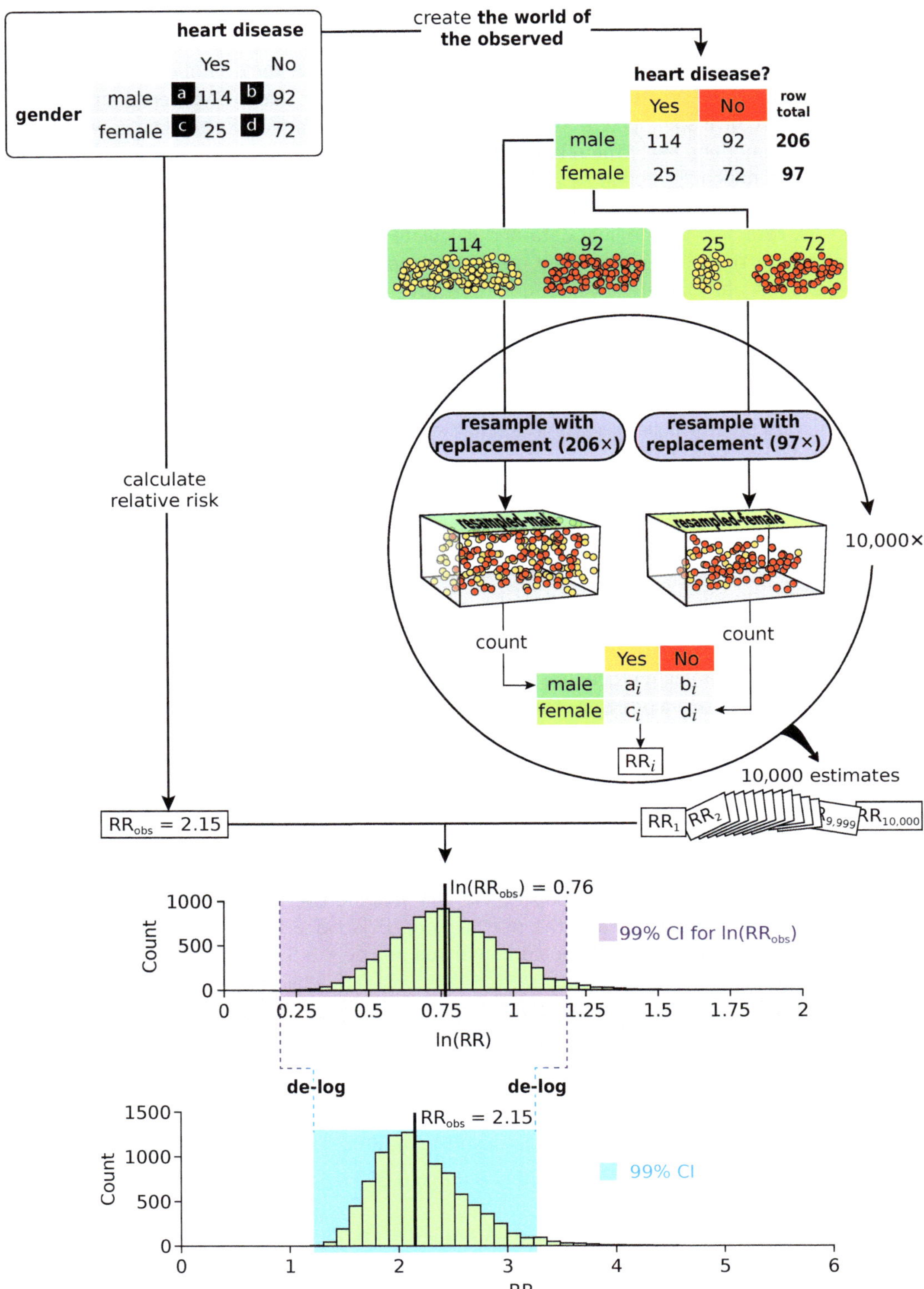

Figure 8.58 Constructing a 99% confidence interval for relative risk. In this illustration, we use the relative risk of being a male for heart disease as an example.

and define the 99% confidence interval as $[2 \times \ln(RR_{obs}) - \ln(RR)_{upper}, 2 \times \ln(RR_{obs}) - \ln(RR)_{lower}]$.

Then we de-log the calculated the 99% confidence interval limits. We present the result by overlaying the 99% confidence interval on top of the resampling distribution, with the observed RR indicated.

The resulting 99% confidence interval for the relative risk of being a male for heart disease in this small study is [1.2, 3.3]. We can interpret this as saying that, in this sample, males are more than twice as likely as females to have heart disease (RR = 2.15), with a 99% confidence interval ranging from 1.2 to 3.3 fold.

Example 5 Diabetes and gender: confidence interval for RR

In Example 2 on page 413, we calculated the relative risk of being a male for the outcome of diabetes to be RR = 1.0018. We reprint the 2 × 2 design here.

		diabetes	
		Yes	No
gender	male	a 2,105	b 19,500
	female	c 2,335	d 21,673

Question Calculate the 99% confidence interval for the relative risk.

Answer The resulting 99% confidence interval for the relative risk of being a male for diabetes is (0.93, 1.08). We can interpret the result as gender is not associated with the prevalence of diabetes (RR = 1.0018), with a 99% confidence interval ranging from 0.93 to 1.08.

We present the result by overlaying the 99% confidence interval on top of the resampling distribution with the observed RR indicated.

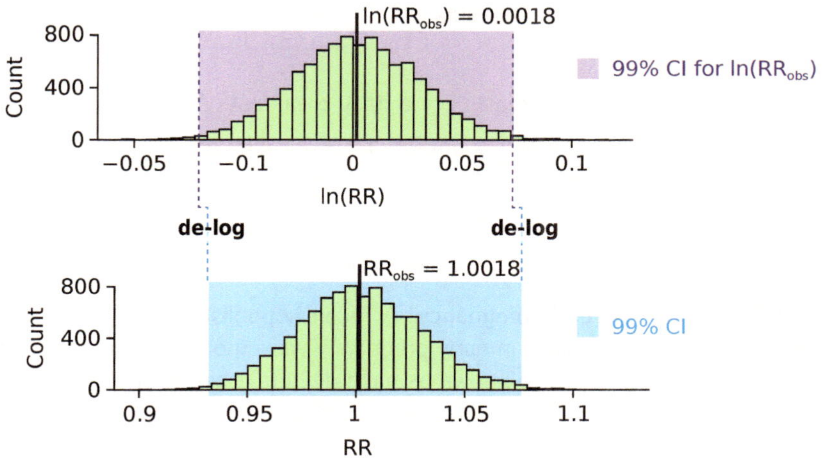

FURTHER EXERCISES 8.11

1. **Sterile surgery.** Joseph Lister published a landmark study in 1867 investigating the benefits of sterile technique in surgery. At the time, it was not customary for surgeons to wash their hands or instruments prior to operating on patients. Lister developed a new operating

procedure in which surgeons were required to wash their hands, wear clean gloves, and disinfect surgical instruments with carbolic acid. This new procedure was compared to the old, non-sterile procedure, and Lister recorded the number of patients in each group that lived or died. Of the 40 patients receiving the sterile procedure, 34 survived and 6 died. Of the 35 patients receiving the control procedure, 16 survived and 19 died. Determine whether sterile operating procedure affected the number of patients who survived surgery.

a) Write a 2 × 2 table for observed counts.

b) State the appropriate Null Hypothesis.

c) Calculate the relative risk of dying for sterile procedure versus control procedure in the observed data.

d) Calculate a p-value for the observed relative risk. Include the null distribution with the observed result indicated.

e) Calculate a 95% confidence interval for the observed relative risk.

f) Interpret the result.

g) Explain why we might want to use a greater than usual α for this particular study.

2. **Have you eaten?** Students who are taking a statistics course surveyed a random sample of students at their college, and one of the questions asked was whether or not they eat breakfast every day. The results they obtained is shown below in a 2 × 2 table. The Null Hypothesis is that male and female students eat breakfast daily in the same proportions.

		Gender?	
		Female	Male
Breakfast?	No	75	24
	Yes	25	6

a. What proportion of males ate a daily breakfast? What about females?

b. Calculate the expected counts in a 2 × 2 table under the assumption that there is no association between gender and whether or not someone eats breakfast daily.

c. Calculate χ_{obs}.

d. Calculate the relative risk of not eating breakfast daily for males compared to females.

3. **COVID vaccines in early pregnancy.** A recently published study examined whether women receiving COVID vaccines in early pregnancy were more likely to give birth to children with congenital anomalies (also known as birth defects) than women who did not receive a vaccine. Of 6,731 babies born to vaccinated mothers, 153 had anomalies. Of 20,193 babies born to unvaccinated mothers, 467 had anomalies. Determine if there is a statistically significant difference in congenital anomaly rate between vaccinated mothers and unvaccinated mothers (see article "A population-based matched cohort study of major congenital anomalies following covid-19 vaccination and sars-cov-2 infection", *Nature Communications*, 2023).

a. Describe the appropriate Null Hypothesis.

b. Write a 2 × 2 contingency table for the observed counts.

c. Write a 2 × 2 contingency table for the expected counts.

d. Calculate the relative risk of anomaly to babies of vaccinated mothers compared to unvaccinated mothers.

e. Calculate a p-value for the relative risk. Include the null distribution with the observed relative risk indicated.

f. Calculate a 95% confidence interval for relative risk.

g. Interpret the result.

h. Explain why we might want to use a greater than usual α for this particular study.

4. Why should we use one-tailed p-values for χ and the F-statistic, but two-tailed p-values for relative risk? *Hint: a sketch would help!*

5. Use the formula for relative risk to explain why the null distribution for relative risk is high near 1.

6. If the relative risk for a particular treatment compared to control is 1.6, then the equally-extreme relative risk that you should use to calculate a two-tailed p-value is 0.625. Explain why 0.625 is equally as extreme as 1.6, instead of 0.4 (which is 1.0 minus 0.6)?

7. **Vitamin E and prostate cancer.** A randomized controlled study investigated whether taking vitamin E reduces the incidence of prostate cancer. Below we provide the 2 × 2 design for this study (see article "Vitamin E and the risk of prostate cancer: the Selenium and Vitamin E Cancer Prevention Trial (SELECT)", *JAMA*, 2011).

		outcome	
		No prostate cancer	Prostate cancer
group	Placebo	8,167	529
	Vitamin E	8,117	620

For any 2 × 2 study, we have two statistical approaches

- a χ-test, in which χ_{obs} is compared to many χ-values under the Null Hypothesis to determine a p-value.
- a relative risk calculation, in which the observed relative risk is compared to the distribution of relative risks under the Null Hypothesis to obtain a p-value. In addition, we can also construct a confidence interval for the observed relative risk.

If we were to use the χ-test

a) State the Null Hypothesis.

b) Calculate the 2 × 2 table for expected results under the Null Hypothesis.

c) What is the observed distance χ_{obs}?

d) Show the distribution of the χ-statistic under the Null Hypothesis.

e) Calculate a p-value.

If we were to use relative risk

a) Calculate the relative risk of prostate cancer for people taking vitamin E vs. placebo.

b) State the Null Hypothesis.

c) Show the distribution of relative risks under the Null Hypothesis.

d) Calculate a p-value.

e) Describe the procedure to construct a confidence interval for relative risk.

f) Construct a confidence interval for relative risk and interpret the result.

8.12 The theoretical approximation (χ^2)

In this chapter, we developed χ as a measure of the distance from an observed histogram to an expected histogram. To our knowledge, this is the only statistics text *ever* that uses this notion! In all other texts and courses, you will encounter a much more widely used notion called "chi-squared" or "chi square" (χ^2).

When we say "more widely used", we aren't kidding: as of this writing, a Google search of "chi-square" returned 424 million hits. Every elementary statistics class and text teaches the doctrine of χ^2. The "chi-squared test" has been the widely accepted approach to n × m contingency tables.

Definition of χ^2

The definition of χ^2 is similar to our χ, with one big difference: as in section 8.2, we used the absolute values of the differences between observed and expected in our definition of χ .

Traditional statistics takes a different path, and uses instead the *squares* of the differences instead of the absolute values. The resulting measure is called "chi squared".

The chi-squared (χ^2) statistic

Assuming there are k bins in each histogram, χ^2 attempts to quantify the **distance between observed and expected counts** by squaring each difference between observed and expected. It uses squaring to avoid positive and negative deviations from cancelling each other out.

$$\chi^2 = \sum_{i=1}^{k} \frac{(\mathrm{O}_i - \mathrm{E}_i)^2}{\mathrm{E}_i} = \sum_{i=1}^{k} \frac{(\mathrm{Observed}_i - \mathrm{Expected}_i)^2}{\mathrm{Expected}_i}$$

It is frequently claimed that the reason for squaring everything is to prevent negative numbers, but, of course, our absolute value does that too, only without the distortions of squaring. For mathematicians, a big motivation for the squaring is that it is possible to prove a mathematical theorem about the distribution of the values of the χ^2 statistic under some very strong assumptions, and using a certain technical formulation of the Null Hypothesis.

The theorem goes like this: suppose we have a total of n observations, divided into k bins. Our Null Hypothesis is that the counts in each of the k bins are numbers drawn from the same Normal distribution. Then, *as n approaches infinity*, the sum

$$\chi^2 = \sum_{i=1}^{k} \frac{(\mathrm{O}_i - \mathrm{E}_i)^2}{\mathrm{E}_i}$$

will approximate a distribution, called the chi-squared (χ^2) distribution. This distribution is a probability density function, giving the probability of a given χ^2 value. It has the following form:

if x is a given χ^2 value, then

$$\text{probability}(x) = \frac{1}{2^{d/2}\Gamma(d/2)} x^{d/2-1} e^{-x/2} \qquad \text{where } \Gamma(z) = \int_0^\infty x^{z-1} e^{-x} dx$$

The quantity d is the "degrees of freedom", which is $k-1$.

These abstruse formulas are buried deep in the realms of mathematical analysis. It is difficult and not particularly useful for practical statistics applications to understand where they came from, or what they mean. The student is never taken through these formulas; the only use that is made of the formulas is that we can use them to calculate how likely it is to have a χ^2 value this large, by consulting this theoretical χ^2 distribution to get its cutoffs for the area remaining under the curve. If the observed value of χ^2 is greater than the 99th percentile of the theoretical χ^2 distribution, then we can declare that the result is statistically significant at the $p < 0.01$ level.

This is what the computer "stats program" is doing when the researcher or student presses the χ^2 button.

We illustrate the χ^2 procedure by an application to a coin-flipping problem.

Calculating χ^2_{obs} for the coin flipping experiment Suppose we have flipped a coin 100 times and recorded the Heads and Tails. Let's say we observed 60 Heads and 40 Tails, so that is our observed histogram. In theory, assuming the coin is a fair coin, we would expect to see 50 Heads and 50 Tails, and so that is the expected histogram, based on theory.

Given the two histograms, observation and expectation, we would use the χ^2 statistic to quantify the distance between them. We simply plug in the observed and expected values of each outcome, to calculate the distance between the observed histogram and the expected histogram (Figure 8.59).

$$\chi^2_{\text{obs}} = \sum_{i=1}^{k} \frac{(\text{O}_i - \text{E}_i)^2}{\text{E}_i} = \frac{(60-50)^2}{50} + \frac{(40-50)^2}{50} = 2 + 2 = 4$$

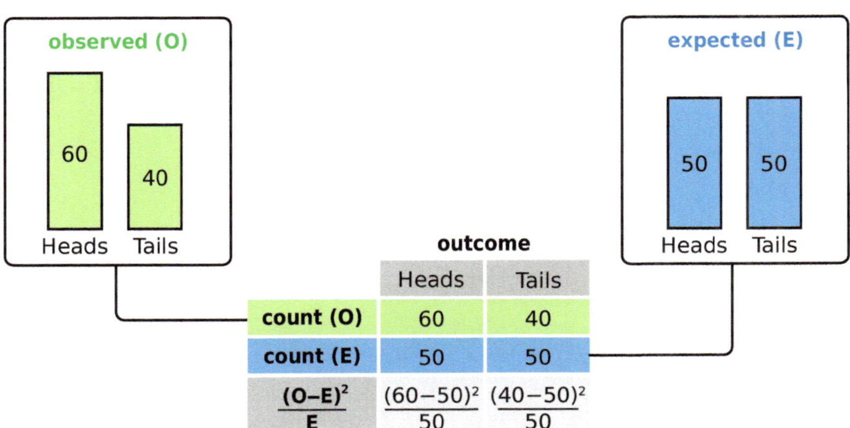

Figure 8.59 Using the χ^2 statistic to quantify the distance between the observed and expected for the 100 coin flip experiment.

How large is "large"? The traditional answer

We've now obtained a value for the "distance" between the two histograms. The χ^2 value for the distance is 4, and the question now is: is 4 large? What is its p-value?

In the traditional, formula-based approach, we now look up (or typically, have a computer look up) "$\chi^2 = 4$" in the relevant theoretical formula to get the p-value. What the computer is doing is, first, finding the right χ^2 curve for the "degrees of freedom" of our data (Figure 8.60), and then calculating how much area there is under the tail of the curve to the right of the observed value. That's how often we would expect to see a value of χ^2 as large as or larger than the observed, under the very strong technical formulation of the Null Hypothesis (see *Problems applying the χ^2 formula* on page 427).

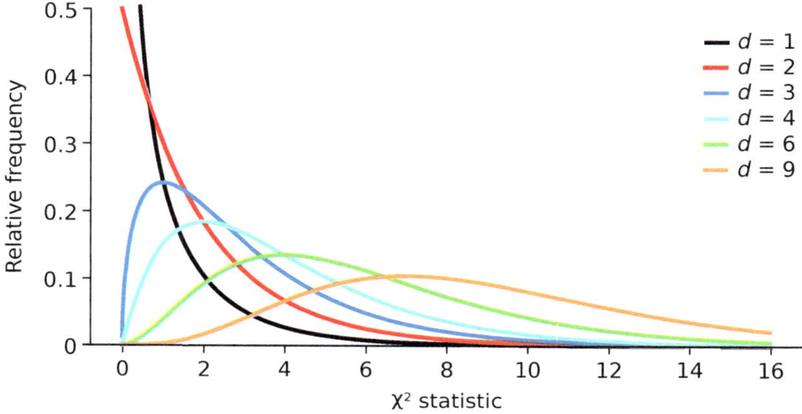

Figure 8.60 χ^2 distributions for various "degrees of freedom".

Calculating the p-value by formula for the coin flipping experiment In the coin-flipping example, there are 2 outcomes, so the number of bins, k, is 2, therefore, the degrees of freedom is $d = k - 1 = 1$. When we calculate the area under the $d = 1$ curve that lies to the right of "4", the answer is 0.0455. Since the total area under the curve is 1 (because this is a probability density function), the fractional area is 0.0455/1, giving us a p-value of 0.0455. Therefore, an outcome of 60 of one face and 40 of the other is not statistically significant at the $p < 0.01$ level (Figure 8.61). (We would have needed $\chi^2 > 6.6$ to have $p < 0.01$ for $d = 1$.)

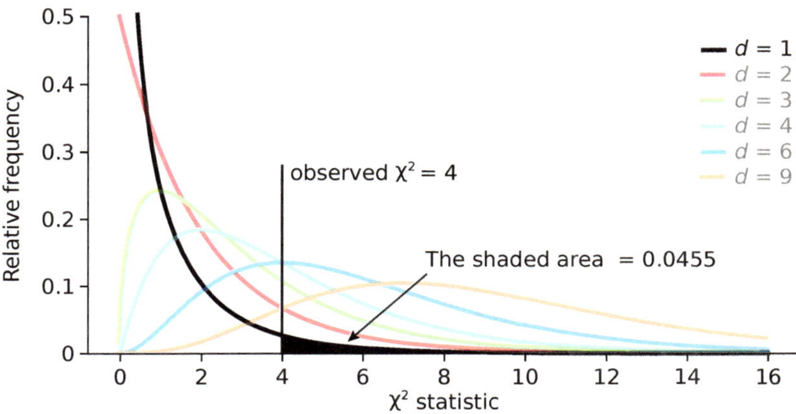

Figure 8.61 **Formula-based** calculation of a p-value for the coin flipping experiment. The theoretical χ^2 curve for $d = 1$ is shown in black. The area to the right of the observed result ($\chi^2 = 4$) is 0.0455.

Problems applying the χ^2 formula

But the traditional approach has major drawbacks.

First of all, the Null Hypothesis in the χ^2 test has a different and technical formulation. In our χ approach, we were able to frame the Null Hypothesis in a particularly straightforward and intuitive way, that the observed histogram was a random sample from the expected histogram. In the χ^2 formula approach, on the other hand, we have to make a narrow technical formulation of the Null Hypothesis, that each of the bin counts is drawn from the same Normal distribution.

This introduces the always-questionable assumption that values are drawn from the same Normal distribution. Why do we need to bring Normal distributions into the discussion? The answer is that in those days before computers, probabilities could only be estimated through theorems and their associated formulas. In order to prove theorems and derive formulas, heavy assumptions had to be made, for example, that quantities follow Normal distributions. In real life, these assumptions are rarely true.

In addition, importantly, we have to assume that n, the total number of observations, is very large, because the theorem says, "... as n approaches infinity, the distribution approaches the probability density function for χ^2" with the appropriate degrees of freedom.

Now we are in the computer age where probabilities can be easily simulated based on realistic assumptions. This was the point of the ground-breaking article "*Statistical data analysis in the computer age*" by Efron and Tibshirani (1991).

In practice, the "n approaches infinity" clause results in the restriction that, to have any validity, there must be a certain minimum number (10 is a frequently cited number) of data points in each bin of the expected histogram. Otherwise, the assumption of smoothness will not be met. However, even in that case, there are discrepancies introduced by the assumption in the χ^2 formula that n, the total number of observations, is infinite.

A better method: a simulation-based approach

The easiest, most intuitive, and most accurate way to answer the question in the coin-toss experiment, is $\chi^2 = 4$ large?, is to start with a more intuitive Null Hypothesis, that the coin is fair (prob of H = prob of T = 0.5). Then we flip the coin 100 times and record the outcomes, then repeat that 10,000 times and see how often 60/40 or more extreme happens.

Isn't this the most natural interpretation, to ask: how often would a χ^2 value of 4 or bigger come up in simulations of a random 100-flip coin toss? who needs the theoretical χ^2 probability distribution for $n = \infty$? just count how often it happens in randomizations of our (finite) data!

Notice that in the simulation-based approach, the Null Hypothesis ("fair coin") is easily understood and makes sense. By contrast, the Null Hypothesis in the formula-based approach is "The values are drawn from the same Normal distribution". But is that really your Null Hypothesis? Why do we have to assume *anything* about Normal distributions?

And secondly, in the formula-based approach, we now have to compare our observed value to an abstruse formula with infinite integrals of products of powers and exponentials, that makes little or no sense to us. Can we really be happy with this?

Calculating the p-value from 10,000 simulations for the coin flipping experiment By contrast, when we carry out the 10,000 simulations of the Null Hypothesis, we see that the observed value $\chi^2 = 4$ was equalled or exceeded in 574 of the 10,000 simulations, so the p-value of the observed result is p = 0.0574 (Figure 8.62).

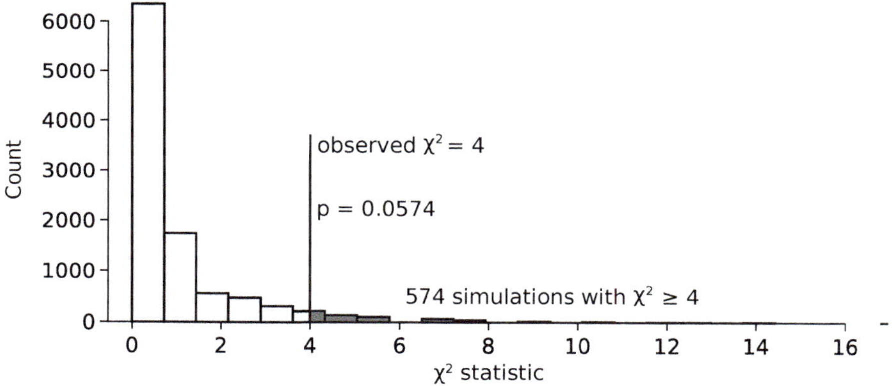

Figure 8.62 Simulation-based approach to calculate a p-value for the coin-flipping example. Of the 10,000 simulations, 574 simulations have a χ^2 value equal to or larger than the observed ($\chi^2 = 4$), resulting in a p-value of 0.0574.

Who is approximating whom? The two answers, from formula and simulation, are somewhat close but definitely not equal: 0.0455 and 0.0574. When we compare the histogram of the 10,000 simulations to the theoretically-computed χ^2 curve for $d = 1$, we see that they are somewhat close, but that there are definite discrepancies (Figure 8.63).

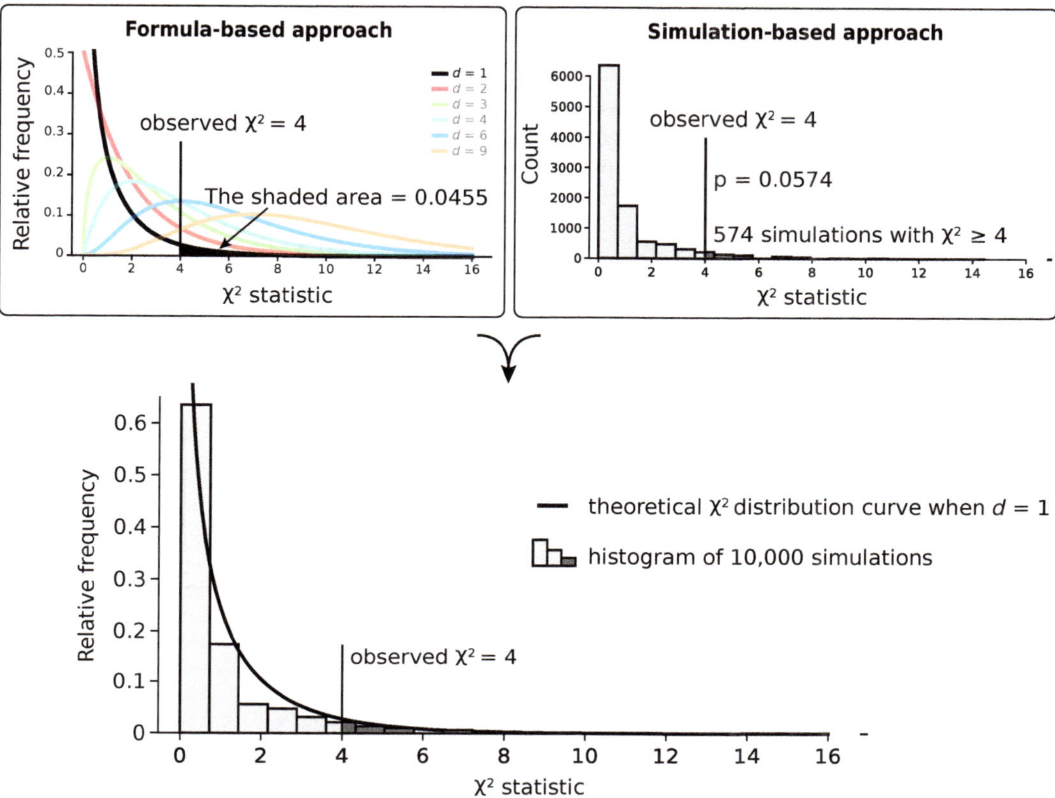

Figure 8.63 Comparing theory and simulation-based approaches.

It is natural to think, "Well, the finite computer simulations are an approximation to the True Theory as given by χ^2". But this would be a mistake. **The discrepancies are due to the $n = \infty$ assumption in the χ^2 approximation, not to the 10,000 simulations; in fact, when we increased the number of simulations to 100,000, the p-value was 0.05708, very close to the value for 10,000 simulations and still not very close to the theoretically calculated value.**

The truth is that for $n = 100$ flips, as in this case, **the simulations are the exact answer, and the "theoretical" value is an approximation, based on the false assumption that 100 = ∞.**

In math, it's often the case, for example in differential equations, that computer simulations are an approximation to the "real truth" which is a theoretical continuous function. But here the opposite is the case.

For a finite sample of n, the theoretical χ^2 distribution is an approximation based on the false assumption that $n = \infty$. The computer simulations are a more precise and correct estimate of significance than the theoretical χ^2 distribution.

Example 6 rock-paper-scissors: formula vs. resampling

Recall the study of people playing the game "rock-paper-scissors"; the frequency of the 3 hand signs was Rock = 107, Paper = 124 and Scissors = 150, for a total of n = 381 games.

We are going to demonstrate the two approaches to calculate a p-value: the formula-based approach and the resampling-based approach, and compare the results.

First, we calculate the observed χ^2.

$$\chi^2_{\text{obs}} = \sum_{i=1}^{k} \frac{(O_i - E_i)^2}{E_i} = \frac{(107-127)^2}{127} + \frac{(124-127)^2}{127} + \frac{(150-127)^2}{127} = 7.386$$

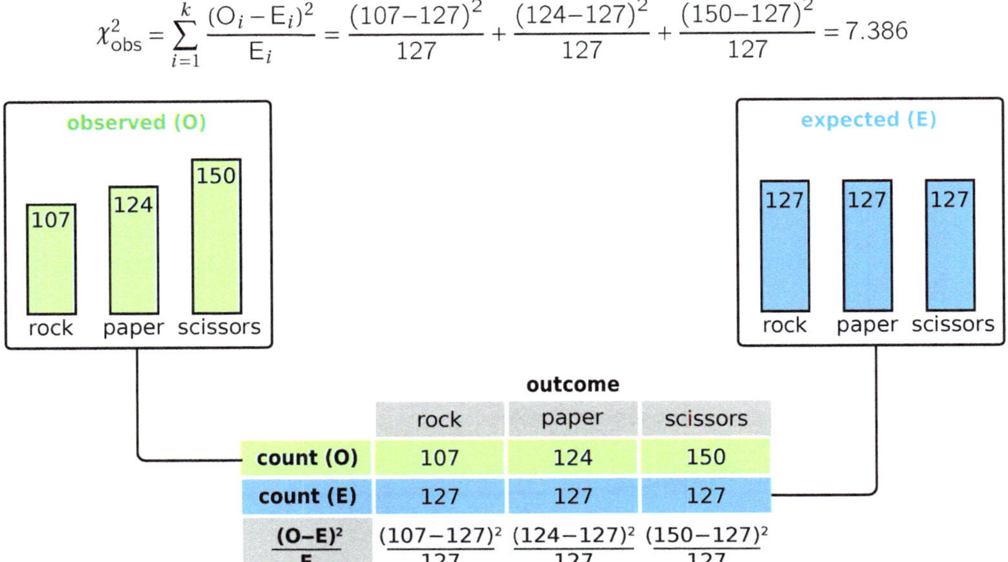

The next question is: is this value of χ^2_{obs} = 7.386 statistically significant?

For the formula-based approach, the "degrees of freedom" is 2, because we have 3 outcomes, and therefore k = 3 bins in our histogram. Looking up the area under the χ^2 distribution curve for d = 2 that is to the right of 7.386, we find that it is 0.025, so that is the theoretical p-value.

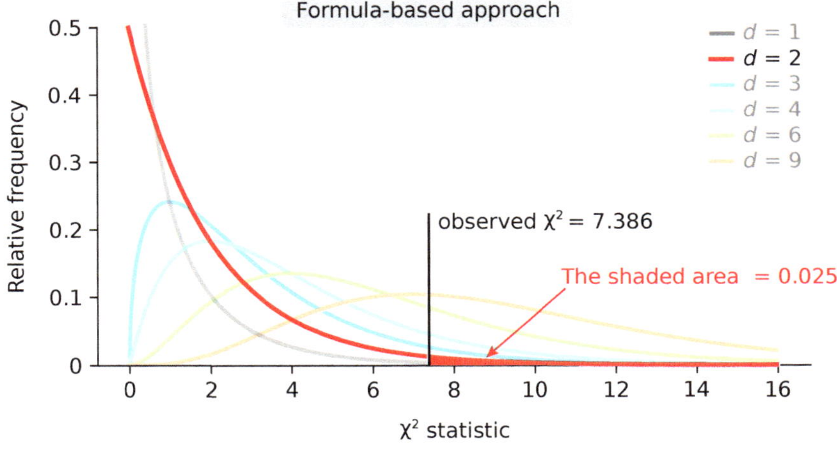

Let's now compare that formula-based result to a simulation-based approach.

To create one simulation, we simply sample $n = 381$ with replacement from the set {R, S, P}, with all three elements having equal chance of being drawn, and record the result into a histogram. This is the first sample from the world of the Null Hypothesis. Then we calculate the χ^2 value for this histogram, compared to the expected histogram, which has $381/3 = 127$ counts in each bin.

We repeat the resampling precess 10,000 times to get 10,000 χ^2 values.

The p-value of the observed $\chi^2 = 7.386$ is then the question of how many simulations ended up with a χ^2 value at least as extreme as the observed.

When we carried out 10,000 simulations under the Null Hypothesis of equal frequency for all signs, the value 7.386 or greater occurred 263 times, giving a p-value of 0.0263.

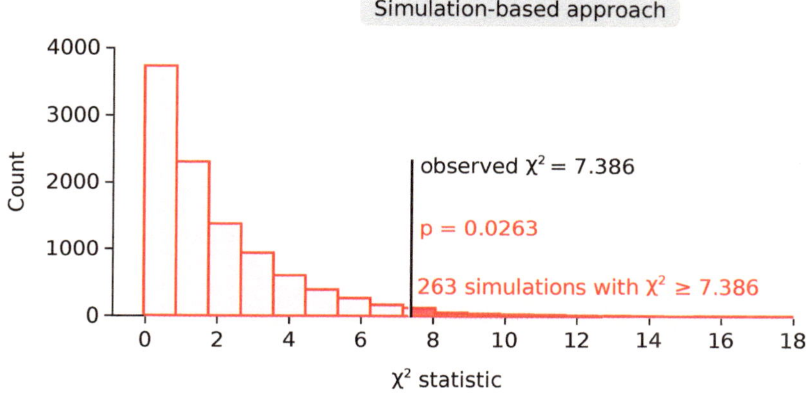

We have *some* evidence against the Null Hypothesis, although not at the $p < 0.01$ level.

When we compare the $d = 2$ theoretical curve to the histogram of 10,000 simulations, we see that there is a rough but not complete agreement: 0.025 vs. 0.0263. The theoretical curve is very close to the simulated histogram, but not exactly equal to it.

We stress again, just as the $d = 1$ case, that **the reason for the discrepancy is the approximations made by the theoretical χ^2 distribution curve, not the finiteness of the 10,000 simulations.** *The simulation result is more correct than the theoretical result.*

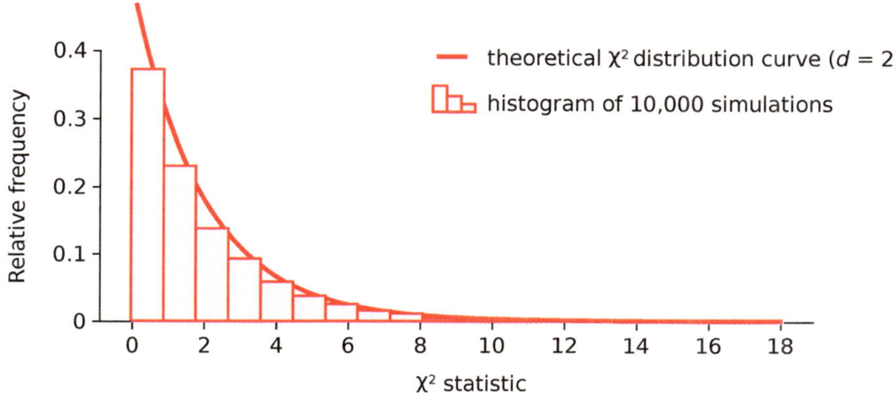

A farewell to χ^2?

As we have seen, computer simulation gives us a much clearer, more intuitive and more valid way of comparing observed histograms to expectations, without having to make unrealistic assumptions, like Normal distributions and infinite samples. Importantly, it also frees us from being forced to square all differences, in order to bring us under the assumptions of the formula; we can use absolute values, and avoid the highly nonlinear distortions introduced by squaring everything.

Efron and Tibshirani entitled their ground-breaking 1992 *Science* paper "Statistical Data Analysis in the Computer Age". Their point was: *"Modern electronic computation has encouraged a host of new statistical methods that require fewer distributional assumptions than their predecessors and can be applied to more complicated statistical estimators."*

Our discussion of χ fits this description perfectly. The resampling approach does indeed "require fewer distributional assumptions", namely, we don't have to assume infinite Normal distributions any more; the resampling procedure embodies the Null Hypothesis. And our χ estimator is "more complicated" than the traditional χ^2 because absolute value can't be written as a single formula; it's *two* formulas tied together by an "if-then" statement. The "if-then" statement means that χ is not a smooth function, which makes it challenging or even impossible to prove theorems about it.

Then, you might ask: why do we have the χ^2 formula at all? Why have we been using and teaching this formula for the last 125 years? The short answer is: because Karl Pearson did not have a computer in the year 1900!

> The excellent text by Freedman, Pisani and Purves, *Statistics*, tells us:
> "In Pearson's time, there were no computers to find the chances. So he developed a method for approximating P by hand ... without having to do a computation that was, — by the standard of his time, quite formidable." (Freedman *et al.*, 2007)

Note that they correctly refer to χ^2 as "approximating P by hand".

And by the standards of *our* time, 10,000 simulations of the Null Hypothesis in a typical case can be done in a matter of minutes on your phone.

So, if we used χ^2 because Karl Pearson did not have a computer in 1900, why does χ^2 continue to be used today?

Chapter 9

Bivariate Data: Correlation and Beyond

LEARNING OBJECTIVES

After studying this chapter, you will be able to:

- apply the concept of correlation to a bivariate data set.
- state what requirements must be met in order to apply the concept of correlation.
- carry out resampling-based approaches to Null Hypothesis Significance Testing and confidence intervals for the correlation coefficient.
- understand how to use the notion of Mutual Information to analyze relations that are not straight lines.

9.1 Bivariate data: two measurements

Experimental design

So far, we've been looking mostly at data where we have one measurement per subject, called a *univariate* data set. But often, we want to look at data sets where we have not one but *two* quantities, call them X and Y, for each subject, and we are interested in exploring whether there is a relationship between a subject's X-value and the corresponding Y-value. This is called a **bivariate data** set. In bivariate data sets, we want to know how those two quantities are related to each other (Figure 9.1).

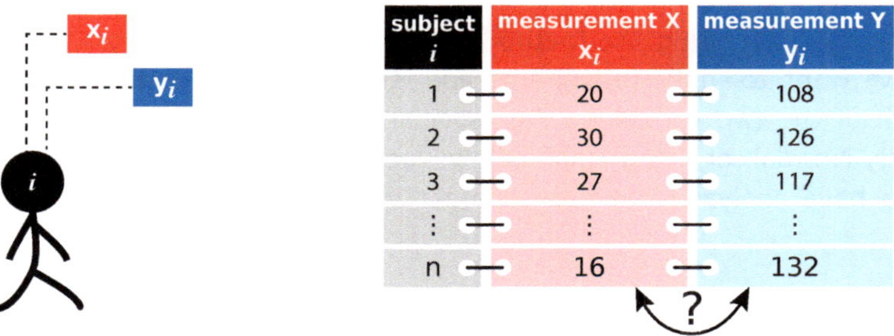

Figure 9.1 A bivariate data set.

A. Garfinkel and Y. Guo, *Understanding Data*,
https://doi.org/10.1007/978-3-032-18600-3_9

So we have not just blood pressure for each subject, but, say, blood pressure and age, or height and weight, or years of experience and annual salary, or the incidence of lung cancer and the number of cigarettes smoked per day, or ..., and we want to know if there is any relation between the X-value of a subject and its Y-value.

Let's call the subjects $\{1, 2, \ldots, n\}$, and call the two quantities X and Y, so our measurements can be represented as $\{(x_1, y_1), (x_2, y_2), \ldots, (x_n, y_n)\}$, giving us n *pairs* of data values. This is different from simply having a pile of X-values and a pile of Y-values: for each subject i, its X-value x_i is linked to its Y-value y_i.

So it is critical for this type of data that a pair of data values, say, the age and blood pressure, **are the age and blood pressure of the same person** (Table 9.1).

		measurement X ⇓	measurement Y ⇓
Subject	Name	Age (years)	Blood Pressure (mmHg)
1	Sally E	$x_1 = 20$	$y_1 = 108$
2	Leander J	$x_2 = 30$	$y_2 = 126$
⋮	⋮	⋮	⋮
n	Kayla W	$x_n = 27$	$y_n = 118$

Table 9.1 A sample bivariate data set.

Of course we are interested, as always, in the two separate variables. What are their distributions? How variable are they? What summary measures are appropriate? etc. But we are especially interested in any *relations* between the X-value of a subject and its Y-value. Is there a relationship between age and blood pressure, or salary and years of experience?

Exercise 9.1.1 Which of the following are bivariate data sets?

a. Left hand and right hand grip strength in a given individual

b. Salary levels of women and men

c. Serum glucose levels in younger and older people

d. Median income and median years of education in each state

How do we approach this kind of data set?

Clearly we want to begin with the analysis of the two separate data sets $\{x_1, x_2, \ldots, x_n\}$ and $\{y_1, y_2, \ldots, y_n\}$, using the methods of Chapter 2 (*Describing and Presenting Data*). In particular, we want to make histograms and Kernel Density Estimates, dot plots, box plots, etc. for each of the two variables. We especially want to note whether their distributions are approximately Normal, or if they markedly violate Normality, for example, by being skewed.

Once we have analyzed the two separate variables, we can go on to ask: how are they related?

Making a picture: the X–Y scatterplot

The key first step in understanding the X–Y relationship is to make a picture of this relationship by making a **2D scatterplot**, in which we simply plot the pairs (x_i, y_i) as points in a 2-dimensional coordinate space, in which the axes are X and Y (Figure 9.2). So each data point is a geometric point in, say, Age–Blood Pressure space, whose axes are Age and Blood Pressure.

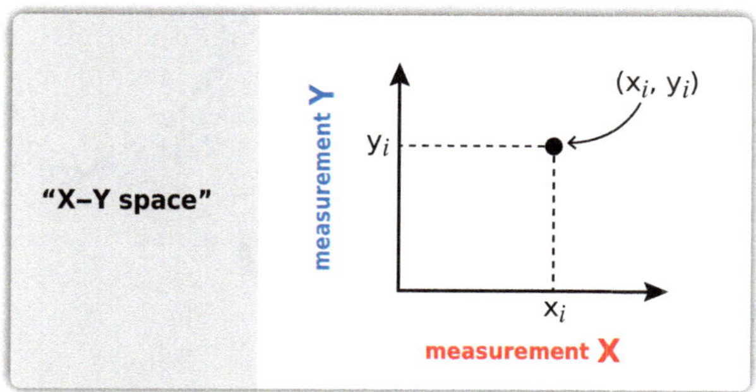

Figure 9.2 In a scatterplot, we plot each pair of measurements (x_i, y_i) as a point in a 2-dimensional coordinate space.

We then want some ways to quantify the relationships between X and Y. But it is important to never forget that **the most important step in the analysis of X–Y data is to make the picture of the scatterplot.** Everything we do depends on what we see in that picture, and *the picture is far more important than any numbers that we use to describe it.*

In analyzing X–Y bivariate data, the critical first step is to make an X–Y scatterplot, in which the data points are plotted on an X–Y axis. All further analyses depend upon what we see in that plot.

Unfortunately, many contemporary research papers lose sight of this, avoiding the scatterplot altogether, while displaying only summary numbers that purport to describe the scatterplot.

Looking at scatterplots

There are many things that can be seen in a scatterplot (Figure 9.3). Nonlinear relationships, like those in panels a, c, d and f, are extremely interesting, scientifically important, and give us major insights into the data. **But they are beyond the scope of the standard concept of correlation, which is solely concerned with one question: *do the data tend to lie on a straight line?* Cases where there is a clear relationship that is not a straight line are simply not addressed by the standard concept of correlation.**

Many people make the mistake of thinking that correlation means "co-relation", but this is not true. The variables in panels a, c, d and f are *co-related,* but they are not *correlated.*

If the data does tend to lie on a straight line, then if Y increases with increasing X, the slope of the line is positive (Figure 9.3 panel b), while if Y decreases with increasing X, the slope of the line is negative (Figure 9.3 panel e).

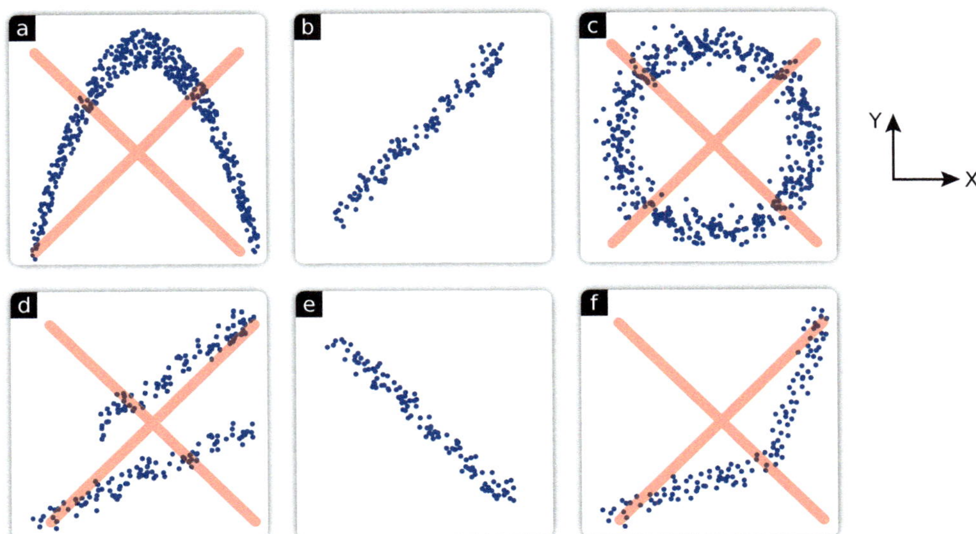

Figure 9.3 Six examples of X–Y scatterplots. The only relationships that can be addressed by the standard concept of correlation are those in panels b and e. Panel f can be addressed with a rank-based correlation (see section 9.3 *Ranks and the Spearman correlation*).

No relation at all Lastly, the scatterplot may indicate *no relation at all* between Y and X. When there is no relationship between Y and X, the scatterplot will resemble a speckled blob, a speckled band or even a speckled box, depending on the distributions of the variables (Figure 9.4).

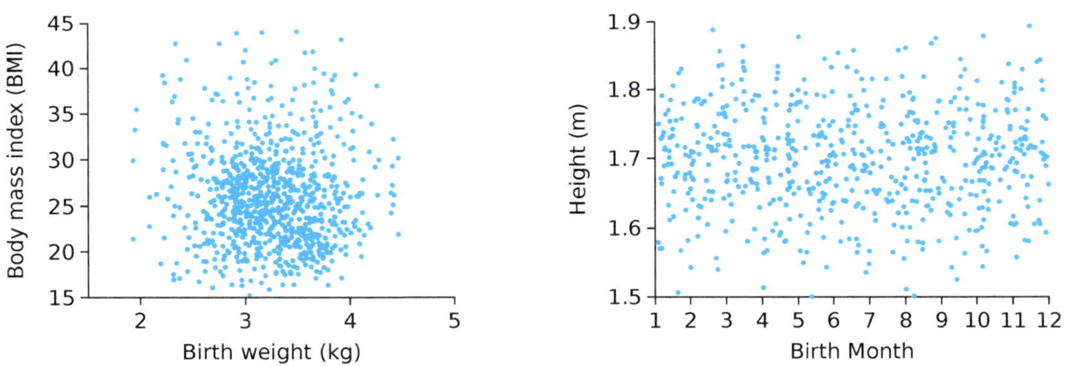

Figure 9.4 Two examples of bivariate data sets that have no relation at all.

For example, an X–Y scatter plot of an adult's Body Mass Index (BMI) with their birthweight resembles a speckled blob, suggesting that an adult's BMI is not related to their birth weight. Given a person's birthweight, we cannot tell anything about the BMI of that person as an adult.

Similarly, the X–Y scatterplot of adult height against birth month is a speckled band, since the birth month variable is uniformly distributed throughout the year. Given a person's birth month, we can say nothing about their adult height. Adult height, in this scatterplot, shows no relation to birth month.

Exercise 9.1.2 Comment on the scatterplots and relationships between the variables.

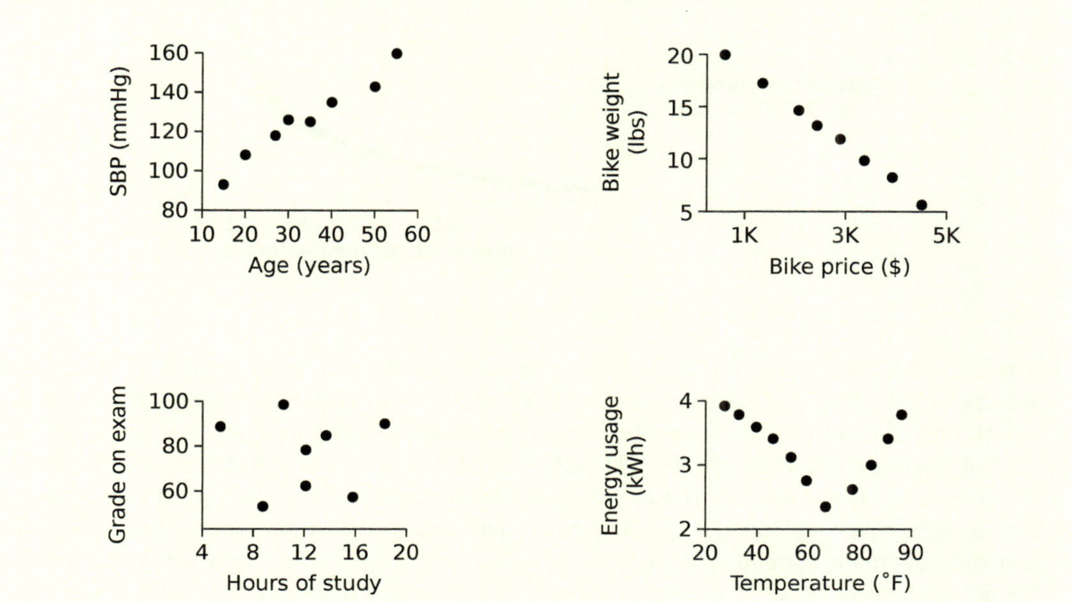

Example 1 Who becomes an inventor?

A study looked at 1.2 million individuals in a patent-tax database ("Who becomes an inventor in America? the importance of exposure to innovation", *The Quarterly Journal of Economics*, 2019).

They defined an "inventor" as someone listed as an inventor on a patent application or a grant. They also had access to the inventors' birth years, and whether they were male or female. This enabled them to calculate, for each birth year, what fraction of inventors born in that year were female. They made a scatterplot of the female fraction against birth year.

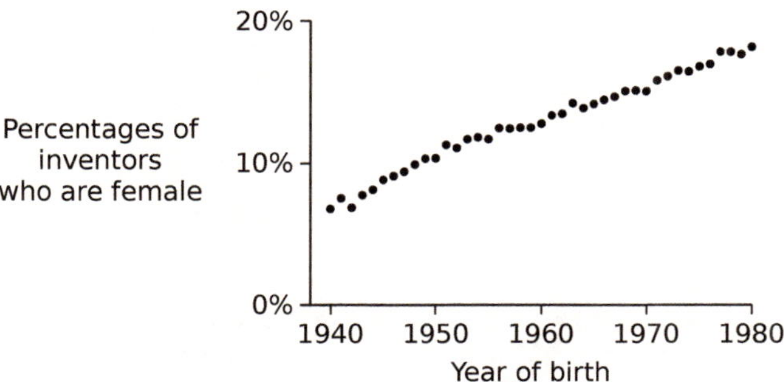

Question Comment on the relationship, if any, between female percentage and year of birth.

Answer The data lies more-or-less on a straight line. Younger generations have a higher percentage of female inventors, and that trend is linear: the percentages grew linearly over time.

The authors also plotted the number of children who became inventors (Y) against their parents' income (X).

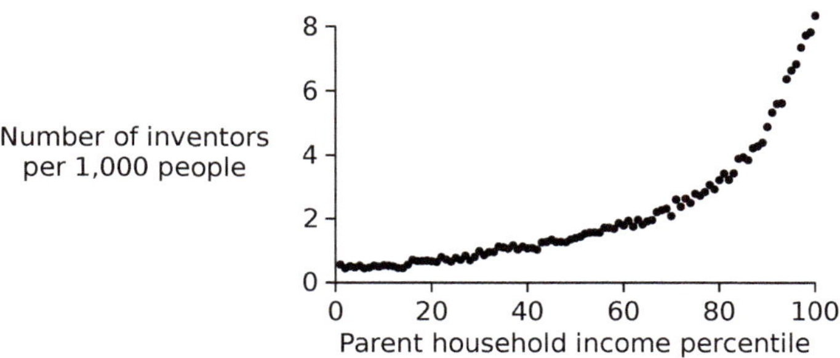

Question Comment on the relationship, if any, between Y and X.

Answer The data are clearly co-related. The data points lie on a curved line. The curve is increasing as a nonlinear function of X, like X^2 or X^3. This certainly suggests that parental wealth has an extremely strong relation to the possibility of becoming an inventor.

But the line is not a straight line. Therefore, it cannot be analyzed by the standard concept of correlation. Data sets like these, where Y always rises (or falls) with X can be analyzed with a more advanced concept of correlation: rank-based correlation (see section 9.3 *Ranks and the Spearman correlation*). A more advanced approach to this data set would include asking *what* curve best fits this data. Is it a square? an exponential? etc. This is the topic of nonlinear regression (see section 10.3 *Nonlinear regression*).

Strength of association If there *is* a linear relationship between Y and X, we can ask how much scatter there is in Y for a given X, and in X for a given Y. We will use the term "strength of association" (or "strength of relationship") to indicate the tightness of the scatter (Figure 9.5).

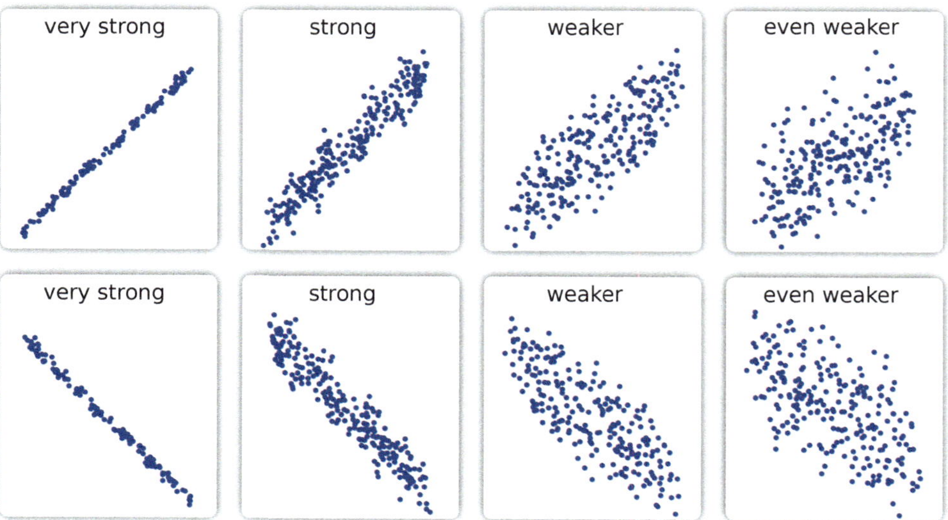

Figure 9.5 If the scatterplot approximates a straight line, then the strength of association depends on how tightly the points cluster around that line: tighter means stronger, more scatter means weaker. The top row shows positive associations; the bottom row shows negative associations.

When the scatter is minimal, so there is very little spread, we will call that a very strong association or a very strong relationship. If the scatter is a little more, we can still call it a strong association, but as the scatter gets larger, the strength of association gets weaker. This applies whether the association is positively sloped or negatively sloped.

Example 2 Change of CO_2 and temperature in the past 800,000 years

The EPICA (European Project for Ice Coring in Antarctica) Dome C ice core from East Antarctica provides the current longest ice core record, going back 800,000 years. From the core, scientists can measure past levels of CO_2 in the atmosphere and estimate temperature changes over the period of the record.

Here is the CO_2–Temperature scatterplot over the past 800,000 years. We down-sampled the temperature record to points for which there was a corresponding CO_2 record, resulting in n = 1768 data points in this scatterplot (see articles "Orbital and millennial antarctic climate variability over the past 800,000 years", *Science*, 2007, and "High-resolution carbon dioxide concentration record 650,000–800,000 years before present", *Nature*, 2008).

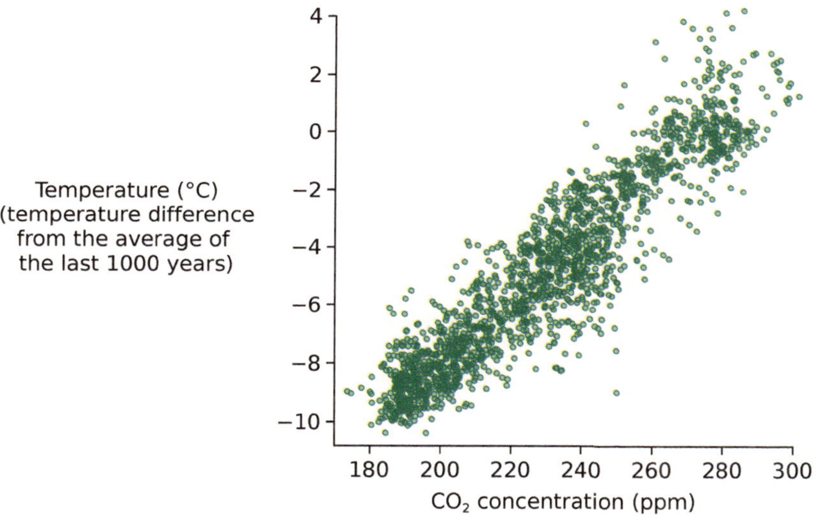

Question Comment on the relationship, if any, between CO_2 (X) and temperature (Y).

Answer Looking at the scatterplot, we ask:

Does it more or less "lie on a line"? The data form a long thin cloud, with a consistent scatter across both X and Y, so yes, the data are "lying approximately on a line".

Linear or not? The long thin cloud does resemble a straight line. The dots in the scatterplot roughly lie on a 45 degree angle. This indicates that the two variables are positively correlated. High concentrations of CO_2 are associated with high temperatures, and low concentrations of CO_2 are associated with low temperatures.

Strength of association. The strength of association here is strong, since the scatter is not large compared to the variation in the variables.

FURTHER EXERCISES 9.1

1. In which of the following scatterplots do the data "lie more or less on a straight line?"

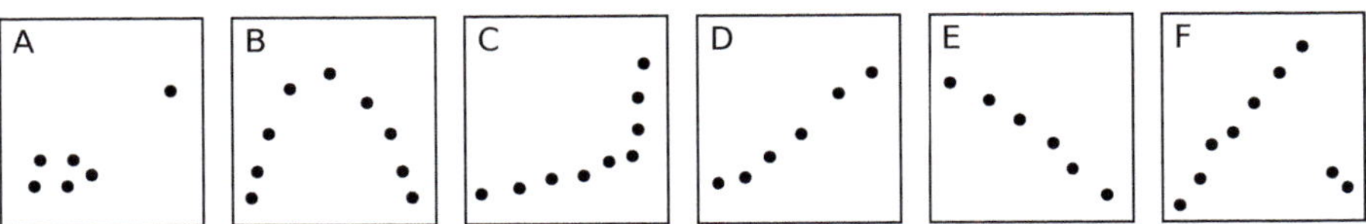

2. **High-altitude cooking.** At elevated altitudes, any cooking that involves boiling or steaming generally needs to be adjusted. This is because, at sea level, water boils at 100 °C (212 °F), and for every 152.4 meter (500 ft) increase in elevation, water's boiling point is lowered by approximately 0.5 °C. Which of the following scatterplots shows the relationship between water's boiling point (B) and elevation (E)?

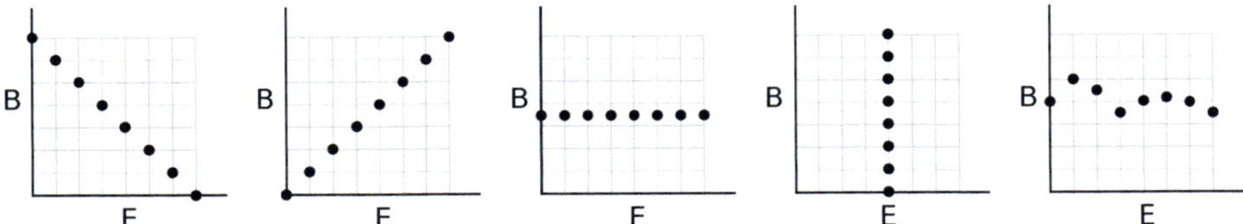

3. **Constructing a scatterplot.** A researcher studying diet and nutrition has collected data on the weights of infants and their corresponding ages. The table below shows the median weights for infants of different ages in the study.

Age (months)	0	1	2	3	4	5	6
Weight (kg)	3.5	4.25	5.0	5.5	6	6.75	7.0

Make a scatterplot of this data set.

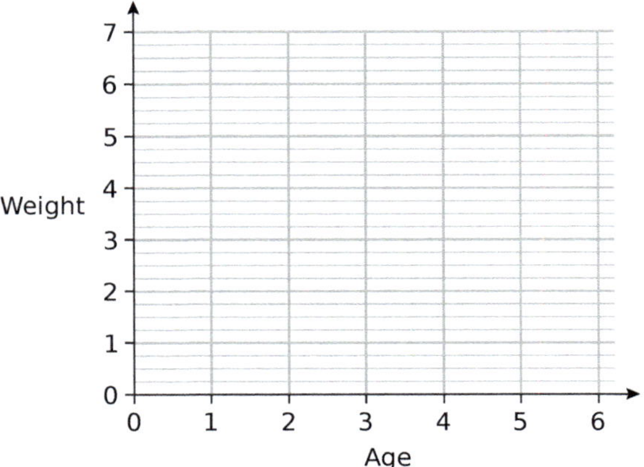

9.2 The correlation coefficient

If our initial visual impression from the scatterplot is that the data *does* lie more or less on a straight line, then we can ask whether we can quantify this. Can we develop a number that reflects the strength of association, the degree to which the data lies on a straight line? The answer is: possibly. If the data and its separate and joint distributions meet some fairly strong assumptions, then there is a number that reflects how well the data lies on a straight line; it's called the **correlation coefficient**.

However, we have to remember that before we proceed to calculate the correlation coefficient, we must have verified in the scatterplot that the data are *not* scattered around some other form that is *not* a line.

If the data lie more or less on a something other than a straight line, the standard concept of correlation does not apply. Do not calculate a standard correlation coefficient in this case; it is not good practice because it mis-represents the relationship in the data.

Covariance

In order to see how to define the correlation coefficient, let's begin with a sample set of data, for illustration purposes (Figure 9.6).

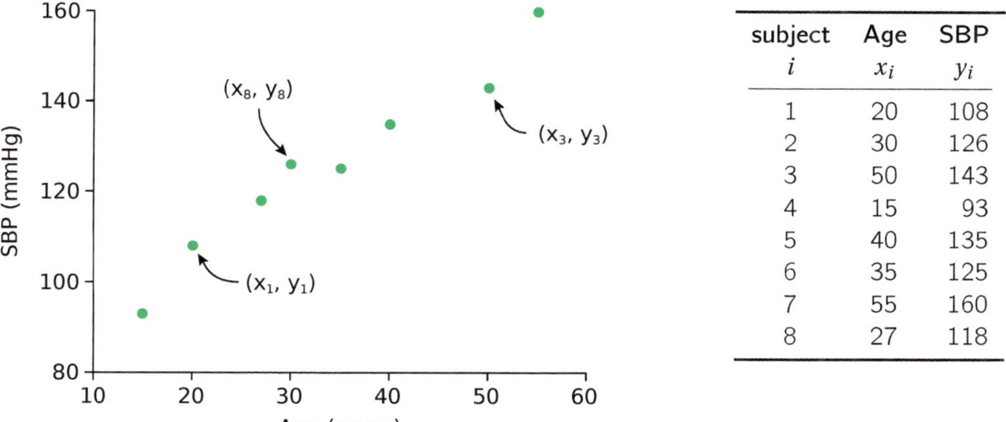

subject i	Age x_i	SBP y_i
1	20	108
2	30	126
3	50	143
4	15	93
5	40	135
6	35	125
7	55	160
8	27	118

Figure 9.6 Ages and systolic blood pressures (SBP) for 8 subjects.

We see that the data does more or less lie on a straight line, and we would like to quantify how much.

To begin, we have to ask: what does it mean to say that points are "lying on a line"? One implication is that larger values of X are paired with larger values of Y, and smaller values of X are paired with smaller values of Y. We want a number that reflects this tendency, that is, a number that will be large when X and Y are both relatively large, and also large when X and Y are both relatively small.

We might think about the products $x_i \cdot y_i$. They are certainly large when both x_i and y_i are large, but the product fails to have the other necessary property, to be large when both x_i and y_i are small.

However, a neat trick will solve this problem (Figure 9.7). In our sample data set, the data values are all positive. But what if we shifted the origin of the axes to the center of the data? Then large numbers would still be relatively large, but small numbers would now become relatively large negative values. This would make the product of two small numbers become the product of two relatively large negative numbers, so the product would again be large and positive, which is what we wanted.

The idea of "moving the origin to the center of the data" can be made precise. First, we need to choose an (x, y) point to be the center of the data. The choice is non-trivial. For now, let's suppose that the data distribution is approximately Normal (see section 2.3 *Summarizing a data set: "measures of central tendency"*), that is, that it's unimodal, symmetric, and has thin tails. In that circumstance, we are justified in using the *mean* as our "center" point. We assume this is true for both the X and Y distributions, so the midpoint of our 2D scatterplot is the point $(\overline{X}, \overline{Y})$.

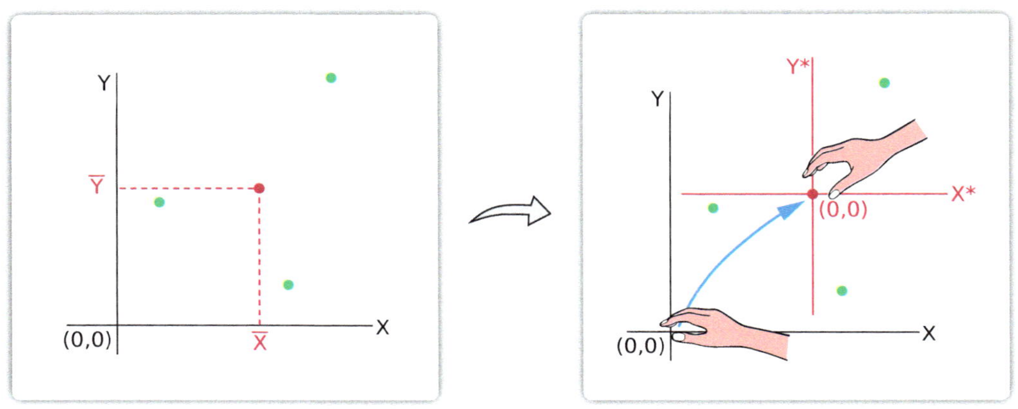

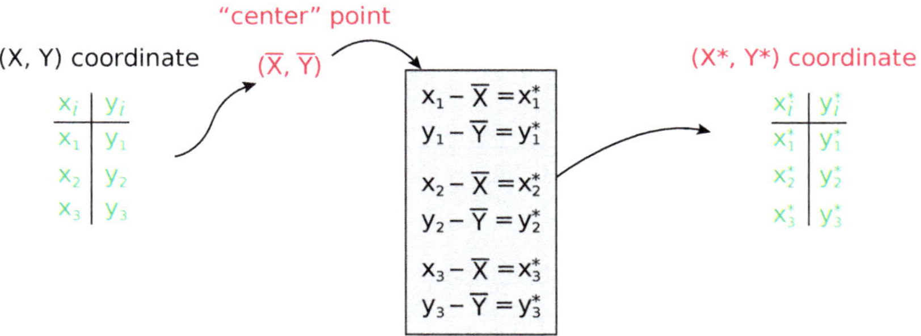

Figure 9.7 Moving the axes to the center of the data.

Moving the origin of the axis from (0, 0) to $(\overline{X}, \overline{Y})$ amounts to subtracting $\overline{X}$ from each X-value x_i, and subtracting $\overline{Y}$ from each Y-value y_i. These new transformed values will be called $(x_i^*, y_i^*) = (x_i - \overline{X}, y_i - \overline{Y})$. The newly-transformed coordinate axes will be called $(X^*, Y^*) = (X - \overline{X}, Y - \overline{Y})$.

Now we can take the products $x_i^* \cdot y_i^*$, and use the fact that small values of x_i (or y_i) get transformed into negative values of x_i^* (or y_i^*). Consequently, we take the products $x_i^* \cdot y_i^*$ and add them up to quantify how much the points lie on a straight line.

But this simple sum-of-products suffers from the drawback that the bigger the data set, that is, the larger n is, the bigger will be the sum of the products, just because there are more terms in the sum. We don't want that, and so we divide the sum of the products by the number of data points to remove that effect.

The sum of the products, divided by n, is then a measure of how much X^* and Y^* (and therefore X and Y) co-vary together in a linear manner. This quantity is called the **covariance of X and Y**.

We can write the definition of covariance mathematically as

covariance $$\text{cov}(X, Y) = \frac{1}{n}\sum(X^* \cdot Y^*) = \frac{1}{n}\sum_{i=1}^{n}(x_i - \overline{X})(y_i - \overline{Y})$$

Notice that the covariance as we defined it is a symmetric quantity: it follows from the definition that $\text{cov}(X, Y) = \text{cov}(Y, X)$. To see this, just substitute X for Y and Y for X in the definition of $\text{cov}(X, Y)$ above. And notice also that if we ask what is the covariance of X with itself, and plug X in for Y in the definition of covariance, we get

$$\text{cov}(X, X) = \frac{1}{n}\sum(X^* \cdot X^*) = \frac{1}{n}\sum_{i=1}^{n}(x_i - \overline{X})^2$$

which is exactly the quantity we called the *variance* of X in section 2.4 *Summarizing a data set: measures of variability.*

Correlation

There is still a problem with the notion of covariance. According to the definition, its magnitude depends on the units with which we measure X and Y. For example, suppose we correlate height and weight, and we decide to measure height in feet and weight in pounds. Suppose subject #1 has a height of 6 feet and a weight of 170 pounds. so (x_1, y_1) is (6, 170). Let's say the average height of our sample data set is $\overline{X} = 5.8$ feet, and the average weight is $\overline{Y} = 145$ pounds.

Then the contribution of subject #1 to the covariance of X and Y is

$$x_1^* \times y_1^* = (x_1 - \overline{X}) \times (y_1 - \overline{Y}) = (6 - 5.8) \times (170 - 145) = 0.2 \times 25 = 5(\text{ft} \times \text{lbs})$$

But if we measure height in inches and weight in ounces, now the data point for subject #1 becomes (72, 2720) and the average value becomes (70, 2320), so the contribution of subject #1 to the covariance now becomes (72 – 70) × (2720 –2320) = 2 × 400 = 800, which is very different from 5!

To put it another way, the covariance has *units*: the value 12.5 has units "ft × lbs", and the value 800 has units "in × oz".

In order to remove this dependence on choice of units, and turn our number into something that is unit-less, we want to divide the covariance by some number or numbers that reflect the choice of units, so that we can cancel out the dependency on units.

In addition, we want to remove the dependency on the spread of the data sets themselves: covariance has the downside that it becomes large whenever the data are highly variable, producing large deviations from the mean.

Relying again on the assumption that the X and Y data are Normally distributed, the traditional approach has been to use the standard deviation of X times the standard deviation of Y as the normalizing factor.

Recall from Chapter 2 that the standard deviation of the X and Y variables is

$$\sigma_X = \sqrt{\frac{1}{n}\sum_{i=1}^{n}(x_i - \overline{X})^2} \qquad \sigma_Y = \sqrt{\frac{1}{n}\sum_{i=1}^{n}(y_i - \overline{Y})^2}$$

It is clear from the formula that the standard deviation of X has the same units as X does (because it is the square root of sums of squares involving X), and that the standard deviation of Y has the same units that Y does.

Therefore, if we divide the covariance, which has the units "X-units × Y-units" by the product of the standard deviations, which is also "X-units × Y-units", then we get a quantity that is unit-less, and therefore independent of the choice of units.

If we divide the covariance by the product of the standard deviations, we arrive at the concept that is commonly called the **correlation coefficient** for short. Its full name is the Pearson Product-Moment Correlation Coefficient. We will refer to it as the correlation coefficient ρ. Other sources use the symbol r. (Later we will see another correlation coefficient that is based on ranks, called the Spearman correlation coefficient.)

correlation coefficient

$$\rho_{XY} = \frac{\text{cov}(X, Y)}{\sigma_X \cdot \sigma_Y} = \frac{\sum_{i=1}^{n}(x_i - \overline{X})(y_i - \overline{Y})}{\sqrt{\sum_{i=1}^{n}\left(x_i - \overline{X}\right)^2}\sqrt{\sum_{i=1}^{n}\left(y_i - \overline{Y}\right)^2}}$$

What does it mean to say that "ρ is independent of the choice of units"? A different choice of units amounts to taking the original measurements and multiplying them by some constant k, which is the unit conversion factor. For example, if the original measurement is in feet, and we are converting the unit to inches, then $k = 12$. Therefore, "ρ is independent of the choice of units" means that we get the same value for ρ if we substitute $k \cdot x_i$ for x_i (because that's what a change of units is).

It is also independent of choice of location, that is, independent of what we choose as the origin. We get the same answer if we move the axes by adding the same constant c_X (positive or negative) to all X-values and the same constant c_Y to all Y-values.

Note that the correlation coefficient is symmetric, $\rho_{XY} = \rho_{YX}$. Because the correlation coefficient is symmetric, we cannot use a correlation to suggest anything causal, such as X causes Y. On the same evidence, we might as well say that Y causes X. (And indeed, it could be neither, see section 9.6 *The interpretation of correlation*.)

z-scores We can look at the correlation coefficient in a useful way by invoking the notion of a z-score. If $X = \{x_i\}$ is a Normally distributed data set, $\overline{X}$ is its mean (average) value, and σ is its standard deviation, then for each value x_i, we can associate its **z-score**, which is defined as

$$z_i = \frac{x_i - \overline{X}}{\sigma}$$

In other words, the z-score of a data point x_i is how far x_i is from the average $\overline{X}$, as measured in units of σ, which is the standard deviation of the data set X. A positive z-score means that the data point is above the mean and a negative z-score means that the data point is below the mean. For example, x_i with a z-score of 2.2 means that the data point x_i is 2.2 standard deviations above the mean; x_i with a z-score of −1.8 means that the data point x_i is 1.8 standard deviations below the mean.

Using the notion of z-score, we can now give another definition of the correlation coefficient: it's the average value of the products of the z-scores of X- and Y-values of each data point

$$\rho_{XY} = \frac{1}{n}\sum_{i=1}^{n}\left(\frac{x_i - \overline{X}}{\sigma_X} \cdot \frac{y_i - \overline{Y}}{\sigma_Y}\right) = \frac{1}{n}\sum_{i=1}^{n}\left(z_{x_i} \cdot z_{y_i}\right)$$

We emphasize that the notion of z-score, and the notion of correlation coefficient, depend critically on the data being Normally distributed.

Exercise 9.2.1 Verify that ρ has the same value if $k_1 \cdot x_i + c_1$ is substituted for every X-value x_i, and $k_2 \cdot y_i + c_2$ is substituted for every Y-value y_i (assuming $k_1 \neq 0$ and $k_2 \neq 0$).

Calculating the correlation coefficient

Positive correlation Let's do a calculation of ρ for a sample data set of $n = 8$ points (data presented in Figure 9.6). Here, the X-values are ages, and the Y-values are systolic blood pressures. In this data set, blood pressure rises with age.

We will carry out the calculation of the correlation coefficient in a table (Table 9.2). Each row is the calculation of the key quantities for subject i ($i = 1, 2, \ldots, n$). The values are then added up according to the formula, and the result is shown at the bottom: the correlation coefficient of this data is 0.98.

① subject ⇓	② age (X) ⇓	③ blood pressure (Y) ⇓	④ recenter $x_i^* = x_i - \overline{X}$	⑤ recenter $y_i^* = y_i - \overline{Y}$	⑥	⑦	⑧
i	x_i	y_i	x_i^*	y_i^*	$x_i^* \cdot y_i^*$	$(x_i^*)^2$	$(y_i^*)^2$
1	20	108	−14	−18	252	196	324
2	30	126	−4	0	0	16	0
3	50	143	16	17	272	256	289
4	15	93	−19	−33	627	361	1089
5	40	135	6	9	54	36	81
6	35	125	1	−1	−1	1	1
7	55	160	21	34	714	441	1156
8	27	118	−7	−8	56	49	64
Σ	272	1008			1974	1356	3004
$\frac{1}{n}\Sigma$	$\overline{X}$ = 34	$\overline{Y}$=126					

$$\rho = \frac{1974}{\sqrt{1356}\sqrt{3004}} \approx 0.98$$

Table 9.2 Calculation of the correlation coefficient for the age and blood pressure bivariate data set (scatterplot shown in Figure 9.6).

A correlation coefficient of 0.98 is very high, indicating a very strong association between age and blood pressure in this data set. The maximum value of a correlation coefficient is 1, so 0.98 is extremely close to lying perfectly on a straight line.

This is what we call a positive correlation, but there are also negative correlations.

Negative correlation Consider the data set in Figure 9.8, which describes the relationship between the weight of cars (in pounds) and their gas mileage in the city (MPG).

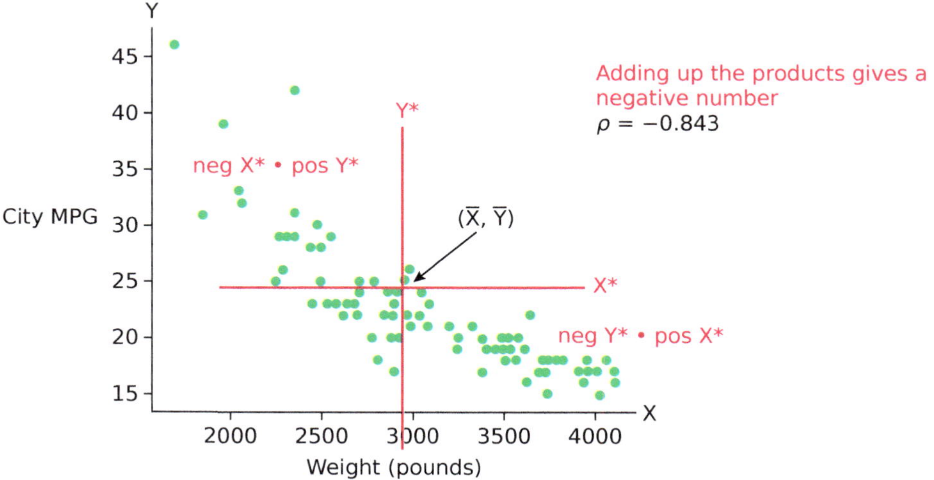

Figure 9.8 An example of negative correlation.

The actual axes of the data are the black coordinates. As in Figure 9.7, moving the axes to center at the average weight ($\overline{X}$) and the average mileage ($\overline{Y}$) amounts to subtracting $\overline{X}$ and $\overline{Y}$ from each data point. This moves the original black axes to the new red axes, (X^*, Y^*), where $X^* = X - \overline{X}$, $Y^* = Y - \overline{Y}$.

In the new coordinate system, in the top left region, there are a lot of data points where X^* is negative and Y^* is positive. This is going to produce a lot of negative numbers when we take the product $X^* \times Y^*$.

In the bottom right region, this data set has a lot of data points that have positive X^* values and negative Y^* values. This is also going to produce a lot of negative numbers.

Summing up all the product terms, we get a negative correlation $\rho = -0.843$. The minimum value of the correlation coefficient is −1, which means lying perfectly on a line with negative slope. So a correlation coefficient of −0.843 indicates a very strong negative association between car weight and gas mileage.

Gallery of correlation coefficients We showcase a gallery of scatterplots with different shapes and their correlation coefficients (Figure 9.9).

- As data approach a straight line, the absolute value of the correlation coefficient, $|\rho|$, approaches 1.
- As the data plot gets fuzzier and fuzzier and the line gets vaguer and vaguer, $|\rho|$ goes down.
- When ρ is close to zero, data does not lie on a straight line at all.

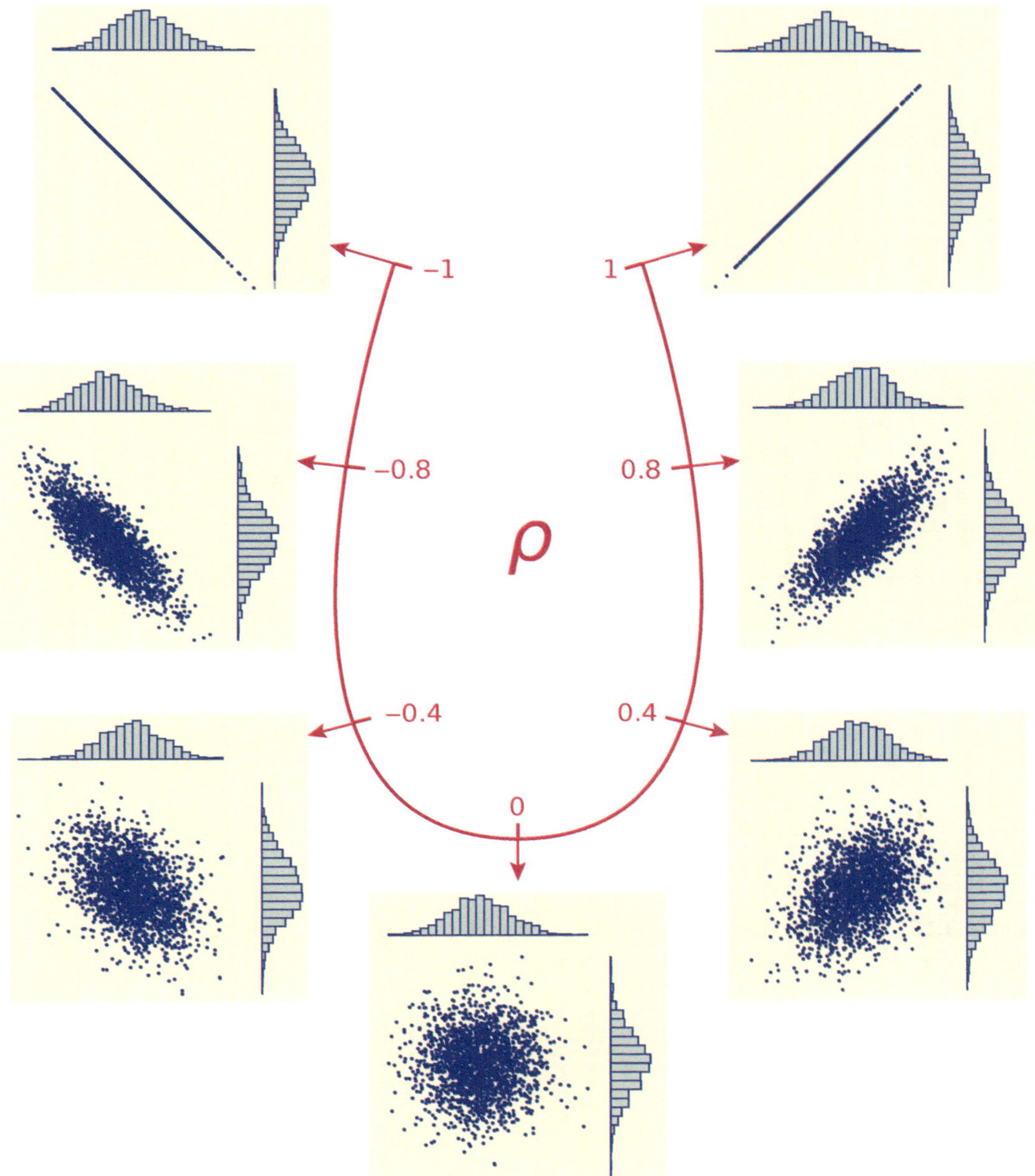

Figure 9.9 Different shapes of scatterplots and their correlation coefficients.

A special case: $\rho = 0$ There is one important special case, when ρ is approximately 0. We typically say in that case that X and Y are *uncorrelated*, or "independent". **But when we say this, we have to be careful that we have followed the instruction at the beginning of this chapter, to verify that the data does *not* lie on a line that is other than a straight line.**

Note that when the data points form a nonlinear shape, such as those shown in Figure 9.10, if we (foolishly) calculated a correlation coefficient, its value would be 0, though X and Y are definitely **not** independent.

When $\rho = 0$, there are two possibilities:

1. X and Y are not linearly correlated (but possibly highly nonlinearly dependent).
2. X and Y are truly independent of each other, in which case their scatterplot will show no relation at all, that is, it will be a featureless point-cloud (Figure 9.4).

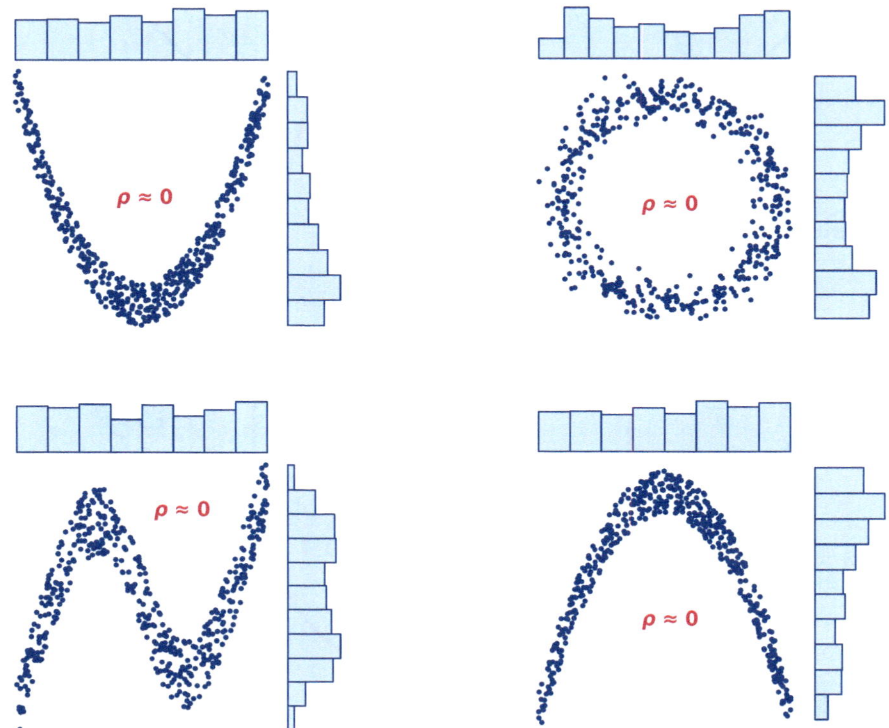

Figure 9.10 **Zero correlation**. There are strong relations between X and Y in these data sets, but the correlation coefficients are all approximately zero.

$\rho \sim 0$ does **NOT** mean "no relation", it means "no **LINEAR** relation".

strength of association ≠ strength of effect

Here is a very important point: *the correlation coefficient does not reflect the strength of the effect, it reflects the strength of the association.*

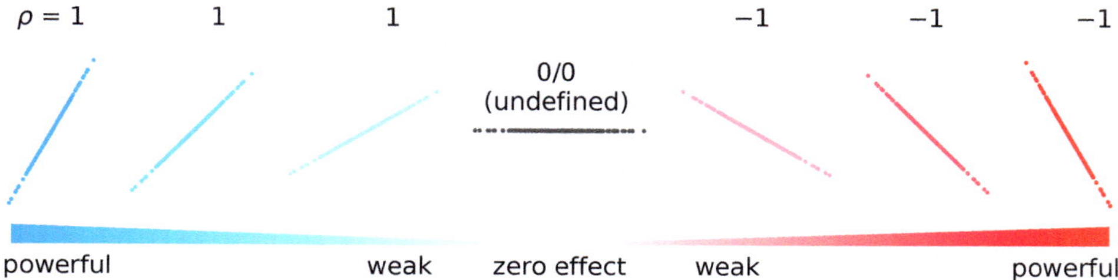

We can have an extremely tight association, with a correlation coefficient $\rho = 1.0$. The data exactly lies on a line, but the effect could be very small. The effect of X on Y is given by ΔY over ΔX. It is how much effect, ΔY, is associated with a change, ΔX. **The slope of the line is the strength of the effect.** In a scatterplot,

- shallow slopes denote a weak effect: a change in X is associated with a small change in Y.
- steep slopes denote a strong effect: a change in X is associated with a big change in Y.

But the correlation coefficient misses strength of effect completely.

In order to quantify the strength of effect, we need to determine the slope of the line, using linear regression, which will be the topic of the next chapter.

Exercise 9.2.2 Verify that ρ has value 1 when $Y = kX$ (line of positive slope) and -1 when $Y = -kX$ (line of negative slope).

Exercise 9.2.3 Which of the following situations describes a negative correlation?

a) The size of a car's gas tank and the number of gallons it fills
b) The number of miles driven and the amount of gas used
c) Horsepower of a car and the time it needs to reach 50 miles/hour speed
d) The size of a car's gas tank and the time it needs to reach 50 miles/hour speed

Exercise 9.2.4 The correlation coefficient for the data below is closest to:

a) -1 b) -0.3 c) 0 d) 0.3 e) 1

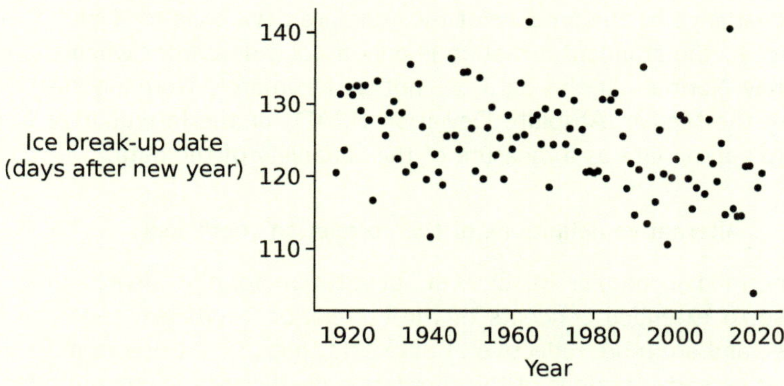

Exercise 9.2.5 Which value of ρ represents data with the strongest linear relationship between two variables?

a) -0.95 b) 0.25 c) 0.84 d) -0.5 e) 1.03

The correlation coefficient concept depends on Normally distributed quantities

The correlation coefficient is married to the Normal distribution. It is defined in terms of means and standard deviations, whose scientific validity depends on the distributions being Normal.

Assumptions of the correlation coefficient If the scatterplot of a data set passes the "no obvious nonlinearity" test, then, given the definition, we can calculate the correlation coefficient for any data set $\{(x_1, y_1), (x_2, y_2), \ldots, (x_n, y_n)\}$.

But just because we can physically calculate it, **it does not mean that the computed value has any meaning or scientific value.** Whether the calculated number has any meaning or scientific value depends on the distribution of the data set. **The correlation coefficient depends on several strong assumptions.**

In developing the formula for the correlation coefficient, we used two concepts (the mean and the standard deviation) that depend for their validity on the assumption of "approximately Normal distributions".

1. First, we subtracted the *mean* value of X from the X-values, and the *mean* value of Y from the Y-values, to get the covariance. But, as we saw in Chapter 2, the mean is only a good indicator if the data are approximately Normal, that is, symmetric, single-humped, and thin-tailed. If these assumptions on X and Y are not met, then the mean value is problematic as a measure of central tendency, and therefore, so are the covariance and the correlation coefficient, which are defined in terms of the mean value. If the data set is not approximately Normally distributed, for example, if it is heavily skewed, a median value will be more appropriate to use for the central tendency.
2. In addition to the limitation of restriction to the mean, there are also limitations related to the use of the *standard deviation*. To get the correlation coefficient of X and Y, we divided the covariance by the product of the standard deviations of X and Y. But, as we saw in Chapter 2, the standard deviation is only a good descriptor when the distribution is approximately Normal. If the data are not approximately Normally distributed, then something like the Median Absolute Deviation (MAD) or the Interquartile Range (IQR) might be more appropriate as a measure of the variability of the data.

Alternative definitions of the correlation coefficient

The use of the median and/or MAD as a substitution for the mean and/or standard deviation gives rise to extended concepts of the correlation coefficient that can be better scientific measures, and are more reflective of data sets that are not approximately Normal.

But, while these extended notions of the correlation coefficient may be the right scientific choice for a given data set, there is no analytical theory of the distributions of these extended notions, like there is for the Normal distribution. Consequently, there are no formulas or equations derived from analytic theory to determine statistical significance or confidence intervals for these extended measures when the Normal distribution assumption is not met. Therefore, **simulation-based approaches are the only way to determine statistical significance and confidence intervals that would be valid for all situations** (see sections 9.4 and 9.5).

Requirements for the analytical theory of statistical significance and confidence intervals for the correlation coefficient

Suppose we have calculated a correlation coefficient on some data, using any of the above definitions (Pearson, median-based, etc.). The next questions are: is my result statistically significant? and what is the confidence interval for my result?

Many scientists think they can look up the statistical significance and confidence intervals for a given value of ρ, knowing n, the size of the data set. There are such tables on the Internet; they tell you that, for example, if $n = 25$, then a ρ of 0.396 is considered statistically significant at the 0.05 level.

The same tables are built into computer programs that "do statistics". Those programs are making calculations based on analytic theory.

But before we look up anything, we have to understand the logic that went into those calculations. The calculations are all based on a mathematical theorem that says that *if* the distributions of X and Y meet certain requirements, then their ρ values will follow a known distribution, and therefore the cutoffs for significance or confidence intervals can be calculated from this theoretical distribution.

The theorem has many limitations:

1. First of all, it applies only to the standard Pearson product-moment correlation coefficient, and not to the non-standard variants using medians, MADs or IQRs. There is no theory for these non-standard variants that can be applied to arbitrary distributions.

2. But even for the Pearson correlation coefficient, the requirements for applying the theorem are stringent. And since it is the theorem that is being used to make all the calculations in statistics software, the validity of those calculations depends on meeting the requirements of the theorem.

Requirements for statistical significance and confidence intervals for the correlation coefficient ρ_{XY} for two variables X and Y

- For analytical theory-based calculations:
  - X is Normally distributed.
  - Y is Normally distributed.
  - Variability in X is constant across the range of Y.
  - Variability in Y is constant across the range of X.
  - No clear nonlinearity between X and Y.
- For resampling-based calculations, we have the freedom to choose Pearson if appropriate, or any of the extended notions of the correlation coefficient. No matter what our choice, the calculation of statistical significance and confidence intervals will be valid.

The classic paper on the subject argues that the correlation coefficient is not robust to violations of the Normal distribution assumption (Kowalski, 1972). Kowalski's general conclusion is that "the distribution of r may be quite sensitive to non-Normality and that normal correlation analyses should be limited to situations in which (X, Y) is (at least very nearly) Normal."

The final requirement for the application of the correlation coefficient is that there is no clear nonlinearity. In Figure 9.11, only the top left data set passes this requirement. In the other three data sets, there are obvious nonlinearities, as well as failures to conform to a Normal distribution.

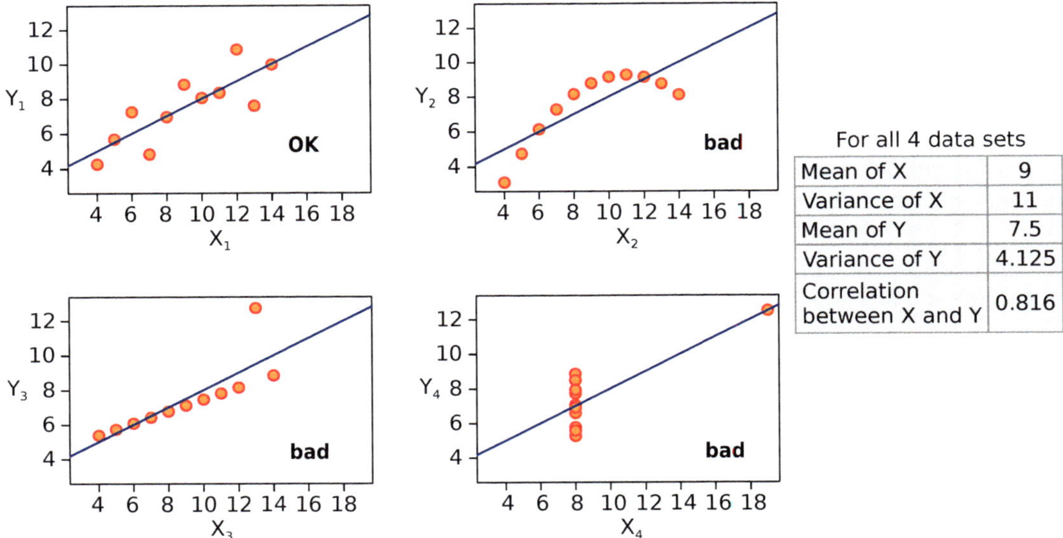

Mean of X	9
Variance of X	11
Mean of Y	7.5
Variance of Y	4.125
Correlation between X and Y	0.816

Figure 9.11 **Anscombe's quartet.** All four data sets have the same means, variances, and correlation coefficients, but they are wildly different.

Of course, we can plug in the data values into the correlation coefficient formula and get 0.816 for all four data sets, but we should not do it; we should stop and see whether there is any obvious nonlinearity, and ask whether correlation is the right tool. If correlation is inappropriate, we need to address that fact first.

The misuse of the Pearson correlation coefficient

Unfortunately, the failure to address this issue is widespread, and can be seen in many publications.

A paper in a major medical journal featured the following figure, claiming that "$R = 0.464$ and $P < 0.05$" (R is the correlation coefficient, P is the p-value).

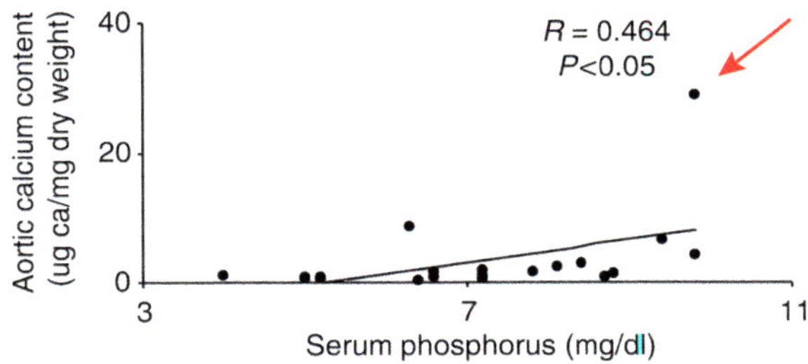

This is a misuse of the correlation coefficient. If we look at the data, we see one subject with very high X- and Y-values. The entire correlation coefficient is driven by that one data point, and if we remove it and look at the rest, there is absolutely no relationship between X and Y.

Recall that to apply the correlation coefficient, the data must be Normally distributed in both X and Y. But Y is clearly *not* Normally distributed. Almost all of the Y-data is very low, and there is one Y-value which is very high. This fails the tests for the use of the correlation coefficient, and should have stopped the authors from plugging in the numbers and pressing the "run" button!

But more deeply, the problem lies not just in the technical execution of the correlation coefficient, but in the scientific claim that the Y (aortic calcium content) and X (serum phosphorus) are related. There is no evidence for that in this data set.

The graph below was in another prestigious medical journal.

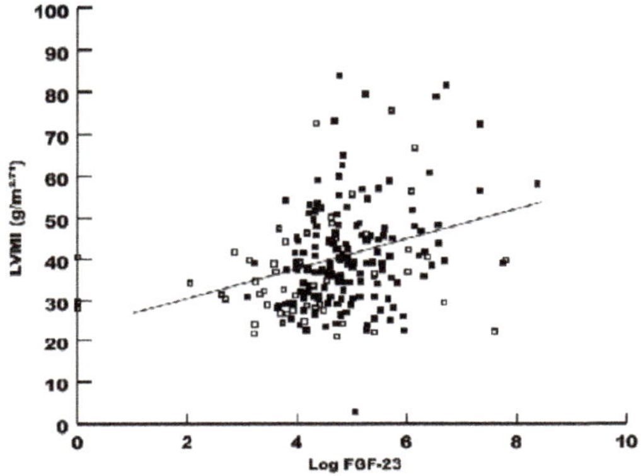

Figure 2. Correlation between log FGF-23 and LVMI (r 0.27, $P<0.001$). □ Indicates non-CKD subjects; ■, subjects with CKD.

The authors are claiming a correlation between Fibroblast Growth Factor-23 levels (X) and Left Ventricular Mass Index (Y), stating that "$r = 0.27$ and $P < 0.001$" (r is the other symbol for Pearson's product-moment correlation coefficient). To the eye, there is no relationship: the scatterplot is a blob. However, there are a handful of points with high X- and Y-values.

Note that the data fails to meet the requirements for applying the correlation coefficient. X and Y are clearly not Normally distributed. The X-values are highly skewed to the right and so are the Y-values. The Pearson correlation coefficient is not the right tool for this data set.

Here is another result, published in a respected journal.

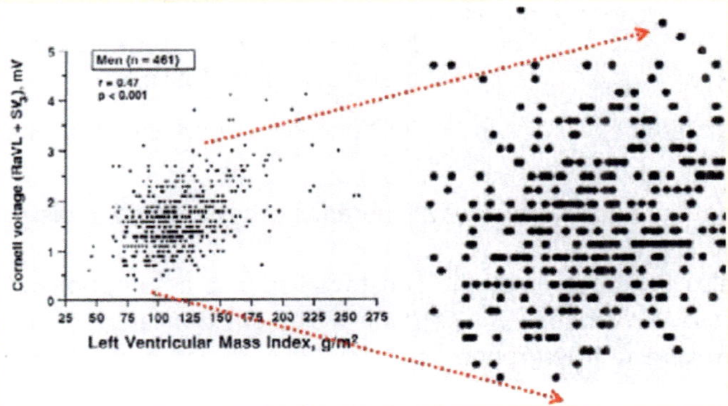

This data set is markedly *not* Normally distributed. Similar to the previous case, there is a mass of data on the left, and a handful of points at large X- and Y-values. The distribution is highly skewed to the right in X and highly skewed to the right in Y. And again, in the vast majority of the data, when blown up, it's clear that there is no relationship at all. The handful of points with very high X- and Y-values are driving the correlation coefficient to 0.461. However, it is poor scientific practice to report this correlation coefficient: for the overwhelming majority of patients, there is no relationship between LVMI (X-value) and the voltage criterion for Left Ventricular Hypertrophy (Y-value). The low p-value ($p < 0.001$) is simply due to the fact that this is a large data set with a handful of extreme (X, Y) values.

Overall, the problem is that the formulas for calculating the Pearson correlation coefficient and statistical significance don't know or care whether the data violates the requirements.

When the requirements are not met, as in the situations above, the basic problem is not the calculation of the Pearson correlation coefficient, but the use of the Pearson correlation coefficient at all. It is simply not the right scientific tool for this data set. As Kowalski (1972) said, "we may be in a situation in which ρ is a poor measure of association."

Countries' per capita energy use and per capita GDP data

As a real-world example, consider the data on a country's *per capita* Gross Domestic Product (X) as related to its *per capita* energy use (Y) (Figure 9.12).

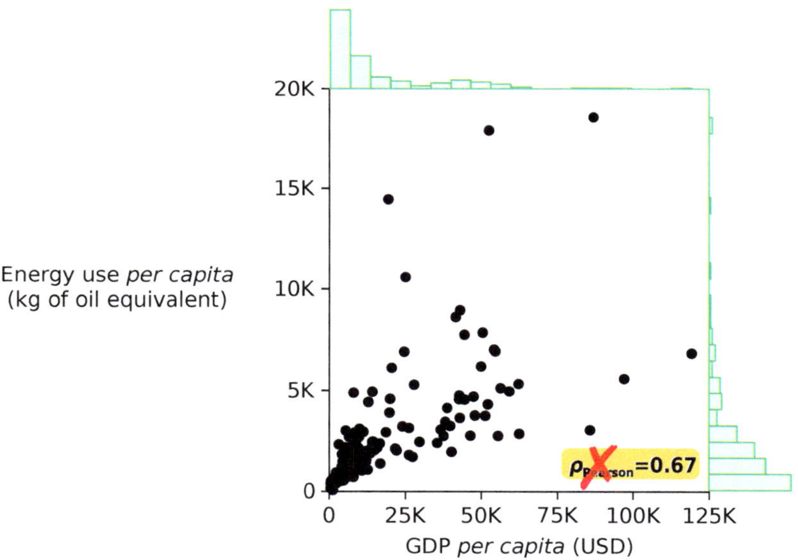

Figure 9.12 Scatterplot of *per capita* GDP and *per capita* energy use, country by country.

Note that the data are not Normally distributed: both the X and Y distributions are markedly skewed to the right.

The data set also does not meet the constant variance assumptions: the variance in Y at low X is low, whereas the variance in Y at high X is high. Similarly, the variance in X at low Y is low; the data are very tightly grouped, but for medium and high Y, the X-variance is much larger, and the data are more spread out.

In cases like these, we can physically calculate a Pearson correlation coefficient. That is, we can plug the values into the formula for the Pearson correlation coefficient, and get back some number, **but the number has no scientific validity.**

As we will see later in next section, this data *is* analyzable, by the rank correlation coefficient (see section 9.3 *Ranks and the Spearman correlation*).

CO_2 and temperature data

Let's consider the data set linking CO_2 levels to temperature (Figure 9.13). We already met this data set in Example 2 on page 439.

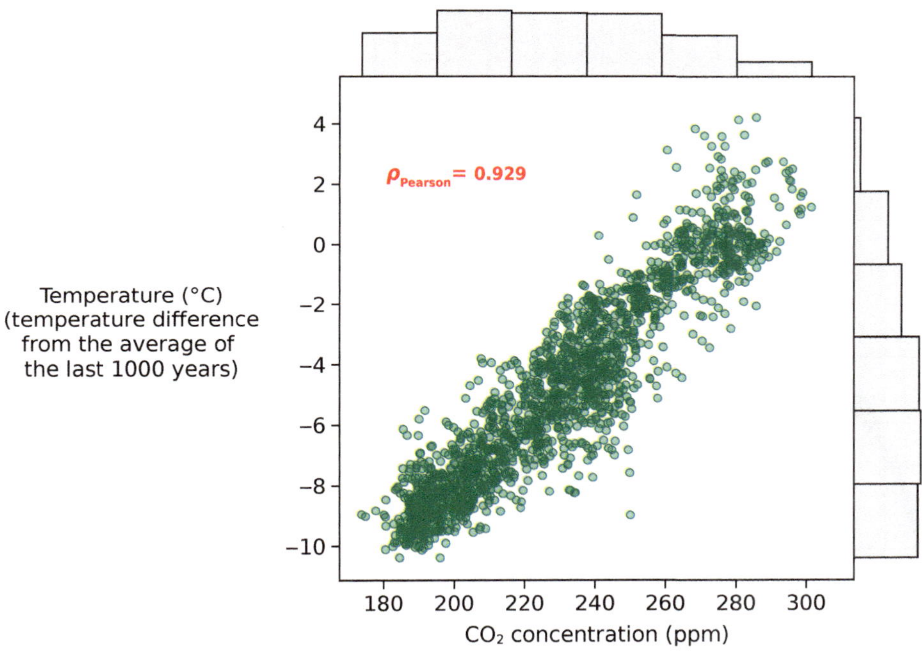

Figure 9.13

Here, the data *do* lie roughly on a straight line. Both X (CO_2) and Y (Temperature) are fairly close to being Normal. The data also meet the constant variance requirements: the variance in Y is roughly constant along X and the variance in X is roughly constant along Y. So a standard correlation coefficient should be suitable for this data. We computed the Pearson correlation coefficient, $\rho = 0.929$, confirming our visual impression that there is a strong association.

FURTHER EXERCISES 9.2

1. List three characteristics of data that would prohibit the use of Pearson's correlation coefficient.

2. Which value of ρ represents the strongest positive linear relationship between two variables?

   a) 0.91 b) 0.38 c) 1.03 d) 0.01

3. Looking at the scatterplot below between rear seat room and maximum horsepower, guess its correlation coefficient. It is closest to:

   a) −1 b) −0.3 c) 0 d) 0.3 e) 1

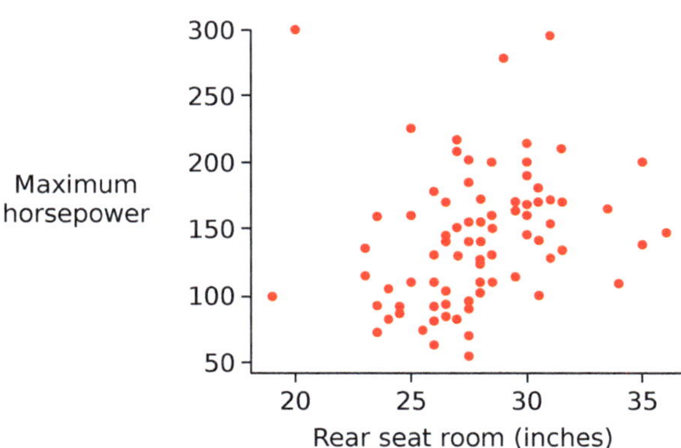

4. Select all scatterplots below for which it is appropriate to use a Pearson correlation coefficient.

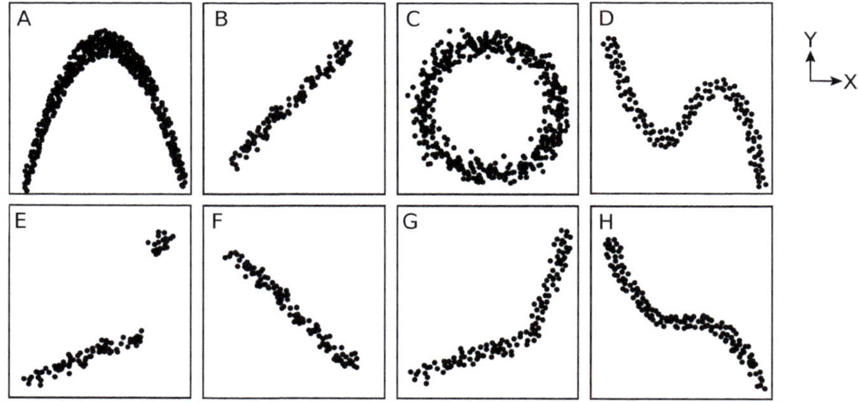

5. Match each of the Pearson correlation coefficients below with the corresponding scatter plot.

   a) 0.99 b) 0.81 c) 0.61 d) −0.38 e) −0.006 f) −0.89

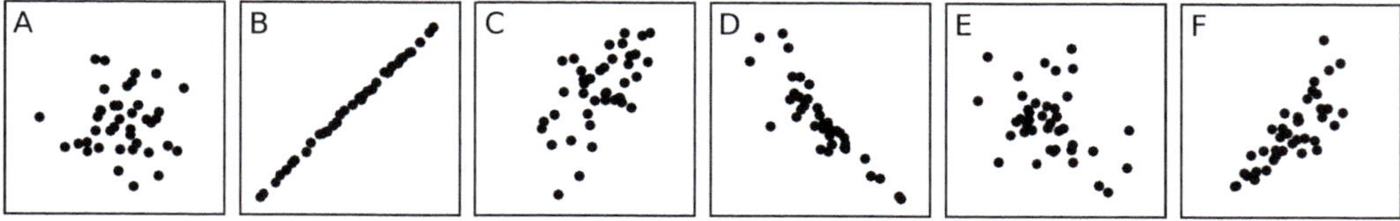

9.3 Ranks and the Spearman correlation

If the data is not Normally distributed, for example, if there is marked skewness or outliers in the data, or, and this is important, *if the data set is too small to say whether or not it is Normally distributed*, there is another strategy that is widely used, which is to consider not the data values, but only their relative *ranks*.

The thinking is that if we don't really know the underlying distribution of the data, or if it is seriously skewed, we don't really know how to compare two data values.

For example, if the X-values are $\{2, 3, 15\}$, then, as we said in section 6.6 (*Ranks*), if we don't know the variability of the data, we can say that 15 is 5 times 3, but if we don't know the range of the data, we don't know if 5 times is a big difference. If the range of the data is $\sim 10^{12}$, then 5 times bigger isn't a big difference. On the other hand, if the standard deviation of the underlying distribution was 1, then 5 times bigger is indeed a big difference.

The idea is that when we have very small samples, we can't really trust the raw data values, because we can't say what distribution they were drawn from; all we can trust is their relative position to one another. We can say that 15 is 5 times 3, but without knowing the underlying distribution, we don't know how big a difference that is, so all we can say is that 15 is larger than 3.

The ranking procedure In this way, we look not at the data values, but their relative **ranks** with regard to each other. To rank the data, we sort all the X-values from smallest to largest, and then replace the X-values with their corresponding ranks. Then we do the same for the Y-values. In case of ties, we assign the average rank, so, for example, if there are two values tied for rank #10, they are both assigned rank #10.5. If the next value is unique, then it will be rank #12;

Calculation of the rank correlation coefficient Then we calculate a (Pearson) correlation coefficient for the new data set consisting of the ranks only. This is called the **rank correlation**, or the **Spearman correlation coefficient**, which we will call ρ_{spearman} or ρ_s (Figure 9.14).

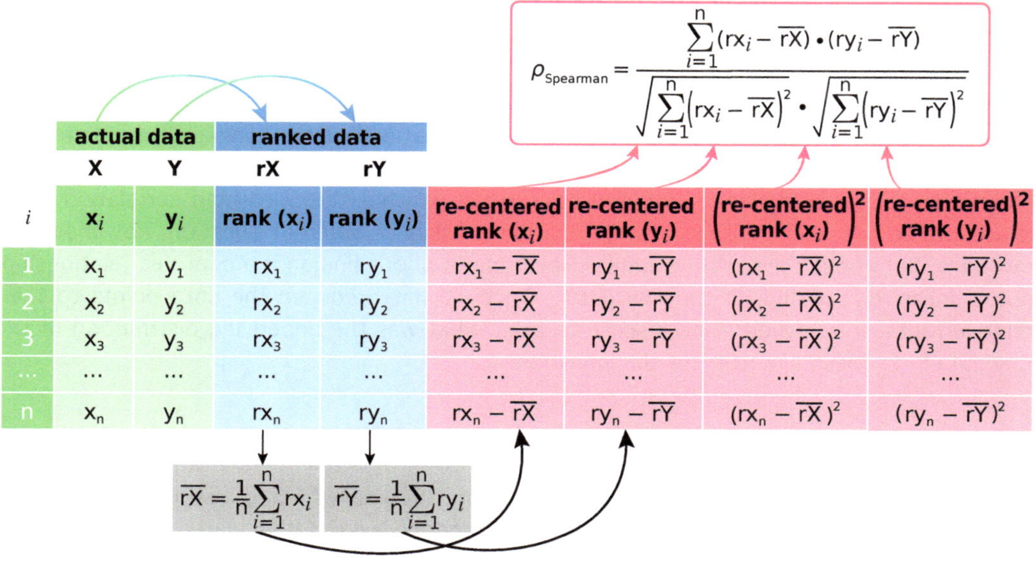

Figure 9.14

When to use the Spearman correlation coefficient While the standard (Pearson) correlation coefficient reflects the degree to which the data lies on a straight line, the (Spearman) rank correlation coefficient answers a different question.

Let's define a **positive monotonic** relation as one where Y never decreases when X increases, and a **negative monotonic** relation as one where Y never increases with increasing X (Figure 9.15).

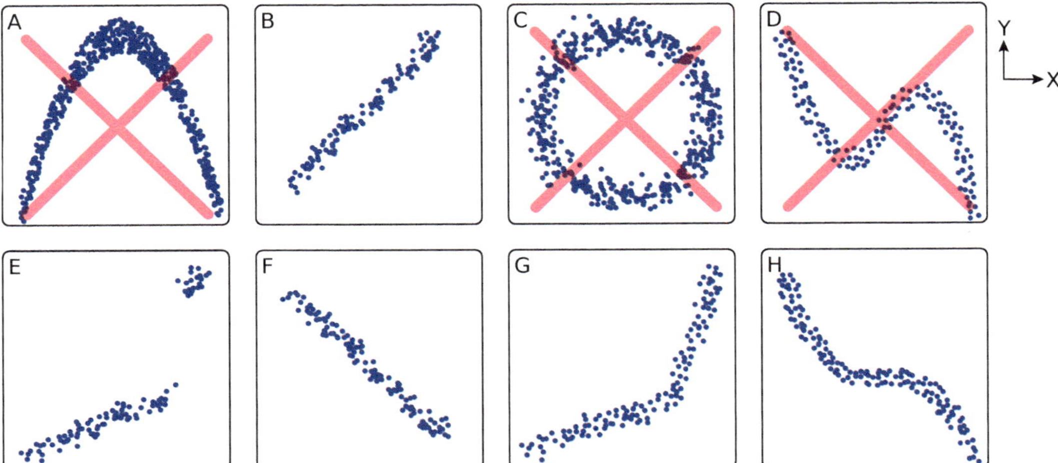

Figure 9.15 Monotonic and non-monotonic relations. Panels A, C and D depict non-monotonic relations: In panel A, the Y-value is first increasing, then decreasing with increasing X. In panel C, the data lies on a circle, and in panel D, the Y-value first decreases, then increases, then decreases again with increasing X. The rest are all monotonic, either increasing or decreasing.

Why is it OK to use the Pearson correlation coefficient on ranks?

The rank correlation coefficient reflects the degree to which the data have a monotonic relationship. Unlike Pearson's correlation, there is no requirement of Normality for the ranked (Spearman) correlation.

When we convert raw data into ranks, the distributions of the ranks will, by definition, not be Normal. In fact, ranks are distributed uniformly. Nevertheless, applying a Pearson correlation to this uniformly distributed data, which is what the Spearman correlation does, is acceptable because the Spearman correlation is asking a different question, not about the raw data, but about their relative ranks. Spearman is about finding a monotonic relationship, not a linear relationship, in the raw data. Once we have reduced the data points to their relative ranks, it no longer makes sense to ask, "what was the underlying distribution of the raw data?" It doesn't matter for this question!

Example 3 Bacteria growth

Bacteria reproduce by a means of asexual reproduction called binary fission. Binary fission creates an exact clone of itself. For example, *Escherichia coli* (abbreviated as *E. coli*), the most widely studied bacteria, double every 20 mins under favorable conditions.

As an example, let's consider a simple case of the population growth of *E. coli* bacteria starting from an initial population of 1. Assume the ideal growth conditions are met and each bacterium doubles every 20 mins, after recording the bacteria population every 20 mins for about 400 mins, we get a simple data set with 21 pairs of time points and bacteria populations. In less than 7 hours time, 1 single *E. coli* has managed to reproduce over a million copies of itself.

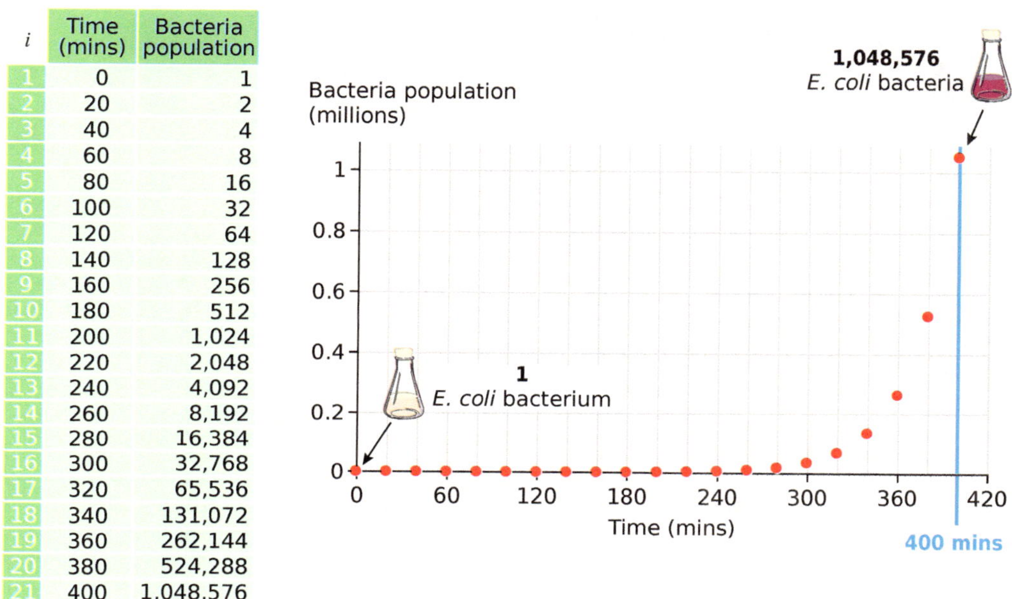

i	Time (mins)	Bacteria population
1	0	1
2	20	2
3	40	4
4	60	8
5	80	16
6	100	32
7	120	64
8	140	128
9	160	256
10	180	512
11	200	1,024
12	220	2,048
13	240	4,092
14	260	8,192
15	280	16,384
16	300	32,768
17	320	65,536
18	340	131,072
19	360	262,144
20	380	524,288
21	400	1,048,576

Question How would you describe the relationship between the passing of time and the *E. coli* population?

Answer From the scatterplot, we see that the passing of time and the bacteria population are positively associated, and that the strength of association is extremely strong. As a matter of fact, the 21 data points perfectly "lie a line", only it is not a *straight* line. This kind of growth is called exponential growth, and the line is $Y = 2^{t/20}$, where t is time in minutes, and Y is the bacteria population. The presence of a nonlinear relationship rules out applying the Pearson correlation coefficient. Instead, the variables have a perfectly increasing monotonic relationship.

But best of all would be to realize from theory that the data might be showing exponential growth, and find the best fit exponential curve using nonlinear regression (section 10.3 *Nonlinear regression*).

Question If we went ahead and calculated the Pearson correlation coefficient for this data, it gives a value of 0.61. What does that mean? What's your interpretation?

Answer The Pearson correlation coefficient measures how well the data points "line on a straight line". It's a measure of the linear relationship between two variables. From our visual inspection, the data clearly do not "line on a straight line", therefore, the correlation coefficient of 0.61 does not reflect that there is indeed a perfect relationship between time and population, *only not a linear relationship*. Therefore, it would be a misuse of the correlation coefficient.

Question When we carry out the ranking procedure for each variable, and calculate a (Spearman) rank correlation coefficient for the new data set consisting of the ranks only, it gives a value of 1. What does that mean? What's your interpretation?

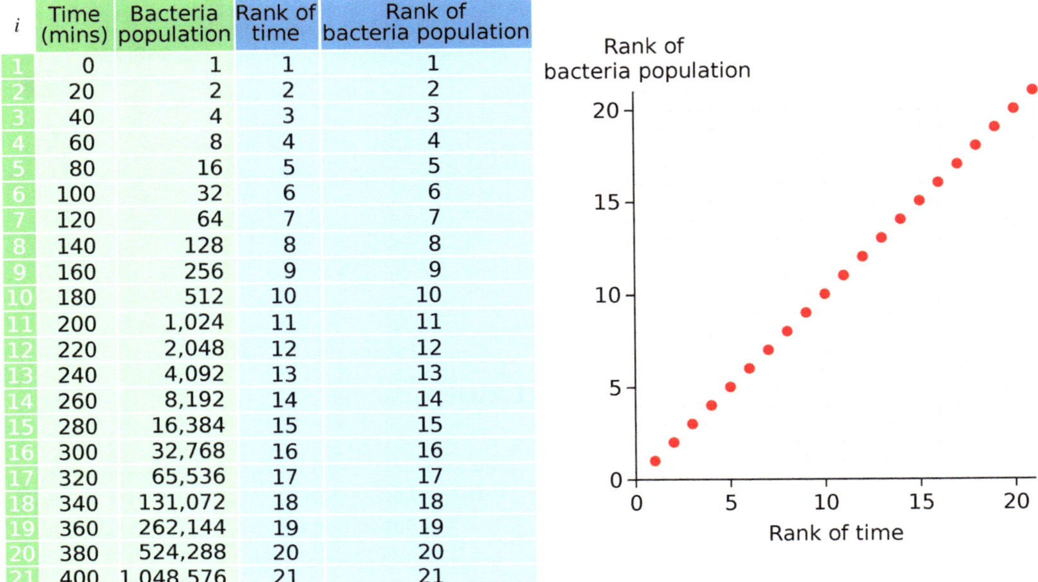

i	Time (mins)	Bacteria population	Rank of time	Rank of bacteria population
1	0	1	1	1
2	20	2	2	2
3	40	4	3	3
4	60	8	4	4
5	80	16	5	5
6	100	32	6	6
7	120	64	7	7
8	140	128	8	8
9	160	256	9	9
10	180	512	10	10
11	200	1,024	11	11
12	220	2,048	12	12
13	240	4,092	13	13
14	260	8,192	14	14
15	280	16,384	15	15
16	300	32,768	16	16
17	320	65,536	17	17
18	340	131,072	18	18
19	360	262,144	19	19
20	380	524,288	20	20
21	400	1,048,576	21	21

Answer Unlike Pearson's correlation coefficient, Spearman's rank correlation coefficient uses the ranked data set, and is therefore a measure of how well the original data monotonically relate to each other. A Spearman value of 1 says that there is a perfect monotonic relationship between the passing of time and the bacteria population. If we look at the scatterplot of the ranked data, we see that they are perfectly linearly related.

Rank correlation for the countries' energy/GDP data Now let's compute the rank correlation for countries' *per capita* energy use and *per capita* GDP data set we had encountered earlier (data presented in Figure 9.12). We first convert the raw numbers into their corresponding ranks.

By looking at the scatterplot of the ranks, we see that the new, ranked data lies more or less on a straight line, and meet the requirements for the Pearson correlation. There is a very strong correlation between the *rank* of a country's GDP *per capita* and the *rank* of its energy use *per capita* (Figure 9.16).

If we carry out the (Spearman) rank correlation coefficient for this data, it is $\rho_{\text{spearman}} = 0.9$.

The high value of the Spearman correlation coefficient, which reflects the straight line arrangement of the ranked data, shows that the ranking procedure has uncovered a strong relationship that was less clear in the raw numbers.

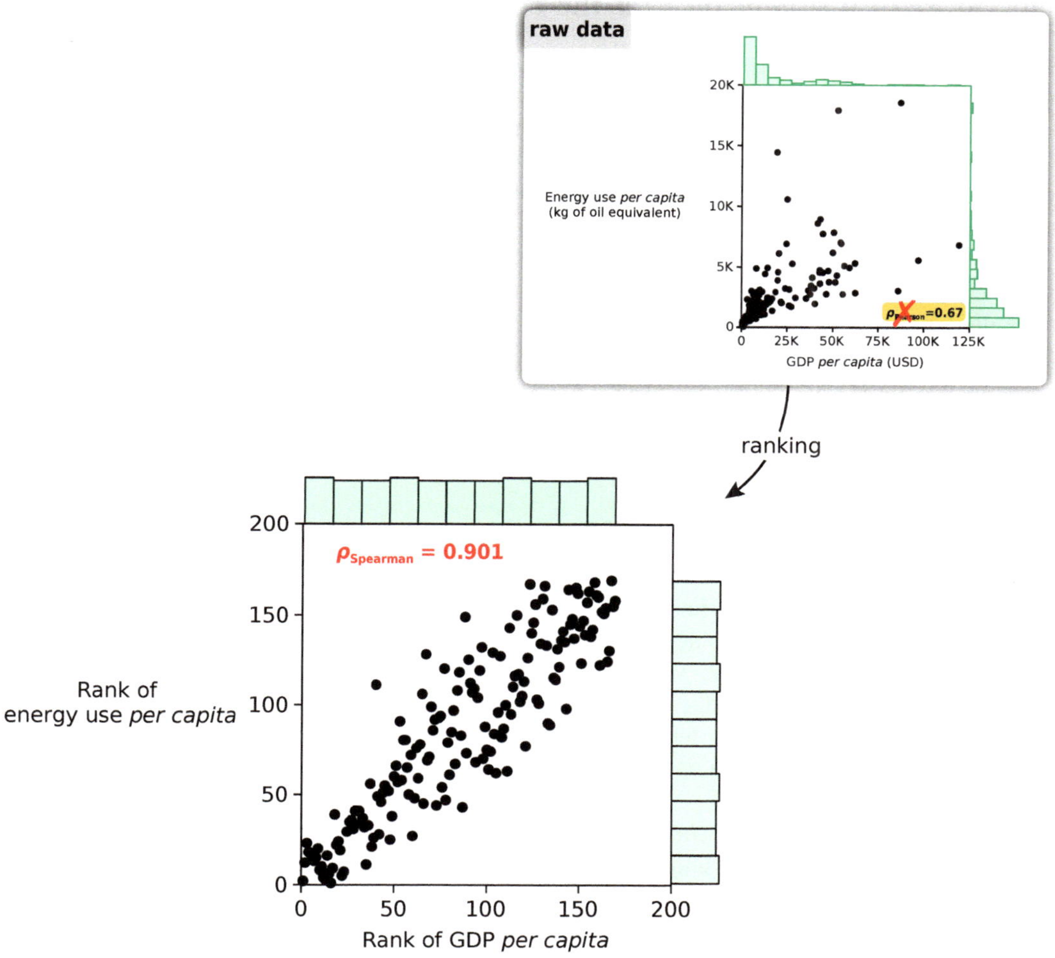

Figure 9.16 Coverting raw data to ranks.

The Spearman correlation coefficient is a measure of a monotonic relationship between two variables. Hence a value of 0 does not imply there is no relationship between the variables. For example, the following scatterplot has no monotonic relationship, which is confirmed by the calculation of a Spearman's correlation, giving a value of 0. But it has a perfect quadratic relationship.

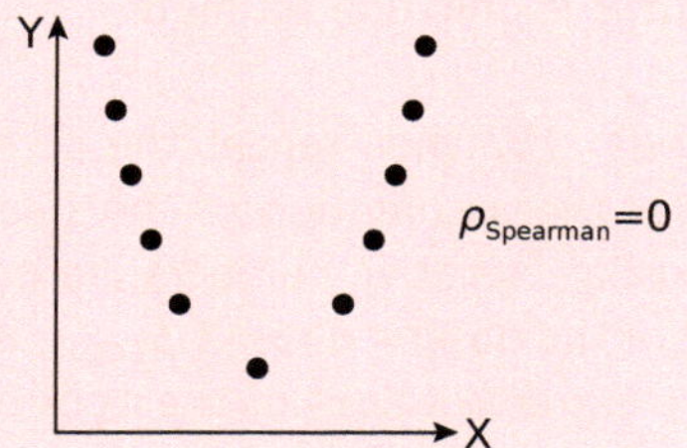

9.4 Statistical significance for the correlation coefficient

For the standard (Pearson) correlation coefficient, we said that if the data meets the heavy requirements of Normal distributions, constant variance and no nonlinearity, then there is an analytical theory to calculate its statistical significance and confidence intervals.

But since these requirements are often not met in practice, we need an alternative to the formula-based method for calculating statistical significance and confidence intervals for the correlation coefficient.

This need for an alternative becomes mandatory when we are using a non-standard definition of correlation, such as one based on medians rather than means. For these extended concepts, there generally are no theoretical formulas.

The alternative, as before, is to use simulation-based methods to generate the distribution of values "under the Null Hypothesis", and then compare the actual, observed value to that distribution to get a p-value.

Statistical significance by simulation

If the deviation from the requirements is small, we can use the Pearson correlation, plus a simulation-based approach to finding statistical significance. If the deviation from Normality/constant variance/straight line relationship is large, we can't use the Pearson correlation. If the data are roughly monotonic, we can use a rank correlation coefficient, and if they are not, then no correlation can be used.

In each case where we can legally use one form of correlation or another, we can *always* estimate statistical significance and calculate confidence intervals using simulation methods.

The world of the Null Hypothesis In order to calculate the significance of the correlation coefficient, however defined, using simulation-based techniques, we need to create a "Null Hypothesis" world, and then simulate the Null Hypothesis many times to get the null distribution.

The Null Hypothesis is that X and Y are unrelated, so we will create a null world that has exactly the same X and Y distributions as ours, but where there is, by construction, *no relation* between X and Y. This is easily created by taking the X-values and permuting (scrambling or shuffling) them, and then reattaching these randomized x_is to the y_is.

Let's form such a world of the Null Hypothesis (Figure 9.17).

One simulation We randomly permute the X-values and assign them to Y-values. We also could have done this by scrambling only the Y-values, or both. All are valid.

For this world, we compute the new correlation coefficient. The result is ρ_1. This is one estimate of how large a correlation coefficient we would expect from this data if X and Y had no relation to each other. But it is only one estimate.

10,000 simulations: p-values We then repeat the permutation procedure 10,000 times to get 10,000 values of ρ_i under the assumption of "no relation". We rank these ρ_is from the smallest to the largest. (As usual, negative numbers are considered to be smaller than positive numbers.) Then, the p-value of the observed result ρ_{actual} is the fraction of the total simulations that have as extreme or more extreme values of their correlation coefficient ρ_i.

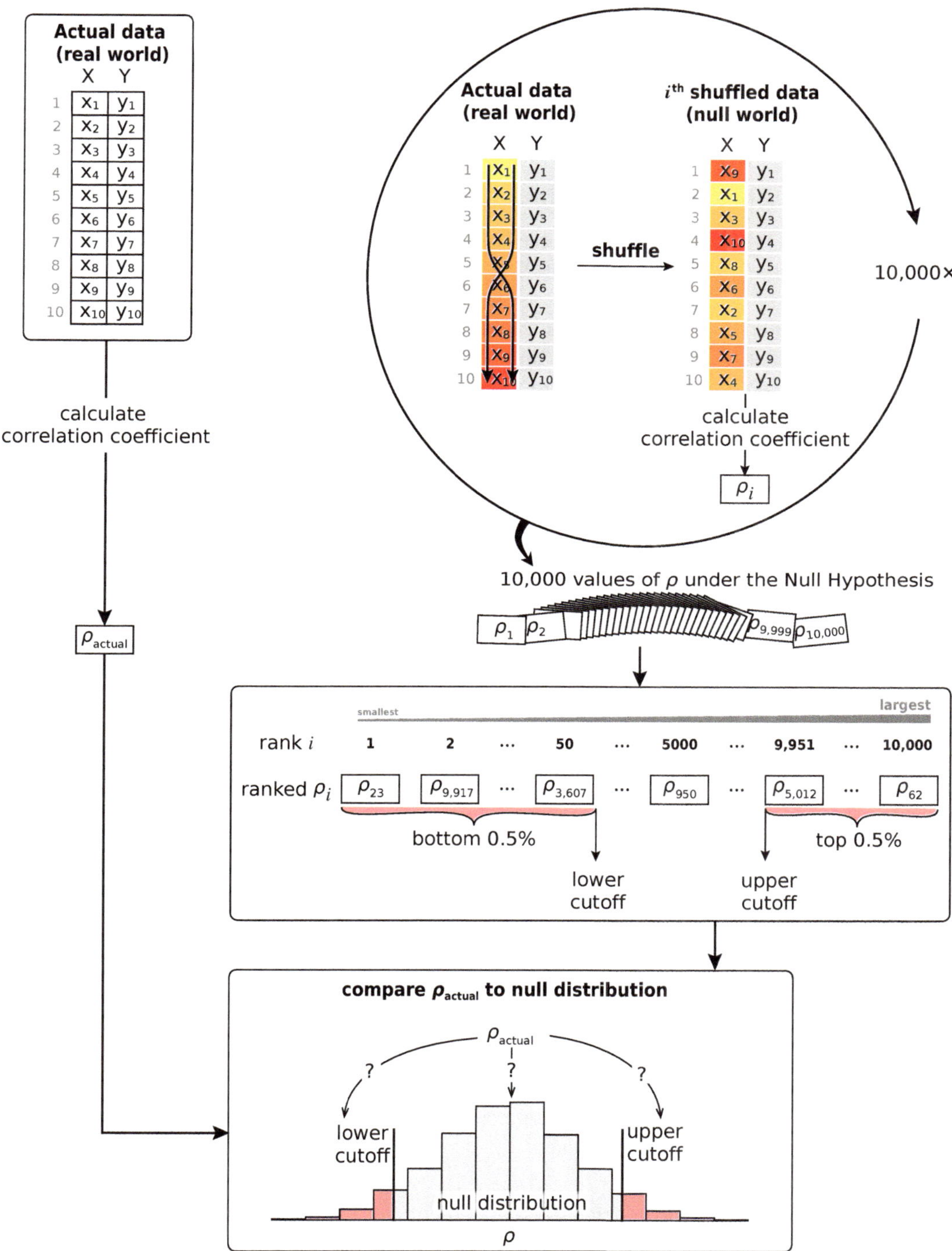

Figure 9.17 NHST for ρ_{actual}. The actual data (left) is used to calculate the observed correlation coefficient, ρ_{actual}. Then the X-values are shuffled (permuted) and reattached to the Y-values to create the ith randomized (null) world. The correlation coefficient for the ith world is ρ_i. We then repeat the permutation 10,000 times, and the 10,000 values of ρ_i from the simulations are used to estimate how extreme our observed ρ_{actual} is under the Null Hypothesis.

Statistical significance If we want the cutoff values for $p < 0.01$, count up to value #50, and count down to value #9,951, those are the bottom 0.5% and top 0.5%, called the lower and upper cutoffs.

If the correlation coefficient of the actual data is larger than the upper cutoff, it is statistically significantly positive, and if it is less than the lower cutoff, it is statistically significantly negative.

NHST for Pearson correlation coefficient: CO_2 and temperature

We carried out NHST for the CO_2 and temperature data that we introduced in Example 2 to determine the statistical significance of the observed correlation coefficient (data presented in Figure 9.13).

We randomly permuted the X-values to create a "world of the Null Hypothesis". We then calculated the correlation coefficient for that randomized data set. This procedure was carried out 10,000 times, and a histogram was made of the results, representing the null distribution. The 0.5% and 99.5% boundaries are the cutoffs for statistical significance at the $p < 0.01$ level. In fact, in 10,000 simulations of the Null Hypothesis, a ρ-value as large of 0.929 was never seen, so we can report that $p < 0.0001$ (Figure 9.18).

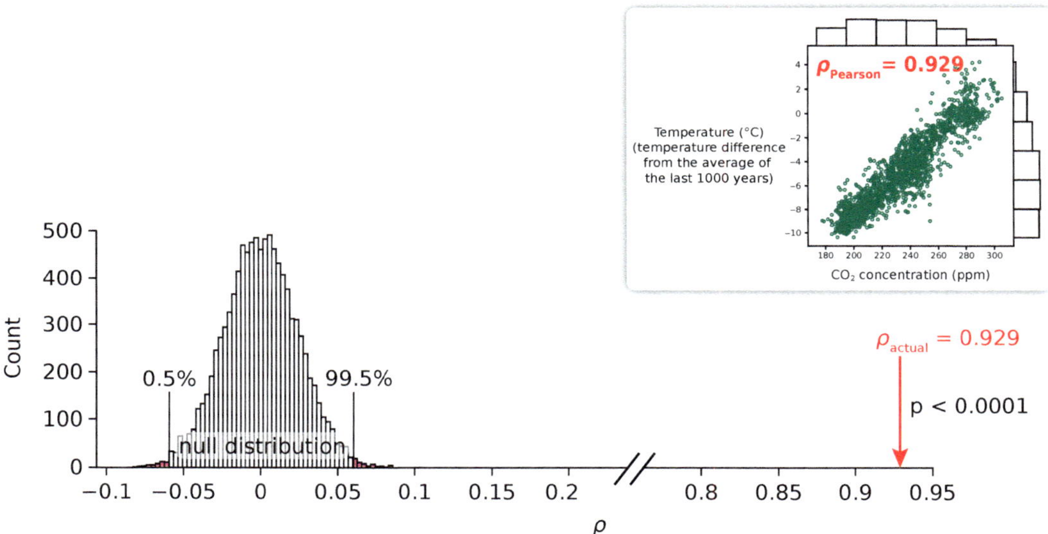

Figure 9.18 NHST for CO_2 and temperature data. The standard (Pearson) correlation coefficient for this data $\rho_{\text{actual}} = 0.929$. The histogram shows the 10,000 correlation coefficients obtained under the Null Hypothesis. Also shown are the cutoffs for statistical significance at the $\alpha = 0.01$ level. Note that the observed correlation coefficient of the actual data set is well outside all of the 10,000 simulated values.

Because this data set largely meets the requirements for the theoretical calculation of statistical significance, we used an online calculator to estimate the p-value for the observed correlation coefficient. The calculator returned $p < 0.00001$, which agrees with our resampling result.

We see that when the data set largely meets the strict requirements for the theoretical calculation, the theoretical approximation closely agrees with resampling.

However, when the data does not meet the theoretical assumptions, which is frequently, only resampling techniques can give us a valid answer.

NHST for Spearman correlation coefficient: countries' energy usage and GDP

We also carried out NHST for the Spearman rank correlation coefficient of the data relating countries' *par capita* energy use to *per capita* GDP (Figure 9.19).

For each simulation in the NHST, we took the raw X–Y data and randomly permuted the X-values. This gave us a randomized data set, and we computed its Spearman correlation coefficient. We repeated this shuffling procedure 10,000 times, and made a histogram of the results. The 0.5% and 99.5% cutoffs for that histogram give us the boundaries of statistical significance at the $p < 0.01$ level. In fact, values as large as $|0.901|$ never occurred in the 10,000 simulations, so we can say that the rank correlation is 0.901 and that that result is statistically significant at the $p < 0.0001$ level.

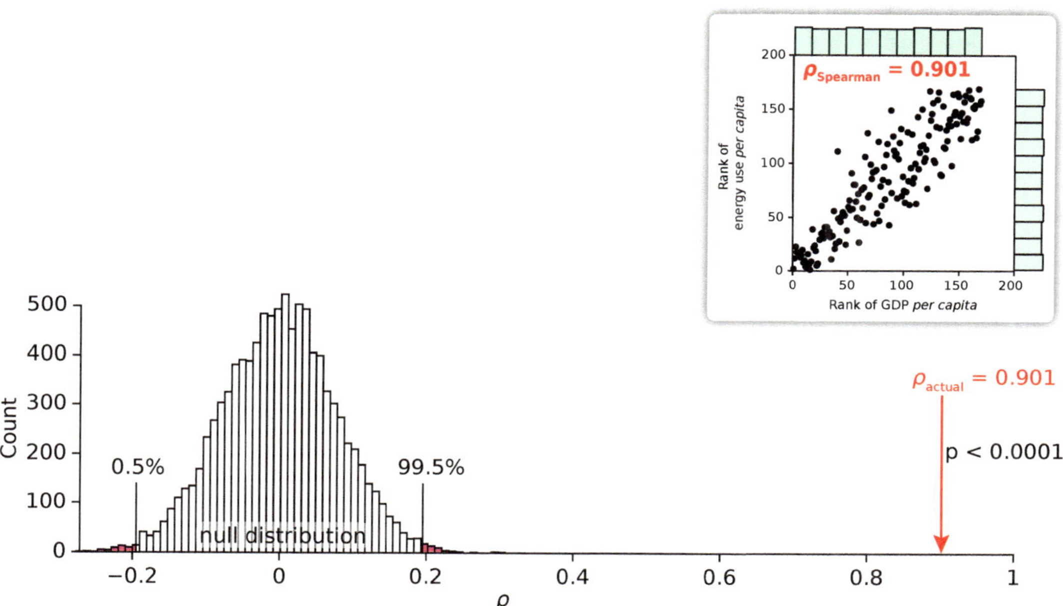

Figure 9.19 NHST for countries' *per capita* energy use/GDP data. The rank (Spearman) correlation coefficient for this data, $\rho_{actual} = 0.901$, was compared to the distribution of correlation coefficients under the Null Hypothesis. The actual value lies well outside the cutoff for statistical significance at the 0.01 level.

Note that the histogram of null results in this case is much wider than for the previous example. Here, 99% of the simulation results lie approximately within the interval [−0.2, 0.2], whereas in the climate change case, 99% of the null results fell within [−0.05, 0.05], a much smaller interval, due to the much larger data set in the CO_2 and temperature data.

FURTHER EXERCISES 9.4

1. Find a research paper in a science journal that has a correlation coefficient in it, and also shows a scatterplot of the data. Do the data meet the requirements for calculating its statistical significance using the traditional formula method?

2. **Water temperature and dissolved oxygen.** Researchers monitored 35 river sites across the Pacific Northwest and recorded water temperature (°C) and dissolved oxygen level (mg/L) at each site. The data are plotted below.

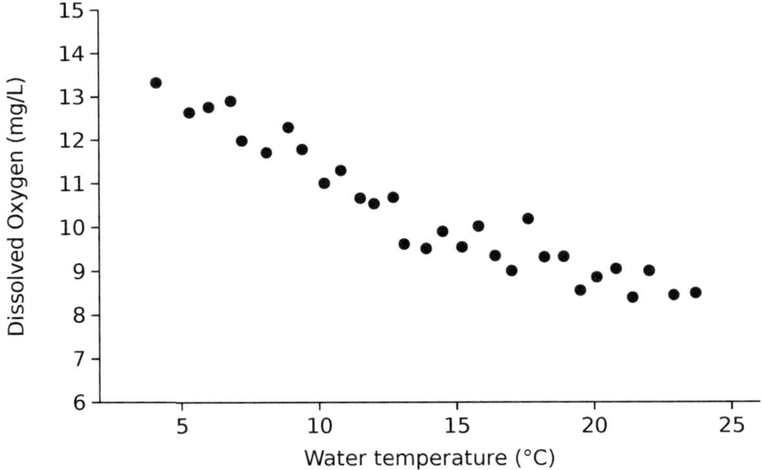

a. Estimate the value of the standard (Pearson) correlation coefficient for this data set.

a) 0.9 b) 0.5 c) −0.5 d) −0.9 e) −1

b. Based on the scatterplot, which correlation coefficient is the most appropriate?

a) Standard correlation coefficient (Pearson)
b) Ranked correlation coefficient (Spearman)
c) No correlation coefficient

c. What is the Null Hypothesis for this study?

a) There is no relationship between water temperature and dissolved oxygen.
b) Warmer rivers cause dissolved oxygen levels to decline.
c) Dissolved oxygen decreases as water temperature increases.
d) There is no difference in dissolved oxygen between warm and cold rivers.

d. To simulate the null hypothesis, what should you do in each simulation?

a) Shuffle the temperature values while keeping dissolved oxygen values fixed.
b) Randomly sample new temperature and dissolved oxygen pairs with replacement.
c) Keep all pairs intact and resample with replacement.
d) Keep all pairs intact and resample without replacement.

9.5 Confidence intervals for the correlation coefficient

We've seen that we can do p-values and Null Hypothesis Significance Testing (NHST) for the correlation coefficient: to test the Null Hypothesis that there is no relation between X and Y, we break the connection between X and Y by randomly shuffling the X-values. Then we compare the observed correlation coefficient to the 10,000 correlation coefficients calculated from the 10,000 randomized data sets. If the actual correlation coefficient is as or more extreme than 99% of the correlation coefficients in the randomized data sets, we can "reject the Null Hypothesis, $p < 0.01$" and claim that there is evidence for "an association" between X and Y.

But we saw in Chapter 3 that NHST has limitations as the only model for how we do statistics. All of the drawbacks of NHST are present when we do NHST on correlation coefficients.

NHST does not give us the strength of the evidence for an association, it only tells us the Yes/No fact that there is a significant association (of some degree or other). It has reduced the data to one bit of information: Yes (it is significantly correlated) or No (it is not).

Even if we provide the p-value, which is more informative than NHST, the p-value still does not give us the strength of the association, only the strength of the evidence that there is some association or other.

We saw that it is much more powerful to state an *effect size* and a *confidence interval* for that effect size.

- **The effect size in this case *is* the correlation coefficient**; it tells us the strength of the association, and not just whether or not there is one.
- The confidence interval for the correlation coefficient then tells us an additional piece of useful information: **how precise is this estimate of the strength of association?**

Confidence intervals by simulation

So how to calculate a confidence interval for a correlation coefficient? Statistical theory tells us that **IF** X and Y are two independent quantities that are distributed Normally, **AND** both satisfy the constant variance assumption, **THEN** there is a theoretical distribution that can be derived for the confidence interval for the correlation coefficient ρ.

But if these strict assumptions are not met, (and even when they are), we can *always* calculate confidence intervals by resampling. We follow the same strategy as for all other confidence intervals: we resample the original data. Here we resample the data by choosing, with replacement, n data points out of the original n. Note that we keep the original pairs intact. So for example, if $n = 5$, and the data points are pairs (x_i, y_i), then our first resample might be $\{(x_1, y_1), (x_2, y_2), (x_3, y_3), (x_2, y_2), (x_4, y_4)\}$.

For each resampled data set, we calculate its correlation coefficient. When we repeat this 10,000 times, we get 10,000 correlation coefficients, roughly centered around the actual correlation. The histogram of the 10,000 correlation coefficients is the resampling distribution (Figure 9.20).

If we rank these from smallest to largest, and count up to #50, that is the value ρ_{lower}, and if we count down from the largest to #9951, that is the value ρ_{upper}. We then calculate the 99% confidence interval, following the method in Chapter 4, as

$$99\% \text{ CI} = [2 \cdot \rho_{\text{actual}} - \rho_{\text{upper}},\ 2 \cdot \rho_{\text{actual}} - \rho_{\text{lower}}]$$

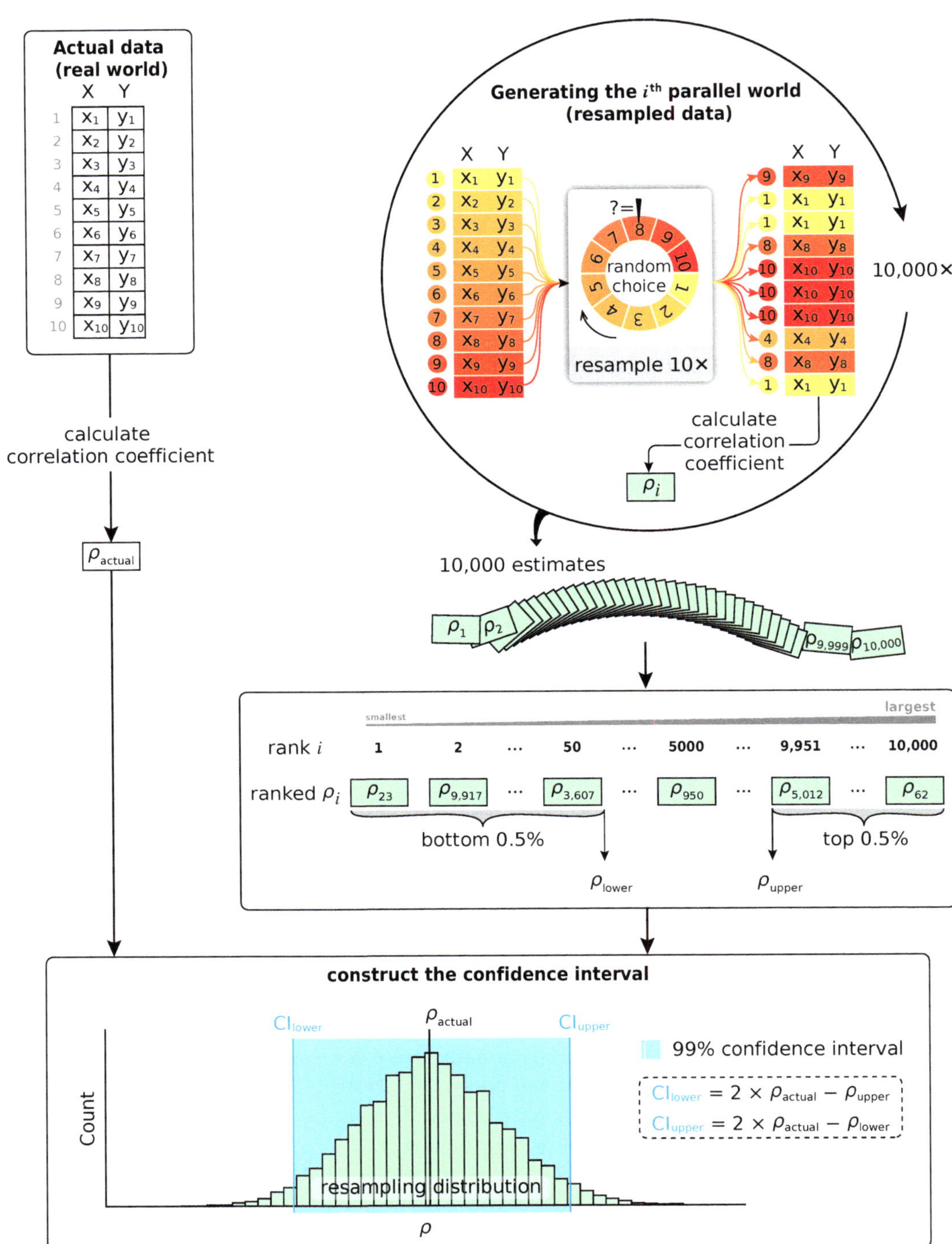

Figure 9.20 **A confidence interval for the correlation coefficient.** First, we calculate the observed correlation coefficient for the actual data, ρ_{actual}. Then, we create 10,000 parallel worlds; for each of the 10,000 parallel worlds, the data pairs are kept intact, and are resampled as pairs (sampled with replacement; note that the same pair may occur several times) to generate ρ_i, a new estimate of the correlation coefficient. Finally, we construct a confidence interval.

Confidence interval: CO_2 and temperature

We calculated a confidence interval, using the resampling method, for the CO_2 and temperature data. The 99% confidence interval for the actual correlation coefficient was 99% CI = [0.921, 0.938] (Figure 9.21).

Notice how much more information this gives us; we know the approximate strength of the effect, which is extremely high at ρ = 0.929, and we can also say that there is a 99% chance that intervals constructed in this way will contain the true correlation coefficient between CO_2 and temperature.[1] This 99% confidence interval is very tight because the data set is very large, and the data points really do tend to lie on a straight line.

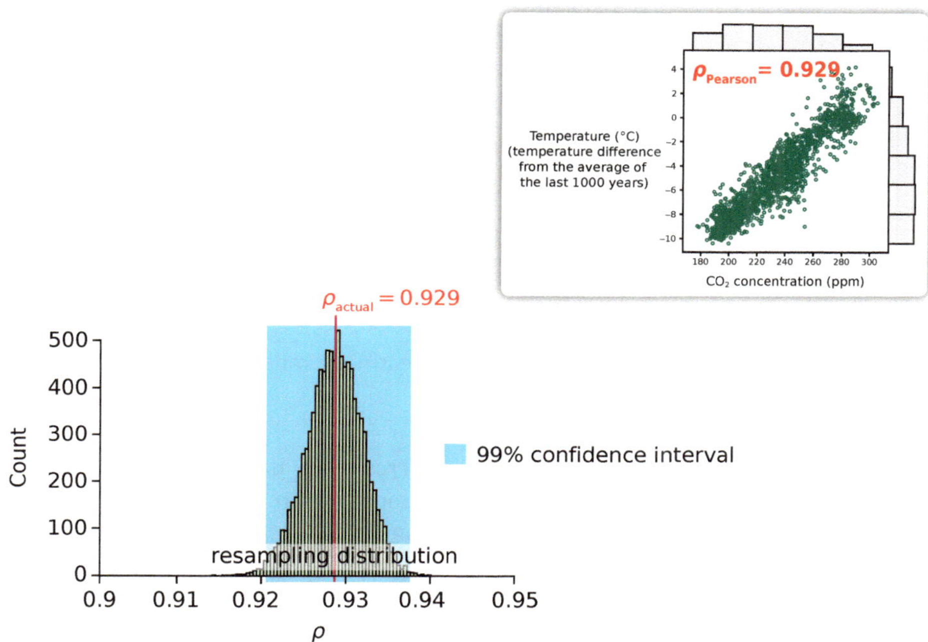

Figure 9.21 99% confidence interval for the CO_2/temperature data using resampling.

Because this data set largely meets the requirements for the theoretical calculation of confidence intervals, we compared the formula-based confidence interval to the confidence interval calculated from resampling. An online calculator that makes a formula-based calculation returned the estimate 99% CI = [0.920, 0.937] which agrees closely with our resampling result.

We see that when the data set largely meets the strict requirements for the theoretical calculation, the theoretical approximation is close to the resampling result.

Confidence interval: countries' energy use and GDP

We also calculated a confidence interval for the rank correlation coefficient for the data on countries' energy usage and GDP. We resampled the original raw data, and then ranked the X- and Y-values to calculate a new Spearman rank correlation. This was done 10,000 times, and a confidence interval was calculated in the usual manner (Figure 9.22).

[1] Note that what we would like to be able to say is: it is highly unlikely to be as low as 0.921 or as high as 0.938. But we cannot say that; that statement is calling for a Bayesian credible interval, and not a Frequentist confidence interval. See *Bayesian Credible Interval* on page 634.

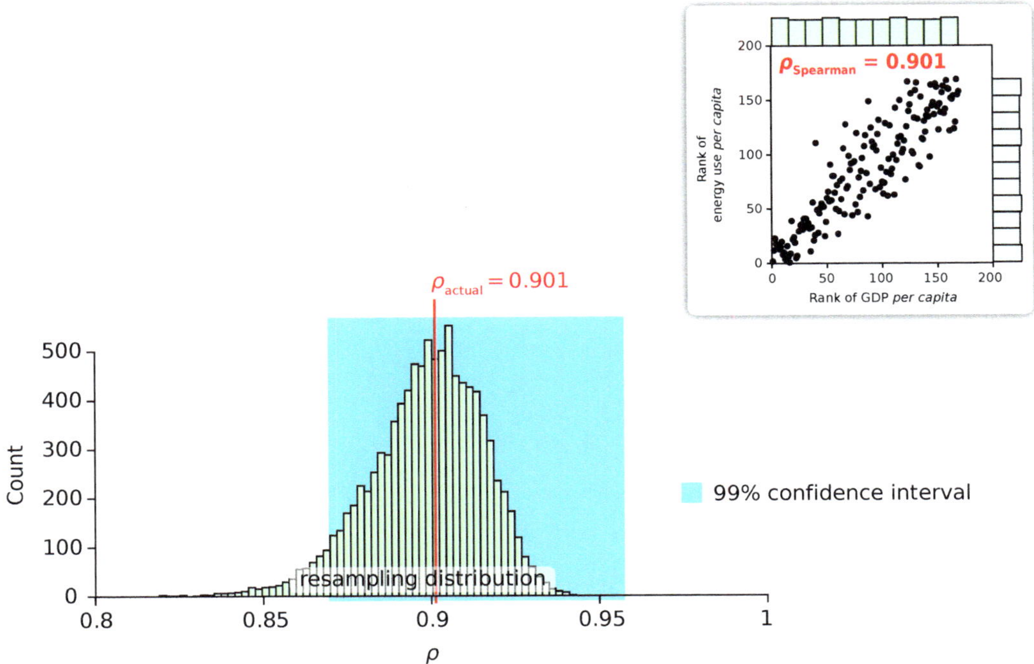

Figure 9.22 99% confidence interval for the (Spearman) rank correlation coefficient for the countries' energy use/GDP data.

Note that the resampling distribution in the energy/GDP data set is much wider than for the CO_2 and temperature data. Here, the histogram's width is around 0.1, whereas in the CO_2 and temperature case, the width was less than 0.02, due to the much larger data set in the CO_2 and temperature data.

NHST and CI

It is instructive to plot the null distribution (NHST) and the resampling distribution (CI) for the correlation coefficient side-by-side.

Here is the plot for the CO_2/temperature data set.

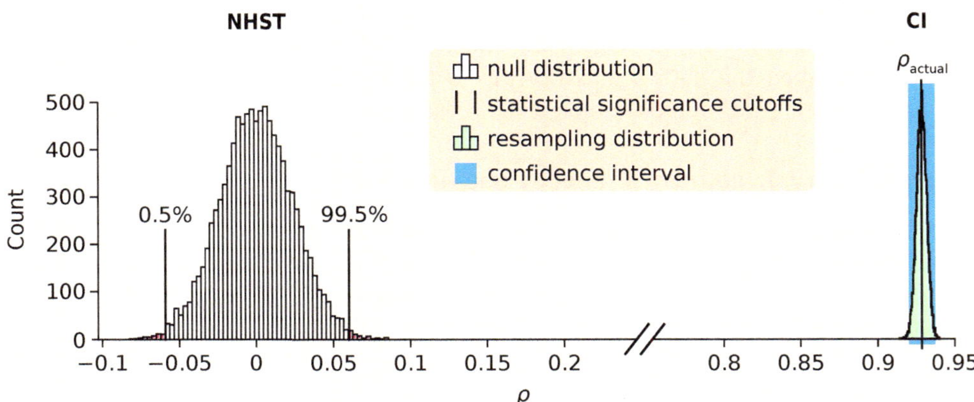

Note that we have constructed two different kinds of intervals. The first, on the left, is defined by Null Hypothesis Significance Testing. It centers around zero, which is the Null Hypothesis, and the cutoffs are the cutoffs for statistical significance. The second kind of interval, on the right, is the confidence interval for the actual (observed) value. It's important not to confuse these two!

Similarly, when we plot the NHST result and the confidence interval for the correlation coefficient in the countries' energy usage and GDP data set, the distribution on the left shows the 99% cutoff for the 10,000 simulations of the Null Hypothesis. It naturally centers around 0. It establishes that the actual value is statistically significant at the 0.01 level. The distribution on the right shows the 99% confidence interval around the actual value. It also implies that the observed result is highly significant.

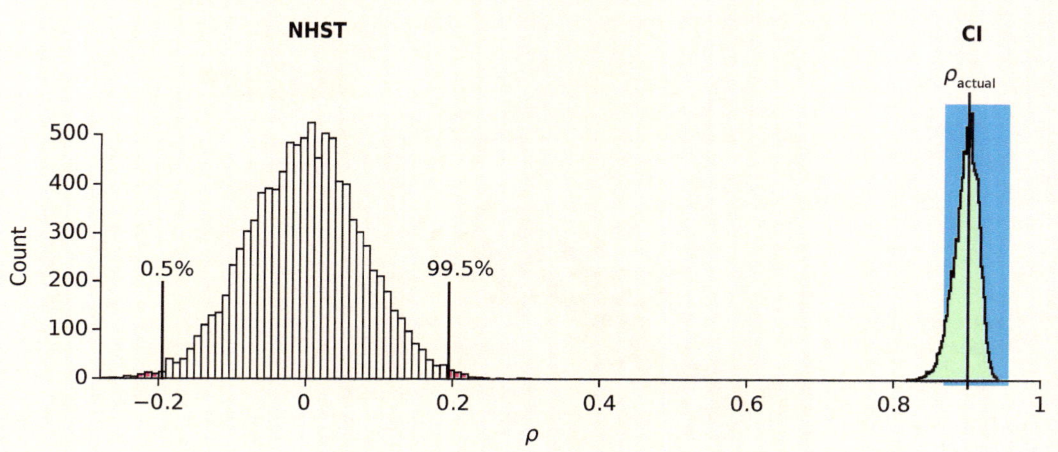

Technical note: the log transform

The raw data on countries' GDP and energy use, as we saw, does not meet the requirement that X and Y be Normally distributed, and also does not meet the requirements of constant variance in X and Y. For example, it has low Y-variance at low X, and high Y-variance at high X. This is common in data, because variability is often a fraction of the value, so if the values are, for example, $Y \pm 10\%Y$, then bigger Y means bigger variability in Y.

There is a data transform that is often used to transform data sets into a new coordinate system, attempting to reshape the data so that the transformed data meets the requirements for using the standard correlation coefficient. The mostly frequently used is the log transform. Transforming from "Y" to "log Y" means that small differences will be enhanced and large differences will be diminished. For example, the difference between \$1 and \$10 is only \$9, whereas the difference between \$10,000 and \$100,000 is \$90,000. However, the difference between log(\$10) and log(\$1), which is (1 – 0), or 1 log(\$), is the same as the difference between log(\$100,000) and log(\$10,000), which is (5 – 4), or 1 log(\$). Note the unit here is "log(\$)". So a difference of \$9 at the low end is the same difference on a log scale as a difference of \$90,000 at the high end.

In the countries' GDP vs. energy data, if we log transform both variables, the result is a distribution that has roughly constant variance. Moreover the distribution of the X and Y variables becomes much closer to Normal (see the log transformed plot below).

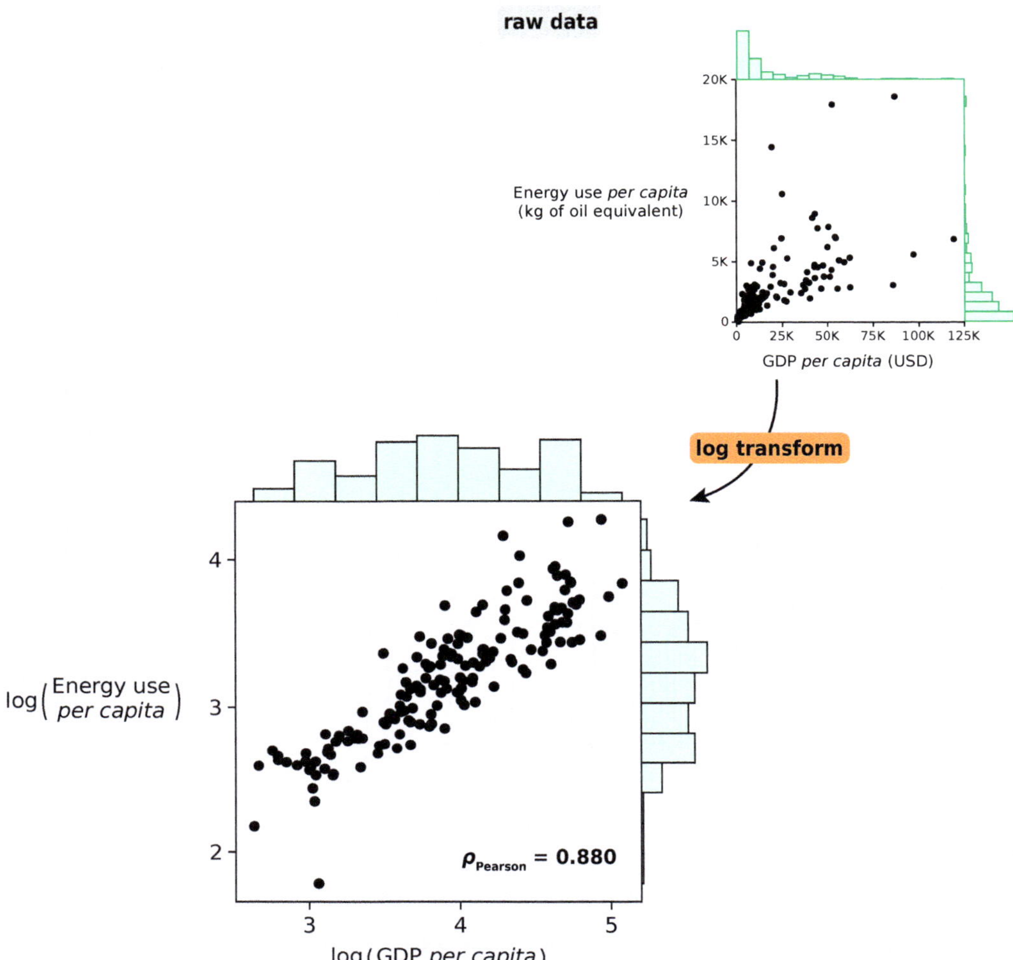

We might then be tempted to calculate a standard (Pearson) correlation coefficient for this logged data, and determine its statistical significance by theoretical formula. The coefficient would be quite high (in this case, $\rho = 0.880$). However, there are two problems.

1. First, the X and Y data are still not Normally distributed. The log transform took the small values and spread them out, and also bunched up the large values; the result is still a skewed distribution, especially in the vertical axis.

2. The second problem is much bigger: even if we were to establish that a correlation coefficient between the logged values was significant, then all we have done is shown that there is a significant correlation **between the logs of the two variables**. We can't really interpret this as a relation between the original quantities, only between their logs.

Then what interest is there in showing a significant correlation among logged variables? The answer is that it is interesting **only if the logs of the variables are a scientifically important quantity in their own right.**

This is sometimes the case. Logs of variables can be scientifically valid concepts. The best-known example is pH, the degree of acidity of a solution in chemistry. Recall that pH is defined as

$$\text{pH} = \log_{10}\left(\frac{1}{[H^+]}\right)$$

where $[H^+]$ is the hydrogen ion concentration, measured in moles/liter.

It would be reasonable to correlate, say, the pH values of many samples of ocean water with, say, the number of plant species observed in the sample.

As another example, sound is generally measured in decibels, which is a log scale, as opposed to linear scale. This is because the human perception of sound follows the logarithm of intensity, not a linear relationship.

FURTHER EXERCISES 9.5

1. **NHST and CI.** When working with a correlation coefficient, you can compute a p-value or a confidence interval. How is the resampling procedure different in the two cases?

2. **Rocks' weight and smoothness.** You decide to study the correlation between the weight and smoothness of rocks. Of your 10 randomly chosen rocks, you notice that the top 3 heaviest rocks are smooth whereas the two lightest rocks are rough. You tentatively conclude that heavier rocks are smoother than lighter rocks. Describe two ways in which you could improve this study.

9.6 The interpretation of correlation

It is commonly said that "correlation does not imply causation". This is very true, and bears repeating! However, as often as it is said, it is frequently then violated, often in the same paragraph. In fact, the deliberate confusion of correlation and causation has dogged discussions since the invention of correlation by Galton and Pearson in the mid 19th century.

We will meet this topic again in Chapter 10 (*Regression*), see especially section 10.5 *The ugly history of correlation, regression, p-values and NHST*.

In general, if X is correlated with Y, there are 4 distinct possibilities.

X causes Y X may cause Y, to be sure.

Consider an experiment where a metal is being heated to a variety of temperatures (X-axis) and its thermal expansion is then measured (Y-axis) (Figure 9.23). There is a very high correlation, and we can say that this is causal, because we know some physics. We know that if you heat a metal, the atoms jiggle faster and that makes the metal expand. This causal relationship between metal being heated and its expanding is reflected in the correlation coefficient.

We also know that this is a causal relationship because it is a result of a controlled experiment: *we*, the experimenter, set the temperatures, and the resulting elongation was then measured (see section 9.7 *Determining causation with an experiment*).

This is one possibility when X is correlated with Y.

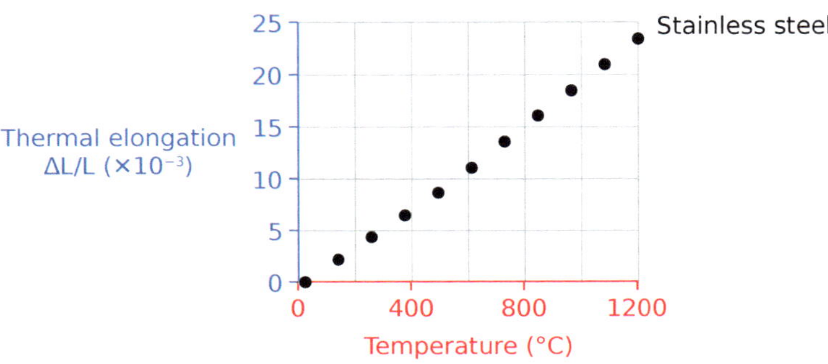

Figure 9.23

Y causes X But it also may be true that Y causes X.

We can also plot this data with the elongation on the X-axis and temperature on the Y-axis (Figure 9.24).

The correlation coefficient is exactly the same. That's a very high correlation between elongation (X) and temperature (Y), but we certainly don't want say that elongation causes temperature changes. Why don't we want to say that? Because we didn't set the elongations directly, and we also know some physics, and we know what can cause what, so we know the answer.

Could you conclude which one of these is true *from looking at the data alone, without knowing how the study was conducted*? Absolutely not. From the data alone, the two possibilities, X causes Y, and Y causes X, are equally likely.

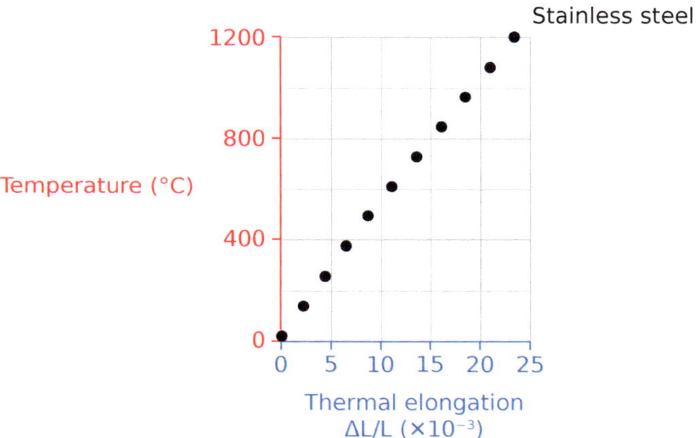

Figure 9.24

We don't want to fall into the error of the person who wrote a letter to the newspaper complaining about windmill farms, and wrote: "I've noticed that when they turn the windmills off, the wind stops."

Some Z causes both X and Y A very frequent possibility is a third kind of relation: it is true neither that X causes Y nor that Y causes X, but rather that **both X and Y are caused by the same factor Z.**

Here the classic example is the well-known correlation between ice cream sales and drowning deaths, month by month (Figure 9.25). When ice cream sales are low, drowning deaths are low, too, and when ice cream sales are high, so are drowning deaths. Do ice cream sales cause drowning deaths? Probably not.

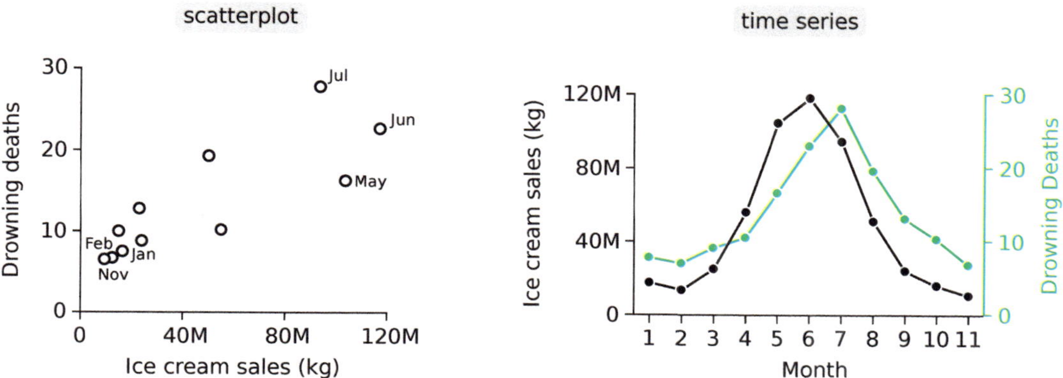

Figure 9.25 Drowning deaths and ice cream sales.

Do drowning deaths cause ice cream sales? Certainly not.

Then why the correlation? It's because both ice cream sales and drowning deaths go up in the summer, when temperatures are high, and go down in the winter, when temperatures are low. It's the season, and the *temperature*, that is driving both ice cream sales and drowning deaths.

Another classic example is the high correlation, in children ages 4–10, of shoe size and reading ability (Figure 9.26).

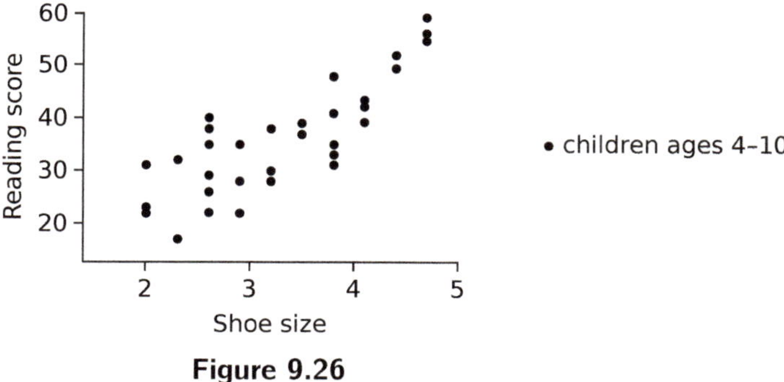

Figure 9.26

People are often surprised at this correlation, but the real factor causing both increases in shoe size and also reading ability is *age*.

In these cases, the third variable that is causing both X and Y is often referred to as a **confounding variable**. So age is a confounding variable if we are considering the relation between shoe size and reading ability in children.

p-hacking: nothing causes both X and Y

And then there is a fourth possibility. X doesn't cause Y, Y doesn't cause X, and there is no Z that causes both X and Y. Rather, the correlation is coincidental, and has been "discovered" by intense data-dredging, a form of p-hacking.

There is a brilliant website devoted to exposing this phenomenon, `tylervigen.com`. The author has also collected these into a book: *Spurious Correlations* (Vigen, 2015).

For example, there is a 99.3% correlation, nearly perfect, between the annual divorce rate in Maine over the years 2000-2009, and the annual *per capita* consumption of margarine over the same years! (Figure 9.27).

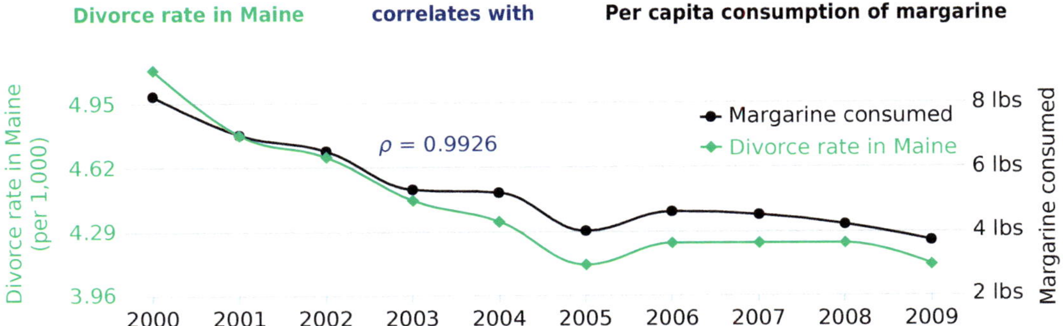

Figure 9.27 This stunning correlation is just the result of data dredging and p-hacking (source: `http://www.tylervigen.com`).

What could possibly account for this correlation? The answer is: nothing. Vigen simply gave his computer access to thousands and thousands of data bases, and then used the computer to play each one against all the others. He waited for occasional hits, mimicking true p-hacking style. So is there a correlation between the divorce rate in *Minnesota* and margarine consumption? No. What about Alabama? No. But if you run each of the 50 states against the *per capita* consumption rates of 50 basic foods, you will be making 2500 tests, and some of them are guaranteed to turn out highly significant, due simply to the play of chance. **This is the "green jelly beans" phenomenon!** (section 7.2 *The perils of multi-group testing*).

The site and the book have many good examples, some of them quite funny. For example, "*per capita* consumption of mozzarella cheese" correlates with "civil engineering doctorates awarded" with a correlation coefficient a stunning 95.9%, and there is an impressive 89.99% correlation between "*per capita* consumption of chicken" and "total US crude oil imports"!

Vigen has now gone even further, and, for each spurious correlation, he now provides an AI-generated (nonsense) causal explanation for the spurious correlation! His entry for the margarine-divorce rate example is:

"Perhaps as people used less margarine, they became less slippery in their relationships. The lack of artificial spread may have kept the couples from buttering each other up, leading to a decrease in overall marital strife. That's the reality when you can't believe it's not butter - it's a recipe for marital success...."

Exercise 9.6.1 Given that "correlation does not imply causation", why do we even care about correlation?

Exercise 9.6.2 We conducted a study of professor's salaries at multiple time points over the last 25 years and attempted to correlate them to the price of alcohol at those times. We find a highly significant correlation coefficient (ρ = 0.7, p = 0.001). What is your interpretation of this finding? Imitate Vigen's AI, and invent a phony causal explanation for each causal direction.

FURTHER EXERCISES 9.6

1. **SAT and family income.** Here is some data on mean SAT scores and family annual income from the College Board, with the first column giving the income bracket.

Income Bracket (in $1000s)	Math	Verbal	Writing
0 – 10	457	429	427
10 – 20	465	445	440
20 – 30	474	462	454
30 – 40	488	478	470
40 – 50	501	493	483
50 – 60	509	500	490
60 – 70	515	505	496
70 – 80	521	511	502
80 – 100	534	523	514
100+	564	549	543

   a. Visualize the data and describe what you see.
   b. What's your interpretation?

2. **CO_2 and plant growth.** Photosynthesis is the process by which a plant uses carbon dioxide (CO_2) to make food (sugars). An experiment showed how plant growth rate varied across carbon dioxide levels for a certain plant species. Is growth rate correlated with CO_2 levels? Do CO_2 levels affect growth rate?

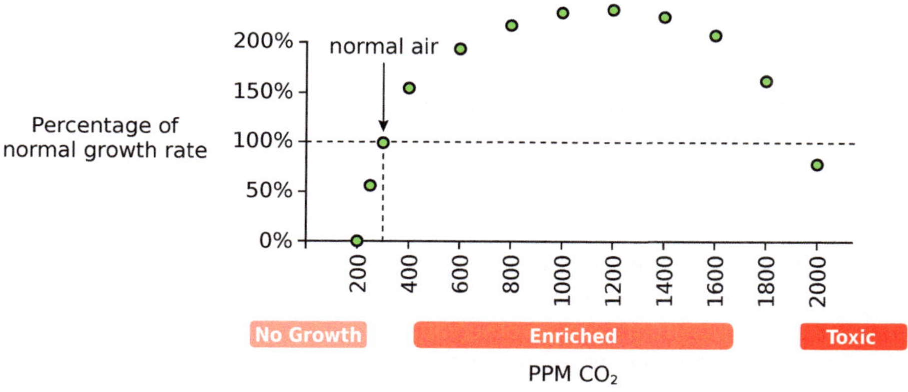

3. **Beef and cholesterol.** A recent study showed that people who eat beef have higher cholesterol than vegetarians.

   a. What are the four possible explanations of this associations?
   b. How could you tell whether eating beef causes higher cholesterol?

4. **The Nenana Classic.** The Tanana River in Alaska freezes over during the winter, and the ice breaks up in late April or early May. In 1917, locals began betting on the exact time the ice would break up and have continued ever since in a contest known as the Nenana Classic. Is there a correlation between break-up dates and years?

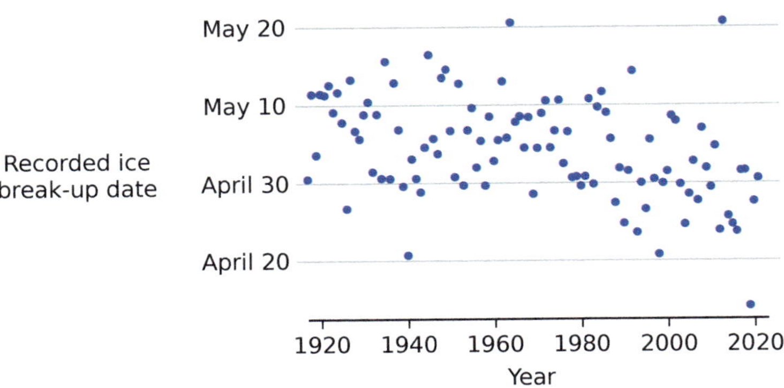

5. **Temperature and the rate of reaction.** Temperature affects the rate of chemical reactions. A study investigated the effect of temperature (°C) on the rate of reaction (μmol per min) of the enzyme acid phosphatase. What measure would you use to determine whether temperature affects the rate of reaction of acid phosphatase?

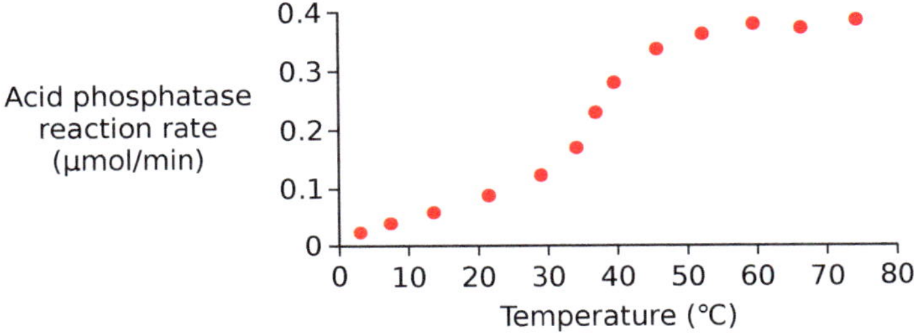

9.7 Determining causation with an experiment

The primary way to obtain causality is by conducting an experiment. In this regard, there is a very important distinction between observational studies and experimental studies.

Observational studies In observational studies, we look at a group of subjects, write down their X- and Y-values, and then look for a relationship between those X- and Y-values.

Can we get causality from such an observational study? Usually not. It's frequently going to be ice cream production and drowning deaths, where weather is the cause, or the correlation between shoe size and reading ability in children under 10, where age is the confounding factor that's causing both. So causality is hard to conclude in an observational study.

Experimental studies In an experimental study, we don't just find subjects and record their X- and Y-values. Instead we find subjects and we actively intervene and *create* their X-values, and see what their resulting Y-value is. This is important **because we, the experimenters, determine the X-value, not the subject itself. Now if there is a correlation, it is likely that X causes Y.**

DUI stops Here is hypothetical data from a series of DUI stops (Figure 9.28). The police stop a number of subjects, and measure their blood alcohol content (BAC) and reaction time.

We see that reaction time is highly correlated with BAC level. So can we infer causality, that drinking increases reaction time?

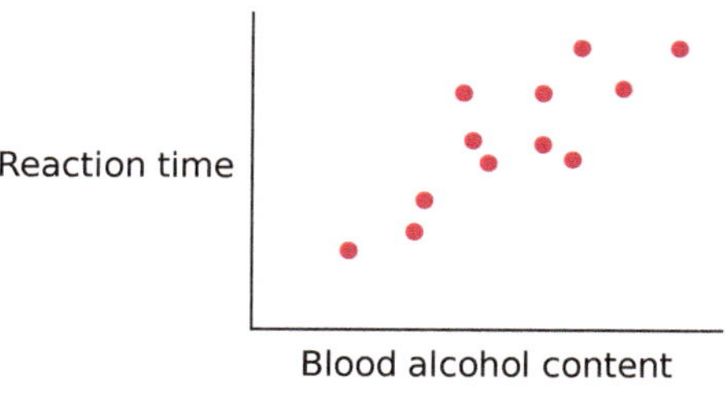

Figure 9.28

We might think that this is an experiment, so the relationship is evidence of causality. The police stopped people and measured their BAC and their reaction time. **But it's not an experimental study, because the experimenter (the police) did not determine how much alcohol people had.** It's an observational study. It may simply be the case that people with long reaction times drink more. We can't rule that out on the basis of this data because it's only an observational correlation.

Running on an elliptical Does running on an elliptical trainer cause pulse rate to increase? The horizontal axis is the running speed on the elliptical and the vertical axis is the pulse rate (Figure 9.29). Is that a causal relationship, that running on an elliptical causes pulse rate to increase?

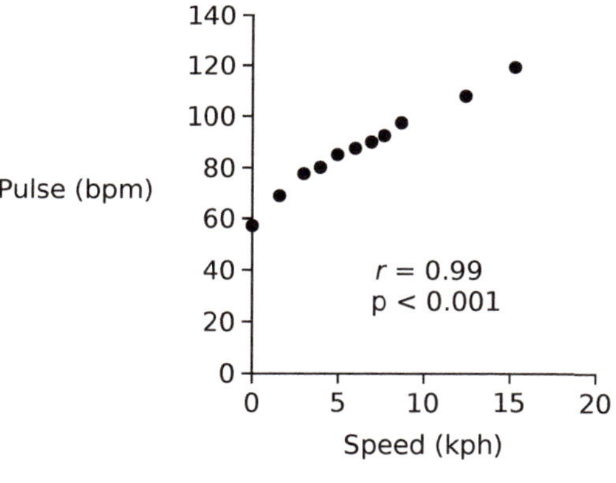

Figure 9.29

If you think it is causal, you are assuming something which has not yet been given. What if people just showed up in the lab and chose a running speed, and then we measured their pulse rates? On the basis of this data, we can't rule out the possibility that people with higher pulse rates run faster.

This is still an observational study. Now, suppose we are given an extra piece of information, that "subjects are required to run at various speeds". With that piece of information, that the experimenter specifies for each subject the speed at which they'll run, this has become an experimental study design, and we **can** now infer a causal relationship based on the correlation.

Causality from an experimental protocol

In an experimental protocol, the subject doesn't set their own X-values, the experimenter does. Therefore, we know Y can't cause X, because we know the cause of X: the cause of X is that this subject was randomly assigned to the 5 km/h group, etc.

If the X assignment is random, then we also know that there cannot be a Z causing both X and Y.

We also know that it can't be due to data dredging (p-hacking).

Therefore, having ruled out 3 of the 4 possibilities, we can conclude that X causes Y.

1. X causes Y
2. ~~Y causes X~~
3. ~~Z causes both X and Y~~
4. ~~p-hacking~~

Example 4 Do higher latitudes cause fewer birds species?

Here is a scatterplot showing the number of bird species at different latitudes. As we go further north, the number of species declines.

Does this enable us to make a causal statement? Is there a causal relationship between latitude and the variety of bird species?

A. Yes, because it's an experiment and there's a relationship.

B. Maybe, because it's an observational study and there is a relationship.

C. No, because there's no relationship stronger than chance would predict.

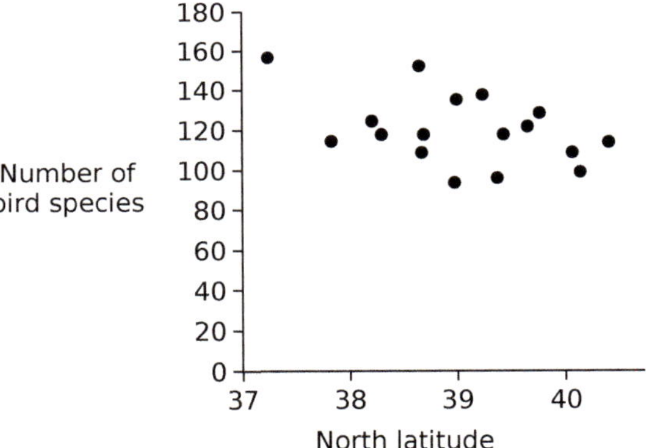

Answer B is the correct answer.

Independent variables and dependent variables Once we have a truly experimental design, we can then distinguish independent and dependent variables.

An **independent variable** is a variable that is controlled by the experimenter. In the example in Figure 9.30, the experimenter decides how much water each plant gets, therefore, the amount of water is an independent variable. On the other hand, measures like the size of the plant, the number of leaves, even whether the plant is alive or dead, are outcomes of the independent variable, and are called **dependent variables**.

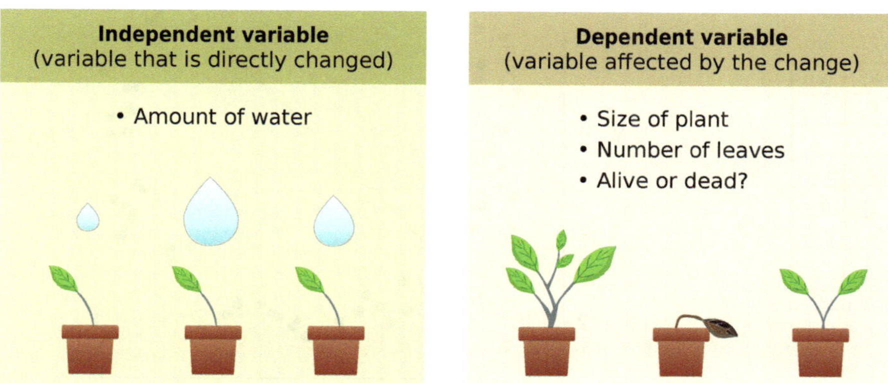

Figure 9.30 Independent and dependent variables.

FURTHER EXERCISES 9.7

1. Which of the following must happen in a study to determine cause and effect?
   a. Taking a random sample from a population
   b. Randomly assigning the observational units to different treatment groups
   c. Simulation-based inference techniques
   d. Theory-based inference techniques
2. You are given a data set and find a very strong correlation between variables A and B. What can you conclude?

9.8 Mutual Information: a nonlinear improvement on the correlation coefficient

It's surprising how many people do not understand that the correlation coefficient is a measure of the ***linear*** association between X and Y. Most people, when shown (X, Y) data that lie on a circle, or that form an upside down "V", think there must be a significant correlation between X and Y. Of course, for these examples, the correct answer to "what is the correlation coefficient?" is "zero".

The correlation coefficient is NOT a measure of nonlinear association. If the data are arranged in a "V" shape, then the answer to the question: how much does the data lie on a straight line? is: zero.

In order to go beyond linear relations and develop a measure that will capture nonlinear relations, we need to rethink (X, Y) data.

Let's take a data set that lies roughly on a circle (Figure 9.31).

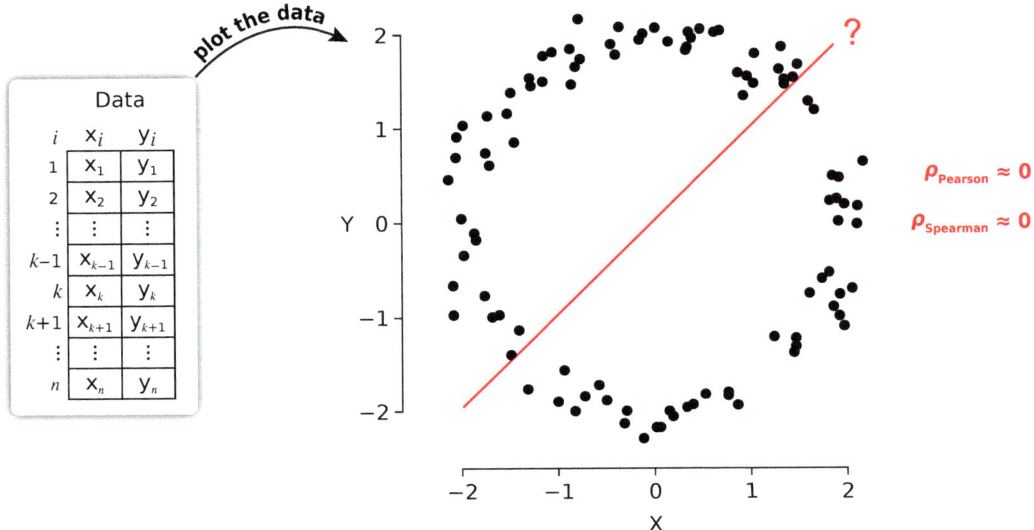

Figure 9.31 A circle-shaped data set.

Here, there is an obvious nonlinear relationship, $X^2 + Y^2 \approx 4$, so the concept of the correlation coefficient does not apply here at all. But if we went ahead anyway and calculated it, the Pearson correlation coefficient ρ_{pearson} is ≈ 0 and the Spearman correlation coefficient ρ_{spearman} is also ≈ 0. But there is clearly *some* relation between the X- and Y-values. How can we capture this?

Rethinking independence One way to capture this relationship is to ask a somewhat different question. Starting from a fresh angle, let's state the idea of **independent** as: **if X is independent of Y, then knowing the X coordinate of a pair tells us nothing about the Y-value.** To put it another way, if we know the X-value, our ignorance of the Y-value is total.

We have just appealed to a new concept: **degree of ignorance**. We would like to quantify this notion, so that we could say that our ignorance is zero if we know the exact value, partial if we know something about Y but not its exact value, and total when we have no idea what the value of Y is.

A formal measure of "degree of ignorance" is the concept of **entropy**. Suppose we have a system that can be in any one of n states. Let $p_1, p_2, \ldots, p_n$ be the probabilities of the system being in state 1, state 2, ..., state n (so $p_1 + p_2 + \cdots + p_n = 1$). Then the entropy of the system at any given moment is

$$\sum_{i=1}^{n} p_i \times \ln(p_i)$$

This quantity is at a maximum when $p_1 = p_2 = \cdots = p_n$. When all probabilities are equal, we are at maximum cluelessness.

Going back to the circle-shaped data, we ask ourselves the following question.

Q: Does knowing the X-value of an (X, Y) pair enable us to (roughly) calculate the Y-value?

A: No.

Because, for example, if the X-value is zero, the Y-value could be highly negative, it could also be highly positive. So knowing X does not enable us to calculate the Y-value, even approximately.

Now we have another question.

Q: Does knowing the X-value of an (X, Y) pair tell us *anything* about the Y-value?

A: Yes.

We know that if the X-value is near zero, the Y-value is either going to be very low or very high. Does that enable us to calculate the Y-value? No. Does it reduce our cluelessness about the Y-value, yes, it does! We also know that if the X-value is very negative, very low, then the Y-value is going to be middling.

So now we are going to define the concept "reduces my cluelessness".

Let's take the circle-shaped X–Y scatterplot with n data points. First, we coarse-grain the space into rows R_1, R_2, ..., R_m, and columns C_1, C_2, ..., C_k. The rows represent the levels of the Y-value and the columns represent the levels of the X-value. The intersection of row R_i and column C_j is the bin $B_{i,j}$ (Figure 9.32).

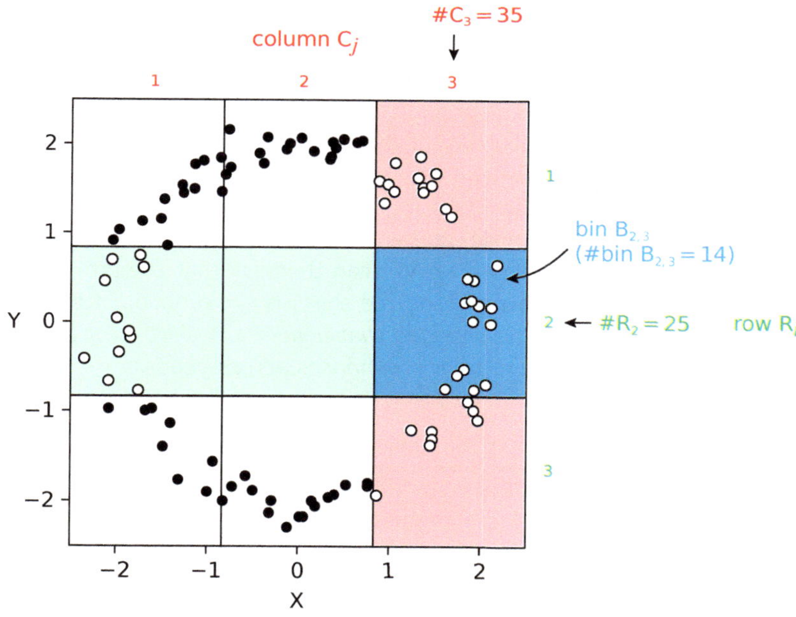

Figure 9.32 Coarse-graining the circle-shaped data.

Let's define the probability of a data point landing in a row, in a column and in a bin cell as:

$$\text{prob}(R_i) = \frac{\#R_i}{n} = \frac{\text{number of data points in } i^{\text{th}} \text{ row}}{\text{total number of data points}}$$

$$\text{prob}(C_j) = \frac{\#C_j}{n} = \frac{\text{number of data points in } j^{\text{th}} \text{ column}}{\text{total number of data points}}$$

$$\text{prob}(B_{ij}) = \frac{\#B_{i,j}}{n} = \frac{\text{number of data points in } (i,\ j)^{\text{th}} \text{ bin}}{\text{total number of data points}}$$

Then we can use a very basic idea from probability theory.

How many data points should we expect in bin cell $(i,\ j)$? We know that the probability of landing in the ith row is prob(R_i), and we know that the probability of landing in the jth column is prob(C_j).

> Let the probability of X be prob(X), and the probability of Y be prob(Y), then
>
> ***IF***
>
> X and Y are independent events
>
> ***THEN***
>
> the probability of X and Y both happening is
> prob(X and Y) = prob(X) × prob(Y)

The principle we will use from probability theory is: if X and Y are "independent", then prob(X and Y) = prob(X) × prob(Y). The probability of drawing a red queen is equal to the probability of drawing a red card × the probability of drawing a queen, because redness and queenness are independent. If they are not independent, we could not make that product statement.

What is the probability of finding a male student in the class who is wearing a red shirt? Let's say that the class is 50% male, and we know overall that roughly 10% of all students will be wearing a red shirt on any given day. We can then say that the probability of finding a male wearing a red shirt, that is prob(male and red shirt) is the product of prob(male) and prob(red shirt), that is, 50% × 10% = 5%, *provided that gender and shirt color are independent, that is, that there is no difference in shirt color preferences across genders.*

Deviation from independence The key idea is then to consider, for each cell, the ratio

$$\frac{\text{prob}(B_{ij})}{\text{prob}(R_i) \times \text{prob}(C_j)}$$

We just said that if rows and columns are independent, then the numerator and denominator will be roughly equal, and the ratio will be 1. **Then we can use the deviation of this quantity from 1 as a measure of the non-independence of row and column, that is, of the non-independence of X and Y for this cell.**

$$\frac{p_{ij}}{p_i \cdot p_j} \approx 1 \iff$$ rows and columns are independent, implying that X and Y are independent, remembering that i is the row value and j is the column value.

So, for each cell, **we want to calculate this ratio $\frac{p_{ij}}{p_i \cdot p_j}$, which captures each cell's contribution to the deviation from independence. We would like to somehow then add up all the cells' contributions to get a total measure of deviation from independence.** This would be a measure of the reduction in our cluelessness.

We could just take this sum as our measure of independence, but there are two technical improvements that will greatly improve this measure.

1. First, the simple sum has the drawback that if the cell does not deviate from independence, then the ratio $\frac{p_{ij}}{p_i \cdot p_j}$ is approximately equal to 1. If we just added up these values, then we would be adding up a lot of 1's, which would increase the total value. But that isn't what we want; we want such a cell to contribute *zero* to the total deviation from independence. Therefore, we use not the ratios, but the logarithm of the ratios (typically using log base 2). The log of 1 is 0, so that works out as we want it.

2. The second improvement to our measure is to introduce a kind of weighting: some cells are more important than others. Cells with many data points in them are more valuable, and should be weighted highly, whereas cells with few data points (that is, cells having a low probability) should be down-weighted. We accomplish this by multiplying each cells' contribution by the probability of a data point being in that cell, that is, by the number of data points in the cell divided by the total number of data points. If that number is low, the cell's contribution is weighted low.

So the weighted contribution of each cell (i, j) is calculated by weighting $\log\left(\frac{p_{ij}}{p_i \cdot p_j}\right)$ by the probability p_{ij}.

$$\mathrm{MI}_{ij} = p_{ij} \times \log\left(\frac{p_{ij}}{p_i \cdot p_j}\right)$$

If we then sum up each cell's contribution, we arrive at the concept called **Mutual Information**.

Mutual Information $$\mathrm{MI} = \sum_{i,j} \mathrm{MI}_{ij} = \sum_{i,j} \left(p_{ij} \times \log \frac{p_{ij}}{p_i \cdot p_j}\right)$$

Example 5 Calculating mutual information

Calculate the mutual information for the following data set:

{ (0.8, 2.8), (1.8, 1.8), (2.8, 0.8), (3.8, 1.8), (4.8, 2.8) }

Answer

Step 1 ► Plot the data points.

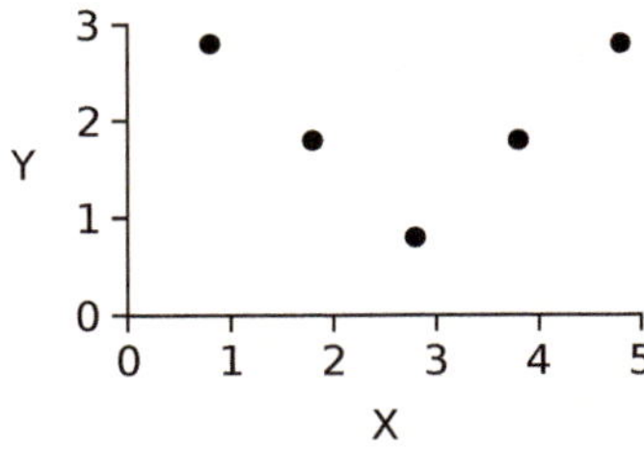

Step 2 ► Choose a mesh size to coarse-grain the data set.

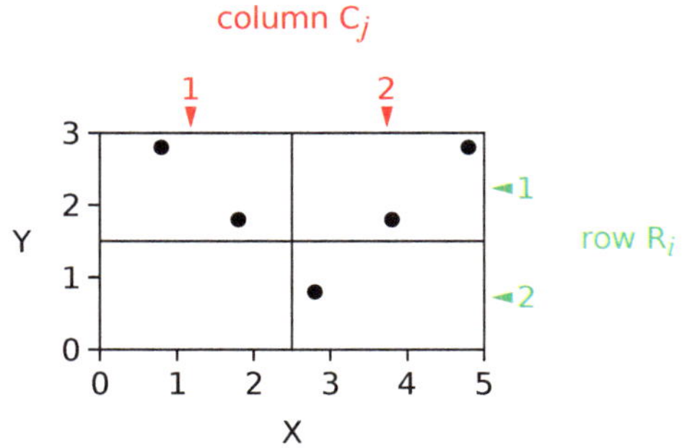

Step 3 ► Calculate the contribution of each bin to the total mutual information. Here, there are 3 bins that contain data points.
For bin (1, 1), its contribution can be calculated as:

$$p_{B_{1,1}} = \frac{\#B_{1,1}}{n} = \frac{2}{5} \qquad p_{R_1} = \frac{\#R_1}{n} = \frac{4}{5} \qquad p_{C_1} = \frac{\#C_1}{n} = \frac{2}{5}$$

$$\text{MI}_{1,1} = p_{B_{1,1}} \times \log\left(\frac{p_{B_{1,1}}}{p_{R_1} \times p_{C_1}}\right) = \frac{2}{5} \times \log\left(\frac{\frac{2}{5}}{\frac{4}{5} \times \frac{2}{5}}\right) = \frac{2}{5} \times \log\left(\frac{5}{4}\right) = 0.129$$

For bin (1, 2), its contribution can be calculated as:

$$p_{B_{1,2}} = \frac{\#B_{1,2}}{n} = \frac{2}{5} \qquad p_{R_1} = \frac{\#R_1}{n} = \frac{4}{5} \qquad p_{C_2} = \frac{\#C_2}{n} = \frac{3}{5}$$

$$\text{MI}_{1,2} = p_{B_{1,2}} \times \log\left(\frac{p_{B_{1,2}}}{p_{R_1} \times p_{C_2}}\right) = \frac{2}{5} \times \log\left(\frac{\frac{2}{5}}{\frac{4}{5} \times \frac{3}{5}}\right) = \frac{2}{5} \times \log\left(\frac{5}{6}\right) = -0.105$$

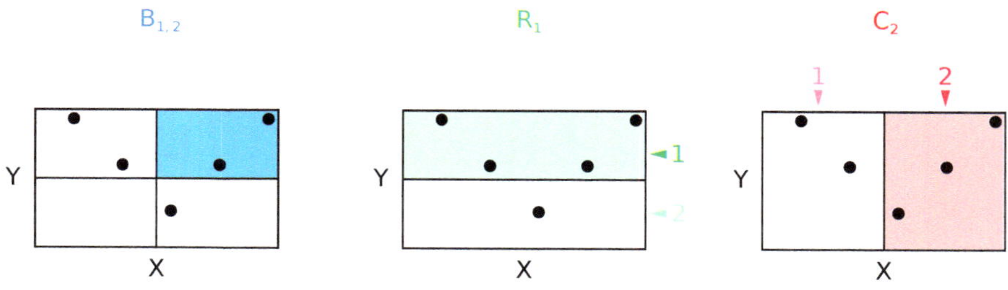

For bin (2, 2), its contribution is calculated as:

$$p_{B_{2,2}} = \frac{\#B_{2,2}}{n} = \frac{1}{5} \quad , \quad p_{R_2} = \frac{\#R_2}{n} = \frac{1}{5} \quad , \quad p_{C_2} = \frac{\#C_2}{n} = \frac{3}{5}$$

$$\mathrm{MI}_{2,2} = p_{B_{2,2}} \times \log\left(\frac{p_{B_{2,2}}}{p_{R_2} \times p_{C_2}}\right) = \frac{1}{5} \times \log\left(\frac{\frac{1}{5}}{\frac{1}{5} \times \frac{3}{5}}\right) = \frac{1}{5} \times \log\left(\frac{5}{3}\right) = 0.147$$

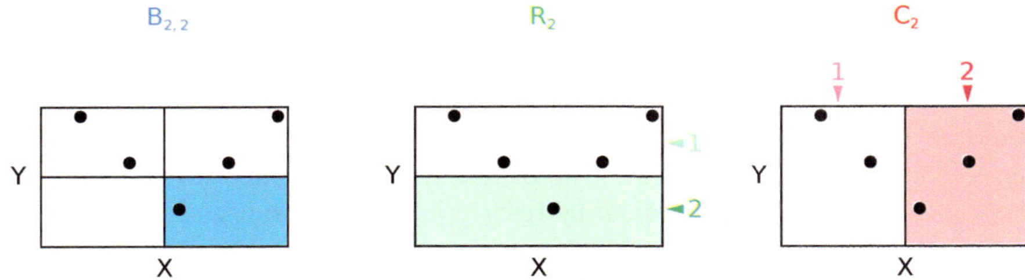

Step 4 ► Summing up each bin's contribution, we get the mutual information for this data set.

$$\begin{aligned}
\mathrm{MI} &= \sum_{i,j} \mathrm{MI}_{i,j} \\
&= \mathrm{M}_{1,1} + \mathrm{M}_{1,2} + \mathrm{M}_{2,2} \\
&= 0.129 + (-0.105) + 0.147 \\
&= 0.171
\end{aligned}$$

Statistical significance and confidence intervals for mutual information

It is straightforward to calculate statistical significance, and even better, confidence intervals, for mutual information. In fact, the procedure is exactly the same as what we did for the correlation coefficient.

Note that in this book we are basically learning one method that's going to work over and over again. Instead of separate theories for the t-test, ANOVA, Welch's test, correlation coefficient, etc, we have one method: resampling.

So we can form our best hypothesis as scientists, enjoying freedom of measure, calculate those measures for our data, and not have to worry about statistical theory that may or may not exist; we know that we can find statistical significance and confidence intervals through resampling.

Statistical significance Returning to the circle-shaped data set, we calculated its mutual information: $\mathrm{MI}_{\mathrm{actual}} = 0.18$ (Figure 9.33). The next question is: is 0.18 large? And of course, "large" means "relative to what we would expect under the Null hypothesis."

The Null Hypothesis is that there is no relationship between the X-value and the corresponding Y-value for each data point. Therefore, to simulate this Null Hypothesis, but keep everything else constant, such as the distribution of the data, etc.

We break up any relation there might be between X and Y. We shuffle the X-values to break this relation (equivalent to shuffling only the Y-values or shuffling both X-values and Y-values).

We then plot the data, recalculate the mutual information for this randomized data set (where we broke the link between X and Y, which embodies the Null Hypothesis) and repeat it, say, 10,000 times.

Then we compare $\text{MI}_{\text{actual}}$ to the 10,000 values of MI computed under the Null Hypothesis. In the case of the circle-shaped data, none of the 10,000 MI-values even come close to $\text{MI}_{\text{actual}}$, therefore, we can say $p < 0.0001$.

For mutual information, we do not use a "0.5% at the top and 0.5% at the bottom" method, because MI can't really be significantly low. MI is non-negative, and the null distribution has MI ≈ 0. So we have to use a one-sided test.

Confidence intervals As we have said throughout this book, the better practice is to report the observed effect size (in this case, the observed MI value), and give confidence intervals.

We derive the confidence interval by resampling: if our original data set is $\{(x_1, y_1), (x_2, y_2), \ldots, (x_n, y_n)\}$, then a resampling of this data set is a set of n of the original (x_i, y_i) pairs, chosen at random with replacement from the original set of pairs. MI for this resampled data set is calculated, and the procedure is repeated 10,000 times, keeping track of the MIs for each new resample.

Then the cutoffs for the 99% confidence interval are constructed from the 0.5^{th} percentile and 99.5^{th} percentile cutoffs of the simulated MIs, and the actually observed MI (Figure 9.34).

In the case of the circled-shaped data set, the 99% confidence interval is [0.03, 0.25].

As introduced in Chapter 4, when we construct a confidence interval, if MI_{lower} and MI_{upper} are the 0.5% and 99.5% cutoffs, then the boundaries of the 99% confidence interval are

$$\text{CI}_{\text{lower}} = 2 \times \text{MI}_{\text{actual}} - \text{MI}_{\text{upper}}$$
$$\text{CI}_{\text{upper}} = 2 \times \text{MI}_{\text{actual}} - \text{MI}_{\text{lower}}$$

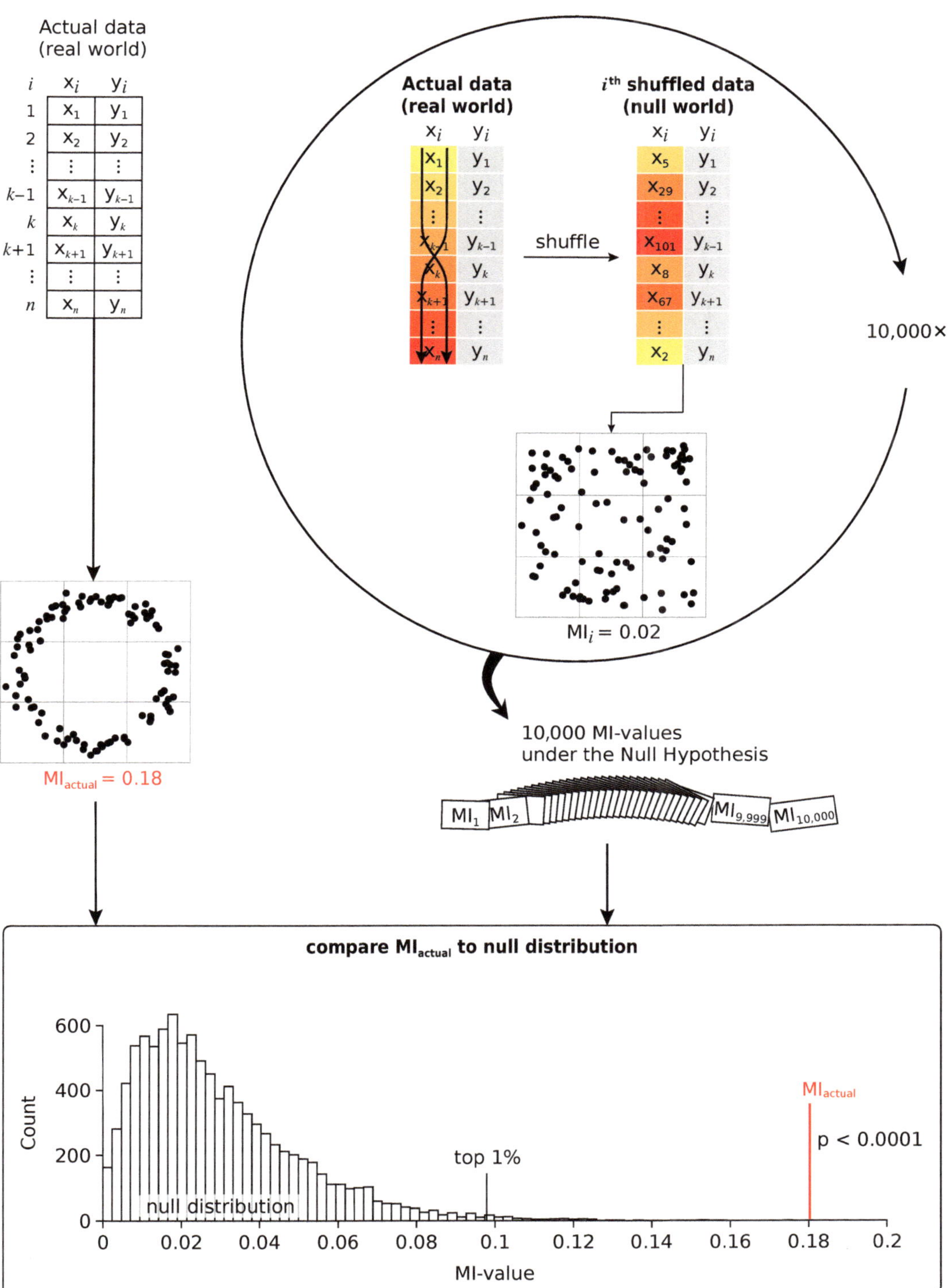

Figure 9.33 NHST for mutual information. We use the circle-shaped data set to illustrate the procedure of calculating a p-value for $\mathrm{MI}_{\mathrm{actual}}$.

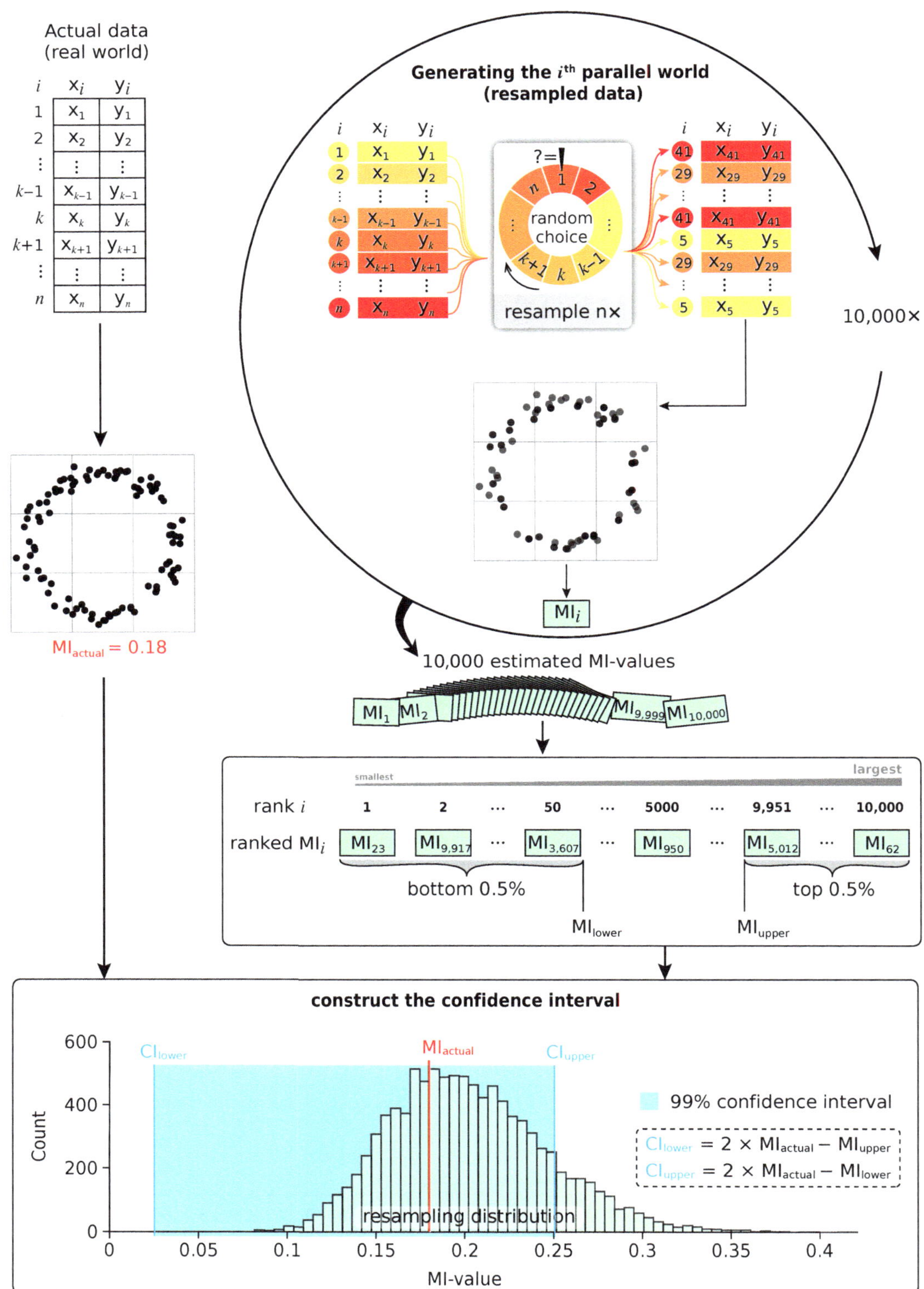

Figure 9.34 Confidence interval for mutual information.

Dependence on mesh size and the concept of Maximal Information Content

When we introduced the concept of Mutual Information, we began by saying "choose a mesh that coarse-grains the (X, Y) space". We did not say *which* mesh to choose. Looking at the circle-shaped data, we can imagine a number of different meshes, of increasing fineness.

Each grid represents a scientific hypothesis, that there is a relation between X and Y at the resolution of that grid. So, for example, if we choose a 2×2 grid across (X, Y)-space, we are asking whether there is a relationship between X and Y when X and Y are considered to be 2-valued (X is either 'high' or 'low', and so is Y). A finer grid would be looking for a relationship between X and Y that existed at those finer resolutions.

Recently, a suggestion was made that would eliminate the dependence on the choice of grid. Reshef *et al.* (2011), writing in the journal *Science*, suggested using the computational power that is now routinely available to define a new concept: **Maximal Information Content** (MIC). In essence, MIC is the maximum value of MI over *all* grids, with a penalty for over-refining the grid, and an upper limit on the number of rows and columns, determined by the size of the data sample.

The resulting MIC has been called "a correlation for the 21st century" (Speed, 2011): "A novel statistical approach has been developed that can uncover nonlinear associations in large data sets."

Measures for bivariate data

We can now summarize all the measures for bivariate data (Figure 9.35).

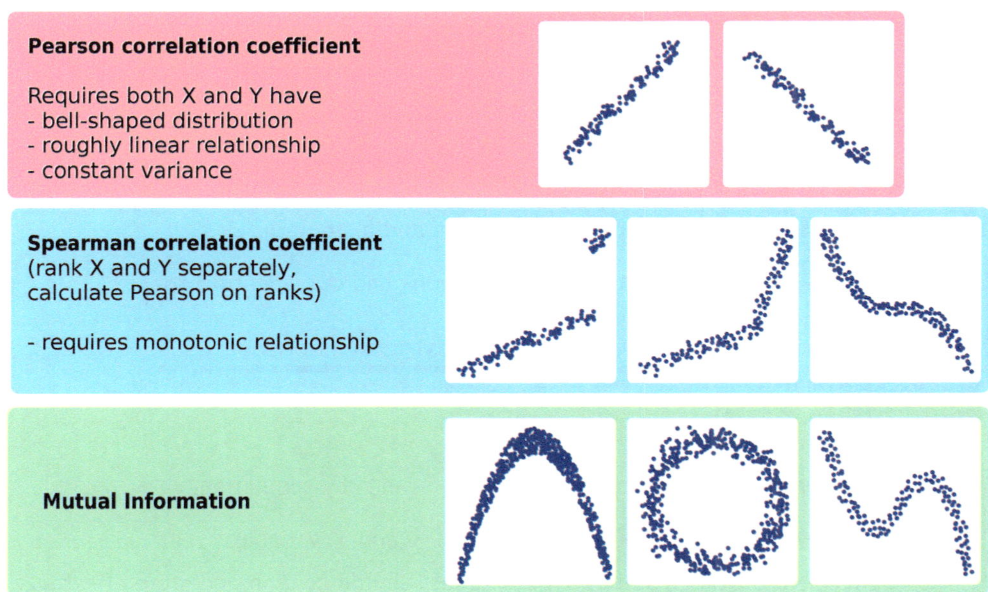

Figure 9.35 Measures for bivariate data.

The first is the Pearson correlation coefficient. Notice the strong requirements on X and Y.

If the data doesn't pass those requirements, but is still monotonic, we can use the weaker form of the correlation coefficient, called the Spearman correlation coefficient. The use of ranks makes it a weaker test, so if the data supported a Pearson correlation coefficient, we should prioritize it, rather than using a Spearman correlation coefficient.

And if the data are not monotonic, the only choice is to use mutual information, which is a powerful detector of nonlinear relationships.

The decision tree to handle bivariate data is shown in Figure 9.36.

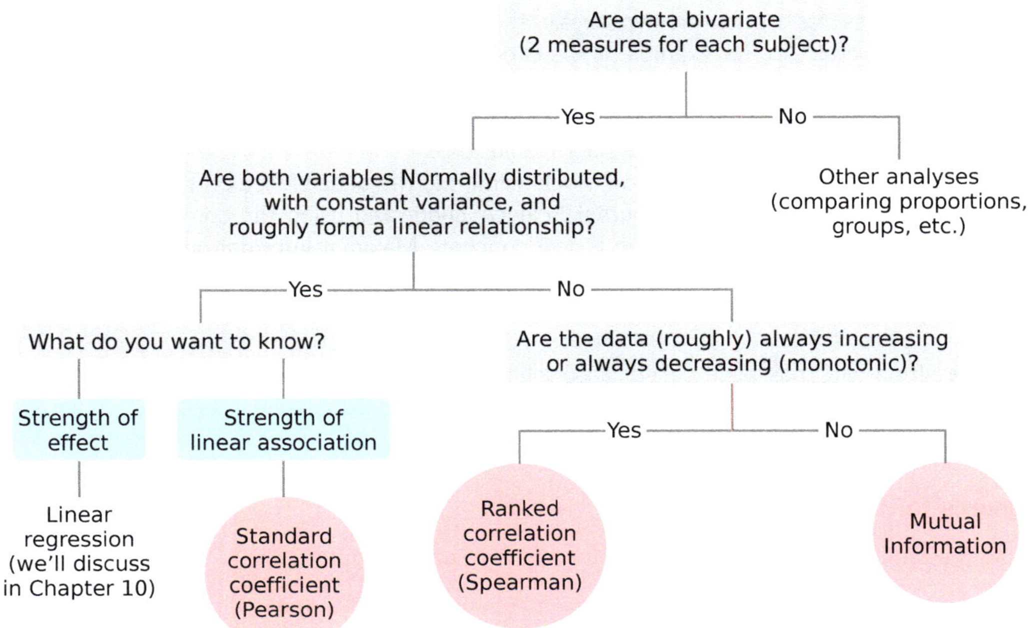

Figure 9.36 Decision tree for analyzing bivariate data.

Analyzing bivariate data

1. Visualize data ("make a picture"): scatterplots and histograms.
2. Choose measures based on the data distributions and our scientific goal.
   - Pearson
   - Spearman
   - Mutual information
3. Calculate p-value and evaluate statistical significance by simulation.
   a. Null Hypothesis: no relationship between X and Y variables.
   b. Simulate the Null Hypothesis many times: shuffle column and store effect size.
   c. Calculate p-value.
4. Construct confidence interval for the observed effect size.
   a. Sample (X, Y) pairs with replacement many times and store effect size.
   b. Find lower and upper cutoffs and construct the confidence interval.

FURTHER EXERCISES 9.8

1. The biologist R. A. Fisher considered a data set, gathered from 3 different plant species. For each plant, there are 4 measures: Sepal Length, Sepal Width, Petal Length and Petal Width. He made scatterplots for each pair of traits. Is correlational analysis a good approach for this data set? How would you proceed with this data set?

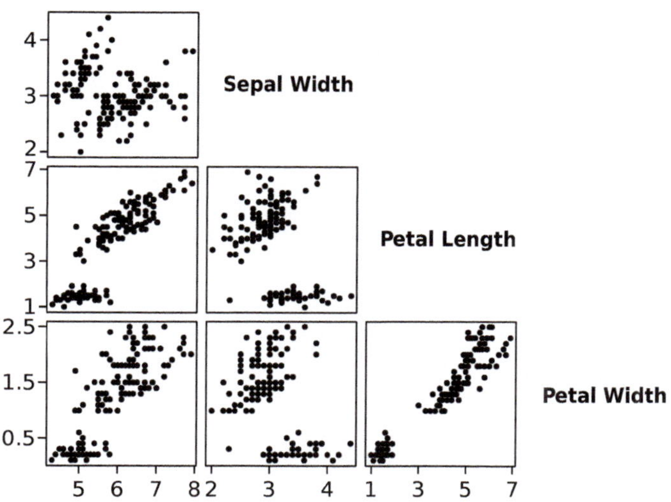

2. Which type of analysis should we use for each of the data sets shown below?

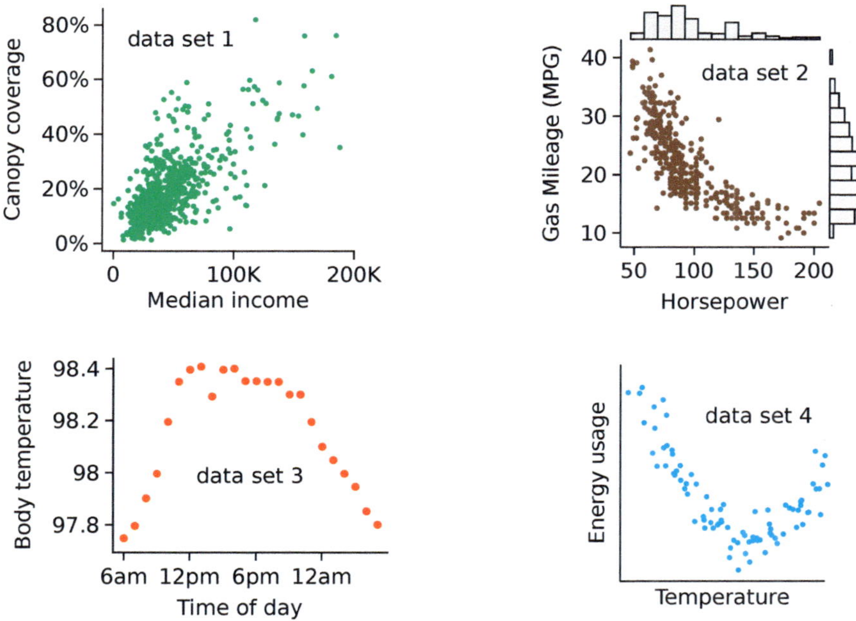

Chapter 10

Regression

LEARNING OBJECTIVES

After studying this chapter, you will be able to:

- explain how linear and nonlinear regression is used in science.
- distinguish among the different methods of regression (ordinary least squares, absolute value, orthogonal), and know their advantages and drawbacks.
- calculate and interpret the statistical significance of linear regression slopes and intercepts using resampling.
- calculate and interpret confidence intervals for linear regression slopes and intercepts using resampling.

10.1 The purpose of regression

As we saw, the correlation coefficient is the answer to the question: **to what degree does this data lie on a straight line?** Possible answers range from -1, meaning "lying exactly on some straight line of negative slope" to 0, meaning "no straight line at all" to $+1$, meaning "lying exactly on some straight line of positive slope".

We saw that if the correlation coefficient is high, positively or negatively, then we can conclude that **there is a straight line that is approximating the data.**

The next question is: <u>what</u> is the line?

Now, when we say "the" line, we are supposing that there is a unique "best" line that fits the data. **In order to decide whether that is true, we have to give a criterion for "best": what makes one line a better fit than another line?** For the time being, let's hold that question off to one side. We will return to it in section 10.2. Let's assume right now that we know what we mean by "best fit line to the data".

Correlation : <u>is there</u> a line?
Regression : <u>what</u> is the line?

Now we want to ask another question: **let's assume there is a "best" line. Why would we**

A. Garfinkel and Y. Guo, *Understanding Data*,
https://doi.org/10.1007/978-3-032-18600-3_10

want to know it?

To put those questions into context, let's look at an example.

Climate change in a nontraditional data set The Tenana River in Alaska usually freezes over during the winter and breaks up in late April or early May. In 1917, locals in the town began a betting contest about the exact time the ice would break up, and have continued ever since (Figure 10.1).

Figure 10.1 The Tenana River in Alaska.

This time series can be used to indicate climate change in the region. If the ice breaks up in early April, the climate is warmer than if the ice doesn't break up until late May. Locals have been recording data for 110 years (Figure 10.2).

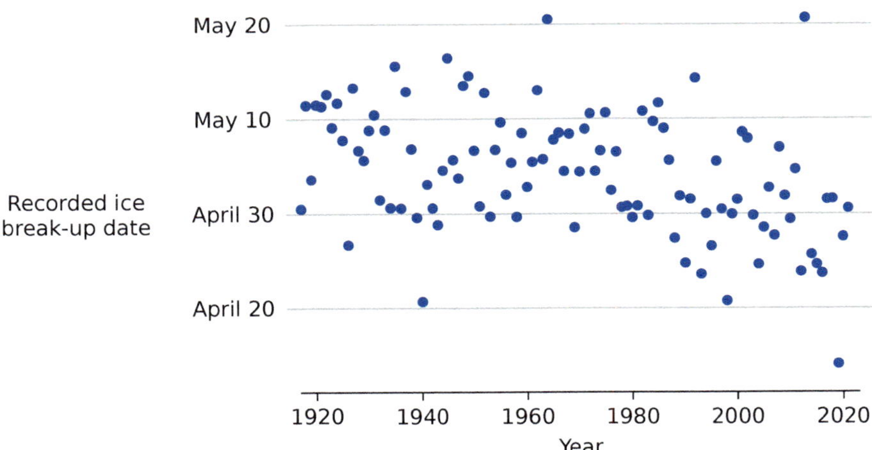

Figure 10.2 Records of each year's ice breakup date on the Tenana River since 1917.

We see that the data does appear to lie more or less on a straight line, and it looks OK to do a correlation coefficient. We calculated the correlation coefficient, $\rho = -0.39$, and verified that it is highly statistically significant. This confirms our visual observation.

But now we ask a further question, which is *what* is the line that the data points all seem be clustering around (Figure 10.3)? This is the question that is answered by linear regression.

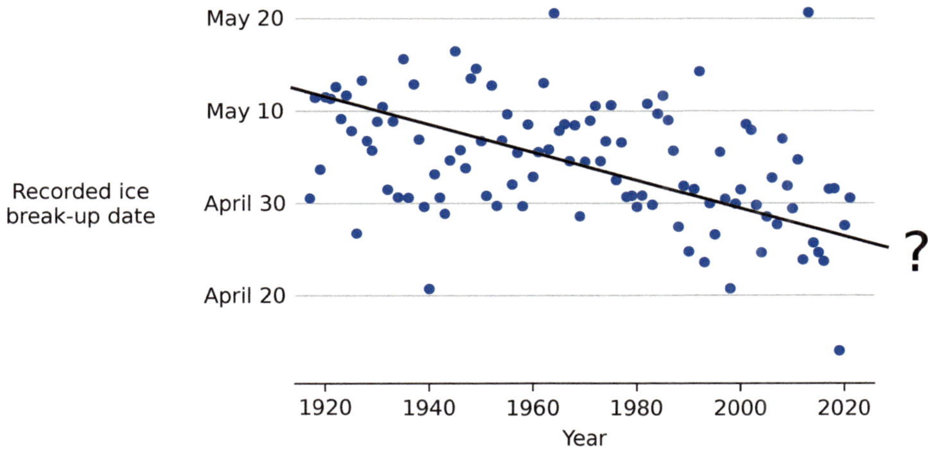

Figure 10.3

Let's first ask what <u>is</u> a line?

What is a line?

The equation for a straight line is $Y = mX + b$ (Figure 10.4).

- **m is the slope of the line**, which is $\frac{\Delta Y}{\Delta X}$. We don't have to ask $\frac{\Delta Y}{\Delta X}$ *where*, since $\frac{\Delta Y}{\Delta X}$ is constant *everywhere* on the line. Indeed, that is the definition of a straight line.
- **b is the Y-intercept**, which is the Y-value when X is 0.

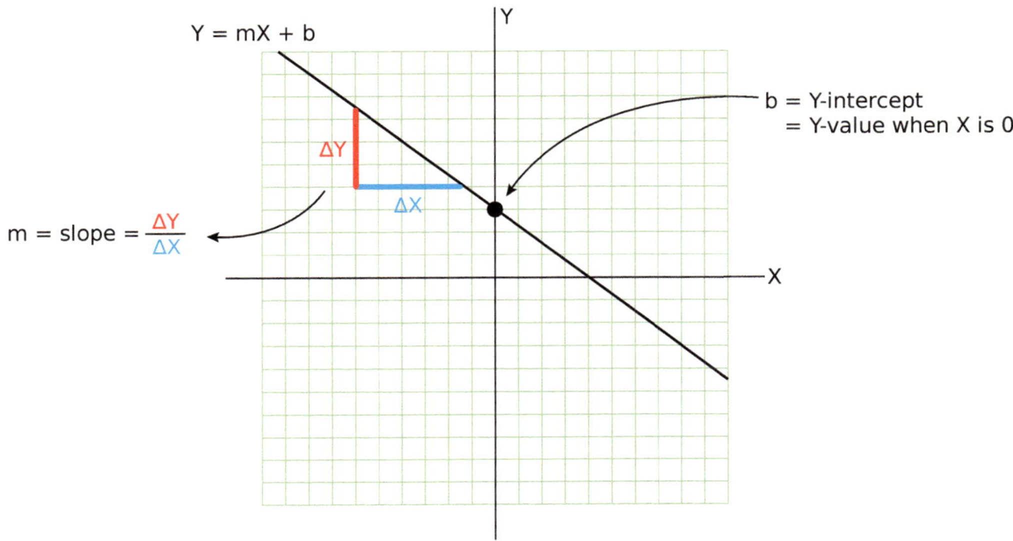

Figure 10.4

Exercise 10.1.1 What is the equation of this line? Each square is 1 × 1.

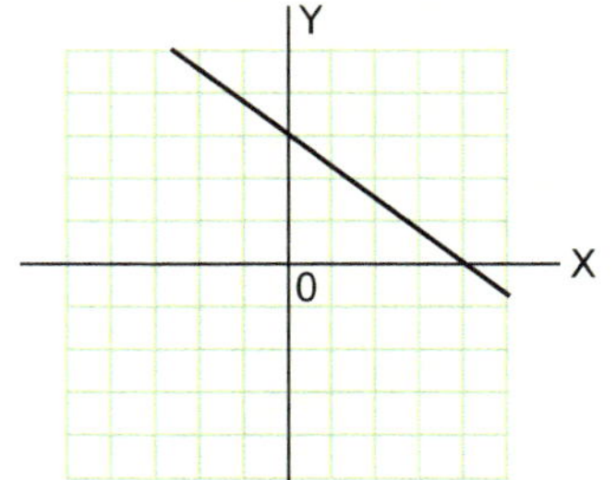

a) $Y = 3X + 3$

b) $Y = 4/3X$

c) $Y = -4/3X + 3$

d) $Y = -3/4X - 3$

e) $Y = -3/4X + 3$

Why do we want to know the line?

We want to know the line because we can give physical meaning to m and b, and use this information to gain understanding and to make predictions.

Interpretation of m The slope of the line, m, is $\frac{\Delta Y}{\Delta X}$.

It tells us, for a given change in the X-value, how big a change in the Y-value is associated with it.

- If (if!) X can be interpreted as a cause and Y as an effect, then m can be interpreted as the strength of the effect of X on Y; when m is small, a change in X produces only a small change in Y; but when m is large, a change in X produces a big change in Y.
- In the many cases in which X and Y are both observational, no causal statement can be made. Instead, we can only say that when m is small, a change in X is associated with a small change in Y; but when m is large, a change in X is associated with a big change in Y.

For example, if we are given a data set like Figure 10.5, we can make a statement that in this data, the incidence of liver cancer goes up markedly with the number of drinks per day. We can quantify that, for every additional drink per day, the incidence of liver cancer increases by 8.3 cases per 100,000.

$$m = \text{strength of effect of } X \text{ on } Y = \frac{60-35}{3} = 8.3 \ \frac{\text{cases per 100,000}}{\text{drinks per day}}$$

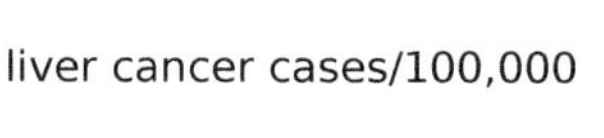

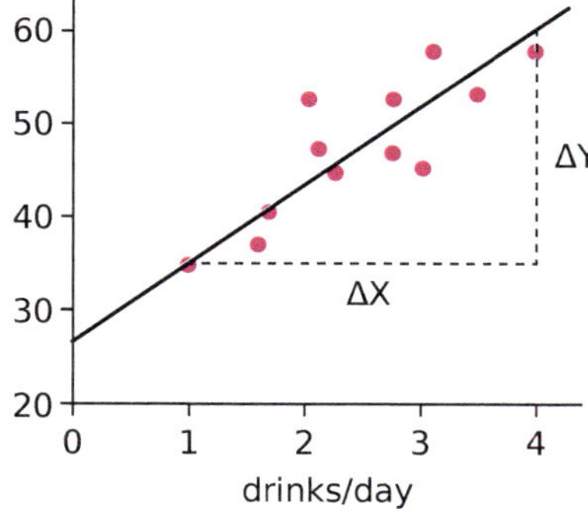

Figure 10.5

However, note although we are calling m the "strength of effect", the word "effect" is really a causal word, and we should only use it when a true causal relation has been established. Otherwise, we are in danger of using it in a faux-causal way (see "Faux-causal words" on page 541).

The interpretation of m is very important and in most cases, the reason why we do linear regression is because we can interpret m as the amount of change in Y that is associated with a given change in X.

For the Tenana River data set, the slope from the linear regression would give us an estimate of the *rate* of climate change.

Strength of effect

Remember that when we were talking about the correlation coefficient in Chapter 9, we said that the correlation coefficient ρ does not reflect the strength of the effect, it reflects the strength of the association between X and Y. **The strength of effect is reflected in the slope m.**

Here are several data sets, all of which have correlation coefficients 1 or −1, meaning that the data are perfectly correlated. In each data set, all data points lie exactly on a line, so the strength of association is perfect.

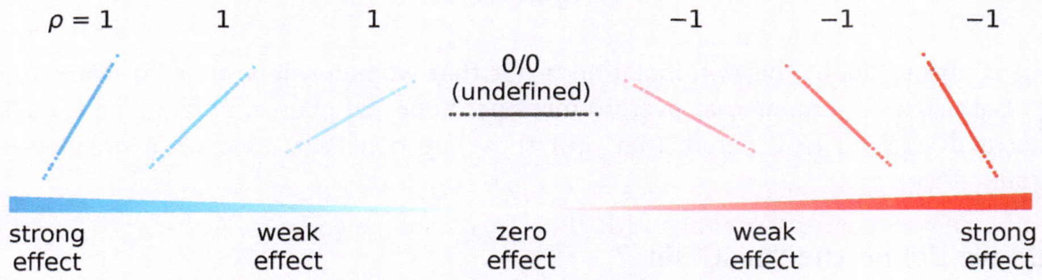

However, they have different strengths of effect. The steeper the slope, the stronger the effect. So for the shallowly-sloped relationships, the association is perfect, but the effect is weak.

Interpretation of b We can also interpret b, the Y-intercept. It is the value of Y when X is 0. That means, for example, that when X is 0, meaning no drinks per day, we still have a 27 per 100,000 incidence. That represents the background incidence of liver cancer, even without drinking.

Predicting new values Another use of the regression line is to predict a Y-value given a new X-value not in the original data set.

If we are given a new X-value, X_{new}, then the predicted Y-value is $Y_{\text{new}} = mX_{\text{new}} + b$.

However, it is important to understand that *the predictions are only valid within the original range of X-values; extrapolations outside the range of the original data are highly dangerous and should be avoided.*

For example, an article called "Momentous sprint at the 2156 Olympics" considered the winning times in the men's and women's 100-meter sprint, over the years 1900-2012. They carried out linear regressions on these two data sets, shown in Figure 10.6. By an absurd extrapolation, they drew the conclusion that women will be running faster than men after 2156.

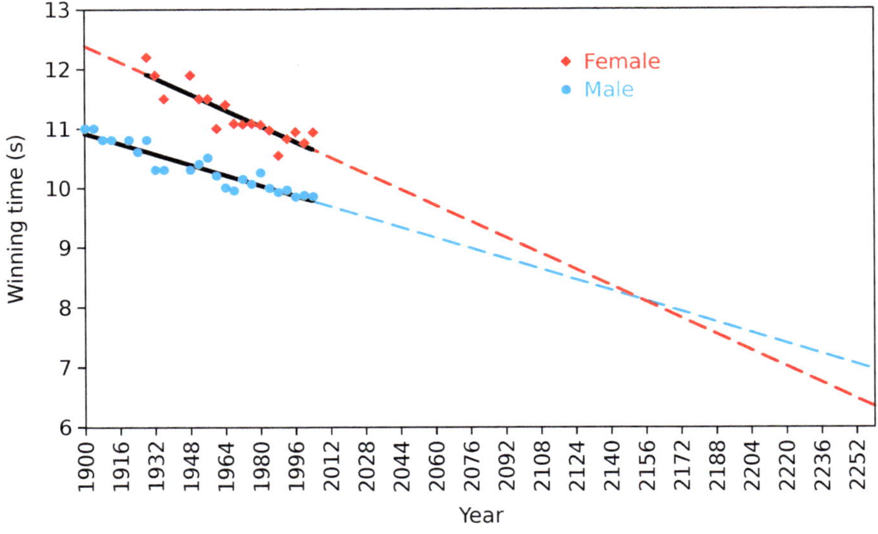

Figure 10.6

What is absurd about this extrapolation is not that women will be running faster than men, but that both sexes are running at over 30 miles per hour. Of course, in fact, the winning times will taper and level off long before that, but these are nonlinear effects not grasped by linear regression.

How do we define the "best" line?

If we imagine some data points, and a straight line that is trying to capture the trend in the data (Figure 10.7), we can ask how good a fit is the line?

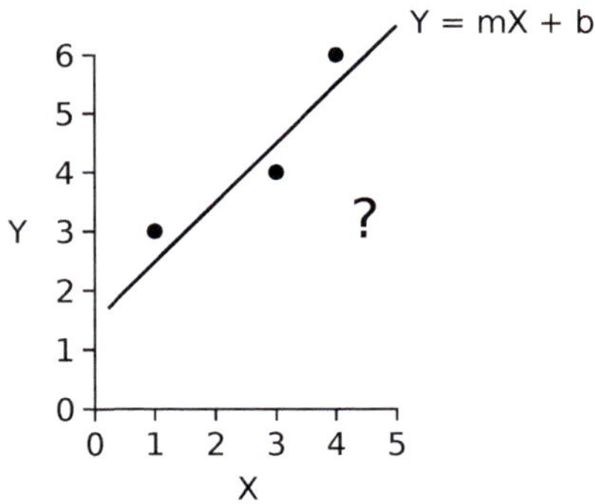

Figure 10.7 A hypothetical data set and a proposed straight line fit.

Let's say we could measure the distance from each data point to the proposed line. Then it would make sense to say that **the best fit line will minimize the total distance**, that is, will minimize the sum of the distances from each data point to the line. But how to measure the distance?

Vertical distances One definition of the distance from a data point to the line is the vertical distance (Figure 10.8 left).

Orthogonal distances Another definition of the distance from a data point to the line is the orthogonal distance (Figure 10.8 right).

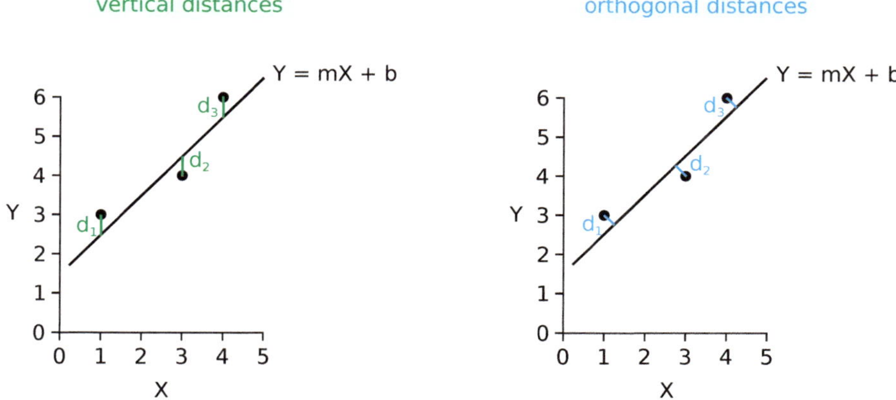

Figure 10.8 Left: the vertical distances from each data point to a proposed line fit. Right: the orthogonal distances from each data point to a proposed line fit.

Both of these definitions have advantages and disadvantages, which we will discuss in the next section.

But there is also another problem. If we tried to simply add up the d_is, we face the familiar problem that positive and negative values would cancel each other out. There are two remedies for this problem.

1. **squaring the distances** so that the total distance equals $(d_1)^2 + (d_2)^2 + (d_3)^2$.
2. **using absolute values of the distances** so that the total distance equals $|d_1| + |d_2| + |d_3|$.

We'll now discuss these possibilities in turn.

10.2 Linear regression

When we fit a straight line to data, by any method, the result is called **linear regression**.

Major methods of linear regression

- **Ordinary Least Squares regression**, in which we use vertical distances ("deviations in Y only"), square them, and add them up to give the total distance from the data set to the proposed line.
- **Absolute value regression**, in which we use vertical distances, take their absolute values, and add them up to give the total distance from the data set to the proposed line.
- **Orthogonal regression**, in which we use orthogonal (perpendicular) distances, whether squared or absolute values, and add them up to give the total distance from the data set to the proposed line.

Ordinary Least Squares regression (OLS) is the standard technique that is taught in most statistics courses, and is used in 90% of publications that use regression lines.

To find the regression line using OLS, we draw a random line, and look at the *vertical* distances from the data points to the proposed line. **This is a very important limitation of OLS regression: we are only looking at the Y-distances.**

What we want to do is find the line that minimizes the total distances (Figure 10.9).

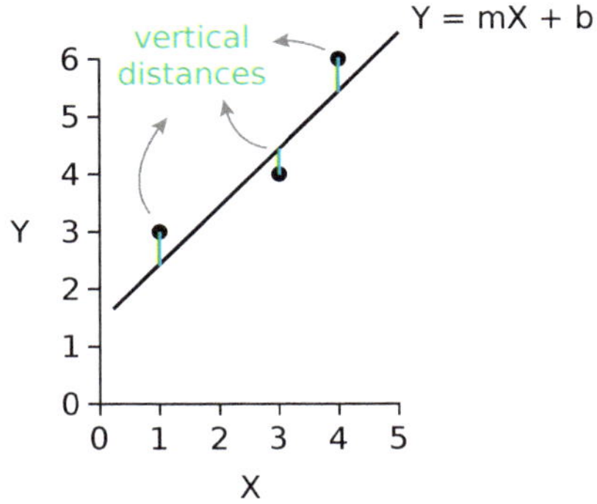

Figure 10.9

OLS then makes another critical assumption: the vertical distances are squared, and the squares are added up to give the total deviation of the data from the proposed line. The line with the minimal total deviation is defined as the "best" line.

When is it OK to use vertical distances?

The use of vertical distances as our measure of deviation says that we only care about variability in the Y-value. The variability in X is ignored. Why is this OK? The answer is that it is appropriate in studies that are *experimental* as opposed to *observational*.

In Chapter 9, we talked about observational and experimental studies (section 9.7 *Determining causation with an experiment*). In an experimental study, the experimenter sets the X-value, and observes the resulting Y-value. *Therefore, the variability in X is not data, it's the decision of the experimenter.* In these cases, we can talk about an "independent" variable (which is X) and a "dependent" variable (which is Y).

OLS is set up for this situation, and is appropriate for experimental studies. The X-value is set by the experimenter, so the variability in X is not on the table as data.

In an observational study, by contrast, we observe both the X- and Y-values, and both are subject to biological variabilities.

So, for example, suppose we are looking at the blood alcohol content (BAC) of subjects and their reaction times. If this is an observational study, then each subject took as much alcohol as they wished, and then, for each subject, we measured their BAC (X-value) and their reaction time (Y-value). In this case, the variability in X is biological data; this person chose to drink this much alcohol, and we recorded their BAC and their reaction time. On the contrary, in an experimental study, we assign each subject a dose of alcohol, this much for one subject, that much for another, etc. Then there is no biological variability in X, *because the subjects didn't choose their alcohol levels, we, the experimenter, did.*

Of course, a great deal of biological data is observational and not experimental; we often see one measurement compared to another using regression, such as height and weight, or two different methods for measuring the same feature of an object. These are two biologically varying quantities, and OLS is the wrong method for this kind of data.

Cases in which there are two biologically varying quantities cannot be analyzed by regression that uses only vertical distances, such as OLS.

Squaring distances vs. absolute values

In OLS, the vertical distances are squared, and then added up to produce the total deviation from the proposed line. There are several advantages to using squaring. They are:

#1 we can find the best fit line using calculus (see immediately below).

#2 we can define a quantity called "R^2", which is a measure of the goodness of fit.

But squaring distances also comes with inherent drawbacks (see *Squaring the residuals* on page 508). These drawbacks can be overcome by using absolute value regression. However, in absolute value regression, we have to give up #1 completely, and also rethink #2.

Ordinary Least Squares regression

Finding the best line in OLS using calculus In OLS, the total deviation of a proposed line is the sum of the squares of the lengths of the vertical green lines (Figure 10.10).

total deviation (OLS) = sum of the squares of the lengths of the vertical green lines = $\sum(\,|\,)^2$

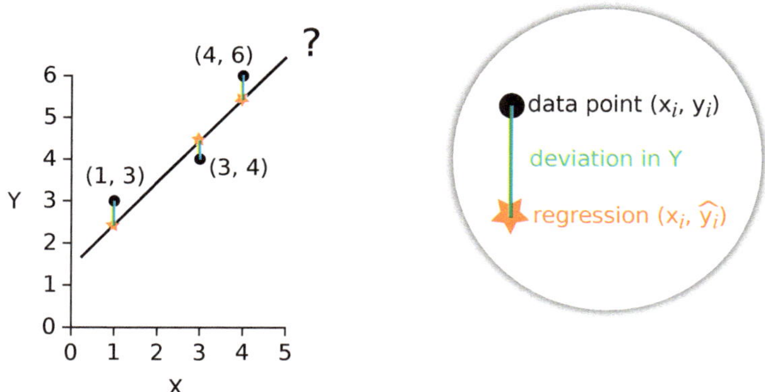

Figure 10.10 Deviations in OLS.

Now let's calculate what those deviations are. The random line we drew is going to define three new points, which are marked by the gold stars.

Let's call this random line $Y = mX + b$. We want to find the m and b that gives the line with the smallest total deviation from the data points, which we are calling the best fit line.

The gold stars are the estimates of Y created by the proposed regression line. These estimated Y-values are called $\widehat{y_1}$, $\widehat{y_2}$, $\widehat{y_3}$, pronounced "y 1 hat" etc. They are the regression estimates of y_1, y_2 and y_3, which were the actual data values.

Let's say the data points (●) are

$$(x_1, y_1) = (1, 3)$$
$$(x_2, y_2) = (3, 4)$$
$$(x_3, y_3) = (4, 6)$$

Then the regression estimated points (★) are

$$(x_1, \widehat{y_1}) = (x_1, mx_1 + b) = (1, m + b)$$
$$(x_2, \widehat{y_2}) = (x_2, mx_2 + b) = (3, 3m + b)$$
$$(x_3, \widehat{y_3}) = (x_3, mx_3 + b) = (4, 4m + b)$$

We then calculate the deviation as the vertical distance from the regression estimate $(x_i, \widehat{y_i})$ to the actual data point (x_i, y_i), squared: $(\widehat{y_1} - y_1)^2$, $(\widehat{y_2} - y_2)^2$, and $(\widehat{y_3} - y_3)^2$.

The total deviation, which is often called the "error" e, is then the sum of the squared vertical distances:

$$\begin{aligned} e &= (\widehat{y_1} - y_1)^2 + (\widehat{y_2} - y_2)^2 + (\widehat{y_3} - y_3)^2 \\ &= (m + b - 3)^2 + (3m + b - 4)^2 + (4m + b - 6)^2 \end{aligned}$$

Once we define the total deviation e, we then want to minimize e with respect to m and b, which are the possible slopes and intercepts. **The m and b that give the minimal total deviation define the "best line".**

We are going to achieve this optimization using calculus.

To see how calculus can be used to find the best fit line that minimizes the total deviation, we construct a 2D surface for e as a function of m and b (Figure 10.11).

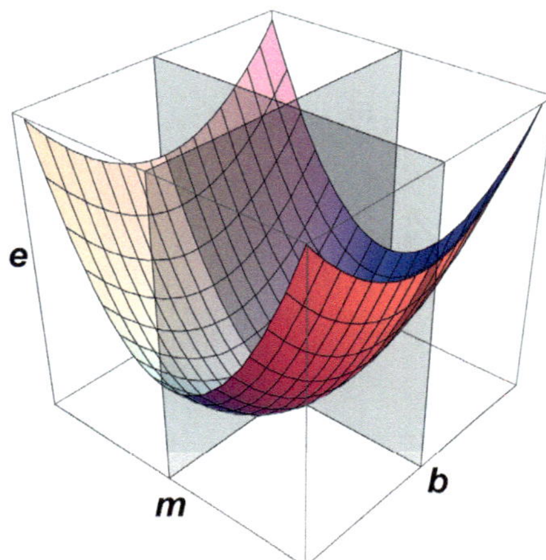

Figure 10.11 Total deviation e in OLS as a function of m and b.

Here, the vertical height is the total deviation e, and the two horizontal axes are m and b. If we look closely at the expression for e, we see that m and b occur in the expression as squares m^2, b^2 and as products $m \times b$. Therefore, squaring the distances makes the 2D surface for e to be bowl-shaped, and the minimum will lie at the bottom of the bowl. But this only applies to OLS. For other types of regression, the minimum has to be found by numerical experiment.

We know from calculus that if the continuous variable e is minimized, then the partial derivatives must be equal to zero. Therefore, we find the minimum total deviation by taking the partial derivatives of e with respect to m and b and setting them equal to zero.

$$\begin{cases} \dfrac{\partial e}{\partial m} = 0 \\ \dfrac{\partial e}{\partial b} = 0 \end{cases}$$

Plugging in the expression for e

$$e = (m+b-3)^2 + (3m+b-4)^2 + (4m+b-6)^2$$

We get

$$\begin{cases} \dfrac{\partial e}{\partial m} = 0 = 2\times(m+b-3) + 6\times(3m+b-4) + 8\times(4m+b-6) \\ \dfrac{\partial e}{\partial b} = 0 = 2\times(m+b-3) + 2\times(3m+b-4) + 2\times(4m+b-6) \end{cases}$$

We now have two linear expressions in two unknowns, and we can calculate m and b

$$\begin{cases} 0 = 52m + 16b - 78 \\ 0 = 16m + 6b - 26 \end{cases}$$

Using an online calculator (search "solving two linear equations calculator"), the answer is

$$m = 0.93 \qquad b = 1.9$$

Therefore, the regression line in OLS, the best fit line that minimizes the total deviation is $Y = \mathbf{0.93}X + \mathbf{1.9}$, shown in Figure 10.12.

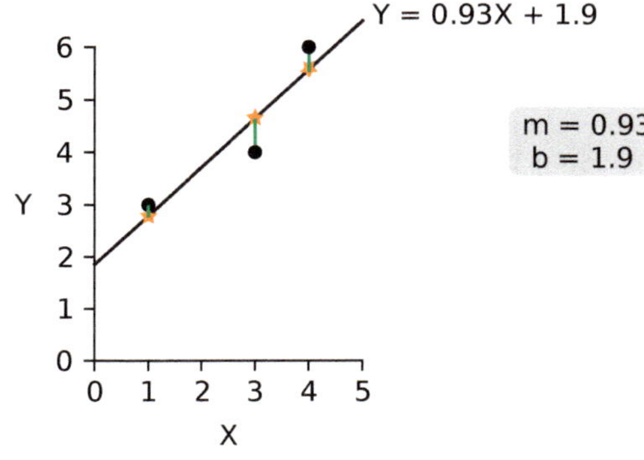

Figure 10.12 The best fit line using OLS for this exemplary data set.

This exercise of using calculus to find the best line is for conceptual illustration of what the computer program is doing. In actual practice, we'll never have to actually carry out this procedure by hand.

Quantifying the goodness of fit As we mentioned earlier, the second advantage of using squares of vertical distances is that it gives us a straightforward definition of the **goodness of fit** for the best fit line, the quantity called R^2, which quantifies how well the best fit line enables us to predict Y from X.

Consider a set of data points, $(x_1, y_1), (x_2, y_2), \ldots, (x_n, y_n)$, and an OLS best fit line $Y = mX + b$. Then we can use the regression to predict y_i from x_i.

Suppose we were asked "here's an X-value, x_i, what is its Y-value y_i?". Once we have the regression, we would predict $y_i = \widehat{y}_i$. Then the total "error" in our prediction is

$$\text{total error (with regression)} = \sum_{i=1}^{n} (y_i - \widehat{y}_i)^2$$

and the average error is

$$\text{average error (with regression)} = \frac{1}{n} \sum_{i=1}^{n} (y_i - \widehat{y}_i)^2$$

Now suppose we didn't have the regression line, and we are asked "here is x_i, what is y_i?". Without the regression, all we can predict (guess) is "y_i equals $\overline{Y}$" (if the y_is are distributed Normally, then $\overline{Y}$ is the most likely value) (Figure 10.13).

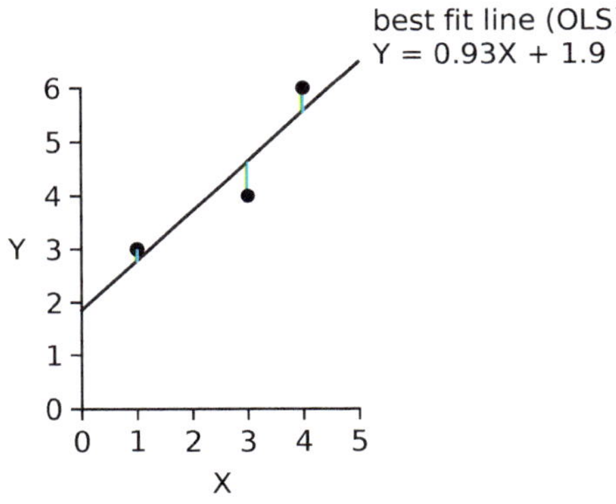

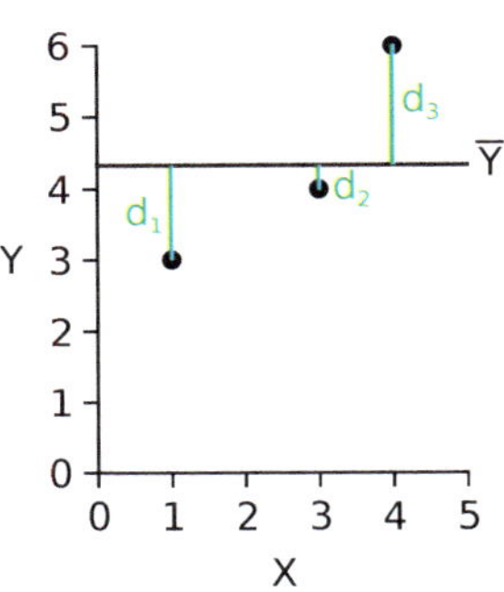

Figure 10.13 Deviations using the regression line, and for the clueless guess "$Y = \overline{Y}$".

Now the total "error" in our prediction (guessing) becomes

$$\text{total error (without regression)} = \sum_{i=1}^{n} (y_i - \overline{Y})^2$$

and the average error is

$$\text{average error (without regression)} = \frac{1}{n} \sum_{i=1}^{n} (y_i - \overline{Y})^2$$

Note that this expression is simply the variance of the y_is.

We can then define a quantity that reflects how much knowing the regression line improves our ability to predict y_i from x_i, that is, improves the prediction as opposed to the clueless guess $y_i = \overline{Y}$.

Consider the quantity that is the ratio of the two average errors, with and without the regression line:

$$\frac{\frac{1}{n}\sum_{i=1}^{n}(y_i - \widehat{y}_i)^2}{\frac{1}{n}\sum_{i=1}^{n}(y_i - \overline{Y})^2}$$

If this fraction is close to 1, then using the regression line didn't help much; the clueless guess is about as good as the prediction using regression. But if the fraction is small, then using the regression has greatly improved our estimate. We therefore define the quantity

$$\mathrm{R}^2 \stackrel{\text{def}}{=} 1 - \frac{\frac{1}{n}\sum_{i=1}^{n}(y_i - \widehat{y}_i)^2}{\frac{1}{n}\sum_{i=1}^{n}(y_i - \overline{Y})^2} = 1 - \frac{\text{average error by using the regression estimates}}{\text{average error by just using the average value}}$$

R^2 is therefore a measure of how good the regression line is as a fit to the data.

Explaining the variance in Y?

There is another interpretation of R^2, which is more problematic. Many textbooks talk about R^2 as the "fraction of variance in Y *explained by* X". It is also often described as "the fraction of the variance in Y that is *due to* X", or as "the fraction of the variance in Y that is *attributable to* variations in X".

We have to be extremely careful about these words. "Explaining" is a causal word, and all we have here is correlational evidence. However, as we said in section 9.7 (*Determining causation with an experiment*), sometimes we can make causal statements, when the X variable is purely experimental. So the entire method of vertical regression, including OLS and the above definition of R^2, is only valid in the experimental context, not the observational one.

Perhaps the worst mistake is to apply vertical regression and the above defined R^2 to a purely observational (variability in both X and Y) data set, which can give rise to the dangerous mis-interpretation that the variation in the Y variable is *due to* the X variable, or that "R^2 percent of the variation in Y is *explained by* X". But "explaining" and "due to" are faux-causal words in this context. Unfortunately, this mistake is quite common in the literature. Beware!

Data requirements for OLS regression In order for OLS regression to be valid, the data set has to meet a number of strict requirements.

- **Linear relationship.** First of all, the data have to lie approximately on a straight line; there can't be any obvious nonlinear relation visible in the X–Y scatterplot.

The other three requirements are about "residuals". Residuals are the (un-squared) distances between the predicted Y-values ($\widehat{y}_i$) and the actual Y-values of the data (y_i).

Figure 10.14 shows a "residuals plot". In a residuals plot, we graph the residual for each X-value against the X-value itself. The residuals will be scattered on either side of the horizontal line representing "residual = 0". If the residual for a particular data point lies on "residual = 0", this means the regression line is a perfect fit for that one data point.

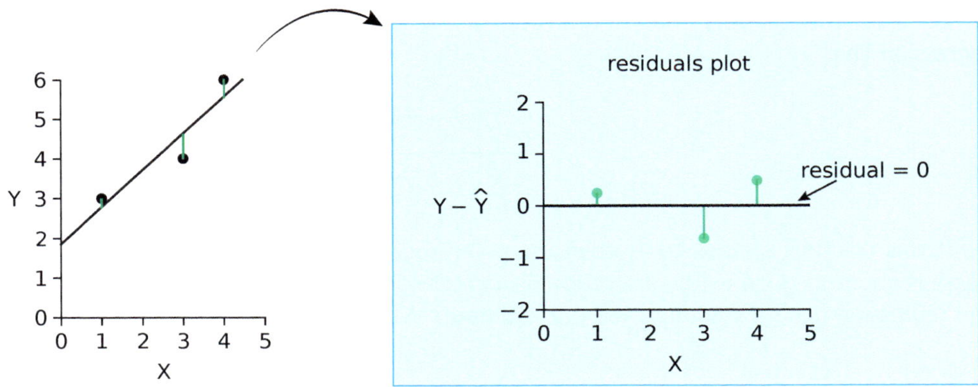

Figure 10.14 A residuals plot.

- **No auto-correlation of residuals.** A diagnostic to detect nonlinearity is to consider the residuals plot. If neighboring residuals are correlated, for example, when a succession of points all lie above the 0 line, or all lie below it, then these "correlated residuals" are a sign that the linear model has missed nonlinear features of the data.
- **Constant variance.** The residuals must have constant variance, that is, their magnitude must be roughly constant across all values of X. If the residuals are much bigger in one part of the X-axis than they are in another, OLS regression is rejected.
- **Residuals are Normally distributed.** A critical requirement is that the residuals be roughly Normally distributed.

A famous article by Anscombe (1973), entitled "Graphs in statistical analysis", makes the case for always making X–Y scatter plots for diagnosing whether the data meet the requirements for OLS regression. He gave four examples ("Anscombe's Quartet"). All four data sets have exactly the same regression line (Figure 10.15 left column). We calculated residuals plots for the four data sets (Figure 10.15 right column).

In the top row, the data set meets the requirements for linear regression.

In the second row, the data set shows a clear example of auto-correlated residuals: the first two data points have negative residuals, then the next seven data points have positive residuals, and the last two data points have negative residuals.

In the third row, the residuals fail to be Normally distributed and are also auto-correlated.

In the fourth row, the variance in the residuals is not constant across X: low X-values have high residual variance, high X-values have low residual variance.

Squaring the residuals In OLS, the best fit line is defined as the line that minimizes the sum of the squared vertical residuals. But remember that squaring things makes big things huge, and makes small things tiny. So if there is an outlying value in the data set, that will singlehandedly pull the OLS regression line in its direction, with a force proportional to the square of its distance from the line.

Consider two data sets, one with and the other without an outlying data point (Figure 10.16).

The blue line is the OLS regression for the full data set. Is it a good fit for that data? Not really. The line is too high for medium-to-large values of X. The fit line has been pulled upward by the outlying value, giving a bad representation of the data. **This is because the OLS regression has to respect the squared distance, so outlying values become over-weighted in OLS**

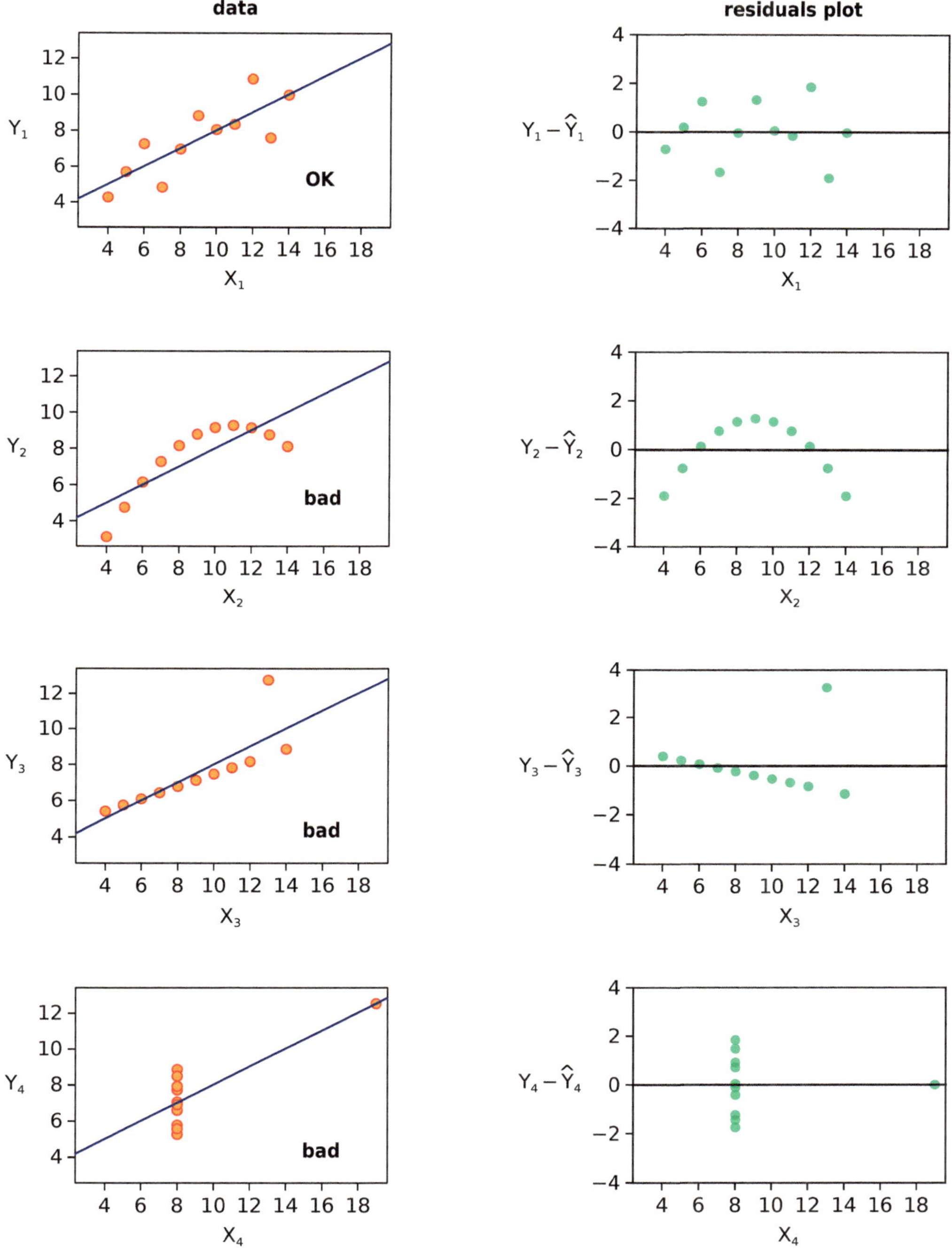

Figure 10.15 Left column: OLS regression line for Anscombe's quartet is shown in blue. Right column: residuals plots for Anscombe's quartet.

regression. The practice of squaring distances gives outlying data points a dominant voice, which enables them to out-vote all of the other data points. This can be seen by the black line which is the OLS regression after removing the outlying value.

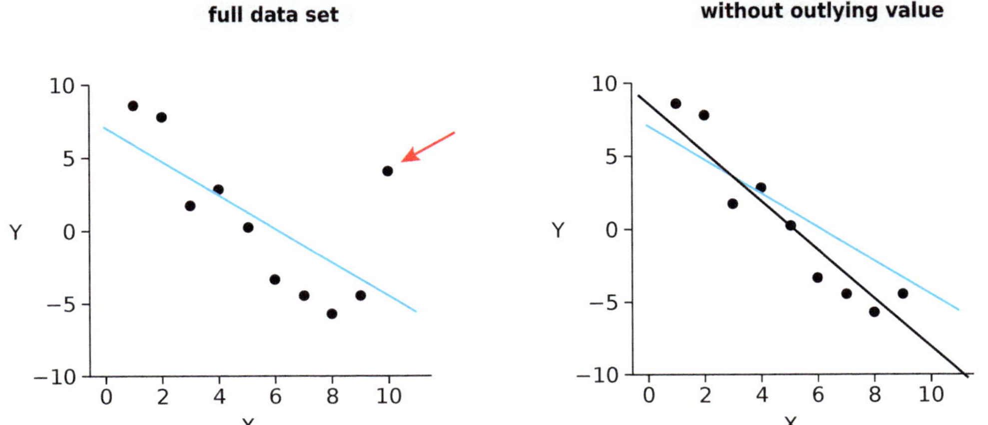

Figure 10.16 Left: OLS regression (blue line) on a data set with an outlying value (arrow). Right: OLS regression (black line) on the same data set with the outlying value removed.

Absolute value regression

The cure for the distortions introduced by squaring is to use the absolute values of the distances instead of the squares. This technique is called **absolute value regression** or **absolute deviation regression**. In absolute deviation regression, the best line is defined as the one that minimizes the MAD, where MAD stands for either "median absolute deviation" or "mean absolute deviation". In a given case, which one to use, as always, depends on the distribution of the data. Whichever we use, absolute deviation regression produces a fit line that doesn't give exaggerated weight to outlying values.

We can compare OLS regression and MAD regression on the same data set with an outlying value (Figure 10.17). We see that when we use MAD regression, we get a much more representative fit than if we had used mean squares (OLS). Note that the mean absolute deviation regression (red line) is a much better approximation to the data than the OLS regression (blue line).

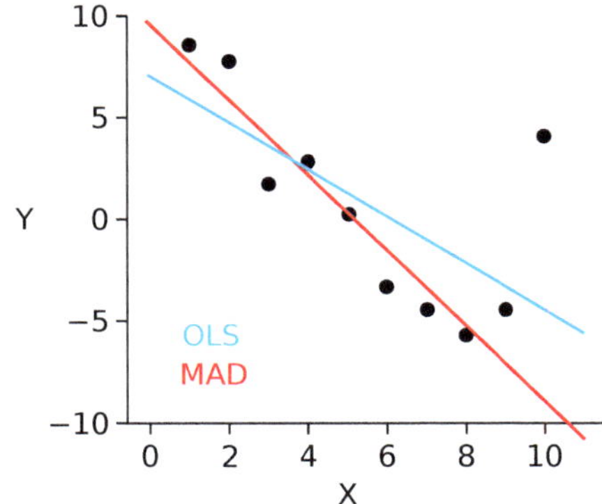

Figure 10.17 Absolute value regression (MAD regression) produces a better fit line than OLS when the data has outlying values.

Finding the best fit line in absolute deviation regression

In absolute deviation regression, finding the best fit line must be done differently than we did in OLS. In OLS, we could use calculus to find the line that minimizes the total error. We wrote the total error as the sum of the squares of the deviations, then took the derivatives of the total error (e) with respect to the slope (m) and intercept (b) parameters, and set those derivatives equal to zero to find the minimum.

But if we are using absolute values, we can't use calculus-based methods. The reason is straightforward.

In OLS, using calculus to find the minimum works because the error function forms a smooth-bottom bowl (Figure 10.11), so calculus can be used to find the m and b that minimize the error. But in absolute deviation regression, since absolute values are being used, the error surface will not be smooth, but will have corners.

Consider a simple 1D example, $f(x) = x^2$. If we are asked to find the minimum value of this function, we can take the derivative df/dx, and set it equal to zero, to get x = 0 as the minimum. Now consider $g(x) = |x|$. At the point (0, 0), this function has no derivative; the corner means that the function is not differentiable there, so we can't take the derivative!

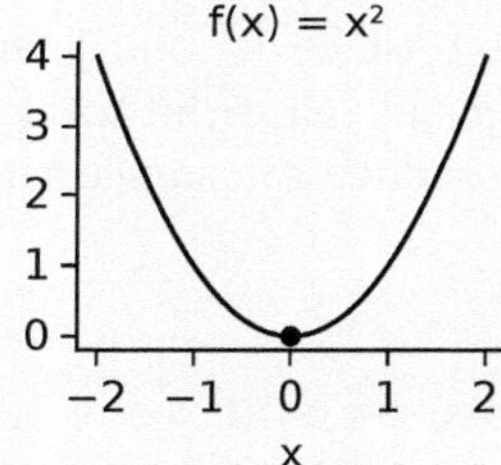

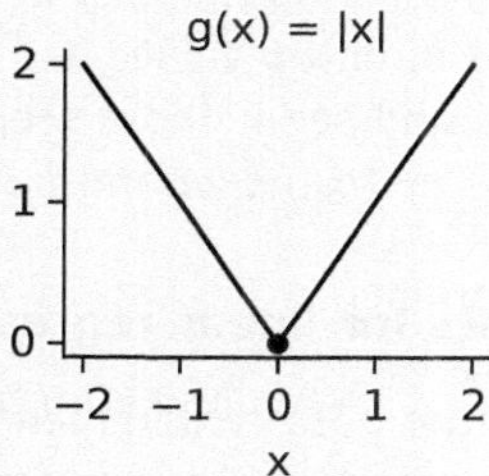

Instead, when we work with absolute values, we have to find the minimum, as it were, experimentally. The procedure would be to pick a random point x_0 on the X-axis, and calculate $g(x_0)$. Then take a small step to the left, and calculate the value of g at that point. Did the value of g decrease? No? Then we are heading in the wrong direction. Go back to x_0 and take a small step to the right. Did the value of g decrease? Yes? Then repeat this action and evaluate again. In this way, after possibly many iterations, we can arrive at the minimum.

The way to find the best fit line using absolute value regression is to use a similar numerical method to pick an m and b through many iterations to find the values that minimize the total error.

Orthogonal regression

Vertical regression, whether OLS or absolute value, cannot deal with situations where both X and Y are observational data, such as height vs. weight. If we are plotting height vs. weight, there is biological variability in both quantities. Vertical regression is only set up to deal with problems of finding the best line that minimizes the total Y error, but in this case, we need to minimize "errors" in both X and Y.

If X and Y are both observational data, which we often see in biological studies, then how should we measure the deviation? It has to be the orthogonal distance, not the vertical distance, because now we have to respect the variability in both X and Y. So now we have a new concept of regression, called **orthogonal regression**, or colloquially, regression with errors in both X and Y (Figure 10.18).

Calculating the orthogonal deviations is a straightforward, if somewhat messy task of repeated application of the Pythagorean Theorem.

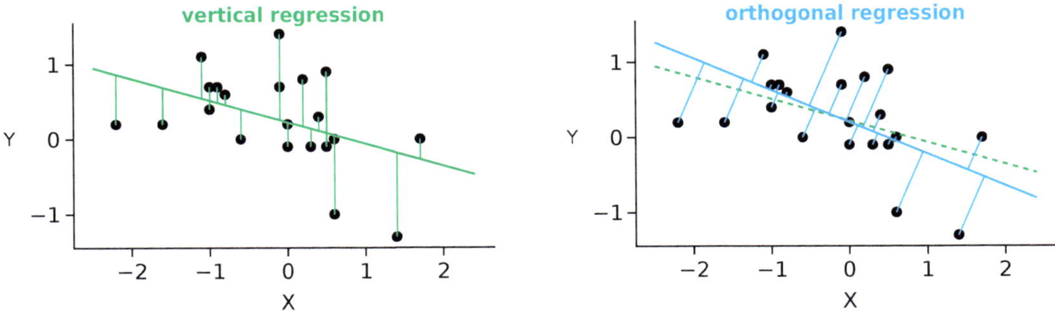

Figure 10.18 Vertical regression looks at errors in Y only, while orthogonal regression considers errors in both variables. In the orthogonal regression, we overlay the vertical regression as dashed.

The approach to find the best fit line using orthogonal regression is also through an experimental method (as in *Finding the best fit line in absolute deviation regression* on page 511). We pick an initial pair of m and b values and calculate its total deviation. Then we take small steps in the neighborhood and see if those steps decrease or increase the total deviation. Repeating this process will lead us to the m and b values that produce the smallest total error.

Data requirements for linear regression

We have discussed the data requirements for OLS regression on page 507. If we use instead one of the more advanced regression methods, such as absolute value regression, or orthogonal regression, the assumptions that refer to Normal distributions can be relaxed. However, the requirements of "no non-linear relationship" and "no auto-correlation of residuals" still apply.

FURTHER EXERCISES 10.2

1. Select all assumptions of OLS regression.
    a. The relationship between X and Y must approximate a straight line.
    b. "Errors" are squared vertical distances between observed and predicted Y.
    c. "Errors" have constant variance.
    d. Observations must be independent.
    e. "Error" in Y is Normally distributed.
    f. There is variability in Y only.
2. Which of the following is a true statement about OLS regression and orthogonal regression?
    a. They always produce the same slopes but may have different intercepts.
    b. They are simply two names for the same analysis.
    c. Orthogonal regression can fit nonlinear relationships, while OLS cannot.
    d. They differ in how residuals are defined and calculated.
    e. Orthogonal regression is always more accurate because it accounts for error in both variables.

10.3 Nonlinear regression

So far, we have been talking about fitting straight lines to data. But the subject of regression can also include fitting curved lines. The logic is exactly the same as it is for linear regression: choose a family of curves, and then ask which member of that family minimizes the total distance from the curve to the data.

In a simple example, which we can find in many chemistry texts, we believe from theory that for certain reactions, the reaction rate V increases with substrate concentration S as

$$V = \frac{V_{max}S}{K_M + S}$$

A chemical reaction following this equation is said to be displaying "Michaelis-Menten kinetics", and K_M is called the Michaelis constant. This equation defines a 2-parameter family of functions depending on the values of V_{max} and K_M. Here, for the purpose of illustrating the procedure of fitting a nonlinear regression to data, we fix $V_{max} = 35$. The equation then becomes a 1-parameter family (Figure 10.19 left).

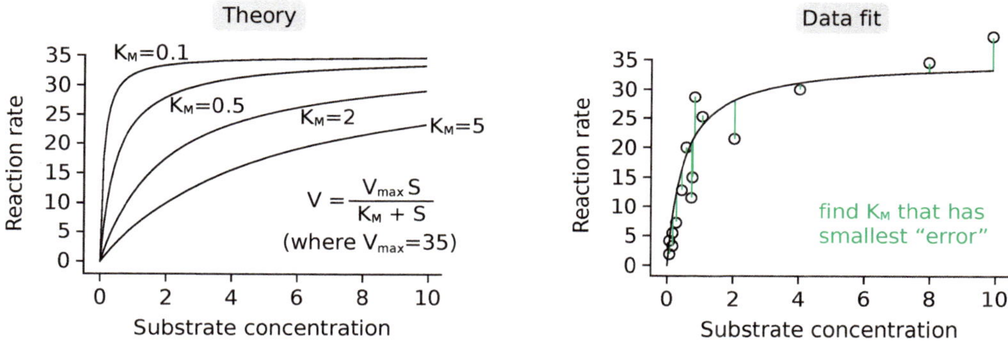

Figure 10.19

Let's say we want to estimate the Michaelis constant of a chemical reaction from data. We will interpret this as finding which value of K_M minimizes the total distance from the curve to the data. Let's suppose it's a controlled experiment, in which we choose various substrate concentrations and measure the corresponding reaction rates, so a regression using vertical deviations can be used (Figure 10.19 right).

Then the best fit curve is the one that minimizes the total "error", that is the K_M that produces the smallest total deviation.

Logistic regression

Logistic regression is heavily used in medicine and in Artificial Intelligence.

Medicine In medical applications, we have a certain yes/no condition, say lung cancer or diabetes, and we want to identify "risk factors" for this condition. Let's say we have some risk factor as our X variable, for example X = number of cigarettes per day, as a risk factor for lung cancer. The condition (lung cancer) is a yes/no, so what we will plot on the Y-axis is whether the subject had the condition ($Y = 1$) or not ($Y = 0$) (black dots) (Figure 10.20).

Then, we fit a sigmoidal ("logistic") curve to this data (blue curve). We can view the curve as estimating the probability that a subject with a given X-value will have the condition. The

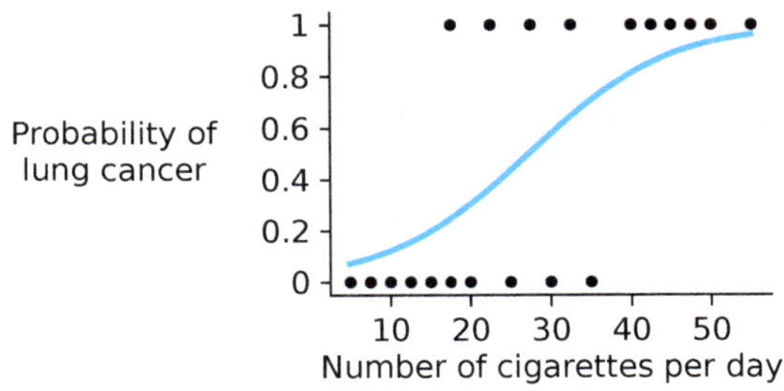

Figure 10.20 A logistic fit (blue curve) to a hypothetical data set (black dots).

best logistic fit will be the one that minimizes the total error, which is the total deviation from the data points to the curve.

Logistic curves

The form of the logistic curve that is usually used is

$$p(t) = \frac{e^t}{1+e^t}$$

Here, p is the probability of having the disease, and t has the form

$$t = \beta_0 + \beta_1 X$$

where X is the amount of the risk factor.[a] We can interpret the two parameters β_0 and β_1:

- β_0 determines the background risk, that is, the probability when $X = 0$.
- β_1 determines the slope of the curve.

[a] In the typical medical application, the quantity X must be non-negative. However, the formula for logistic regression accepts negative values of X and assigns them a non-zero probability. In practice, the curve is shifted to the right, so that negative values of X are assigned extremely low probabilities. The parameter that shifts the curve to the left or right is β_0: by setting β_0 to a negative number, we can shift the curve to the right, so that almost all of the curve is in positive X.

To understand the role of β_1, we will consider not the probability p of contracting the disease, but rather the **odds** of contracting the disease. The transform from probability to odds is simple: if p is the probability of something happening, then the odds on its happening are $\frac{p}{1-p}$. So if the probability of an event is 75%, then the odds on its happening are (0.75/0.25) = 3:1.

If we convert the Y-axis from p to $\frac{p}{1-p}$, we get

$$p = \frac{e^t}{1+e^t}$$

$$\Downarrow \textit{ convert } Y\textit{-axis from } p \textit{ to } \frac{p}{1-p}$$

$$\frac{p}{1-p} = \frac{\frac{e^t}{1+e^t}}{1-\frac{e^t}{1+e^t}} = \frac{\frac{e^t}{\cancel{1+e^t}}}{\frac{1}{\cancel{1+e^t}}}$$

that is,

$$\frac{p}{1-p} = e^t$$

then taking the natural log on both sides gives us

$$\ln\frac{p}{1-p} = t$$

Since $t = \beta_0 + \beta_1 X$, we derive a linearized version of the logistic equation:

Logistic curves (linearized version)

$$\ln\frac{p}{1-p} = \beta_0 + \beta_1 X$$

In other words, the logistic equation for the probability p is mathematically equivalent to the linearized equation for $\ln\frac{p}{1-p}$ (Figure 10.21).

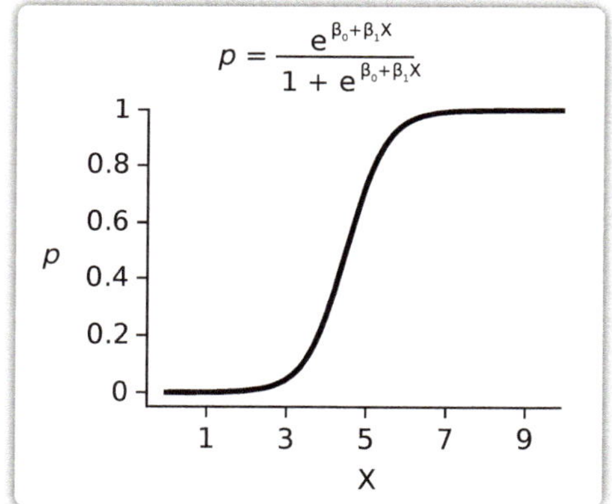

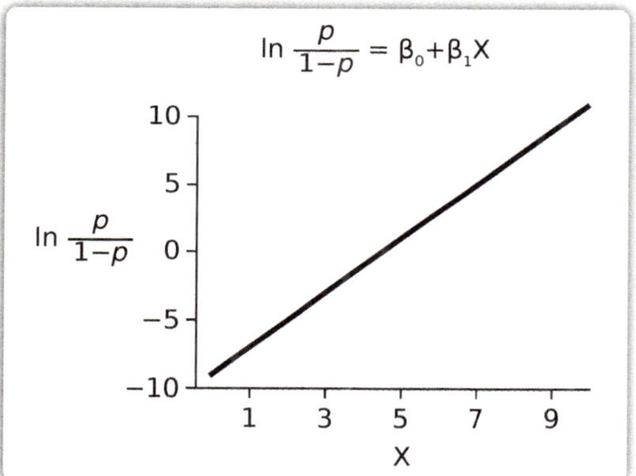

Figure 10.21 Two representations of the logistic relation. The two forms are mathematically equivalent but not statistically equivalent (see *Mathematical vs. statistical equivalence* on page 516). In this example, $\beta_0 = -9$, $\beta_1 = 2$.

Then we can interpret β_1 as the effect on the log of the odds, ln(odds), that would be associated with a unit change in X. In other words, β_1 is the "strength" of the risk factor. For example, when we fix β_0 and vary β_1, we see that as β_1 increases, the steepness of the logistic curve increases, and the slope of the linearized version of logistic regression increases as well, which means that the strength of the effect of the risk factor (X-value) on the outcome variable (p or ln(odds)) increases (Figure 10.22).

Artificial Intelligence Logistic regression is also heavily used in Artificial Intelligence. In these applications, we are interested in determining whether an object has or does not have a certain property. For example, in self-driving cars, the system must examine an image, and decide whether the image contains a car or not. Or we may want to determine whether an object on a radar screen is hostile or not, given features such as speed, direction, etc.

Here, X is some feature of the object, and p is the probability of the object belonging to some binary category, such being a car or not. (If p is close to 1, then this object is very likely

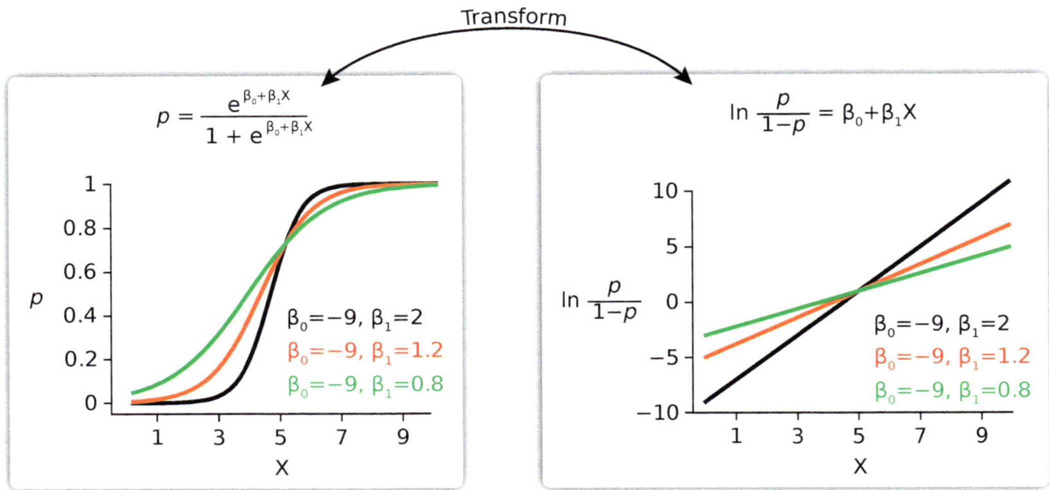

Figure 10.22 Left: logistic curves for three different values of β_1, 2 (black), 1.2 (orange), and 0.8 (green). We fix $\beta_0 = -9$. Right: linearized logistic curves for the same β_1 values.

to be a car, and if p is close to 0, this object is probably not a car.) When we use logistic regression, β_1 is the strength of this feature, how decisive it is in determining the outcome.

Mathematical vs. statistical equivalence

The two forms of the logistic relation, the nonlinear function for probability and the linearized function for ln(odds), are *mathematically* equivalent, but they are *not statistically equivalent*.

In the statistical practice that is called logistic regression, we fit logistic curves to data to find "the best fit curve". It might be thought that to avoid the difficulties of nonlinear functions, we can just linearize the logistic function, use linear regression to find the best straight line, and then "delinearize" the straight line to give us a best fit logistic function. However, this is not true, as we will now show by example.

Effect of Transforms on Regression

Let's generate a data set by adding a little bit of noise onto a logistic function:

$$p = 0.1 + 0.8\,\frac{e^{\beta_0+\beta_1 X}}{1+e^{\beta_0+\beta_1 X}} + \text{noise}$$

where

$$\beta_0 = -9$$
$$\beta_1 = 4$$
$$\text{noise} = \text{uniform distribution } [-0.01,\ 0.01]$$

Here is the scatter plot of the data set (blue dots) and the logistic curve as a function of X. The data points lie very closely to the logistic curve, indeed, that's how they were generated. The amount of noise is so small that you might have to look twice to see that it wasn't just a line.

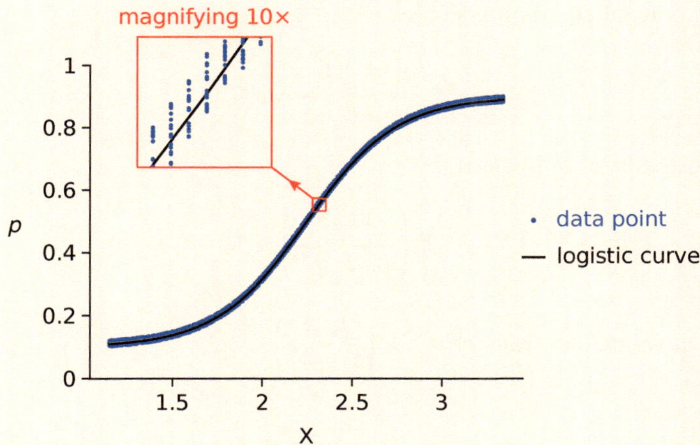

Now we will see if we can recover the original logistic curve from this very slightly noisy data using logistic regression.

Using least squares regression to fit nonlinear logistic curves, we then estimate the parameters β_0 and β_1 that minimize the total error from the curve to the data. By using resampling, we construct a 99% confidence interval for each parameter:

Parameter	True value	Estimated value	99% CI
β_0	−9	−8.991	[−9.007, −8.974]
β_1	4	3.996	[3.988, 4.003]

As we see, the results are satisfactory. The estimates are very close to the true values and both true values are well within their respective CIs.

Now let's linearize the logistic equation by a log transform.

First, we rearrange terms,

$$p = 0.1 + 0.8\,\frac{e^{\beta_0+\beta_1 X}}{1+e^{\beta_0+\beta_1 X}}$$

$$\Updownarrow$$

$$\frac{p-0.1}{0.8} = \frac{e^{\beta_0+\beta_1 X}}{1+e^{\beta_0+\beta_1 X}}$$

$$\Updownarrow$$

$$\frac{p-0.1}{0.8} + \frac{p-0.1}{0.8}e^{\beta_0+\beta_1 X} = e^{\beta_0+\beta_1 X}$$

$$\Updownarrow$$

$$\frac{p-0.1}{0.8} = (1 - \frac{p-0.1}{0.8})e^{\beta_0+\beta_1 X}$$

$$\Updownarrow$$

$$\frac{\frac{p-0.1}{0.8}}{1-\frac{p-0.1}{0.8}} = e^{\beta_0+\beta_1 X}$$

so the original equation becomes

$$\frac{\frac{p-0.1}{0.8}}{1-\frac{p-0.1}{0.8}} = e^{\beta_0+\beta_1 X}$$

Taking the ln of both sides,

$$\ln\left(\frac{\frac{p-0.1}{0.8}}{1-\frac{p-0.1}{0.8}}\right) = \beta_0 + \beta_1 X$$

Then, if we define Transform = $\ln\left(\frac{\frac{p-0.1}{0.8}}{1-\frac{p-0.1}{0.8}}\right)$
We get

$$\text{Transform} = \beta_0 + \beta_1 X$$

But note that the transform, since it involves a log, does not treat the uniform noise uniformly.

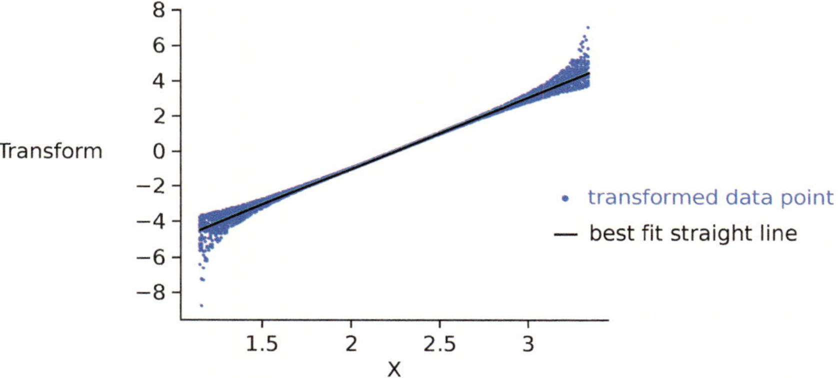

In the log transformed data set, noise is no longer uniform in the vertical direction, and the distribution of noise is different at different X-values: noise is relatively compact at middling X-values, is spread wider and pulled downward at lower X-values, and is spread wider and pulled upward at higher X-values.

This has a significant effect on the estimate of both parameters β_0 and β_1. The logged data now tilts the regression line upward.

Term	Parameter	True value	Estimated value	99% CI
Intercept	β_0	−9	−9.157	[−9.216, −9.095]
Slope	β_1	4	4.070	[4.043, 4.095]

Notice that the 99% confidence interval actually excludes the correct answer $\beta_0 = -9$ and $\beta_1 = 4$.

The log transform has altered the regression and has returned an incorrect estimate, rejecting the correct answer with $p < 0.01$.

Spline regression

In another type of nonlinear regression, called **spline regression**, we fit not one, but two lines, joined together, to the data.

This type of regression came up, in a critical way, in a paper entitled "Systolic blood pressure and mortality", discussing hypertension and its relation to mortality (Port *et al.*, 2000). (The first author of this text was senior author in that study.) Here we follow the argument of that paper.

Let's ask ourselves a question: if we were to graph Systolic Blood Pressure (SBP) on the X-axis, and probability of death on the Y-axis, it's clear that the curve is going to rise: high blood pressures carry an increased risk of death. The question is: *how* does it rise?

There are two competing theories (Figure 10.23):

- **Smoothly rising curve.** One model holds that the risk of death increases smoothly with SBP. This model then fits a (smoothly rising) logistic curve to the data.
- **Hockey stick.** As an alternative, a logistic spline model postulates that the risk of death can be described by two logistic curves, joined together at a connecting point. This type of model allows for two different slopes in distinct ranges of the data, on either side of the connecting point. In one type of spline, the left hand portion is much shallower than the right hand portion, which makes the joint curve look like a hockey stick.

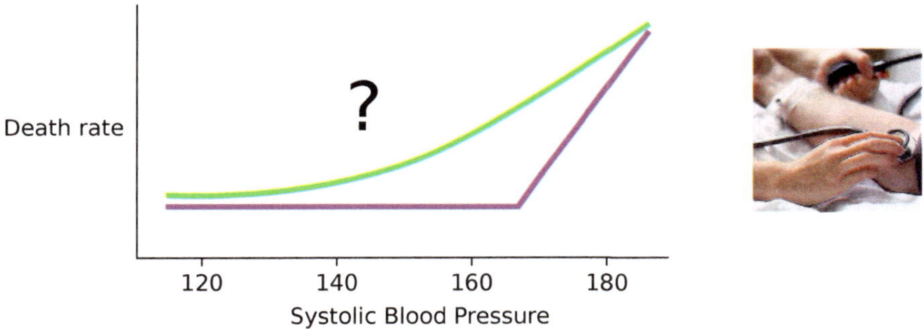

Figure 10.23 Two competing theories of how blood pressure affects mortality: a smoothly rising logistic curve (green) and a hockey stick model (magenta).

Seven Countries Study There was a well documented study, called the Seven Countries Study, led by Ancel Keys (see *Seven Countries: A Multivariate Analysis of Death and Coronary Heart Disease*, Harvard University Press, 1980).

If we look at their plot of "all deaths" or "coronary deaths" against decile (1 decile = 10%) of SBP, we see that the data resembles a hockey stick: flat in one part and rising in another, as indicated by the magenta lines that we overlaid on the data (Figure 10.24).

The choice of a hockey stick model is important because it means that increased blood pressure is not associated with increased death until the 7th decile, that is the 70th percentile of this distribution.

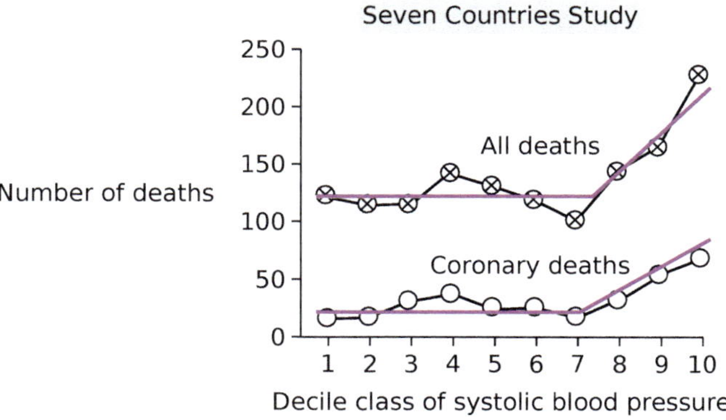

Figure 10.24 Raw data from Seven Countries Study, with our overlay (magenta).

The Framingham Study Despite the seeming presence of a hockey stick relation, the model was not adopted by the medical community. Instead, the community adopted another model, that fit a smoothly rising curve to the data. The clearest advocate of the smoothly rising curve model was the Framingham Study (The Framingham Study, NIH 74-599. Bethesda: USHEW NIH, 1968).

We plotted the results from the study (Figure 10.25). Death as a function of systolic blood pressure does indeed appear to be a smoothly rising curve.

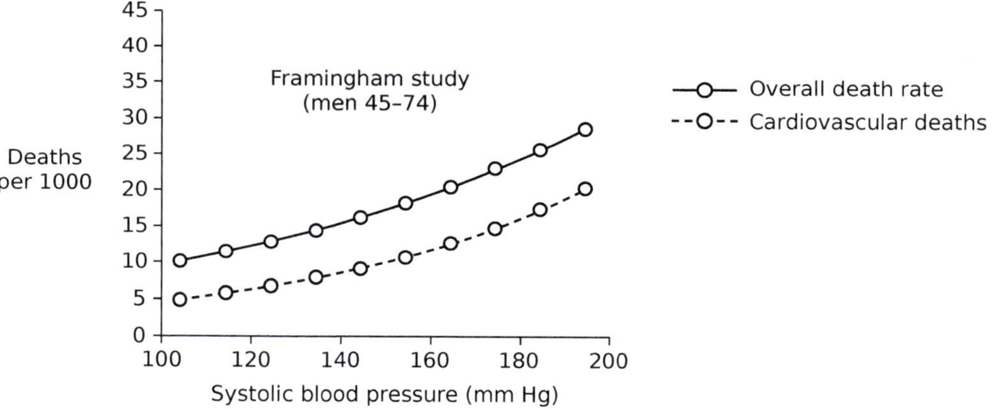

Figure 10.25 When the "results" of the Framingham study are plotted, they appear to support the smoothly-rising curve model (Reprinted from Port *et al.* (2000)).

However, note that the data points lie *exactly* on a smoothly rising curve. We asked how could this be the case?

When we examined the supplementary tables of the Framingham study, we saw that these "data points" are *not* the raw data; they are the "regression coefficients of the logistic function." In other words, the points are lying right on the regression line because these are not the data points, these are the fit points from the logistic regression.

But there is a deep circularity here. If the question is: is this a smoothly rising relation?, then that question cannot be answered by fitting only smoothly rising curves to the data!

So that leaves open the question: what is the shape of the relationship between blood pressure and mortality?

We were able to obtain the raw data, and when we plotted it, there was some reason to think that a hockey stick model was, after all, the preferred model (Figure 10.26).

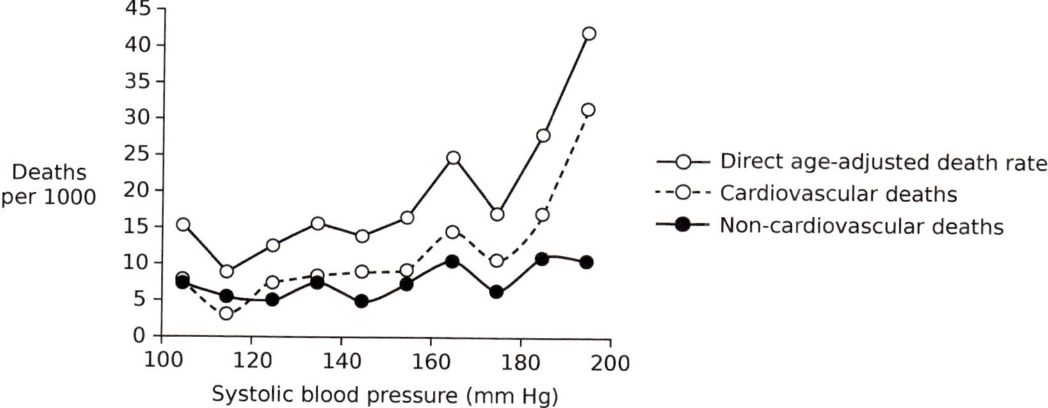

Figure 10.26 Raw data from the Framingham study (Port *et al.*, 2000).

And indeed, if we take the Framingham data, and plot it by percentiles, we see a hockey stick relationship (Figure 10.27), the same relationship that Ancel Keys saw in his Seven Countries study with completely different data (Figure 10.24).

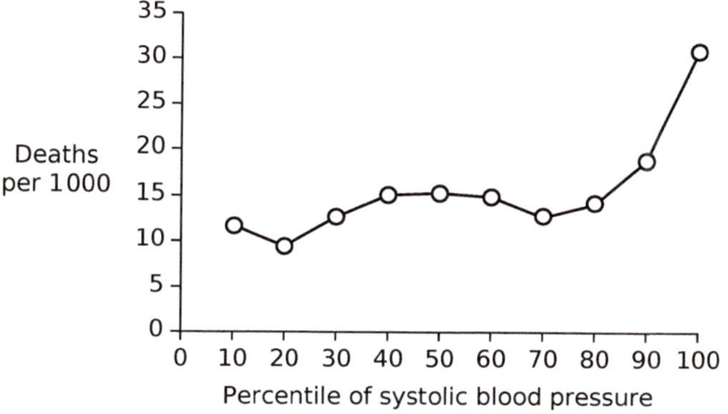

Figure 10.27 Percentile plot of the raw data from the Framingham study (Port *et al.*, 2000).

Note that for the first 70% of the data, risk does look more or less flat, and then risk rises sharply afterwards.

A better fit to the Framingham data In fact, we were able to show that the first 70% of the data has no significant slope at all, even at the modest $p < 0.05$ level. In other words, we cannot rule out the Null Hypothesis that the line is flat.

So what we did was fit a more complex function to the data, called a **logistic spline** (Figure 10.28). In a logistic spline, we join two distinct logistic curves, with the join at the 70th percentile of the data. Then, we can ask

- does the left hand curve have a slope different from zero?

- does the right hand curve have a slope different from zero?
- do the two curves have different slopes?

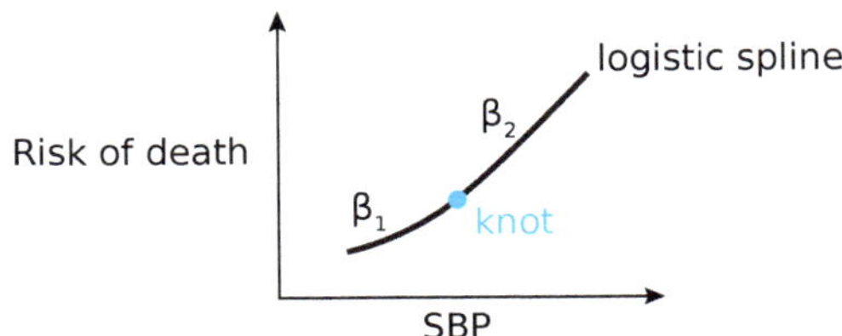

Figure 10.28 A logistic spline model.

When we fit a logistic spline to the data, we found that β_1 was not significantly different from 0, and that β_2 was significantly greater than 0. We were also able to reject the Null Hypothesis that the two curves have the same slope. In other words, the best fit is a flat line followed by a smoothly rising curve that has a much steeper slope.

We also compared this hockey stick model to a logistic regression model with the same number of parameters, and found that the hockey stick is a better fit to the data.

A new definition of hypertension from a better fit to data

Notice that in this new model, there is a natural definition of hypertension: hypertension is the point where risk starts to rise significantly.

By comparison, what is hypertension on the smoothly rising curve model? The answer is that it's completely arbitrary, and panels of cardiologists meet regularly to argue the virtues of one arbitrary definition versus another.

We disaggregated the data by age and sex, and found that the point at which risk starts to rise, that is, the proposed definition of hypertension, is dependent on both age and sex. In particular, older people can tolerate higher blood pressures without increased risk of death, and women can tolerate higher blood pressures than can men of the same age (Figure 10.29).

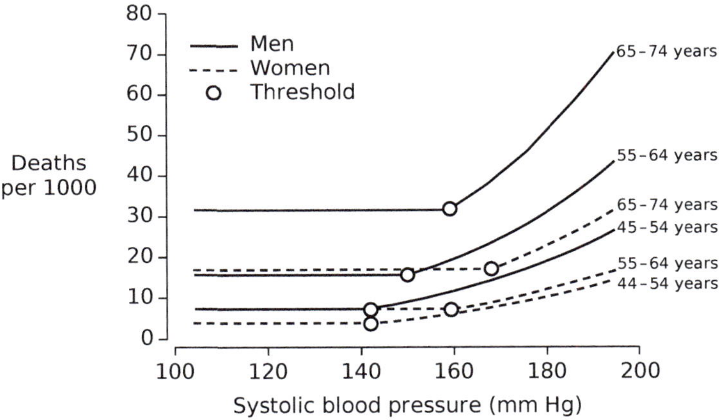

Figure 10.29 The definition of hypertension should depend on age and sex (Port *et al.*, 2000).

10.4 Significance Testing and Confidence Intervals

In general, significance testing and confidence intervals can be done for any form of regression using resampling methods. Here, we will illustrate the procedure using the example of linear regression.

Suppose that we have some data and we found the best fit line $Y = mX + b$. How do we do significance testing and confidence intervals for the parameters m and b?

In fact, we can do all of the following:

- NHST for slope (m)
- NHST for Y-intercept (b)
- Confidence interval for slope (m)
- Confidence interval for Y-intercept (b)

The Null Hypothesis is that there is no relationship between X and Y, that is, that the slope is zero ($m = 0$), and the Y-intercept b is equal to the average value, $\overline{Y}$. We then do NHST as follows (Figure 10.30).

NHST on regression

Step 1 ► Take original raw data pairs. Find regression line, yielding slope m and intercept b.

Step 2 ► Break the connection between X and Y by shuffling the Y-values and re-attaching them to the X-values, generating a new data set, representing the Null Hypothesis that there is no relation between X and Y.

Step 3 ► Calculate the "best fit" regression line for this new data set. Store the slope and intercept for this line, calling them m_1 and b_1.

Step 4 ► Repeat the shuffling procedure 10,000 times to generate 10,000 regression lines. Store the 10,000 values of m_i and b_i. Make histograms of the m_i values and b_i values, which are the null distribution for m and b from the 10,000 simulations.

Step 5 ► Calculate p-values for m and/or b as the fraction of the 10,000 simulations with more extreme results.

Confidence intervals are calculated in the standard method, by resampling the data (Figure 10.31).

Confidence intervals for regression coefficients

Step 1 ► Take original raw data pairs. Find regression line, yielding slope m and intercept b.

Step 2 ► Resample the raw data in pairs (do not break the X–Y connections).

Step 3 ► Compute a new regression line on this resampled data, store m_1 and b_1.

Step 4 ► Repeat the resampling procedure 10,000 times to get 10,000 m_i and b_i values.

Step 5 ► Use the 10,000 simulated values to construct a confidence intervals for m and/or b.

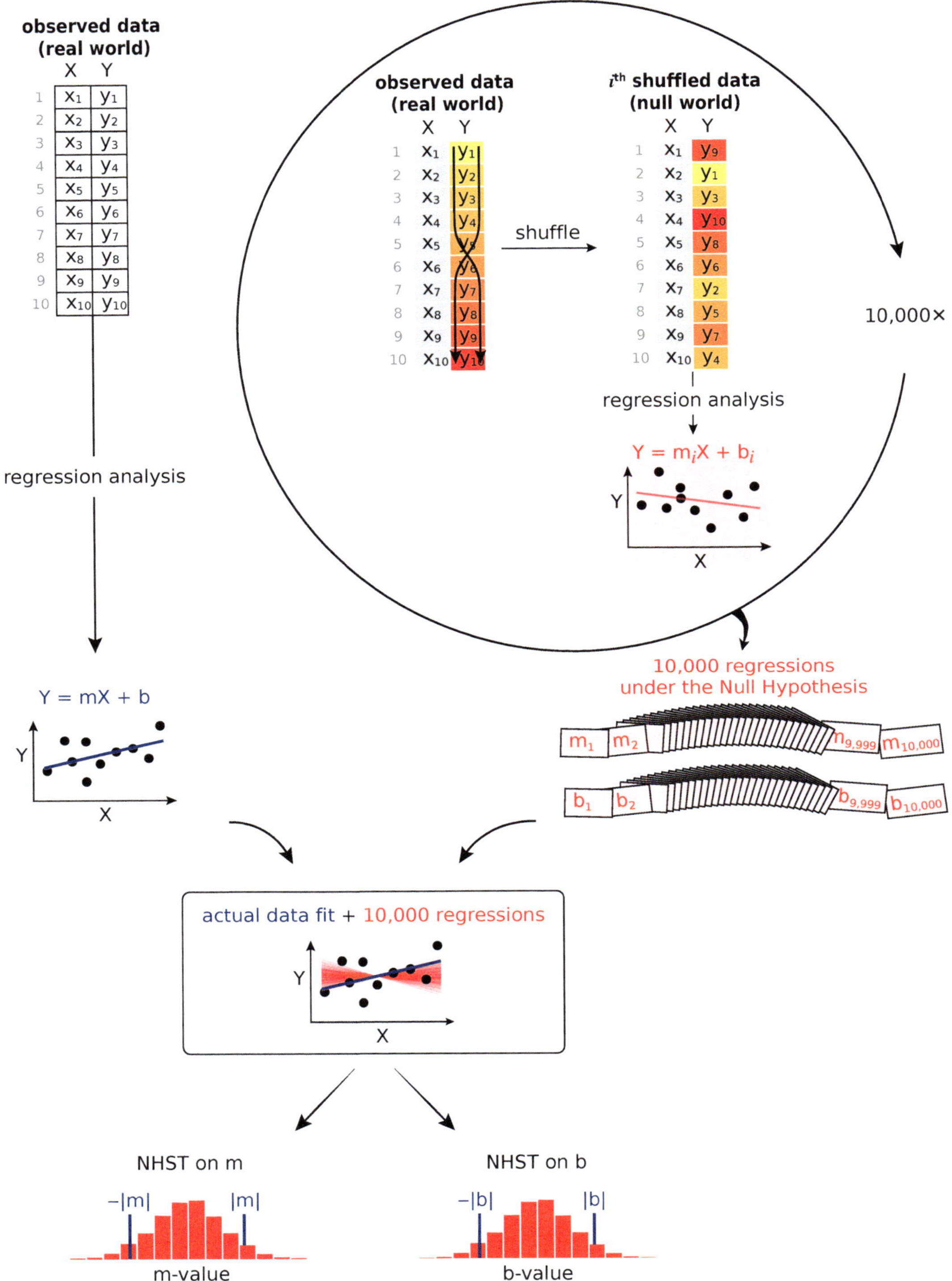

Figure 10.30 NHST for regression coefficients.

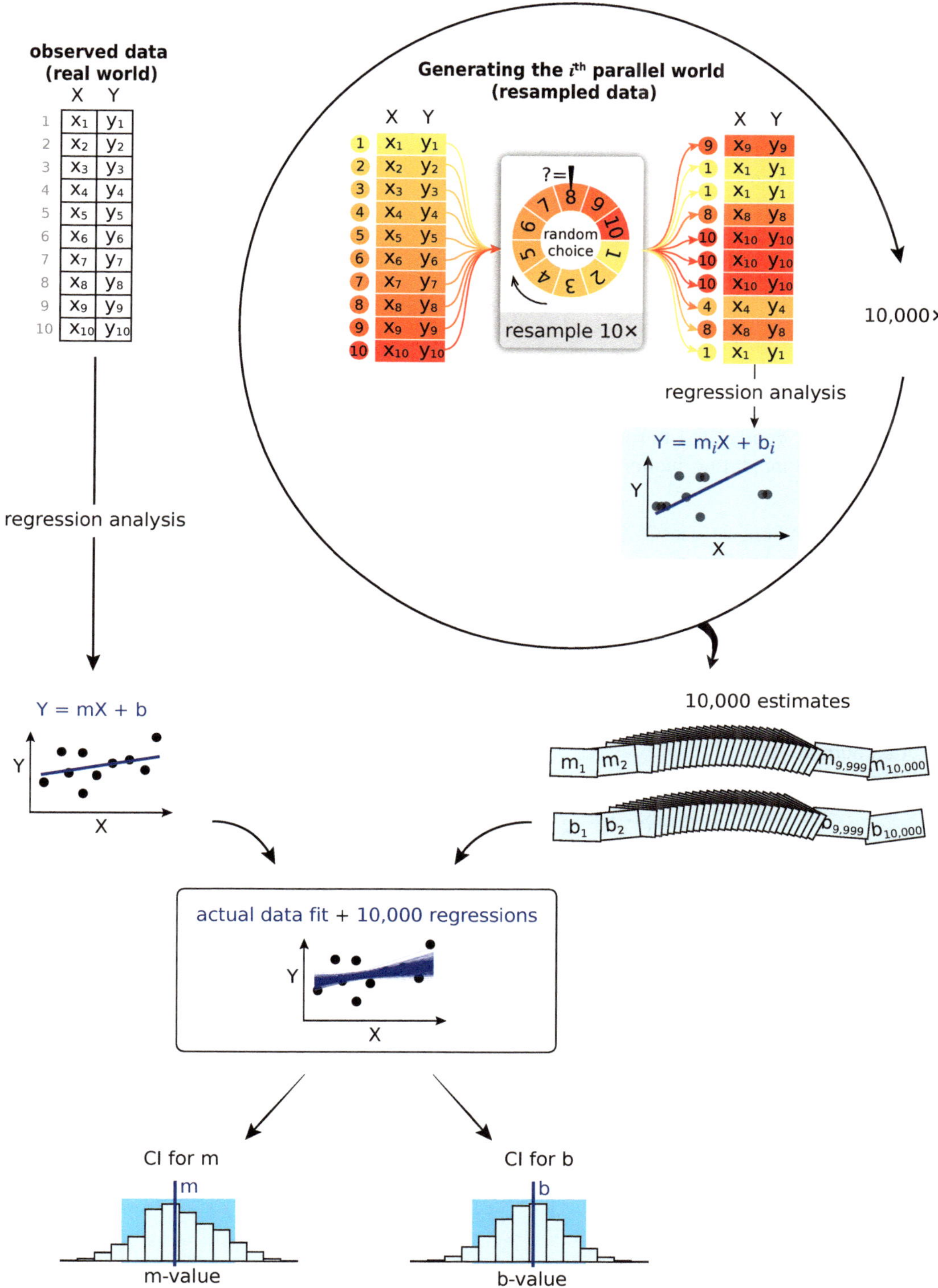

Figure 10.31 Confidence intervals for regression coefficients.

Are wildfires increasing in CA?

Let's consider a data set detailing the area burned in California by wildfires each year, to see whether there is a significant trend over time (Figure 10.32).

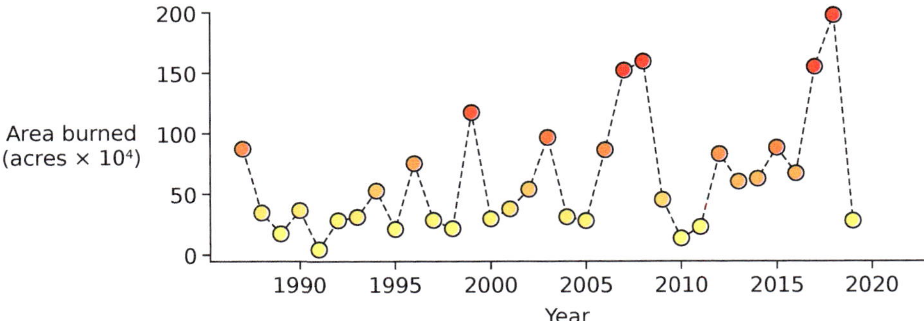

Figure 10.32 Yearly area burned in California wildfires,1987-2019. Colors encode the annual totals, with red indicating higher values.

Regression When we use OLS regression to fit a line to this wildfires data, we get a slope of 2.14 (blue line). The red line is the Null Hypothesis, which is that the slope is equal to 0 (Figure 10.33).

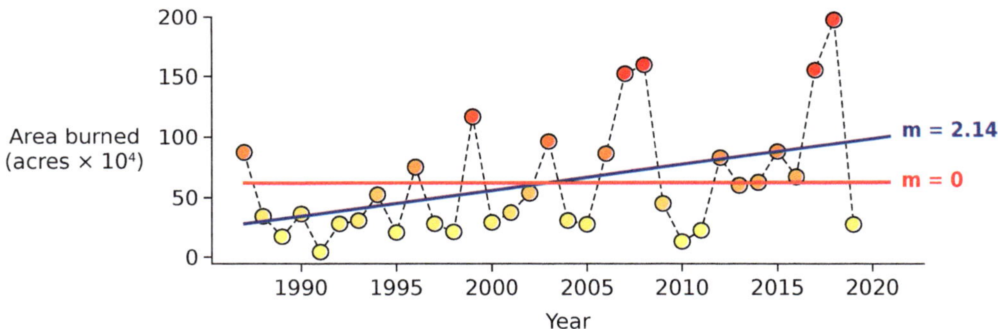

Figure 10.33 The OLS best fit line (blue) and the Null Hypothesis (red).

NHST for slope m The Null Hypothesis is that the annual area burned in California is not changing over time. So our first question is: is our regression slope, m, significantly different from zero?

To do NHST, we break the X-Y connection by shuffling the Y-values and re-attaching them to the X-values. For each shuffled data set, we compute a regression slope.

For the 1st shuffle, we calculate the best fit line, and store its slope as m_1 (Figure 10.34).

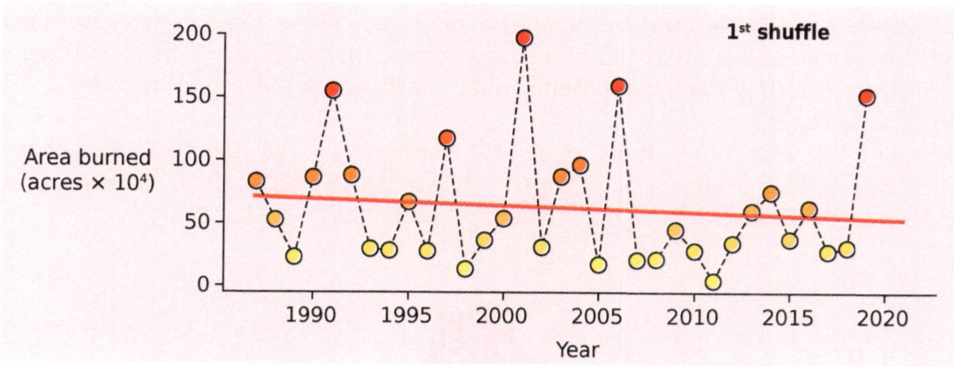

Figure 10.34 The 1st shuffle and its best fit line.

For the 2nd shuffle, similarly, we calculate the best fit line and store its slope as m_2 (Figure 10.35).

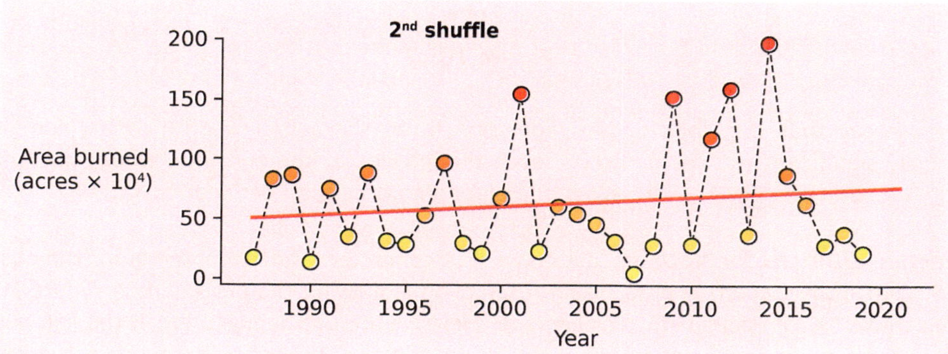

Figure 10.35 The 2nd shuffle and its best fit line.

Repeating the shuffling procedure 10,000 times gives us 10,000 regression lines, with 10,000 estimates of the slope under the Null Hypothesis. Notice that all the regression lines pass through the point $(\overline{X}, \overline{Y})$, the average value of X and Y (Figure 10.36).

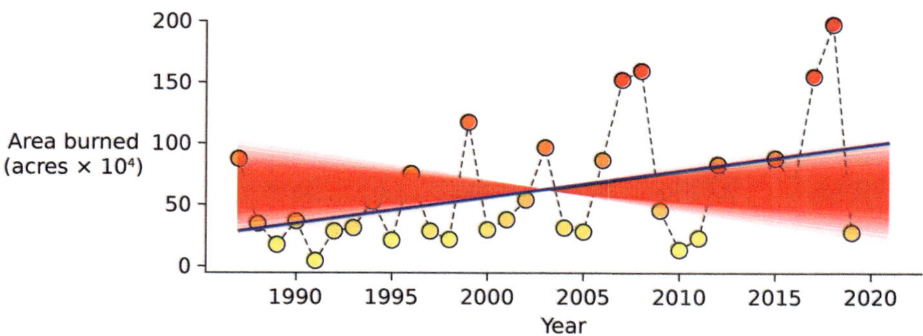

Figure 10.36 10,000 regression estimates under the Null Hypothesis. The actual data (dots) are shown together with the best fit lines for each of the 10,000 shuffled data sets (red lines) and the best fit of the actual data (blue line).

If we compare the actual data fit (blue line) to the 10,000 regression lines generated under the Null Hypothesis, we can see that the blue line is outside the envelope of the 10,000 regression lines under the Null Hypothesis, suggesting that the slope of the actual data, $m = 2.14$, is statistically significant.

To confirm this visual observation, we plot the distribution of the 10,000 slopes, and compute the statistical significance numerically: only 61 + 67 simulations out of 10,000 have slopes equally or more extreme than the actual slope, therefore, p = 0.0128 (Figure 10.37).

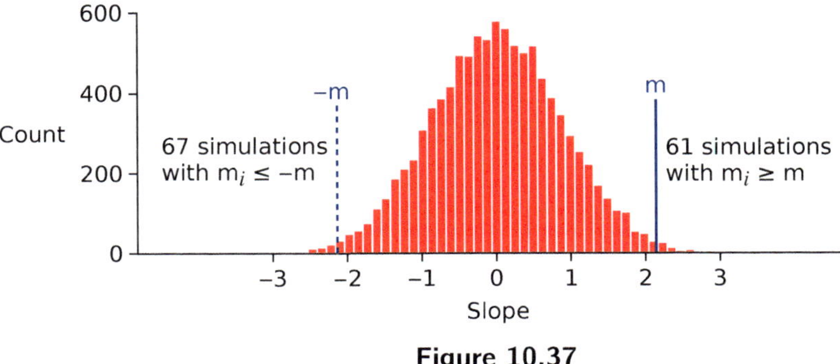

Figure 10.37

We can say that the slope of the regression fit of the actual wildfire data (blue line) is statistically significant (at the $p < 0.05$ level, though not at the $p < 0.01$ level), and that on average, an additional 2.14×10^4 acres of California are burning per year ($p < 0.05$).

Confidence intervals for slope m Next, we'll calculate a confidence interval for the observed slope of 2.14 from the actual data, using resampling (following the procedure in Figure 10.31).

We sample (X, Y) pairs with replacement. Here we are not going to break the link between X and Y because we only want to know the effect of sampling variability on the actual slope.

In the first resample, we obtained a slope $m_1 = 3.30$ (Figure 10.38).

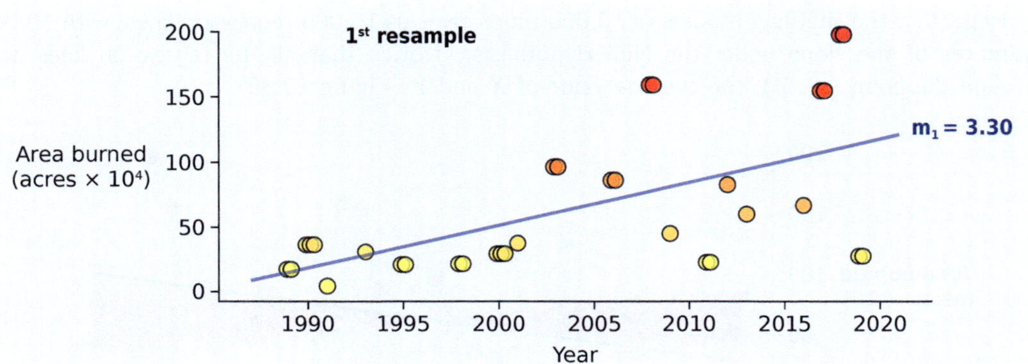

Figure 10.38 The 1st resampling from the actual data and its best fit line. Note the occurrence of multiple copies of the same data points.

In the next resample, we calculated a slope of $m_2 = 1.47$ (Figure 10.39).

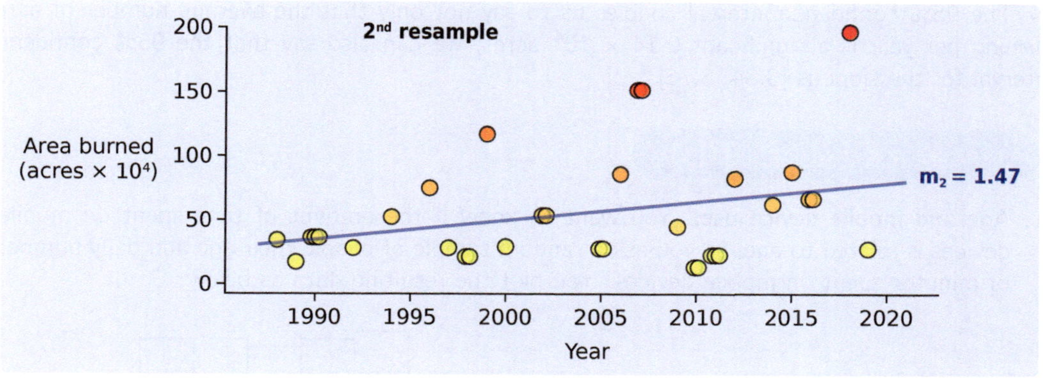

Figure 10.39 The 2nd resampling from the actual data and its best fit line.

We then repeated the resampling 10,000 times, and plotted the 10,000 regression lines. Note that the actual regression line (slope $m = 2.14$) is right in the middle, and now we have a good sense of what the variability is of our estimate (Figure 10.40).

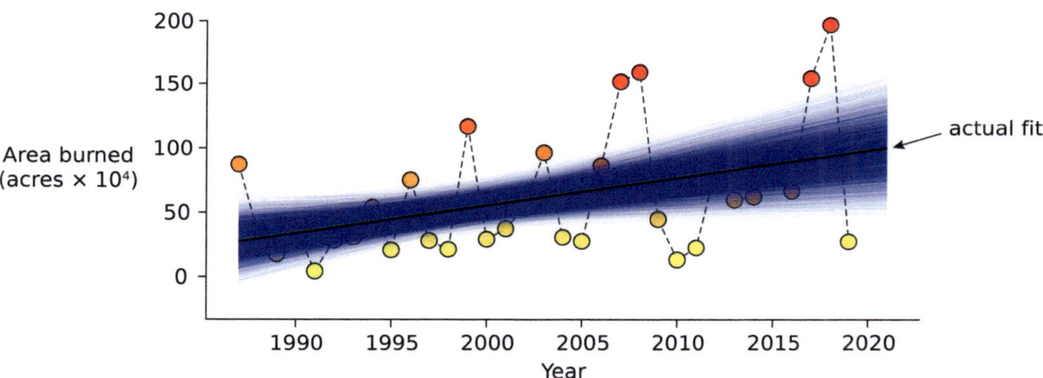

Figure 10.40 The best fit lines for the 10,000 resamples (blue lines) are overlaid on top of the actual data points (dots). The best fit line for the actual data is the darker line (arrow).

From the 10,000 slopes, we constructed a confidence interval for the slope of the regression fit to the actual wildfire data, following the method in Chapter 4 (Figure 10.41).

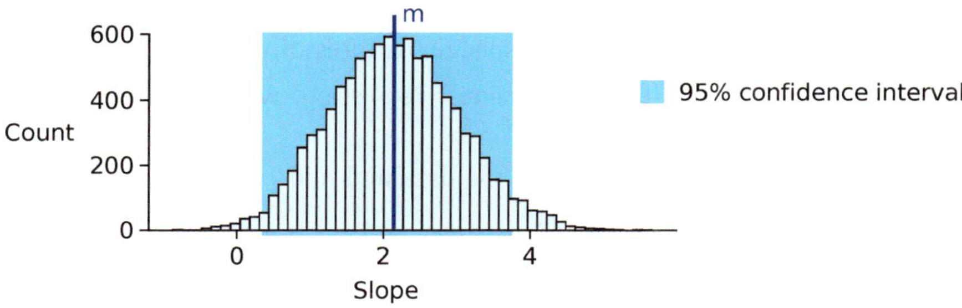

Figure 10.41 95% confidence interval (blue shaded area) for the slope of the wildfires data and the distribution of the 10,000 slopes from the resamples.

The 95% confidence interval enables us to say not only that the average number of acres burning per year is a significant 2.14×10^4 acres; we can also say that the 95% confidence interval for the slope is [0.34, 3.75].[1]

FURTHER EXERCISES 10.4

1. **Age and mobile device use.** You want to know if the amount of time spent on mobile devices is related to age, so you ask a random sample of people their age and daily number of minutes spent on mobile devices. You plot the resulting data as below.

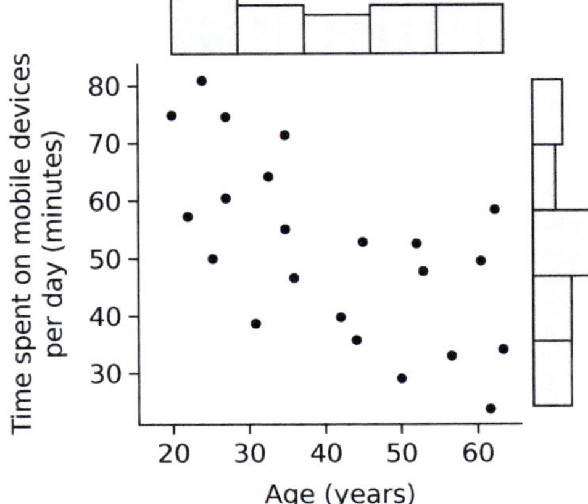

a. Which of the following is a true statement about Ordinary Least Squares regression and orthogonal regression?

   a) The only difference is that orthogonal regression can be used to fit nonlinear relationships.

   b) They always produce the same slopes but different intercepts.

   c) They differ in how residuals are defined and calculated: OLS minimizes vertical distances while orthogonal regression minimizes perpendicular distances to the line.

   d) They are two alternative names for the same analysis.

   e) OLS is always more accurate than orthogonal regression regardless of the data.

b. Looking at the data displayed, is it appropriate to conduct linear regression and NHST here? Explain by referring to specific features of what you see.

c. If it's appropriate to conduct linear regression, which type of regression should you use for this data set?

d. You then fit a straight line to the data and find:

$$Y = -1.3X + 104.9$$

[1] For further discussions of wildfires and their implications for biological species, see the work of Prof. Morgan Tingley (UCLA), for example, "Cross-scale occupancy dynamics of a postfire specialist in response to variation across a fire regime", *Journal of Animal Ecology*, 2018, and "Evolutionary implications of trait–fire mismatches for animals", *Global Change Biology*, 2025.

where Y is time spent on mobile devices per day (minutes) and X is age (years). What does the slope tell you?

a) For each year older, people generally spent about 105 minutes more per day on mobile devices.

b) For each year older, people generally spent about 1.3 minutes less per day on mobile devices.

c) For each year older, people generally spent about 1.3 minutes more per day on mobile devices.

d) A person less than a year old would be predicted to spend about 1.3 minutes per day on mobile devices.

e) A person less than a year old would be predicted to spend about 105 minutes per day on mobile devices.

e. You decide to estimate a confidence interval for the regression slope. Which simulation procedure is correct?

a) Keep the data pairs intact and resample them with replacement in each simulation.

b) Randomly shuffle at least one of the variables in each simulation.

c) Pool both variables together and randomly sample for each group.

d) Keep the data pairs intact and resample them without replacement in each simulation.

f. You calculate a 99% confidence interval for the regression slope as $[-1.5, -1.1]$. What can you conclude about the NHST for the regression slope using $\alpha = 0.01$?

a) Since the interval does not contain 0, we can reject the Null Hypothesis and conclude the slope is statistically significantly different from 0, though running the NHST explicitly would confirm this.

b) There is no way to determine the result of the NHST from the confidence interval alone.

c) Since the interval does not contain 0, we can reject the Null Hypothesis and conclude the slope is statistically significantly different from 0.

d) Since the interval contains only negative values, we cannot draw any conclusions about the Null Hypothesis.

e) The confidence interval and NHST always agree, so we can reject the Null Hypothesis without further analysis.

2. **Penguins.** A scientist measures the body mass and flipper length of penguins at Palmer Station, Antarctica and produces the scatterplot below.

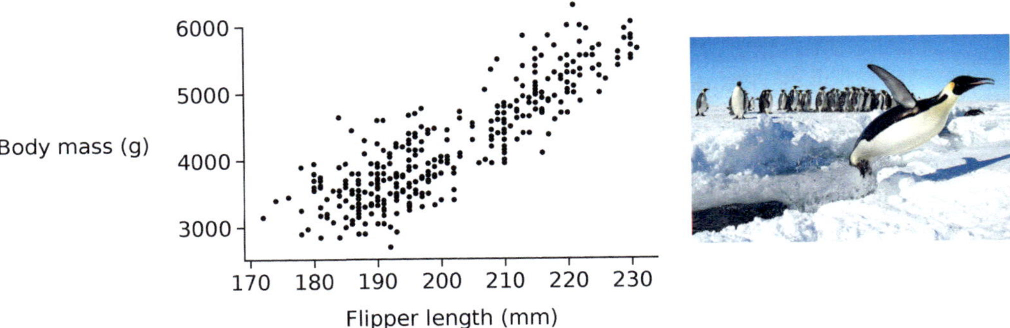

a. What statistic would you use to quantify the relationship between flipper length and body mass, and why?
b. Should you use Ordinary Least Squares (OLS) or orthogonal regression for this data? Explain your reasoning.
c. What is the key difference between OLS and orthogonal regression, and when is each appropriate?
d. Using the correct regression method, a line is fitted to the data with equation $Y = 65X - 8862$, where X = flipper length (mm) and Y = body mass (g). Interpret the slope in the context of this study.

3. **City rents.** An urban economist investigates whether distance from the city center predicts monthly apartment rent. Data were collected from 40 randomly selected apartments across the city, and the scatterplot is shown below.

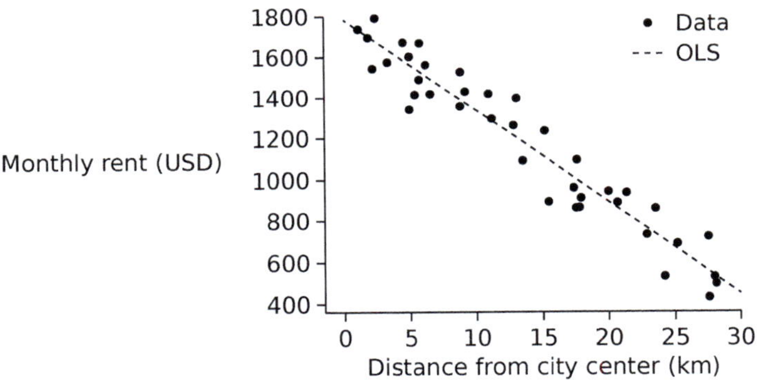

The equation of the least-squares regression line for predicting monthly rent (Y) from distance (X) is:

$$Y = -44X + 1783$$

where X is distance from the city center (km) and Y is monthly rent (USD).

a. Is this an experiment or an observational study? Give your reasoning.
b. Why is OLS more appropriate here than orthogonal regression? What assumption about X makes this valid?

c. Interpret the slope in the context of this problem.
d. Use the regression line to predict the rent of an apartment 10 km from the city center. How confident are you in this prediction?
e. A friend asks you to predict rent for an apartment 60 km from the city center. How do you respond?
f. You decide to use NHST to test whether distance is a statistically significant predictor of rent. State the Null Hypothesis in plain language and mathematically. What test statistic will you use, and why is the slope a natural choice here?
g. Choose the histogram below that is the most likely result of your NHST. On it, indicate the bins that you would use to calculate the p-value.

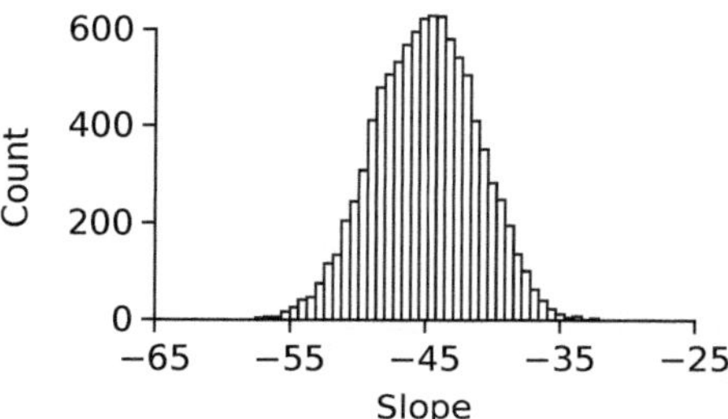

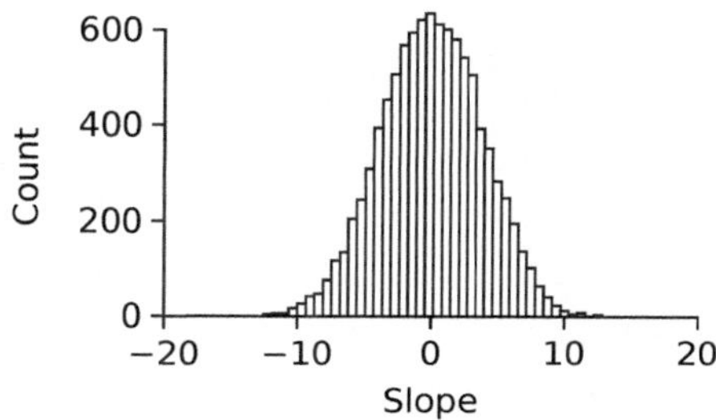

4. **Life expectancy and per-capita GDP.** A study examines the relationship between a country's life expectancy and its economic output. As the first step, the analysis produces the scatterplot below, along with marginal histograms for each variable.

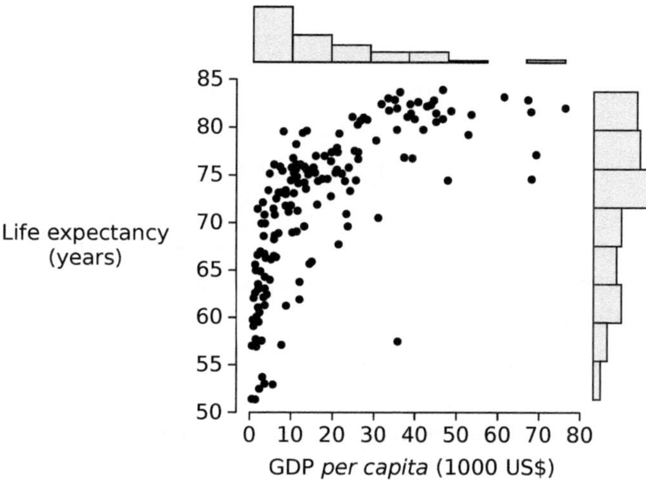

a. What statistic would you use to quantify the relationship between these two variables, and why?
b. State the null hypothesis for this study both in plain language and in terms of the test statistic.
c. Assuming you have calculated the statistic of interest, describe how you would use simulation to conduct NHST. In your answer, specify the box model, the sample size,

whether sampling should be done with or without replacement, and how the p-value would be calculated.

d. Based on the information provided, is linear regression appropriate for this data set? Why?

5. **Wildfires in the US.** You want to know whether the annual area burned by wildfires in the United States has been increasing over time. The data are visualized below.

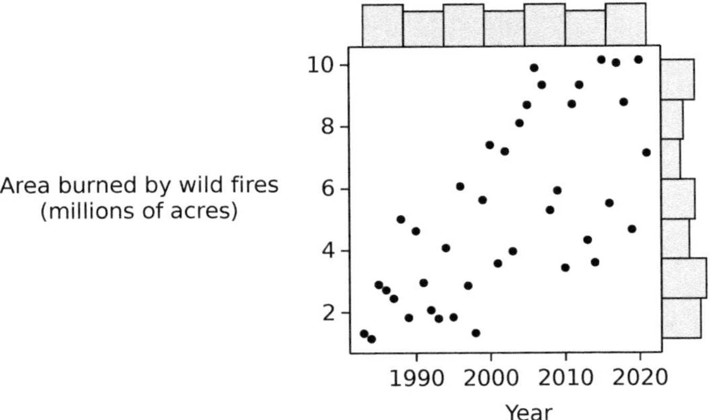

a. Is this an experimental or an observational study? Explain your reasoning.

b. Looking at the plots, is it appropriate to conduct linear regression and NHST here? What features of the plots inform your answer?

c. What statistic would you use to quantify whether area burned has been increasing over time, and why?

d. Based on the plots, circle the value below that you expect your sample statistic to be closest to.

$-0.6 \quad -0.2 \quad -0.05 \quad 0 \quad 0.05 \quad 0.2 \quad 0.6$

e. You calculate the statistic of interest and want to construct a 99% confidence interval for it using resampling. Describe how you would do this. Include the box model, sample size, whether to sample with or without replacement, and how to calculate the confidence interval.

f. If a 99% confidence interval for the statistic of interest is [0.100, 0.242], what can you conclude?

g. Based on this confidence interval, what can you conclude about the NHST for this statistic at $\alpha = 0.01$?

h. Why is reporting a confidence interval more informative than reporting only a p-value?

i. Your regression line has an R^2 of 0.52. Interpret what R^2 means.

10.5 The ugly history of correlation, regression, p-values and NHST

The big three

Most of what is thought of as statistics: the correlation coefficient, the Normal distribution, statistical significance, NHST, χ^2, ANOVA, etc., was developed in the 19th and early 20th centuries largely by three people: Francis Galton, Karl Pearson and Ronald Fisher.

Each of them were world leaders in the movement that was called Eugenics, and each of them, as we'll see in the following pages, were eye-popping racists and White Supremacists. As we will see, the association between these flawed statistical concepts and the racial views of the people who invented those concepts was not accidental: they needed the flawed concepts to advance their social agenda.

What is Eugenics?[2]

The term itself was coined by Galton. The Oxford Dictionary defines Eugenics as

> "The study of how to arrange reproduction within a human population to increase the occurrence of heritable characteristics regarded as desirable. Developed largely by Sir Francis Galton as a method of improving the human race, eugenics was increasingly discredited as unscientific and racially biased during the 20th century, especially after the adoption of its doctrines by the Nazis."

Eugenics, developed by Galton, was embraced enthusiastically by the Nazis. Hitler cited the Eugenics movement in the US as people who were leading the way to his concept of racial purity.

After World War II, the term "Eugenics" became highly unpopular, and prewar Eugenicists professed to be horrified by its consequences in Nazi hands (except that Fisher continued to

[2] All quotes in this section from Galton, Pearson, and Fisher are taken from Clayton (2020).

cling to the concept of Eugenics and defended the Nazis, even after the war, as we'll see later on).

"Heritable characteristics"?

In the Oxford Dictionary definition, note the word "heritable". It's a word that is used frequently in discussions of the genetic bases for various conditions, such as diseases or traits. In one meaning of the word, the meaning used by the Oxford Dictionary, "heritable" means "transmitted by genetic inheritance". In this meaning, heritability is a causal concept: if something is heritable, then you inherited it through genetics.

However, there is a second, purely statistical concept of heritability. In this meaning, heritability is essentially the strength of the regression estimate of children's trait from parent's trait.

This second meaning of heritability masquerades as a causal concept, but it isn't. "Heritable" is a faux-causal word, probably the most important faux-causal word, but in fact, it's a correlation-based notion and has no causality in it at all. The intentional confusion of the two concepts of "heritable" is responsible for a great deal of invalid social theorizing. In an important paper called *The heritability fallacy*, Moore and Shenk (2017) said

> "The term 'heritability,' as it is used today in human behavioral genetics, is one of the most misleading in the history of science."

Galton Galton's eugenics was based on his claim, for which he had no real evidence, that both socially desirable traits and socially undesirable traits were determined largely by heredity. He argued that people disfavored by society should not be allowed to reproduce.

> "Reproduction should be discouraged for people afflicted by lunacy, feeble-mindedness, habitual criminality, and pauperism."
>
> — Sir Francis Galton

What he means by "feeble-mindedness" is unclear, "habitual criminality" is a judgement call, and of course, "pauperism" means "poor people".

Galton's social views were not limited to his assessments of the poor; he was also a fanatic believer in the concept of "race" and in the idea that war between races was desirable and necessary. He was the founding editor of *The Eugenics Review*, and wrote there:

> "There exists a sentiment, for the most part quite unreasonable, against the gradual extinction of an inferior race."
>
> — Sir Francis Galton

Of course we know who those inferior races are. They were exactly the people that the British Empire was colonizing at the time in Africa, in India, and in South East Asia. Those were the "inferior races" who were being marked for extinction.

Galton also promoted the idea that human abilities like "mental capacities" must follow a Normal distribution, or a bell curve. What is his argument? First, he claims that there is a Normal distribution for height. As we saw, that's not true. There are separate distributions for men and women, and even if they were both Normal distributions (which they were not), the mixture of two different Normal distributions is never a Normal distribution.

He goes on to make a stupendous leap, and claims that therefore *all* physiological quantities must be Normally distributed.

> "if this [the Normal distribution] be the case with stature, it will be true as regards every other physical feature — as circumference of head, size of brain, weight of grey matter, number of brain fibres, etc."
>
> — Sir Francis Galton

Note that Sir Francis has absolutely no idea what he is talking about. "weight of grey matter"? "number of brain fibers"? These are concepts completely outside the science of his time, and indeed, outside even the science of our time. He is simply pontificating on a subject he could not possibly know anything about.

But then he goes on to do something even worse. He invents the concept of "mental capacity", whatever that could possibly mean, and claims that this mystical quantity must also follow a Normal distribution.

> "and thence, by a step on which no physiologist will hesitate, as regards mental capacity. . . that the deviations from that average — upwards toward genius, and downwards toward stupidity — must follow the law that governs deviations from all true averages."
>
> — Sir Francis Galton

His argument is so wild and speculative that he feels the need to add the assurance that "no physiologist will hesitate", whereas in fact, no decent physiologist would ever dream of making an unscientific statement like this.

This is what passes for scientific argument if you were promoting racism in 19th century Britain. Not only was he not ridiculed for the absurdity of this statement, he was awarded a knighthood for "services to the Empire." Services to the Empire indeed!

He then invoked the Normal distribution to express his personal opinions on race, using the pseudo-scientific language of the Normal distribution to lend credibility to his racist beliefs. He invented the notion of "grades", which are essentially standard deviations of a Normal distribution, and said

> "the average intellectual standard of the Negro race is some two grades below our own. . . The average ability of the Athenian race is, on the lowest possible estimate, very nearly two grades higher than our own"
>
> — Sir Francis Galton (*The Comparative Worth of Different Races*)

There are several complete absurdities in this statement. First of all, Galton has absolutely zero data on "the Negro race", and admits that he is basing this on his subjective impressions of a handful of personal encounters. The use of the phrase "two grades" gives a pseudo-scientific cover to his personal prejudices.

Second, what is the "Athenian race"? It's Greek people who lived in and around Athens during the 5th and 4th centuries BC! The whole concept of race is completely confused. But in no concept of race do "Greek people who lived in and around Athens during the 5th and 4th centuries BC" constitute a race.

How can Galton, sitting in London in 1880, possibly make a judgement about the average "ability" of the "Athenian race"? The answer is that it's impossible, and this is again Galton's personal prejudices, pontificated in a pseudo-scientific language.

Pearson Galton's student and protege, Karl Pearson, shared his racist agenda. Pearson developed the correlation coefficient that is named after him, the notion of standard deviation, χ^2, and regression.

> My view — and I think it may be called the scientific view of a nation — is that of an organized whole, kept up to a high pitch of internal efficiency by insuring that its numbers are substantially recruited from the better stocks . . . and kept up to a high pitch of external efficiency by contest, chiefly by way of war with inferior races.
>
> . . .
>
> History shows me one way, and one way only, in which a high state of civilization has been produced, namely the struggle of race with race, and the survival of the physically and mentally fitter race.
>
> — Karl Pearson

He was proposed for a knighthood in 1935 for "services to the Empire" but refused it for philosophical reasons.

Fisher The third element of the trio was Sir Ronald Fisher, who further developed the notion of χ^2, who coined the term "$p < 0.05$", developed NHST and the F-distribution.

Fisher was a vocal supporter of eugenics from an early age, helping to found the Cambridge University Eugenics Society as an undergraduate. He became editor of the *Annals of Eugenics* (since renamed *Annals of Human Genetics*).

University College London, Fisher's alma mater, used to have buildings and lectures named after Fisher. They had to rethink this and published a self examination:

> Fisher held onto his views about eugenics long after the Second World War, when eugenics had fallen into disrepute. In the years immediately after the war, Fisher remained on friendly terms with former Nazi geneticist Otmar Freiherr Verschuer. Fisher also expressed sympathy towards the eugenics policies of the Nazis."
>
> . . .
>
> During the war, Verschuer had worked directly alongside Josef Mengele, using biological samples obtained from Jews murdered in concentration camps.
>
> — University College London (Fisher's alma mater)

In World War II, both Axis powers were motivated by theories of racial superiority. The Nazis glorified themselves as champions of the "Aryan race", and the Japanese military government popularized the concept of Japan as the "Yamato race", which they believed to be destined to rule over the inferior Asian races.

After World War II, racist views like these were forced to take a step backward. The newly established United Nations formed an Educational Scientific and Cultural Organization (UNESCO), which made a strong postwar statement condemning racism. Fisher was opposed to this anti-racist statement.

> Another notable expression of Fisher's political values occurred in 1950, when UNESCO was assembling a statement on the nature of race. Their view, which is mainstream in science today, was that race is a socially constructed categorisation, and has little basis in genetic variation. Fisher refused to sign this declaration, arguing that human groups profoundly differ 'in their innate capacity for intellectual and emotional development'.
>
> — University College London

Fisher was knighted in 1952 for "service to the Empire."

Sir Ronald Fisher: cancer causes smoking

To give an idea of what Fisher's intellectual standards were, he became quite active in the 1950s, trying to deny the relation between smoking and cancer. He said:

> Unfortunately, considerable propaganda is now being developed to convince the public that cigarette smoking is dangerous...
>
> Such results suggest that an error has been made of an old kind, in arguing from correlation to causation.

In other words, he is attacking the view that smoking causes cancer, by accusing those people of arguing from correlation to causation, arguing instead for

> something else causing both. Is it possible, then, that lung cancer — that is to say, the pre-cancerous condition which must exist and is known to exist for years in those who are going to show overt lung cancer — is one of the causes of smoking cigarettes? I don't think it can be excluded.

So lung cancer could cause smoking, argued Sir Ronald Fisher.

Statistician Paul Stoltey, writing in the *American Journal of Epidemiology* (*When Genius Errs: R. A. Fisher and the Lung Cancer Controversy*), points out that Fisher didn't look at the evidence and simply pontificated his position.

> he was unwilling to seriously examine the data and to review all the evidence before him to try to reach a judicious conclusion.
>
> ...
>
> in addition, he took a fee from the tobacco industry.

But is their science wrong?

We can conclude that these three held horrible social views. But does that mean that their science is wrong? Is their science also corrupt?

After all, the correlation coefficient, the Normal curve, χ^2, Null Hypothesis Significance Testing, and "$p < 0.05$" are the foundation of traditional statistics, and are enshrined in traditional statistics classes and texts.

But what you've been hearing about in this book is what's wrong with the correlation coefficient, what's wrong with the Normal curve, what's wrong with χ^2, what's wrong with NHST, and what's wrong with "$p < 0.05$".

Indeed, all of the current crises in traditional statistics can be traced to precisely these traditional statistical concepts, as *what's wrong with traditional statistics!* So, on the one hand, some people want to celebrate Pearson and Fisher for inventing "statistical significance", but, as we saw in Chapter 1 (*The Problem with Statistics*), in 2019, 800 signatories in *Nature* said: "we call for the entire concept of statistical significance to be abandoned." Some people celebrate Galton and Person as the inventors of correlation and linear regression, but correlation and linear regression have frequently been badly misused to make false inferences about cause and effect, especially with regard to racial groups.

So we have gone from "great men, great science" to "bad guys, great science" to "bad guys, bad science". The final question is: is the connection between "bad guys" and "bad science" purely coincidental? Or is there some link between the two?

In an important essay *How Eugenics Shaped Statistics*, Clayton (2020) argued for a link between these traditional statistical concepts and their racist origins.

> Statistical thinking and eugenicist thinking are, in fact, deeply intertwined, and many of the theoretical problems with methods like significance testing — first developed to identify racial differences — are remnants of their original purpose, to support eugenics.
>
> ...
>
> It's no coincidence that the method of significance testing and the reputations of the people who invented it are crumbling simultaneously. Crumbling alongside them is the image of statistics as a perfectly objective discipline, another legacy of the three eugenicists.
>
> — Aubrey Clayton (*How Eugenics Shaped Statistics*)

So the big three, in their quest to somehow justify racism, developed the tools that became traditional statistics. Subsequently, the eugenics origin got whitewashed out of the picture, and statues of the big three were erected.

But now, we are finding deep problems with these tools. The problem of irreproducibility and the flood of false results are forcing sharp criticisms of these statistical tools. In fact, one major root cause of the problem of irreproducibility lies in these flawed statistical tools.

In this book, we have argued that one of the biggest problems with correlation, regression, p-values, NHST, and χ^2, is their reliance on extreme assumptions about Normal distributions, which are necessary to derive formulas for statistical significance. We have suggested that the resampling approach rescues these concepts from their dependence on Normal distributions. *But resampling cannot fix the fundamental problem that NHST and p-values are answers to the wrong question.*

What Clayton is suggesting is that the flawed concepts stem from the need to support racist conclusions. We can't get to racist conclusions with good, causal science, and so the big three needed to invent some bad science, which would have to avoid causality.

The biggest obstacle that they faced was that up to this point, mathematics in science was completely causal. Before the big three, the accepted path to applications of math was via differential equations, which are statements of causality. Newton, Euler, Faraday, Maxwell, and Lagrange all dealt with differential equations.

Faux-causal words

But if you want a math to draw racist conclusions, causality is not your friend. Better find a math that avoids causality, and will enable you to use faux-causal words, like

- "Y is affected by X"
- "the effect of X on Y"
- "X explains Y"
- "X determines Y"
- "X reduces/increases Y"
- "Controlling for X"

This was explicitly acknowledged by Pearson himself. As cited by Clayton (2021):

> There was a category broader than causation, namely correlation,. . . this new conception of correlation brought psychology, anthropology, medicine and sociology in large parts into the field of mathematical treatment. *It was Galton who first freed me from the prejudice that sound mathematics could only be applied to natural phenomena under the category of causation.*
>
> — Karl Pearson

Now, with this new mathematics, they were able to make statements that membership in racially marginalized groups was *correlated* with low "mental capacity", and that that correlation was "statistically significant". Then, having established this, they conjured the fallacious reasoning that

- Parents' height is correlated with children's height.
- Height is genetically inherited.
- Parents' 'achievement' is correlated with children's 'achievement'.
- Therefore 'achievement' is genetically inherited.

> We are forced, I think literally forced, to the general conclusion that the physical and psychical characters in man are inherited within broad lines in the same manner, and with the same intensity.... We inherit our parents' tempers, our parents' conscientiousness, shyness and ability, even as we inherit their stature, forearm and span.
>
> — Karl Pearson, "On the Inheritance of the Mental and Moral Characters in Man: II," Biometrika (1904).

In other words, after warning that "correlation is not causation", Pearson went on to claim just that, pretending to pull the causal rabbit (inheritance) out of the correlational hat!

In contemporary applications of correlation and regression, this fallacy is committed frequently. Scientists use regression techniques on two observational variables X and Y, and then use faux-causal words to describe the relationship, for example, saying that "eating more corn increases the risk of cancer" or "higher test scores increase future income prospects".

As scientists and as consumers of science, we need to be vigilant and critical of these misuses of correlation and regression.

Chapter 11

Power

LEARNING OBJECTIVES

After studying this chapter, you will be able to:

- explain why statistical power is important when declaring "negative" findings.
- use resampling methods to calculate the power of any study design.
- explain how power calculations can help in the prospective design of a study.
- explain why low power means that even a positive finding may be suspect.

11.1 Statistical Power

Positive claims and negative claims

Positive claims We know that if we want to make a positive claim that, say, "X is associated with Y", or "there is a difference between group A and group B", we need to have statistically valid evidence for the claim. As we saw in earlier chapters, this evidence requirement is expressed by the concept of *statistical significance*: if you don't have statistical significance, you can't really claim that you have a positive result.

Of course, there were many problems with the idea of statistical significance and Null Hypothesis Significance Testing, but the general idea is valid that you need some level of evidence to make a claim that "X is associated with Y", or "there is a difference between group A and group B".

Negative claims But what about negative claims? Suppose we wanted to say that "X is not associated with Y", or "group A is no different from group B". Do we need statistical evidence for these kinds of "negative results"?

A very naïve, and very wrong, point of view would be to say "no, you don't need statistical evidence to make a negative claim; you just say that you failed to find a positive association, and that supports the negative statement. You are merely saying what is true, that you did not find a relation". What could be wrong with that?

A. Garfinkel and Y. Guo, *Understanding Data*,
https://doi.org/10.1007/978-3-032-18600-3_11

Viral infection and MRI To understand why this is wrong, imagine a med student, looking at a patient with a possible viral infection. Various tests have been done on this patient. The senior attending doc says to the med student "Can you rule out viral infection?", and the med student says "I looked at the MRI carefully, and I saw no viruses."

Attending: "Rule out viral infection"

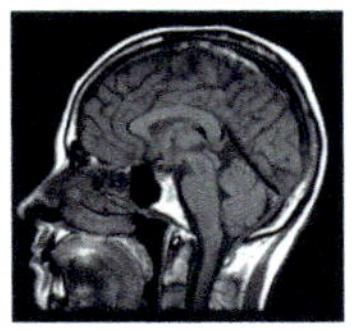

Med student: "I saw no viruses on MRI"

There is something deeply wrong with the med student's reply. The problem is that the lower limit of spatial resolution of MRI is about 1 mm, and the size of a virus is about 10^{-4} mm, 4 orders of magnitude too small to be seen on MRI!

But then how can we say what is wrong with the med student's reply? After all, notice that the statement "I saw no viruses on MRI" **is a true statement**: there really weren't any viruses seen on the MRI! So how can we say that it is wrong to make a true statement?

Letter of recommendation To appreciate why it is wrong, here's another example: imagine a student asking for a letter of recommendation from one of their professors. The professor, who knows the student from a lecture class, writes "I have never seen this student act competently in lab." Note that this is a true statement, and it is true because the professor has never seen the student in a lab setting at all. Can the professor ethically make this true statement?

Do you have the power to make a negative claim? In both of these cases, a true-yet-wrong negative statement was made. The source of the error is the fact that the person making the statement was not in a position to make the positive statement, had the positive statement been true! The med student *could not* have seen viruses *had they been there*, and the evil professor *could not* have seen competent lab performance *had it been there*.

You can't make a negative statement unless you were in a position to make the positive finding had it been there.

A study cannot conclude "there is no association" or "there is no difference" unless it had the ability to find a difference if there had been one.

In the case of "negative conclusions" in statistical studies, a similar requirement holds: negative findings do indeed require some kind of statistical evidence, namely, evidence that the test in question **could** have found a difference **had one actually been there.**

The ability of a study to find a difference that is there is called the power of the study.

> **Power**: how often could the study design find a difference if there really was a difference?

We can now state our principle about "negative findings".

> "Negative results" without power estimates are meaningless.
> It is unethical, statistical malpractice, to draw a negative conclusion from a low power study.

Let's look further into this concept of power.

The two types of errors

As we said in Chapter 3, whenever we have a "test" and a "reality", then there are four possibilities. The reality (say, pregnancy) could be there or not there, and the test can say that the reality is there or say that it is not there.

There are two types of errors we can make. We can make a Type I error, which is a false positive: in reality, the condition was not there, but the test came back positive. The other kind of mistake is a Type II error, which is a false negative: in reality, the condition was there, but the test came back negative.

To illustrate the four possibilities and the two types of errors, we use pregnancy tests as an example and summarize these in a table format (Figure 11.1).

		In reality	
		Result is *really* "there"	Result is *really* not "there"
Test result	Test says: "Result is there"	**correct conclusion**	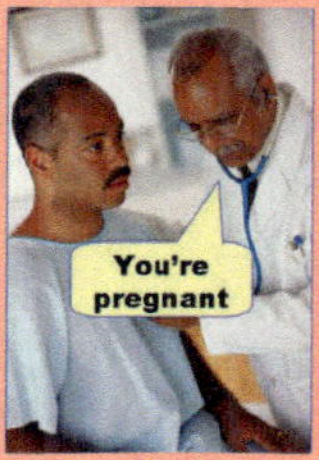 You're pregnant — **false positive** — **Type I error**
	Test says: "Result is not there"	 You're not pregnant — **false negative** — **Type II error**	**correct conclusion**

Figure 11.1 The four possibilities for a "test" and a "reality".

A similar situation holds in the case of the statistical testing of a Null Hypothesis: the Null Hypothesis can, in reality, be either true or false, and the statistical test can either claim that the Null Hypothesis has been rejected or that it has not been rejected. Here, the two types of error we can make are called: False Positive (Type I) error, and False Negative (Type II) error (Figure 11.2) (see also section 3.3 *What's wrong with Null Hypothesis Significance Testing?*).

False Positive (Type I) error is the error of rejecting a Null Hypothesis that is really true.

False Negative (Type II) error is the error of not rejecting a Null Hypothesis that is really false.

		In reality	
		Null Hypothesis *really* false	Null Hypothesis *really* true
Study result	Study rejects Null Hypothesis	**True Positive**	**False Positive** (Type I error)
	Study does not reject Null Hypothesis	**False Negative** (Type II error)	**True Negative**

Figure 11.2 The four possibilities for a Null Hypothesis and a study.

Example 1 Is this kidney cancer?

A CT scan shows a small smudge in the kidney (red arrow). The radiologist must decide whether this smudge is a sign of kidney cancer or not.

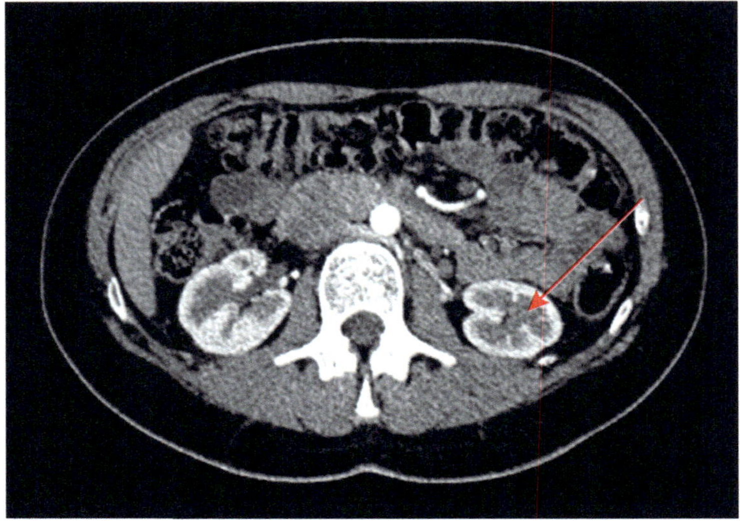

Question Given the CT scan of a patient, what would be the scenarios for a false positive and a false negative?

Answer A false positive would say "it's cancer" when it isn't, and a false negative would say "it's not cancer" when it is.

What determines the likelihood of Type I (false positive) error? In statistical testing of a Null Hypothesis, we set the criterion α, which is the p-value cutoff for "statistical significance". If our p-value is less than α, we say that our result is statistically significant.

For example, if we set α to be 0.05, and our actual p-value was p = 0.03, then p = 0.03 would be statistically significant. However, if we had set the α to be 0.01, a result with p = 0.03 would not be considered statistically significant.

But what is the meaning of the p-value? Recall that saying "p = 0.03" *means* "results this big or bigger happen by chance 3% of the time under the Null Hypothesis". So if we set α = 0.05 as our criterion for significance we are saying that we will accept the probability that a null finding will be called significant to be less than 5%. **But things that happen by chance 5% of the time happen by chance 5% of the time!** Therefore if we set α = 0.05, we automatically accept a 5% false positive rate.

Similarly, if we set α to be 0.01, which means we only accept results that could happen by chance less than 1% of the time, then we are allowing false results up to 1% of the time, and therefore setting the false positive rate to be 1% (Figure 11.3).

		In reality	
		Null Hypothesis *really* false	Null Hypothesis *really* true
Study result	Study rejects Null Hypothesis	**True Positive**	**False Positive** (Type I error) (probability = α)
	Study does not reject Null Hypothesis	**False Negative** (Type II error)	**True Negative** (correct conclusion) (probability = $1 - \alpha$)

Figure 11.3 Type I error and false positive rate α. Type I error occurs when the Null Hypothesis is really true, but is mistakenly rejected. The rate of Type I error is α.

The choice of α controls the false positive rate: the lower the α, the lower the false positive rate. However, there will then be more false negatives, because now there is a more stringent test for statistical significance, which can miss results that are really "there".

What determines the likelihood of Type II (false negative) error? Is there an equivalent "false negative rate", just like α, the false positive rate? Yes, it's called β! In order to see how to define β, we need to examine the concept of the **power** of a statistical test.

As we said, the power of a test is the answer to the question: **how likely** is this test to have *found* a difference **had a difference actually been there?** On the other hand, β, the false negative rate, is then **how likely** is this test to have *missed* a difference **had a difference actually been there**, which is clearly (100% – power) (Figure 11.4).

Study result		In reality: Null Hypothesis *really* false	In reality: Null Hypothesis *really* true
	Study rejects Null Hypothesis	**True Positive** (correct conclusion) (probability = $1 - \beta$)	**False Positive** (Type I error)
	Study does not reject Null Hypothesis	**False Negative** (Type II error) (probability = β)	**True Negative** (correct conclusion)

Figure 11.4 Type II error and false negative rate β. Type II error occurs when the Null Hypothesis is really false, but is mistakenly not rejected. The rate of Type II error is β.

The definition of power

Given that a difference is really there, that is, that the Null Hypothesis is false, there are two possibilities for our study result: either our study will correctly reject the Null Hypothesis (a True Positive), or our study will mistakenly fail to reject the Null Hypothesis (a False Negative). The fraction that is correctly rejected is the power of the study.

$$\text{power} = \frac{\text{TP}}{\text{TP} + \text{FN}} = 1 - \beta$$

How to determine the power of an MRI machine

Let's consider the example of the MRI machine. How do radiologists calculate the power of an MRI machine? The answer is, that since power is the ability to find a difference that is really there, we have to create some examples where a difference really is there and see if the MRI can spot it.

For this purpose, radiologists use what they call "phantoms", which are manufactured objects with gridlines of various known sizes. They put the phantom under the MRI and simply ask: what is the smallest spacing where we can actually detect two lines instead of one? That's the power of our MRI.

The power of an MRI machine is determined by using phantoms

The figure below shows an MRI scan of a series of phantoms for testing the power of MRI machines.

The phantom has panels of vertical lines, with indicated spacings in mm. We can see that the MRI easily distinguishes the bottom two rows, with the widest spacings. Even in the second row, the machine can distinguish the separate lines, but in the first row, the machine cannot distinguish the separate lines that are spaced less than 0.3 mm.

We have therefore determined that the power of this MRI machine is about 0.3 mm.

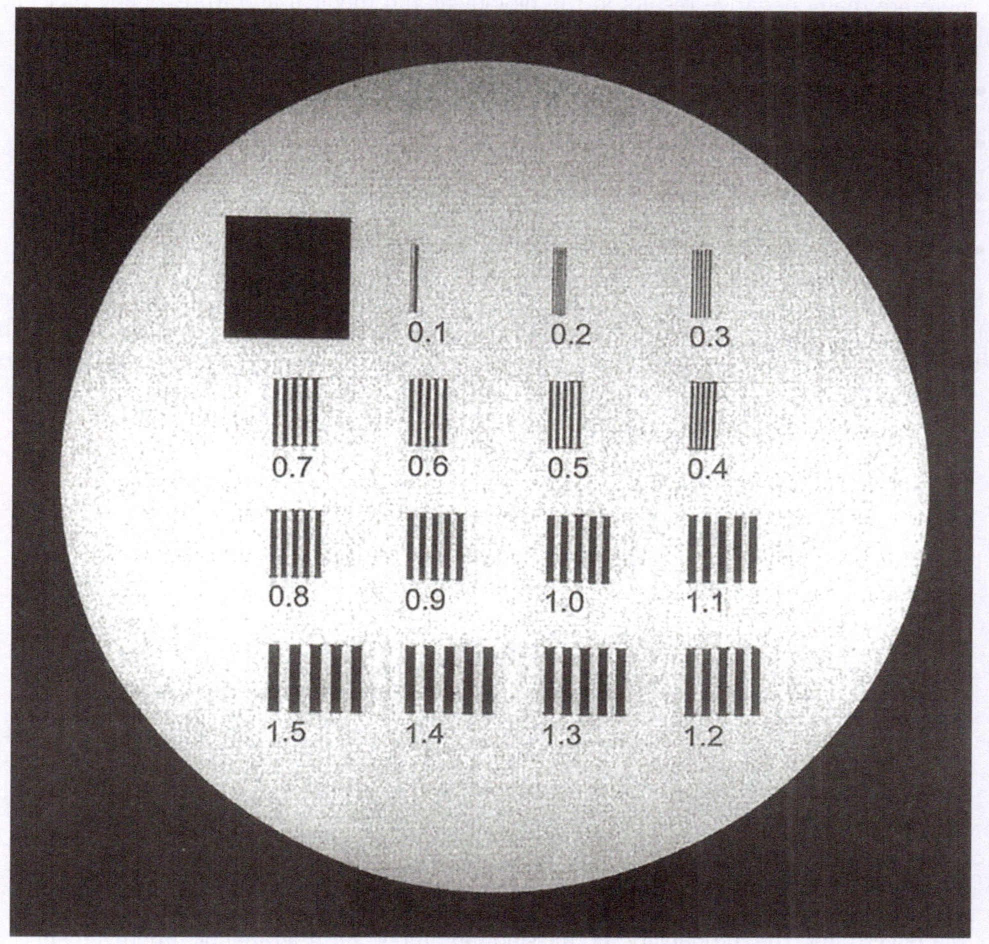

Power is the power to find a *given* difference

When we talk about power, it's not the power of a study design or a test, it's the power of a study design or a test *to find a given difference*. Looking at the MRI phantoms, we see that the machine has a high power to resolve differences of 1.0 mm; if many MRI machines looked at the 1.0 mm spacing, and we forced an answer to the yes/no question "is there a difference?", close to 100% of them would be able to detect the difference. But for the 0.2 mm spacing, maybe 10% of them would correctly call a difference. So the power of the MRI machine to resolve the 0.2 mm spacing would be around 10%.

Why is power important?

Power in negative findings If a study has low power, then there is a big chance of a false negative, missing something that is really there. For example, the med student's response to the attending's question "rule out viral infection" is a false negative due to low power. The evil professor's non-recommendation is a false negative due to low power. So we have to beware negative findings from low-power studies: they could well be false negatives.

There are many negative results in the biological and medical literature. They typically are found in claims like:

- "X is not associated with Y"
- "X is independent of Y"
- "There is no relation between X and Y"
- "There is no *significant* relation between X and Y"

Negative results from low power studies are a major problem. There is a famous paper, published in *The New England Journal of Medicine* in 1978, entitled "The importance of beta, the type II error and sample size in the design and interpretation of the randomized control trial".

SPECIAL ARTICLE

The Importance of Beta, the Type II Error and Sample Size in the Design and Interpretation of the Randomized Control Trial

Survey of 71 "Negative" Trials

Jennie A. Freiman, A.B. Thomas C. Chalmers, M.D.
Harry Smith, Jr., Ph.D. Roy R. Kuebler, Ph.D.

Abstract Seventy-one "negative" randomized control trials were re-examined to determine if the investigators had studied large enough samples to give a high probability (>0.90) of detecting a 25 per cent and 50 per cent therapeutic improvement in the response. Sixty-seven of the trials had a greater than 10 per cent risk of missing a true 25 per cent therapeutic improvement, and with the same risk, 50 of the trials could have missed a 50 per cent improvement. Estimates of 90 per cent confidence intervals for the true improvement in each trial showed that in 57 of these "negative" trials, a potential 25 per cent improvement was possible, and 34 of the trials showed a potential 50 per cent improvement. Many of the therapies labeled as "no different from control" in trials using inadequate samples have not received a fair test. Concern for the probability of missing an important therapeutic improvement because of small sample sizes deserves more attention in the planning of clinical trials.

The NEW ENGLAND JOURNAL OF MEDICINE 1978

They surveyed 71 trials that drew a negative conclusion, and found that 57 of these negative trials had such low power that they would have missed a 25% improvement, and 34 of those 71 had such low power that they could not have found even a 50% improvement. So almost half of the published "negative" findings could not have found a 50% improvement.

The conclusion of the paper is that 'many of the therapies labeled as "no different from control" in trials using inadequate samples have not received a fair test.'

More recent studies have followed up on the conclusions of this paper, and found a continuing and troubling pattern of negative claims from low power studies.

Power in positive findings Power is necessary in claiming negative findings, but as we will see in section 11.5 (*Role of power in positive results*), power is also important in supporting positive findings: a low power study drawing a positive conclusion has low trustworthiness.

11.2 How to calculate the power of a study?

To calculate the power of a statistical study, we have to basically replicate all the conditions of the radiological phantoms. We have to create a phantom world with *a given difference*, and then see if our study design can find the difference that we know is there because we put it there.

Imagine a "mini-me": a model of our study with the same n, same methods, same statistical measures etc. (Figure 11.5).

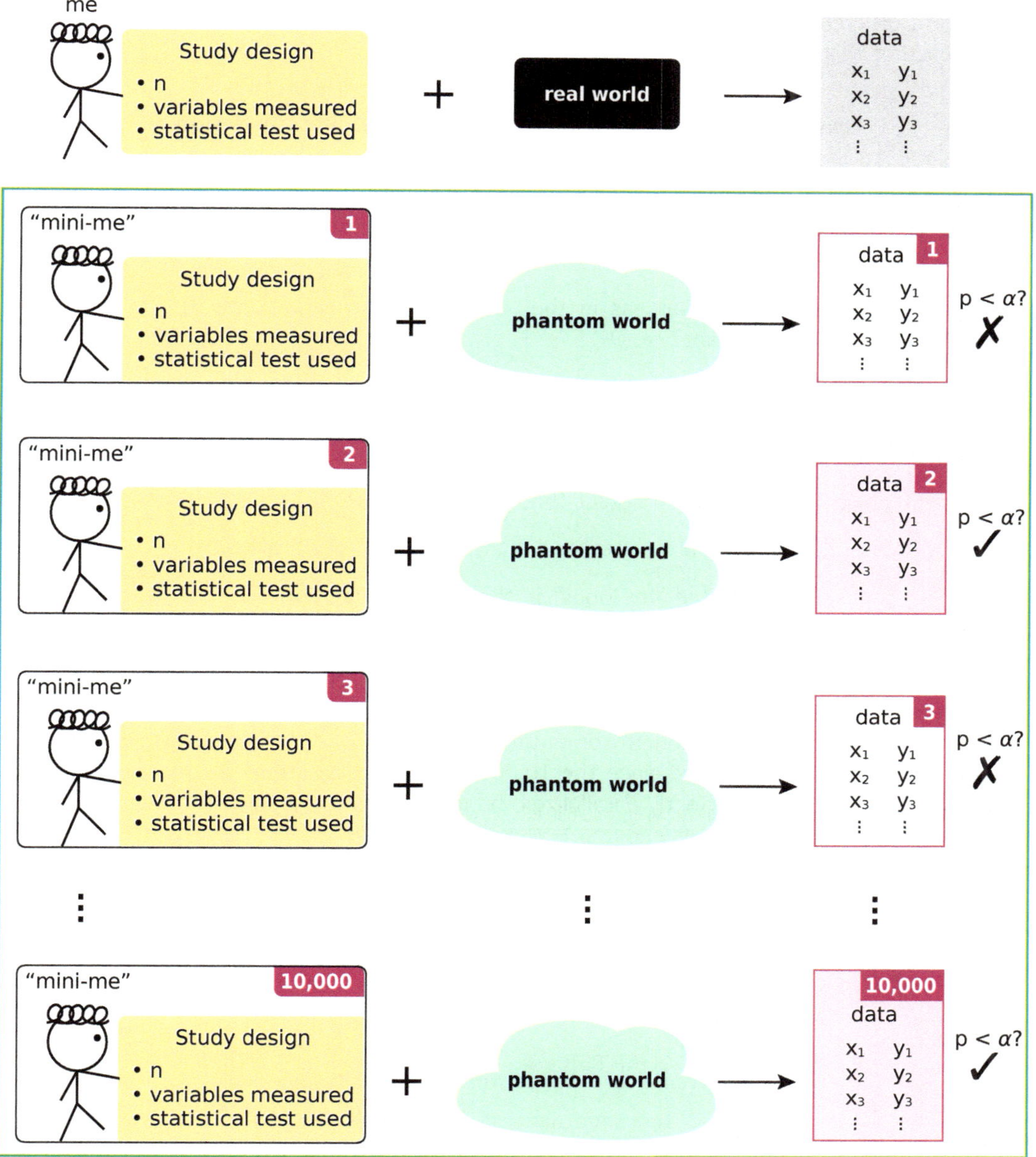

Figure 11.5 Calculating the power of a study using a phantom world. First, we decide on the phantom world and the difference we hope to detect. Then we throw the "mini-me" at the phantom world many times and record how often the "mini-me" detected a significant difference.

A "mini-me" is a simu-study. A phantom world is a simulated environment where we create an effect size and data variability. By replicating the study design in this controlled setting, we can see if the study can reliably detect the effect.

Each time, the mini-me goes fresh into the phantom world, and uses the same study design to generate a data set, then uses the same statistical methods to determine statistical significance. We write down the number of times the mini-me found a significant difference. If the mini-me found a significant difference x% of the time, then the power of this study to detect *this difference* with that amount of within-group variation is x%. For example, if the mini-me found a significant difference 85% of the time, then the power of that study to find this difference, in the presence of this much variability, is 85%.

Using simulations to calculate the power of a study design

Step 1 ► Create a phantom world and give the difference we hope to detect.

Step 2 ► Randomly sample from the phantom world exactly as the actual study would have done, generating simu-study #1.

Step 3 ► Apply the same measure as in the actual study to simu-study #1. Is the result statistically significant?

Step 4 ► Repeat steps 2 and 3 for a total of 10,000 times, generating simu-studies #1, #2, ..., #10,000. How many times was a significant result found? Divide that number by 10,000.

That is the power of the study to detect that given difference.

Nested "for loops" in the power calculation

Notice that, when we use a resampling-based test for significance, then we will have nested "for loops", because we need to have 10,000 mini-mes, and each mini-me has to do 10,000 simulations to determine statistical significance. So this will involve 10,000 × 10,000, or 100 million simulations. But with modern computation, this is easily done. It would take a modern laptop a few hours or less. And since the 10,000 mini-me calculations are independent, the power calculation can be perfectly parallelized, bring the total time down to minutes.

```
for j in range(num_mini_mes)
        # Generate jth simu-study
        ......
        ......
        ......
        # Calculate p-value for jth simu-study
        for i in range(num_simulations)
                ......
                ......
```

In theoretical statistics, power is a fearsome subject and takes up many pages. In a traditional statistics textbook, there are separate chapters on calculating the power of a two group analysis, calculating the power of a regression, calculating the power of a χ^2 test, etc. In each case, radical assumptions must be made of Normal distributions, and means and standard deviations, in order to derive any formulaic results about power.

By contrast, the resampling-based method gives us one procedure to calculate the power of any methodology whatsoever.

What is a good "given difference"? Recall the example of testing the power of an MRI machine using radiological phantoms: radiologists created markings that were separated by various given spacings, in order to see which spacings the MRI could resolve (see page 548). Similarly, to test the power of a study, we need to create groups that have given across-group differences. Where do we get these across-group differences? One place is the literature: how much of a change could we expect with drugs of this type? Another source of across-group differences is preliminary pilot experiments.

The power of a study is determined by using a phantom world Here we illustrate the procedure of generating a phantom world with a 2-group comparison.

Creating a phantom world for a 2-group comparison

We create a phantom world by specifying:

- **Across-group difference.** The first decision we have to make is: how small a difference are we hoping to be able to detect? If we're expecting to see huge effects, then pretty much any study size will do to find a significant effect. If you think that all of the subjects who take the drug will live, and all that don't will die, then there's no point in having $n = 100$ patients. A much smaller study would suffice to establish a significant difference. So first we have to answer: how big an effect size are we expecting?
- **Within-group difference.** Once we have specified an effect size, an across-group difference, we then have to specify how much within-group variability we expect to see. After all, if within-group variability is small, it is easy to detect across-group differences.

Study design After creating the phantom world, the next step is to specify the structure of the simu-study, and specify what exactly the "mini-me" is going to do. First, we need to specify the sample size n, that is, how many subjects the "mini-me" will draw for each group. Then we have to specify the statistical test we will use to determine statistical significance. Let's say we decide to do a median-based test, and will use a Re-centered Two-Box resampling method to assess the p-value, with a p-value of $p < 0.01$ being considered statistically significant.

To illustrate, let's say we are testing an anti-cancer drug. We are going to get n mice in the control group and n in the drug group. All of the mice will be given the relevant form of cancer, and we will use "days of survival" as our outcome measure.

Now we have to make two decisions: how much of an across-group difference we are hoping to see, and how much within-group variation we might see. Where could we possibly get those numbers? The answer is: we have to make them up! (Figure 11.6).

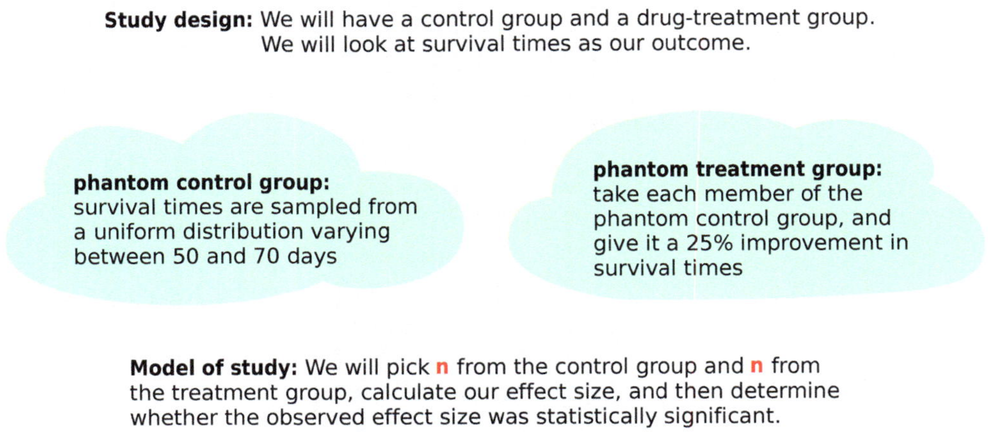

Study design: We will have a control group and a drug-treatment group.
We will look at survival times as our outcome.

phantom control group:
survival times are sampled from a uniform distribution varying between 50 and 70 days

phantom treatment group:
take each member of the phantom control group, and give it a 25% improvement in survival times

Model of study: We will pick **n** from the control group and **n** from the treatment group, calculate our effect size, and then determine whether the observed effect size was statistically significant.

Figure 11.6 Structure of a simu-study.

Phantom control group We might be guided by previous studies. With this type of cancer, how long does the typical mouse live? That kind of data can often be found in the literature. If not, then we make an educated guess. Let's say we estimate that the survival time of mice without the drug will be around 60 days, plus or minus 10 days. That means we have n phantom mice in the control group, and for each mouse, we pick a number of days of survival by sampling randomly (with replacement) from the set {50, 51, 52,..., 70}. This is our phantom control group.

Phantom treatment group Then we have to make our phantom treatment group, and for this, we take the control group and decide how big a difference we are going to see. Let's say we are hoping to get a 25% improvement due to the drug. Then for our treatment group, we take the control group data, and multiply each data point by 1.25. This is our phantom treatment group.

Calculating power

Power **We test the power of this study design by applying the study to the phantom world 10,000 times, and seeing how often the study design found and declared significant the 25% difference that we put there. That number divided by 10,000 is the power of this study design to detect a 25% increase** (Figure 11.7).

False negative rate β We saw that choosing a value for α, the standard for statistical significance, is equivalent to accepting a false positive rate of α. And we then asked whether there was an equivalent false negative rate. We can now define β, the false negative rate, as (1 − power), assuming power is being put as a fraction.

$$\text{false negative rate} \qquad \beta = 1 - \text{power}$$

So if a power test found that 82% of the simu-studies found significance, that gives us a power of 0.82, and a false negative rate = β = (1 − 0.82) = 0.18.

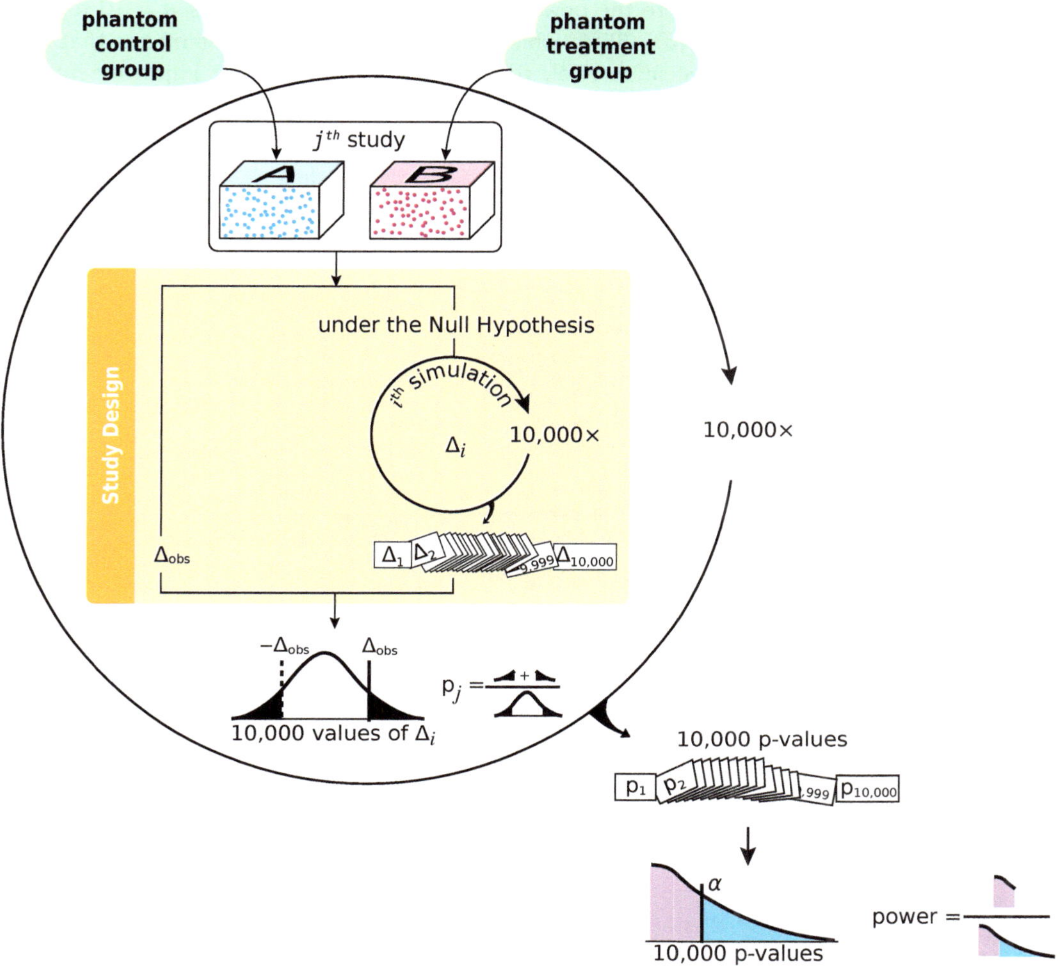

Figure 11.7 How to calculate the power of a study design to detect a predefined difference. Here we use a 2-group comparison to illustrate.

Phantoms in action

The anti-cancer drug study Let's go back to our study in mice of an anti-cancer drug: we postulated that the phantom control group will have survival times varying uniformly between 50 and 70 days, and that the phantom treatment group will have a 25% improvement. Note that we have not yet mentioned the size of the study.

Now we need to determine how large a study we propose to do: how many mice will we have in each group?

10 mice? For our first study design, we chose an $n = 10$ (that is, 10 mice) in each group, and we chose for our summary measure the median of each group, so the effect size is the median of the experimental group minus the median of the control group. As our statistical test, we used a Re-centered Two-Box resampling method.

When we applied this study design to the phantom world 10,000 times, we found that there were 3518 simu-studies, 3518 mini-me simulations, that found statistical significance at the p

< 0.01 level (Figure 11.8). **Therefore the power of this study design, with its $n = 10$ for each group, to detect a 25% increase is $\approx$ 35%, which is not very impressive. It means that even when there is a 25% improvement in survival, this study design found that improvement only 35% of the time.**

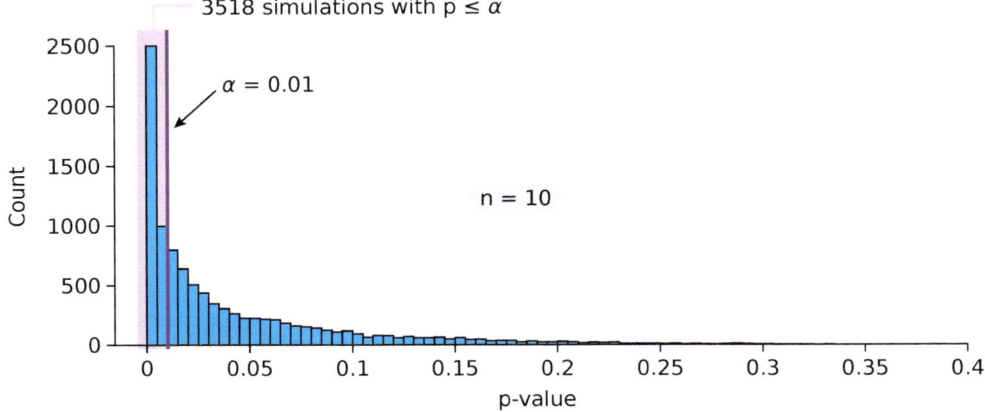

Figure 11.8 Power of the study when $n = 10$.

A bigger study Then, for comparison, we considered a bigger study, that had $n = 25$ animals in each group. Now, when we computed the power, we found that 8230 of our 10,000 trials found significance at $p < 0.01$, so the power of *this* study design is a much better 82% (Figure 11.9).

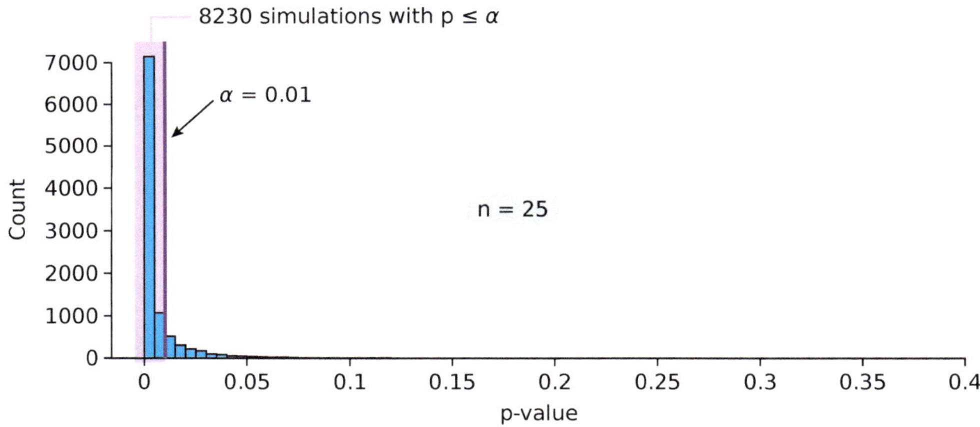

Figure 11.9 Power of the study when $n = 25$.

We see that sample size greatly affects the power of the study design. But sample size is not the only factor that we need to consider.

Exercise 11.2.1 For the small study design, we chose the sample size $n = 10$, and calculated the power of the study to be around 35% (Figure 11.8). How would the power of the study change if we were to increase α from 0.01 to 0.05? 0.2?

Factors affecting the power of a study

As we design a study, we obviously want it to have a good chance of success, that is, we want it have sufficient power. There are a number of ways to do this: sample size, across-group difference, within-group differences and study design (Figure 11.10).

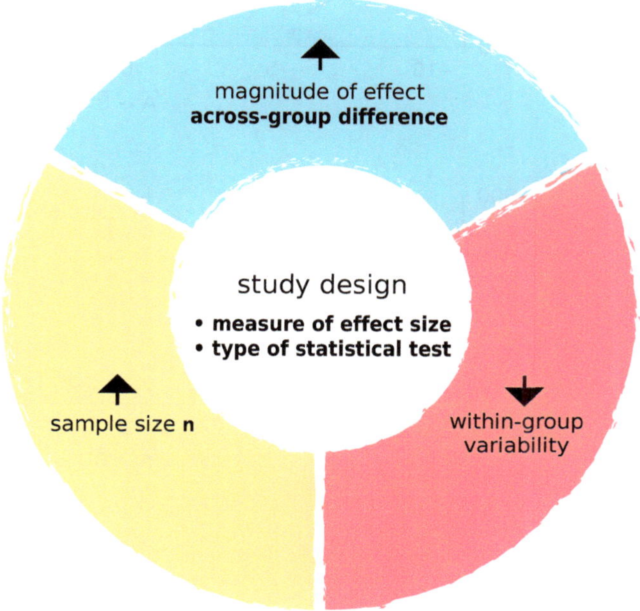

Figure 11.10 Factors that influence the power of a study.

Sample size As our example above makes clear, sample size is a large determinant of statistical power. In the eyes of many, it is *the* principal contributor to power. The reason for this is that the larger the sample size, the more likely it is to find a significant difference.

As a further example of the importance of sample size for finding statistical significance, we created a data set that had two samples drawn from Normal distributions (Figure 11.11). Group A was drawn from the Normal distribution of Mean = 16 and Standard Deviation = 5 and group B was produced by adding 4 to every data point in group A.

When we set our $n = 7$, the Big Box resampling test on the difference in the means was not significant. The p-value was 0.1691. When we took $n = 12$ in each group, the Big Box resampling returned a p-value of 0.0518, and when we took $n = 23$ in each group, the p-value was 0.0148. If the sample size n increases to 50, the p-value becomes < 0.0001.

Sample size has a major effect on power, by making it easier to detect statistical significance in each simu-study, hence increasing power.

Effect size The bigger the effect size, the easier it is to find a significant difference, which translates to greater power. Thus, if our study power is low, one way we can increase its power is by trying to get a bigger effect size, for example, by giving larger doses of the drug. Of course this can have its own drawbacks.

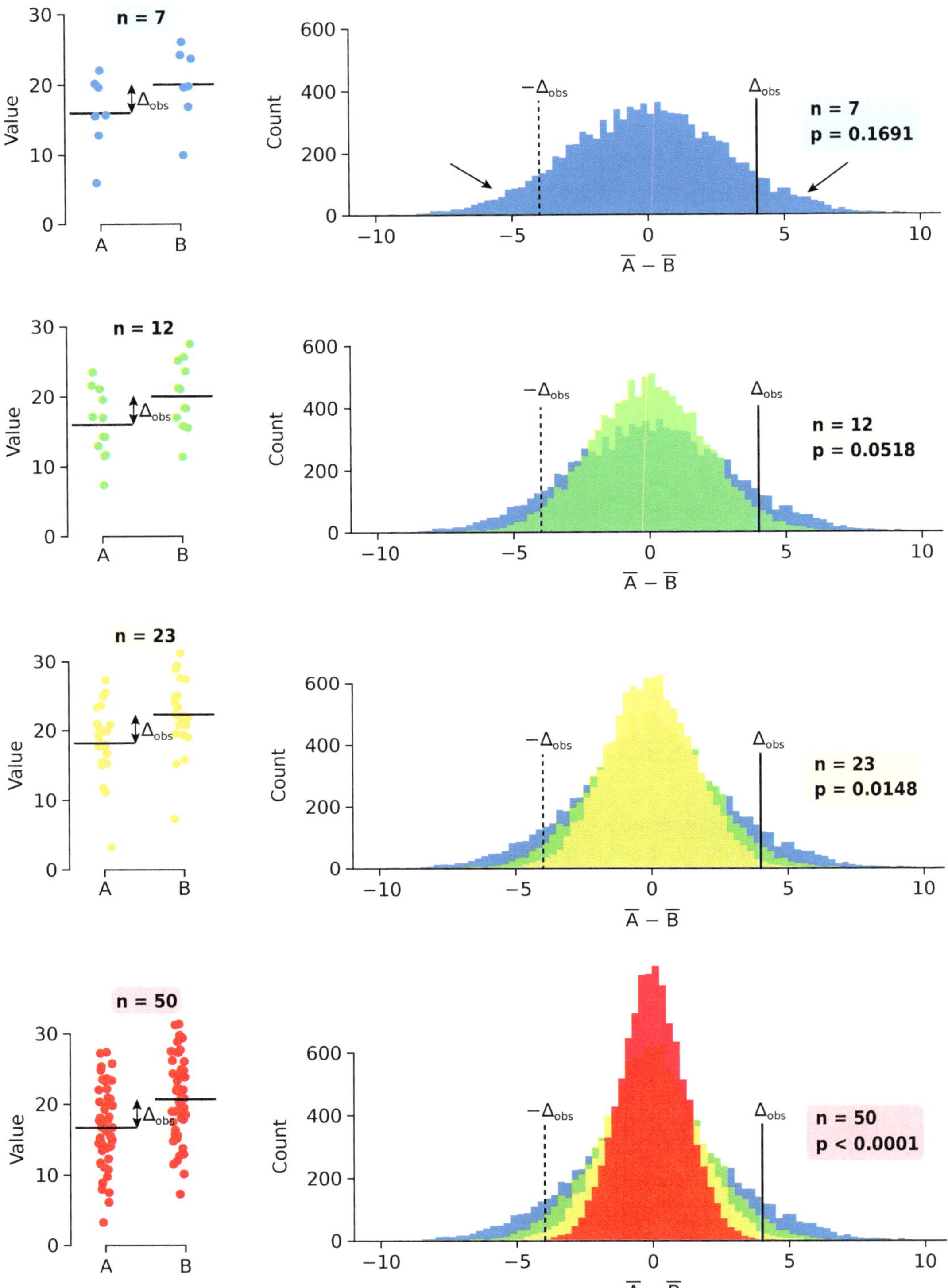

Figure 11.11 The impact of sample size on statistical significance. The histograms show the null distributions for the various data sets. The larger the sample size, the tighter the null distribution and the higher the chance of finding statistical significance in each simu-study, hence raising the power of the study.

Within-group variation The smaller the within-group variation, the easier it is to find statistical significance. Remember that, in general, statistical significance is the battle between across-group differences and within-group differences. If across-group differences are much bigger than within-group differences, the across-group differences are likely to be statistically significant (Figure 11.12).

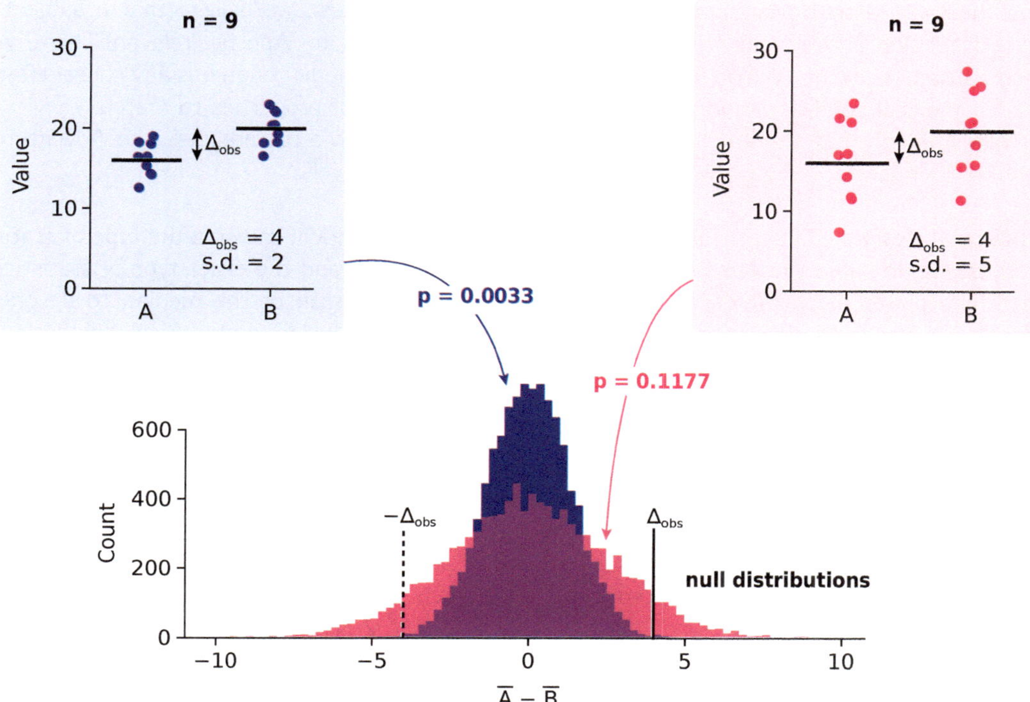

Figure 11.12 The impact of within-group variation on statistical significance. For a given across-group difference, the smaller the within-group variation, the tighter the null distribution and the higher the chance to find statistical significance in each simu-study, hence raising the power of the study.

But how to keep within-group variation small? This is difficult. In the past, many study designers tried to limit within-group variation by making their study populations as homogeneous as possible.

In the worst examples, which happened frequently, study designers excluded women from the study! This is a terrible idea, since, obviously, it means that your results are only applicable to half the population. What's more, the drug in question could have distinct effects in women vs. men (this is common), and then your study would be absolutely invalid for, and wrongly applied to, women. But these study designers made a devil's calculation, that malpractice against women was a small price to pay for getting a good p-value, and with it, publication, funding and promotion.

One ethical method of reducing within-group variation is to create a paired methodology. Remember that if we have a truly paired structure, then we can use the δs, the differences between the two members of each pair. We can try to capture this by finding for each subject, another subject who is closely matched in relevant properties, such as age, gender, cholesterol level, etc., and then using a paired design.

But where did we get the idea that age, gender, and cholesterol level are the "relevant" matching properties? We must be thinking about some kind of medical outcome, like blood pressure or cardiovascular events. What about astrological signs? Should the matched subject have the same astrological sign? We want to say "No", because we believe that a subject's astrological sign is irrelevant to the medical outcome in question. And by "relevant" here, we mean *causally* relevant. We do not believe that a subject's astrological sign has any causal effect on their medical condition, and so, we exclude it from the list of properties to match.

Now imagine that the study was actually to determine movie preferences, like Adventure, Comedy, Crime, etc., why we would try to match cholesterol levels?

Statistical design The last factor that strongly impacts statistical power is the type of statistical test we do, which includes both the measure of effect size and the exact type of statistical test. The power of a test depends on whether we use the mean or the median to describe the groups, and depends on the choice of α, the standard for declaring statistical significance, among other factors. It also depends on whether we test for significance by formula-based tests versus resampling-based tests.

Exercise 11.2.2 The paper of Drummond and Vowler (2012) "Not different is not the same as the same: How can we tell?" has a great graph showing the various factors that affect the power of a test.

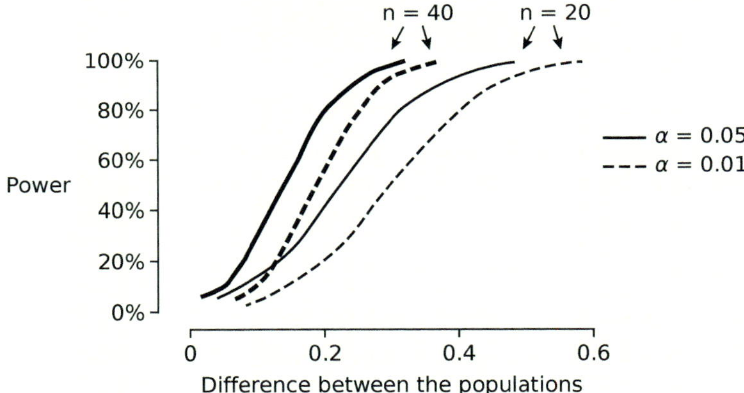

a) What are the factors illustrated in the graph that affect the power of a statistical test?

b) Use the graph to demonstrate how each of the above factors affects the power of a statistical test.

c) What factors are not mentioned in the graph that also significantly affect the power of a statistical test.

d) If the difference between the populations is around 0.2, what sample size would you choose? Would it help to conduct a bigger study?

e) If the difference between the populations is around 0.5, what sample size would you choose? Does it make sense to conduct a bigger study?

Exercise 11.2.3 Below are random samples from populations under two different treatment conditions, Hot and Cold. In the study on the left (black dots) the difference in means is bigger than the difference in means in the study on the right (green dots). And yet the smaller difference was found statistically significant, while the larger one was not. Explain how this is possible.

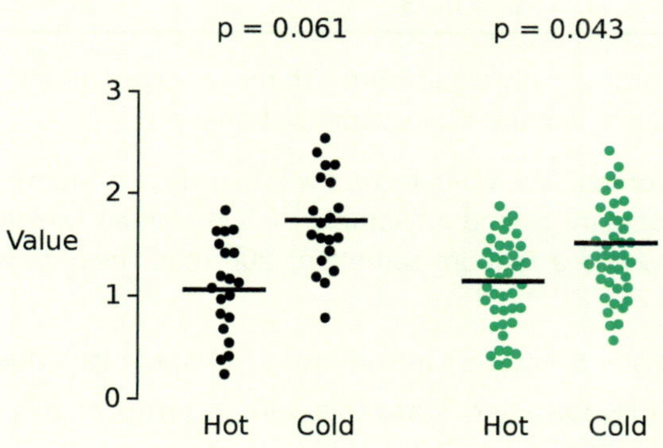

FURTHER EXERCISES 11.2

1. Select all changes that would increase the power of a statistical test:

   a) Increase sample size
   b) Increase α from 0.01 to 0.05
   c) Increase within-group variation
   d) Increase effect size (across-group variation)
   e) Use a paired design
   f) Increase Type I error
   g) Increase variability of the data
   h) Decrease Type II error

2. A recent study with 30 participants reported that drinking tea doesn't affect stress levels, while another study with 300 participants reported that drinking coffee doesn't affect stress levels. Assume that the studies were performed similarly. Which are you more likely to believe? Explain your answer using statistical concepts.

3. **Power study for wetlands and plant species richness.** In FE 6.5.4 on page 225, we found no statistically significant difference between ephemeral and permanent wetlands for plant species richness ($p \approx 0.23$) using $\alpha = 0.01$.

   a. Calculate the power for this analysis to detect the observed difference.
   b. Should you report "there is no difference between ephemeral and permanent wetlands in plant species richness"?
   c. What are two possible ways you could modify this study design to increase its statistical power?

4. **Why a different result for macroinvertebrates?** In FE 6.5.4 on page 225, we did not detect a statistically significant difference between ephemeral and permanent wetlands for plant species richness, but the same analysis of macroinvertebrate species richness for the exact same wetlands finds a statistically significant difference as shown below.

	Plant species richness			Macroinvertebrates		
	median	MAD	sample size	median	MAD	sample size
ephemeral	30	8	33	26	3	33
permanent	35.5	7.5	24	32	2	24
		p = 0.23			p = 0.0033	

Why did we detect a statistically significant difference for the macroinvertebrate data set, but not the plant data set for the exact same wetlands?

5. **Heights of men and women.** We want to test whether men or women are taller in a certain town. We want to compare a random sample of 200 women (heights with median 172.1 cm and MAD 4.6 cm) and a random sample of 200 men (heights with median 175.1 cm and MAD 11.3 cm).

   a. What is the statistic of interest in this study? What is its value in this sample?
   b. What is the Null Hypothesis? State this both in terms of this study and in terms of the statistic of interest.
   c. We conduct NHST using 10,000 simulations. Which of the three distributions below should the NHST produce as a null distribution?

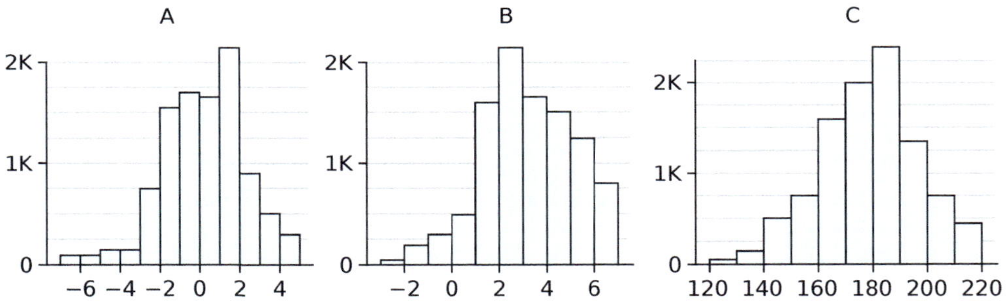

   d. Label the axes for the null distribution.
   e. Circle the regions of the distribution that we should use to calculate the p-value.
   f. Using the information you're given, calculate the approximate p-value.
   g. Using your calculated p-value, could you make the claim that "there is a difference between the heights of men and women" or "there is no difference between the heights of men and women?" Explain your reasoning. What additional information would you need to make that statement?
   h. You have been told you should do a power analysis, so you plan out how you're going to do this using simulations. Break down the power analysis into steps with short sentences for each. Use the observed data (including effect size and sample size) in this study.

6. **Tailgating.** When you analyzed the effects of several drugs on tailgating distance, you found no significant effect at the $\alpha = 0.01$ level (p = 0.157). The groups investigated are No Drug (N), Alcohol (A), Marijuana (M), and Ecstasy (E).

a. If you find that the statistical power of the study is 0.2, how should you interpret the results of your study?

b. How could you change the design for a follow-up study to increase statistical power? Describe one way that you can ethically use and one way that could work but should be avoided.

c. Your analysis of the re-designed study produces the visualization below and finds a statistically significant effect of drug type on tailgating distance (p = 0.002) with statistical power of 90%. Can you conclude that drug type affects tailgating distance? Explain your reasoning.

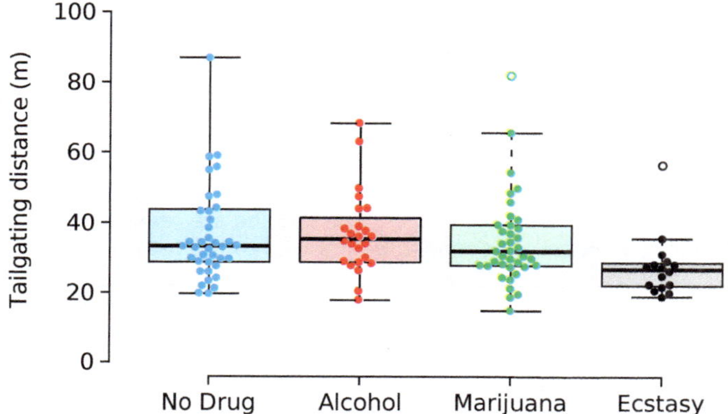

d. You compare tailgating distance for each pair of treatments and calculate p-values:

	N vs. A	N vs. M	N vs. E	A vs. M	A vs. E	M vs. E
p-value	0.45	0.32	0.0001	0.01	0.02	0.003

Plug the p-values into the Benjamini-Hochberg multiple testing correction to determine statistical significance. Can you conclude that Ecstasy users have a statistically significantly lower tailgating distance than the other groups?

11.3 A low-power study wrongly drawing a negative conclusion

As we saw in the beginning of this chapter, it is unethical to draw a negative conclusion from a low power study. We will now consider a historical case of this error.

Was Bendectin use by pregnant women associated with an increased risk of birth defects?

Bendectin was a drug given to pregnant women to treat pregnancy-related nausea ("morning sickness"). There was some concern about the safety of the drug, and so a study was done to determine whether Bendectin use was associated with an increased risk of birth defects. The primary paper reporting the outcomes was published in the *American Journal of Obstetrics and Gynecology* in 1977 (Shapiro *et al.*, 1977). So this was a paper written by epidemiologists for practicing OBG doctors. This is important, because the audience would not be statistical experts; they are practicing doctors who cannot critique the statistical methods that were used, and must rely on the epidemiologists for their take-home conclusions.

What did the epidemiologists say? The take-home conclusions that the epidemiologists made were quite striking: they very strongly and repeatedly stated the negative conclusion that the drug "does not appear to be harmful to the fetus". They repeated this three times in their paper:

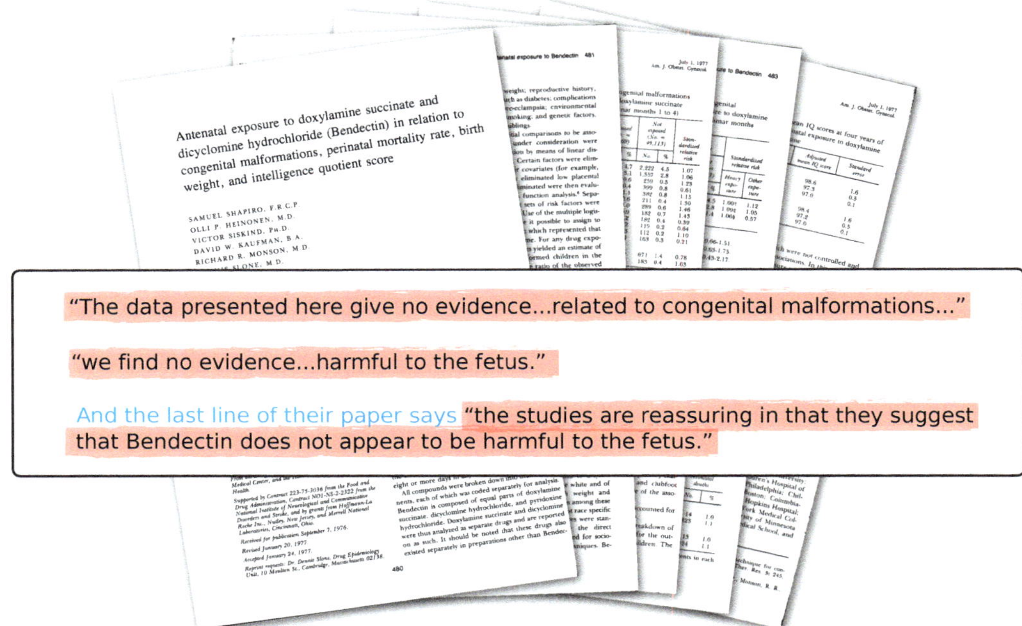

An OBG doc would therefore feel justified in prescribing the drug, given these findings.

Analysis of the Bendectin study In their paper, the authors state that their statistical method was "multiple logistic risk function analysis", run on a mainframe computer. This is an example of something we have seen in previous chapters: a highly technical method that has many hidden assumptions and many drawbacks.

Because we did not want to make these assumptions and accept the heavy drawbacks, we used a straightforward "relative risk" analysis, as defined in Chapter 8 (*Independence, Proportions, and Relative Risk*).

Here, we will focus specifically on "Musculoskeletal Malformations" as reported in their study. We chose this category because there were a number of articles in the popular press, and a number of lawsuits, charging that the drug had caused limb reduction disorder, a birth defect in which children are born with shortened or missing limbs.

The relevant data from their study is given below (Figure 11.13).

		outcome?	
		Musculoskeletal Malformations	No Musculoskeletal Malformations
took the drug?	Bendectin	a 13	b 1,156
	No Bendectin	c 382	d 48,731

Figure 11.13 The data from the Bendectin study (Shapiro *et al.*, 1977), here presented in a 2 × 2 design.

The observed relative risk Let's calculate the relative risk of Musculoskeletal Malformations for children of mothers who took Bendectin compared to those who did not.

First we calculate the risk with Bendectin

$$\text{risk with Bendectin} = \frac{a}{a+b} = \frac{13}{13+1156} = 0.011$$

and the risk without Bendectin

$$\text{risk without Bendectin} = \frac{c}{c+d} = \frac{382}{382+48731} = 0.0078$$

Therefore, the relative risk is

$$\text{relative risk} = \frac{\text{risk with Bendectin}}{\text{risk without Bendectin}} = \frac{\frac{a}{a+b}}{\frac{c}{c+d}} = 1.43$$

So the observed relative risk is

$$RR_{obs} = 1.43$$

In other words, in this study, *women who took the drug were 43% more likely to have an infant with a Musculoskeletal Malformation.*

Is the observed result significant? Is the observed relative risk statistically significant? The Null Hypothesis is that Bendectin has no effect on the risk of birth defects. Under the Null Hypothesis, we would ignore the group identifiers, mothers who took Bendectin and those who didn't, to create the Big Deck: 13 + 382 cards of "Musculoskeletal Malformation" and 1156 + 48731 cards of "no Musculoskeletal Malformation". Then, in the null world, we shuffle the Big Deck and deal the shuffled deck out into two piles, the "drug" pile ($n = 1169$) and the "No drug" pile ($n = 49{,}113$). Then we count the number of "Musculoskeletal Malformation" cards and "no Musculoskeletal Malformation" cards to construct a 2 × 2 table, and calculate the relative risk for that shuffled sample (Figure 11.14).

Then we repeat the "shuffle, deal and count each pile" procedure 10,000 times, generating 10,000 values for the relative risk under the Null Hypothesis, which forms the null distribution (Figure 11.15). (see Figures 8.56 and 8.57 for another illustration of how the "shuffle and deal, then count each pile" procedure is used to conduct NHST for relative risk.)

Because relative risk is a ratio, in order to do a two-sided test, we have to look at results equally or more extreme than RR_{obs} or $1/RR_{obs}$. From this we conclude that the p-value for our observed result of relative risk of 1.43 is 0.289.

Therefore, if we adopt a "$p < 0.05$" standard for statistical significance, this result would be declared "not statistically significant".

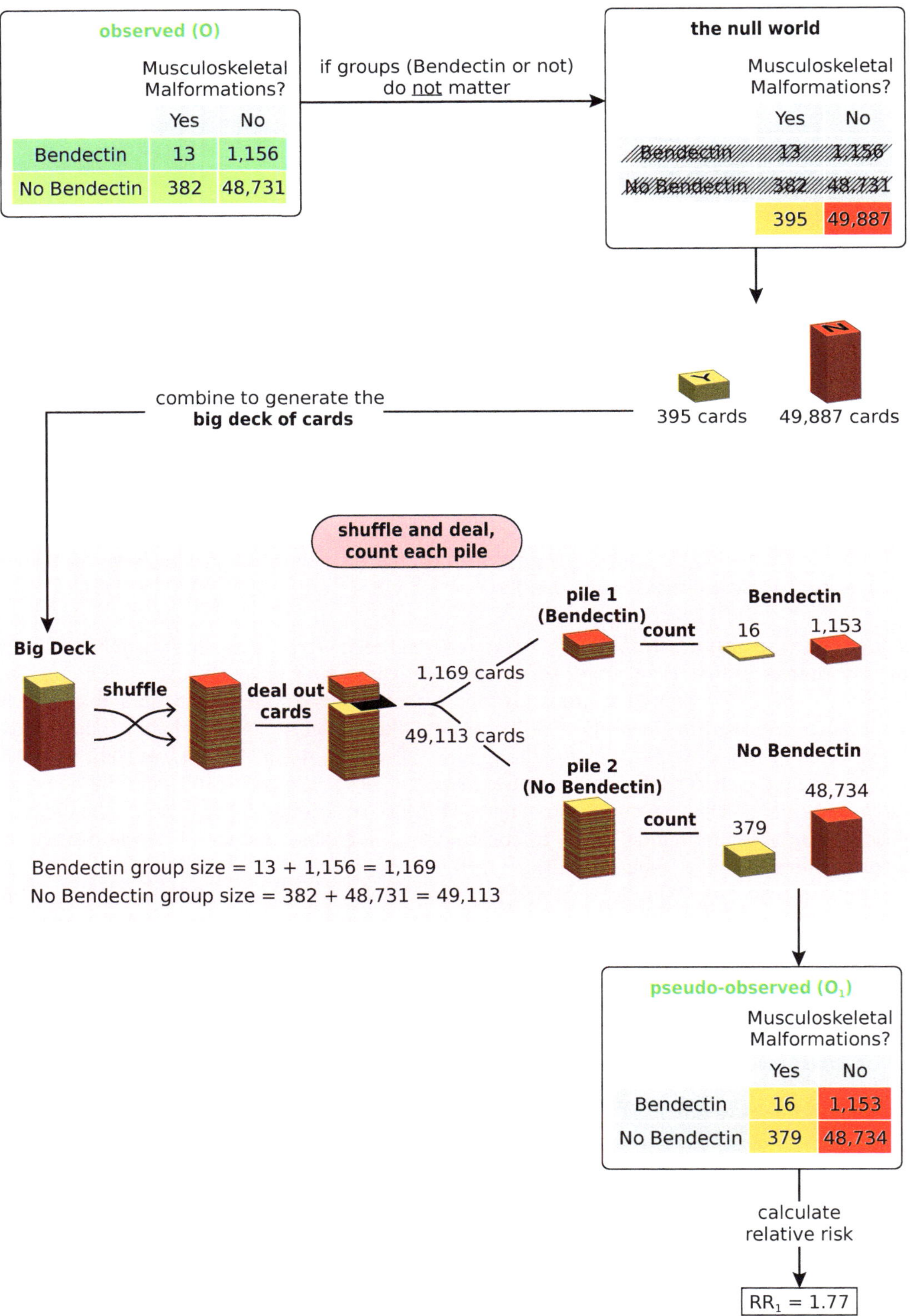

Figure 11.14 Generating one estimate of the relative risk under the Null Hypothesis.

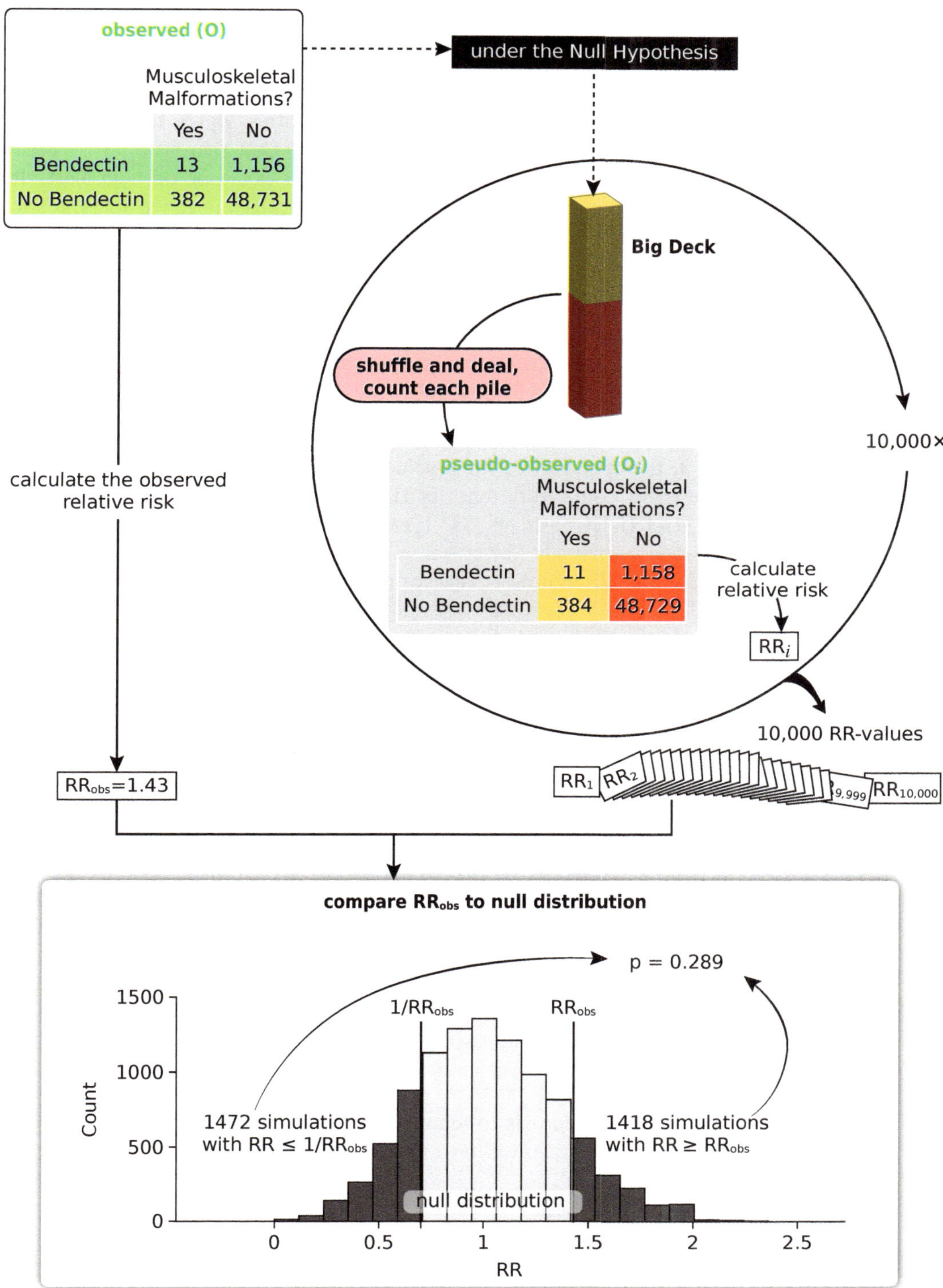

Figure 11.15 NHST for relative risk. Distribution under the Null Hypothesis of the relative risk of taking Bendectin for the outcome of having Musculoskeletal Malformations. The p-value of the observed relative risk RR_{obs} is the fraction of the 10,000 simulations that generated an RR-value more extreme than either RR_{obs} or $1/RR_{obs}$.

Is this cause for concern? We see two causes for concern.

1. First, the raw risk of having a Musculoskeletal Malformation is 43% higher in the drug-yes group. You would *certainly* want to know if a drug you are taking has a 43% increase in the likelihood of these serious birth defects! In cases like this, where the stakes are so high, and a back-of-the-envelope calculation, as we did, says that there is a 43% increase in risk, if you were at all concerned about the patients, you would be on high alert. It's extremely important to distinguish between scientific or medical significance and statistical significance. A 43% increase in risk of a serious birth defect is a medically significant risk.

2. The second cause for concern is the fact that the authors have drawn a negative conclusion from their failure to find a statistically significant relation. Our calculation of the p-value, using relative risk, was p = 0.289. The relatively high p-value is due to the small sample size in the Bendectin group. The p-value is certainly larger than 0.01 or even 0.05, and therefore fails the standard "$p < 0.05$" criterion for statistical significance.

But is this the right criterion?

As we saw in Chapter 3, the choice of α, the p-value cutoff for "statistical significance", is a *social policy decision*. It doesn't come from math or statistics, it comes from our valuation of the social costs of Type I error (false positives) vs. Type II error (false negatives). For example, suppose that we were studying a socially low-stakes question, like whether left-handed people have more musical skills. Here, there are no catastrophic social costs associated with either Type I or Type II error. The cost of a Type I error is that a false result will be enshrined into the literature, while the cost of a Type II error is the failure to publish a real positive finding. The scientific community holds that, in this case, Type I error is worse than Type II error: we would much rather *not* publish a borderline study than publish a false positive. So we set α low, and require strong evidence to publish.

Lowering α, that is, raising our standards, will result in fewer false positives, but at the cost of increasing the number of false negatives. Higher standards means that we will miss more true results.

But when the case in question is whether a drug causes birth defects, then the situation is, or should be, reversed: it is much better socially to publish a warning, risking a false positive, and much worse to fail to warn patients of a danger, risking a false negative. **Therefore, in this kind of case, the alpha value should be set much higher, say 0.3 or even 0.5**.

In the engineering literature, Type I error (calling a good product bad) is called "Producer's Risk", and Type II error (calling a bad product good) is called "Consumer's Risk". By choosing $\alpha = 0.05$, the authors have covertly decided that Producer's Risk is much more important to them than Consumer's Risk.

Therefore what the authors have done is to report a negative conclusion because they failed to find a statistically significant result. So we have to ask: what was the power of this study?

The power of the Bendectin study

Because the study makes a strong negative conclusion, and no power estimates were supplied, we set out to calculate the power of their study.

However, as we said, we do not accept the assumptions and drawbacks of the authors' "multiple logistic risk function analysis" of the "standardized relative risk". Instead, we treated the problem as a 2 × 2 contingency table (as in Figure 11.13), with relative risk as our effect size and Big Deck shuffling for significance.

The last parameter that has to be chosen is the desired effect size: how small is the difference that we hope to be able to see?

If we were formulating this study design, we would ask ourselves: what would be an increase that we should warn people about? a 1% increase? 10%? Let's be generous and say that if there really was a 50% increase in birth defects, we would definitely want the study design to have the power to find that.

So we asked: what is the power of this study design to find a 50% difference?

To answer this question, we created a phantom world in which the drug-yes group had a 50% higher rate of defects than the drug-no group (Figure 11.16). We did this by moving cases in the "Bendectin" group so that it had a 1.5 relative risk compared to the "No Bendectin" group, while preserving the same group sizes.

		outcome?	
		Musculoskeletal Malformations	No Musculoskeletal Malformations
took the drug?	Bendectin	a 14	b 1,155
	No Bendectin	c 382	d 48,731

Figure 11.16 A phantom world, based on the data in the Bendectin study, but altered to produce a 50% increase in relative risk.

Exercise 11.3.1 Verify that in the phantom world shown in Figure 11.16, there is indeed at least a 50% higher rate of defects in the drug-yes group than the drug-no group.

For the first simu-study, we sampled a new drug-yes group and a new drug-no group out of the respective groups in the phantom world exactly as the original study did. In particular, the mini-me goes into the phantom world, consisting of a phantom Bendectin group and a phantom No Bendectin group.

- The phantom Bendectin group consists of 14 cases of malformation-yes and 1,155 cases of malformation-no.
- The phantom No Bendectin group consists of 382 cases of malformation-yes and 48,731 cases malformation-no.

The mini-me selects, with replacement, 1,169 data points from the phantom Bendectin group and selects, with replacement, 49,113 data points from the phantom No Bendectin group, forming the data set for simu-study #1. Then mini-me calculates the relative risk in this simu-study, which we will call RR_1, and tests it for statistical significance by the "Big Deck, shuffle and deal" method.

We repeated this 10,000 times, and calculated what fraction of simu-studies were able to find statistical significance. This fraction is the power of this study design to find a 50% increase in relative risk.

The answer was gloomy: this study design had a power of 0.3984, which means it has less than a 40% chance of finding a large, 50% increase in risk (Figure 11.17). Then the power to detect the 43% increase in risk that was observed must be still less.

Thus, **the authors' negative finding was highly likely given the low-power study design**.

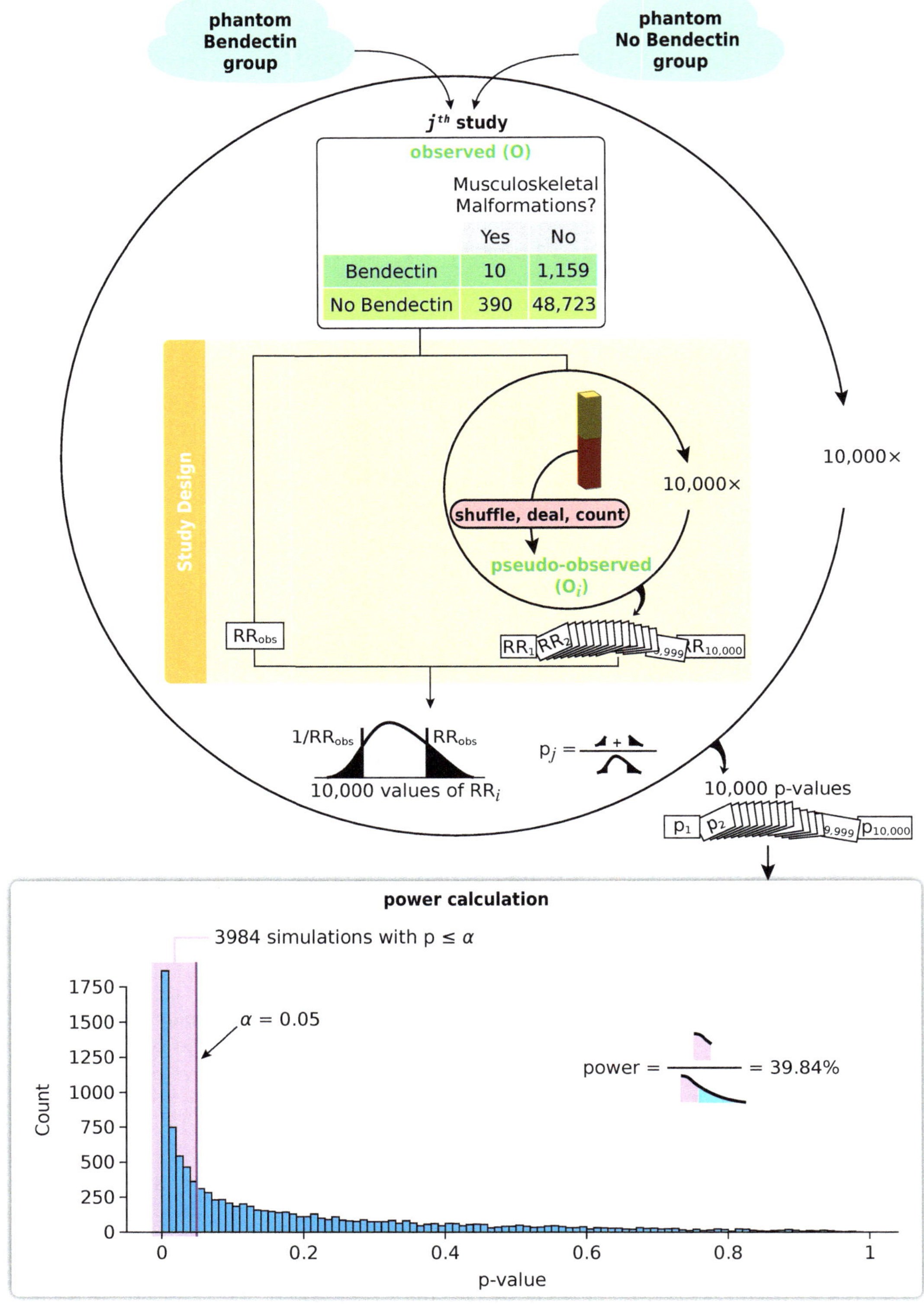

Figure 11.17 The power of the "2 × 2 relative risk" study design to find a 50% increase in relative risk, given the numbers from the actual Bendectin study.

Exercise 11.3.2 Consult Figures 11.15 and 11.17. If the cutoff for statistical significance α is set at 0.01, what would be the consequence for the power of the study design, and how would that affect the outcome of the Null Hypothesis Significance test? What if α is set at 0.3?

Exercise 11.3.3 How should the Bendectin study have presented their result?

Conclusions So we have seen an example of a highly dangerous recommendation that a drug is safe, based on a finding of "no statistical significance". The authors did not provide a power calculation, and in their published paper, there is not enough information about the statistical methodology to permit us to calculate its power. But when we attempted a statistical analysis, using a straightforward relative risk measure, we found a relative risk of 1.43 with a p-value of 0.289, that is, not statistically significant at the 0.05 level. But when we calculated the power of this analysis to detect an even larger, and therefore easier to detect, relative risk of 1.5, we found that the power was less than 40%. Therefore, it would seem that the authors have made a negative finding, based on a low-power study. As we said in the beginning of this chapter, it is unethical, statistical malpractice, to draw a negative conclusion from a low power study.

What could possibly have led to this dangerous recommendation? Perhaps the answer lies in the Acknowledgment that their study was

> "Supported by...grants from Hoffman-LaRoche Inc., Nutley New Jersey and Merrell National Laboratories, Cincinnati, Ohio"

Merrell National Laboratories was the manufacturer of the drug.

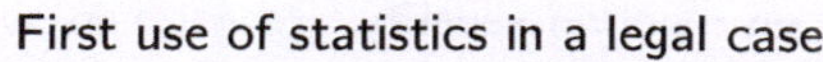

First use of statistics in a legal case

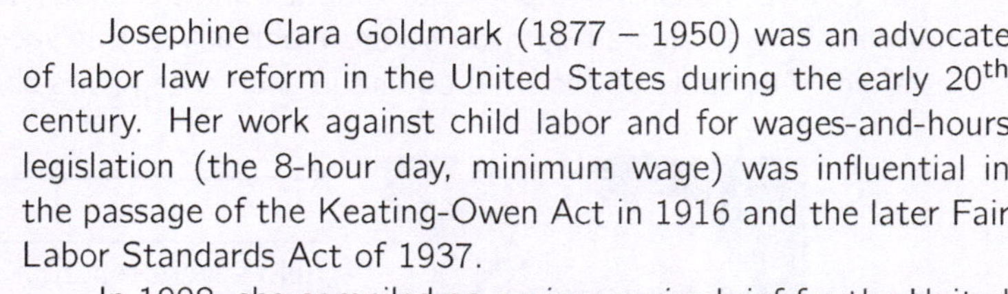

Josephine Clara Goldmark (1877 – 1950) was an advocate of labor law reform in the United States during the early 20th century. Her work against child labor and for wages-and-hours legislation (the 8-hour day, minimum wage) was influential in the passage of the Keating-Owen Act in 1916 and the later Fair Labor Standards Act of 1937.

In 1908, she compiled an *amicus curiae* brief for the United States Supreme Court case Muller v. Oregon. The brief was instrumental in getting the Supreme Court to declare that state maximum-hours laws were constitutional.

Nowadays, statistics in the law is considered a major subject. There are statistics courses in Law Schools, and many texts on statistics for lawyers.

For example, if we wanted to prove employment discrimination, that's a statistical claim. We're claiming that there is a significant difference between pay rates in group A vs. group B. If we wanted to prove housing discrimination, that's a statistical claim. We're claiming that group A gets significantly better treatment than does group B.

FURTHER EXERCISES 11.3

1. **Helmet wearing and head injuries in bicycle accidents.** Researchers wanted to know whether wearing a helmet really does reduce the likelihood of head injury in bicyclists involved in collisions with motor vehicles. The table of observed counts in each category from their study in Australia is shown below (see article "The effectiveness of helmets in bicycle collisions with motor vehicles: a case-control study", *Accident Analysis & Prevention*, 2013).

	Wearing helmet	Not wearing helmet
Head injury	372	267
No head injury	4715	1391

   a. Calculate the relative risk of head injury for bicyclists not wearing helmets compared to bicyclists wearing helmets.

   b. What is the Null Hypothesis for this study?

   c. After simulating the Null Hypothesis 10,000 times, you found that results equally or more extreme than the observed never happened. What is the p-value?

   d. Your friend wanted to see if people wearing helmets are also less likely to experience head injury in the U.S., so he asked 20 people who had been involved in a motor vehicle collision while bicycling if they were wearing a helmet at the time and if they experienced head injury. Do you expect this study to show a statistically significant effect? If it doesn't, does that mean that people in the U.S. shouldn't wear helmets? Explain your reasoning.

2. **Effect of vaccine location on reaction severity.** Researchers wanted to know whether it is better to give the DTaP (diphtheria, tetanus, and pertussis) vaccine in either the thigh or the arm. They collected data on severe reactions to this vaccine in children aged 3 to 6 years old. The table of observed counts in each category is shown below.

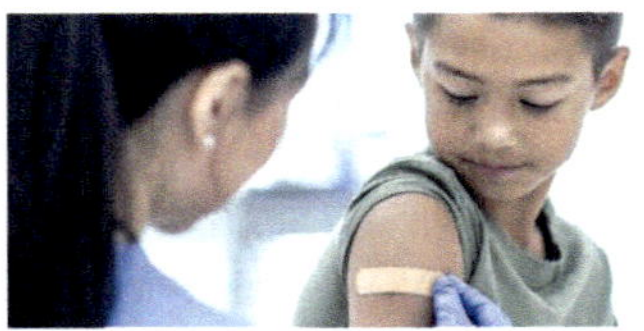

	Thigh	Arm
No severe reaction	4760	8840
Severe reaction	30	80

   a. Among children who received the vaccine in the thigh, what is the observed probability of a severe reaction?

   b. Calculate the relative risk of a severe reaction for children who received the vaccine in the thigh relative to those who received it in the arm.

   c. What is the Null Hypothesis for this study?

   d. Generate a 2 × 2 contingency table of the expected counts of children if there was no association between vaccination location and the presence of severe reaction.

   e. Suppose the researchers simulated the Null Hypothesis 10,000 times. The computer output reported "p = 0". Based on this result, which of the following could we conclude?

a) The probability of observing results at least as extreme as the observed result is less than 0.0001 under the Null Hypothesis.

b) The probability of observing results at least as extreme as the observed results is zero under the Null Hypothesis.

c) The probability of observing results less extreme than the observed result is 1 under the Null Hypothesis.

f. Select all situations in which it is appropriate to calculate power.

a) After a study with a significant result

b) When designing a study to determine the necessary sample size

c) After a study with a non-significant result

g. Outline the steps of a power analysis for this study, using the observed relative risk and sample size in this study.

h. Select all changes that would increase the power of this statistical test.

a) Use paired design

b) Decrease α from 0.05 to 0.01

c) Decrease within-group variation

d) Decrease sample size

e) Decrease effect size (across-group variation)

3. **Do Statins help?** A research article claimed that "Statins were not associated with a reduction in atherosclerotic cardiovascular disease (CVD) or all cause mortality in primary prevention in people without diabetes older than 74 years independently of age subgroup." and concluded their results "do not support the widespread use of statins in old and very old populations."

   Their analysis result shows that "In participants without diabetes, the hazard ratios for statin use in 75-84 year olds were 0.94 (95% confidence interval 0.86 to 1.04) for atherosclerotic CVD and 0.98 (0.91 to 1.05) for all cause mortality, and in those aged 85 and older were 0.93 (0.82 to 1.06) and 0.97 (0.90 to 1.05), respectively." (see article "Statins for primary prevention of cardiovascular events and mortality in old and very old adults with and without type 2 diabetes: retrospective cohort study", *BMJ*, 2018).

   a. Explain how the researchers reached their conclusion. (Note: hazard ratio = 1 means no difference.)

   b. What would the researchers need to do to back up their negative claims?

4. **Calcium and vitamin D supplements: do you need them?** In a meta-analysis of 33 randomized clinical trials that included 51,145 participants, the use of supplements that included calcium, vitamin D, or both was not associated with a significant difference in the risk of hip fractures compared with placebo or no treatment (calcium: RR, 1.53 [95%CI, 0.97 to 2.42]; vitamin D: RR, 1.21 [95% CI, 0.99 to 1.47]; combined: RR, 1.09 [95% CI, 0.85 to 1.39]) (see article "Association Between Calcium or Vitamin D Supplementation and Fracture Incidence in Community-Dwelling Older Adults", *JAMA*, 2017).

   a. Explain how the researchers reached their conclusion.

   b. What would the researchers need to do to back up their negative claims?

11.4 The prospective importance of power

So far, we have been discussing the use of power retrospectively, to look back on a study and ask whether its negative conclusions were justified. But perhaps the most important use of power calculations is in the prospective design of a study, to answer the question: is this study design likely to succeed? (Figure 11.18).

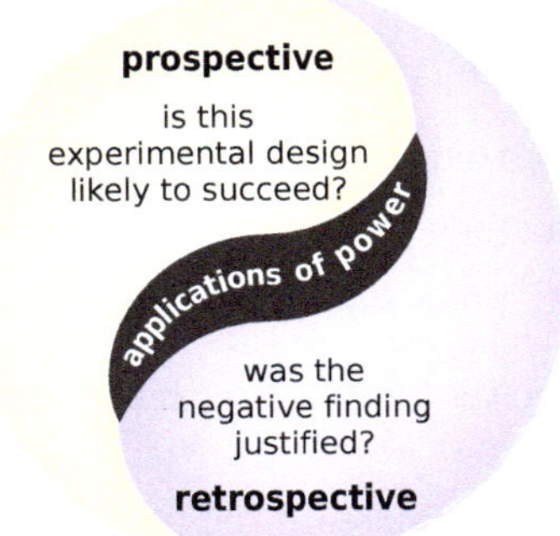

Figure 11.18

In particular, one of the critical questions in prospective power calculations is: "is my sample size big enough"?

There is a lot of confusion in biology about choosing the right sample size. If we are studying an anti-cancer drug in mice, how many mice do we need? A large n costs money, effort, and time, but is more likely to find significance, while a small n is cheaper, faster, and easier, but risks failing to find anything significant, even when there's something there. Where do we draw the line? (Figure 11.19).

Figure 11.19

The answer is to create a phantom world that represents your best guess as to what reality looks like, and create a model of your proposed study, including the sample size. **Then apply that study design to the phantom world 10,000 times, and see how often the study found significance.**

What you are doing, in effect, is making a model of the world, and then fashioning a "mini-me", a model of your study. When you throw the mini-me at the world thousands of times, you are in effect carrying out your proposed study thousand of times, which will give you a good sense of what you are likely to find.

Then we vary the sample size n, and see how big a study we need to get various levels of power. Granting agencies like the U.S.'s NIH, when they are considering funding a proposed study, typically require a power calculation to show them that you will not be wasting taxpayers' money by carrying out an under-powered study. In these contexts, after having defined a suitable effect size, we would typically like to see a power of at least 80%.

FURTHER EXERCISES 11.4

1. We can use power analysis to estimate the minimum sample size necessary to get sufficient power for a desired effect size. When comparing two groups, we can do this by considering increasingly larger sample sizes and running power analyses for each until we find sufficient power. What minimum sample size would we need in order to have at least 80% power for the effect size that was found in the plant species richness study in FE 6.5.4 on page 225?

11.5 Role of power in positive results

So far, we've been talking about the need for power when drawing "negative conclusions". It might be thought that power is limited to backing negative conclusions, and that power plays no role when we have positive conclusions, that is, findings of statistical significance. But this is not true. As a matter of fact, considerations of power play an important role in the interpretation of positive findings.

Kidney cancer A vivid illustration of the dangers of claiming positive findings from low power studies can be seen in an excellent study by epidemiologist Howard Wainer (Wainer, 2007).

A study was done of the US, county-by-county, tabulating the rates of kidney cancer in each county.

In Figure 11.20, we see in teal the *lowest* 10% of American counties by kidney cancer rate, that is, the *best* 10% of counties for kidney cancer.

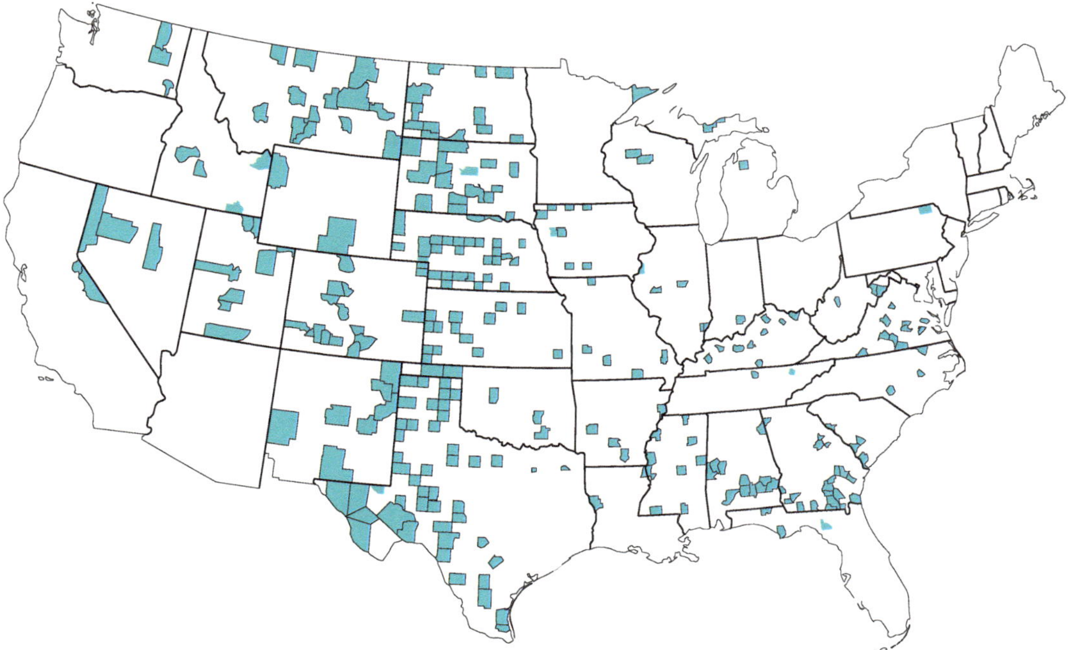

Figure 11.20 Counties in the US that have the lowest kidney cancer rates (adapted from Wainer (2007)).

Looking at the map, we are immediately struck by the fact that no big cities have *any* county that is among the best 10%. Not one in Boston, or New York, or Miami, or Dallas, or LA, or San Francisco, or Seattle, etc. Instead, the best 10% of counties for kidney cancer are *all* rural.

So, for example, if we constructed a 2 × 2 contingency table with the two factors as "best 10% for kidney cancer?" (Yes/No), and "rural?" (Yes/No), then we would find a highly significant association between "best county" and "rural".

What are we to make of this statistically significant fact? We immediately start to think: is there something about rural life that protects against kidney cancer? or on the other hand, could there be something about city life that promotes kidney cancer? It would be very tempting to start speculating about the benefits and deficits of city vs. country life.

But this would all be a mistake. In Figure 11.21, we see in orange the *worst* 10% of US counties, that is, the counties that have the *highest* rates of kidney cancer. And they are in the same rural places! The worst 10% of US counties are all in rural areas, which disproves our previous speculations about the virtues of rural life and/or the drawbacks of city life.

We would find a highly significant association between "worst counties" and "rural"!

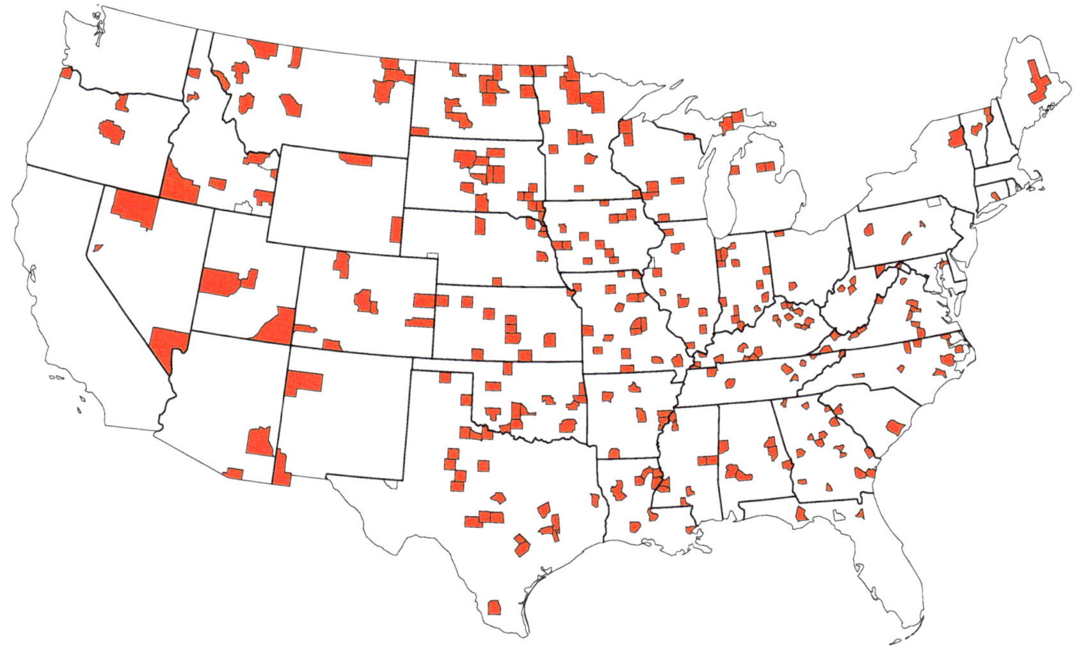

Figure 11.21 Counties in the US that have the highest kidney cancer rates (adapted from Wainer (2007)).

And even more surprisingly, if we superimpose maps of the best and the worst counties, in many if not most cases, the "worst" counties are directly adjacent to the "best" ones! (Figure 11.22).

What is going on here has nothing to do with cancer at all, and nothing to do with city life vs. rural life either.

The explanation is that the rural counties have the smallest populations, and **small samples produce highly variable outcomes**, as a rule. To take an extreme example, if there are only 10 people in a county, and one of them has kidney cancer, then that county has a kidney cancer

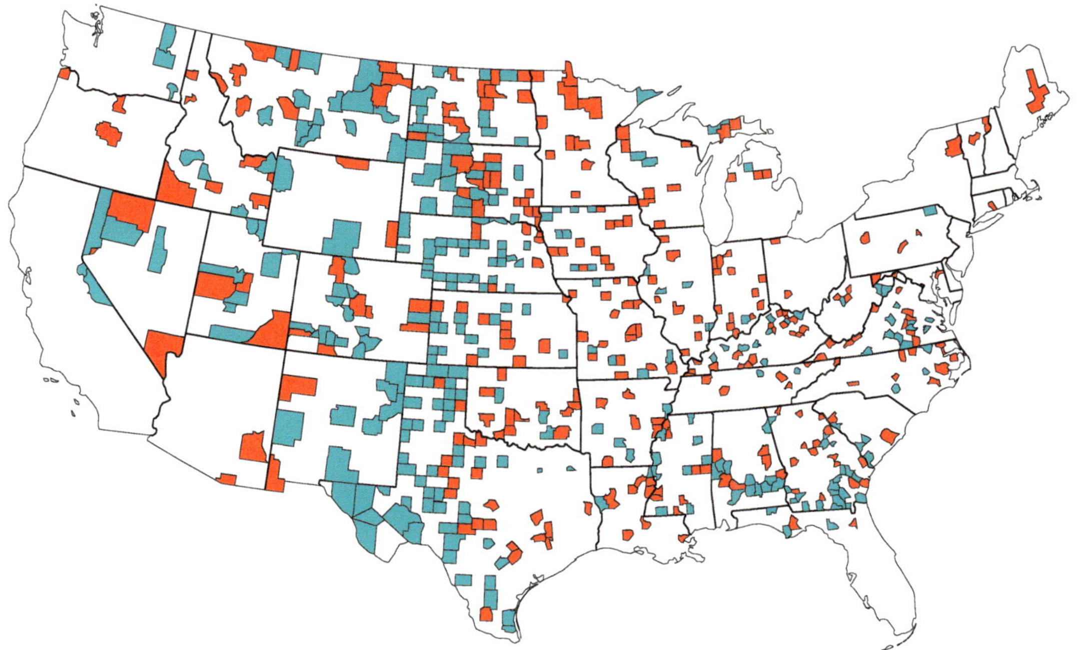

Figure 11.22 The highest and the lowest kidney cancer rate counties in the US (adapted from Wainer (2007)).

rate of 10%, which is extremely high and makes it one of the worst counties. But if that person moves away, the county rate drops to 0%, making it one of the best counties. In small sample sizes, randomness can result in the most extreme values. Hence, small samples have large fluctuations, and are unreliable as estimates.

These statistically significant positive findings are unreliable, in this case, simply wrong, because the studies had small sample sizes, and therefore, low power.

Low power makes a positive finding unreliable. If power is low, a positive finding could well be a false positive.

Exercise 11.5.1 Why Positive Results from Low-Powered Analyses Are Unreliable: The Case of Eli Lilly's Solanezumab. Solanezumab was a monoclonal antibody designed to clear amyloid-beta plaques from the brain — a leading theory for treating Alzheimer's disease. Eli Lilly invested over a decade and billions of dollars in its development.

Two Initial Trials: EXPEDITION 1 & 2 (2012).

Neither EXPEDITION 1 nor EXPEDITION 2 showed significant benefit on primary cognitive or functional outcomes in mild-to-moderate Alzheimer's disease. The company concluded: "Solanezumab, a humanized monoclonal antibody that binds amyloid, failed to improve cognition or functional ability." (see "Phase 3 Trials of Solanezumab for Mild-to-Moderate Alzheimer's Disease", *New England Journal of Medicine*, 2014).

Subsequent Pooled Subgroup Analyses.

The company responded by restricting the previous analysis to a smaller subgroup: only **mild** Alzheimer's patients.

In this smaller sample, they found that "A pre-specified secondary analysis of pooled data in patients with mild Alzheimer's disease showed a slowing of cognitive decline (p=.001) compared with placebo ... this finding represented a 34 percent reduction in decline."

In other words, a smaller n yielded a substantial effect size that was highly statistically significant.

Findings from the subsequent pooled subgroup analysis:

- Solanezumab showed a 34% slowing of cognitive decline
- p = 0.001 (statistically significant!)

The question is: were those results reliable?

Doubling Down: EXPEDITION 3.

Based on these statistically significant findings in the subsequent analyses of only **mild** Alzheimer's patients, Eli Lilly launched EXPEDITION 3 (2013-2016) (see article "Trial of Solanezumab for Mild Dementia Due to Alzheimer's Disease", *The New England Journal of Medicine*, 2018).

The expectation was to replicate the 34% slowing effect and its statistical significance. The cost was estimated to be $500M-1B for this trial alone.

But in the actual trial, solanezumab showed an 11% reduction in cognitive decline compared to placebo, not the 34% seen in the pooled subgroup analysis.

Findings from EXPEDITION 3:

- Difference in cognitive decline: 11% (not the 34% seen before)
- p = 0.10 (not significant)
- No meaningful clinical benefit

a. Explain why the subgroup analysis (mild Alzheimer's patients) was more vulnerable to unreliable results.

b. Why are statistically significant results from low-powered studies likely to overestimate the true treatment effect?

Power failure This fact that positive results from low-powered analyses are unreliable was explained in a paper in *Nature Reviews Neuroscience* called "Power failure: why small sample size undermines the reliability of neuroscience".

NATURE REVIEWS | NEUROSCIENCE 2013 ANALYSIS

Power failure: why small sample size undermines the reliability of neuroscience

Katherine S. Button[1,2], John P. A. Ioannidis[3], Claire Mokrysz[1], Brian A. Nosek[4], Jonathan Flint[5], Emma S. J. Robinson[6] and Marcus R. Munafò[1]

Abstract I A study with low statistical power has a reduced chance of detecting a true effect, but it is less well appreciated that low power also reduces the likelihood that a statistically significant result reflects a true effect. Here, we show that the average statistical power of studies in the neurosciences is very low. The consequences of this include overestimates of effect size and low reproducibility of results. There are also ethical dimensions to this problem, as unreliable research is inefficient and wasteful. Improving reproducibility in neuroscience is a key priority and requires attention to well-established but often ignored methodological principles.

We know that if a study has low power, it can easily miss a true result, and produce a false negative. This paper argues that if a study has low power, and makes a positive finding, then even that positive finding has low trustworthiness and credibility.

Let's say we have a positive finding (statistically significant). Let's ask the question, what is the probability that the positive finding is really, actually, true?

Remember the terminology p(X | Y) that we introduced to denote "the probability of X given Y". We can express what we are asking as the probability, given a positive finding, that the positive finding is really true. This can be expressed as:

what is p(really true | positive finding)?

This is equivalent to asking: what fraction of all positive findings are true positives?

$$\text{p(really true} \mid \text{positive finding)} = \frac{\text{TP}}{\text{TP} + \text{FP}}$$

If true positives (TP) are almost all of the positives (TP + FP), then our positive is probably a true positive, because most positives are true positives. On the other hand, if that fraction is low, then our positive has a good chance of being a false positive, and therefore, not really true.

Since power is the likelihood of correctly identifying a positive result as a true positive, it follows that the lower the power, the fewer the TPs. Therefore, low power means low $\frac{\text{TP}}{\text{TP+FP}}$, which means that there is a lower chance that this positive is a true positive!

The quantity $\frac{\text{TP}}{\text{TP+FP}}$ is also called the Positive Predictive Value (PPV). It will be the first thing we study in the following chapter, on Bayesianism, and will enable us to make precise the idea that low power studies generate lots of false positives.

In the *Nature Reviews Neuroscience* paper, the authors reviewed 49 meta-analyses in neuroscience that were published in 2011. A meta-analysis is a study of published studies. Meta-analyses are very useful to summarize publications in a particular research area.

The authors calculated, for each meta-analysis, the median power of the studies reviewed in that meta-analysis. The distribution of median powers in the 49 meta-analyses is shown in (Figure 11.23).

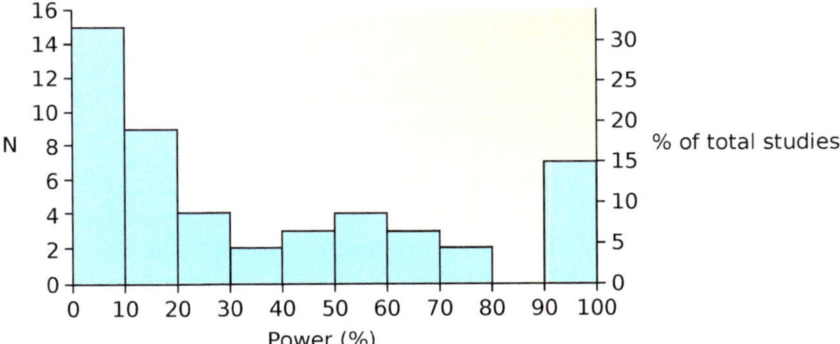

Figure 11.23 Distribution of median powers in the 49 meta-analyses reviewed in Button *et al.* (2013).

When we are considering the power of a study, 80% is considered good. 50% is considered somewhat low power, and 20% is considered a poor study. Their finding, therefore, is a quite gloomy. Only 15% of the studies had high power, two thirds of them had low power (under 50%), and half of them had extremely low power (under 20%).

These findings give us insights into the root cause of the reproducibility crisis. **Low power studies create irreproducible results.**

This is not just limited to neuroscience. A similar pattern can be found across the biomedical sciences.

FURTHER EXERCISES 11.5

1. Briefly explain how low statistical power affects the interpretation of results from a study that finds a statistically significant effect.

2. A trial tests a new antihypertensive in 24 patients (12 drug, 12 placebo). The drug group has systolic BP 8 mmHg lower, but $p = 0.09$. An 8 mmHg reduction could be clinically meaningful, but the result was not statistically significant. The authors conclude: "The drug had no significant effect." A senior colleague responds: "With only 24 patients, the study was likely underpowered. It may never have had a reasonable chance to detect a clinically meaningful effect." Which of the following statements about this trial are FALSE?

   a. Power is an important consideration when interpreting this negative finding, as the trial may have been too small to detect a true effect.

   b. The senior colleague's criticism is misplaced. Power only matters when interpreting positive findings, not negative ones.

   c. Had the researchers aimed for 80% power, they would likely have needed a larger sample. A negative result from that larger trial would be more informative.

   d. Calculating the required sample size before the study, the researchers would have needed to specify an expected effect size, the significance level, and the desired power.

Chapter 12

Bayesian Statistics

LEARNING OBJECTIVES

After studying this chapter, you will be able to:

- explain why p-values are the answer to the wrong question.
- use the concept of Positive Predictive Value in medicine.
- use Bayes' Theorem to derive $p(A \mid B)$ from $p(B \mid A)$, with additional information.
- use Bayes' rule to update prior probabilities into posterior probabilities.

"The Bayesian approach is now the preferred method of comparing scientific theories."

— Richard Feynman

12.1 Answering the wrong question

When we were talking about p-values in Chapter 3 (*Introduction to Probabilities and Hypothesis Testing*), we discussed the fact that the p-value is simply the answer to the wrong question.

To see why, consider a Null Hypothesis, say, that the coin we're flipping is unbiased and that flips are independent. Then we flipped the coin 10 times, and got 8 Heads in the 10 flips. Now, what we want to know is: what is the likelihood that the Null Hypothesis is false, that is, that the coin is not fair? And there we have a problem.

$$\text{Null Hypothesis } H_0 : \text{the coin is fair and flips are independent.}$$
$$\text{observed result } D : \text{we observed 8 Heads out of 10 flips.}$$

Recall the notation we introduced there, that "$p(X \mid Y)$" means "the probability of X, given Y". For example, p(female | pregnant) means the probability of being female given that you are pregnant. This probability is 1. On the other hand p(pregnant | female) is the probability of being pregnant given that you are female, which is obviously much less.

If H_0 is the Null Hypothesis, and D is the observed result, that is, this particular outcome, say, 8 heads out of 10 tosses, then the p-value gives us the probability of getting a result at

A. Garfinkel and Y. Guo, *Understanding Data*,
https://doi.org/10.1007/978-3-032-18600-3_12

least as extreme as our observed result (8 Heads out of 10), assuming the Null Hypothesis that the coin is fair and flips are independent. The at least as extreme cases are: at least 8 Heads or at least 8 Tails in the 10 flips:

$$\begin{aligned}\text{p-value} &= p(8 \text{ Heads} \mid H_0) + p(9 \text{ Heads} \mid H_0) + p(10 \text{ Heads} \mid H_0) \\ &\quad + p(8 \text{ Tails} \mid H_0) + p(9 \text{ Tails} \mid H_0) + p(10 \text{ Tails} \mid H_0)\end{aligned}$$

Following Chapter 3, we do thousands of simulations of the Null Hypothesis, and it turns out that in only 1,108 of the 10,000 do we get a result at least as extreme as 8 of one face (note that we are doing a two-sided test). We then say that the p-value of the observed result is $\frac{1{,}108}{10{,}000} = 0.11$ (that is, somewhat unlikely).

But is that the answer to the question "how likely is the Null Hypothesis given the data?"

No, it is not. *It is not even in the correct grammatical form to be an answer!*

The p-value is the answer to the question "how likely is this data, given the Null Hypothesis?" (Strictly speaking, the p-value is the sum of a number of terms, all of which have the form $p(\text{data} \mid H_0)$.) And those are not the same question!

What we really wanted was $p(H_0 \mid D)$, the probability that the coin is fair given our observed result. This is a fundamentally different question, and as we just showed in the example about pregnancy, the two probabilities are not equal.

In this chapter, we are going to provide an answer to the general question: what is $p(H \mid D)$?

In general, in science, we have a hypothesis H, and some data D.
Our scientific question is

$$\text{What is } p(H \mid D)?$$

Whereas the p-value answers questions of the form

$$\text{What is } p(D \mid H)?$$

and

$$p(H \mid D) \neq p(D \mid H)$$

The confusion of the two probabilities $p(D \mid H)$ and $P(H \mid D)$ is the p-value fallacy, and has led to disastrous results in science, including the epidemic of false positives and irreproducible research.

It is also the root of the Prosecutor's Fallacy, which wrongly convicted the Collins couple (see section 3.3 *What's wrong with Null Hypothesis Significance Testing?*) and wrongly convicted the British lawyer Sally Clark of murdering her two sons.

FURTHER EXERCISES 12.1

1. What's wrong with Null Hypothesis Significance Testing? ("$p < 0.05$")?

   a. There's nothing wrong with it and it's used everywhere.
   b. 0.05 is not strong enough.
   c. "$p <$ anything" is binary, just yes or no, no clue as to effect size.
   d. No sense of precision, even if you do give an effect size.
   e. The p-value is the answer to the wrong question.

12.2 Positive Predictive Value in medicine and social science: an introduction to Bayesianism

Disease testing

Our first example will be not about Null Hypothesis Significance Testing, but a different subject, disease testing in medicine.

Sensitivity and specificity Suppose we have a test for some disease, say a blood test for HIV, an EKG for detecting a heart condition, or a mammogram for detecting breast cancer. A given person, in reality, either has the disease ("Dis") or does not ("noDis"). On our test, people either test positive ("+") or negative ("−") for the disease.

We can then define a number of quantities that reflect the accuracy of the test. Using the terminology "p(X | Y)", the probability of X, given Y, we can define:

Conditional probabilities in disease testing

Terminologies	
p(X) = probability of X	
p(X | Y) = probability of X given Y	
+	test is positive
−	test is negative
Dis	actually has disease
noDis	actually doesn't have disease

Then we can define 2 very important medical concepts:

$$\begin{aligned}\text{sensitivity} &= p(+ \mid \text{Dis})\\ &= \text{correctly diagnosing presence of disease}\\ &= \text{probability of the test being positive if the disease is there}\end{aligned}$$

$$\begin{aligned}\text{specificity} &= p(- \mid \text{noDis})\\ &= \text{correctly diagnosing absence of disease}\\ &= \text{probability of returning a negative finding when the disease is not there}\end{aligned}$$

Let's consider a test for which the sensitivity and specificity of the test are both extremely high, at 99%.

Question: on this test, if a patient has tested +, what is the likelihood that the patient has the disease?

a) 99%

b) um, 99%

c) I said, 99%

d) is this a trick question?

e) not enough information to answer this question.

The answer is: This is a trick question, and the correct answer is e)!

What's missing? We would like to know: ***given the positive test, what is the probability that the patient actually has the disease?*** This is a very natural question. It's the question the doctor, and the patient, really want to know the answer to.

But what is this question? In our new terminology, note that this is asking for $p(\text{Dis} \mid +)$, the probability of actually having the disease, given that the test was positive.

But $p(\text{Dis} \mid +)$ isn't either the sensitivity or the specificity. It's a new and very important quantity, which is called the **Positive Predictive Value (PPV)**.

How can we calculate it? Here's one way. Remember back in Chapter 3, we drew up a table to represent the four kinds of judgments you could make after looking at a particular MRI scan for kidney cancer (Figure 12.1).

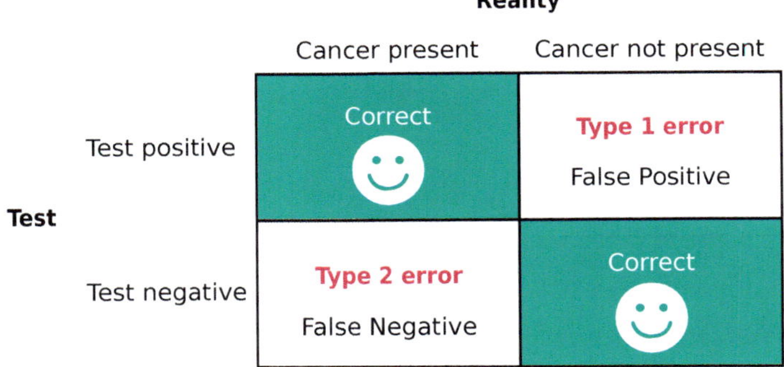

Figure 12.1

We also introduced another terminology, which is also helpful here: we talked about the four kinds of judgments as "True Positive (TP)", "False Positive (FP)", "True Negative (TN)" and "False Negative (FN)" (Figure 12.2).

Test?	**Disease?** Yes	No
+	**TP**	**FP**
−	**FN**	**TN**

Figure 12.2 Four kinds of judgements that can be made in disease testing.

Given the notions of TP, FP, TN, and FN, we can now define the Positive Predictive Value (PPV) of a test, administered to a given group. Since it's the probability of a person's actually having the disease given that they tested positive, that is equal to the proportion of positive tests that are true positives. If that fraction is low (few positives are true positives), then the PPV is low, and if it is high (most positives are true positives), the PPV is high.

Positive Predictive Value $$\text{PPV} = p(\text{Dis} \mid +) = \frac{\text{TP}}{\text{TP} + \text{FP}}$$

TP and FP are the number of true positives and the number of false positives in the study group to which the test was administered. But we do not know TP or FP, because we cannot calculate them from the test results, since all we have is the number of positives (TP + FP) and the number of negatives (TN + FN). We don't know how many of them are true and how many are false.

- **Sensitivity** is the proportion of all positives that are true positive. If we let TP = the number of true positives and FN = the number of false negatives, then the sensitivity of a test can be defined as the fraction of all people with the condition that the test called correctly.

$$\text{sensitivity} = \frac{\text{TP}}{\text{TP}+\text{FN}}$$

  Note that the denominator TP + FN is the total number of diseased people. If, at the extreme, the test pronounced every result to be cancer, the sensitivity of the test would be 100%. The number of False Negatives would be zero, because the number of Negatives would be zero!

- **Specificity** is the proportion of all negatives that are true negative. The specificity of a test is defined as its ability to say "negative" when the condition is not there. We can define it as the fraction of negatives that are true negatives,

$$\text{specificity} = \frac{\text{TN}}{\text{TN}+\text{FP}}$$

  Note that the denominator TN + FP is the total number of non-diseased people, those truly diagnosed plus those who produced false positive tests. So the specificity of the test that says "all results indicate cancer" is zero.

Now, TP, the number of true positives, is equal to the sensitivity of the test × the number of diseased people in our test group, which we will call #Dis.

$$\text{TP} = \text{sensitivity of the test} \times \#\text{Dis}$$

FP can also be calculated. We know that TN, the number of true negatives, is equal to $p(- \mid \text{noDis}) \times$ the number of not-diseased people in our test group, which we will call #noDis.

$$\text{TN} = \text{specificity of the test} \times \#\text{noDis}$$

Then what is the number of false positives, FP? A false positive must be a not-diseased person who has tested positive, so FP is then the number of not-diseased minus the number of true negatives.

$$\begin{aligned}\text{FP} &= \#\text{noDis} - \text{TN}\\ &= \#\text{noDis} - \text{specificity of the test} \times \#\text{noDis}\\ &= (1 - \text{specificity of the test}) \times \#\text{noDis}\end{aligned}$$

Then the PPV of the test is the fraction of positives that are true positives.

$$\text{PPV of the test} = p(\text{Dis} \mid +) = \frac{\text{TP}}{\text{TP} + \text{FP}}$$
$$= \frac{\text{sensitivity} \times \#\text{Dis}}{\text{sensitivity} \times \#\text{Dis} + (1 - \text{specificity}) \times \#\text{noDis}}$$

If #Dis = number of diseased people, and #noDis = number of not-diseased people, then the total number of people, #total, is #Dis + #noDis. Then we divide the numerator and denominator by #total.

$$\text{PPV} = \frac{\text{sens} \times (\#\text{Dis})}{\text{sens} \times (\#\text{Dis}) + (1 - \text{spec}) \times (\#\text{noDis})}$$

$\Downarrow$ *dividing top and bottom by (#total)*

$$= \frac{\text{sens} \times (\frac{\#\text{Dis}}{\#\text{total}})}{\text{sens} \times (\frac{\#\text{Dis}}{\#\text{total}}) + (1 - \text{spec}) \times (\frac{\#\text{noDis}}{\#\text{total}})}$$

$$= \frac{\text{sens} \times (\frac{\#\text{Dis}}{\#\text{total}})}{\text{sens} \times (\frac{\#\text{Dis}}{\#\text{total}}) + (1 - \text{spec}) \times (\frac{\#\text{total} - \#\text{Dis}}{\#\text{total}})}$$

$$= \frac{\text{sens} \times (\frac{\#\text{Dis}}{\#\text{total}})}{\text{sens} \times (\frac{\#\text{Dis}}{\#\text{total}}) + (1 - \text{spec}) \times (1 - \frac{\#\text{Dis}}{\#\text{total}})}$$

Now, let's define the **prevalence** of the disease as

$$\text{prevalence} = \frac{\#\text{Dis}}{\#\text{total}}$$

That is, the prevalence of a disease is the fraction of the population that actually has the disease.

Since we don't know the $\frac{\#\text{Dis}}{\#\text{total}}$ in our test group, we use the prevalence in the general population from which our test group was drawn. For example, the prevalence of breast cancer in a 20-30 year old cohort is much lower than the prevalence of breast cancer in a 60-70 year old cohort.

In addition to knowing the sensitivity and specificity of a test, the missing information required to calculate PPV is prevalence.

So, we have actually derived the positive predictive value (PPV) of the test, p(Dis | +), in terms of the sensitivity and specificity of the test AND ONE MORE QUANTITY, THE PREVALENCE.

PPV as a function of sensitivity, specificity and prevalence

$$\text{PPV of the test} = p(\text{Dis} \mid +) = \frac{\text{TP}}{\text{TP}+\text{FP}}$$
$$= \frac{\text{sensitivity} \times \#\text{Dis}}{\text{sensitivity} \times \#\text{Dis} + (1-\text{specificity}) \times \#\text{noDis}}$$
$$= \frac{\text{sens} \times \text{prev}}{\text{sens} \times \text{prev} + (1-\text{spec}) \times (1-\text{prev})}$$

sensitivity: The sensitivity of the test is the probability of the test picking up the disease if it is there, that is, $p(+ \mid \text{Dis})$.

specificity: The specificity of the test is the probability of returning a negative finding when the disease is not there, that is, $p(- \mid \text{noDis})$.

prevalence: The prevalence of a disease is the fraction of the population that has the disease, that is, $\frac{\#\text{Dis}}{\#\text{total}}$.

Does the patient actually have the disease? Let's work this out for our original disease testing. We were asking about the patient who tested positive: what is the probability that this patient actually has the disease?

We stipulated that the sensitivity and specificity of the test are both 99%. But as we just saw, in order to calculate PPV, we need one more quantity, the prevalence, $\frac{\#\text{Dis}}{\#\text{total}}$, that is, the fraction of the population that is diseased. Is the disease a very rare one ($\frac{\#\text{Dis}}{\#\text{total}}$ is low), or is it common ($\frac{\#\text{Dis}}{\#\text{total}}$ is high)? **We must have an estimate of this quantity in order to calculate PPV.**

So let's say that the prevalence of this disease is $\frac{1}{200}$; in the general population, we expect 1 in 200 people to be diseased. Then

$$\text{PPV} = \frac{\text{sens} \times \text{prev}}{\text{sens} \times \text{prev} + (1-\text{spec}) \times (1-\text{prev})}$$
$$\Downarrow \quad \textit{sens} = 0.99 \quad \textit{spec} = 0.99 \quad \textit{prev} = \frac{1}{200} = 0.005$$
$$\text{PPV} = \frac{0.99 \times 0.005}{0.99 \times 0.005 + (1-0.99) \times (1-0.005)}$$
$$\text{PPV} = \frac{0.00495}{0.00495 + 0.00995} = 0.332$$

In other words, the true probability that the patient has the disease, given sensitivity = 0.99 and specificity = 0.99 and prevalence = 0.005, is ONE THIRD!

This is an absolutely stunning conclusion! It is completely different from our naïve expectations.

If your non-believing friends are unwilling to go through the calculation above, here's an even simpler way to see the shocking but correct answer: if the specificity is 99%, there is a 1% false positive rate. Then, in 200 people there will be 2 false positives, on average.

But in 200 people, there is, on average, one true positive. So the chance that person X is the true positive and not one of the false ones is 1 in 3, = 33%!!

In other words, false positives outnumber true positives, because the condition is rare. Prevalence has a strong effect on PPV; low prevalence lowers the PPV of a test.

Exercise 12.2.1 Consider a test whose sensitivity and specificity are both 99%.
Suppose the prevalence of the disease is 1 in 1000 people.
A patient tests +.
What is the likelihood that the patient actually has the disease?

Exercise 12.2.2 A new type of blood test is meant to detect cancer in dogs. The test has a sensitivity of 56.1% and a specificity of 97.5%. Suppose the cancer incidence is 1/1000.

a. If a dog tests positive, what's the probability they have cancer?
b. If a dog tests negative, what's the probability they don't have cancer?

Exercise 12.2.3 Different breeds of dogs have different cancer susceptibilities. In what kind of breed is a positive screening test more meaningful?

a. A more susceptible breed
b. A less susceptible breed
c. The same

Breast cancer screening The idea that when prevalence is low, most positives are false positives, is extremely important. Alas, most people, including most doctors, simply do not understand this point.

An example similar to ours, concerning screening women for breast cancer, was posed to groups of European and American doctors by the psychologist Gerd Gigerenzer.

The New York Times

When Gigerenzer asked 24 other German doctors the same question, their estimates whipsawed from 1 percent to 90 percent. Eight of them thought the chances were 10 percent or less, 8 more said 90 percent, and the remaining 8 guessed somewhere between 50 and 80 percent. Imagine how upsetting it would be as a patient to hear such divergent opinions. As for the American doctors, 95 out of 100 estimated the woman's probability of having breast cancer to be somewhere around 75 percent.

The right answer is 9 percent.

From ''Chances Are'' by Steven Strogatz (2010)

Graphical approaches

There's an intuitive way of visualizing the concepts above; we call it the "shaded area method".

Let's make a graphical representation of the number of diseased people and the number of not-diseased people (two columns) (Figure 12.3). The height of the bar represents the number of people in each category.

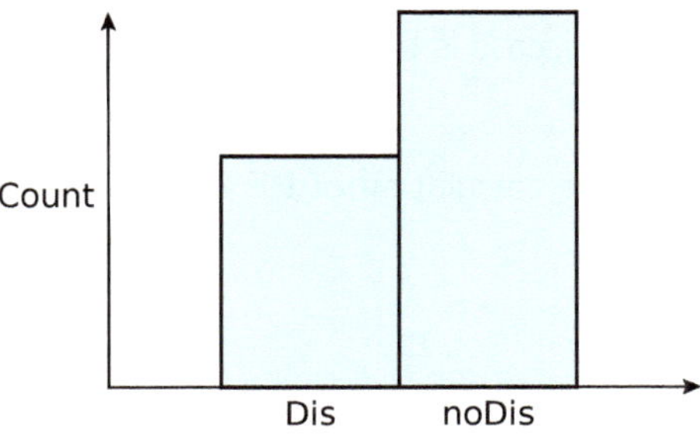

Figure 12.3

Then we will use a dark shade to indicate the number of people in each group who tested positive (Figure 12.4).

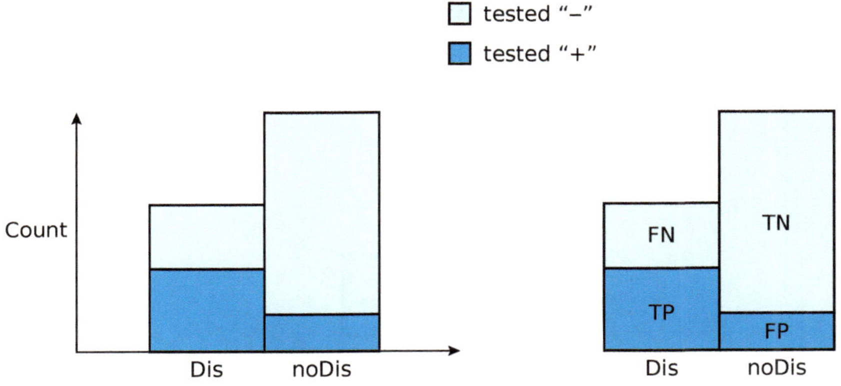

Figure 12.4

We can now label the 4 subrectangles. Since the left hand column is the disease-yes group, the dark-shaded group, that has tested positive, is the true positives, and the other area, light-shaded, is the false negatives (since they tested negative but truly had the disease). Similarly, in the right hand column (noDis), the darkly shaded area is the false positives (do not have the disease but tested positive) and the other area is the true negatives (do not have the disease and tested negative).

Positive Predictive Value: shaded area method

Positive Predictive Value (PPV) is the fraction of positives that are true positives. In the case of disease testing, PPV answers the question: given the positive test, what is the probability that the person actually has the disease?

$$\text{PPV} = p(\text{Dis} \mid +) = \frac{\text{TP}}{\text{TP} + \text{FP}} = \frac{\text{sens} \times \text{prev}}{\text{sens} \times \text{prev} + (1 - \text{spec}) \times (1 - \text{prev})}$$

sensitivity: The sensitivity of the test is the probability of the test picking up the disease if it is there, that is, $p(+ \mid \text{Dis})$.

specificity: The specificity of the test is the probability of returning a negative finding when the disease is not there, that is, $p(- \mid \text{noDis})$.

prevalence: The prevalence of a disease is the fraction of the general population that has the disease, that is, $\frac{\#\text{Dis}}{\#\text{total}}$.

Graphically, we can present the calculation of PPV using the shaded area method.

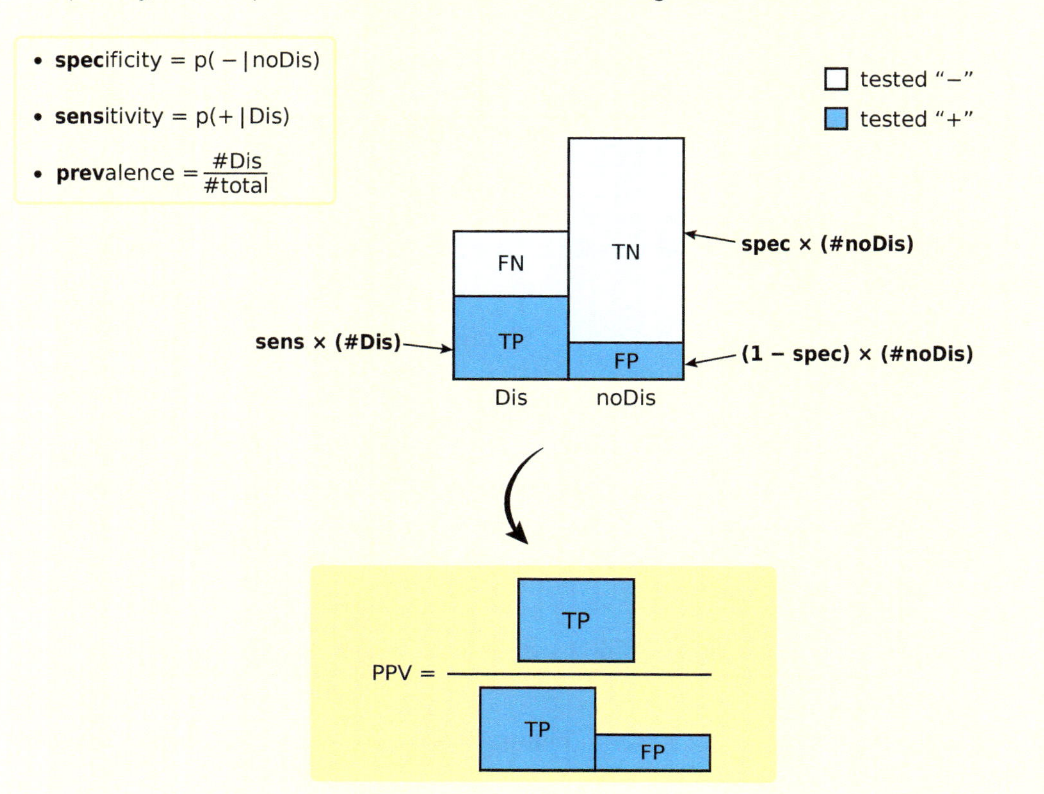

Example 1 Calculating PPV using shaded area method

A patient takes a screening test for a disease and it comes back positive. The test has a 99% sensitivity and a 99% specificity. The overall prevalence of the disease in the population is 1/100.

Question Suppose 10,000 people take this test. Fill in the number of people expected in each of the four judgements.

	Diseased	Not diseased
Test +		
Test −		

Answer Since the overall prevalence of the disease in the population is 1/100, then out of 10,000 people, we would expect to see 100 diseased people and 9,900 not-diseased people.

The sensitivity of the screening test is 99%, meaning it would test positive for 99 out of the 100 diseased people, and test negative for 1 out of the 100 diseased people.

Similarly, the specificity of the screening test is 99%, meaning it would test negative for 9801 out of the 9,900 not-diseased people and test positive for 99 out of the 9,900 not-diseased people.

	Diseased	Not diseased
Test +	99	99
Test −	1	9,801
	100	9,900

Question What is the probability that the patient who tested positive has the disease?

Answer We see that 50% of the people who test positive have the disease. In other words, the probability that the patient who tested positive has the disease is 50%, that is, the Positive Predictive Value of the test is 50%.

	Diseased	Not diseased
Test +	99	99
Test −	1	9,801
	100	9,900

50% of the people who test + have the disease

Question How can we represent our analysis using the shaded area method?

Answer

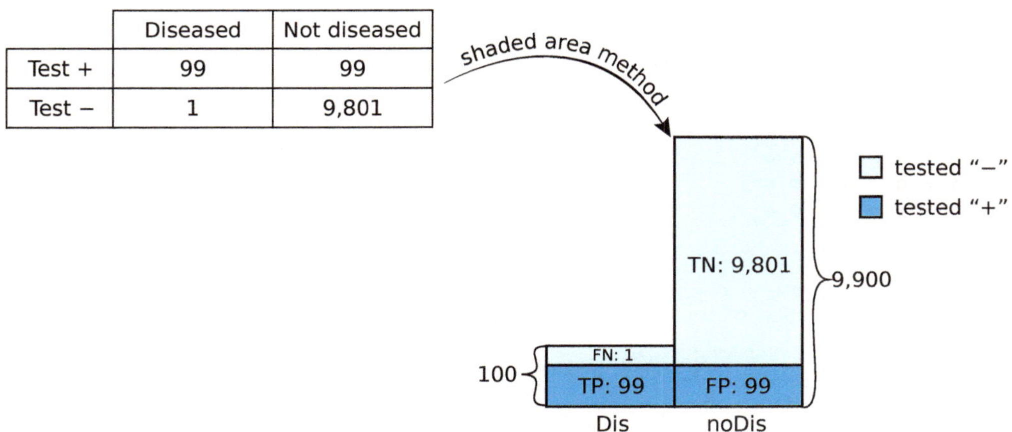

	Diseased	Not diseased
Test +	99	99
Test −	1	9,801

A geometric approach

Loong (2003) (*Understanding sensitivity and specificity with the right side of the brain*) introduces another visual approach that is useful.

Image a theoretical population, in which a filled dot means "has disease", and an empty circle means "no disease". The green regions mean "tested −", and the pink regions mean "tested +".

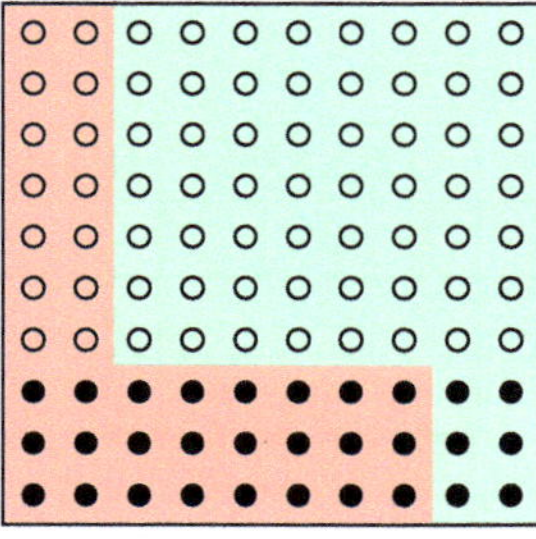

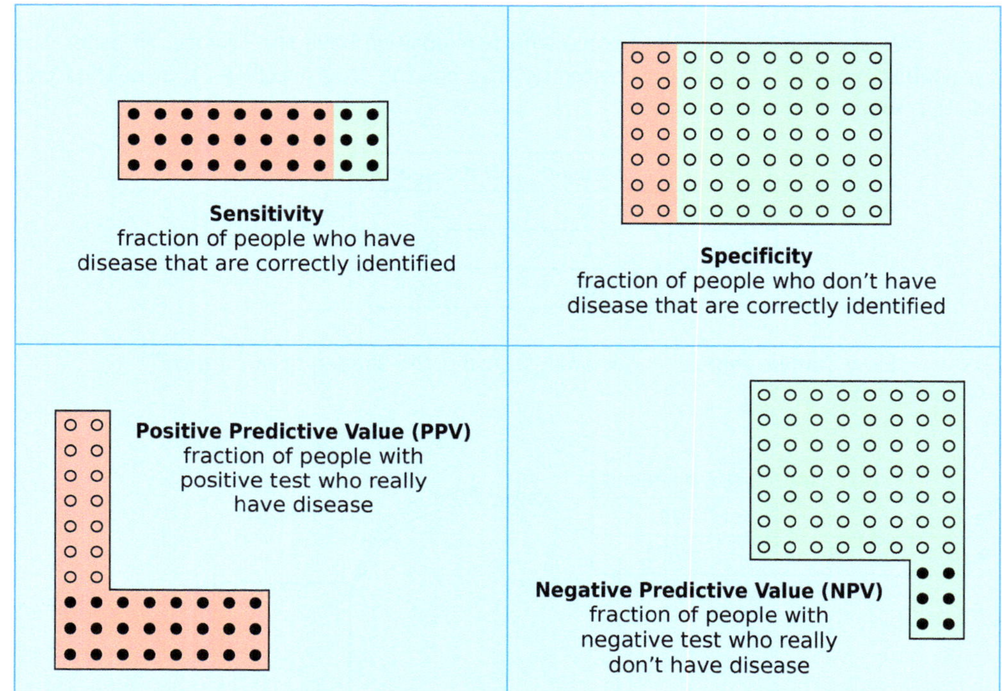

Now let's revisit each of the following concepts geometrically.

- **Sensitivity.** We need to look at only the diseased population (filled dots). Of the filled dots, some of them tested positive (pink). Sensitivity is the fraction of the people who have the disease that are correctly identified, in other words, the total number of filled dots in the pink area divided by the total number of filled dots.

- **Specificity.** We now have to look at only the empty circles. Of the empty circles, that is the people with no disease, some of them tested positive (pink). So specificity is the fraction of the people who don't have the disease that are correctly identified, in other words, the green empty circles over all the empty circles.

- **PPV.** Positive Predictive Value is the fraction of all positives (pink) that are true positives (filled dots in pink area).

- **NPV.** Negative Predictive Value is an equally important concept. It's the answer to the question "if I tested negative, what is the likelihood that I don't have the disease". Therefore, NPV is the fraction of all negatives (green) that are true negatives (empty circles in green area).

  For example, in a test for cystic fibrosis, the PPV is very low, around 3%, yet it's given as a standard test for newborns. The question was why would you use a test that has a PPV of 3%? The answer is that the test has a high NPV. It says that if the baby tests negative, we can be fairly certain that it doesn't have the disease.

PPV and NPV as a decision tree

Another way to look at PPV and NPV is with this tree structure.

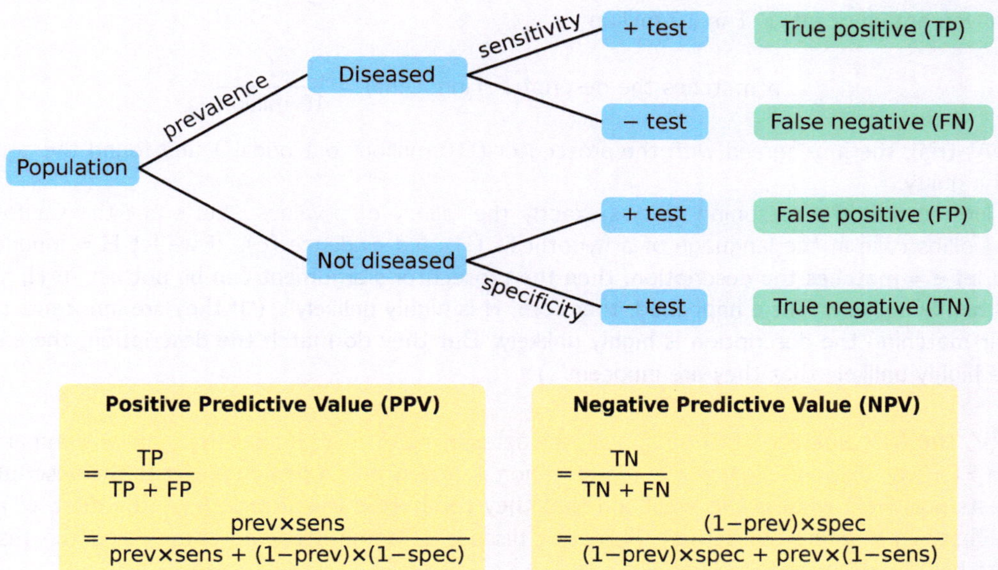

The total population is divided into the "Diseased" and the "Not diseased". The prevalence is

$$\text{prevalence} = \frac{\text{Diseased}}{\text{Diseased} + \text{Not diseased}}$$

In the Diseased group, a person can test positive or negative. Sensitivity is the probability that if a person is diseased, they'll have a positive test. In the Not diseased group, specificity is the probability that if a person is not diseased, they'll have a negative test.

Following the tree, we can then reach each leaf with the four possibilities: TP, FN, FP, TN.

$$\begin{aligned}
TP &= \text{prev} \times \text{sens} \times \text{population} \\
FN &= \text{prev} \times (1 - \text{sens}) \times \text{population} \\
FP &= (1 - \text{prev}) \times (1 - \text{spec}) \times \text{population} \\
TN &= (1 - \text{prev}) \times \text{spec} \times \text{population}
\end{aligned}$$

With this tree structure in mind, we can easily derive the expressions for PPV and NPV.

- PPV is the fraction of all positives (TP + FP) that are true positives (TP).
- NPV is the fraction of all negatives (TN + FN) that are true negatives (TN).

The prosecutor's fallacy

As an illustration of Positive Predictive Value, let's apply it to a case we first saw in Chapter 3, California vs. Collins. Recall that it concerned the Collins couple, who were arrested and tried after a brutal assault had been committed.

Reviewing the prosecutor's argument The prosecution argued that they matched a description that fit 1 in 10 million couples. That is, the probability of matching the description, given that you are innocent, is 1 in 10 million[1].

$$\text{p(matches the description} \mid \text{innocent)} = \frac{1}{10 \text{ million}}$$

At trial, the jury agreed with the prosecutor ("10 million to 1 odds!") and found the couple to be guilty.

Notice that the reasoning here is exactly the fallacy of p-values. Let's put the California vs. Collins case in the language of a hypothesis (H) and evidence (e). **If we let H = innocent, and let e = matches the description, then the prosecutor's argument can be put as: "If H, then e is highly unlikely. But e happened, therefore, H is highly unlikely". ("If they are innocent, then their matching the description is highly unlikely. But they do match the description, therefore, it is highly unlikely that they are innocent".)**

PPV, the true answer But on appeal, the defense made an argument that can be summarized as a PPV calculation, because the real question is: **given that a person matches the description ("tests positive") what is the likelihood that they committed the crime ("have the disease").**

First, let's define our terms. Here, the disease is "committed the crime", and the test is "matches the description". Then

$$\begin{aligned} \text{sensitivity} &= \text{p}(+ \mid \text{Dis}) \\ &= \text{p(matches the description} \mid \text{committed the crime)} \\ &= 1 \end{aligned}$$

The prevalence of "committed the crime", that is, the fraction of the population that committed the crime, is 1 in 30 million, because the search range included 30 million people.

$$\text{prevalence} = \frac{1}{30 \text{ million}}$$

Therefore, the numerator in PPV, sens × prev = 1 × 1/30 million = 1/30 million

The other concept we need is specificity, the likelihood of getting a negative test, that is, doesn't meet the description, given that the person did not have the disease, that is, did not commit the crime.

---

[1] For clarity of exposition, we have rounded off some of the actual numbers. The overall conclusion is not changed.

$$
\begin{aligned}
\text{specificity} &= p(- \mid \text{noDis}) \\
&= p(\text{does not match the description} \mid \text{innocent}) \\
&= 1 - p(\text{matches the description} \mid \text{innocent}) \\
&= 1 - \frac{1}{10 \text{ million}}
\end{aligned}
$$

a number very close to 1.

We now have all the pieces necessary to calculate the PPV for "matches the description", that is, to calculate the answer to the question: given that the couple matches the description, what is the likelihood that they committed the crime? (Analogous to "given that they tested positive, what is the likelihood that they actually have the disease?")

$$
\text{PPV} = \frac{\text{sens} \times \text{prev}}{\text{sens} \times \text{prev} + (1 - \text{spec}) \times (1 - \text{prev})}
$$

$$
= \frac{1 \times \frac{1}{30 \text{ million}}}{1 \times \frac{1}{30 \text{ million}} + \frac{1}{10 \text{ million}} \times \left(1 - \frac{1}{30 \text{ million}}\right)}
$$

$\Downarrow$ *multiply top and bottom by 30 million*

$$
= \frac{1 \times \frac{1}{30 \text{ million}} \times 30 \text{ million}}{1 \times \frac{1}{30 \text{ million}} \times 30 \text{ million} + \frac{1}{10 \text{ million}} \times \left(1 - \frac{1}{30 \text{ million}}\right) \times 30 \text{ million}}
$$

$$
= \frac{1 \times \frac{1}{\cancel{30 \text{ million}}} \times \cancel{30 \text{ million}}}{1 \times \frac{1}{\cancel{30 \text{ million}}} \times \cancel{30 \text{ million}} + \frac{1}{\cancel{10 \text{ million}}} \times \left(1 - \frac{1}{30 \text{ million}}\right) \times \cancelto{3}{30 \text{ million}}}
$$

$$
= \frac{1}{1 + \left(1 - \frac{1}{30 \text{ million}}\right) \times 3}
$$

$$
\approx \frac{1}{1 + 1 \times 3}
$$

$$
= \frac{1}{4}
$$

So the PPV of "matches the description", the likelihood that the couple committed the crime, given that they match the description, is 1/4! This means that there is only a 25% chance that any couple matching the description is a true positive, that is, there is only a 25% chance that a couple matching the description actually committed the crime. The Collins couple are probably innocent! (Figure 12.5).

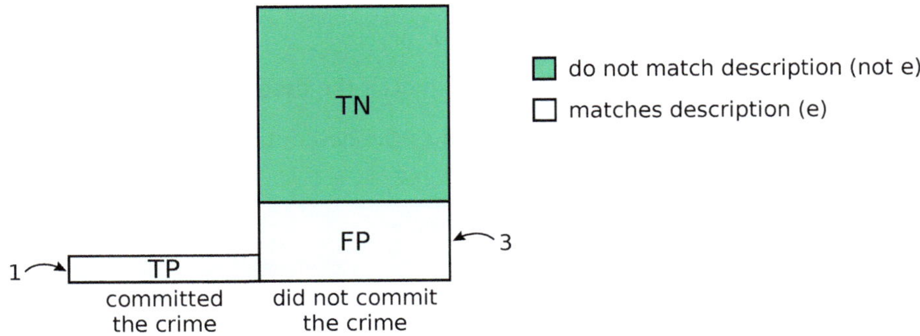

Assume H = couple committed the crime and e = matches description

$$\text{PPV} = p(H \mid e) = \frac{\text{TP}}{\text{TP} + \text{FP}} = \frac{1}{1+3} = 1/4$$

Figure 12.5 Shaded areas method for the PPV calculation of the California vs. Collins case.

The lessons here are the same as in the case of disease testing: when the prevalence is very low, most positives are false positives.

Racial profiling by law enforcement

Racial profiling by law enforcement and security forces has been sharply criticized as unjust discrimination. This criticism has merit, but an objection could be made that the policy is justified by its results. But is that the case?

In an article in the *Proceedings of the National Academy of Sciences*, the author demonstrated that racial profiling simply doesn't work mathematically.

PNAS PNAS PNAS

Strong profiling is not mathematically optimal for discovering rare malfeasors 2009

William H. Press

Department of Computer Science and School of Biological Sciences, University of Texas, Austin, TX 78703; and Los Alamos National Laboratory, Los Alamos, NM 87545

The use of profiling by ethnicity or nationality to trigger secondary security screening is a controversial social and political issue. Overlooked is the question of whether such actuarial methods are in fact mathematically justified, even under the most idealized assumptions of completely accurate prior probabilities, and secondary screenings concentrated on the highest-probablity individuals. We show here that strong profiling (defined as screening at least in proportion to prior probability) is no more efficient than uniform random sampling of the entire population, because resources are wasted on the repeated screening of higher probability, but innocent, individuals.

Why does racial profiling not work? For the same reason that when we test for a rare condition, most positives are false positives. So we would be spending the bulk of our time on the repeated screening of innocent people, because the positive predictive value is very low.

A new take on prevalence

To summarize, and to link this example to two forthcoming topics, we should note that we have defined a new concept, *prevalence.*

- Key new quantity: **prevalence** = $\frac{\texttt{\#Dis}}{\texttt{\#total}}$
- Another way to think of this is as the **prior probability** that a person was infected, prior to knowing that they tested +. Learning that they tested + then ***updated*** our probability from the prior to a new, **posterior probability**.

This new take enables us to link prevalence to two upcoming topics:

1) the role of the prior probability, and
2) the concept of updating probabilities, that are at the heart of Bayesianism.

FURTHER EXERCISES 12.2

1. Consider the following test. What are the sensitivity and specificity of this test? What are the Type I and Type II errors?

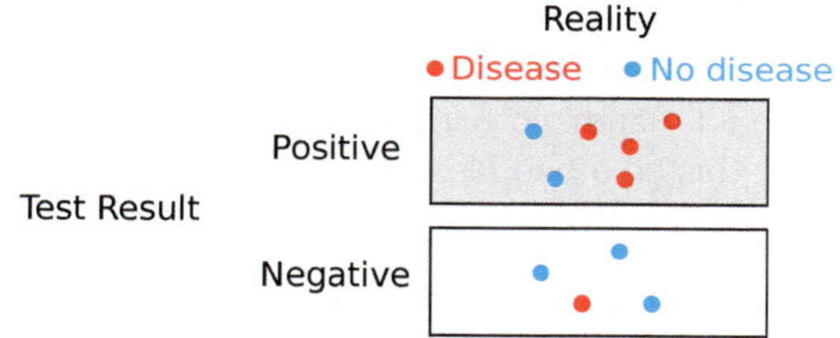

2. **Rapid test.** Several rapid tests have been developed to produce faster results for COVID-19. One test has been found to have a sensitivity of 87.1% and a specificity of 91.2%.

   a. In a population where 1 in 10,000 people have COVID, if this test is used for mass screening (testing people regardless of whether they have symptoms), what fraction of people who test positive actually have the disease?
   b. A person has a fever and cough but tests negative for the seasonal flu. When given this rapid COVID test, the person tests positive. Without making calculations, would you say the probability that this person has COVID is higher, lower or about the same as someone from the general population who tested positive on the COVID test. Briefly explain why.
   c. Suppose this test was given to a different population where the prevalence of COVID is 1 in 50. What's the probability that a person testing positive actually has COVID?
   d. When do you recommend using this test? What should you do if you test positive or test negative with a test like this?

3. **Breast cancer and the mammogram test.** Suppose that, in a population, the probability of a woman having breast cancer is 0.8 percent. If a woman has breast cancer, the probability

is 90 percent that she will have a positive mammogram. If a woman does not have breast cancer, the probability is 7 percent that she will have a positive mammogram.

a. For a mammogram test, what are the sensitivity, specificity and prevalence according to the description above?
b. If a woman has a positive mammogram test, what is the probability that she actually has breast cancer?

4. **Prostate cancer screening.** Screening for prostate cancer in men involves detecting prostate-specific antigen (PSA) in a blood test. PSA levels are higher in men with prostate cancer, men who develop prostate cancer later, and men with other (benign) conditions. If a man receives a positive PSA test, he may be advised to have a biopsy (invasive procedure) to confirm or deny the existence of prostate cancer. Studies have shown that about 40% of men in their 50s have prostate cancer (although they may not show symptoms or be aware of it).

   a. Using a cutoff of 3 ng/mL, the PSA blood test correctly identifies 32% of men with prostate cancer and correctly identifies 85% of men that do not have prostate cancer. If a man in his 50s receives a positive PSA test, what is the likelihood that he actually has prostate cancer? Give the numeric value of each of these quantities: specificity, sensitivity, prevalence and Positive Predictive Value.
   b. Younger men are less likely to have prostate cancer. Do you expect the likelihood that a man in his twenties with a positive test actually has prostate cancer to be greater, smaller or about the same as the likelihood for men in their fifties?
   c. We sometimes see ads on TV and in other places promoting prostate cancer screening for all men. Why might this be a bad idea?

5. **Van Herick test.** We want to determine how well the Van Herick test diagnoses primary angle close glaucoma (PACG). We examined 100 people that have PACG, as diagnosed by the "gold standard" test (gonioscopy). The Van Herick test diagnosed 75 of them as having the condition and 25 as not having the condition. We also examined 100 people that do not have PACG, as determined by the "gold standard" test. The Van Herick test diagnosed 15 of them as having PACG and 85 of them as not having PACG.

   a. Calculate the specificity and sensitivity of this test.
   b. Would you recommend using this test to diagnose PACG? Why or why not?

6. **Cystic fibrosis.** In order to provide early diagnosis and therefore delay or prevent serious, lifelong health problems related to cystic fibrosis (CF), all newborns in the United States are screened for CF with a blood test for the protein immunoreactive trypsinogen (IRT). If the test comes back positive, the child is tested further, such as by genetic screening, that can determine whether they actually have CF. The IRT blood test correctly identifies 87% of newborns with CF and correctly identifies 99% of newborns that do not have CF. About 1 in 3,000 newborns actually has CF (see article "Lessons learned from 20 years of newborn screening for cystic fibrosis". *Medical Journal of Australia*, 2012).

   a. If a newborn has a positive IRT blood test, what is the probability that they actually have CF? Give the numeric value of each of these quantities for the IRT blood test: specificity, sensitivity, prevalence and Positive Predictive Value.
   b. Explain why it makes sense that every newborn is given an IRT blood test.

12.3 Bayes' Theorem

In the disease-testing example, we saw that the key quantity we are looking for is $p(\text{Dis} \mid +)$, the positive predictive value of the test, where "Dis" means "has the disease" and "+" means "tested positive". We saw that, in order to calculate $p(\text{Dis} \mid +)$, we needed three quantities: the sensitivity $p(+ \mid \text{Dis})$ and specificity $p(- \mid \text{noDis})$ of the test, and also the prevalence of the disease in the population $p(\text{Dis})$.

To put it another way, PPV gives us a method for going from sensitivity $p(+ \mid \text{Dis})$ to positive predictive value $p(\text{Dis} \mid +)$, using the two additional factors specificity and prevalence.

We now want to generalize this to a bigger question: given *any* A and B, is it possible to go from $p(A \mid B)$ to $p(B \mid A)$?

The answer is yes, and the general method for going from $p(A \mid B)$ to $p(B \mid A)$ for arbitrary A and B is called Bayes' Theorem.

General Question: is $p(A \mid B)$ equal to $p(B \mid A)$?

Answer: no!

General Question: can we go from $p(A \mid B)$ to $p(B \mid A)$?

Answer: yes, if we supply two additional pieces of information: $p(B)$ and the "prior" probability $p(A)$.

$$p(A \mid B) \neq p(B \mid A)$$

Equating these two probabilities is a mistake that has infected and distorted hundreds of years of scientific research, according to Aubrey Clayton's excellent book, *Bernoulli's Fallacy* ("Bernoulli's fallacy" is his name for the mistake of thinking $p(A \mid B)$ must be equal to $p(B \mid A)$).

We've seen several examples illustrating this fallacy. Here's another. Let's say

A = the patient suffers from abdominal pain
B = the patient has inflammatory bowel disease (IBD)

and we know that

- most patients with IBD will experience abdominal pain, so $p(A \mid B)$ is high.
- $p(B \mid A)$, however, is much lower because only around 1% of the U.S. population has IBD, although many individuals suffer from abdominal pain symptoms at some point in their lives. This stomach pain may come from many other causes, such as indigestion, food poisoning, hernias, pancreatitis, etc.

We said that the big problem with p-values is that p-values are the answer to the wrong question. Now we can state this clearly, and understand what the wrong question is. This fallacy of thinking $p(A \mid B)$ must be equal to $p(B \mid A)$ *is* the p-value fallacy.

In order to understand how to go from the conditional probability $p(A \mid B)$ to the conditional probability $p(B \mid A)$, we first need a definition of the concept of conditional probability.

Conditional probabilities

We've been talking about "the probability of A given B", and we even introduced a notation for it, "$p(A \mid B)$", but we haven't defined it.

Let's imagine a Universe (the points in the pink oval in Figure 12.6). If X is any property, then the probability of X, $p(X)$, is defined as the area of the set of points that have property X. By definition, the probability of a point being in the Universe is 1, so the area of the pink oval is 1.

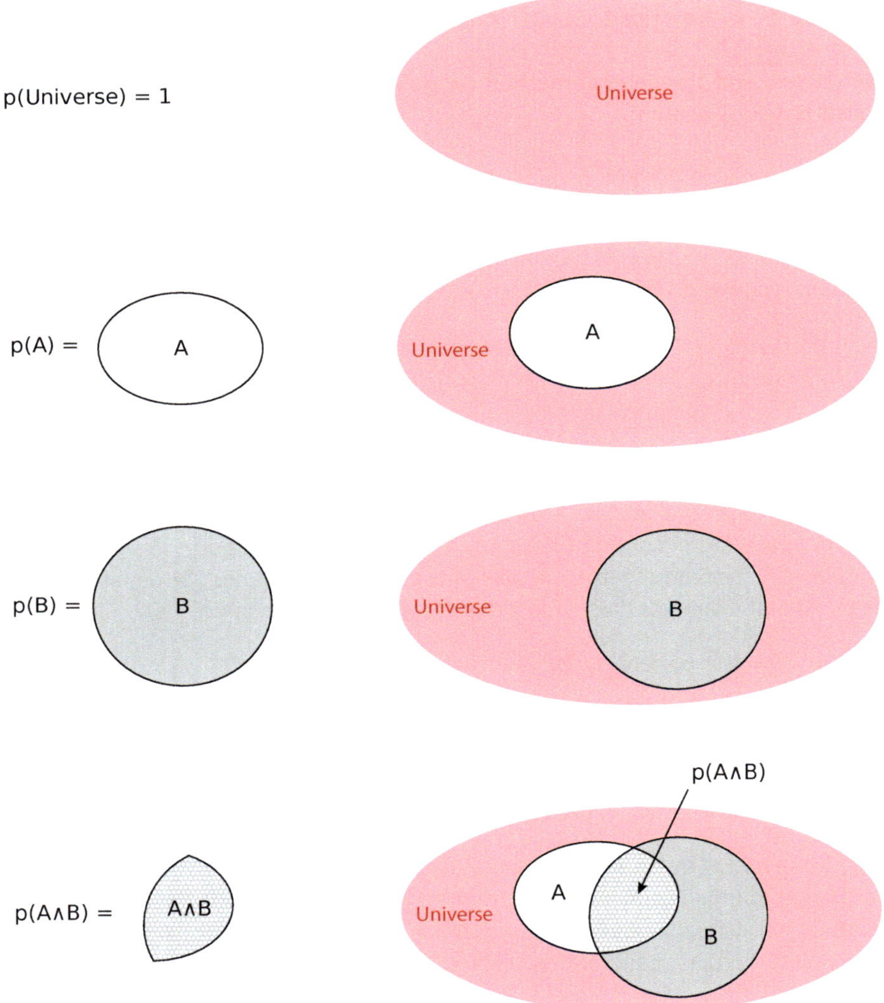

Figure 12.6 The Universe, A, B and A∧B, and their corresponding probabilities.

Let's consider a property A, and let the white oval represent the points in the Universe that have A. The probability of being an A, $p(A)$, is equal to *the area of A divided by the total area of the pink oval including A itself.* But the total area of the pink oval, including A itself, is 1, so $p(A)$ is the area of the white oval. $p(A)$ represents the fraction of the Universe that has A, which is indeed the probability of a point in the Universe being an A.

Similarly, let the gray disk B represent all the points in the Universe that have property B. So the probability of B, p(B), is *the area of the gray disk divided by the total area of the pink oval (including B)*. Again, the area of the pink oval is 1, so p(B) = the area of the gray disk.

In this example, note that there is an overlap between the white oval A and the gray disk B. This is the wedge-shaped region where both A and B occur. In set theory, the wedge-shaped region is called the intersection of A and B, and is written "A∧B", which is read as "A intersect B". In logic and probability theory, the wedge symbol ∧ is also used to mean "and". The double meaning is intentional: the probability of A and B, p(A∧B), is equal to the area of the region "A intersect B", that is, p(A∧B).

p(A | B) We can now define the probability of A given B, p(A | B) (Figure 12.7 top). Since B has happened, we can forget the pink Universe and the white segment. p(A | B) is the fraction of B's who are also A's, which is the area of the wedge-shaped intersection (gray dotted area) of the A oval and the B disk, divided by the area of the B disk.

$$p(A \mid B) = \frac{p(A \wedge B)}{p(B)}$$

p(B | A) Similarly, we can also define the probability of B given A, p(B | A) (Figure 12.7 bottom). Since A has happened, we can forget the pink Universe and the gray segment. p(B | A) is the fraction of A's who are also B's, the area of the wedge-shaped intersection (gray dotted area) of the A oval and the B disk, divided by the area of the A oval.

$$p(B \mid A) = \frac{p(A \wedge B)}{p(A)}$$

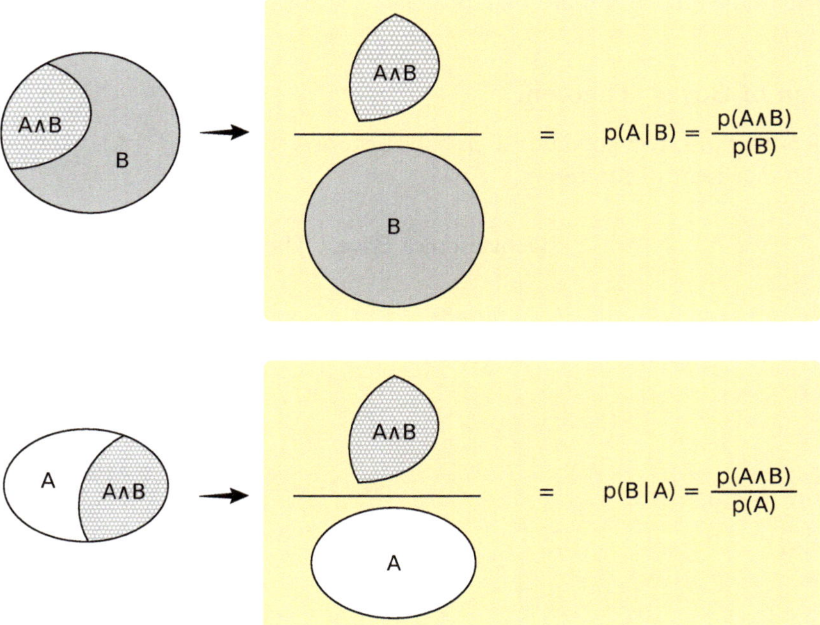

Figure 12.7 Graphical presentation of the two conditional probabilities, p(A | B) and p(B | A).

p(A | B) ≠ p(B | A) To answer the question "is p(A | B) equal to p(B | A)", looking at the two definitions, the answer is clearly no. Conditional probabilities are not commutative, the reversed order in a conditional probability does not yield the same result.

Exercise 12.3.1 The conditional probability p(A | B) is read "probability of A given B". A is what we are trying to find and B is what we know.

trying to find → p(A | B) ← know

read "probability of A given B"

Here is a table showing the number of times students visited tutoring, for full time students and part time students.

	Number of time students visited tutoring			
	never	1 to 5 times	more than 5 times	total
full time student	12	25	8	45
part time student	2	5	6	13
total	14	30	14	58

Calculate the following conditional probabilities. For each conditional probability, what is the "trying to find" and what is the "know"? Highlight those two components in the table.

- p(more than 5 times | full time student)
- p(full time student | more than 5 times)
- p(never | full time student)
- p(full time student | never)

Derivation of Bayes' Theorem

Having defined those two conditional probabilities, p(A | B) and p(B | A), we can then do some simple rearrangement of the terms.

Derivation of Bayes' Theorem

① $$p(A \mid B) = \frac{p(A \wedge B)}{p(B)} \Rightarrow p(A \wedge B) = p(A \mid B) \times p(B)$$

② $$p(B \mid A) = \frac{p(A \wedge B)}{p(A)} = \frac{p(A \mid B) \times p(B)}{p(A)}$$

③ $$p(B \mid A) = \frac{p(A \mid B) \times p(B)}{p(A)}$$

① Starting from the definition of $p(A \mid B)$ on the left, we can multiply both sides by $p(B)$, to get $p(A \wedge B) = p(A \mid B) \times p(B)$.

② Then we take the definition of $p(B \mid A)$, and substitute for $p(A \wedge B)$ using line 1.

③ That gives us line 3, which is Bayes' Theorem.

From $p(A \mid B)$ to $p(B \mid A)$ We see that we can get $p(B \mid A)$ from $p(A \mid B)$ by multiplying $p(A \mid B)$ by the ratio of two new quantities: the numerator $p(B)$ and the denominator $p(A)$.

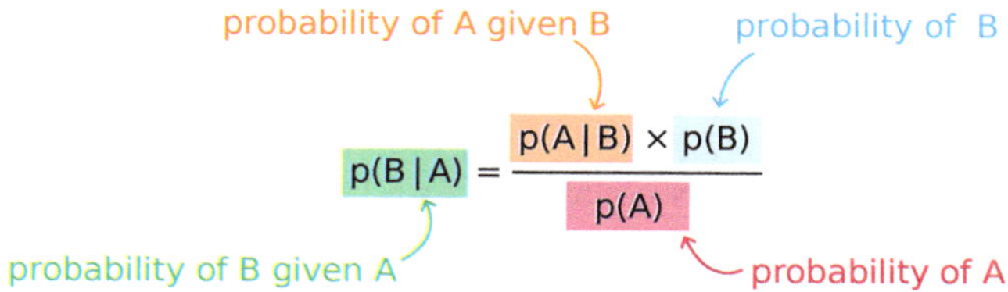

Figure 12.8 Bayes' Theorem, from $p(A \mid B)$ to $p(B \mid A)$.

The understanding of these two quantities, and how to apply them in particular cases, is the heart of Bayesian statistics. We'll now go on to discuss several different applications of Bayes' Theorem, each with their own interpretation of $p(B)$ and $p(A)$.

Bayes as updating

An important insight into Bayes' Theorem is that Bayes' Theorem is all about *updating* probabilities based on new information. In traditional statistics, a probability is a static number, something you calculate once and for all.

Update $p(B)$ to $p(B|A)$ by learning A In particular, we imagine that we have a "prior" probability of property B to happen, $p(B)$. Then we get a new piece of data, A, and we want to update the prior probability $p(B)$ to a new probability, $p(B \mid A)$, the "posterior" probability of B given the additional information that A has happened (Figure 12.9).

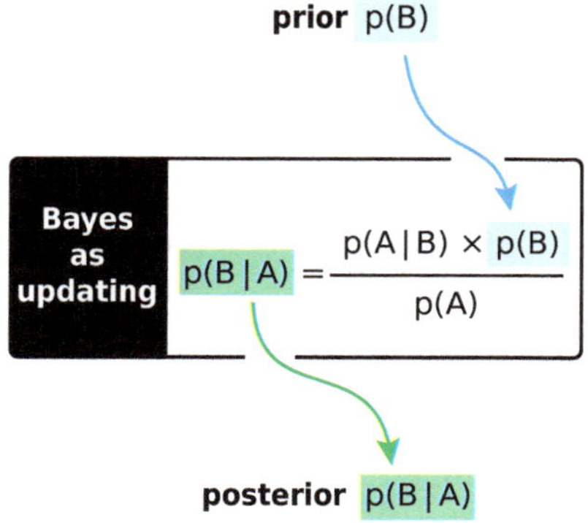

Figure 12.9 Bayes' Theorem as probability updating using new data A.

Bayes' Theorem tells us that we can update the prior probability p(B) to get the posterior probability p(B | A) if we know two additional quantities, p(A) and p(A | B).

- p(A) is the overall probability of A happening.
- p(A | B) is the probability of A happening when B is present.

The terms in Bayes' Theorem

We can gain a lot of insight into Bayes' Theorem by thinking about the terms in the numerator and denominator in the fraction on the right-hand side.

The abstract form In the abstract form of Bayes' Theorem (Figure 12.9),

- The **numerator** is p(A | B) × p(B). Consulting Figure 12.10, we see that it is the area of the wedge-shaped region, which is equal to p(A∧B).
- The **denominator** is p(A). But what is p(A)? If we refer to the diagram (Figure 12.10), p(A) is the area of the white oval A. We see that the white oval A is the sum of two pieces, one of which is inside the gray disk B and the other of which is outside B. The first piece is A∧B and the second piece is A∧~B. (We use ~B to denote not-B.)

  But if

  $$A = A \wedge B + A \wedge {\sim}B$$

  then

  $$p(A) = p(A \wedge B) + p(A \wedge {\sim}B)$$

  and if we substitute from the definitions of conditional probabilities (Figure 12.7), we get

  $$p(A) = \underbrace{p(A \mid B) \times p(B)}_{\text{the B case}} + \underbrace{p(A \mid {\sim}B) \times p({\sim}B)}_{\text{the} \sim\text{B case}}$$

  That is, the total probability of A happening is equal to the (probability of A given B) times the probability of B (the B case) plus the (probability of A given not-B) times the probability of not-B (the not-B case). For example, the probability of seeing a traffic jam in a given week is equal to the probability of seeing a traffic jam on a weekday times the probability of it being a weekday (5/7) plus the probability of seeing a traffic jam on a weekend times the probability of it being a weekend (2/7).

  Thus, we have "split the denominator" p(A) into the sum of two mutually exclusive and jointly exhaustive cases, the "B case" and the "not-B case". We can graphically illustrate "splitting the denominator" (Figure 12.10).

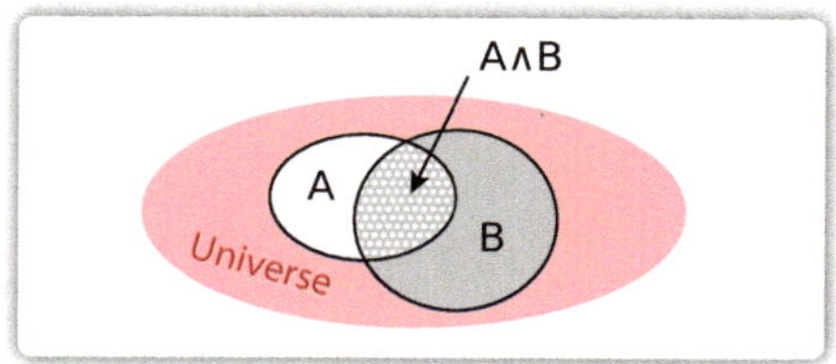

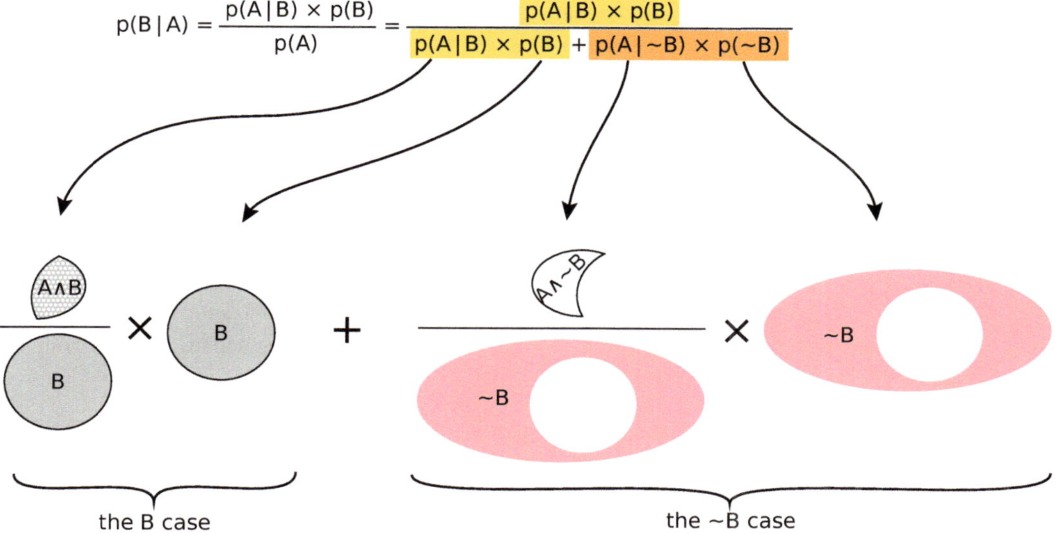

Figure 12.10 Splitting the denominator in Bayes' Theorem.

Disease testing Positive Predictive Value, in its simplest and most straightforward form, can be written as

$$\text{PPV} = \frac{\text{TP}}{\text{TP} + \text{FP}}$$

- The **numerator**, TP, is the number of true positives, which is the sensitivity of the test × the prevalence of the disease × the total number of people tested. If we call the total number of people tested N, then,

$$\text{TP} = \text{sens} \times \text{prev} \times \text{N}$$

- The **denominator**, TP + FP, is the number of all positives. It has already been split into the sum of the two types of positives: the diseased case (True Positives) and the not-diseased case (False Positives).

  TP is the same as in the numerator, that is sensitivity × prevalence × N, while FP is the number of false positives, that is, the probability of the test saying "positive" when the subject does not have the disease × the total number of people who do not have the disease. In other words, FP is (1 − specificity) × the probability of not having the disease × N. The probability of not having the disease is equal to (1 − the probability of having the disease).

$$\text{TP} + \text{FP} = \text{sens} \times \text{prev} \times \text{N} + (1 - \text{spec}) \times (1 - \text{prev}) \times \text{N}$$

Therefore,

$$\begin{aligned}\text{PPV} &= \frac{\text{TP}}{\text{TP} + \text{FP}} \\ &= \frac{\text{sens} \times \text{prev} \times \cancel{\text{N}}}{\text{sens} \times \text{prev} \times \cancel{\text{N}} + (1 - \text{spec}) \times (1 - \text{prev}) \times \cancel{\text{N}}}\end{aligned}$$

We see that the Ns cancel out. So we can refer to TP as either the number of true positives or the probability of being a true positive in the test population. Similarly, we can refer to FP as either the number of false positives or the probability of being a false positive in the test population.

Let's now consider PPV by applying Bayes' Theorem. What we are looking for, PPV, can be put as p(Dis | +), where "Dis" is "having the disease", and "+" is "tested positive". Bayes' Theorem then tells us that

$$\text{PPV} = \text{p}(\text{Dis} \mid +) = \frac{\text{p}(+ \mid \text{Dis}) \times \text{p}(\text{Dis})}{\text{p}(+)}$$

- The **numerator** is p(+ | Dis) × p(Dis). We recognize that this is the sensitivity × prevalence, which is the number of True Positives.
- The **denominator** is p(+). But what is p(+)? We split the denominator p(+) into two terms, the diseased case and the not-diseased case. The diseased case is the True Positives, and the not-diseased case is the False Positives.

$$\text{p}(+) = \underbrace{\text{p}(+ \mid \text{Dis}) \times \text{p}(\text{Dis})}_{\text{the diseased case}} + \underbrace{\text{p}(+ \mid \text{noDis}) \times \text{p}(\text{noDis})}_{\text{the not-diseased case}}$$

Therefore, the concept of PPV is a direct application of Bayes' Theorem.

Hypothesis and evidence Let's consider the application of Bayes' Theorem to the case where we have a hypothesis H and evidence e. What we really want to know is p(H | e): given this evidence, how likely is the hypothesis? That's the real scientific question.

Bayes' Theorem tells us that

$$\text{p}(\text{H} \mid \text{e}) = \frac{\text{p}(\text{e} \mid \text{H}) \times \text{p}(\text{H})}{\text{p}(\text{e})}$$

- The **numerator** is p(e | H) × p(H). The first term p(e | H) is the probability of this evidence given the hypothesis. The hypothesis could be anything; when the hypothesis is the Null Hypothesis, then $\text{p}(\text{e} \mid \text{H}_0)$ is part of the calculation (sometimes the whole calculation) of the p-value; the p-value is the sum of a number of terms of the form $\text{p}(\text{e} \mid \text{H}_0)$.

  The second term, p(H), is the prior probability of the hypothesis being true. In order to do Bayesian updating, there has to be a prior to get updated.

  This need to supply a prior probability has been held to be both a strength and a weakness of Bayesianism.

On the one hand, the strength of the need for a prior means that each new study is treated as information to update our prior collective knowledge. In Bayesianism, knowledge accumulates through the updating process.

And it is certainly true that we cannot evaluate a hypothesis in a vacuum, without any context or general knowledge or previous experiments. For example, if a hypothesis has a very low prior probability, it *should* take more evidence to establish it.

In an important and widely-discussed example, Cornell psych prof Darryl Bem claimed to have found evidence for the existence of Extra-Sensory Perception (ESP), in the form of "precognition", the mystical ability to "perceive" events before they happen. His evidence was statistical; he used Null Hypothesis Significance Testing. For example, he asked 100 participants to guess the outcome of yes/no events, a guess that should have split 50/50, and found that 53% had guessed right. Using a (one-sided!) t-test, he found this result to be statistically significant ($p = 0.01$), and concluded that this is evidence for ESP.

Critics responded that with a low prior probability for "ESP exists", you need better than a $p < 0.01$ level of evidence before you reject the hypothesis of random chance. As Wagenmakers *et al.* (2011) wrote,

> "... in order to convince a skeptical audience of a controversial claim, one needs to conduct strictly confirmatory studies and analyze the results with statistical tests that are conservative rather than liberal. We conclude that Bem's p values do not indicate evidence in favor of precognition; instead, they indicate that experimental psychologists need to change the way they conduct their experiments and analyze their data."

On the other hand, the need to supply a prior has been held to be a weakness of Bayesianism. Critics cite the prior as potentially simply embodying previous prejudices. If we decided, based on prejudice or bad science, that the existence of witchcraft has a very high probability, then new evidence against the existence of witchcraft will not be able to alter the probability very much.

Where to get the priors?

There are a number of places to get a prior probability:

- **Literature.** The scientific literature is a good source of priors. Use the findings from other studies in the field as the prior. That way, you are viewing the present study in the context of previous studies, not in isolation.
- **Pilot study.** Do a small trial experiment and see what you get.
- **Background knowledge.** "In my clinical experience, this type of cancer occurs x% of the time in this type of situation".
- **Flat prior.** Assume all possibilities are equally likely. For example, if we don't know the percentage cured of a drug, we can assume that all percentages, from 0 to 100%, are equally likely. This is sometimes referred to as a "flat prior"; we sometimes call it "IHNI" (I Have No Idea).
- **Guess and iteration.** Studies have shown that in many cases, if we have enough data, it doesn't really matter what the prior is, because it will be "washed out"

by the subsequent stream of data. This is especially true for situations that are inherently updating with lots of fresh data. This happens in clinical trials, in cryptography, in gambling, in target identification, etc. In these applications, we can start with a flat prior, and let the incoming data update it. For an example of this process in the context of clinical trials, see section 12.4 *Bayes and Clinical trials*.

- The **denominator** is p(e). What does this mean? It's "the probability of this evidence occurring". What that means is the total probability of this evidence occurring in any situation whatever. We need to know the probability of e happening if H is true, but we also need to know the probability of e happening if H is false!

$$p(e) = \underbrace{p(e \mid H) \times p(H)}_{\text{probability of seeing this evidence when H is true}} + \underbrace{p(e \mid {\sim}H) \times p({\sim}H)}_{\text{probability of seeing this evidence when H is false}}$$

The idea here is that $p(e \mid {\sim}H)$ is like a false positive rate. It is the likelihood of seeing this evidence when the hypothesis is false.

The importance of that second term, $p(e \mid {\sim}H) \times p({\sim}H)$, in judging guilt or innocence given a positive drug test, was critical in overturning the doping charge against the US Olympic runner Mary Decker Slaney, who tested positive for testosterone. She was initially barred from the sport, in a classic example of the prosecutor's fallacy.

At the Olympics, testing for the steroid drug testosterone is done by measuring the ratio of two substances: testosterone T and epitestosterone E. If the ratio T/E is greater than 6, the athlete is suspected of doping, that is, cheating by administering performance enhancing drugs.

To see the fallacy, let's agree that if an athlete is actually doping, then their T/E ratio will be greater than 6. In the language of Bayesianism, p(high T/E | doping) is high. **But what we really need to know to convict is not p(high T/E | doping) but p(doping | high T/E).** In other words, we need to know the Positive Predictive Value of a positive doping test.

$$p(\text{doping} \mid \text{high T/E}) = \frac{p(\text{high T/E} \mid \text{doping}) \times p(\text{doping})}{p(\text{high T/E})}$$

$$= \frac{p(\text{high T/E} \mid \text{doping}) \times p(\text{doping})}{\underbrace{p(\text{high T/E} \mid \text{doping}) \times p(\text{doping})}_{\text{the doping case}} + \underbrace{p(\text{high T/E} \mid \text{no doping}) \times p(\text{no doping})}_{\text{the no doping case}}}$$

If there is a high probability of a positive doping test when the athlete is innocent, that radically lowers the Positive Predictive Value of a positive test.

As *Science* magazine summed it up:

Science NEWS FOCUS | 1999

The Reverend Bayes Goes to Court

Bayesian clarity was also critical to undermining a doping accusation against U.S. long-distance runner and former record holder Mary Decker Slaney. Slaney was barred from the sport after failing a drug test for performance-enhancing steroids at the 1996 Olympic games. But the test, which measures the ratio of the steroids testosterone to epitestosterone (T/E) in urine, is statistically impossible to interpret as evidence of guilt or innocence, says statistician Don Berry of the University of Texas M. D. Anderson Cancer Center in Houston, who helped Slaney's defense team win a reversal from U.S. track officials.

Berry used a Bayesian approach to show that prior knowledge is essential to conclude anything from the test. For one, he says, testers need to know "the probability that you'd get a T/E greater than 6 [the prohibited level] if she used a banned substance, and the same probability if she didn't use it." Doctors who rely on tests to diagnose everything from AIDS to paternity face similar problems, he notes, as do prosecutors trying to pin convictions on other kinds of drug tests. In each case, other statisticians say, the Bayesian framework exposes the range of prior information needed to interpret test results properly.

Exercise 12.3.2 You randomly choose a treasure chest to open, and then randomly choose a coin from that treasure chest. If the coin you choose is gold, then what is the probability that you choose chest A?

100 gold coins

50 gold coins + 50 silver coins

(hint: let H = "chest A was chosen" and e = "a gold coin was chosen". Clearly what we want is p(H | e). Use Bayes' Theorem to calculate p(H | e).)

Exercise 12.3.3 A sign on the local bus says "2/3 of pedestrian fatalities take place in crosswalks". Does this mean you should avoid crosswalks?

Bayes is Positive Predictive Value

We have now developed two concepts: Positive Predictive Value and Bayes' Theorem. They are equivalent statements, and typically we can use either form to get insight into an application.

We also derived several different forms of PPV and Bayes' Theorem:

- an abstract form
- a form applicable to testing for a disease
- a form applicable to a hypothesis and evidence

These are summarized in the table below.

Abstract form

Positive Predictive Value	Bayes' Theorem
	$p(B \mid A) = \frac{p(A \mid B) \times p(B)}{p(A)} = \frac{p(A \mid B) \times p(B)}{p(A \mid B) \times p(B) + p(A \mid \sim B) \times p(\sim B)}$
Fraction of Positives that are True Positives $PPV = \frac{TP}{TP + FP}$	$p(TP \mid +) = \frac{p(+ \mid TP) \times p(TP)}{p(+)} = \frac{p(+ \mid TP) \times p(TP)}{p(+ \mid TP) \times p(TP) + p(+ \mid FP) \times p(FP)} = \frac{1 \times p(TP)}{1 \times p(TP) + 1 \times p(FP)} = \frac{p(TP)}{p(TP) + p(FP)} \times \frac{\text{population size N}}{\text{population size N}} = \frac{TP}{TP + FP}$

Disease and test

PPV of the test	Bayes' Theorem
$p(Dis \mid +) = \frac{TP}{TP + FP} = \frac{p(+ \mid Dis) \times p(Dis)}{p(+ \mid Dis) \times p(Dis) + p(+ \mid \sim Dis) \times p(\sim Dis)} = \frac{sens \times prev}{sens \times prev + (1 - spec) \times (1 - prev)}$	$p(Dis \mid +) = \frac{p(+ \mid Dis) \times p(Dis)}{p(+)} = \frac{p(+ \mid Dis) \times p(Dis)}{p(+ \mid Dis) \times p(Dis) + p(+ \mid \sim Dis) \times p(\sim Dis)}$

Hypothesis and evidence

PPV of the evidence	Bayes' Theorem
$p(H \mid e) = \frac{TP}{TP + FP} = \frac{p(e \mid H) \times p(H)}{p(e \mid H) \times p(H) + p(e \mid \sim H) \times p(\sim H)}$	$p(H \mid e) = \frac{p(e \mid H) \times p(H)}{p(e)} = \frac{p(e \mid H) \times p(H)}{p(e \mid H) \times p(H) + p(e \mid \sim H) \times p(\sim H)}$

Example 2 When drug screening meets Bayes' Theorem

Employment-related drug screening is in wide use, but does this common practice provide credible evidence of illegal substance use? Considering the serious consequence of false accusations, we want to ask a simple question: what are the chances that a potential employee who tests positive for an illegal substance actually uses that substance?

Suppose that a drug test for some illegal substance is 99% accurate (sensitivity) in the case of users of that substance, and 99% accurate (specificity) in the case of non-users. Suppose it is known that 0.5% (prevalence) of the entire population uses the substance. What are the chances that someone who tested positive has not used the substance?

You may recognize this problem: it's identical to the one we just developed in the context of disease-detection (section 12.2 *Positive Predictive Value in medicine and social science: an introduction to Bayesianism* on page 583). Even the numbers are the same; just substitute "drug use" for "disease" and they are identical. In that discussion, we used the concept of Positive Predictive Value to reach a surprising conclusion about the probability of having the disease given that you have tested positive.

Now, let's do the same problem, but *purely by using Bayes' Theorem, and not relying on the concept of Positive Predictive Value.*

Question What parameters about this drug test are presented in the above description?

Answer Sensitivity, specificity and prevalence.

Question Let

$$\mathrm{H} = \text{person uses the substance}$$
$$\mathrm{e} = \text{person tests positive}$$

How to present the test parameters mathematically?

Answer We would present the parameters as

- sensitivity = $p(e \mid H) = 99\%$
- specificity = $p(\text{not-}e \mid \text{not-}H) = 99\%$
- prevalence = $p(H) = 0.5\%$

Question What is the probability that someone who tested positive actually uses the substance?

Answer $p(H \mid e)$

Question What is the probability that someone who tested positive does not use this substance?

Answer $p(\text{not-}H \mid e)$

Question What is the relationship between $p(H \mid e)$ and $p(\text{not-}H \mid e)$?

Answer $p(\text{not-}H \mid e) + p(H \mid e) = 1$

Question Use Bayes' Theorem to calculate the probability $p(H \mid e)$.

Answer The probability of H is what we are interested in updating upon receiving the information e. Applying Bayes' Theorem, we get

$$p(H \mid e) = \frac{p(e \mid H) \times p(H)}{p(e)}$$

We already know the two terms in the numerator, p(H), the prior probability of H, and p(e | H), the probability of e given H. The only term we need to calculate is the denominator, p(e), the probability of testing positive. To calculate p(e), we notice that there are only two scenarios for a person to test positive, 1) a person who uses the substance tests positive, and 2) a person who does not use the substance but falsely tests positive.

$$\begin{aligned} p(e) &= \overbrace{p(e \mid H) \times p(H)}^{\substack{\text{person who}\\ \text{uses the substance}\\ \text{tests positive}}} + \overbrace{p(e \mid \text{not-}H) \times p(\text{not-}H)}^{\substack{\text{person who does not}\\ \text{use the substance}\\ \text{but falsely tests positive}}} \\ &= p(e \mid H) \times p(H) + \left(1 - p(\text{not-}e \mid \text{not-}H)\right) \times \left(1 - p(H)\right) \end{aligned}$$

Substituting this expression for p(e) back into the denominator of p(H | e), we get

$$\begin{aligned} p(H \mid e) = \frac{p(e \mid H) \times p(H)}{p(e)} &= \frac{p(e \mid H) \times p(H)}{p(e \mid H) \times p(H) + \left(1 - p(\text{not-}e \mid \text{not-}H)\right) \times \left(1 - p(H)\right)} \\ &= \frac{99\% \times 0.5\%}{99\% \times 0.5\% + (1 - 99\%) \times (1 - 0.5\%)} \\ &= 33.22\% \end{aligned}$$

So the probability that someone who has tested positive actually uses the substance, p(H | e), is 33.22%.

We have just reached the same conclusion as in the disease testing case, purely by using Bayes' Theorem and not relying explicitly on the concept of PPV.

Question How to use Bayes as updating to understand the probability update from the prior to the posterior?

Answer We had a prior probability that a person uses the substance: 1 in 200.

prior p(H) = 0.5%

Then we updated that prior probability with the new data that the person tested positive (e).

Now, the updated, posterior probability, is

posterior p(H | e) = 33.22%

Question Recall the expression for PPV (Positive Predictive Value) in the context of disease testing in section 12.2, the probability of someone testing positive for a disease who actually has the disease,

$$\text{PPV} = p(\text{Dis} \mid +) = \frac{\text{sens} \times \text{prev}}{\text{sens} \times \text{prev} + (1 - \text{spec}) \times (1 - \text{prev})}$$

Compare this PPV expression with the p(H | e) expression and find the three terms in the p(H | e) expression: sensitivity, specificity, and prevalence.

Answer If we consider the expression of p(H | e) deduced purely from Bayes' Theorem, and compare it to the expression for PPV, we find that the three main components of PPV: sensitivity, specificity, and prevalence are manifested in Bayes' Theorem as

$$p(H \mid e) = \frac{\overbrace{p(e \mid H)}^{\text{sensitivity}} \times \overbrace{p(H)}^{\text{prevalence}}}{\underbrace{p(e \mid H)}_{\text{sensitivity}} \times \underbrace{p(H)}_{\text{prevalence}} + \left(1 - \underbrace{p(\text{not-}e \mid \text{not-}H)}_{\text{specificity}}\right) \times \left(1 - \underbrace{p(H)}_{\text{prevalence}}\right)}$$

Question What is the probability that someone who tested positive does not use the substance?

Answer p(not-H | e) is the probability that someone who tested positive does not use the substance. Based on the relationship between p(H | e) and p(not-H | e), we can calculate

$$p(\text{not-}H \mid e) = 1 - p(H \mid e) = 1 - 33.22\% = 66.78\%$$

This drug screen is extremely accurate, with 99% sensitivity and 99% specificity. However, our calculation using Bayes's Theorem has shown that this supposedly reliable drug test would result in a 2 out of 3 chance of a false accusation, and therefore has a low positive predictive value!

What's more alarming is that from an individual person's perspective, if an individual tests positive, it's more likely than not that that individual does not use that illegal substance. *In cases like this, most positives are False Positives.*

Question To see this with actual numbers, fill in the blanks below.

> If 1,000 individuals are tested, we expect ____ non-users and ____ users. Among the non-users, ____ false positives are expected. Among the users, ____ true positives are expected. Out of ____ positive results, only ____ are genuine. (Round your calculations to integers.)

Answer If 1,000 individuals are tested, we expect 995 non-users and 5 users. Among the non-users, 10 ($\approx$ 995 × 1%) false positives are expected. Among the users, 5 ($\approx$ 5 × 99%) true positives are expected. Out of 15 positive results, only 5 are genuine, and 10 are false positives.

Question: higher sensitivity Now suppose there is an updated version 2.0 of the previous drug test and it has increased the sensitivity from 99% to 100%. Its specificity remains the same 99%, and population prevalence stays at 0.5% level. What are the chances that someone who tested positive actually uses that substance?

Answer The new drug test has the following parameters:

- sensitivity = p(e | H) = 100%
- specificity = p(not-e | not-H) = 99%

- prevalence = p(H) = 0.5%

Using the formula p(H | e) with the new parameters, we get

$$\textbf{Drug test 2.0: } p(H \mid e) = \frac{p(e \mid H) \times p(H)}{p(e)} = \frac{p(e \mid H) \times p(H)}{p(e \mid H) \times p(H) + \left(1 - p(\text{not-}e \mid \text{not-}H)\right) \times \left(1 - p(H)\right)}$$
$$= \frac{100\% \times 0.5\%}{100\% \times 0.5\% + (1 - 99\%) \times (1 - 0.5\%)}$$
$$= 33.44\%$$

In drug test 2.0, the perfect sensitivity of 100% has little effect on the PPV: it only increases the probability that a person who tests positive actually uses that substance from 33.22% to 33.44%.

Question: higher sensitivity and higher specificity Now suppose there is a further updated version 3.0 of the drug test and it maintains the same sensitivity of 100%. Its specificity increased from 99% to 99.5%, and population prevalence stays at 0.5% level. What are the chances that someone tested positive actually uses that substance?

Answer The further improved new drug test has the following parameters:

- sensitivity = p(e | H) = 100%
- specificity = p(not-e | not-H) = 99.5%
- prevalence = p(H) = 0.5%

Using the formula p(H | e) with the new parameters, we get

$$\textbf{Drug test 3.0: } p(H \mid e) = \frac{p(e \mid H) \times p(H)}{p(e)} = \frac{p(e \mid H) \times p(H)}{p(e \mid H) \times p(H) + \left(1 - p(\text{not-}e \mid \text{not-}H)\right) \times \left(1 - p(H)\right)}$$
$$= \frac{100\% \times 0.5\%}{100\% \times 0.5\% + (1 - 99.5\%) \times (1 - 0.5\%)}$$
$$= 50.1\%$$

By increasing specificity, we have substantially increased the PPV of the test, however, it is still true that half of all the positives detected by drug test 3.0 are false accusations!

Question: higher prevalence To see the effects of disease prevalence on PPV, suppose we give the original test (sensitivity = 99%, specificity = 99%) to a different population. In this population, the prevalence is 13%. In this population, what are the chances that a person who tested positive actually uses that substance?

Answer In this new population, the drug test has the following parameters:

- sensitivity = p(e | H) = 99%
- specificity = p(not-e | not-H) = 99%

- prevalence = p(H) = 13%

Using the formula p(H | e) with the new parameters, we get

New population with higher prevalence:

$$p(H \mid e) = \frac{p(e \mid H) \times p(H)}{p(e)}$$
$$= \frac{p(e \mid H) \times p(H)}{p(e \mid H) \times p(H) + \big(1 - p(\text{not-e} \mid \text{not-H})\big) \times \big(1 - p(H)\big)}$$
$$= \frac{99\% \times 13\%}{99\% \times 13\% + (1 - 99\%) \times (1 - 13\%)}$$
$$= 93.7\%$$

We conclude that the effect of disease prevalence on PPV is very strong: in a population where prevalence is high, a positive test is a much stronger indicator of the presence of disease.

Example 3 The prosecutor's fallacy: California vs. Collins

In the California vs. Collins example, we applied the concept of positive predictive value and showed that the PPV of "matches the description", that is, the likelihood that the couple committed the crime, given that they match the description, is 1/4 (see section 12.2 *Positive Predictive Value in medicine and social science: an introduction to Bayesianism* on page 594).

In our calculation of the PPV,

$$\text{PPV ("matches the description")} = p(\text{committed the crime} \mid \text{matches the description})$$
$$= \frac{\text{sens} \times \text{prev}}{\text{sens} \times \text{prev} + (1 - \text{spec}) \times (1 - \text{prev})}$$

we identified the three main components,

- sensitivity = p(matches the description | committed the crime) = 1
- specificity = p(does not match the description | innocent)

$$= 1 - p(\text{matches the description} \mid \text{innocent})$$
$$= 1 - \frac{1}{10 \text{ million}}$$

- prevalence = $\frac{1}{30 \text{ million}}$

In that discussion, we used the concept of positive predictive value to reach a surprising conclusion that there is only a 25% chance that this positive finding is a true positive. In other words, the likelihood that the couple committed the crime, given that they match the description, is 25%.

Now, let's do the same problem, but *purely by using Bayes' Theorem, and not relying explicitly on the concept of positive predictive value.*

Question Assume that the hypothesis H and the evidence e are

$$\begin{aligned} \mathrm{H} &= \text{couple committed the crime} \\ \mathrm{e} &= \text{matches the description} \end{aligned}$$

How should you present sensitivity, specificity, and prevalence mathematically?

Answer We would present them as

- sensitivity = $p(e \mid H) = 1$
- specificity = $p(\text{not-e} \mid \text{not-H}) = 1 - \dfrac{1}{10 \text{ million}}$
- prevalence = $p(H) = \dfrac{1}{30 \text{ million}}$

Question What is the probability that a couple who matches the description actually committed the crime?

Answer $p(H \mid e)$

Question What is the probability that a couple matches the description but did not commit the crime?

Answer $p(\text{not-H} \mid e)$

Question What is the relationship between $p(H \mid e)$ and $p(\text{not-H} \mid e)$?

Answer $p(\text{not-H} \mid e) + p(H \mid e) = 1$

Question Use Bayes' Theorem to calculate the probability $p(H \mid e)$.

Answer The probability of H is what we are interested in updating upon receiving the information e. Applying Bayes' Theorem, we get

$$p(H \mid e) = \frac{p(e \mid H) \times p(H)}{p(e)}$$

We already know the two terms in the numerator, $p(H)$, the prior probability of H, and $p(e \mid H)$, the probability of e given H. The only term we need to calculate is the denominator, $p(e)$, the probability of a couple matching the description. To calculate $p(e)$, there are only two scenarios for a couple to match the description, 1) a couple who committed the crime and matches the description, and 2) a couple who did not commit the crime but matches the description.

$$\begin{aligned} p(e) &= \overbrace{p(e \mid H) \times p(H)}^{\substack{\text{a couple who committed} \\ \text{the crime and} \\ \text{matches the description}}} + \overbrace{p(e \mid \text{not-H}) \times p(\text{not-H})}^{\substack{\text{a couple who did not} \\ \text{commit the crime but} \\ \text{matches the description}}} \\ &= p(e \mid H) \times p(H) + \big(1 - p(\text{not-e} \mid \text{not-H})\big) \times \big(1 - p(H)\big) \end{aligned}$$

Substituting this expression for p(e) back into the denominator of p(H | e), we get

$$p(H \mid e) = \frac{p(e \mid H) \times p(H)}{p(e)} = \frac{p(e \mid H) \times p(H)}{p(e \mid H) \times p(H) + \big(1 - p(\text{not-e} \mid \text{not-H})\big) \times \big(1 - p(H)\big)}$$

$$= \frac{1 \times \frac{1}{30 \text{ million}}}{1 \times \frac{1}{30 \text{ million}} + \left(\frac{1}{10 \text{ million}}\right) \times \left(1 - \frac{1}{30 \text{ million}}\right)}$$

$$\approx \frac{1}{4}$$

$$= 25\%$$

So the probability that a couple who matches the description has actually committed the crime, p(H|e), is 25%. We have reached the same conclusion, now purely by using Bayes' Theorem and not relying explicitly on the concept of PPV.

Question How to use Bayes-as-updating to understand the probability update from the prior to the posterior?

Answer We had a prior probability that a couple committed the crime, that is 1 in 30 million.

$$\textbf{prior} \qquad p(H) = \frac{1}{30 \text{ million}}$$

Then we updated that prior probability with the new data that they meet the description (e).

Now, the updated, posterior probability, is

$$\textbf{posterior} \qquad p(H \mid e) = \frac{1}{4}$$

Exercise 12.3.4 Comparing the prior and posterior. For each of the scenarios below, you are given two events, A and B. Determine what is likely to be the relationship between the posterior and prior probabilities of B:

$$p(B \mid A) > p(B),\ p(B \mid A) = p(B),\ p(B \mid A) < p(B)$$

a. A = You just finished watching season 1 of a TV series and you immensely enjoyed it!
B = You will also enjoy watching season 2.

b. A = You wore an N95 mask during a gathering.
B = You will test positive for COVID-19 a week afterwards.

c. A = On average, Los Angeles has 284 sunny days per year, compared with the U.S. average of 205.
B = It will rain tomorrow in Los Angeles.

d. A = From a deck of cards, you randomly picked one and it is a 9.
B = The card that you randomly picked is a red card.

e. A = There are 2 children in a family. You are told that one of them is a girl.
B = Both children are girls.

Cookie plates: the problem that broke the Nazi code

We are now going to discuss a very important problem that has wide application from cryptography to science to clinical trials in medicine. It's called **model selection**. A homespun illustration of model selection involves plates of cookies, and deciding which cookie plate a particular sample of cookies came from. Solving this problem was the key, for Alan Turing, to break the Nazi Enigma code during WW2, and the method is widely used today.

As the physicist Richard Feynman said "The Bayesian approach is now the preferred method of comparing scientific theories" (Ziliak and McCloskey (2008) quoting James Press' *Subjective and Objective Bayesian Statistics*).

Imagine that there are two plates of cookies, A and B. The two plates have different numbers of Sugar and Chocolate Chip cookies: plate A has 10 Sugar cookies and 30 Chocolate Chip, while plate B has 20 and 20 (Figure 12.11).

Figure 12.11 Two cookie plates A and B.

We are told that one of these plates has been selected in advance, with equal probability.

$$p(A) = p(B) = 0.5$$

We do not know which plate, and we have to find that out from the data. The data will consist of sequential samples (with replacement) from that selected plate.

Our problem is then: given repeated samples from the selected plate, can we guess which plate that is?

First data point: Chocolate Chip Suppose that our first data point, the first sampled cookie, was a Chocolate Chip. We started with a prior belief that the two plates were equally likely, that is, that $p(A) = p(B) = 0.5$.

Now, how does that drawing of a Chocolate Chip update our prior belief about which of the two plates was chosen? Thinking about it, we want to say something like: "well, it kind of makes plate A more likely, because plate A has relatively more Chocolate Chip cookies than plate B".

This makes sense, but how can we calculate *how much* more likely it makes plate A?

The calculation is straightforward using Bayes' Theorem. Let S be a Sugar cookie, and CC be a Chocolate Chip cookie. We know that

- the probability of picking an S from plate A is $p(S \mid A) = 1/4$
- the probability of picking a CC from plate A is $p(CC \mid A) = 3/4$
- the probability of picking an S from plate B is $p(S \mid B) = 1/2$
- the probability of picking a CC from plate B is $p(CC \mid B) = 1/2$

What we want to know is $p(A \mid CC)$, the probability of plate A being the selected plate given that we picked a CC. We have the prior as $p(A) = p(B) = 0.5$. After taking into account the first data point (a Chocolate Chip cookie), using Bayes' Theorem, we calculate the posterior probability.

So now, given that CC, we have new values for p(A) and p(B), namely the posterior probabilities, $p(A \mid CC) = 3/5$ and $p(B \mid CC) = 2/5$. We have updated the prior probability given the new data (Figure 12.12).

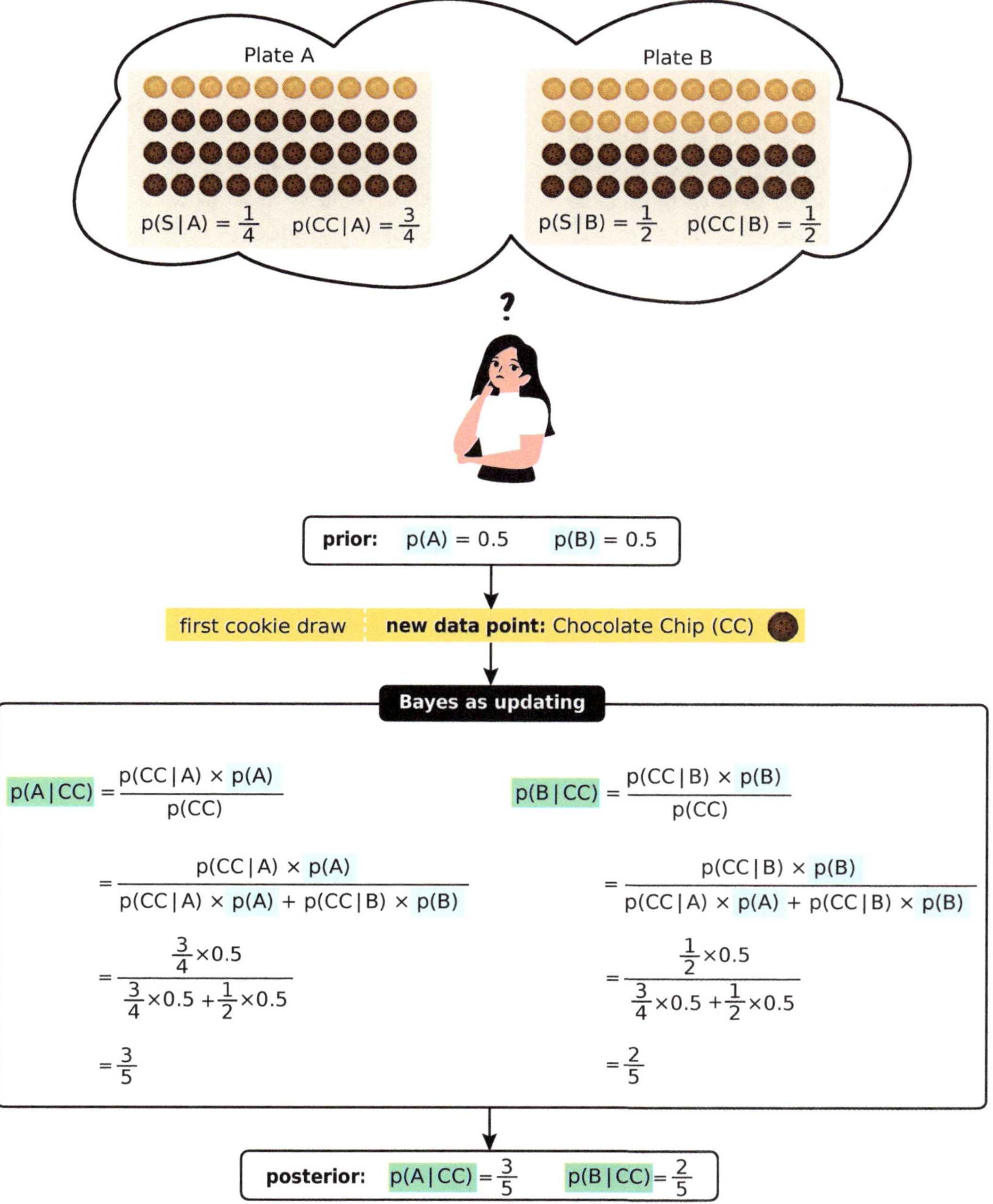

Figure 12.12 After drawing the first cookie from the selected plate, a Chocolate Chip, we invoked Bayes' Theorem to update the probability of either plate being the selected.

Second data point: also Chocolate Chip We put that Chocolate Chip cookie back in the plate whose identity we still don't know. Now we draw another cookie from the plate. It's also a Chocolate Chip. We then apply Bayes' Theorem on this new data (Figure 12.13).

The posterior probability from the last cookie draw becomes the prior probability for the next cookie draw. That's the key step in the updating procedure. Then we plug the new prior and the

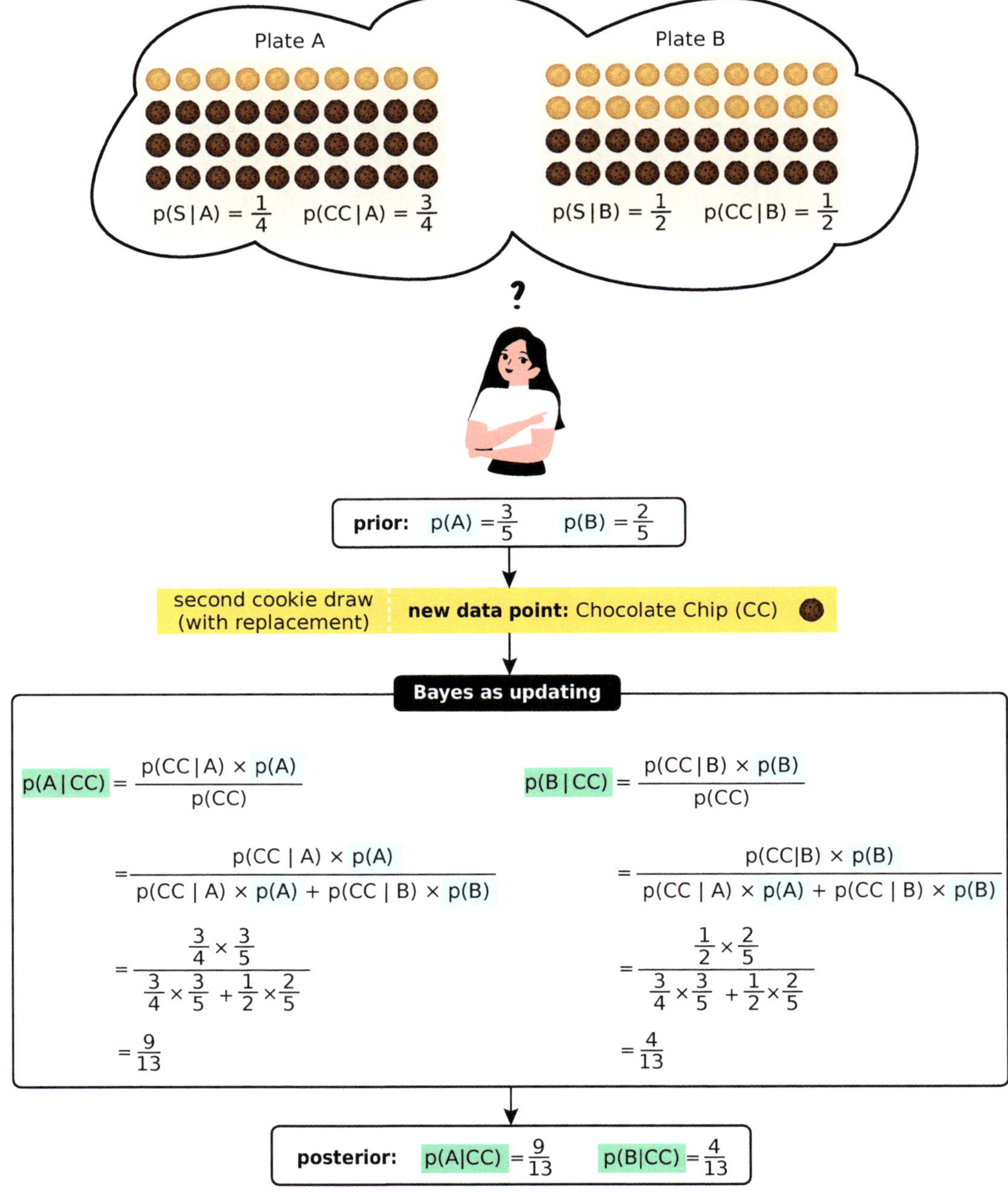

Figure 12.13 After drawing the second cookie from the selected plate, a Chocolate Chip, we invoked Bayes' Theorem to update the probability of either plate being the selected.

next data point into Bayes' Theorem, and update the probability of plate A being the selected plate, given the second Chocolate Chip cookie, which is now 9/13.

Alternative scenario: what if the second cookie was a Sugar?

Supposing that the second cookie had been a Sugar cookie, we use Bayes' Theorem to update the probability of either plate being the selected.

We take the posterior from the first chocolate chip cookie as our prior, p(A) = 3/5, p(B) = 2/5, and apply Bayes' Theorem to update the prior and calculate the new posterior.

Applying Bayes' Theorem, the probability of plate A being the selected one becomes 3/7. It was 9/13 which is more than 3/7.

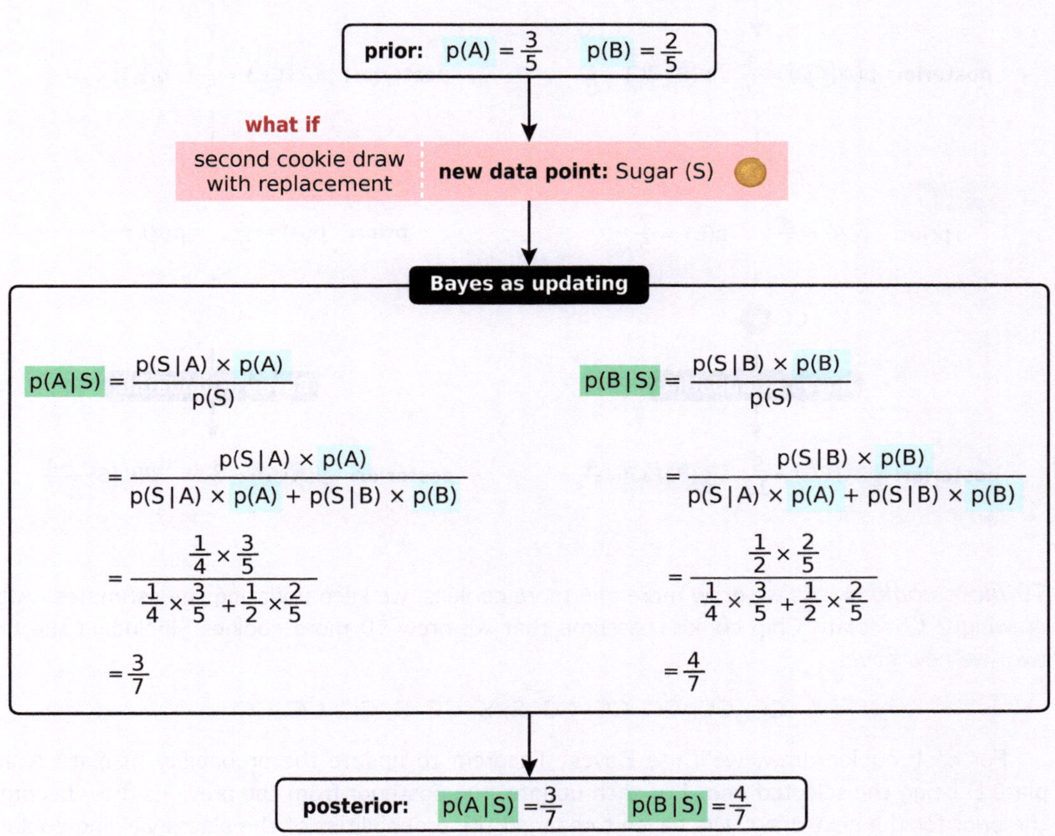

Let's compare the two scenarios.

- In the first scenario, two draws from the selected plate were two Chocolate Chip cookies. After the 1st draw, p(A) increased from 1/2 to 3/5, after the 2nd draw, it further increased the chance of plate A to 9/13.
- In the alternative scenario, when we pulled a Sugar cookie on the second draw, our estimate of p(A) was updated by calculating the probability of A given the Sugar cookie, and it decreased the chance of plate A to 3/7.

The two scenarios illustrate how probabilities are updated by new data.

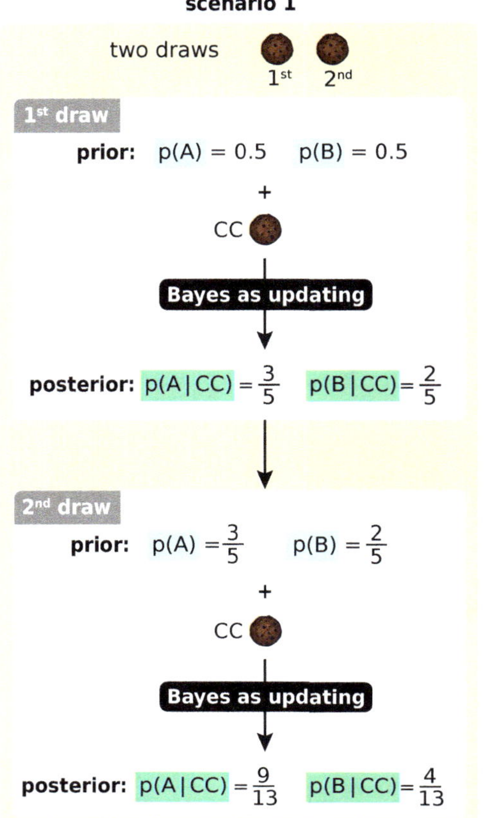

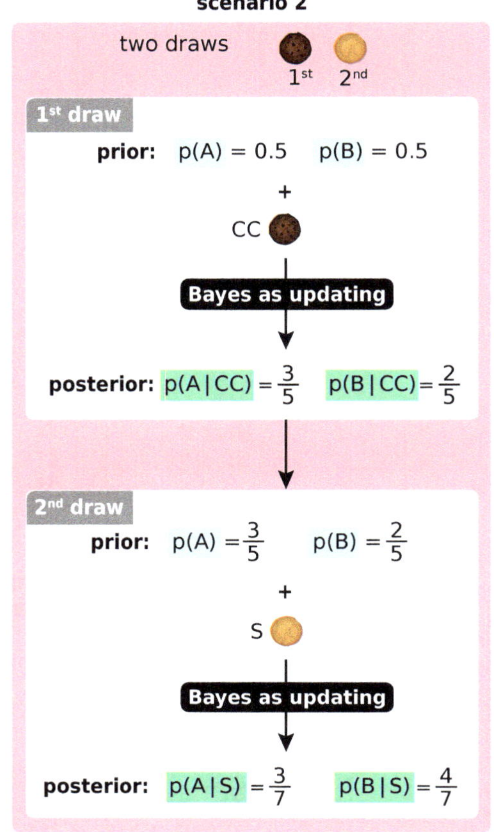

10 more cookies As we draw more and more cookies, we keep updating the estimates. After drawing 2 Chocolate Chip cookies, assume that we drew 10 more cookies. Including the first two, we now have:

CC, CC, CC, CC, CC, S, S, CC, S, CC, CC, CC

For each cookie draw, we'll use Bayes' Theorem to update the probability of plate A and plate B being the selected one. For each update, the posterior from the previous draw becomes the prior for the next draw. We haven't changed the probabilities of the plates yielding cookies, what we've changed is our knowledge, our estimate of the probabilities, p(A) and p(B), of each plate being the selected plate (Table 12.1).

draw	0	1 CC	2 CC	3 CC	4 CC	5 CC	6 S	7 S	8 CC	9 S	10 CC	11 CC	12 CC
p(A)	0.5	0.6	0.69	0.77	0.84	0.88	0.79	0.65	0.74	0.59	0.68	0.76	0.83
p(B)	0.5	0.4	0.31	0.23	0.16	0.12	0.21	0.35	0.26	0.41	0.32	0.24	0.17

Table 12.1 Probabilities of plate A and plate B being the selected plate as each cookie was drawn.

If we convert Table 12.1 to graphics, by plotting p(A) and p(B) against each cookie drawn, we see how each successive cookie draw changes the probability (Figure 12.14).

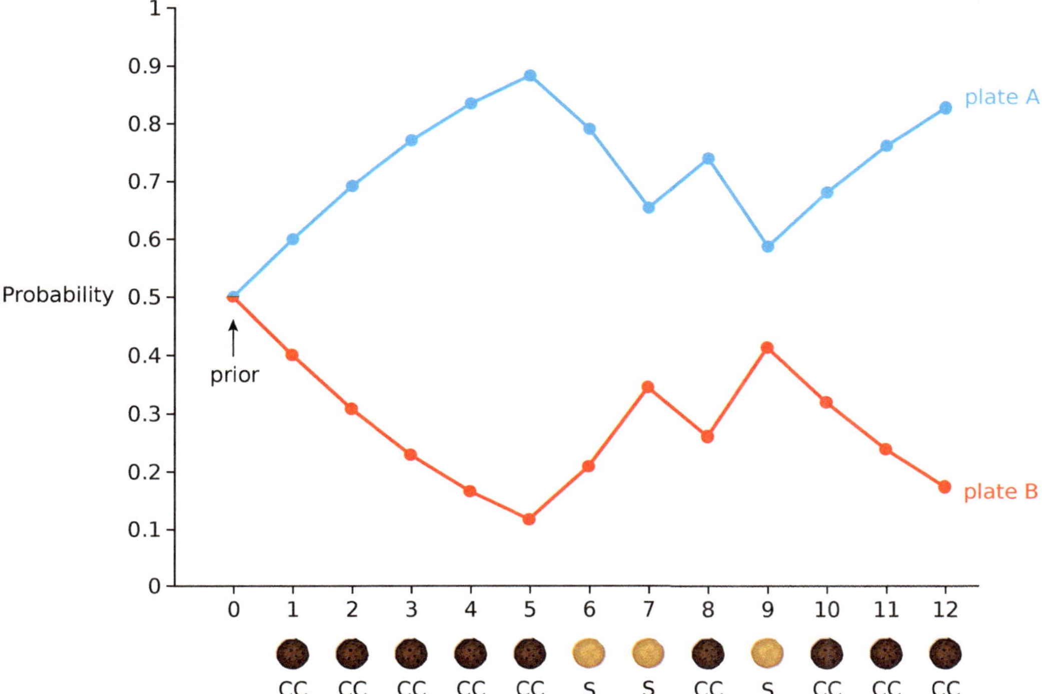

Figure 12.14 Probabilities of plate A and plate B being the selected plate as each cookie was drawn.

We see that, starting from the flat prior p(A) = p(B) = 0.5, each CC drawn nudges the likelihood of plate A being the selected plate upwards, which makes sense since there are more CCs in plate A. Each S drawn lowers the likelihood of model A being the selected plate because plate A has fewer Ss in it.

Exercise 12.3.5 Does the order matter? In the cookie plates alternative scenario (*Alternative scenario: what if the second cookie was a Sugar?* on page 621), before we have any data, the probabilities of either plate being the selected one are equal, p(A) = p(B) = 0.5.

We drew the first cookie and it was a Chocolate Chip (CC) cookie. Using Bayes' Theorem, this first data point gave us an updated probability of either plate being the selected one, p(A) = 3/5, and p(B) = 2/5. Plate A has a higher chance of being the one.

Then we drew a second time and when it turned out to be a Sugar (S) cookie, we used Bayes' Theorem again, updated the probability for either plate being the selected one, and from this second data point we learnt that plate A become slightly less likely to be the selected plate, p(A) = 3/7 and p(B) = 4/7.

What if the first cookie was a Sugar(S) cookie and the second cookie was a Chocolate Chip (CC), how would this reverse of sequence change the probabilities of p(A) and p(B) after each draw?

Model choice and model selection

The cookie plate problem is an archetype for a very broad and important class of problems. We have two distinct models, call them A and B, for a phenomenon, and we have to estimate how likely each model is as a description of reality. So we start sampling from reality, one sample at a time, and update our probability estimates as we go along. The initial prior probabilities of the two models could be equal (the so-called "flat prior") or they could be initially unequal if we have some prior knowledge.

In this general form, the cookie plate problem can be thought of as choosing between two models of reality: model A, which predicts distribution A, and model B, which predicts distribution B. As data comes in the door, we update the likelihood of model A being true vs. model B. The word "model" here could be a hypothesis, a theory, a deterministic model, a statistical model, etc.

Cookie plates in the language of model selection Now we can see the reasoning in the cookie plate problem applied in **model choice** or **model selection**.

Let's imagine that we have two different models (or two different theories) for a given process. The two models have different predictions about the distribution of outcomes. Model A predicts that the outcome will happen 75% of the time, while model B predicts that it will happen 50% of the time (Figure 12.15).

Figure 12.15

For example, model A has the consequence that the ratio of Chocolate Chip cookies to Sugar cookies should be 3 to 1, model B has the prediction of a 50:50 split. We, the scientist, have to choose between model A and model B, based on the data, which is coming in sequentially. Bayes' Theorem, cast in the language of model choice, asks: what is the probability of model A given this data? That's what we want to know as a scientist.

But of course, while this example uses Chocolate Chip and Sugar cookies, in less homespun applications, the outcome could be different antibody concentrations, different levels of neural activity, etc.

Then we start collecting data. In our first trial, this outcome happens, in the second trial it does not, in the third....

We immediately recognize that this is identical to the cookie plates problem, down to the same numbers. Thus, the cookie plates problem is an archetype for the model selection or model choice problem. And, in general, we can say the probability of model A given the data is:

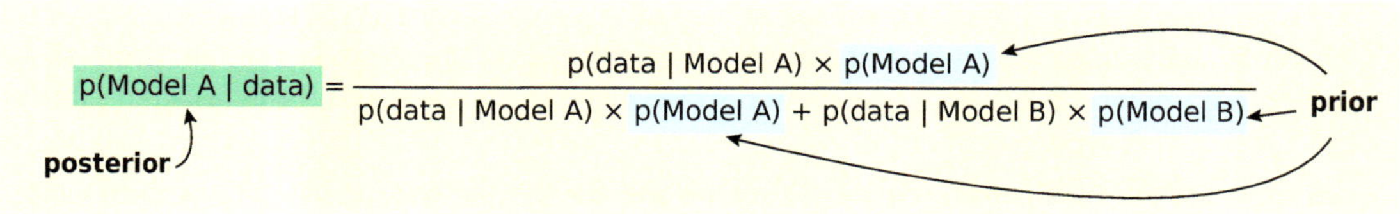

Model selection is an important method in the application of statistics to serious scientific questions.

Breaking the Enigma code

"Nearly all applications of probability to cryptography depend on ... Bayes' Theorem"

– Alan Turing

You might be wondering what this has to do with Alan Turing and his group breaking the Nazi code in WW2.

As Hitler and his staff began to build up the German army in the 1930s, they realized that they needed a more sophisticated system to code messages. Instead of hand coding, they turned to a new invention: cryptographic machines.

To code a message, the message was typed into the cryptographic machine, and each letter entry caused a rotor to turn by a predetermined amount, which caused another rotor to turn, again by a predetermined amount, which caused a third rotor to turn, again by a predetermined amount. When the rotors stopped, the letter at which they stopped was the encoding of the letter that was originally typed.

The number of possible combinations in the machine, called the Enigma, was staggering, making the code seemingly unbreakable.

When World War II began in Europe in 1939, British cryptographers, led by the legendary Alan Turing, inventor of the digital computer, among other things, took on the problem of decoding the Enigma. They had the advantage of brilliant earlier cryptographic work by Polish mathematicians who had solved earlier versions of the Enigma, and even supplied Turing with a stolen actual Enigma machine itself.

But there was still an obstacle. To make things more difficult, there was a three-letter key, giving the starting point for the rotors, that was changed every day at midnight. Without knowing that day's three-letter rotor setting key, the problem was just too hard.

So Turing turned to Bayesian updating. The question was: what is today's three-letter rotor key? Initially, before any messages are intercepted for the day, he had to assume a flat prior: all three-letter combinations were equally likely {AAA, AAB, AAC, . . . , ZZZ}.

The data coming in the door every hour was new, freshly intercepted coded messages. Turing and his group built an automated working model of the Enigma, which could carry out thousands of encodings automatically, to help tackle the problem of finding today's rotor key setting. They called the working model of the Enigma the "bombe". Turing and the code breakers were able to run the bombe continuously, to determine how likely various letter sequences (coded messages) were under different rotor key settings (Figure 12.16). But, as Turing realized, the real question was *not*:

> "If today's key setting is XDW, then the following sequence is unlikely" (that is, p(this sequence | XDW) is small).

The real question is about "how likely is XDW?":

> "If we see this sequence, how likely is XDW?" (that is, p(XDW | this sequence)).

And this is a Bayesian question!

Each rotor key setting gave a slightly different probability distribution for the occurrence of the letters A, B, C, . . . , Z in the coded message. These probability distributions could be worked out by the bombe. Then as coded messages came in the door, Turing's team used Bayesian

Figure 12.16 Left: an Enigma machine. When the plain-text letter was typed into the machine, rotors turned, and the encoding letter was lit. Middle: Turing devised a mechanical device, called a "bombe", that essentially duplicated the rotor structure. But the machine alone was unable to deal with all the combinations without knowing the daily 3-letter rotor key. Right: these women, working at the British cryptanalysis center, Bletchley Park, are using Bayesian updating to work out the most likely prospects for today's 3-letter key.

updating to update the probabilities for all three-letter combinations as candidates for today's rotor setting.

This can be viewed as a model selection problem. Only now we have not 2 possible models as in the cookie plates problem, but 26 × 26 × 26, or around 18,000 possible models.

The technique worked, more often than not, and resulted in many major military victories. In many cases, the Allied commanders were reading German High Command signals on the same day they were sent out.

Dwight Eisenhower, the Supreme Allied Commander (and later President of the US), said

> "The intelligence which has emanated from [the Enigma project] before and during this campaign has been of priceless value to me. It has simplified my task as a commander enormously. It has saved thousands of British and American lives and, in no small way, contributed to the speed with which the enemy was routed and eventually forced to surrender."

Winston Churchill said that the decoding of the Enigma probably shortened the war by 1 to 2 years, which means several million lives.

Bayesian methods were quickly adopted by the military after the war. The US Navy used Bayesian methods to locate the lost nuclear submarine Scorpion, and the US Air Force used Bayesian methods to find the hydrogen bomb that was lost at sea off the coast of Palomares, Spain.

FURTHER EXERCISES 12.3

1. **Venn diagram.** Consider two properties A and B such that p(A∧B) = 0. If p(A) = 0.4 and p(A∨B) = 0.95, draw a diagram like Figure 12.6 (these are sometimes called "Venn diagrams"). What is p(B)? *(A∧B means "A and B" and A∨B means "A or B".)*

2. **Airline bags.** You're waiting at a baggage carousel after a flight. How many bags do you have to see that are not yours before it's more likely than not that your bag has been lost? Assume there are 100 bags on the airplane and the background rate of luggage loss by airlines is 1%.

3. **Race and death penalty.** A study looked at 326 murder cases tabulated by the victim's race, the defendant's race, and whether the defendant was sentenced to death (see article "Racial Characteristics and the Imposition of the Death Penalty", *American Sociological Review*, 1981).

Did white defendant receive death penalty?		
	Yes	No
White victim	19	132
Black victim	0	9
Did black defendant receive death penalty?		
	Yes	No
White victim	11	52
Black victim	6	97

In this data set,

a. What is the probability of a white defendant receiving the death penalty?
b. What is the probability of a black defendant receiving the death penalty?
c. What is the probability of a defendant receiving the death penalty given that the victim is white?
d. What is the probability of a black defendant receiving the death penalty given that the victim is white?
e. What is the probability of a white defendant receiving the death penalty given that the victim is black?

4. **Revisiting breast cancer and the mammogram test.** In the previous section, when we talked about the concept of positive predictive value, we used the mammogram test for detecting breast cancer as an example FE 12.2.3 and calculated the positive predictive value of a positive mammogram test. We said that:

"The probability that one of these women has breast cancer is 0.8 percent. If a woman has breast cancer, the probability is 90 percent that she will have a positive mammogram. If a woman does not have breast cancer, the probability is 7 percent that she will still have a positive mammogram."

Assume that D = "patient has breast cancer", + = "patient tests positive"

a. Use Bayes' Theorem to derive the probability that the patient has breast cancer when testing positive. What is the prior and the posterior?
b. You are a doctor and one of your patients has tested positive. Before sending the patient to get a biopsy, you decide to just repeat the test. It comes back negative. Now what is the probability that the patient has the disease? What is the prior and the posterior?

5. **Cormorants and seagulls.** Imagine you are studying birds in a coastal region. Two models describing the abundance of seagulls and cormorants have been proposed. Model A predicts that there will be 75% cormorants and 25% seagulls, while model B predicts that each

species will make up 50%. Heading out to the shore, you will collect data to determine which model better reflects your observations. Let's use C to denote "observing a cormorant" and S to denote "observing a seagull".

a. Before you collect any data, if you know nothing that would favor one model over the other, what is the probability that model A describes the actual bird population? What is the probability that model B describes the actual bird population? What is the prior?
b. What is p(C)? (*hint: p(C) is the overall probability of observing a cormorant, regardless of model, therefore, using the concept of "splitting the denominator" in Figure 12.10, it is the sum of the probabilities of observing a cormorant predicted by each model, weighted by the relative probabilities of the models.*)
c. Similarly, what is p(S)?
d. You observed a cormorant. Use Bayes' Theorem to update the prior. What are the probabilities of model A and model B after your first observation?
e. The next bird observed is also a cormorant. Now what is the probability of model A?
f. Then you observed a seagull. Using Bayes' Theorem, what are your beliefs in model A and model B?
g. Given the next 10 birds you observed: C, S, C, C, C, S, C, C, S, C, how does the probability of model A change as you collect each new evidence? What about the probability of model B, how does it change after each of your observations?

6. **HIV in-home test.** An in-home test to detect HIV infection, by swabbing gums to detect antibodies, is available. If the in-home test is positive, it needs to be further confirmed by a clinical blood test. If not treated, an HIV infection can develop into the potentially life-threatening condition AIDS. The test's package insert reports that clinical studies have shown:

   - the percentage of results that will be negative when HIV is absent is 99.98% (1 false positive result is expected out of every 5,000 test results in uninfected individuals).
   - the percentage of results that will be positive when HIV is present is 92% (1 false negative result is expected out of every 12 test results in HIV infected individuals).

   In 2019, the estimated incidence of HIV infections in the U.S. was 34,800 infections (12.6 per 100,000 people). Use this information to answer the following questions:

   a. What are the sensitivity, specificity, and prevalence for this test?
   b. If someone receives a positive test result, what is the probability that they are actually infected with HIV?
   c. A patient is having symptoms of HIV infection and knows they were exposed to an HIV-positive person. How would the quantities in parts a) and b) change?
   d. Should all people in populations that are at high risk of HIV get tested regularly with this test?
   e. Explain why Bayesian analysis (in this case, PPV) is superior to just providing the sensitivity and specificity of the test.

7. **Eye colors.** We have two models for the frequency of the recessive allele B for blue eyes and the dominant allele D (for other colors) in a particular population: 50% B and 50% D (Model X) and 20% B and 80% D (Model Y). We don't have any information about the prior probability of the two models. If you observe a D allele, what are the new values of the probability of Model X and Model Y? If you next observe a B, what are the new values of the probability of Model X and Model Y?

8. **Seat belts.** According to data collected by the Florida Department of Highway Safety and Motor Vehicles on people involved in traffic accidents in 1998, 0.37% suffered a fatal accident, 28.4% did not wear a seat belt, and 0.28% did not wear a seat belt and suffered a fatal accident.

   a. What is the probability that an accident involved a person not wearing a seat belt, given that it was a fatal accident?

   b. What is the probability that an accident involved a fatal injury, given that the person was not wearing a seat belt?

   c. What is the probability that an accident involved a fatal injury, given that the person was wearing a seat belt? How does this compare to (b)?

12.4 Bayes and Clinical trials

An important application of Bayes' Theorem in medical practice is in the evaluation of clinical trials, say, of the efficacy of a new drug. In order to understand how to apply Bayes' Theorem in these cases, we need to dive a little more deeply into the denominator in Bayes' Theorem.

Expanding the denominator

So far, we've focused on examples of Bayes' Theorem where the denominator p(A) is the sum of two terms, the "B case" and the "~B case". The probability of A happening in the B case is $p(A \mid B) \times p(B)$, and the probability of A happening in the ~B case is $p(A \mid {\sim}B) \times p({\sim}B)$.

For example, in the disease testing case, the denominator p(+), the probability of a positive test, was the sum of two terms

$$p(+) = p(+ \mid \text{Dis}) \times p(\text{Dis}) + p(+ \mid \text{noDis}) \times p(\text{noDis})$$

And in the case of a hypothesis H and evidence e, the denominator p(e), the probability of obtaining this evidence, was the sum of two terms

$$p(e) = p(e \mid H) \times p(H) + p(e \mid {\sim}H) \times p({\sim}H)$$

It's important to expand this idea, and to realize that in other cases, there may be many terms in the denominator, because there are many alternative hypotheses.

For example, if we wanted to calculate the probability of a student in our class wearing a red shirt, we might have approached it in a binary model, as

$$p(\text{red shirt}) = p(\text{red shirt} \mid \text{male}) \times p(\text{male}) + p(\text{red shirt} \mid \text{female}) \times p(\text{female})$$

But it's important to realize that this binary splitting of the denominator is not required by Bayesianism. If we want to be more inclusive, and split the denominator into three possibilities: male, female and non-binary, Bayes' Theorem easily accommodates this new model, and would write the denominator as

$$\text{p(red shirt)} = \text{p(red shirt} \mid \text{male)} \times \text{p(male)} + \text{p(red shirt} \mid \text{female)} \times \text{p(female)} + \text{p(red shirt} \mid \text{non-binary)} \times \text{p(non-binary)}$$

3 terms in the denominator Let's consider a cookie plate case, but now there are three plates of cookies A, B and C, instead of two. Bayes' Theorem would then have the form

$$\text{p(A} \mid \text{data)} = \frac{\text{p(data} \mid \text{A)} \times \text{p(A)}}{\text{p(data} \mid \text{A)} \times \text{p(A)} + \text{p(data} \mid \text{B)} \times \text{p(B)} + \text{p(data} \mid \text{C)} \times \text{p(C)}}$$

Or consider a case in disease testing where there are three kinds of people: diseased, healthy, and carriers. Here the positive predictive value of a positive test would have the form

$$\text{p(Dis} \mid +) = \frac{\text{p(+} \mid \text{Dis)} \times \text{p(Dis)}}{\text{p(+} \mid \text{Dis)} \times \text{p(Dis)} + \text{p(+} \mid \text{Healthy)} \times \text{p(Healthy)} + \text{p(+} \mid \text{Carrier)} \times \text{p(Carrier)}}$$

n terms in the denominator In general, if there are n hypotheses, $H_1, H_2, \ldots, H_n$, then after receiving a new data point, the form of Bayes' Theorem we would use to update the probability of H_i would be

$$\text{p}(H_i \mid \text{data}) = \frac{\text{p}(\text{data} \mid H_i) \times \text{p}(H_i)}{\text{p}(\text{data} \mid H_1) \times \text{p}(H_1) + \text{p}(\text{data} \mid H_2) \times \text{p}(H_2) + \cdots + \text{p}(\text{data} \mid H_n) \times \text{p}(H_n)}$$

Using the $\sum$ notation, we can rewrite Bayes' Theorem as

Bayes' Theorem: n hypotheses

$$\text{p}(H_i \mid \text{data}) = \frac{\text{p}(\text{data} \mid H_i) \times \text{p}(H_i)}{\sum_{k=1}^{n} \text{p}(\text{data} \mid H_k) \times \text{p}(H_k)}$$

Since the n hypotheses are mutually exclusive and exhaust all the possibilities, we know that

$$\sum_{k=1}^{n} \text{p}(H_k) = \text{p}(H_1) + \text{p}(H_2) + \cdots + \text{p}(H_n) = 1$$

Exercise 12.4.1 You randomly choose a treasure chest to open, and then randomly choose a coin from that treasure chest.

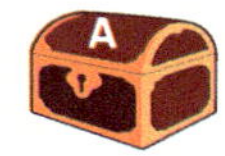

100 gold coins

50 gold coins +
50 silver coins

20 gold coins +
80 silver coins

If the coin you choose is gold, let

- H_1 = "chest A was chosen"
- H_2 = "chest B was chosen"
- H_3 = "chest C was chosen"
- data = "a gold coin was sampled"

What is the probability that you chose chest A? What is the probability that you chose chest B? What about chest C?

Continuous denominator We now know how to do Bayesian updating when the number of alternative hypotheses is 2 or 3 or ... or n. For any finite number of hypotheses H_1, H_2, ..., H_n, upon receiving new data, we can update their probabilities by

$$p(H_i \mid \text{data}) = \frac{p(\text{data} \mid H_i) \times p(H_i)}{\sum_{k=1}^{n} p(\text{data} \mid H_k) \times p(H_k)}$$

The final step is now to extend this discussion to the case of *infinitely* many alternative hypotheses.

The classical example of a case with infinitely many hypotheses is the *coin flipping problem*. We imagine that we're given a coin. The coin is *not* necessarily fair. In fact, we have *no idea* what the true probability of Heads (the "bias" of the coin) is for this coin. We'll denote the bias of the coin as θ, where θ can take any value from $\theta = 0$ (all Tails) to $\theta = 1$ (all Heads), including possibly a fair coin $\theta = 0.5$.

Our task is then to estimate θ, the coin's true probability of Heads, from a sequence of flips of the coin.

Since θ can take any value between 0 and 1, we have infinitely many hypotheses H_θ, each saying "the true bias of the coin is θ". The number of hypotheses here becomes a continuous infinity: there's an H_θ, saying that "the true bias of the coin = θ", for every possible value of θ between 0 and 1. We will use $p(\theta)$ to denote the probability of the hypothesis H_θ. The shape of $p(\theta)$ can take many forms (Figure 12.17).

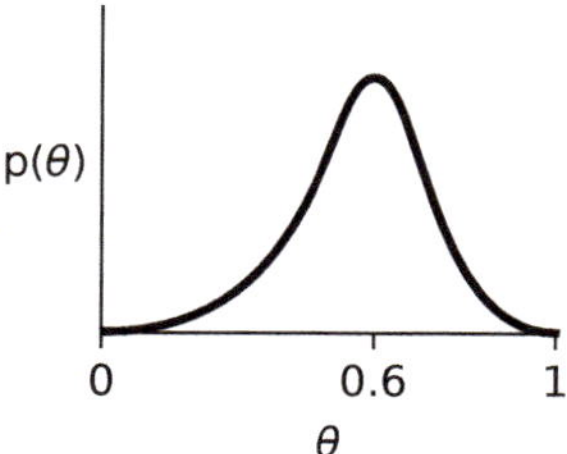

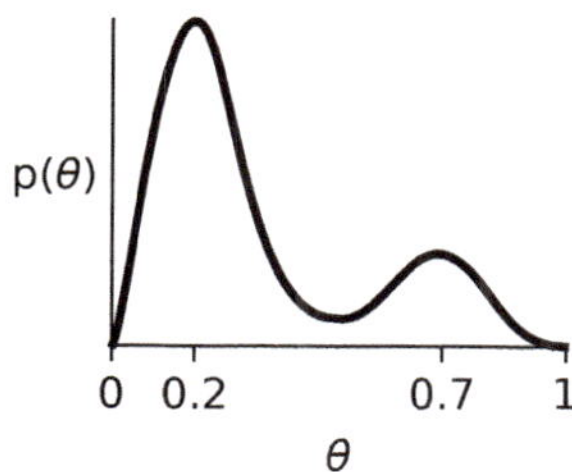

Figure 12.17 Examples of what $p(\theta)$ might look like. On the left, the most probable value for the bias of the coin is $\theta = 0.6$, with nearby values of θ also having higher likelihood. On the right, the most likely value is $\theta = 0.2$. There is a secondary peak at $\theta = 0.7$, but notice that many values around 0.2 are more likely than any of the values around 0.7.

Note that $p(\theta)$ is a *probability density distribution*. It tells us that the probability of the true bias of the coin falling in the interval $[\theta, \theta + d\theta]$ is equal to the area under the curve between θ and $\theta + d\theta$.

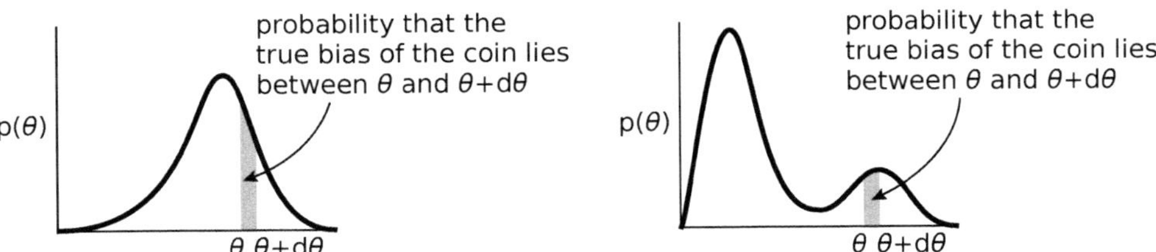

Figure 12.18 The shaded area is the probability that the true bias of the coin lies between θ and $\theta + d\theta$.

Now let's suppose that we start with a prior probability density distribution $p(\theta)$, and we obtain new data. We want to update the probabilities according to this new data, giving us a posterior probability density distribution. If there were finitely many cases, $\theta_1, \theta_2, \ldots, \theta_i, \ldots, \theta_n$, we would update each $p(\theta_i)$ by multiplying it by $p(\text{data} \mid \theta_i)$ and dividing it by $p(\text{data})$, where $p(\text{data}) = \sum_{k=1}^{n} p(\text{data} \mid \theta_k) \times p(\theta_k)$.

$$p(\theta_i \mid \text{data}) = \frac{p(\text{data} \mid \theta_i) \times p(\theta_i)}{\sum_{k=1}^{n} p(\text{data} \mid \theta_k) \times p(\theta_k)}$$

For the continuous case, we let n approach infinity, giving us

Bayes' Theorem: continuous form

$$p(\theta \mid \text{data}) = \frac{p(\text{data} \mid \theta) \times p(\theta)}{p(\text{data})}$$

$$\text{where } p(\text{data}) = \int_0^1 p(\text{data} \mid \theta) \times p(\theta)\, d\theta$$

Since the infinitely many hypotheses are mutually exclusive and exhaust all the possibilities, we know that

$$\int_0^1 p(\theta)\, d\theta = 1$$

In the case of n hypotheses, Bayes' Theorem requires that the probabilities of each of the n hypotheses must collectively add up to 1. Similarly, in the continuous case, the requirement that the probabilities of all possible hypotheses must add up to 1 is reflected by $\int_0^1 p(\theta)d\theta = 1$, that is, that the total area under the probability density curve $p(\theta)$ must be equal to 1.

In the continuous form, note that the denominator p(data) is an integral. Occasionally, it may be possible to calculate this integral in closed form. But most of the time, in actual practice, we will approximate the continuum of θ values with a finite number of values, say, $\theta_1, \theta_2, \ldots, \theta_{100}$. This amounts to **Riemann integration** of the continuous integral.

This continuous form of Bayes' Theorem has many applications.

Clinical trials

In the evaluation of a new drug in a clinical trial, our question is: what is the cure rate of this drug? Notice that this is not an NHST question: does the drug have an effect (yes/no)?, and this is also not a p-value question: how likely would the observed cure rate be if the drug had no effect? This is a different question: *how* effective is the drug?, which means: what is our best estimate of the true cure rate?

In a clinical trial, the basic situation is that patients who have taken the drug come through the door each day, and each patient is either cured or not. On the basis of each new piece of evidence, we then update our estimate of the cure rate.

So let θ = the true cure rate of the drug. This is what we would like to estimate.

In other words, estimating the true cure rate of a drug from successive patient outcomes is an exact mathematical analogue of estimating the true bias of a coin, based on successive coin flips.

At the outset of the trial, before any patient has been seen, let's say we are completely ignorant about the value of θ: it could be zero, it could be 100% (that is $\theta = 1$), or it could be any number in between, with equal probability. In other words, the initial prior $p(\theta)$ is a flat line.

After collecting each data point and updating the prior distribution, we'll have an updated probability density distribution $p(\theta)$, that tells us, for any interval $[\theta, \theta + d\theta]$, what is the probability that the true cure rate lies in that interval (Figure 12.19).

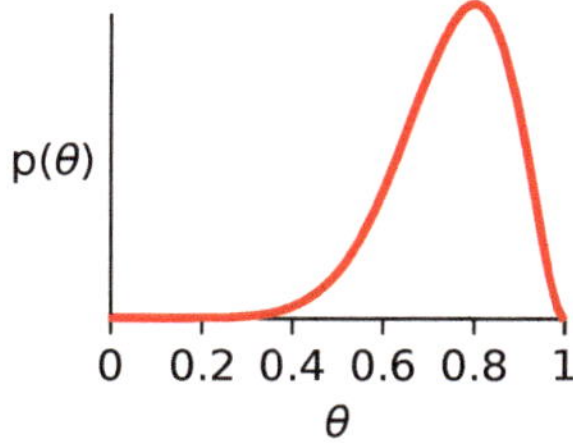

Figure 12.19 An example of what $p(\theta)$ might look like. Here the 80% cure rate ($\theta = 0.8$) is the most likely value.

Reading a probability density distribution

A curve like Figure 12.19 contains a great deal of information.

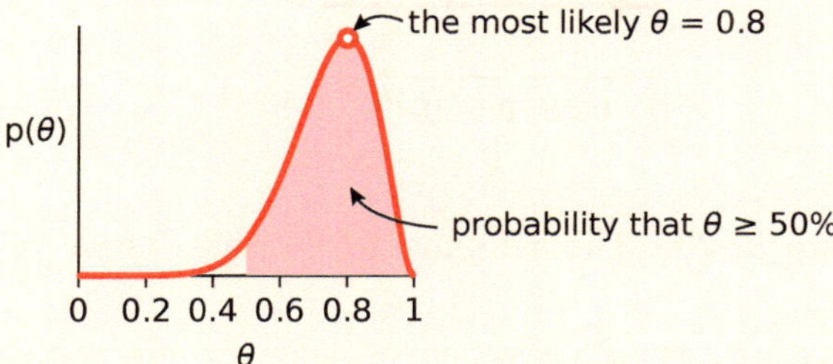

- First, the most likely value for θ is around 0.8.
- But **it also tells us the likelihood of every other θ value.** This is very valuable additional information.
- With this additional information, we can estimate *how reliable* the estimate of 0.8 is. If other θ values are also quite likely, then our estimate is weak and has low precision.

Wider and flatter probability distributions indicate low-precision estimates. On the other hand, sharply peaked probability distributions give high-precision estimates.

- The curve also enables us to answer many other questions. For example, if we wanted to know what is the likelihood that the true cure rate θ is greater than 50%, we would calculate the area under the curve that is to the right of 0.5.

Bayesian Credible Interval

From the final posterior probability density distribution, after all the data, we can then define the 99% **credible interval** as the central 99% of the posterior probability density distribution. This credible interval is the Bayesian answer to quantifying the uncertainty of a measure, which is the straightforward and direct approach, in contrast to the convoluted and indirect Frequentist concept of "confidence interval".

Recall that when we discussed so-called confidence intervals (section 4.9 *The interpretation of the confidence interval*), we wished we could say "we are 99% confident that the true measure lies in *this* interval [x, y]", but we couldn't. We could only say the awkward and somewhat backwards "99% of confidence intervals constructed in this manner will contain the true value". We could not say "values outside this confidence interval are highly unlikely". Only the Bayesian credible interval allows to say "values outside this credible interval are highly unlikely".

Where to begin? So let's consider a hypothetical clinical trial of a new drug.[2] We begin with a prior distribution that is flat: we have no idea what θ is, and all possibilities are equally likely. This is not the only possible prior. We might be in a different situation, and have some prior belief that 0.3 is the most likely value, etc. But here we will assume a flat prior (Figure 12.20).

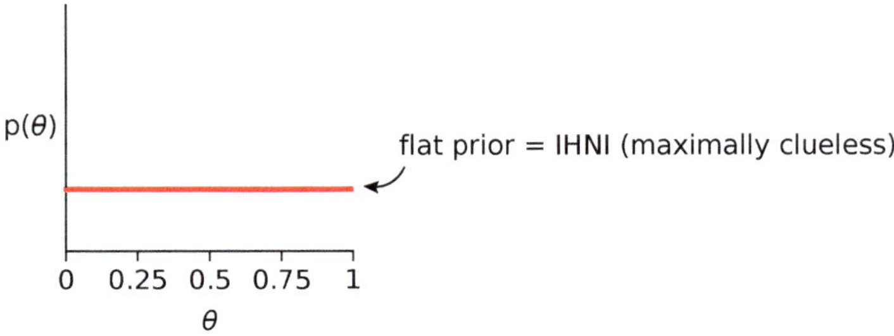

Figure 12.20 The flat prior. IHNI means "I have no idea".

Then we will update $p(\theta)$ as each new patient, cured or not, comes through the door. The curve will change its shape as new data keeps coming in.

Approximating the continuous distribution In order to apply Bayes' Theorem to update $p(\theta)$, a continuous function, we need to calculate integrals. Since we know how to apply Bayes' Theorem when the denominator is a sum of a large number of possibilities, we will use a discrete approximation to the continuous distribution, that is, Riemann integration.

[2] We follow the excellent exposition of Donald Berry's *Bayesian clinical trials* (Berry, 2006).

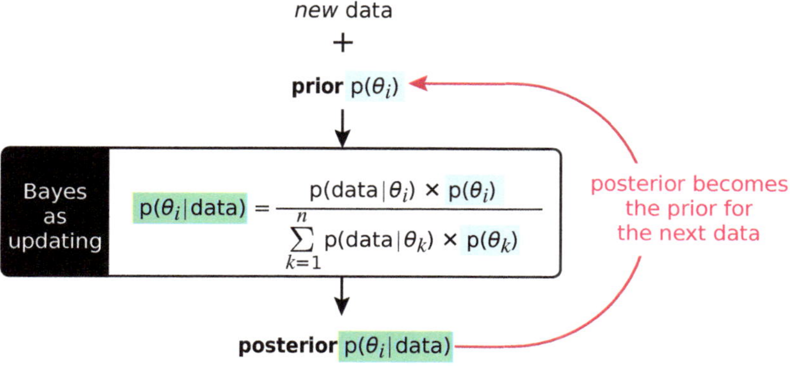

Figure 12.21

Flat prior In this discrete approximation, we'll assume that θ has 100 possible values, 0.01, 0.02, 0.03, ..., 1. That means we have 100 hypotheses,

Hypothesis #1: $\theta = \theta_1 = 0.01$

Hypothesis #2: $\theta = \theta_2 = 0.02$

...

Hypothesis #100: $\theta = \theta_{100} = 1$

We will begin with a flat prior $p(\theta_1) = p(\theta_2) = \cdots = p(\theta_{100}) = \frac{1}{100}$. Each hypothesis is equally likely, and the total probability must add up to 1, so each of the hundred hypotheses has initial probability 0.01.

Now we can apply Bayesian updating directly to the 100 hypotheses (Figure 12.21).

First update Suppose patient #1 walks in the door, and has been cured. How does this change the probabilities of the various hypotheses? Clearly, if a hypothesis said that the cure rate was very low, then 1 patient cured out of 1 should lower the probability of that hypothesis. But by how much? That's the job of Bayesian updating.

We have 100 hypotheses that need updating. Let's start with θ_1, which says that $\theta = 0.01$. We can now update $p(\theta_1)$ given that the first patient was cured.

$$
\begin{aligned}
p(\theta_1 \mid \text{cure}) &= \frac{p(\text{cure} \mid \theta_1) \times p(\theta_1)}{\sum_{k=1}^{100} p(\text{cure} \mid \theta_k) \times p(\theta_k)} \\
&= \frac{0.01 \times 0.01}{0.01 \times 0.01 + 0.02 \times 0.01 + \cdots + 0.99 \times 0.01 + 1 \times 0.01} \\
&= 0.00019802
\end{aligned}
$$

Note that the update takes $p(\theta_1)$ from the prior value of 0.01 to 0.00019802, a very large reduction. This is saying that if we got 1 cure out of 1, then the probability that the true cure rate is 0.01 is very low.

Similarly, we update $p(\theta_2), \ldots, p(\theta_{100})$

$$\begin{aligned} p(\theta_2 \mid \text{cure}) &= \frac{p(\text{cure} \mid \theta_2) \times p(\theta_2)}{\sum_{k=1}^{100} p(\text{cure} \mid \theta_k) \times p(\theta_k)} \\ &= \frac{0.02 \times 0.01}{0.01 \times 0.01 + 0.02 \times 0.01 + \cdots + 0.99 \times 0.01 + 1 \times 0.01} \\ &= 0.00039604 \end{aligned}$$

which tells us that the probability that the true cure rate is 0.02 has also been downgraded from 0.01 to 0.00039604 by this data point.

At the other end of the spectrum, the hypotheses that guess a high cure rate are upgraded in probability. For example,

$$\begin{aligned} p(\theta_{100} \mid \text{cure}) &= \frac{p(\text{cure} \mid \theta_{100}) \times p(\theta_{100})}{\sum_{k=1}^{100} p(\text{cure} \mid \theta_k) \times p(\theta_k)} \\ &= \frac{1 \times 0.01}{0.01 \times 0.01 + 0.02 \times 0.01 + \cdots + 0.99 \times 0.01 + 1 \times 0.01} \\ &= 0.01980198 \end{aligned}$$

Note that $p(\theta_{100})$ has been *increased* by the update: one cure out of one increases the likelihood that the cure rate of the drug is 100%.

In a similar fashion, we get 100 updated probabilities, which are the posterior probabilities

$$\begin{aligned} p(\theta_1 \mid \text{cure}) &= 0.00019802 \\ p(\theta_2 \mid \text{cure}) &= 0.00039604 \\ &\vdots \\ p(\theta_{100} \mid \text{cure}) &= 0.01980198 \end{aligned}$$

If we graph those 100 probabilities $p(\theta_1 \mid \text{cure})$, $p(\theta_2 \mid \text{cure})$, ..., $p(\theta_{100} \mid \text{cure})$, we get an updated probability distribution $p(\theta)$ (Figure 12.22).

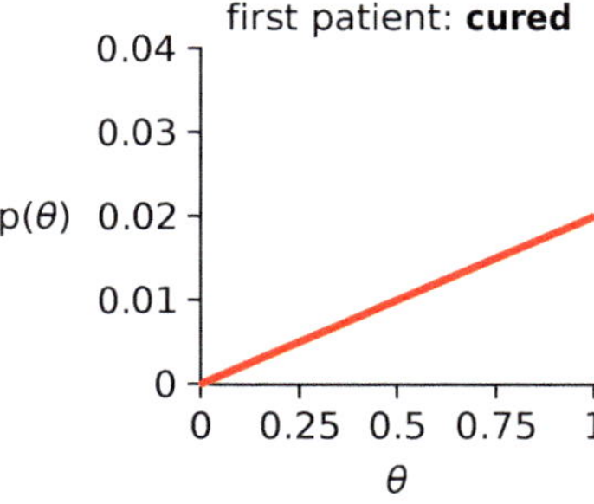

Figure 12.22 Probabilities of different values of cure rate θ after the first update.

We see that the posterior distribution of probabilities, given 1 cured patient, increases with θ. **This posterior distribution will become the prior distribution for the next patient.**

$$p(\theta_1) = 0.00019802$$
$$p(\theta_2) = 0.00039604$$
$$\vdots$$
$$p(\theta_{100}) = 0.01980198$$

Second update We are now ready for the second patient. It turns out that the second patient was also cured.

We can then update the probabilities $p(\theta_i)$, using the previous posterior as our new prior.

$$p(\theta_1 \mid \text{cure}) = \frac{p(\text{cure} \mid \theta_1) \times p(\theta_1)}{\sum_{k=1}^{100} p(\text{cure} \mid \theta_k) \times p(\theta_k)}$$
$$= \frac{0.01 \times 0.00019802}{0.01 \times 0.00019802 + 0.02 \times 0.00039604 + \cdots + 1 \times 0.01980198}$$
$$= 0.0000029555$$

$$p(\theta_2 \mid \text{cure}) = \frac{p(\text{cure} \mid \theta_2) \times p(\theta_2)}{\sum_{k=1}^{100} p(\text{cure} \mid \theta_k) \times p(\theta_k)}$$
$$= \frac{0.02 \times 0.00039604}{0.01 \times 0.00019802 + 0.02 \times 0.00039604 + \cdots + 1 \times 0.01980198}$$
$$= 0.000011822$$

$$\vdots$$

$$p(\theta_{100} \mid \text{cure}) = \frac{p(\text{cure} \mid \theta_{100}) \times p(\theta_{100})}{\sum_{k=1}^{100} p(\text{cure} \mid \theta_k) \times p(\theta_k)}$$
$$= \frac{1 \times 0.01980198}{0.01 \times 0.00019802 + 0.02 \times 0.00039604 + \cdots + 1 \times 0.01980198}$$
$$= 0.029555$$

If we graph these 100 new posterior probabilities, $p(\theta_1 \mid \text{cure})$, $p(\theta_2 \mid \text{cure})$, ..., $p(\theta_{100} \mid \text{cure})$, we see that the second data point has **morphed the posterior distribution** into a rising curve (Figure 12.23). **Each new data point will then further morph the curve**.

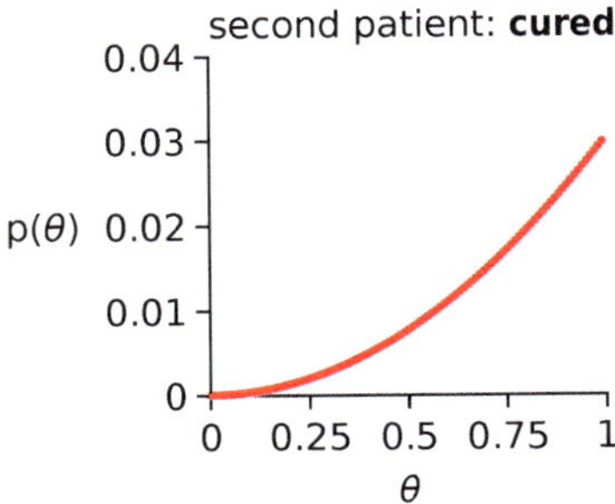

Figure 12.23 Probabilities of different values of cure rate θ after the second update.

Exercise 12.4.2 How has the second update changed the probabilities of the drug having a high cure rate, and a low cure rate?

This posterior distribution will become the prior distribution for the next patient.

$$p(\theta_1) = 0.0000029555$$
$$p(\theta_2) = 0.000011822$$
$$\vdots$$
$$p(\theta_{100}) = 0.029555$$

Third update Now a third patient walks in, and that patient is not cured. The update calculation for the posterior probabilities for H_1 now takes the form

$$p(\theta_1 \mid \text{no cure}) = \frac{p(\text{no cure} \mid \theta_1) \times p(\theta_1)}{\sum_{k=1}^{100} p(\text{no cure} \mid \theta_k) \times p(\theta_k)}$$

What is $p(\text{no cure} \mid \theta_1)$? It's $1 - p(\text{cure} \mid \theta_1)$.

$$p(\theta_1 \mid \text{no cure}) = \frac{p(\text{no cure} \mid \theta_1) \times p(\theta_1)}{\sum_{k=1}^{100} p(\text{no cure} \mid \theta_k) \times p(\theta_k)} = \frac{\big(1 - p(\text{cure} \mid \theta_1)\big) \times p(\theta_1)}{\sum_{k=1}^{100} \big(1 - p(\text{cure} \mid \theta_k)\big) \times p(\theta_k)}$$

We then update the probabilities $p(\theta_i)$ for each of the 100 hypotheses, using the previous posterior as our new prior.

$$\begin{aligned} p(\theta_1 \mid \text{no cure}) &= \frac{p(\text{no cure} \mid \theta_1) \times p(\theta_1)}{\sum_{k=1}^{100} p(\text{no cure} \mid \theta_i) \times p(\theta_i)} = \frac{\big(1 - p(\text{cure} \mid \theta_1)\big) \times p(\theta_1)}{\sum_{k=1}^{100} \big(1 - p(\text{cure} \mid \theta_k)\big) \times p(\theta_k)} \\ &= \frac{(1-0.01) \times 0.0000029555}{(1-0.01) \times 0.0000029555 + (1-0.02) \times 0.000011822 + \cdots + (1-1) \times 0.029555} \\ &= 0.000011881188 \end{aligned}$$

$$\mathrm{p}(\theta_2 \mid \text{no cure}) = \frac{\mathrm{p}(\text{no cure} \mid \theta_2) \times \mathrm{p}(\theta_2)}{\sum_{k=1}^{100} \mathrm{p}(\text{no cure} \mid \theta_k) \times \mathrm{p}(\theta_k)} = \frac{\left(1 - \mathrm{p}(\text{cure} \mid \theta_2)\right) \times \mathrm{p}(\theta_2)}{\sum_{k=1}^{100} \left(1 - \mathrm{p}(\text{cure} \mid \theta_k)\right) \times \mathrm{p}(\theta_k)}$$

$$= \frac{(1-0.02) \times 0.000011822}{(1-0.01) \times 0.0000029555 + (1-0.02) \times 0.000011822 + \cdots + (1-1) \times 0.029555}$$

$$= 0.0000470447$$

$$\vdots$$

$$\mathrm{p}(\theta_{100} \mid \text{no cure}) = \frac{\mathrm{p}(\text{no cure} \mid \theta_{100}) \times \mathrm{p}(\theta_{100})}{\sum_{k=1}^{100} \mathrm{p}(\text{no cure} \mid \theta_k) \times \mathrm{p}(\theta_k)} = \frac{\left(1 - \mathrm{p}(\text{cure} \mid \theta_{100})\right) \times \mathrm{p}(\theta_{100})}{\sum_{k=1}^{100} \left(1 - \mathrm{p}(\text{cure} \mid \theta_k)\right) \times \mathrm{p}(\theta_k)}$$

$$= \frac{(1-1) \times 0.029555}{(1-0.01) \times 0.0000029555 + (1-0.02) \times 0.000011822 + \cdots + (1-1) \times 0.029555}$$

$$= 0$$

A graph of these values shows that the probability distribution, $\mathrm{p}(\theta)$, now turns downward at high values of θ (Figure 12.24). The not-cured patient has substantially lowered the probabilities of cure rates that are very high, which is as it should be.

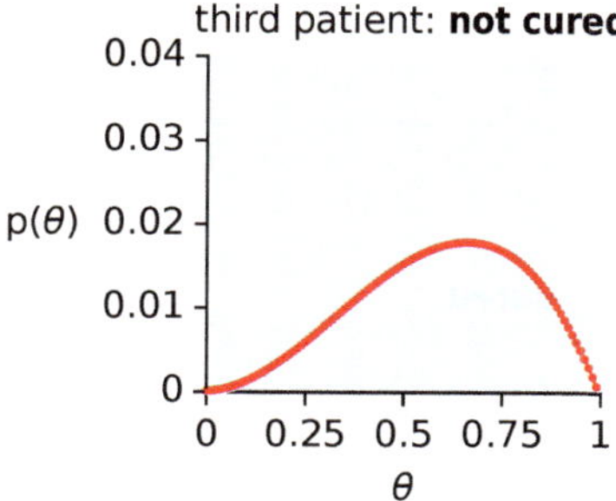

Figure 12.24 Probability of different values of cure rate θ after the third update.

These posterior probabilities then become the priors for the next patient.

$$\mathrm{p}(\theta_1) = 0.000011881188$$
$$\mathrm{p}(\theta_2) = 0.0000470447$$
$$\vdots$$
$$\mathrm{p}(\theta_{100}) = 0$$

All patients Let's imagine that we had 7 more patients, for a total of 10 patients in our trial, and let's suppose that the 10 patients were, in the following order.

cured, cured, not cured, cured, cured, not cured, cured, cured, cured, cured

The successive updates would look like panels E-K in Figure 12.25.

Then, if tomorrow, a new patient walks in the door, that patient would morph the curve as shown in panel L in Figure 12.25.

The output of our Bayesian analysis for this hypothetical clinical trial is panel K.

From panel K, we can draw a number of conclusions. First, the highest point, the most likely value for θ, is 80%, which was the overall cure rate. But the Bayesian curve gives us a great deal more information, as we said at the beginning. It tells us how precise this estimate is, by showing us how likely other values are. The width of that curve is a readout of our precision in saying $\theta = 0.8$, because after all, it could also had been 0.5, 0.6, or 0.9, with slightly lower probability. The thinner the peak of the curve, the more precise is the estimate.

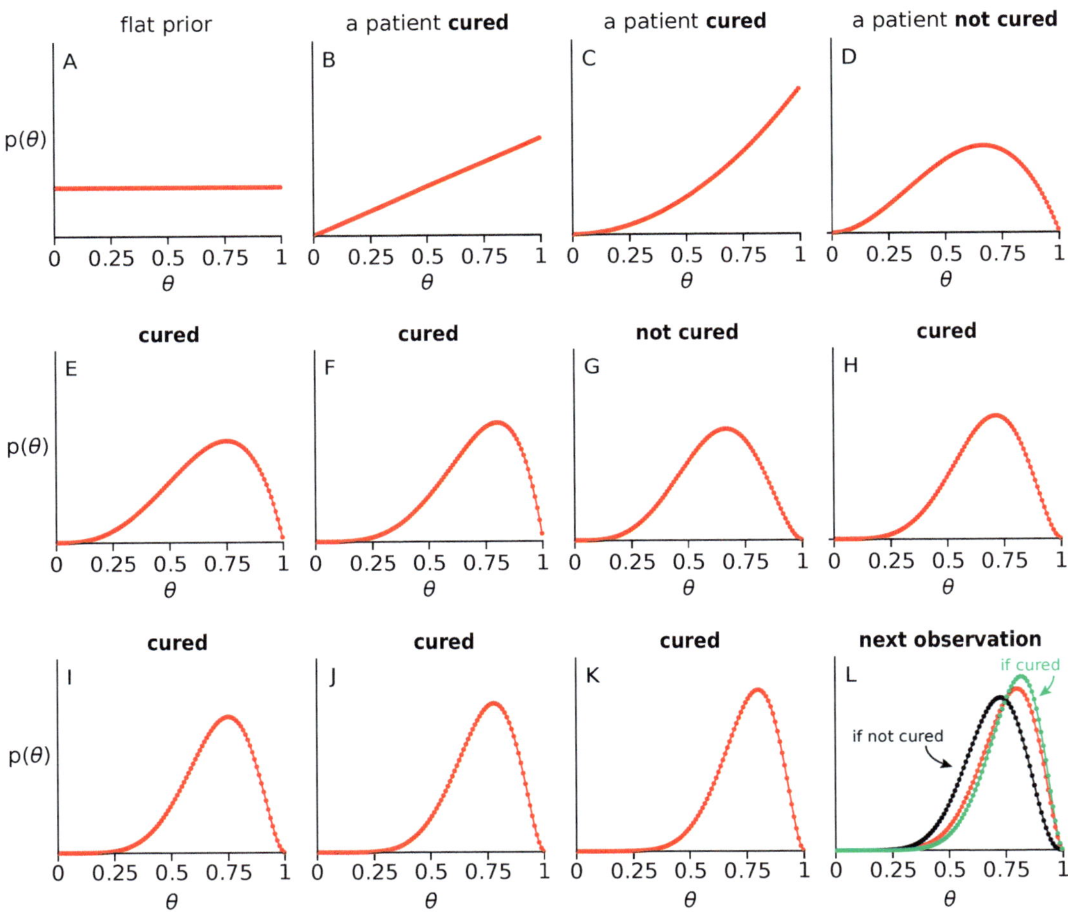

Figure 12.25 Panels A–K: successive updates of the probabilities as 10 patients come in the door. Panel L: how different results of the next observation (the 11th patient in this case) would morph the distribution curve depending on the outcome, cured (green) or not not cured (black), overlaying the prior probability curve (red), copied from panel K.[3]

[3] This figure looks identical to Figure 2 in Berry (2006), however, our figure was computed by Riemann Integration, whereas Berry's figure was calculated using analytically derived integrals.

Additionally, the Bayesian procedure enables us to update probability distributions. The fact we can deal with probability *distributions* means that we can answer questions like: what is the probability that the cure rate is higher than 75%? (Answer: it's the area under the curve to the right of $\theta = 0.75$.)

Order doesn't matter Although the Bayesian updating is done sequentially, it would be a drawback if the final answer depended on the order in which the patients walked in the door. Fortunately, this is not the case: the final answer does not depend on the order in which the patients present themselves.

Bayes vs. p-value

What would it look like if we were to use the p-value approach on the same study result?

In the hypothetical clinical trial, we observed that, out of 10 patients, 8 were cured and 2 were not.

If we were to calculate a p-value, using Null Hypothesis Significance Testing, we would have to provide a Null Hypothesis, and then ask: what is the probability, under the Null Hypothesis, of results equally or more extreme than our observed? The typical choice for the Null Hypothesis would be that the drug has no effect, that is, that the true cure rate is the same as the background rate of spontaneous cures in the population. This is an additional fact; let's say we look that up, and the fraction of cases that have spontaneous cures is 0.35. Then our Null Hypothesis is that $\theta = 0.35$.

We can then calculate the likelihood (that is, the p-value) of getting 8 or more cures out of 10 if the Null Hypothesis is true. A simple simulation tells us that the probability is 0.0034 (Figure 12.26). This is a one-sided p-value calculation. Berry (2006) suggests that we can obtain a two-sided p-value by "including probabilities in the opposite tail of the distribution for observations with probabilities smaller than for the actual observation". Consulting Figure 12.26, we see that there are no outcomes on the left whose probability is smaller than the observed 0.029. Therefore, our two-sided p-value is 0.034.

Count	152	729	1745	2532	2337	1523	742	206	29	5	0
Probability	0.0152	0.0729	0.1745	0.2532	0.2337	0.1523	0.0742	0.0206	0.0029	0.0005	0

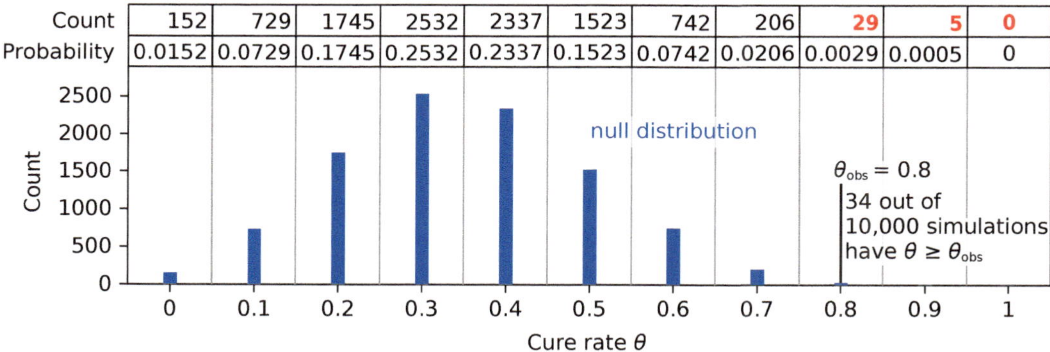

Figure 12.26 Distribution of cure rates θ under the Null Hypothesis that the background cure rate is 35%.

What can we conclude from this? That, if we follow the usual backwards logic of NHST, and also specify a weak "$p < 0.05$" standard, then we can rule out the Null Hypothesis, that the drug has no effect, that is, that $\theta = 0.35$. But this is a very weak conclusion. By contrast, the Bayesian approach tells us the probability of *any* cure rate: the most likely is $\theta = 0.8$, but we also have probabilities for all the other cure rates. Moreover, the p-value approach is still giving us the answer to the wrong question, telling us p(e | H) rather than p(H | e).

The advantages of the Bayesian approach to clinical trials were succinctly summarized by Donald Berry.

NATURE REVIEWS

Bayesian clinical trials

Donald A. Berry

Abstract | Bayesian statistical methods are being used increasingly in clinical research because the Bayesian approach is ideally suited to adapting to information that accrues during a trial, potentially allowing for smaller more informative trials and for patients to receive better treatment. Accumulating results can be assessed at any time, including continually, with the possibility of modifying the design of the trial, for example, by slowing (or stopping) or expanding accrual, imbalancing randomization to favour better-performing therapies, dropping or adding treatment arms, and changing the trial population to focus on patient subsets that are responding better to the experimental therapies. Bayesian analyses use available patient-outcome information, including biomarkers that accumulating data indicate might be related to clinical outcome. They also allow for the use of historical information and for synthesizing results of relevant trials. Here, I explain the rationale underlying Bayesian clinical trials, and discuss the potential of such trials to improve the effectiveness of drug development.

12.5 Bayes' Theorem and the problem of unreliable results

As we saw in Chapter 1, a major problem facing modern science is the *crisis of irreproducibility*, the fact that many studies that have appeared in the published literature cannot be replicated. Bayes' Theorem can help us understand this problem, as explained in Button *et al.* (2013).

The basic question is: **given that a published paper has a result that is statistically significant, what is the likelihood that the result is actually true?** In other words, should you believe this result? That's a fundamental question, and it can only be approached through Bayes' Theorem.

To answer this, let's get the terms of Bayes' Theorem set up correctly. Let's call the positive significant finding "+ finding". Then, in reality, either 1) there is actually an effect ("effect"), and so the positive finding was reliable, or 2) there is actually no effect ("no effect"), in which case the positive finding was not reliable.

Then what we want to know is: what is p(effect | + finding)? That is exactly asking for the Positive Predictive Value of this + finding.

If we apply Bayes' Theorem

$$p(\text{effect} \mid + \text{finding}) = \frac{p(+ \text{finding} \mid \text{effect}) \times p(\text{effect})}{p(+ \text{finding})}$$

But what do these terms mean? Let's start with the numerator.

- p(+ finding | effect) is the likelihood that the study will have returned a positive finding given that there really is an effect. That is just the **power** of the study.
- p(effect) is the **prior** probability that the effect is really there, the probability before this study was undertaken. This may seem like an odd request, since what we wanted to know

was what is the likelihood that this effect is actually there, and we are now being asked to supply an initial guess, that will be updated by the new finding.

This can be tricky, since the value we assign to this prior probability could be decisive. For example, if we give an extremely low prior probability for the effect, then we would need very strong evidence to update it to a status of "likely".

Here, we have to make a guess as to what is the prior likelihood of this effect actually being there. We will deal with this requirement to supply a prior probability by trying several estimates, $p = 0.5$, $p = 0.1$ etc.

For the denominator, which is the total probability of a + finding, we split the denominator into two cases, one where the effect is really there, and the other where it isn't.

$$\mathrm{p(effect \mid + finding)} = \frac{\mathrm{p(+ finding \mid effect)} \times \mathrm{p(effect)}}{\mathrm{p(+ finding)}}$$

$$\Downarrow \textit{ splitting the denominator}$$

$$= \frac{\mathrm{p(+ finding \mid effect)} \times \mathrm{p(effect)}}{\mathrm{p(+ finding \mid effect)} \times \mathrm{p(effect)} + \mathrm{p(+ finding \mid no\ effect)} \times \mathrm{p(no\ effect)}}$$

Now let's look at the denominator.

- what is p(+ finding | no effect)? That is, what is the probability of a positive finding when there is actually no effect? That is just the **False Positive Rate**, which is α (see Figure 3.16 for the definition of α).

Bayes' Theorem for published studies

$$\mathrm{p(effect \mid + finding)} = \frac{\overbrace{\mathrm{p(+ finding \mid effect)}}^{\text{power}} \times \overbrace{\mathrm{p(effect)}}^{\text{prior}}}{\underbrace{\mathrm{p(+ finding \mid effect)}}_{\text{power}} \times \underbrace{\mathrm{p(effect)}}_{\text{prior}} + \underbrace{\mathrm{p(+ finding \mid no\ effect)}}_{\text{False Positive Rate } \alpha} \times \underbrace{\mathrm{p(no\ effect)}}_{\text{1-prior}}}$$

By visually inspecting this equation, several facts immediately stand out:

- **Power.** If we hold α and the prior fixed, then p(effect | + finding) is essentially

$$\mathrm{p(effect \mid + finding)} = \frac{\text{power}}{\text{power + constant}}$$

  This fraction goes down as power goes down. Therefore, the lower the power of the study, the lower is the likelihood that the + finding represents a real effect. This is very significant: it says that in a low power study, the + finding is likely to be a false positive, that is, a statistical fluke.

- **Prior.** If we hold everything else constant, it is also clear that the lower the prior, the lower is the likelihood that the + finding represents a real effect. *Positive results from studies of hypotheses with low inherent plausibilities are likely to be false positives.* When Cornell prof Darryl Bem claimed to find statistically significant evidence for Extra-Sensory Perception (the mystical ability to "perceive" events before they happen), critics objected that for such an implausible conclusion, much stronger evidence is required than simply doing Null Hypothesis Significance Testing. (And indeed, some even suggested that Bem intended his study to be a critique of these methods, saying in effect "look at the silly things I can prove using NHST".)

- **False Positive Rate α.** We could try to increase our power by just increasing α, that is, by loosening our standard for statistical significance. That would certainly increase the power, but at the expense of increasing the false positive rate. Is the tradeoff worth it? We can estimate an answer by dividing both the numerator and the denominator by power × prior. Rearranging terms,

$$\text{p(effect} \mid + \text{finding)} = \frac{1}{1 + \text{constant} \times \frac{\alpha}{\text{power}}}$$

Generally, increasing α yields diminishing returns in power but increases false positives linearly. Therefore, doubling α (e.g., from 0.01 to 0.02) generally reduces PPV and makes findings less reliable, even if power increases somewhat. For example see Figure 11.8, and note that when we increase α from 0.01 to 0.02, power increases less than by a factor of 2. **It follows (and makes sense) that the higher the false positive rate α, the lower is the likelihood that the + finding represents a real effect.**

Calculating the probability that a published result is reliable

The prior What is p(effect)? The probability that there really is an effect. Note that this is not the probability after this study, but the **prior probability** before this study appeared. But where do we get the prior probability of there being a real effect? Do we guess?

In the following few examples, we will try a few possible numbers for the prior and calculate the probability that a published positive finding is reliable. Later, we'll address the question of priors more systematically (see further discussion on *The Prior* on page 645).

Bayes' Theorem approach Let's say the prior probability of a given effect was 10%, the power of our study is 80%, and we set α to be 0.05. Applying Bayes' Theorem

$$\begin{aligned}
\text{p(effect} \mid + \text{finding)} &= \frac{\text{p(+ finding} \mid \text{effect)} \times \text{p(effect)}}{\text{p(+ finding)}} \\
&= \frac{\text{p(+ finding} \mid \text{effect)} \times \text{p(effect)}}{\text{p(+ finding} \mid \text{effect)} \times \text{p(effect)} + \text{p(+ finding} \mid \text{no effect)} \times \text{p(no effect)}} \\
&= \frac{\text{power} \times \text{prior}}{\text{power} \times \text{prior} + \alpha \times (1 - \text{prior})} \\
&= \frac{0.8 \times 0.1}{0.8 \times 0.1 + 0.05 \times (1 - 0.1)} \\
&= 0.64
\end{aligned}$$

Therefore, setting a False Positive Rate of 5% doesn't mean that a + finding has a 95% chance of being a real effect. On the contrary, in this case (that is, power = 80%, prior = 10%), the probability that the + finding represents a real effect is only 64%, not 95%!

PPV approach We get the same result if we look for the positive predictive value of the positive finding. Since the prior likelihood of there being a real effect is 10%, then, in 1000 studies there will be 100 real effects. Power = 80% means 80 of those will be called +. In the 900 other cases, in which the effect is absent, $\alpha = 0.05$ means we will call 5% of the 900, or 45 out of 900, positive. These will be false positives.

Then the fraction of all positives that are true positives is:

$$\text{Positive Predictive Value of + findings} = \frac{\text{TP}}{\text{TP} + \text{FP}} = \frac{80}{80 + 45} = 64\%$$

So, in this setting, if a paper has a + finding, there's a 64% chance it's true.

In reality The calculation above was for a fairly high power (80%) study. In reality, most published studies have much lower power. If the power of a study is 40%, then the Positive Predictive Value of a $p < 0.05$ statistically significant + finding drops to less than 50%.

$$\begin{aligned}
p(\text{effect} \mid + \text{finding}) &= \frac{p(+\text{ finding} \mid \text{effect}) \times p(\text{effect})}{p(+\text{ finding})} \\
&= \frac{p(+\text{ finding} \mid \text{effect}) \times p(\text{effect})}{p(+\text{ finding} \mid \text{effect}) \times p(\text{effect}) + p(+\text{ finding} \mid \text{no effect}) \times p(\text{no effect})} \\
&= \frac{\text{power} \times \text{prior}}{\text{power} \times \text{prior} + \alpha \times (1 - \text{prior})} \\
&= \frac{0.4 \times 0.1}{0.4 \times 0.1 + 0.05 \times (1 - 0.1)} \\
&= 0.47
\end{aligned}$$

In other words, in this lower power (40%) study, a + finding is more likely than not to be a false positive.

It is interesting to look at the published literature to see what kinds of power published studies actually have. Button *et al.* (2013) reviewed studies in the subject of Neuroscience, and was quite gloomy: "Our results indicate that the median statistical power in neuroscience is 21%". and "Our results indicated that the median statistical power of [neuroimaging] studies was 8% across 461 individual studies contributing to 41 separate meta-analyses, which were drawn from eight articles that were published between 2006 and 2009".

Exercise 12.5.1 We conduct a study on whether a particular drug reduces the pain level of migraines, and we find a statistically significant result. We used $\alpha = 0.05$, and the power of our study was 40%. We use a pilot study and previous literature to find that the prior probability of our hypothesis that the drug reduces pain is 0.5. What is the probability that our statistically significant result represents a real effect?

The Prior

It is essential to the Bayesian approach that we start with a prior; fresh data then updates the prior. As we mentioned in section 12.3 (*Where to get the priors?* on page 607), there are a number of ways to get the priors.

In the case of published findings, the updating takes the form of: we have a prior probability of an effect, and the + finding of this study updates the probability. For published findings, the most important source of priors is the published literature.

Previous literature The fact that Bayesian analyses allow the previous literature to be used as the prior is not a drawback of Bayesianism, it's an advantage: it enables us to evaluate the present study in the context of our knowledge from previous studies. Therefore, Bayesianism allows us to aggregate our knowledge.

In contrast, in the p-value approach, every study stands alone with its own p-value, calculated in isolation; the p-values don't speak to each other, and there is no accumulation of knowledge. For example, let's say we're a p-value theorist studying the efficacy of a new drug. We learn that there are three papers already on this drug that found the efficacy to be x, y, and z. How do we take account of those previous studies? The answer is: we can't. The p-value approach does not enable us to incorporate other studies.

Attempting to overcome this "every study stands alone" limitation, statistical theorists developed a technique called "meta-analysis". We discussed meta-analysis briefly in Chapter 6, but we left that discussion at simply plotting the results of the surveyed studies, more or less as a list (such as a forest plot); there was no means to *combine* these results.

Meta-analysis has proposed some methods for aggregating results, for example, by taking the weighted average of all the effect sizes, weighted by the size of each study. But such a meta-analysis is different from a Bayesian approach. In the Bayesian approach, we incorporate the results of other studies into our study to form our conclusion. By contrast, in the p-value approach, our study stands alone with its own effect size and p-value, and then, later on, our study might or might not become incorporated into someone else's meta-analysis.

Why do p-values persist?

The p-value approach cannot, by itself, combine studies to produce a socially and scientifically useful conclusion. So why do p-values persist?

One reason is that, post WW2, the scientific and medical establishment grew rapidly, which required evaluators to make many thousands of judgements of the form: should we fund this study? should we publish this paper? This required a somewhat mechanized process to evaluate each study, and assign a number to *this* result, *this* paper, *this* researcher. Is this study a good one? If so, we'll give this person tenure, grants, etc. Whereas if it's not a good study, we will not. But then you say, the Bayesian point of view enables us to incorporate all of the other studies on the same subject so it gives an answer not to how good is this study/person, but an answer to: what is our collective social knowledge of the efficacy of this drug, combining everybody's result? Therefore, the p-value approach sacrifices the important question "what is our overall knowledge?" for "how do I evaluate this researcher/study?" because that's what the scientific and medical establishment needed.

This point is made well in *The Cult of Statistical Significance* by Ziliak and McCloskey (2008). They cite "what Robert Merton called the 'the Bureaucratization of Knowledge'...statistical significance, originally conceived as a means to substantive significance, became transformed by Fisher and then by bureaucracies of science into an end in itself".

A good example of the use of Bayesianism to incorporate prior studies was a Bayesian reanalysis of the cardiac GUSTO trial. The background of the trial is: if a cardiologist detects coronary artery blockage in a patient, the doctor and patient may decide on a minimally invasive procedure, in which "clot-busting" drugs are injected. The drug in standard use had been Streptokinase (SK), but a new drug, Tissue Plasminogen Activator (t-PA) was developed and introduced. t-PA is 10 times more expensive than SK, so a natural question was: is t-PA clinically superior to SK?

The GUSTO study found a clinical superiority for t-PA, but when this study was re-analyzed using Bayesian methods, the positive finding disappeared (Brophy and Joseph, 1995).

> **Special Communications** *Journal of the American Medical Association 1995*
>
> **Placing Trials in Context Using Bayesian Analysis**
> GUSTO Revisited by Reverend Bayes
>
> James M. Brophy, MD, Lawrence Joseph, PhD
>
> Standard statistical analyses of randomized clinical trials fail to provide a direct assessment of which treatment is superior or the probability of a clinically meaningful difference. A Bayesian analysis permits the calculation of the probability that a treatment is superior based on the observed data and prior beliefs. The subjectivity of prior beliefs in the Bayesian approach is not a liability, but rather explicitly allows different opinions to be formally expressed and evaluated. The usefulness of this approach is demonstrated using the results of the recent GUSTO study of various thrombolytic strategies in acute myocardial infarction. This analysis suggests that the clinical superiority of tissue-type plasminogen activator over streptokinase remains uncertain.

There couldn't be a paper entitled "Placing Trials in Context Using p-value Analysis" because the p-value approach does not enable you to incorporate other trials.

Situations that are inherently updating Often, the choice of a prior does not play a critical role, because the difference between one prior and another is washed out by the updating process: it doesn't really matter what prior you start with, because with enough data, the updating processes will converge to the same answer.

Several good examples of this process, showing how new data washes out differences in the initial choice of priors, can be found in the excellent text by Bolstad and Curran (2016), *Introduction to Bayesian Statistics*. For example, they consider a case of a Bayesian analysis of the paired differences of two groups. Three investigators, Aleece, Brad and Curtis, have three very different priors (Figure 12.27 left). But after 15 data points (15 pairs), the three posteriors are almost identical (Figure 12.27 right).

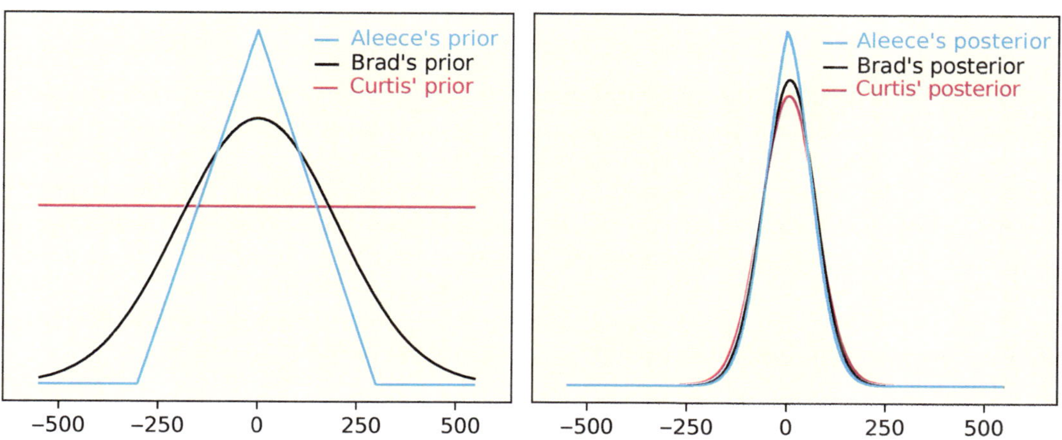

Figure 12.27 New data can wash out differences in initial choice of priors. Three investigators with radically different priors come to similar posteriors after Bayesian updating. Adapted from Bolstad and Curran (2016).

Bayesian methods are therefore especially applicable to situations that inherently involve updating. For example, in clinical trials, we are trying to estimate the probability that this drug is successful. On day one, patient #1 walks in the door, took the drug and got cured. This gives us a probability. Then patient #2 walks in the door, and they took the drug, and they are also cured. So we are going to update the probabilities. Then patient #3 walks in the door, took the drug, and were not cured. We then update the probability again.

Other situations that are inherently updating occur in code breaking, and also in gambling.

In code breaking, as we saw, Turing and the group at Bletchley park, started each midnight with a flat, I-Have-No-Idea prior: all three-letter rotor key sequences were equally likely. Then they updated that prior with the first intercepted message to get the posterior, which became the prior for the next message.

In gambling in Blackjack, we would be interested in determining the fraction of high cards outstanding in the deck. Whether that fraction is high or low matters to the bettor (see "Sampling with replacement in casino Blackjack" on page 99). And so, after each deal, a Bayesian blackjack player updates the probability that the remaining deck is rich in high cards.

12.6 Bayesianism as a paradigm shift in statistics

Bayesianism vs. Frequentism

We have now seen the Bayesian approach, and we can reflect on the deep differences between Bayesianism and the p-value approach. (The p-value approach is also called Frequentism.)

Answering the wrong question We've already seen the fundamental difference that is rooted in the fact that Frequentism and Bayesianism are simply answering different questions.

- **Frequentism is answering the question: what is the probability of this data, given this hypothesis?**

- **Bayesianism is answering the question: what is the probability of this hypothesis, given this data?**

Obviously, the Bayesian question is the right question; it's what we really want to know as scientists.

By contrast, if we adopt the frequentist approach, we have to make backward and convoluted statements that are difficult to interpret. For example, in Chapter 3, it was important to insist that "$p < 0.01$" does NOT mean "the probability of the Null Hypothesis being true is < 0.01". Instead, we had to insist on the convoluted wording: "if this experiment were run thousands of times, a result at least this extreme would occur less than 1% of the time". And when we considered the notion of confidence intervals, we had to insist that the 99% confidence interval cannot be read as "we are 99% confident that the true value is in this interval". We can only say "if this experiment were run thousands of times, generating thousands of 99% confidence intervals, the true value would be in the 99% confidence intervals 99% of the time".

Frequentism is about what you might expect to see in the long run. Bayes is about *this* Hypothesis and *this* data.

Thus, Bayesianism can make statements like: given this data, the probability of candidate X winning the election is 70%, whereas frequentism can only make the philosophically problematic statement "if this election were held 1000 times, in 700 of them, candidate X will win". *But what can that possibly mean?*

Because Bayesianism is about the Hypothesis, the Bayesian result can tell us, as in the clinical trial problem, not just what is the probability of *this* value of the parameter, for example, the cure rate of the drug, but what is the probability *distribution* over *all* values of the parameter. The frequentist approach can only tell us that one particular value of the parameter, usually, the Null Hypothesis, rarely produces this data at least this extreme. This is because **Bayesianism assigns probabilities to hypotheses, frequentism assigns probabilities to data.**

Probability as frequency vs. probability as subjective belief Lying at the heart of the disagreement is a fundamental difference in how the two views treat the concept of probability.

Suppose I flip a coin, and hide the result from you. I then ask you: **what is the probability that the coin came up Heads?** Because Frequentism is about frequencies in the long run, the frequentist answer is that there *is no* probability of that coin having come up Heads or Tails. Either it was Heads (100% probability) or it was Tails (0% probability). For the Bayesian, the question is about your subjective belief, and the answer is that given the state of your knowledge, you can only assign a 50% probability to each outcome.[4]

We might say that Frequentism is about the coin, and Bayesianism is about your belief.

Another corollary of the fact that Bayesianism is about our beliefs is the fact that the Bayesian approach allows us to incorporate prior knowledge, which we then update. Bayesianism therefore produces a unified synthesis of our knowledge, whereas in Frequentism, it's every study for itself, and so Frequentism always faces a problem in combining information from different studies.

The "logic" of NHST In many statistics texts and courses, it's unfortunate that a completely fallacious "logic" of NHST continues to be presented. The ill-logic consists of two separate fallacies, both of which have been thoroughly exposed and refuted by philosophers of science in the mid-20th century.

The fallacy of falsification In the first fallacy, a doctrine is announced called "falsificationism", which is the idea that scientific theories can only be falsified, and that a single contrary observation falsifies and invalidates the theory.

But, as philosophers such as W.V.O. Quine and Hilary Putnam have pointed out, well-established scientific theories are never refuted by a single observation; rather, the scientific community will question what Putnam called the "auxiliary hypotheses" that are necessary to derive any observational consequences from theory (see "The 'corroboration' of theories" by Putnam (1974)).

For example, Newton's theory of gravitation produced many successful calculations of planetary orbits, but when it came to calculating the orbit of the planet Uranus, the observed trajectory did not agree with the theoretical calculations. Did astronomers discard Newton's theory? No. Instead, they realized that the problem might lie in one of the auxiliary hypotheses, that the planets up to Uranus were all the planets there were. They thought: what if *that* weren't true? So they asked themselves: if there were another planet perturbing the orbit of Uranus, what would the planet and its trajectory have to look like to have this effect? They calculated the trajectory of such an object, and observational astronomers pointed their telescopes where the theoreticians told them to look, and discovered the planet Neptune.

---

[4] Cassie Kozyrkov, former Chief Decision Scientist at Google.
`https://towardsdatascience.com/statistics-are-you-bayesian-or-frequentist-4943f953f21b`

Similarly, when an experimenter reports observational data that's in conflict with a theory that has been perviously corroborated, we need to look at the exact conditions under which the data was collected, and the assumptions that the data-gatherers had to make. For example, was the sample biased?

The contrapositive fallacy And then, having embraced the false doctrine of falsificationism, the NHST "logic" commits another fallacy, which is even more critical. It takes the logically valid inference "if A then not-B, but B, therefore not-A" and tries to probabillize it, which turns it into a totally invalid form: "if A then *probably* not-B, but B, therefore *probably* not-A". In NHST, this fallacy takes the form of "If the Null Hypothesis is true, then this data is unlikely. But this data was observed, therefore the Null Hypothesis is unlikely."

In Chapter 3, we explained this fallacy, and gave a convincing counter-example, that by using this principle, we can easily "prove" that your senator is not an American! The logic is a perfect instance of the NHST argument: if a person is American, it is highly unlikely that they are a senator (100 senators in 350 million people, $p < 0.0001$), but your senator is a senator, therefore your senator is probably not an American ($p < 0.0001$) (see section 3.3 *The p-value is the answer to the wrong question* on page 127).

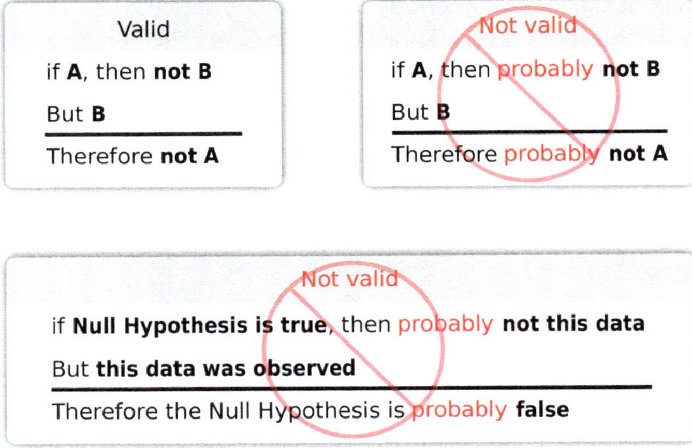

Thus, the fundamental "logic" of NHST rests on two major logical fallacies.

Shifting the question

Scientific revolutions often involve fundamental changes in the questions that are being asked (*Forms of Explanation* (Garfinkel, 1981)).

For instance, when Descartes looked at an object in motion, he asked the question "why does the object keep moving?" He postulated a physical quantity he called "impetus" as the factor that keeps the object moving. This led to a poor physics.

The revolution that Newton introduced began with a fundamental shift in the explanatory framework: we no longer try to explain why an object keeps moving. If asked what keeps the object moving, the answer is: nothing, an object in motion tends to remain in motion unless acted on by outside forces. Rather, we explain *changes* in the motion of an object. This shifts the question to: what causes the *change* in the motion of an object? The answers to this question Newton called "forces".

The shift from frequentism to bayesianism involves an analogous shift in the question being asked. Frequentism is trying to explain why this probability? whereas bayesianism is answering a different question: why this *change* in probability?

Update your priors Two concepts that are unique to Bayesianism are: update and priors.

The slogan "update your priors" (often seen on t-shirts in Silicon Valley) is actually an excellent three-word summary of Bayesianism. We have a prior, we get some new data, we update our prior. That's Bayesianism.

FURTHER EXERCISES 12.6

1. What's the difference between a p-value in frequentism and a posterior probability in Bayesian statistics? Include the definitions of each (in words and as a conditional probability) in your answer.

Afterword

The crisis of irreproducibility Medical, biological, and social science research is in a crisis of irreproducibility, which is having the negative effect of impeding scientific progress and undermining public confidence in science.

As we mentioned from the very beginning, many published medical and scientific studies that used flawed statistical methodologies are being questioned. In particular, two subject areas in which there are epidemics of false positives are studies of the genetic bases for diseases and other conditions, and studies that employ imaging methodologies like fMRI. The rate of false claims is relatively greater in these areas because both of them have Big Data. When we have Big Data, we can get lots of false positives.

For example, statistical methods. especially NHST, had found some 85 genetic variants that were statistically linked to acute coronary syndrome. But a follow-up study found that *none* of the 85 genetic variants was actually associated with the disease! So the 85 published "genes for" articles were all false positives. A view that is emerging is that "genes for" is a problematic idea to begin with, and that most of the "findings" are false positives.

Odds Are, It's Wrong

Science fails to face the shortcomings of statistics

By Tom Siegfried

Nowhere are the problems with statistics more blatant than in studies of genetic influences on disease. In 2007, for instance, researchers combing the medical literature found numerous studies linking a total of 85 genetic variants in 70 different genes to acute coronary syndrome, a cluster of heart problems. When the researchers compared genetic tests of 811 patients that had the syndrome with a group of 650 (matched for sex and age) that didn't, only one of the suspect gene variants turned up substantially more often in those with the syndrome — a number to be expected by chance.

How could so many studies be wrong? Because their conclusions relied on "statistical significance," a concept at the heart of the mathematical analysis of modern scientific experiments.

SCIENCE NEWS | 2010

A similar situation happened in the search for genetic bases for depression. A number of studies had gained headlines by announcing the discovery of "genes for" depression. But a well-done follow-up study, using a large data base, found that "the large number of associations reported in the depression candidate gene literature are likely to be false positives".

A. Garfinkel and Y. Guo, *Understanding Data*,
https://doi.org/10.1007/978-3-032-18600-3

ARTICLES

No Support for Historical Candidate Gene or Candidate Gene-by-Interaction Hypotheses for Major Depression Across Multiple Large Samples

Richard Border, M.A., Emma C. Johnson, Ph.D., Luke M. Evans, Ph.D., Andrew Smolen, Ph.D., Noah Berley, Patrick F. Sullivan, M.D., Matthew C. Keller, Ph.D.

Objective: Interest in candidate gene and candidate gene-by-environment interaction hypotheses regarding major depressive disorder remains strong despite controversy surrounding the validity of previous findings. In response to this controversy, the present investigation empirically identified 18 candidate genes for depression that have been studied 10 or more times and examined evidence for their relevance to depression phenotypes.

Methods: Utilizing data from large population-based and case-control samples (Ns ranging from 62,138 to 443,264 across subsamples), the authors conducted a series of preregistered analyses examining candidate gene polymorphism main effects, polymorphism-by-environment interactions, and gene-level effects across a number of operational definitions of depression (e.g., lifetime diagnosis, current severity, episode recurrence) and environmental moderators (e.g., sexual or physical abuse during childhood, socioeconomic adversity).

Results: No clear evidence was found for any candidate gene polymorphism associations with depression phenotypes or any polymorphism-by-environment moderator effects. As a set, depression candidate genes were no more associated with depression phenotypes than noncandidate genes. The authors demonstrate that phenotypic measurement error is unlikely to account for these null findings.

Conclusions: The study results do not support previous depression candidate gene findings, in which large genetic effects are frequently reported in samples orders of magnitude smaller than those examined here. Instead, the results suggest that early hypotheses about depression candidate genes were incorrect and that the large number of associations reported in the depression candidate gene literature are likely to be false positives.

Am J Psychiatry 2019

The epidemic of false positives is also seen in many studies that rely on imaging modalities like fMRI. We discussed the fMRI study of the "dead salmon" in Chapter 3 (page 123), which found several "statistically significant" voxels, implying active brain activity in those regions of the very dead salmon. The dead salmon study, of course, was intended as a satirical rebuke, but there have been many studies that seriously suggested "findings" that were false positives.

What got these studies into publication? NHST plus multiple testing!

The fundamental principle at work here is the "green jelly beans" phenomenon: given multiple tests at $p < 0.05$, or $p < 0.01$, or even $p < 0.001$ if we are doing thousands of tests, we are bound to see many false positives. It is more important to understand the green jelly beans phenomenon than to memorize any formulas in any textbooks.

That's why over 800 signatories in a lead article in *Nature* called for the retirement of statistical significance.

COMMENT **Retire statistical significance**

Valentin Amrhein, Sander Greenland, Blake McShane and more than 800 signatories call for an end to hyped claims and the dismissal of possibly crucial effects.

21 MARCH 2019 | VOL 567 | NATURE | 305

We ... call for the entire concept of statistical significance to be abandoned.

Where to now? Our goal is to educate future life scientists to do honest, reliable and reproducible science. We urge:

- **Know your data.** Use visualization techniques to capture the key features of the data. Make dot plots, beeswarm plots, histograms and KDEs to visualize the distribution.
- **Break the NORM(AL distribution).** Avoid the default assumption that distributions must be Normal. Based on the visualized actual distribution, choose appropriate measures. Understand the Flaw of Averages; be ready to use measures beyond the mean.
- **Understand p-values.** Understand the meaning of a p-value and the Null Hypothesis. What are the limitations of Null Hypothesis Significance Testing? Why does using a "$p < 0.05$" standard lead to a flood of false positives?
- **p-hacking.** Understand the perils of multiple testing and how to correct for multiple testing.[5]
- **Focus on effect sizes, not just p-values.** *Never* say "drug X reduces heart attacks ($p < 0.05$)", instead say "drug X reduces heart attacks *by how much* (effect size)", and give a confidence interval for that effect size.
- **Use resampling-based computer simulations, based on your data, instead of formulas, based on theoretical distributions.** All statistics tests, such as p-values, effect sizes, confidence intervals and power analyses can be studied and calculated using resampling, and applied to two-group comparisons, multi-group comparisons, χ, χ^2, relative risk, correlation, regression, etc.
- **Know your power.** Compute the statistical power of a study design using simulation-based techniques. Understand how statistical power is used.
  - Prospective study design (For example, is my sample size n large enough?)
  - Retrospective analysis (Did the study have the power to make the negative statement? Did it have sufficient power to ensure that a positive finding is not a false positive?)
- **Bayesianism.** Focus on positive and negative predictive value and on the probability of hypotheses given data. Understand how to update prior probabilities with new evidence, by using Bayes' Theorem.

[5] In this book, we have focused on a few forms of p-hacking, especially, multiple testing. However, there are many forms of p-hacking which we did not discuss. HARKING (Hypothesizing After Results are Known), pseudoreplication, sample size increments, variable definitions, and outcome shifting are also common and hard-to-detect p-hacking practices.

Acknowledgments

In writing this text, we've had enormous help from a number of sources. We benefitted from strong institutional support from UCLA, particularly from Eric Deeds, Vice Chair for Mathematics Education in the Life Sciences Core. We also acknowledge the support of the University of Oxford, where parts of this work were written while the first author enjoyed a sabbatical as the Newton-Abraham Professor.

Morgan Tingley has shared his insights from teaching an earlier draft of this text, as well as helping on several theoretical points.

Jane Shevtsov supplied a large number of exercises. Many of the examples and exercises that discuss ecological systems are due to her. We also benefitted from theoretical discussions with her about resampling methods.

Adrian Soto-Mota provided wonderful strategic advice and encouragement, as well as making a number of highly useful suggestions about our exposition.

Eric Demer supplied a number of theoretical points to clarify our exposition.

Helen Ringley was a superb TA for this course, and worked through the text sentence-by-sentence, giving us dozens of helpful suggestions to improve our presentations and arguments.

Kristin McCully gave us extensive feedback on the classroom teaching: what worked and what didn't. She also supplied a very large number of exercises. Many of the exercises that discuss public health, epidemiology, disease testing, educational policy and other social science issues are due to her.

We want to thank the thousands of UCLA students from Life Sciences 40 who field-tested our earlier versions and gave many helpful corrections, ranging from stylistic suggestions to pointing out gaps in the exposition.

A. Garfinkel and Y. Guo, *Understanding Data*,
https://doi.org/10.1007/978-3-032-18600-3

Bibliography

Amrhein, V., Greenland, S., and McShane, B. (2019). Scientists rise up against statistical significance. *Nature*.

Anscombe, F. J. (1973). Graphs in statistical analysis. *The American Statistician*.

Austin, P. C., Mamdani, M. M., Juurlink, D. N., and Hux, J. E. (2006). Testing multiple statistical hypotheses resulted in spurious associations: a study of astrological signs and health. *Journal of Clinical Epidemiology*.

Bass, T. A. (1985). *The eudaemonic pie*. Houghton Mifflin.

Benjamin, D. J., Berger, J. O., Johannesson, M., Nosek, B. A., Wagenmakers, E.-J., Berk, R., Bollen, K. A., Brembs, B., Brown, L., Camerer, C., *et al.* (2018). Redefine statistical significance. *Nature Human Behaviour*.

Benjamini, Y. and Yekutieli, D. (2001). The control of the false discovery rate in multiple testing under dependency. *Annals of Statistics*.

Berry, D. A. (2006). Bayesian clinical trials. *Nature Reviews Drug Discovery*.

Bolstad, W. M. and Curran, J. M. (2016). *Introduction to Bayesian statistics*. John Wiley & Sons.

Bono, R., Blanca, M. J., Arnau, J., and Gómez-Benito, J. (2017). Non-normal distributions commonly used in health, education, and social sciences: a systematic review. *Frontiers in Psychology*.

Brophy, J. M. and Joseph, L. (1995). Placing trials in context using Bayesian analysis: GUSTO revisited by Reverend Bayes. *JAMA*.

Button, K. S., Ioannidis, J., Mokrysz, C., Nosek, B. A., Flint, J., Robinson, E. S., and Munafò, M. R. (2013). Power failure: why small sample size undermines the reliability of neuroscience. *Nature Reviews Neuroscience*.

Clayton, A. (2020). How eugenics shaped statistics. *Nautilus*.

Clayton, A. (2021). *Bernoulli's fallacy: Statistical illogic and the crisis of modern science*. Columbia University Press.

Cohen, J. (1994). The earth is round (p< .05). *American Psychologist*.

Colquhoun, D. (2014). An investigation of the false discovery rate and the misinterpretation of p-values. *Royal Society Open Science*.

A. Garfinkel and Y. Guo, *Understanding Data*,
https://doi.org/10.1007/978-3-032-18600-3

De Lucas, R. D., Balikian, P., Neiva, C. M., Greco, C. C., and Denadai, B. S. (2000). The effects of wet suits on physiological and biomechanical indices during swimming. *Journal of Science and Medicine in Sport.*

Drummond, G. and Vowler, S. (2011). Show the data, don't conceal them. *British Journal of Pharmacology.*

Drummond, G. B. and Vowler, S. L. (2012). Not different is not the same as the same: how can we tell? *Advances in Physiology Education.*

Efron, B. and Tibshirani, R. (1991). Statistical data analysis in the computer age. *Science.*

Fieberg, J. R., Vitense, K., and Johnson, D. H. (2020). Resampling-based methods for biologists. *PeerJ.*

Fisher, R. A. (1936). The use of multiple measurements in taxonomic problems. *Annals of Eugenics.*

Freedman, D., Pisani, R., and Purves, R. (2007). *Statistics.* WW Norton & Company.

Garfinkel, A. (1981). *Forms of explanation: rethinking the questions in social theory.* Yale University Press.

Halsey, L. G., Curran-Everett, D., Vowler, S. L., and Drummond, G. B. (2015). The fickle P value generates irreproducible results. *Nature Methods.*

Head, M. L., Holman, L., Lanfear, R., Kahn, A. T., and Jennions, M. D. (2015). The extent and consequences of p-hacking in science. *PLOS Biol.*

Herder, E. A., Spence, A. R., Tingley, M. W., and Hird, S. M. (2021). Elevation correlates with significant changes in relative abundance in hummingbird fecal microbiota, but composition changes little. *Frontiers in Ecology and Evolution.*

Ioannidis, J. P. (2005). Why most published research findings are false. *PLOS Medicine.*

Janosi, Andras, S. W. P. M. and Detrano, R. (1988). Heart Disease. UCI Machine Learning Repository. DOI: https://doi.org/10.24432/C52P4X.

Johnson, V. E. (2013). Revised standards for statistical evidence. *Proceedings of the National Academy of Sciences.*

Jones, M. C., Marron, J. S., and Sheather, S. J. (1996). A brief survey of bandwidth selection for density estimation. *Journal of the American Statistical Association.*

Kent, D. and Hayward, R. (2007). When averages hide individual differences in clinical trials: analyzing the results of clinical trials to expose individual patients' risks might help doctors make better treatment decisions. *American Scientist.*

Kowalski, C. J. (1972). On the effects of non-normality on the distribution of the sample product-moment correlation coefficient. *Applied Statistics.*

Loong, T.-W. (2003). Understanding sensitivity and specificity with the right side of the brain. *BMJ.*

Moore, D. S. and Shenk, D. (2017). The heritability fallacy. *Wiley Interdisciplinary Reviews: Cognitive Science.*

Morgan, T. M., Krumholz, H. M., Lifton, R. P., and Spertus, J. A. (2007). Nonvalidation of reported genetic risk factors for acute coronary syndrome in a large-scale replication study. *JAMA*.

Nakagawa, S. and Cuthill, I. C. (2007). Effect size, confidence interval and statistical significance: a practical guide for biologists. *Biological Reviews*.

Port, S., Demer, L., Jennrich, R., Walter, D., and Garfinkel, A. (2000). Systolic blood pressure and mortality. *The Lancet*.

Putnam, H. (1974). The 'corroboration' of theories. *Philosophy of Science*.

Reshef, D. N., Reshef, Y. A., Finucane, H. K., Grossman, S. R., McVean, G., Turnbaugh, P. J., Lander, E. S., Mitzenmacher, M., and Sabeti, P. C. (2011). Detecting novel associations in large data sets. *Science*.

Ringwalt, C., Paschall, M. J., Gorman, D., Derzon, J., and Kinlaw, A. (2011). The use of one- versus two-tailed tests to evaluate prevention programs. *Evaluation & the Health Professions*.

Rose, T. (2016). When the US Air Force Discovered the Flaw of Averages. *Toronto Star*.

Ruxton, G. D. and Neuhäuser, M. (2010). When should we use one-tailed hypothesis testing? *Methods in Ecology and Evolution*.

Schoenfeld, J. D. and Ioannidis, J. P. (2013). Is everything we eat associated with cancer? A systematic cookbook review. *The American Journal of Clinical Nutrition*.

Shapiro, S., Heinonen, O. P., Siskind, V., Kaufman, D. W., Monson, R. R., and Slone, D. (1977). Antenatal exposure to doxylamine succinate and dicyclomine hydrochloride (Bendectin) in relation to congenital malformations, perinatal mortality rate, birth weight, and intelligence quotient score. *American Journal of Obstetrics and Gynecology*.

Siegfried, T. (2010). Odds are, it's wrong: Science fails to face the shortcomings of statistics. *Science News*.

Snow, J. (1855). *On the mode of communication of cholera*. John Churchill.

Speed, T. (2011). A correlation for the 21st century. *Science*.

Sullivan, G. M. and Feinn, R. (2012). Using effect size—or why the p value is not enough. *Journal of Graduate Medical Education*.

Tukey, J. W. *et al.* (1977). *Exploratory data analysis*. Pearson.

Viechtbauer, W. (2010). Conducting meta-analyses in R with the metafor package. *Journal of Statistical Software*.

Vigen, T. (2015). *Spurious correlations*. Hachette UK.

Wagenmakers, E.-J., Wetzels, R., Borsboom, D., and Van Der Maas, H. L. (2011). Why psychologists must change the way they analyze their data: the case of psi: comment on bem (2011).

Wainer, H. (2007). The most dangerous equation. *American Scientist*.

Ziliak, S. and McCloskey, D. N. (2008). *The cult of statistical significance: How the standard error costs us jobs, justice, and lives*. University of Michigan Press.

Index

A. Garfinkel and Y. Guo, *Understanding Data*,
https://doi.org/10.1007/978-3-032-18600-3

GPSR Compliance

The European Union's (EU) General Product Safety Regulation (GPSR) is a set of rules that requires consumer products to be safe and our obligations to ensure this.

If you have any concerns about our products, you can contact us on ProductSafety@springernature.com

In case Publisher is established outside the EU, the EU authorized representative is:

Springer Nature Customer Service Center GmbH
Europaplatz 3
69115 Heidelberg, Germany

Batch number: 10371079

Printed by Printforce, the Netherlands